Advances in Intelligent Systems and Computing

Volume 320

About this Series

The series "Advances in Intelligent Systems and Computing" contains publications on theory, applications, and design methods of Intelligent Systems and Intelligent Computing. Virtually all disciplines such as engineering, natural sciences, computer and information science, ICT, economics, business, e-commerce, environment, healthcare, life science are covered. The list of topics spans all the areas of modern intelligent systems and computing.

The publications within "Advances in Intelligent Systems and Computing" are primarily textbooks and proceedings of important conferences, symposia and congresses. They cover significant recent developments in the field, both of a foundational and applicable character. An important characteristic feature of the series is the short publication time and world-wide distribution. This permits a rapid and broad dissemination of research results.

More information about this series at http://www.springer.com/series/11156

El-Sayed M. El-Alfy · Sabu M. Thampi
Hideyuki Takagi · Selwyn Piramuthu
Thomas Hanne
Editors

Advances in Intelligent Informatics

Editors
El-Sayed M. El-Alfy
King Fahd University of Petroleum and Minerals
Dhahran
Saudi Arabia

Sabu M. Thampi
Indian Institute of Information Technology and Management - Kerala (IIITM-K)
Kerala
India

Hideyuki Takagi
Faculty of Design
Kyushu University
Fukuoka
Japan

Selwyn Piramuthu
Department of Information Systems and Operations Management
Warrington College of Business Administration
University of Florida
Florida
USA

Thomas Hanne
Institute for Information Systems
University of Applied Sciences
Olten
Switzerland

ISSN 2194-5357 ISSN 2194-5365 (electronic)
ISBN 978-3-319-11217-6 ISBN 978-3-319-11218-3 (eBook)
DOI 10.1007/978-3-319-11218-3

Library of Congress Control Number: 2014948178

Springer Cham Heidelberg New York Dordrecht London

Printed on acid-free paper

Springer is part of Springer Science+Business Media (www.springer.com)

Preface

The third International Symposium on Intelligent Informatics (ISI-2014) provided a forum for sharing original research results and practical development experiences among experts in the emerging areas of Intelligent Informatics. This edition was co-located with third International Conference on Advances in Computing, Communications and Informatics (ICACCI-2014), September 24–27, 2014. The Symposium was hosted by Galgotias College of Engineering & Technology, Greater Noida, Delhi, India.

ISI-2014 had two tracks: 'Advances in Intelligent Informatics' and a special track on 'Intelligent Distributed Computing'. In response to the call for papers, 134 papers were submitted for the Intelligent Informatics Track and 63 submissions for the Intelligent Distributed Computing track. All the papers were evaluated on the basis of their significance, novelty, and technical quality. Each paper was rigorously reviewed by the members of the program committee. This book contains a selection of refereed and revised papers of Intelligent Informatics Track originally presented at the symposium. In this track, 58 were accepted. The peer-reviewed papers selected for this Track cover several intelligent informatics and related topics including signal processing, pattern recognition, image processing, data mining and intelligent information systems.

Many people helped to make ISI-2014 a successful event. Credit for the quality of the conference proceedings goes first and foremost to the authors. They contributed a great deal of effort and creativity to produce this work, and we are very thankful that they chose ISI-2014 as the place to present it. Thanks to all members of the Technical Program Committee, and the external reviewers, for their hard work in evaluating and discussing papers. We wish to thank all the members of the Steering Committee and Organising Committee, whose work and commitment were invaluable. Our most sincere thanks go to all keynote speakers who shared with us their expertise and knowledge. The EDAS conference system proved very helpful during the submission, review, and editing phases.

We thank the Galgotias College of Engineering & Technology, Greater Noida, Delhi for hosting the conference. Sincere thanks to Suneel Galgotia, Chairman, GEI,

Dhruv Galgotia, CEO, GEI, R. Sundaresan, Director, GCET and Bhawna Mallick, Local Arrangements Chair for their valuable suggestions and encouragement.

We wish to express our thanks to Thomas Ditzinger, Senior Editor, Engineering/Applied Sciences Springer-Verlag for his help and cooperation.

September 2014

El-Sayed M. El-Alfy
Sabu M. Thampi
Hideyuki Takagi
Selwyn Piramuthu
Thomas Hanne

Organization

ISI'14 Committee

General Chair

Kuan-Ching Li	Providence University, Taiwan

TPC Chairs

El-Sayed M. El-Alfy	King Fahd University of Petroleum and Minerals, Saudi Arabia
Selwyn Piramuthu	University of Florida, USA
Thomas Hanne	University of Applied Sciences, Switzerland

TPC Members

A.B.M. Moniruzzaman	Daffodil International University, Bangladesh
A.F.M. Sajidul Qadir	Samsung R&D Institute-Bangladesh, Bangladesh
Abdelmajid Khelil	Huawei European Research Center, Germany
Aboul Ella Hassanien	University of Cairo, Egypt
Adel Alimi	REGIM, University of Sfax, National School of Engineers, Tunisia
Afshin Shaabany	University of Fasa, Iran
Agostino Bruzzone	University of Genoa, Italy
Aitha Nagaraju	CURAJ, India
Ajay Jangra	KUK University, Kurukshetra, Haryana, India
Ajay Singh	Multimedia University, Malaysia
Akash Singh	IBM, USA
Akhil Gupta	Shri Mata Vaishno Devi University, India
Akihiro Fujihara	Fukui University of Technology, Japan
Alex James	Nazarbayev University, Kazakhstan

Ali Yavari	KTH Royal Institute of Technology, Sweden
Amit Acharyya	IIT HYDERABAD, India
Amit Gautam	SP College of Engineering, India
Amudha J.	Anrita Vishwa Vidyapeetham, India
Anca Daniela Ionita	University Politehnica of Bucharest, Romania
Angelo Trotta	University of Bologna, Italy
Angelos Michalas	Technological Education Institute of Western Macedonia, Greece
Anirban Kundu	Kuang-Chi Institute of Advanced Technology, P.R. China
Aniruddha Bhattacharjya	Amrita School of Engineering Bangalore, India
Anjana Gosain	Indraprastha University, India
Antonio LaTorre	Universidad Politécnica de Madrid, Spain
Arpan Kar	Indian Institute of Management, Rohtak, India
Ash Mohammad Abbas	Aligarh Muslim University, India
Ashish Saini	Dayalbagh Educational Institute, India
Ashraf S.	IITMK, India
Athanasios Pantelous	University of Liverpool, United Kingdom
Atsushi Takeda	Tohoku Gakuin University, Japan
Atul Negi	University of Hyderabad, India
Azian Azamimi Abdullah	Universiti Malaysia Perlis, Malaysia
B.H. Shekar	Mangalore University, India
Belal Abuhaija	University of Tabuk, Saudi Arabia
Bhushan Trivedi	GLS Institute Of Computer Technology, India
Bilal Gonen	University of West Florida, USA
Bilal Khan	University of Sussex, United Kingdom
Chia-Hung Lai	National Cheng Kung University, Taiwan
Chia-Pang Chen	National Taiwan University, Taiwan
Chien-Fu Cheng	Tamkang University, Taiwan
Chiranjib Sur	ABV-Indian Institute of Information Technology & Management, Gwalior, India
Chunming Liu	T-Mobile USA, USA
Ciprian Dobre	University Politehnica of Bucharest, Romania
Ciza Thomas	Indian Institute of Science, India
Constandinos Mavromoustakis	University of Nicosia, Cyprus
Dalila Chiadmi	Mohammadia School of Engineering, Morocco
Daniela Castelluccia	University of Bari, Italy
Deepak Mishra	IIST, India
Deepti Mehrotra	AMITY School of Engineering and Technology, India
Demetrios Sampson	University of Piraeus, Greece
Dennis Kergl	Universität der Bundeswehr München, Germany
Dhananjay Singh	Hankuk University of Foreign Studies, Korea
Dimitrios Stratogiannis	National Technical University of Athens, Greece

Durairaj Devaraj	Kalasalingam University, India
Emilio Jiménez Macías	University of La Rioja, Spain
Evgeny Khorov	IITP RAS, Russia
Farrah Wong	Universiti Malaysia Sabah, Malaysia
Fikret Sivrikaya	Technische Universität Berlin, Germany
G. Thakur	MANIT Bhopal, India
Gancho Vachkov	The University of the South Pacific (USP), Fiji
Gorthi Manyam	IIST, India
Gregorio Romero	Universidad Politecnica de Madrid, Spain
Grienggrai Rajchakit	Maejo University, Thailand
Gwo-Jiun Horng	Fortune Institute of Technology, Taiwan
Habib Kammoun	University of Sfax, Tunisia
Habib Louafi	École de Technologies Supérieure (ETS), Canada
Haijun Zhang	Beijing University of Chemical Technology, P.R. China
Hajar Mousannif	Cadi Ayyad University, Morocco
Hammad Mushtaq	University of Management & TEchnology, Pakistan
Hanen Idoudi	National School of Computer Science - University of Manouba, Tunisia
Harikumar Sandhya	Amrita Vishwa Vidyapeetham, India
Hemanta Kalita	North Eastern Hill University, India
Hideaki Iiduka	Kyushu Institute of Technology, Japan
Hossam Zawbaa	Beni-Suef University, Egypt
Hossein Malekmohamadi	University of Lincoln, United Kingdom
Huifang Chen	Zhejiang University, P.R. China
Igor Salkov	Donetsk National University, Ukraine
J. Mailen Kootsey	Simulation Resources, Inc., USA
Jaafar Gaber	UTBM, France
Jagdish Pande	Qualcomm Inc., USA
Janusz Kacprzyk	Polish Academy of Sciences, Poland
Javier Bajo	University of Salamanca, Spain
Jaynendra Kumar Rai	Amity School of Engineering and Technology, India
Jia-Chin Lin	National Central University, Taiwan
Jose Delgado	Technical University of Lisbon, Portugal
Jose Luis Vazquez-Poletti	Universidad Complutense de Madrid, Spain
Josip Lorincz	University of Split, Croatia
Jun He	University of New Brunswick, Canada
Junyoung Heo	Hansung University, Korea
K. Majumder	West Bengal University of Technology, India
Kambiz Badie	Iran Telecom Research Center, Iran
Kandasamy SelvaRadjou	Pondicherry Engineering College, India
Kanubhai Patel	Charotar University of Science and Technology (CHARUSAT), India

Kaushal Shukla	Indian Institute of Technology, Banaras Hindu University, India
Kenichi Kourai	Kyushu Institute of Technology, Japan
Kenneth Nwizege	University of SWANSEA, United Kingdom
Kuei-Ping Shih	Tamkang University, Taiwan
Lorenzo Mossucca	Istituto Superiore Mario Boella, Italy
Lucio Agostinho	University of Campinas, Brazil
Luis Teixeira	Universidade Catolica Portuguesa, Portugal
M. Manikandan	Anna University, India
M. Rajasree	IIITMK, India
Mahendra Dixit	SDMCET, India
Malika Bourenane	University of Senia, Algeria
Manjunath Aradhya	Sri Jayachamarajendra College of Engineering, India
Mantosh Biswas	National Institute of Technology-Kurukshetra, India
Manu Sood	Himachal Pradesh University, India
Marcelo Carvalho	University of Brasilia, Brazil
Marco Rospocher	Fondazione Bruno Kessler, Italy
Marenglen Biba	University of New York, Tirana, USA
Martin Randles	Liverpool John Moores University, United Kingdom
Martin Zsifkovits	University of Vienna, Austria
Massimo Cafaro	University of Salento, Italy
Melih Karaman	Bogazici University, Turkey
Mikulas Alexik	University of Zilina, Slovakia
Mohamad Noh Ahmad	Universiti Teknologi Malaysia, Malaysia
Mohamed Ba khouya	University of Technology of Belfort Montbeliard, France
Mohamed Dahmane	University of Montreal, Canada
Mohamed Moussaoui	Abdelmalek Esaadi UniversitY, Morocco
Mohammad Monirujjaman Khan	University of Liberal Arts Bangladesh, Bangladesh
Mohammed Mujahid Ulla Faiz	King Fahd University of Petroleum and Minerals (KFUPM), Saudi Arabia
Mohand Lagha	Saad Dahlab University of Blida - Blida - Algeria, Algeria
Mohd Ramzi Mohd Hussain	International Islamic University Malaysia, Malaysia
Monica Chis	Frequentis AG, Romania
Monika Gupta	GGSIPU, India
Mukesh Taneja	Cisco Systems, India
Mustafa Khandwawala	University of North Carolina at Chapel Hill, USA
Muthukkaruppan Annamalai	Universiti Teknologi MARA, Malaysia

Naveen Aggarwal	Panjab University, India
Nestor Mora Nuñez	Cadiz University, Spain
Nico Saputro	Southern Illinois University Carbondale, USA
Nisheeth Joshi	Banasthali University, India
Noor Mahammad Sk	Indian Institute of Information Technology Design and Manufacturing (IIITDM) , India
Nora Cuppens-Boulahia	IT TELECOM Bretagne, France
Olaf Maennel	Loughborough University, United Kingdom
Omar ElTayeby	Clark Atlanta University, USA
Oskars Ozolins	Riga Technical University, Latvia
Otavio Teixeira	Centro Universitário do Estado do Pará (CESUPA), Brazil
Pedro Gonçalves	Universidade de Aveiro, Portugal
Peiyan Yuan	Henan Normal University, P.R. China
Petia Koprinkova-Hristova	Bulgarian Academy of Sciences, Bulgaria
Philip Moore	Lanzhou University, United Kingdom
Praveen Srivastava	Indian Institute of Management (IIM), India
Pravin Patil	Graphic Era University Dehradun, India
Qin Lu	University of Technology, Sydney, Australia
Rabeb Mizouni	Khalifa University, UAE
Rachid Anane	Coventry University, United Kingdom
Rafael Pasquini	Federal University of Uberlândia - UFU, Brazil
Rajeev Kumaraswamy	Network Systems & Technologies Private Ltd, India
Rajeev Shrivastava	MPSIDC, India
Rajib Kar	National Institute of Technology, Durgapur, India
Rakesh Nagaraj	Amrita School of Engineering, India
Rama Garimella	IIIT Hyderabad, India
Ranjan Das	Indian Institute of Technology Ropar, India
Rashid Ali	College of Computers and Information Technology, Taif University, Saudi Arabia
Raveendranathan Kl C.	University of Kerala, India
Ravibabu Mulaveesala	Indian Institute of Technology Ropar, India
Rivindu Perera	Auckland University of Technology, New Zealand
Rubita Sudirman	Universiti Teknologi Malaysia, Malaysia
Ryosuke Ando	Toyota Transportation Research Institute (TTRI), Japan
Sakthi Balan	Infosys Ltd., India
Salvatore Venticinque	Second University of Naples, Italy
Sameer Saheerudeen Mohammed	National Institute of Technology Calicut, India
Sami Habib	Kuwait University, Kuwait
Sasanko Gantayat	GMR Institute of Technology, India
Satish Chandra	Jaypee Institute of Information Technology, India

Satya Ghrera	Jaypee University of Information Technology, India
Scott Turner	University of Northampton, United Kingdom
Selvamani K.	Anna University, India
Shalini Batra	Thapar University, India
Shanmugapriya D.	Avinashilingam Institute , India
Sheeba Rani	IIST Trivandrum, India
Sheng-Shih Wang	Minghsin University of Science and Technology, Taiwan
Shubhajit Roy Chowdhury	IIIT Hyderabad, India
Shuping Liu	University of Southern California, USA
Shyan Ming Yuan	National Chiao Tung University, Taiwan
Siby Abraham	University of Mumbai, India
Simon Fong	University of Macau, Macao
Sotiris Karachontzitis	University of Patras, Greece
Sotiris Kotsiantis	University of Patras, Greece
Sowmya Kamath S.	National Institute of Technology, Surathkal, India
Sriparna Saha	IIT Patna, India
Su Fong Chien	MIMOS Berhad, Malaysia
Sujit Mandal	National Institute of Technology, Durgapur, India
Suma V.	Dayananda Sagar College of Engineering, VTU, India
Suryakanth Gangashetty	IIIT Hyderabad, India
Tae (Tom) Oh	Rochester Institute of Technology, USA
Teruaki Ito	University of Tokushima, Japan
Tilokchan Irengbam	Manipur University, India
Traian Rebedea	University Politehnica of Bucharest, Romania
Tutut Herawan	Universiti Malaysia Pahang, Malaysia
Usha Banerjee	College of Engineering Roorkee, India
V. Vityanathan	SASTRA Universisty, India
Vatsavayi Valli Kumari	Andhra University, India
Veronica Moertini	Parahyangan Catholic University, Bandung, Indonesia
Vikrant Bhateja	Shri Ramswaroop Memorial Group of Professional Colleges, Lucknow (UP), India
Visvasuresh Victor Govindaswamy	Concordia University, USA
Vivek Sehgal	Jaypee University of Information Technology, India
Vivek Singh	Banaras Hindu University, India
Wan Hussain Wan Ishak	Universiti Utara Malaysia, Malaysia
Wei Wei	Xi'an University of of Technology, P.R. China
Xiaoya Hu	Huazhong University of Science and Technology, P.R. China
Xiao-Yang Liu	Columbia University, P.R. China
Yingyuan Xiao	Tianjin University of Technology, P.R. China

Yong Liao	NARUS INC., USA
Yoshitaka Kameya	Meijo University, Japan
Yuming Zhou	Nanjing University, P.R. China
Yu-N Cheah	Universiti Sains Malaysia, Malaysia
Zhenzhen Ye	IBM, USA
Zhijie Shen	Hortonworks, Inc., USA
Zhuo Lu	Intelligent Automation, Inc, USA

Additional Reviewers

Abdelhamid Helali	ISIMM, Tunisia
Adesh Kumar	UPES, India
Adriano Prates	Universidade Federal Fluminense, Brazil
Amitesh Rajput	Sagar Institute of Science & Technology, Bhopal, India
Antonio Cimmino	Zurich University of Applied Sciences, Switzerland
Aroua Hedhili	SOIE, National School of Computer Studies (ENSI), Tunisia
Azhana Ahmad	Universiti Tenaga Nasional, Malaysia
Behnam Salimi	UTM, Malaysia
Carolina Zato	University of Salamanca, Spain
Damandeep Kaur	Thapar University, India
Devi Arockia Vanitha	The Standard Fireworks Rajaratnam College for Women, Sivakasi, India
Divya Upadhyay	Amity School of Engineering & Technology, Amity University Noida, India
Gopal Chaudhary	NSIT, India
Hanish Aggarwal	Indian Institute of Technology Roorkee, India
Indrajit De	MCKV Institute of Engineering, India
Iti Mathur	Banasthali University, India
Kamaldeep Kaur	Guru Gobind Singh Indraprastha University, India
Kamyar Mehranzamir	UTM, Malaysia
KOTaiah Bonthu	Babasaheb Bhimrao Ambedkar University, Lucknow, India
Laila Benhlima	Mohammed V-Agdal University, Mohammadia School of Engineering, Morocco
M. Karuppasamypandiyan	Kalasalingam University, India
Mahalingam Pr.	Muthoot Institute of Technology and Science, India
Manisha Bhende	University of Pune, India
Mariam Kassim	Anna University, India

Mohammad Abuhweidi	University of Malaya, Malaysia
Mohammad Hasanzadeh	Amirkabir University of Technology, Iran
Muhammad Imran Khan	University of Toulouse, France
Muhammad Murtaza	University of Engineering and Technology Lahore, Pakistan
Muhammad Rafi	FAST-NU, Pakistan
Mukundhan Srinivasan	Indian Institute of Science, India
Naresh Kumar	GGSIPU, India
Nasimi Eldarov	Saarland University, Germany
Nattee Pinthong	Mahanakorn University of Technology, Thailand
Nouman Rao	Higher Education Commission of Pakistan, Pakistan
Omar Al Saif	Mosul University, Iraq
Orlewilson Maia	Federal University of Minas Gerais, Brazil
Pankaj Kulkarni	Rajasthan Technical University, India
Paulus Sheetekela	MIPT SU, Russia
Pooja Tripathi	IPEC, India
Prakasha Shivanna	RNS Institute of Technology, India
Pratiyush Guleria	Himachal Pradesh University Shimla, India
Preetvanti Singh	Dayal Bagh Educational Institute, India
Prema Nedungadi	Amrita University, India
Rathnakar Achary	Alliance Business Academy, India
Rohit Thanki	C U Shah Unversity, India
Roozbeh Zarei	Victoria University, Australia
Saida Maaroufi	Ecole Polytechnique de Montréal, Canada
Saraswathy Shamini Gunasekaran	Universiti Tenaga Nasional, Malaysia
Sarvesh Sharma	BITS-Pilani, India
Senthil Sivakumar	St. Joseph University, Tanzania
Seyedmostafa Safavi	Universiti Kebangsaan Malaysia, Malaysia
Shanmuga Sundaram Thangavelu	Amrita Vishwa Vidyapeetham, India
Shruti Kohli	Birla institute of Technology, India
Sudhir Rupanagudi	WorldServe Education, India
Thenmozhi Periasamy	Avinashilingam University, India
Tripty Singh	Amrita Vishwa Vidyapeetham, India
Umer Abbasi	Universiti Teknologi PETRONAS, Malaysia
Vaidehi Nedu	Dayananda Sagar College of Engineering, India
Vijender Solanki	Anna University, Chennai, India
Vinita Mathur	JECRC, Jaipur, India
Vipul Dabhi	Information Technology Department, Dharmsinh Desai University, India
Vishal Gupta	BITS, India

Vrushali Kulkarni	College of Engineering, Pune, India
Yatendra Sahu	Samrat Ashok Technological Institute, India
Yogesh Meena	Hindustan Institute of Technology and Management, India
Yogita Thakran	Indian Institute of Technology Roorkee, India
Zhiyi Shao	Shaanxi Normal University, P.R. China

Steering Committee

Antonio Puliafito	MDSLab - University of Messina, Italy
Axel Sikora	University of Applied Sciences Offenburg, Germany
Bharat Bhargava	Purdue University, USA
Chandrasekaran K.	NITK, India
Deepak Garg, Chair	IEEE Computer Society Chapter, IEEE India Council
Dilip Krishnaswamy	IBM Research - India
Douglas Comer	Purdue University, USA
El-Sayed M. El-Alfy	King Fahd University of Petroleum and Minerals, Saudi Arabia
Gregorio Martinez Perez	University of Murcia, Spain
Hideyuki Takagi	Kyushu University, Japan
Jaime Lloret Mauri	Polytechnic University of Valencia, Spain
Jianwei Huang	The Chinese University of Hong Kong, Hong Kong
John F. Buford	Avaya Labs Research, USA
Manish Parashar	Rutgers, The State University of New Jersey, USA
Mario Koeppen	Kyushu Institute of Technology, Japan
Nallanathan Arumugam	King's College London, United Kingdom
Nikhil R. Pal	Indian Statistical Institute, Kolkata, India
Pascal Lorenz	University of Haute Alsace, France
Raghuram Krishnapuram	IBM Research - India
Raj Kumar Buyya	University of Melbourne, Australia
Sabu M. Thampi	IIITM-K, India
Selwyn Piramuthu	University of Florida, USA
Suash Deb, President	Intl. Neural Network Society (INNS), India Regional Chapter

ICACCI Organising Committee

Chief Patron

Suneel Galgotia, Chairman	GEI, Greater Noida

Patrons

Dhruv Galgotia	CEO, GEI, Greater Noida
R. Sundaresan, Director	GCET, Greater Noida

General Chairs

Sabu M. Thampi	IIITM-K, India
Demetrios G. Sampson	University of Piraeus, Greece
Ajith Abraham	MIR Labs, USA & Chair, IEEE SMCS TC on Soft Computing

Program Chairs

Peter Mueller	IBM Zurich Research Laboratory, Switzerland
Juan Manuel Corchado Rodriguez	University of Salamanca, Spain
Javier Aguiar	University of Valladolid, Spain

Industry Track Chair

Dilip Krishnaswamy	IBM Research Labs, Bangalore, India

Workshop and Symposium Chairs

Axel Sikora	University of Applied Sciences Offenburg, Germany
Farag Azzedin	King Fahd University of Petroleum and Minerals, Saudi Arabia
Sudip Misra	Indian Institute of Technology, Kharagpur, India

Special Track Chairs

Amit Kumar	BioAxis DNA Research Centre, India
Debasis Giri	Haldia Institute of Technology, India

Demo/Posters Track Chair

Robin Doss	School of Information Technology, Deakin University, Australia

Keynote/Industry Speakers Chairs

Al-Sakib Khan Pathan	IIUM, Malaysia
Ashutosh Saxena	Infosys Labs, India
Shyam Diwakar	Amrita Vishwa Vidyapeetham Kollam, India

Tutorial Chairs

Sougata Mukherjea	IBM Research-India
Praveen Gauravaram	Tata Consultancy Services Ltd., Hyderabad, India

Doctoral Symposium Chairs

Soura Dasgupta	The University of Iowa, USA
Abdul Quaiyum Ansari	Dept. of Electrical Engg., Jamia Millia Islamia, India
Praveen Ranjan Srivastava	Indian Institute of Management (IIM), Rohtak, India

Organizing Chair

Bhawna Mallick	Galgotias College of Engineering & Technology (GCET), India

Organizing Secretaries

Sandeep Saxena	Dept. of CSE, GCET
Rudra Pratap Ojha	Dept. of IT, GCET

Publicity Chairs

Lucknesh Kumar	Dept. of CSE, GCET
Dharm Raj	Dept. of IT, GCET

Contents

Pattern Recognition, Signal and Image Processing

Artificial Immune System Based Image Enhancement Technique

Susmita Ganguli, Prasant Kumar Mahapatra*, and Amod Kumar

Abstract. Artificial immune system (AIS) inspired by immune system of vertebrates can be used for solving optimization problem. In this paper, image enhancement is considered as a problem of optimization and AIS is used to solve and find the optimal solution of this problem. Here, image enhancement is done by enhancing the pixel intensities of the images through a parameterized transformation function. The main task is to achieve the best enhanced image with the help of AIS by optimizing the parameters. The results have proved better when compared with other standard enhancement techniques like Histogram equalization (HE) and Linear Contrast Stretching (LCS).

Keywords: AIS, Image enhancement, parameter optimization, Histogram Equalization, Linear Contrast Stretching.

1 Introduction

The purpose of image enhancement is to improve the interpretability of information in images. The main aim of image enhancement is to modify parameters of an image to make it more relevant for a given task (Bedi & Khandelwal, 2013) (Maini & Aggarwal, 2010). According to (Gonzalez, Woods, & Eddins, 2009), image enhancement techniques can be divided into four main categories: point operation,

Susmita Ganguli · Prasant Kumar Mahapatra · Amod Kumar
V-2(Biomedical Instrumentation Division), CSIR-Central Scientific Instruments Organisation, Sector-30, Chandigarh 160030, India
e-mail: {gsush19,csioamod}@yahoo.com,
{gangsush,prasant22}@gmail.com, prasant22@csio.res.in

Susmita Ganguli
Department of Electronics and Communication, Tezpur University, Tezpur 784028, India

* Corresponding author, Member IEEE.

El-Sayed M. El-Alfy et al. (eds.), *Advances in Intelligent Informatics*,
Advances in Intelligent Systems and Computing 320, DOI: 10.1007/978-3-319-11218-3_1

spatial operation, transformation, and pseudo-coloring. Here, image enhancement technique is done on the basis of spatial operation i.e. it directly operates upon the gray levels of image which includes gray-scale correction, gray-scale transformation, histogram equalization, image smoothing and image sharpening.

Artificial Immune systems are inspired by theoretical aspects of immunology and observed immune functions (Ji & Dasgupta, 2007). Due to some of its properties like: self-organisation; learning and memory; adaptation; recognition; robustness and scalability, it has become a novel approach for solving complex computation problems. For image segmentation AIS methods such as NSA, Clonal Selection theory and Immune Network model have already been applied. It has been used for pattern recognition and image thresholding in (Wang, Gao, & Tang, 2009) and (Mahapatra et al., 2013), whereas in (Keijzers, Maandag, Marchiori, & Sprinkhuizen-Kuyper, 2013) it is used to find similar images.

Several computational techniques have been applied to enhance images like genetic algorithm (Hashemi, Kiani, Noroozi, & Moghaddam, 2010; Shyu & Leou, 1998) , evolution algorithm (Gogna & Tayal, 2012; Munteanu & Rosa, 2004), PSO (Gorai & Ghosh, 2009), (Braik, Sheta, & Ayesh, 2007), (Wachowiak, Smolíková, Zheng, Zurada, & Elmaghraby, 2004), Firefly algorithm (Hassanzadeh, Vojodi, & Mahmoudi, 2011), etc. In this paper, gray-level image contrast enhancement is proposed using AIS. The algorithm is applied on Lathe tool images and MATLAB inbuilt images. The enhanced images by AIS are found to be better when compared with other automatic enhancement techniques. The algorithm has been developed using MATLAB® software (version: 8.2.0.701 (R2013b)). The detailed approach is discussed in methodology.

2 Methodology

2.1 Overview of Artificial Immune System

Artificial Immune System is basically a bio-inspired technique which is used to develop new methods of computing for solving various problems. It uses learning, memory and associative retrieval to solve recognition and classification tasks. AIS is a diverse field of research which describes an interrelation between immune system, engineering and computer science. There are four models under AIS, viz : Negative selection algorithm (NSA), Clonal Selection Algorithm (CLONALG), immune network model and danger theory. In this paper, NSA is used for image enhancement.

The purpose of negative selection is to provide self-tolerance to body-cells. It detects unknown antigens, without reacting with the self cells (Hormozi, Akbari, Javan, & Hormozi, 2013) (Thumati & Halligan, 2013). The main idea of NSA is to generate a detector set in order to detect anomaly. The algorithm works in the following manner:

It defines a self string which defines the normal behavior of the system. A detector set is then generated which can recognize the elements which does not belong to self string (i.e. non-self string). According to NSA, a self string, which gives the normal behavior of the system, and a randomly generated string are initialized. Then the affinity of all the elements of random string is calculated with respect to all the elements of self string. If the affinity of a random element is greater or equal to a given cross-reactivity threshold (De Castro & Timmis, 2002) then this random element is recognized as self element and is discarded; otherwise it is accepted and introduced to the detector set.

2.2 Functions Used

The image enhancement is done by enhancing the pixel intensities of the image. A parameterized transformation function is used to enhance the images (Gorai & Ghosh, 2009) which uses both local and global information of the images. These local and global information is similar to the statistical scaling given in (Gonzalez et al., 2009). The transformation function is defined as:

$$g(i,j)=\left[\frac{k.D}{\sigma(i,j)-b}\right]\left[f(i,j)-c\times m(i,j)\right]+m(i,j)^{a} \tag{1}$$

where, ***f(i,j)***=the gray value of the $(i,j)^{th}$ pixel of the input image
g(i,j)=the gray value of the $(i,j)^{th}$ pixel of the enhanced image
D=the global mean of the pixel of the input image
σ(i,j)=the local standard deviation of the $(i,j)^{th}$ pixel over an nxn window
m(i,j)=local mean of the $(i,j)^{th}$ pixel over an nxn window
a,b,c,k=parameters to produce large variations in the processed image.

To check the quality of the images an objective function is defined which uses number of edge pixels, intensity of edges and entropy measure of images as parameters. For a good contrast enhanced image the value of these parameters should be high. The objective function is defined as:

$$F(I_e)=\log(\log(E(I_s)))\times\frac{edgels\ (I_s)}{M\times N}\times H(I_e) \tag{2}$$

Here, Ie is the enhanced image and Is is the edge image which is produced by an edge detector. E(Is) is the sum of M×N pixel intensities of the edge image, edgels(Is) is the number of edge pixels, whose intensity value is above a threshold in the edge image. H(Ie) is the entropy of the enhanced image.

2.3 Proposed Method

In this research work, the proposed NSA algorithm at first generates the detector set. By using the transformation function explained in eq. (1), it then processes the HE of the original image. The image is processed by changing the parameters used in the transformation function which are further optimized by the algorithm. A flowchart of the proposed method is shown in Fig. 3.

The self and random sets are initialized at first. The original image is defined as the self set and the HE of the original image is considered as the random set of elements. The affinity between the elements of random in relation to the elements of self is calculated by using eq. (6) (Aickelin & Qi, 2004; Zheng & Li, 2007) which is considered as the reactivity threshold (Rthreshold).

$$Affinity\_measure = \frac{ObservedValue - ExpectedValue}{1 - ExpectedValue} \tag{3}$$

Then the image is enhanced using eq. (1) and its affinity is calculated. If the affinity of the enhanced image is less than reactivity threshold then they are accepted, otherwise discarded. This process is iterated a number of times to get the best solution. To check the quality of the resulted image the objective function value, given in eq. (2), of the images are found. In each and every case the objective function value of the image obtained from the proposed algorithm is found better than histogram equalization and linear contrast stretching.

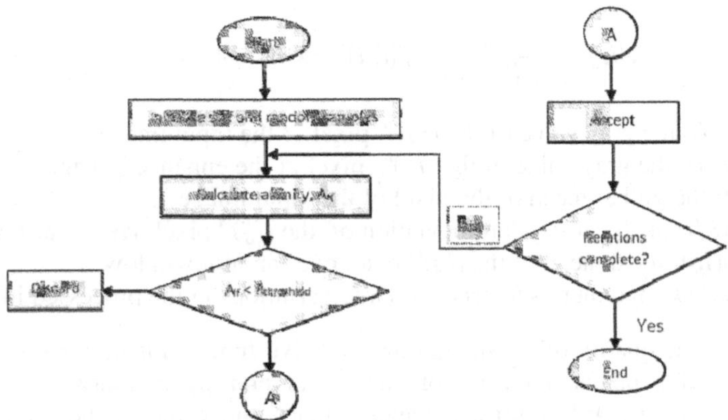

Fig. 1 Flowchart of the proposed NSA method for image enhancement

3 Results and Discussions

In this section, the result obtained from the newly proposed method is discussed. The new algorithm is applied on lathe tool and inbuilt MATLAB images. The lathe tool images were used because the tool is being used by the authors to carry out different experiments on micro and nano scale movements for desktop machining. The ultimate goal of the proposed algorithm is to increase the number of edgels (edge pixels), overall intensity of the edges and entropy values. For comparison purpose the results of standard techniques, i.e., HE and LCS are also considered. It is found that the images obtained from the new technique are much better than the other two.

The gray scale images of lathe tool at different horizontal positions are captured using single AVT Stingray F125B monochrome camera mounted on Navitar lens, and their enhanced images obtained from different enhancement techniques are shown in Fig. 2.

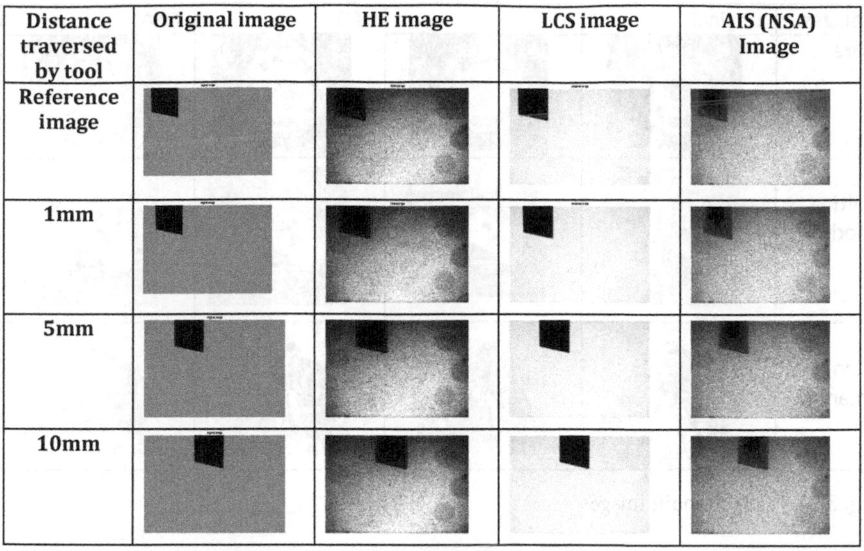

Distance traversed by tool	Original image	HE image	LCS image	AIS (NSA) Image
Reference image				
1mm				
5mm				
10mm				

Fig. 2 Images of lathe tool

In order to evaluate the proposed method number of edgels and fitness function are considered as the criterions. The performances of the resulted images are indicated by the increasing value of these two parameters. Sobel edge detector is used to determine the number of edgels. The value of the objective function and number of edge pixels are shown in Table 1.

Table 1 Fitness Value and number of edge pixels of enhanced images

Images	No of iterations	Parameters	Original	HE	LCS	AIS (NSA)
Reference	52	Value of Objective Function	0.0287	0.2722	0.0301	0.4058
		No of edgels	941	12269	941	12309
1mm	52	Value of Objective Function	0.0296	0.2760	0.0309	0.4276
		No of edgels	973	12475	973	12464
5mm	55	Value of Objective Function	0.0304	0.2627	0.0321	0.4245
		No of edgels	991	11964	991	12286
10mm	55	Value of Objective Function	0.0309	0.2722	0.0324	0.4415
		No of edgels	1005	12356	1006	12482

To check the effectiveness of the algorithm, it has also been applied to MATLAB inbuilt images. Three images are considered viz. tire, liftingbody and cameraman on which the new algorithm is tested. Their enhanced image obtained from different enhancement techniques are shown in Fig. 3.

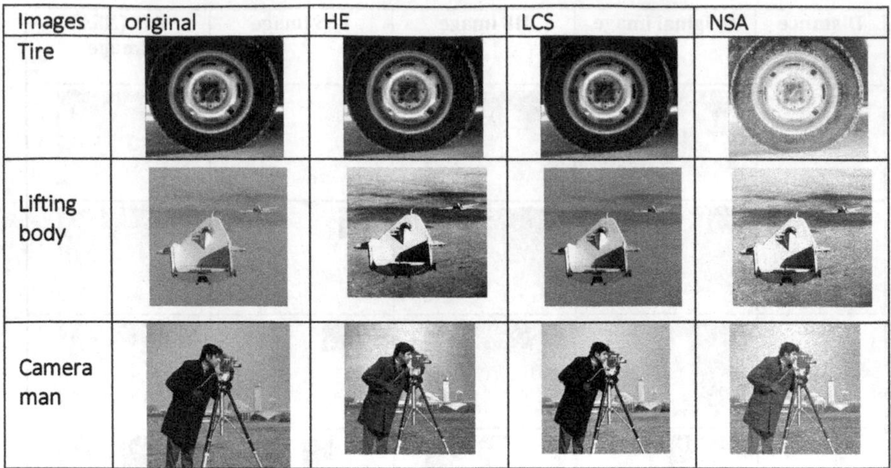

Fig. 3 MATLAB inbuilt images

Table 2 Value of Objective Function and number of edge pixels of enhanced images

Images	No. of iterations	Parameters	Original	HE	LCS	NSA
Tire	55	Value of Objective Function	0.4846	0.4323	0.4805	0.5224
		No of Edgels	1859	1994	1844	2052
Liftingbody	50	Value of Objective Function	0.2239	0.2200	0.2137	0.3040
		No of Edgels	4511	5281	4350	6186
Cameraman	50	Value of Objective Function	0.4826	0.4030	0.4800	0.5075
		No of Edgels	2503	2430	2503	2740

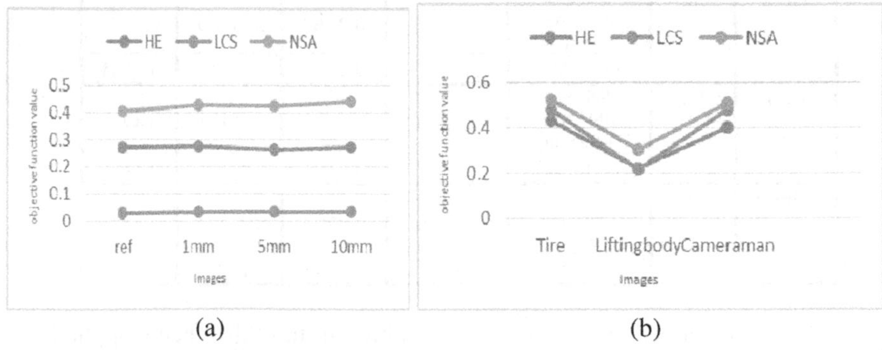

Fig. 4 Chart showing the variation of the value of objective function obtained from HE, LCS and NSA of (a): Lathe tool images (b): MATLAB inbuilt images

From the data shown in Table 1, 2 and the chart shown in Fig. 4, it can be concluded that the value of the parameters i.e. objective function and number of edgels obtained from the proposed technique is better, i.e. their values are higher, than the values obtained from HE and LCS.

4 Conclusions

In this paper, a new method of image enhancement using NSA has been proposed to enhance the contrast of images. The main objective of the algorithm is to enhance the gray intensity of the pixels of images by increasing the number of edgels, edge intensity and entropy of the images. The algorithm is applied on the images of a lathe tool at different horizontal positions and also to MATLAB inbuilt images. The results are compared with that of HE and LCS techniques. In each and every case the result of the proposed algorithm is found better.

Acknowledgment. This work is supported by the Council of Scientific & Industrial Research (CSIR, India), New Delhi under the Network programme (ESC-0112) in collaboration with CSIR-CMERI, Durgapur. Authors would like to thank Director, CSIR-CSIO for his guidance during investigation.

References

Aickelin, U., Qi, C.: On affinity measures for artificial immune system movie recommenders. Paper presented at the The 5th International Conference on: Recent Advances in Soft Computing, Nottingham, UK (2004)

Bedi, S., Khandelwal, R.: Various Image Enhancement Techniques-A Critical Review. International Journal of Advanced Research in Computer and Communication Engineering 2(3), 1605–1609 (2013)

Braik, M., Sheta, A.F., Ayesh, A.: Image Enhancement Using Particle Swarm Optimization. Paper presented at the World congress on engineering (2007)

De Castro, L.N., Timmis, J.: Artificial immune systems: a new computational intelligence approach. Springer (2002)

Gogna, A., Tayal, A.: Comparative analysis of evolutionary algorithms for image enhancement. International Journal of Metaheuristics 2(1), 80–100 (2012)

Gonzalez, R.C., Woods, R.E., Eddins, S.L.: Digital image processing using MATLAB, vol. 2. Gatesmark Publishing, Knoxville (2009)

Gorai, A., Ghosh, A. (2009). Gray-level Image Enhancement By Particle Swarm Optimization. Paper presented at the World Congress on Nature & Biologically Inspired Computing, NaBIC 2009 (2009)

Hashemi, S., Kiani, S., Noroozi, N., Moghaddam, M.E.: An image contrast enhancement method based on genetic algorithm. Pattern Recognition Letters 31(13), 1816–1824 (2010)

Hassanzadeh, T., Vojodi, H., Mahmoudi, F.: Non-linear grayscale image enhancement based on firefly algorithm. In: Panigrahi, B.K., Suganthan, P.N., Das, S., Satapathy, S.C. (eds.) SEMCCO 2011, Part II. LNCS, vol. 7077, pp. 174–181. Springer, Heidelberg (2011)

Hormozi, E., Akbari, M.K., Javan, M.S.: Performance evaluation of a fraud detection system based artificial immune system on the cloud. Paper presented at the 2013 8th International Conference on Computer Science & Education (ICCSE), April 26-28 (2013)

Ji, Z., Dasgupta, D.: Revisiting negative selection algorithms. Evolutionary Computation 15(2), 223–251 (2007)

Keijzers, S., Maandag, P., Marchiori, E., Sprinkhuizen-Kuyper, I.: Image Similarity Search using a Negative Selection Algorithm. Paper presented at the Advances in Artificial Life, ECAL (2013)

Mahapatra, P.K., Kaur, M., Sethi, S., Thareja, R., Kumar, A., Devi, S.: Improved thresholding based on negative selection algorithm (NSA). Evolutionary Intelligence, 1–14 (2013)

Maini, R., Aggarwal, H.: A comprehensive review of image enhancement techniques. Journal of Computing 2(3), 8–13 (2010)

Munteanu, C., Rosa, A.: Gray-scale image enhancement as an automatic process driven by evolution. IEEE Transactions on Systems, Man, and Cybernetics, Part B: Cybernetics 34(2), 1292–1298 (2004)

Shyu, M.-S., Leou, J.-J.: A genetic algorithm approach to color image enhancement. Pattern Recognition 31(7), 871–880 (1998)

Thumati, B.T., Halligan, G.R.: A Novel Fault Diagnostics and Prediction Scheme Using a Nonlinear Observer With Artificial Immune System as an Online Approximator. IEEE Transactions on Control Systems Technology 21(3), 569–578 (2013)

Wachowiak, M.P., Smolíková, R., Zheng, Y., Zurada, J.M., Elmaghraby, A.S.: An approach to multimodal biomedical image registration utilizing particle swarm optimization. IEEE Transactions on Evolutionary Computation 8(3), 289–301 (2004)

Wang, W., Gao, S., Tang, Z.: Improved pattern recognition with complex artificial immune system. Soft Computing 13(12), 1209–1217 (2009)

Zheng, H., Li, L.: An artificial immune approach for vehicle detection from high resolution space imagery. International Journal of Computer Science and Network Security 7(2), 67–72 (2007)

Grayscale to Color Map Transformation for Efficient Image Analysis on Low Processing Devices

Shitala Prasad, Piyush Kumar, and Kumari Priyanka Sinha

Abstract. This paper presents a novel method to convert a grayscale image to a colored image for quality image analysis. The grayscale IP operations are very challenging and limited. The information extracted from such images is inaccurate. Therefore, the input image is transformed using a reference color image by reverse engineering. The gray levels of grayscale image are mapped with the color image in all the three layers (red, green, blue). These mapped pixels are used to reconstruct the grayscale image such that it is represented in a 3 dimensional color matrix. The algorithm is very simple and accurate that it can be used in any domain such as medical imaging, satellite imaging and agriculture/environment real-scene. The algorithm is implemented and tested on low cost mobile devices too and the results are found appreciable.

Keywords: Digital Image Processing, Grayscale to Color Transformation, Image Analysis, Low Processing Devices.

1 Introduction

The technological advancements are contributing a major role in human society and colored cameras and scanners are becoming ubiquitous. There is a demand for better image analysis, in general. The growth of image processing from grayscale

Shitala Prasad
Computer Science and Engineering, IIT Roorkee
e-mail: shitala@ieee.org

Piyush Kumar
Information Technology, IIIT Allahabad
e-mail: piyushkumariiita@gmail.com

Kumari Priyanka Sinha
Information Technology, NIT Patna
e-mail: priyankasinha2008@gmail.com

El-Sayed M. El-Alfy et al. (eds.), *Advances in Intelligent Informatics*,
Advances in Intelligent Systems and Computing 320, DOI: 10.1007/978-3-319-11218-3_2

uni-variant data analysis to a colored multi-variant data analysis is still challenging the researchers. The areas like biomedical imaging are still working with grayscale image and phase problems like segmentation, edge detection, and blob detection, in case of tumor detection. Enhanced medical imaging techniques have attracted attention after the advanced medical equipments used in medical field. Colored images are much easier to distinguish the various shades by human eye.

Computer vision technologies are applied in various domains for image enhancements and automatic image summarization. The traditional process of identification involves lot of technical expert knowledge, lack of which misleads the results, particularly in medical cases. Therefore, the image is enhancement before the data is applied for any decision making analysis which increases the difference between the objects in image [1]. The aim of this paper is to present an algorithm to automatically transform a grayscale image to a colored image, which adds extra information in it making analysis easier.

Barghout and Sheynin [2] used *k*-mean clustering algorithm to segment different objects in real-scene based on the pixel color intensity. The algorithm guaranteed to converge but highly depends on *k*-value. Since the real-scenes include irregular shapes the *k*-value selection may fail to return an optimal solution. Chang et al. [3] agreed with this and thus proposed a novel environment scene image analysis approach based on the perceptual organization which incorporates *Gestalt* law. This algorithm can handle objects which are unseen before and so used for quality image segmentation. Few authors have used object-shape based models to overcome the segmentation problem with fixed objects [4], but it's not the real case. The major problems with such algorithms are that, they results in over segmentation and under segmentation.

A solution to over segmentation, in 2011, Prasad et al. [5], proposed a block-based unsupervised disease segmentation algorithm which segment the sub-images into small classes that is later on combined to form a bigger class. While in 2013, Prasad et al. [6], used a cluster based unsupervised segmentation approach to identify the diseased and non-diseased portion from a colored natural plant leaf image, considering only a single disease attack at a time. This algorithm granted the best result in disease detection in $L^*a^*b^*$ color space. On the other side, Grundland and Dodgson [7] introduced a new contrast enhanced color to grayscale conversion algorithm for real-scenes in real-time. They used image sampling and dimension reduction for the conversion. They transformed a colored image to a high contrast grayscale for better understanding. But again losing information in this dimension reduction which creates problem of under segmentation.

Sharma and Aggarwal [8] discuses the limitations of segmenting CT and MR scanned images due to its image type (i.e. grayscale images) [9]. The problem faced in segmentation may be due to its grayscale nature and so if it was in color space their might not be such case. Therefore, the aim of this paper is to transform grayscale image into colored (pseudo colored) image for better analysis and segment. This pseudo colored images may also be used in controlling various computer operations like operating power point presentation without using any statist input devices, as proposed and used by Prasad et al. [10-11]. Here, different color markers are used to identify the objects; as a pointer and tracking the movements of these pointers they are mapped with the static mouse operations.

Hence, color image processing bit complex but resolves many problems faced with grayscale processing and give users new dimensions to think and analysis the input image. This pseudo color transforms helps to gain color images from devices that are limited with intensity values only such as CT scanner and X-ray machines.

In this paper, we represent a gray pixel with a color pixel value and map it with the color map to form color image with extra information. To be very specific and to the point, the paper is transformed into 4 phase: the section 2 discusses about the proposed grayscale to color transform and in section 3 results are highlighted in different domains. The last phase is the conclusion and future scopes.

2 Proposed Grayscale to Color Transform

As we know that a grayscale digital image, say $I^2_{gray}(x, y)$, is a single dimension pixel intensity matrix whereas a colored image, $I^3_{color}(x, y)$ is a three dimension matrix. The first layer of matrix is red, second is green and third is blue layer because of these layers a gray image I^2_{gray} is transformed to a color image I^3_{color}. Figure 1 explains the concept of gray and color image.

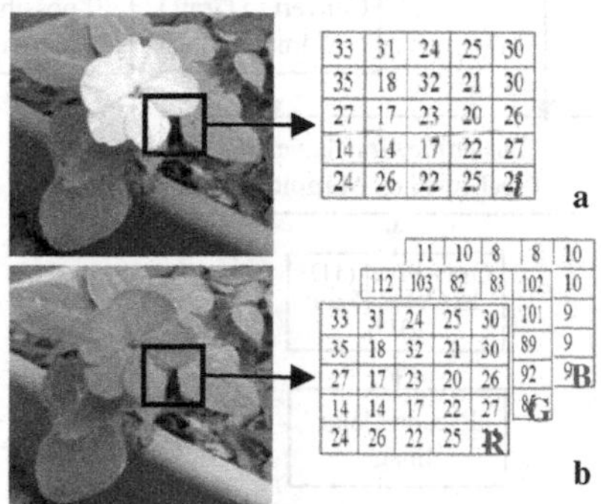

Fig. 1 Image, I: (a) grayscale intensity image matrix, I^2_{gray}, where i=intensity; and (b) colored three dimension (red, green , blue) image matrix, I^3_{color}

There are various devices that capture gray and colored images with different formats in different domain, such as CT scanner and mobile camera respectively. It is very simple and easy to convert a colored image to a single dimension grayscale image by losing any two layers information called single layer grayscale but the visa-versa is bit difficult and challenging. Few the RGB to grayscale image conversion formulas are shown in equations 1-5, they are the most commonly used.

$$gray = (\text{red} + \text{green} + \text{blue})/3 \tag{1}$$

$$gray = red * .3 + g * .59 + b * .11 \tag{2}$$

$$gray = red * .2126 + green * .7152 + blue * .0722 \quad (3)$$

$$gray = (max(red, green, blue) + min(red, green, blue))/2 \quad (4)$$

$$gray = (max\ /\ min)(red, green, blue) \quad (5)$$

Here, equation 1 is the average grayscale image formed from RGB. It is quick but results an inaccurate dirty image. Whereas by grayscale image formed by using equation 2-3, the result is good and it treats each color same as human eye perceive. Thus they are also called as *Luma* or luminance gray image. The 4th and 5th formula results in a de-saturation and decomposition gray image respectively. There are many other methods and approaches to get a grayscaled image but it's very difficult to convert a gray image to a colored one. These grayscale images from grayscale capturing devices are very challenging and involve many hurdles for image processing operations. This paper aims the same. A pseudo color image transformation is proposed for better IP operations. Below is the complete flow graph of the proposed system, see figure 2.

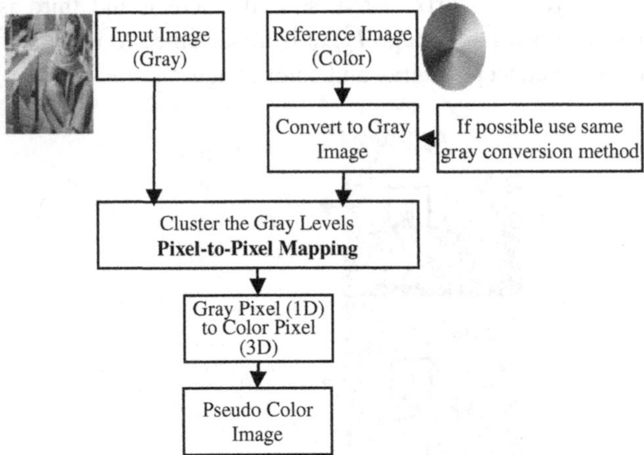

Fig. 2 System flow graph – grayscale to color image conversion

As in figure 2, a grayscale image, $\boldsymbol{I^2_{gray}}$ is taken as an input image to be converted to a color image from a reference colored image, $\boldsymbol{I^3_{refe}}$. The reference image $\boldsymbol{I^3_{refe}}$ is again transformed into a grayscale image, $\boldsymbol{I^3_{refe\_gray}}$ such that the method of grayscale conversion for both $\boldsymbol{I^2_{gray}}$ and $\boldsymbol{I^3_{refe\_gray}}$ images are same, but is not necessary. The next step is to cluster the gray levels of both the images and map it to the reference image, $\boldsymbol{I^3_{refe}}$. That is, a single intensity pixel of 8-bits is represented by a three layer pixel of 24-bits, reversed engineered from $\boldsymbol{I^2_{refe\_gray}}$ to $\boldsymbol{I^3_{refe}}$. In such a way, the single dimension (1D) $\boldsymbol{I^2_{gray}}$ is transformed to a three dimension (3D) colored image, $\boldsymbol{I^3_{pseudo}}$. This colored image can be used in-for segmentation or any other IP operations.

The mapping of a single pixel for 8-bit grayscale image to 24-bit color image. That is, if a pixel from $\boldsymbol{I^2_{gray}}$ with i^{th} gray level then the pixel with same i^{th} gray level from $\boldsymbol{I^2_{refe\_gray}}$ act as a reference pixel. Referring to $\boldsymbol{I^3_{refe}}$ with the same coordinate value, say $(\boldsymbol{x}, \boldsymbol{y})$, as the ith gray level pixel in $\boldsymbol{I^2_{refe\_gray}}$ the ith gray pixel is mapped to this 24-bit pixel from $\boldsymbol{I^3_{refe}}$. The mapping of pixels from 8-bit to 24-bit is shown in figure 3 for better explanation. The complete pseudo algorithm is presented in algorithm 1 below.

***Algorithm 1.* Grayscale Image to Color Image Transformation.**

Assumptions: Input: 1D grayscale image, I^2_{gray} and 3D reference color image, I^3_{refe} of any size.

Output: Pseudo colored image, I^3_{pseudo}

1. Transform the referenced image I^3_{refe} to grayscale using any of the methods from equations (1-5) or simply single layer gray image $I^2_{refe\_gray}$
2. Cluster the gray levels into 256 clusters for $0 - 255$ levels for both $I^2_{refe\_gray}$ and I^2_{gray}. K-Mean clustering algorithm is used here.
3. For every i^{th} gray level in I^2_{gray} we map with the same i^{th} level in $I^2_{refe\_gray}$ such that the coordinate (x, y) of ith of $I^2_{refe\_gray}$ refer to the coordinate (x, y) of I^3_{refe} for the intensity values of different layers (red, green, blue), say (r, g, b)
4. For every layer of I^3_{pseudo} assign values (r, g, b)
5. Repeat till gray level reaches 255

In mapping, there may be a case where many pixels are of ith gray level in $\boldsymbol{I^2_{refe\_gray}}$, as in figure 3, so pixel p(x, y) in $\boldsymbol{I^2_{gray}}$ is mapped with the first pixel p(x', y') in $\boldsymbol{I^2_{refe\_gray}}$.

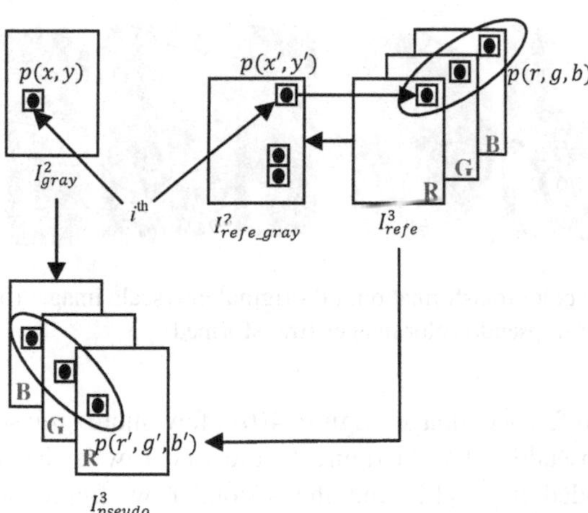

Fig. 3 Pixel mapping from 8-bit to 24-bit, using reverse engineering

The I^3_{pseudo} is the output pseudo color image with various extra details which a colored image has, without losing any other information. I^3_{pseudo} now can be used in digital color image processing operation which was not possible in earlier case for I^2_{gray}. Note that the pseudo color image depends upon the reference color image inputted in the system, i.e. as the reference color image changes the resultant output pseudo color image changes. Therefore, a standard reference color image is chosen where all color shades are present, as in figure 2.

In next section experimental results with different domains are presented and analyzed.

3 Experimental Results and Analysis

Since there are many datasets available in grayscale but are untested with digital color image processing operations and thus fails in proposing a general algorithm for grayscale and colored image, this paper points on this limitation. The algorithm proposed is very simple and thus implemented and tested on low cost computing devices such as mobile phones, $\mathcal{M}_d$. An application is designed for Android operating system version 2.3.x. The algorithm is resolution tolerant, that is, for any resolution of reference color image the output pseudo image has no effect or has very marginal change.

The grayscale to color image transformation in a standard 512x512 grayscale test images [12] is applied and shown in figure 4, below. Here, figure 4(c) and 4(d) are the pseudo images formed by reference image 4(b). Figure 4(c) is formed by using red portion of 4(b) reference image and so output image is red-oriented.

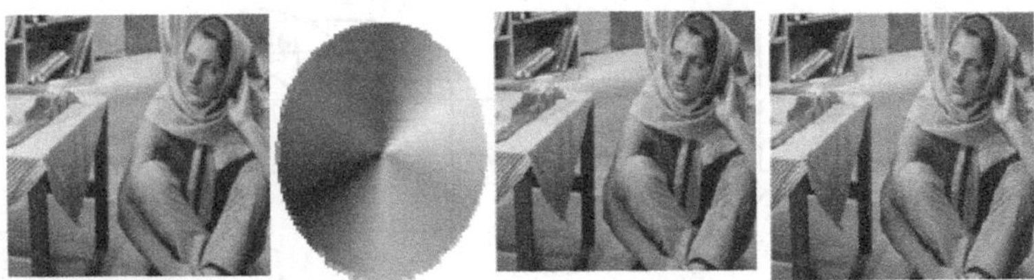

Fig. 4 Grayscale to color transformation: (a) original grayscale image; (b) original reference color image; and (c-d) pseudo color images (transformed)

Using same reference image, figure 4(b), few more grayscaled images are transformed to pseudo color. In figure 5, the first row is the original grayscale images downloaded from [12] and the second row, figure 5(b) is the output colored image mapped to gray values of figure 4(b).

Fig. 5 Color image transformation of standard 512x512 grayscale images [12]: (a) original gray images; and (b) pseudo color images

3.1 k-Means Clustering Algorithm

The gray level of image is clustered using simple *k*-means clustering algorithm. *k*-means is a vector quantization iterative technique to partition an image into k-selected clusters. The pixel intensity is used for computing mean cluster and distance is the squared difference between the intensity and the cluster center. The algorithm may not return optimal result but guaranties to converge and so is used in this paper.

Since *k*-means is very common we must not discuss in details and move to test results of various other datasets and images.

Figure 6(a), is a set of images recorded from satellite which is very difficult to analyze as which is what. The same is then transformed through our proposed system and tried for its pseudo images using same reference images, figure 4(b), we get the second row of figure 6 which is much more clear than the earlier one. Now, images from figure 6(b) can be easily clustered using unsupervised segmentation algorithm [6].

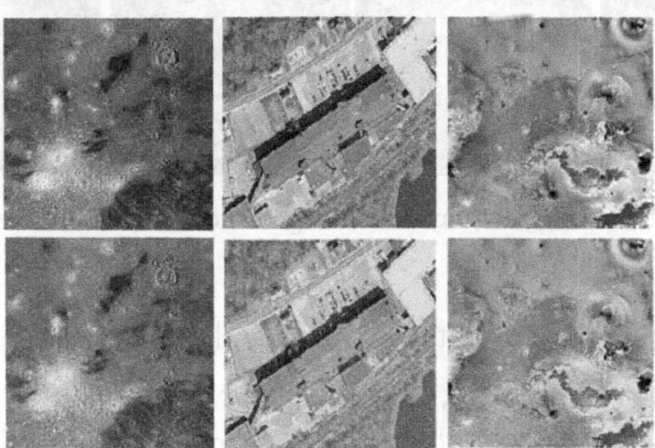

Fig. 6 Satellite sample images: (a) original grayscale image; and (b) pseudo color image (segmented)

Over this, the proposed algorithm plays a better role in biomedical images such as CT scanned images and MRI images. Such images are very complicated and if a colored or pseudo colored images are provided by the physicians, it will be easy and fast way to diagnose the cases. In figure 7, few of the brain MRI images are captured in grayscale and the proposed color transform method is applied over it. The resulted output pseudo color image is much easier to understand and derive and decision. It clearly shows the portion where brain tumors are occurred by just seeing the color changes. It can be visualized by and layman too.

Another CT scan grayscale image with its pseudo color transform is represented in figure 8. It's clear that the proposed algorithm is very simple and accurate and can be used in any type of domain and also can be used to segment the image.

3.2 Applications and Scopes

The algorithm feasibility was tested under various different conditions and found to be positive. The robustness of proposed method is clearly visualized in this paper with the above results. The scope of this paper is mainly in segmentation and image analysis where the texture of an image is not so clear and unique. This method will also be used in biomedical and satellite image processing to classify various patterns and objects in the image. It also enhances the image.

This paper processes grayscale images and represent it in a 3D pseudo color image format. It is not limited to and for any domain and any reference image. Changing the reference image will change the pseudo image but the transformation equation will be the same. Finally, the four concludes this paper with the future scope.

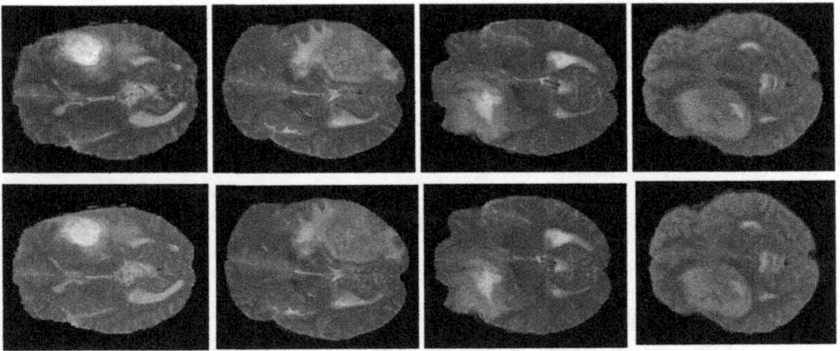

Fig. 7 MRI images: (a) original grayscale image; and (b) pseudo colored transformed images

The interesting point is that, if the reference color image, I_{refe}^3 is same, that is, $I_{refe}^3 \xrightarrow{gray} I_{gray}^2$ then the I_{pseudo}^3 image is approximately the same, that is, $I_{pseudo}^3 \approx I_{refe}^3$. Figure 9 shows the same. Where in figure 9(d) is a different I_{gray}^2image is which uses figure 9(a) as I_{refe}^3 image giving output to pseudo image shown in figure 9(e).

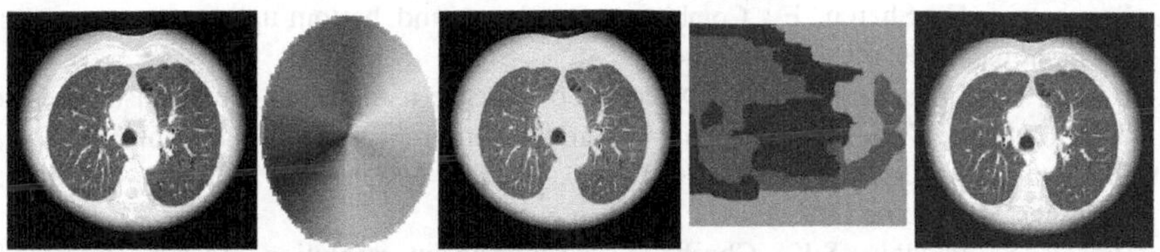

Fig. 8 CT scanned image: (a) original grayscale image; (b) reference color image 1; (c) pseudo color image from reference image 1; (d) reference color image 2; and (e) pseudo color image from reference image 2

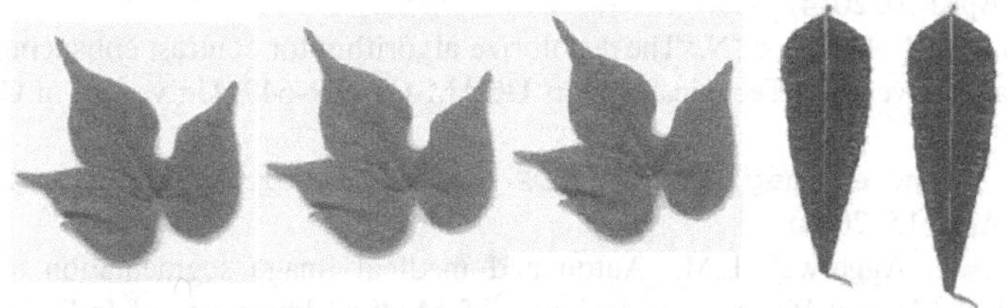

Fig. 9 (a) The original I^3_{refe} image, (b) input I^2_{gray} image from I^3_{refe}, (c) the output I^3_{pseudo} image, (d) input I^2_{gray} image and (e) the output I^3_{pseudo} image using I^3_{refe}.

4 Conclusion and Future

This paper aims to transform a grayscale image with 8-bit pixel representation to a pseudo color image with 24-bit pixel representation. The system proposed is robust and flexible with image resolution and reference color image used. Algorithm is so simple and less computationally cost that it can be deployed on and low processing devices such as mobile phones, $\mathcal{M}_d$. The algorithm is unsupervised. Firstly, the input gray image is mapped with the reference gray image formed by the reference color image and then applying k-mean clustering similar gray levels are clustered to 256 clusters. Then using reverse engineering 8-bit pixel is transformed to 24-bit pixel. This method is used to enhance, segment and analyze the image in better way, even though they are of gray in nature. After some many experiments there is still gap between the output pseudo color image and actual color image. Thus, the work may include more realistic color image transformation and robust output.

References

1. Alparone, L., Wald, L., Chanussot, J., Thomass, C., Gamba, P.: Comparison of pansharpening algorithms. IEEE Trans. Geosci. Remote Sens. 45, 3012–3021 (2007)
2. Barghout, L., Jacob, S.: Real-world scene perception and perceptual organization: Lessons from Computer Vision. Journal of Vision 13(9), 709 (2013)
3. Chang, C., Koschan, A., Page, D.L., Abidi, M.A.: Scene image segmentation based on Perceptual Organization. In: IEEE Int'l Conf. on ICIP, pp. 1801–1804 (2009)

4. Borenstein, E., Sharon, E.: Combining top-down and bottom-up segmentation. In: Workshop. CVPR, pp. 46–53 (2004)
5. Prasad, S., Kumar, P., Jain, A.: Detection of disease using block-based unsupervised natural plant leaf color image segmentation. In: Panigrahi, B.K., Suganthan, P.N., Das, S., Satapathy, S.C. (eds.) SEMCCO 2011, Part I. LNCS, vol. 7076, pp. 399–406. Springer, Heidelberg (2011)
6. Prasad, S., Peddoju, S.K., Ghosh, D.: Unsupervised resolution independent based natural plant leaf disease segmentation approach for mobile devices. In: Proc. of 5th ACM IBM Collaborative Academia Research Exchange Workshop (I-CARE 2013), New York, USA, Article 11, 4 pages (2013), `http://dl.acm.org/citation.cfm?id=2528240&preflayout=tabs` (Cited April 10, 2014)
7. Grundland, M., Dodgson, N.: The decolorize algorithm for contrast enhancing, color to grayscale conversion. Technical Report UCAM-CL-TR-649, University of Cambridge (2005), `http://www.eyemaginary.com/Portfolio/TurnColorsGray.html` (Cited April 15, 2014)
8. Sharma, N., Aggarwal, L.M.: Automated medical image segmentation techniques. Journal of Medical Physics / Association of Medical Physicists of India 35(1), 3–14 (2010)
9. Kumar, P., Agrawal, A.: GPU-accelerated Interactive Visualization of 3D Volumetric Data using CUDA. World Scientific International Journal of Image and Graphics 13(2), 1340003–13400017 (2013)
10. Prasad, S., Peddoju, S.K., Ghosh, D.: Mobile Augmented Reality Based Interactive Teaching & Learning System with Low Computation Approach. In: IEEE Symposium on Computational Intelligence in Control and Automation (CICA), pp. 97–103 (2013)
11. Prasad, S., Prakash, A., Peddoju, S.K., Ghosh, D.: Control of computer process using image processing and computer vision for low-processing devices. In: Proceedings of the ACM International Conference on Advances in Computing, Communications and Informatics, pp. 1169–1174 (2012)
12. Dataset of Standard 512x512 Grayscale Test Images, `http://decsai.ugr.es/cvg/CG/base.htm` (last accessed on May 20, 2014)

Automatic Classification of Brain MRI Images Using SVM and Neural Network Classifiers

N.V.S. Natteshan and J. Angel Arul Jothi

Abstract. Computer Aided Diagnosis (CAD) is a technique where diagnosis is performed in an automatic way. This work has developed a CAD system for automatically classifying the given brain Magnetic Resonance Imaging (MRI) image into 'tumor affected' or 'tumor not affected'. The input image is preprocessed using wiener filter and Contrast Limited Adaptive Histogram Equalization (CLAHE). The image is then quantized and aggregated to get a reduced image data. The reduced image is then segmented into four regions such as gray matter, white matter, cerebrospinal fluid and high intensity tumor cluster using Fuzzy C Means (FCM) algorithm. The tumor region is then extracted using the intensity metric. A contour is evolved over the identified tumor region using Active Contour model (ACM) to extract exact tumor segment. Thirty five features including Gray Level Co-occurrence Matrix (GLCM) features, Gray Level Run Length Matrix features (GLRL), statistical features and shape based features are extracted from the tumor region. Neural network and Support Vector Machine (SVM) classifiers are trained using these features. Results indicate that Support vector machine classifier with quadratic kernel function performs better than Radial Basis Function (RBF) kernel function and neural network classifier with fifty hidden nodes performs better than twenty five hidden nodes. It is also evident from the result that average running time of FCM is less when used on reduced image data.

Keywords: Computer aided diagnosis (CAD), Fuzzy C Means (FCM), Active Contour model (ACM), feature extraction, Neural network, Support vector machine (SVM), Magnetic Resonance Imaging (MRI).

N.V.S. Natteshan · J. Angel Arul Jothi
Department of Computer Science and Engineering, College of Engineering Guindy, Anna University, Chennai, India
e-mail: natteshann.v.s@gmail.com, jothi@cs.annauniv.edu

El-Sayed M. El-Alfy et al. (eds.), *Advances in Intelligent Informatics*,
Advances in Intelligent Systems and Computing 320, DOI: 10.1007/978-3-319-11218-3_3

1 Introduction

Brain tumor is a solid neoplasm of uncontrolled cell division. CAD is a process of using computation capacity effectively to diagnose a condition from medical images or other medical related data. CAD is mainly used to provide a second opinion thereby helping doctors while performing diagnosis. Image processing is a technique where the input is an image and the output being certain parameters related to image. Pattern recognition algorithms aim to assign a label to a given input feature vector. One of the most important applications of pattern recognition is within medical science where it forms the basis for CAD [16]. The process of interpreting a disease from a medical image is not hundred percent accurate. The reason is segmentation error involved in manual segmentation. Even for a highly skilled and experienced radiologist there will be inter-observer and intra-observer variation involved in diagnosis. These problems can be reduced by building a CAD system which performs the process of diagnosis in an automated way and is more accurate and reliable for diagnosis. It can also aid the physician in the diagnosis procedures by providing second opinion. This work aims at developing a CAD system that can classify the given brain MRI image as 'tumor affected' or 'tumor not affected'. The primary objectives of the CAD system are (1) to remove the noise and to enhance the contrast of the image, (2) to effectively and precisely separate the tumor region from the rest of the image, (3) to quantify the features from the tumor segment and (4) to train a classifier to classify the image as 'tumor affected' or 'tumor not affected'.

This paper is organized as follows. Section 2 deals with the literature survey where some of the related works in this area are discussed. Section 3 describes the overall design of the CAD system. Section 4 provides the description about the data set, experiments conducted, results and discussion. Section 5 concludes the work along with future enhancements.

2 Related Works

This section details about the previous works relating to Computer aided diagnosis of brain MRI images. Mahesh et al. have proposed the use of modified cluster center and membership updating function in the conventional FCM algorithm to segment the tumor region from brain MRI images [4]. Shasidhar et al. have proposed a method for modifying the conventional FCM algorithm by using quantization and aggregation as a result of which the dataset is reduced [8]. Fahran et al. have proposed a method for segmenting the given brain MRI image into three tissues gray matter, white matter and cerebrospinal fluid by using the ACM [2].Wang et al. have proposed the use of region based active contour model along with Local Gaussian Distribution (LGD) fitting energy for the contour to evolve [11]. Troy et al. performed gray level manipulation in which first they changed the gray level distribution within the picture and a method for extracting relatively noise free objects from a noisy back ground [15].

Ubeyli et al. proposed the automatic diagnostic system for Doppler ultrasound signals with diverse and composite features and classifiers like multi-layer perception neural network and combined neural network [12]. Qurat-ul-ain et al. extracted histogram features using Gray level co-occurrence matrix. In order to capture the spatial dependence of gray level values, which contribute to the perception of texture, a two-dimensional dependence texture analysis matrix is taken into consideration in this system [10]. Fritz has discussed the GLCM method which is a way of extracting second order statistical texture features [13]. Gebejes et al. in their work analyzed the texture of the samples taken from the KTH-TIPS2 data base through second order statistical measurements based on the GLCM proposed by Haralick [3]. Dong et al. in their work proposes the volumetric texture analysis computation using Gray Level Run Length Matrix (GLRL) features [14].

Kavitha et al. in their work have performed the classification of the given brain MRI image by using neural network to segment the brain tumor into benign or malignant tumor [7]. Sridhar et al. in their work proposed the use of neural network to classify the brain MRI image as tumor affected or not [5]. Gladis et al. in their work proposed a technique by which feature selection is done according to the grouping class variable which reduces the features and increases the classifier accuracy [6]. Ibrahim et al. proposed classification of brain tumor using Magnetic resonance imaging using neural network [1]. Abdullah et al. have used SVM classifier to classify brain MRI images as tumor affected or not [9].

3 Design of the CAD System

This section provides the overall design of the CAD system. The entire architecture is divided into training and testing phase as shown in Fig.1. The system architecture consists of four steps namely preprocessing, segmentation, feature extraction and classification. During the training phase the training brain MRI image is given as input to the system and is subjected to all the steps mentioned above. A classifier model is chosen and the extracted features along with the class labels are used to train the classifier. The classifier now becomes a trained classifier. During the testing phase a test brain MRI image is input to the system and the image passes through all the above mentioned steps. The trained classifier now accepts the features of the test image and with the knowledge it has already gained during the training phase will assign a class label to the test image as either 'tumor affected' or 'tumor not affected'.

3.1 Preprocessing

Preprocessing is the first step in the CAD system. It preprocesses the input image such that there is a change in the output image. This change can be a reduction of noise and/or an enhancement of contrast. In this work wiener filter is

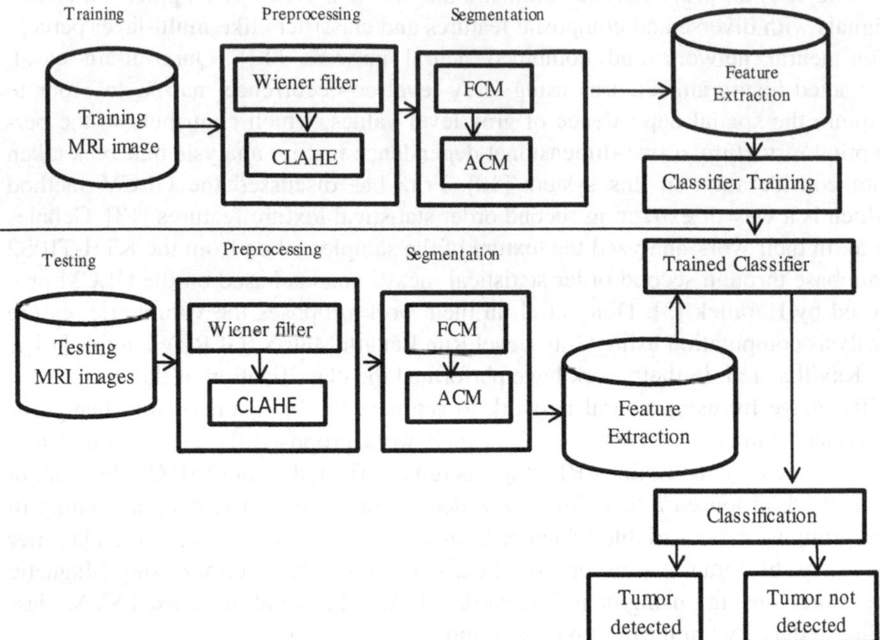

Fig. 1 System Architecture

used to reduce the effect of noise in the image. This is because wiener filter has the highest Peak Signal to Noise Ratio (PSNR) as shown in Table 1 when calculated for the entire data set. The output after applying wiener filter is given as the input to the CLAHE. Contrast limited adaptive histogram equalization is used to enhance the contrast of the given input image. CLAHE does not considers the image in whole but separates it as tiles and enhances the contrast of each tile thereby enhancing the contrast of the entire image.

Table 1 PSNR values of different filters for the entire data set

SNO	Filter Name	PSNR value
1	Median Filter	19.35
2	Wavelet Filter	18.42
3	Weiner Filter	24.44

3.2 *Segmentation*

Segmentation is a process of obtaining homogenous region from the input image. Generally in a brain MRI image the region which is affected by tumor occurs with high intensity. In this work two segmentation algorithms are used in order to locate the tumor segment exactly. This involves a FCM and an ACM algorithm.

3.2.1 Fuzzy C Means Algorithm

FCM is an unsupervised clustering algorithm which partitions the image into desired number of regions. It reduces the burden of providing ground truth data sets. The input to this algorithm consists of dataset and the required number of clusters. In this work the numbers of clusters are four which represents the gray matter, white matter, cerebrospinal fluid and the high intensity cluster. FCM works by assigning membership values to the pixels and assigns it to one of the cluster based on membership value. The membership values are given by the membership matrix which contains values between zero and one. The values zero and one indicate no membership and complete membership respectively and values in between zero and one indicates partial membership. The problem with FCM is that it takes large time to perform the clustering. This is because the FCM algorithm is an iterative algorithm and during every iteration each and every pixel of the input image is compared with every representative cluster center to calculate the distance of that pixel with the cluster center. The data point or pixel is then assigned a membership value based on its distance with the cluster center. Thus the number of comparisons in-order to cluster the input image is in turn proportional to the number of data points in the dataset. In order to reduce these comparisons and to decrease the running time of FCM two techniques quantization and aggregation are used in this work. This in turn helps to complete the clustering quickly.

In quantization every pixel in the input image is converted into its equivalent eight bit binary representation. Usually a selected number of lower order bits are masked by making it as zero [4]. This process of quantization converts the input image in such a way that many of the pixel values in the given image become almost similar and thus instead of comparing all the intensity values with the cluster center, only a representative intensity value is compared. i.e., if three intensity values are same then only one pixel is required for comparison. In this work the last three bits of the binary representation is replaced with zeroes.

Aggregation is performed by selecting a reference pixel. If the absolute difference between the left and top pixel to the reference pixel does not exceed a threshold value, then the reference pixel value is left as such else the mean of the two pixels (i.e., top and left) is computed and the reference pixel is replaced with this value. In this project the threshold value is set to ten. Thus the process of quantization and aggregation reduces the time for the convergence of the FCM algorithm [8].

The concept of aggregation is explained by considering a sample dataset as in Table 2. It can be observed from Table 2 that by masking three lower bits certain intensity values become same in their binary value. The intensity values 11, 13, and 10 after quantization becomes 8 and the intensity values 16 and 18 after quantization becomes 16. The quantization step is illustrated using Table 3 and Table 4. Consider the pixel at second column fourth row which has an intensity value 13. Let this pixel be the reference pixel. The absolute difference between the top pixel and left pixel of this reference pixel is greater than the threshold value (i.e., 10). So the reference pixel value is changed as '19' which is the mean of top pixel and left pixel. If absolute difference between the left and the top pixels is less than the threshold value then there is no change in the reference pixel value. These processes make many intensity values to become similar and hence the image data becomes reduced thereby reducing the number of comparisons during the FCM iterations. The process is explained in fig.2.

Table 2 Example to illustrate quantization of data

Pixel value	Binary value	Quantized value	Decimal value
11	00001011	00001000	8
16	00010000	00010000	16
18	00010010	00010000	16
13	00001101	00001000	8
10	00001010	00001000	8

Table 3 Pixel values before aggregation

12	15	15
16	12	14
16	12	14
25	13	15

Table 4 Pixel values after aggregation

12	15	15
16	12	14
16	12	14
25	19	15

3.2.2 Active Contour Model

Active contour model is a method used to find the boundary of the ROI in an image. Active contour model is of two types namely region based active contour model and edge based active contour model. In a region based active contour model there is a region stopping function which stops the contour evolution when it crosses the region. In an edge based active contour model there will be an edge indicator function which will stop the contour evolution when it crosses the edges. In this work a region based active contour model in a variation level set formulation is used. The contour is initialized by locating the first non-zero pixel in the high intensity cluster and then plotting a circle with a definite radius. This contour is evolved around the tumor for hundred iterations. The contour evolves and fits the tumor segment at the end of hundred iterations.

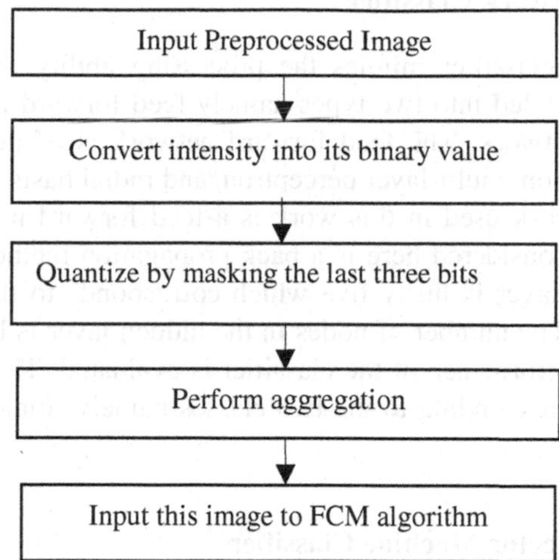

Fig. 2 Process of quantization and aggregation

The next step measures the feature values of the extracted Region of Interest (ROI). This feature extraction is used to train a classifier to predict the condition. Generally for an image color, texture, shape, edge, shadows, temporal details are the features that can be extracted. A total of thirty five features with the following composition are used in this work.

Shape based features: Eccentricity, major axis length and minor axis length.

GLRL features: Short run emphasis, long run emphasis, gray level non uniformity, run percentage, run length non uniformity, low gray level run emphasis and high gray level run emphasis.

GLCM features: Contrast, homogeneity, energy, correlation, inverse difference normalized, inverse difference is homomorphic, info measure of correlation-I, info measure of correlation-II, Difference entropy, Difference variance, sum entropy, sum variance, sum of squares, maximum probability, sum average, dissimilarity, cluster shade and cluster prominence.

Statistical features: Mean variance, standard deviation, median, skew, kurtosis and entropy.

3.3 Classification

Classification is the next step after feature extraction and it is a supervised learning procedure. It involves two steps training and testing. During the training phase, the classifier is trained with features from training images. In testing phase, an unknown image's features are given to the classifier and it has to classify the image as 'tumor affected' or 'tumor not affected'. In this project two classifier models are used namely Support vector machine and neural network.

3.3.1 Neural Network Classifier

A Neural network classifier mimics the processing ability of biological brain. They are mainly divided into two types namely feed forward network and recurrent or feedback network. The feed forward network is of three types' namely single layer perceptron, multi-layer perceptron, and radial basis function nets [17].

The Neural network used in this work is a feed forward neural network. The training algorithm considered here is a back propagation method. The number of nodes in the input layer is thirty five which corresponds to the total number of features extracted. The number of nodes in the hidden layer is kept as twenty five and fifty and the performance of the classifier is evaluated. The output layer contains two nodes corresponding to the two classes namely 'tumor affected' or 'tumor not affected'.

3.3.2 Support Vector Machine Classifier

Support vector machine is a learning method used for classification. In this work a two class SVM classifier is used which finds a hyper plane which separates the data into two classes. A two class SVM contains two classes namely 'tumor affected' and 'tumor not affected' as its two outputs. During the training phase the SVM classifier is trained with a training data set which contains feature vectors extracted from the training images and their respective class labels. During the testing phase if an unknown image's feature vector is given as an input to the trained classifier, it classifies the test image as belonging to one of classes [18]. Experiments are conducted with support vector machine classifier using quadratic kernel function and RBF kernel function.

4 Experimental Results

4.1 Data Set Used

The images for this work are taken from Cancer imaging archive [22]. There are 83 tumor affected and 9 non tumor grayscale images in DICOM format.

4.2 Performance Analysis of FCM on Original and Reduced Image Data

The performance of FCM on original and reduced image data is analyzed by considering the time in seconds for the algorithm to complete. Every image in the dataset is segmented into four clusters using FCM. The time taken to perform the segmentation on original and reduced image data is measured. The average running time is calculated by repeating the segmentation process on every image for five times. Table 5 shows the comparison of average running time for FCM for original and reduced image data for seven sample images. Table 6 shows the comparison of average time for FCM for original and reduced image data for the entire dataset.

4.3 Parameters Used for Classifier Evaluation

The following parameters are used for evaluating the performance of the classification algorithms. They are sensitivity, specificity, precision, and accuracy.

Table 5 Average running time of FCM for seven sample images

SNO	Image id	FCM original image data (time in sec)	FCM reduced image data (time in sec)
1	8.DCM	15.8	12.6
2	9.DCM	33.4	29.4
3	57.DCM	21.1	17.7
4	58.DCM	27.4	22.3
5	68.DCM	13.4	10.5
6	74.DCM	32.7	26.3
7	138.DCM	31.6	25.5

Table 6 Average running time of FCM for the entire dataset

Total no of images	FCM original image data (time in sec)	FCM reduced image data (time in sec)
92	22.96	22.14

Sensitivity or Recall: Sensitivity or recall measures the actual positives which are identified as [20] such and is given by equation 1.

$$\text{Recall} = TP/((TP+FN)) \tag{1}$$

Specificity: Specificity measures the actual negatives which are identified as such [20] and is given by equation 2.

$$\text{Specificity}= TN/((TN+FP)) \tag{2}$$

Precision: Precision measures the positive predictive rate [21] and is given by equation 3.

$$\text{Precision}= TP/((TP+FP)) \tag{3}$$

Accuracy: Accuracy of a classifier can be defined as how well a classifier can predict a condition [21] and is given by equation 4.

$$\text{Accuracy}= ((TP+TN))/((TP+TN+FP+FN)) \tag{4}$$

True Positive (TP): The classification result of CAD system is positive in the presence of the tumor.

True Negative (TN): The classification result of CAD system is negative in the absence of tumor.

False Positive (FP): The classification result of CAD system is positive in the absence of tumor.

False Negative (FN): The classification result of CAD system is negative in the presence of tumor [19].

4.4 Experiments Conducted

Four experiments are conducted using SVM and neural network classifiers. Two third of the total images are taken for training and one third of the images are taken for testing. In first experiment neural network classifier with twenty five hidden nodes is used. Second experiment involves neural network classifier with fifty hidden nodes. Third experiment involves SVM with quadratic kernel function and the fourth experiment involves SVM with RBF as the kernel function. Table 7 shows the accuracy, specificity, precision, and recall values of the classifiers. From the results obtained from Table 5 and Table 6 it is found that there is a reduction in time when the FCM algorithm is run on the reduced image data. The accuracy of the neural network classifier increases when the number of hidden nodes increases. The increase in the accuracy comes at the cost of training time. In case of SVM classifier quadratic kernel function has better accuracy when compared to RBF kernel function. When comparing between the accuracy obtained from neural network classifier with fifty nodes and SVM classifier with quadratic kernel function the accuracy of SVM classifier with quadratic kernel function is better.

Table 7 Performance analysis of neural network and SVM classifiers

Description	Accuracy	Specificity	Precision	Recall
Neural network with 25 hidden nodes	0.8058	0.76234	0.7229	0.8667
Neural network with 50 hidden nodes	0.8433	0.7614	0.6800	1
SVM with quadratic kernel function	0.8540	0.8471	0.819	0.88
SVM with RBF kernel function	0.8361	0.8038	0.7831	0.8757

5 Conclusion and Future Work

In this paper a CAD system that classifies the brain MRI image as 'tumor affected' or 'tumor not affected' is developed with an emphasis to reduce the segmentation time. A two step segmentation process is considered to get the exact tumor segment. This CAD system with the above mentioned preprocessing, segmentation and feature extraction steps can be coupled with a SVM classifier using quadratic kernel function to aid the radiologists in the diagnosis procedures.

As a future enhancement the number of bits masked in quantization step can be further varied depending upon the binary representation of the input image. The process of considering the pixels for the aggregation purpose can also be varied. The features can also be increased by considering other shape and statistical features. The classifier's accuracy can be enhanced by adding the adaptive boosting classifier.

References

[1] Ibrahim, W.H., Osman, A.A.A., Mohamed, Y.I.: MRI Brain Image Classification using neural networks. In: International Conference on Computing, Electrical and Electronics Engineering, pp. 253–258 (2013)

[2] Akram, F., Kim, J.H., Choi, K.N.: Active contour method with locally computed signed pressure force function: An application to brain MR image segmentation. In: Seventh International Conference on Image and Graphics, pp. 154–159 (2013)

[3] Gebejes, A., Huertas, R.: Texture characterization based on GLCM. In: Conference on Informatics and Management Sciences (2013)

[4] Badmera, M.S., Nilawar, A.P., Karawankar, A.R.: Modified FCM approach for Brain MR Image segmentation. In: International Conference on Circuits, Power and Computing Technologies, pp. 891–896 (2013)

[5] Sridhar, M.K.: Brain Tumor classification using Discrete Cosine transform and Probabilistic neural network. In: International Conference on Signal Processing Image Processing & Pattern Recognition (ICSIPR), pp. 92–96 (2013)

[6] Gladis Pushparathi, V.P., Palani, S.: Brain tumor MRI image classification with feature selection and extraction using Linear Discriminant analysis. Computer Vision and Pattern Recognition (2012)

[7] Kavitha, A.R., Chellamuthu, C., Rupa, K.: An efficient approach for brain tumor detection based on modified region growing and Neural Network in MRI images. In: International Conference on Computing, Electronics, Electrical Technologies, pp. 1087–1095 (2012)

[8] Shasidhar, M., Raja, V.S., Kumar, B.V.: Mri brain image segmentation using modified fuzzy c means clustering. In: International Conference on Communication and Network Technologies, pp. 473–478 (2011)

[9] Abdullah, N., Ngah, U.K., Aziz, S.A.: Image classification of brain MRI using support vector machine. In: International conference on Imaging Systems and Techniques, pp. 242–247 (2011)

[10] Qurat-Ul-ain, Latif, G., Kazmi, S.B., Jaffer, M.A., Mirza, A.M.: Classification and segmentation of Brain tumor using texture analysis. In: International Conference on Artificial Intelligence, Knowledge Engineering and Data bases, pp. 147–155 (2010)

[11] Wang, L., Hi, L., Mishra, A., Li, C.: Active contour model driven by Local Gaussian Distribution Fitting Energy. Signal Processing, 2435–2447 (2009)

[12] Ubeyeli, E.D., Goeler, I.: Feature extraction from Doppler ultrasound signals for automated diagnostic system. Computers in Biology and Medicine 7, 678–684 (2007)

[13] Albergtsen, F.: Statistical Feature measures computed from gray level co-occurrence matrices. International Journal on Computer Applications (2005)
[14] Huixu, D., Kurani, A.S., Furst, J.D., Raicu, D.S.: Run length encoding for volumetric texture. In: International Conference on Visualization, Imaging and Image Processing (2004)
[15] Troy, E.B., Deutch, E.S., Rosen Feld, A.: Gray level manipulation experiments for texture analysis. IEEE Transactions on System, Man, Cybernetics smc-3, 91–98 (1973)
[16] http://www.en.wikipedia.org/wiki/Pattern_recognition
[17] http://www.cse.unr.edu/~bebis/MathMethods/NNs/lecture.pdf
[18] http://www.en.wikipedia.org/Support_vector_machine
[19] http://cs.rpi.edu/~leen/miscpublications/SomeStatDefs
[20] http://en.wikipedia.org/wiki/Sensitivity_and_Specificity
[21] http://en.wikipedia.org/wiki/Accuracy_and_Precision
[22] http://www.cancerimagingarchive.net

An Investigation of fSVD and Ridgelet Transform for Illumination and Expression Invariant Face Recognition

Belavadi Bhaskar, K. Mahantesh, and G.P. Geetha

Abstract. This paper presents a wide-eyed yet effective framework for face recognition based on the combination of flustered SVD(fSVD) and Ridgelet transform. To this end we meliorate in the sense of computation efficiency, invariant to facial expression and illumination of [21]. Firstly fSVD is applied to an image by modelling SVD and selecting a proportion of modelled coefficients to educe illumination invariant image. Further, Ridgelet is employed to extract discriminative features exhibiting linear properties at different orientations by representing smoothness along the edges of flustered image and also to map line singularities into point singularities, which improves the low frequency information that is useful in face recognition. PCA is used to project higher dimension feature vector onto a low dimension feature space to increase numerical stability. Finally, for classification five different similarity measures are used to obtain an average correctness rate. We have demonstrated our proposed technique on widely used ORL dataset and achieved high recognition rate in comparision with several state of the art techniques.

Keywords: Face Recognition, fSVD, Ridgelet, Compounding argument, PCA, Similarity Measures.

1 Introduction

Face recognition is one of the most dynamic areas of research with a wide variety of real-world applications, and in recent years a clearly defined face-recognition furcates has egressed. Further, there are still open problems yet to be answered completely such as recognition under different pose, illumination, background, occlusion and/or expressions[6]. Our work mainly concentrates on feature extraction or

Belavadi Bhaskar · K. Mahantesh · G.P. Geetha
SJB Institute of Technology, Bengaluru-560060
e-mail: {bhaskar.brv,mahantesh.sjbit}@gmail.com,
geetha_nivas@yahoo.com

El-Sayed M. El-Alfy et al. (eds.), *Advances in Intelligent Informatics*,
Advances in Intelligent Systems and Computing 320, DOI: 10.1007/978-3-319-11218-3_4

description and recognizing face under illumination and expressions. Vytautas Perlibakas [13] developed a well-known subspace method based on Principal Component Analysis (PCA) with modified sum mean square error (SMSE) based distance. A rather more veritable 2DPCA algorithm was developed by Yang et.al[1], in which covariance matrix is directly constructed using original image matrices. 2DPCA needs more coefficients in representing an image and to overcome this problem, a bidirectional projection based 2DPCA (Generalized 2DPCA) was coined by Hui Kong et.al [2], which employed kernel based eigen faces to improve the recognition accuracy. In counterpart to 2DPCA, DiaPCA [15] retains the correlations between rows and those of columns of images by extracting optimal projective vectors from the diagonal face images. Adaptively weighted Sub-pattern PCA (Aw-SpPCA) [14] extracts the features from the sub patterns of the whole image, thercby enhancing the robustness to expression and illumination variations.

Statistical techniques are used widthways in face recognition because of its sinewy characteristic of extracting features. Kim et.al[8] proposed mixture of eigenfaces to subdue the variations in pose and illumination. Second order mixture of eigenfaces [7] combines the second order eigenface method and the mixtures to take care of retrieval rank and false identification rate. Gaussian Mixture Model (GMM) [16] in the Fourier domain makes the recognition system more robust to illumination changes and thereby evidencing the importance of transforms in face recognition. The use of wavelet transformation with sparse representation [12] of face images slays the surplus data thereby making the algorithm conformable to achieve higher recognition rate under occlusions. The Gabor wavelet transform [10] provides good approximation to the sensitivity profiles of neurons found in visual cortex of higher vertebrates. A Gabor based decorrelation transformation in the eigen space [11] makes the recognition system illumination and expression invariant. Dual Tree-Complex Wavelet Transform (DT-CWT) [9] provides shift invariance and better directionality with oscillations and aliasing eliminated, which are required for effective feature representation and expression/ illumination invariant. Complex wavelet moments with phase information [3] helps to achieve invariance attribute. Discrete Cosine Transform (DCT) [4] helps to extract the required coarse features and there by aiding in reduced feature size. Spectrum based feature extraction [5] uses a combination of Discrete Fourier Transform (DFT), Discrete Cosine Transform (DCT) and Discrete Wavelet Transform (DWT) with Binary Particle Swamp Optimization (BPSO).

In many face recognition tasks, a sparse representation of a face image is used in order to compact the image into a small number of samples. Wavelets are a good example of sparse geometrical image representation. But regardless of the success of the wavelets, they exhibit strong limitations in terms of efficiency when applied in more than one dimension. Candies and Donoho has developed ridgelets: a new method of representation to deal with line singularities of the image in 2D [20]. The basic difference between ridgelets and wavelets is that wavelets relate the scales of the point position and ridgelets relate the scales of the line positions as given by [18, 19].

Wth the facts in hand and due to the intrinsic benifits of using Ridgelets and fSVD, we process the obtained highly discriminative invariant features in compressed domain using PCA to improve the recognition rate. The anatomy of the paper is as follows: Section 2 explains proposed feature extraction and classification techniques. Section 3 confronts experimental results on ORL datasets and performance analysis considering several standard techniques. Conclusion and future work are delineated at the end.

2 Proposed Methodology

In this section, we key out the proposed face image depiction and classification technique. We take into account the primal issues of face recognition and reduce the effect of these in improving the accuracy of recognition. The proposed algorithm derives a new face image by using flustered SVD, which reduces the illumination problem. The features are extracted out of this image by applying ridgelet transform, which are invariant to lighting, facial expressions (open / closed eyes, smiling / not smiling) and facial details (glasses / no glasses)[22]. The features obtained after transformation are projected onto the low dimension subspace using PCA. Classification is done by making use of five different similarity measures.

2.1 flustered SVD (fSVD)

The SVs (Singular Values) of an image have very good stability, i.e. when a small noise is added to an image; its SVs do not vary rapidly. They represent algebraic image properties which are intrinsic and not visual [21]. Making use of these properties, in order to effectively recognize the faces, we derive an image from the original image by flustering the face matrix's singular values. The intensity face image I(x,y) of size (mxn) is initially normalized to reduce the effect of intensity variations as given below

$$I_{norm}(x,y) = \frac{I(x,y)}{max(max(I(x,y)))} \tag{1}$$

Applying Singular Value Decomposition on the normalized face image gives

$$[U,\Sigma,V] = SingularValuedecomp(I_{norm}(x,y)) \tag{2}$$

The image D is derived by modelling SVD defined as

$$D(x,y) = U * \Sigma^{\gamma} * V^{T} \tag{3}$$

Where U and V are (mxm) left and (nxn) right singular matrices respectively and Σ is (mxn) singular diagonal matrix. γ is a real value which takes on between 1 and 2. Finally the derived image is combined with the original image to get an Illumination invariant image data set as given below

$$I_{flustered}(x,y) = \frac{I(x,y) + \xi D(x,y)}{1+\xi} \quad (4)$$

Where ξ is a compounding argument and plays a very crucial role in deciding the recognition accuracy. For train and test set ξ is set to 0.5, so that it tries to minimize any variations in the illumination.

2.2 Ridgelet Transform

Wavelets are very good at representing point singularities. However, when it comes to line singularities wavelets fail and ridgelets are the solution [17]. The idea is to map a line singularity into a point singularity using the Radon transform. A wavelet transform can then be used to effectively handle the point singularity in the Radon domain. Given an integrable bivariate function I(x,y), its Radon transform (RDN) is expressed as:

$$RDN_f(\phi,t) = \int_{R^2} (I(x,y) * \delta(x cos\phi + y sin\phi - t)) dxdy \quad (5)$$

The radon transform operator plots the information from spatial domain to projection domain (ϕ, t), in which one of the directions of the radon transform coincides and maps onto a straight line in the spatial domain. On a contrary each point in the spatial domain turns out to be a sine wave representation in the projection domain.

The Continuous Ridgelet Transform (CRT) is merely the application of 1-D wavelet $\psi_{i,j}(t) = i^{-\frac{1}{2}\psi^{(t-j)/i}}$ the slice of the radon transform.

$$CRT(i,j,\phi) = \int_R \psi_{i,j}(t) RDN_I(\phi,t) dt \quad (6)$$

Where $\psi_{a,b,\theta}(x)$ in 2-D is defined from a wavelet type function $\psi(t)$ and given as

$$\psi_{i,j,\phi}(x,y) = i^{-\frac{1}{2}\psi(\frac{x cos\phi + y sin\phi - j}{i})} \quad (7)$$

Our aim of applying ridgelets to 2-D face image was simplified by Do and Vetterli [19] by exploring Finite Ridgelet Transform (FRIT). FRIT is developed based on Finite Radon Transform (FRAT) and can be defined as summation of image pixel intensity values over certain set of predefined line in Eucledian space.

The FRAT of real discrete function f on the definite grid Z^2P can be defined as

$$FRAT_I(k,l) = \frac{1}{\sqrt{p}} \Sigma_{(m,n)\varepsilon L_{k,l}} I(m,n) \quad (8)$$

Where 'p' is the prime number with modulus operation in finite field Z_p & $L_{k,l}$ denotes the set of points that make up a line on the lattice Z^2P and expressed as

$$L_{k,l} = \begin{cases} (m,n) : n = (km+l)(modp), m\varepsilon Z_p & if \ \ 0 \le k \le p \\ (l,n) : n\varepsilon Z_p & if \ \ k = p \end{cases} \quad (9)$$

A fine number of energy information is found to be accumulated in the low-pass band of the ridgelet decomposed image. It is also worth noting that ridgelet coefficients include the information about the smoothness in t and ϕ of the Radon transform. Radon transform exhibits a certain degree of smoothness in particular; we can instantly see that the ridgelet coefficients decay rapidly as ϕ and/or j moves away from the singularities (kindly refer equation 7).

2.3 *Classification*

Classification also plays a very important role in face recognition [23]. For classification, we use five distance measures such as: Weighted angle based Distance, Weighted Modified Manhattan Distance, Minkowski Distance, Canberra Distance, and Mahalanobis Distance to obtain a mean classification rate.

3 Experimental Results and Performance Analysis

This section presents the results grounded on widely used ORL face database consisting of 40 classes. Fig. 1 shows few samples of ORL face database. We venture into the recognition system by applying flustered SVD to the face image to derive an illumination invariant face image. Ridgelet transformation is applied to extract the salient features out of the derived face image and is projected onto the reduced dimensionality eigen subspace. We have used five different distance measures to validate the query face image.

To assess our proposed technique we espoused the standard experimental procedure [1], where five random images from every class is used for training and rest of the images are entailed for testing purpose. We performed experiments with different γ holding ξ value in both train and test phase. In this case ξ was set to 0.5. Fig. 2 shows the graph of variation in number of feature vectors required to achieve accuracy of 100%. Note that for $\gamma = 1.75$, the no of feature vectors required to achieve 100% accuracy is only 25.

In the next phase, γ was set to 1.75 and ξ of both test and train were varied. Fig. 3 shows the variation of number of feature vectors required to achieve accuracy of 100%. It is found that for ξ between 0.5 and 0.55, the no of feature vectors required to achieve 100% accuracy is again 25.

Fig. 4 shows the variation of recognition accuracy with ξ of test while setting γ=1.5 and ξ of train to 0.1. From the result it is very clear that with the deviation in the value of ξ (test) away from ξ (train), accuracy comes down and more feature vectors are required to achieve good accuracy.

From the preliminary experiments conducted we set the value of ξ to 0.5 and γ to 1.75 and achieved the accuracy of 100% with minimal feature vectors of 25. Table 1 shows the best performance for each method. The results of the experiments show that our proposed method outdoes the other traditional face recognition methods in terms of accuracy and dimension.

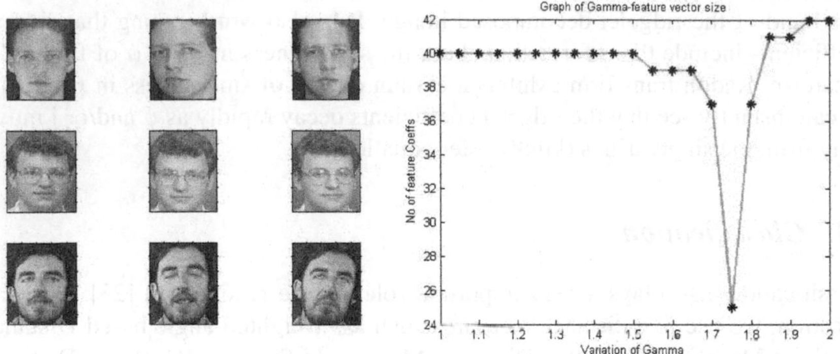

Fig. 1 Sample images of ORL database **Fig. 2** Graph of γ vs Feature Vector size

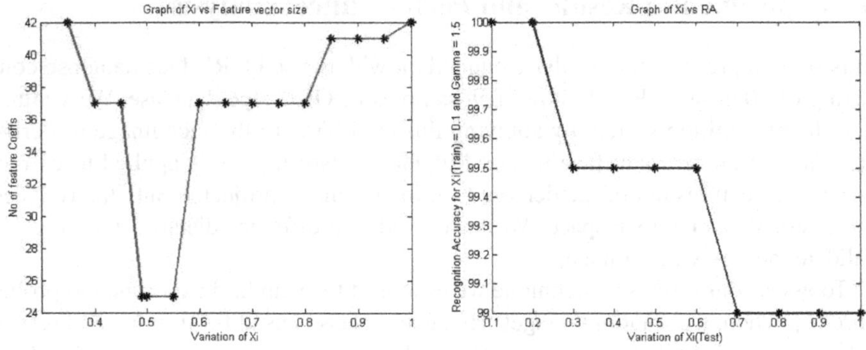

Fig. 3 Graph of ξ vs Feature Vector size **Fig. 4** Graph of ξ vs RA

Table 1 Best Recognition Accuracy on ORL database

Method	No of (train x test) images	Fea Vec Dimension	RA (in %)
(2D)2PCA[24]	5 x 5	27 x 26	90.5
SFA[25]	6 x 4	40	96.07
c Kmeans+NN[26]	6 x 4	-	96.97
PCA+LDA+SVM[27]	5 x 5	9	97.74
2FNN[28]	5 x 5	-	98.5
1D HMM+SVD[29]	5 x 5	-	99
Proposed System	5 x 5	25	100

4 Conclusion

The fSVD + Ridgelet based face recognition has been developed in this paper. The main advantage of this technique is that it derives a new face image by using fSVD, thereby retaining only the important features of the face image. The features are extracted by ridgelet transform, which not only reduces the dimensionality but also extracts salient features which are less affected by pose and expressions. The simulation experiment suggests that the proposed technique outperforms the conventional algorithms.

References

1. Yang, J., Zhang, D., Frangi, A.F., Yang, J.Y.: Two- Dimensional PCA: A New Approach to Appearance - Based Face Representation and Recognition. IEEE Transactions on Pattern Analysis and Machine Intelligence 26(1), 131–137 (2004)
2. Kong, H., Li, X., Wang, L., Teoh, E.K., Wang, J.-G., Venkateswarlu, R.: Generalized 2D Principal Component Analysis. In: Neural Networks, IJCNN 2005, vol. 18(5-6), pp. 585–594 (2005)
3. Singh, C., Sahan, A.M.: Face recognition using complex wavelet moments. Optics & Laser Technology 47, 256–267 (2013)
4. Manikantan, K., Govindarajan, V., Sasi Kiran, V.V.S., Ramachandran, S.: Face Recognition using Block-Based DCT Feature Extraction. Journal of Advanced Computer Science and Technology 1(4), 266–283 (2012)
5. Deepa, G.M., Keerthi, R., Meghana, N., Manikantan, K.: Face recognition using spectrum-based feature extraction. Applied Soft Computing 12, 2913–2923 (2012)
6. Shah, S., Faizanullah, Khan, S.A., Raiz, N.: Analytical Study of Face Recognition Techniques. IJSP, Image Processing and Pattern Recognition 6(4) (2013)
7. Kima, H.-C., Kim, D., Bang, S.Y., Lee, S.-Y.: Face recognition using the second-order mixture-of-eigenfaces method. Pattern Recognition 37, 337–349 (2004)
8. Kim, H.-C., Kim, D., Bang, S.Y.: Face recognition using the mixture-of-eigenfaces method. Pattern Recognition Letters 23, 1549–1558 (2002)
9. Santikaa, D.D., Ang, D., Adriantoc, D., Luwindad, F.A.: DT-CWT And Neural Network For Face Recognition. Procedia Engineering 50, 601–605 (2012)
10. Sharma, P., Arya, K.V., Yadav, R.N.: Yadav, Efficient face recognition using wavelet-based generalized neural network. Signal Processing 93, 1557–1565 (2013)
11. Deng, W., Hu, J., Guo, J.: Gabor-Eigen-Whiten-Cosine: A Robust Scheme for Face Recognition. In: Zhao, W., Gong, S., Tang, X. (eds.) AMFG 2005. LNCS, vol. 3723, pp. 336–349. Springer, Heidelberg (2005)
12. Zhu, Z., Ma, Y., Zhu, S., Yuan, Z.: Research on Face Recognition based onWavelet Transformation and Improved Sparse Representation. Advances in Systems Science and Applications 10(3), 422–427 (2010)
13. Perlibakas, V.: Distance measures for PCA-based face recognition. Pattern Recognition Letters 25, 711–724 (2004)
14. Tan, K., Chen, S.: Adaptively weighted sub-pattern PCA for face recognition. Neurocomputing 64, 505–511 (2005)
15. Zhang, D., Zhou, Z.-H., Chen, S.: Diagonal principal component analysis for face recognition. Pattern Recognition 39, 140–142 (2006)

16. Mitra, S.: Gaussian Mixture Models for Human Face Recognition under Illumination Variations. Applied Mathematics 3, 2071–2079 (2012)
17. Birgale, L., Kokare, M.: Iris Recognition Using Ridgelets. Journal of Information Processing Systems 8(3) (2012)
18. Terrades, O.R., Valveny, E.: Local Norm Features based on ridgelets Transform. In: ICDAR 2005 Proceedings of the Eighth International Conference on Document Analysis and Recognition, pp. 700–704. IEEE Computer Society (2005)
19. Do, N., Vetterli, M.: The Finite Ridgelet Transform for Image Representation. IEEE Transactions on Image Processing 12(1), 16–28 (2003)
20. Candies, E.J.: Ridgelets: Theory and applications. PhD thesis. Department of Statistics, Stanford University (1998)
21. Zhao, L., Hu, W., Cui, L.: Face Recognition Feature Comparison Based SVD and FFT. Journal of Signal and Information Processing 3, 259–262 (2012)
22. Kautkar, S., Atkinson, G., Smith, M.: Face recognition in 2D and 2.5D using ridgelets and photometric stereo. Pattern Recognition 45(9), 3317–3327 (2012) ISSN 0031-3203
23. Liu, C.: Discriminant analysis and similarity measure. Pattern Recognition 47, 359–367 (2014)
24. Zhang, D., Zhou, Z.H. (2D)2PCA: Two - directional two - dimensional PCA for efficient face representation and recognition. Neurocomputing 69, 224–231 (2005)
25. Wang, F., Wang, J., Zhang, C., Work, J.K.: Face recognition using spectral features. Pattern Recognition 40, 2786–2797 (2007)
26. Yen, C.Y., Cios, K.J.: Image recognition system based on novel measures of image similarity and cluster validity. Neurocomputing 72, 401–412 (2008)
27. Zhou, C., Wang, L., Zhang, Q., Wei, X.: Face recognition based on PCA image reconstruction and LDA. Optik 124, 5599–5603 (2013)
28. Jyostna Devi, B., Veeranjaneyulu, N., Kishore, K.V.K.: A Novel Face Recognition System based on Combining Eigenfaces with Fisher Faces using Wavelets. Procedia Computer Science 2, 44–51 (2010)
29. Miar-Naimi, H., Davari, P.: A New Fast and Efficient HMM - Based Face Recognition System Using a 7 - State HMM Along With SVD Coefficients. Iranian Journal of Electrical & Electronic Engineering 4(1-2) (2008)

Coslets: A Novel Approach to Explore Object Taxonomy in Compressed DCT Domain for Large Image Datasets

K. Mahantesh, V.N. Manjunath Aradhya, and S.K. Niranjan

Abstract. The main idea of this paper is to exploit our earlier work of image segmentation [11] and to propose a novel transform technique known as Coslets which is derived by applying 1D wavelet in DCT domain to categorize objects in large multiclass image datasets. Firstly, k-means clustering is applied to an image in complex hybrid color space and obtained multiple disjoint regions based on color homogeneity of pixels. Later, DCT brings out low frequency components expressing image's visual features and further wavelets decomposes these coefficients into multi-resolution sub bands giving an advantage of spectral analysis to develop robust and geometrically invariant structural object visual features. A set of observed data (i.e. transformed coefficients) is mapped onto a lower dimensional feature space with a transformation matrix using PCA. Finally, different distance measure techniques are used for classification to obtain an average correctness rate for object categorization. We demonstrated our methodology of the proposed work on two very challenging datasets and obtained leading classification rates in comparison with several benchmarking techniques explored in literature.

Keywords: Image Retrieval System, k-means, Complex hybrid color space, DCT, Wavelet, Coslets, PCA, Similarity Measures.

1 Introduction

The problem of facilitating machine to learn and recognize objects and its category in images has turned out to be one of the most fascinating and challenging task

K. Mahantesh
Department of ECE, Sri Jagadguru Balagangadhara Institute of Technology, Bangalore, India
e-mail: mahantesh.sjbit@gmail.com

V.N. Manjunath Aradhya · S.K. Niranjan
Department of MCA, Sri Jayachamarajendra College of Engineering, Mysore, India
e-mail: {aradhya.mysore, sriniranjan}@gmail.com

El-Sayed M. El-Alfy et al. (eds.), *Advances in Intelligent Informatics*,
Advances in Intelligent Systems and Computing 320, DOI: 10.1007/978-3-319-11218-3_5

in computer vision and machine learning applications. The recent approaches can perform well for categories ranging from 101 through to 256 (e.g. the Caltech - 101/256 dataset for object), much more to a great extent leftover to be done to reach 30,000+ categories of human visual recognition [1].

In this paper, an effort toward designing highly invariant feature descriptor to identify within the class variations by integrating transform techniques and subspace model is addressed. Being motivated by the fact that none of the feature descriptor has got the same discriminative power for all the classes, in [2] peter et al., used combined set of diverse visual features such as color, shape and texture information for between class discrimination. Since the Bag of Features (BoF) ignore local spatial information, SPM method evaluates histograms of local features after partioning image into progressively finer spatial sub-regions and computes histogram intersection in this space provides better learning and increase in object recognition tasks [3]. Fei-Fei et al. making use of prior information developed incremental Bayesian model in maximum likelihood framework for classifying images in large dataset collection with very few training samples [4]. In [5], Nearest Neighbor classifier is remodeled by picking K nearest neighbors and computed pair wise distances between those neighbors, further distance matrix is converted into a kernel matrix and then multi-class SVM is applied to recognize objects on large scale multiclass datasets outperforming nearest neighbor and support vector machines.

Retaining some degree of position and scale information after applying Gabor and max pooling operations has increased the performance of model on Caltech-101 dataset [6]. A plain but effective non-parametric Nearest Neighbor (NN) classifier proposed without using descriptor quantization and makes use of Image-to-Class distance showed significant improvement compared to conventional classifiers using descriptor quantization and Image-to-Image distances [7]. Discrete Cosine Transforms (DCT) and wavelet transforms are widely used in image and video compression and for still images it is revealed that wavelets outperforms DCT by 1DB in PSNR where as for video coding it is less noticed [8]. DCT & Wavelets with proper thresholding techniques shows significant improvement in compression ratio in terms of MSE and PSNR [10]. Lower frequency DCT coefficients are considered neglecting high frequency components from normalized image patches, and also selecting the number of DCT coefficients has shown the affect on retrieval performance [9]. Malik & Baharudin considered 1 DC with first 3 AC coefficients in DCT transformed matrix at different histogram bins and analyzed distance metrics for classification in Content Based Image Retrieval (CBIR) [14].

Being motivated by the facts that DCT efficiently preserves local affine information and robust to the variant factors such as illumination, occlusion & background clutter present in an image. Wavelet transform provides sparse representation for piecewise-linear signals. Our contribution is to integrate DCT and wavelets generating Coslet coefficients representing significant discriminative features in segmented image, and further processing these features in compressed domain using PCA to achieve better recognition rate. The rest of the paper is structured as follows: Section 2 explains proposed feature extraction and classification techniques.

Section 3 presents experimental results on Caltech-101/256 datasets and performance analysis considering several benchmarking techniques. Conclusion and future work are drawn at the end.

2 Proposed Method

In this section, we describe the proposed Coslet transformation technique for better image representation and classification. Addressing the crucial problem of learning and deriving discriminating image features to reduce the semantic gap between low level semantics and high level visual features, we carried out image analysis in frequency domain by integrating transform & subspace based techniques for better image representation and classifying images in large image datasets. The proposed model extracts features by applying Coslet transforms on segmented image in complex hybrid color space, the feature vectors obtained after transformation are projected onto reduced dimensional feature space using PCA. We make use of four different similarity distance measure techniques for classification to obtain an average classification rate.

2.1 Segmentation in Complex Hybrid Color Space

We put together our earlier work of partioning image into significant disjoint regions based on identifying color homogeneity of neighborhood pixels in complex hybrid colorspace and further k-means clustering is applied for effectively discriminating foreground pixels from background by retaining low frequency components [11]. The crucial steps involved in segmentation process are as given below:

Step 1: Input RGB image is transformed into YCbCr & HSI color spaces.
Step 2: Consider H component of HSI & CbCr components in YCbCr.
Step 3: Augmenting three higher dimensional matrices to generate hybrid color space - HCbCr.
Step 4: Further HCbCr is transformed into LUV color space.
Step 5: Finally k-means clustering is applied.

Above steps are sequentially processed in supervised context to obtain segmented image of highly coherent regions preserving low frequency components which will be further exploited for better image representation.

2.2 Feature Extraction and Classification

In this section we describe the techniques related to frequency domain (DCT) & combined time-frequency domain (Wavelets). DCT (Discrete Cosine Transforms) is used to convert 2-D signal into elementary frequency components of lower frequency and wavelets representing point singularities by isolating edge discontinuities and capturing maximum energy of the given image. Implementation details of combining DCT and wavelets to form Coslet are given in following sections.

2.2.1 Discrete Cosine Transform (DCT)

DCT is a very popular transform technique used in image compression [12] and face recognition [13]. For most of the images, much of the signal strength lies at low frequencies and DCT significantly separates out image spectral sub-bands with respect to image's visual features. It transforms a signal from time domain to frequency domain.

Let $f(x), x = 0, 1, 2, ..., N-1$, be a sequence of length N, Then 1D DCT is defined by the following equation:

$$F(u) = \left(\frac{2}{N}\right)^{\frac{1}{2}} \sum_{x=0}^{N-1} \Lambda(x) \cos\left[\frac{(2x+1)u\pi}{2N}\right] f(x) \tag{1}$$

Unlike 1D signal, the DCT of an m x n image $f(x,y)$ is given by:

$$C(u,v) = \frac{2}{\sqrt{mn}} \alpha(u)\alpha(v) \sum_{x=1}^{m} \sum_{y=1}^{n} f(x,y) \cos\left[\frac{(2x+1)u\pi}{2m}\right] \cos\left[\frac{(2y+1)v\pi}{2n}\right] \tag{2}$$

where $\forall u = 1, 2, ..., m$ and $\forall v = 1, 2, ..., n$ are scaling factors.

Fig.1(a) shows 256 X 256 pixels size color airplane image, Fig.1(b) shows the color map of log magnitude of its DCT signifying high energy compaction at the origin and Fig.1(c) illustrates the scanning strategy of DCT coefficients for an image using conventional zigzag technique.

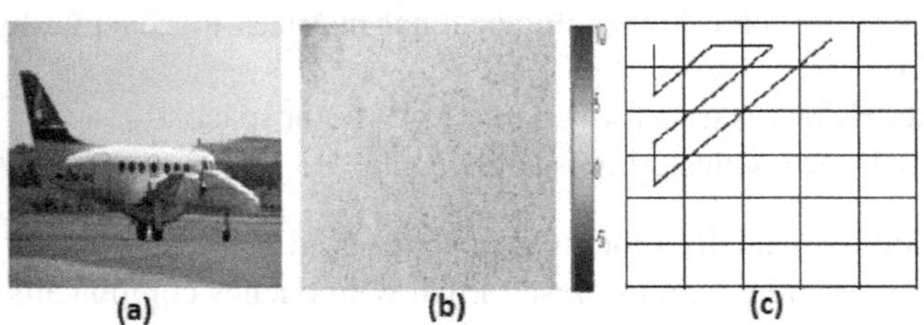

Fig. 1 (a) Original image; (b) Color map of quantized magnitude of DCT; (c) Scanning strategy of DCT coefficients

DCT coefficients preserve low frequency coefficients to identify the local information which is invariant to illumination, occlusion, clutter and almost nullifying the effect of high frequency coefficients.

2.2.2 Wavelets

Wavelets are increasingly becoming very popular due to its significant property of analyzing 2D signal in multi-resolution domain. DCT co-efficients obtained from previous section are analyzed using wavelets & proved very effective for identifying information content of images. Since choice of wavelet bases is highly subjective in nature as it depends on data in hand, we selected 'Haar' as the best wavelet basis for image representation [15]. The information difference between the approximation of signals at 2^{j+1} & 2^j can be obtained by decomposing the signal on orthogonal Haar basis. Let $\psi(x)$ be the orthogonal wavelet, an orthogonal basis of H_{2j} is computed by scaling the wavelet $\psi(x)$ with a coefficient 2^j and translating it on a lattice with an interval 2^{-j}. Haar wavelet is given as:

$$\psi(x) = \begin{cases} 1 & \text{if } 0 \le x < \frac{1}{2} \\ -1 & \text{if } \frac{1}{2} \le x < 1 \\ 0 & \text{otherwise} \end{cases}$$

It is noticed that selecting basis, scaling function and wavelet we can obtain good localization in both spatial and Fourier domains. Let $'x'$ be a given 1D signal, wavelet consists of $\log_2 N$ stages at the most. At the first step, we obtain two sets of coefficients: Approximation coefficients $CA1$ and Detailed coefficients $CD1$. These coefficients are obtained by convolving $'x'$ with low-pass filter for approximation, and with the high-pass filter for detail, followed by dyadic decimation.

The resultant decomposition structure contains the wavelet decomposition vector C and vector length L. Fig.2(a) shows the structure of level-2 decomposition. Fig.2(b) exhibits an example of original signal and its 1-level wavelet decomposition structure. To brief this section, 1-D wavelet is applied to DCT transformed feature space to obtain highly discriminative coslet features.

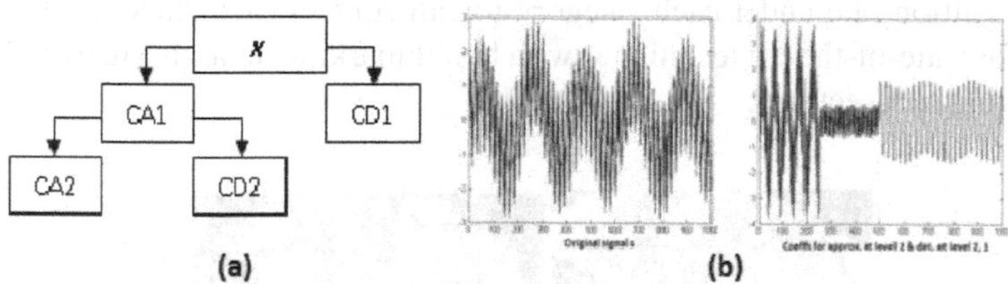

Fig. 2 (a) Level-2 Decomposition of wavelet; (b) Original signal & Wavelet Decomposition structure

2.2.3 Classification

Due to the advantage of considering low frequency coefficients as discriminating features for representing the actual information in an image by nearly eliminating the background information, we obtained huge amount of observed data to be processed during classification. In this regard, observed data is transformed onto the reduced orthogonal feature space using PCA [16]. For further classification, we use four distance measures such as: Manhattan distance, Euclidean distance, Modified squared Euclidean distance, Angle-based distance to improve the average classification rate.

To conclude this section, we introduced a novel transformation technique known as Coslet transform by applying 1D wavelet to DCT transformed feature space and increased the numerical stability of the model using PCA for better image analysis and classification.

3 Experimental Results and Performance Analysis

This section presents the results based on two widely used benchmarking image datasets: Caltech-101 [4] and Caltech-256 [17] consisting of 101 & 256 various object categories respectively. We initialize our system by applying k-means technique for the image in complex hybrid color space into a collection of fuzzy clusters based on the color homogeneity of the pixels. Fig.3 shows few samples of original image and its resultant segmented image. Lower frequency coefficients are collected from segmented image by applying 1-D wavelet in compressed DCT domain and vectorizing each image coefficients (with at least 2500 coefficients from intermediate subbands) from 2-D to 1-D using zigzag scan strategy. Since resultant feature vector is extensively high, PCA is applied to reduce the dimension. Further, for classification four different distance measure techniques are applied in reduced feature space to measure the similarity between train feature dataset and query feature vector.

We have followed the standard experimental procedures mentioned in [17, 18, 19, 20] by dividing the entire dataset into 15 and 30 images/category as training phase and remaining images for testing. We also experimented our proposed method in four stages with above mentioned two different procedural settings. We have considered four different similarity distance measures to obtain an average of per class recognition rate under each category for all runs in each stage. Performance analysis for state-of-the-art techniques with benchmarking datasets are explained in the following sub-sections.

Fig. 3 Samples of Original and Resultant segmented image of Caltech-101 dataset

3.1 Caltech - 101 Dataset

Caltech - 101 dataset comprises of 9,144 images of 101 different categories of natural scenes (animals, butterfly, chandelier, garfield, cars, flowers, human face, etc.). Categories ranging from 31 to 800 with most of the images centered, occluded, affected by corner artifacts, and with large intensity variations have made this dataset most challenging [4]. We demonstrated our method by labeling first 15 & 30 images per category as n_{labels} and generated train feature dataset and remaining images for testing generating query feature vector. Fig. 4 illustrates few samples of Caltech - 101 dataset with high and low classification rates for $n_{labels} = 30$. An experimental results revealed in Table 1 proves that the proposed model with fuzzy segmented image is superior compared with the most popular state-of-the-art techniques found in the literature considering similar dataset and experimental procedures.

With reference to Table.1, we notice that proposed method with k-means has obtained leading classification rates of 46% & 54.7% for 15 & 30 images per category respectively in comparison with conventional methods and classifiers mentioned in [18, 19, 22] and found aggressive with spatial pyramid feature technique [20] & sparse localized features [6].

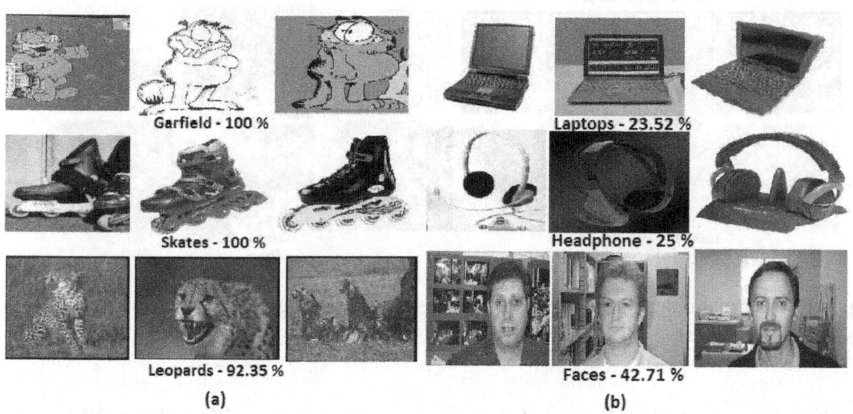

Fig. 4 Few sample images of Caltech - 101 dataset with (a) High classification rates (b) Low classification rates (at $n_{labels} = 30$)

Note: We have considered first 15 and 30 images respectively in all categories for labeling; whereas benchmarking techniques mentioned in [6, 20] has randomly selected them for training dataset.

3.2 Caltech - 256 Dataset

Griffin et al. [17] created challenging set of 256 object categories containing 30,607 images in total by downloading Google images and manually screening out with different object classes. Images possess high variations in intensity, clutter, object size,

Table 1 Performance analysis for Caltech - 101 dataset (Average of per class recognition for four runs in %)

Model	15 Training images/cat	30 Training images/cat
Serre et al. [18]	35	42
Holub et al. [19]	37	43
Berg et al. [5]	45	-
Mutch and Lowe [6]	51	56
Lazebnik et al.[20]	56.4	64.6
Coslets	37	43.1
Coslets + segmentation	46	54.7

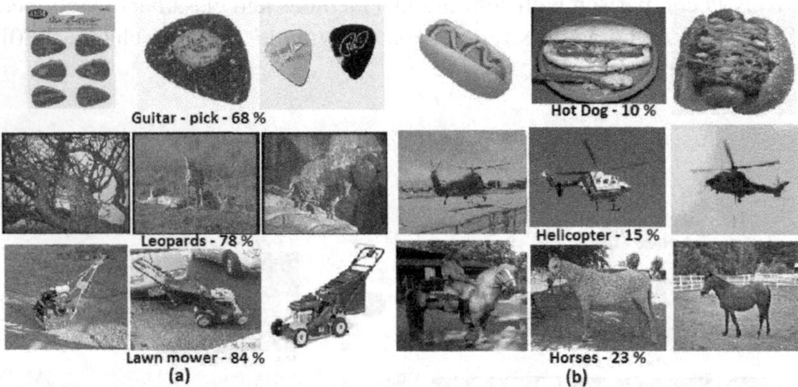

Fig. 5 Few sample images of Caltech - 256 dataset with (a) High classification rates (b) Low classification rates (at $n_{labels} = 30$)

location, pose, and also increased the number of category with at least 80 images per category.

In order to evaluate the performance of proposed method we followed the standard experimental procedure mentioned in [17], labeled first 15 & 30 images per category for generating train feature vector and remaining as test. Fig.5 exhibits few sample images of Caltech - 256 dataset with high and low classification rates and average recognition rates for four different runs are tabulated in Table.2. Proposed method with segmentation leads by 3.43% compared to [21] and achieved highly competitive results in contrast with the techniques mentioned in [17, 22, 23] which has used not more than 12,800 (viz. not exceeding 50 images per category) images for testing.

Table 2 Performance analysis for Caltech - 256 dataset (Average of per class recognition for four runs in %)

Model	15 Training images/cat	30 Training images/cat
Van et al. [21]	-	27.17
Griffin et al. [17]	28.3	34.1
Sancho et al. [22]	33.5	40.1
Jianchao et al. [23]	27.73	34.02
Coslets	16.9	21
Coslets + segmentation	24.3	30.6

Note: We have considered the entire set of images for testing purpose; whereas benchmarking techniques mentioned in [22, 23] uses not more than 50 images per category in testing phase.

4 Discussion and Conclusion

Several benchmarking techniques mentioned in literature failed to connect user's query specification to the image representation. We investigated a novel transform technique of preserving local affine & sparse information for better image representation and gained importance by producing excellent results on large multi-class dataset. Experimental results revealed that Coslet transform is first of its kind in the literature and gained competence by recognizing low frequency components in segmented image and recorded highest classification rate in reduced feature space.

We demonstrated our method on two widely used challenging datasets along with different similarity distance measure techniques in compressed domain and achieved high correctness rate compared to the hybrid model of [5] which mainly concentrated on color, shape & texture features. Our approach exhibits best performances over biologically inspired visual cortex feed forward computer vision model [18], generative semi supervised Fisher score model [19] and kernel visual codebook [21]. It is important to develop methods describing discriminative features taking an advantage of both local and global information in highly variant datasets.

References

1. Datta, R., Joshi, D., Li, J., Wang, J.Z.: Image retrieval: ideas, influences, and trends of the new age. ACM Computing Surveys 40(2), 1–60 (2008)
2. Gehler, P., Nowozin, S.: On Feature Combination for Multiclass Object Classification. In: Computer Vision, pp. 221–228 (2009)
3. Grauman, K., Darrell, T.: The Pyramid Match Kernel: Discriminative Classification with Sets of Image Features. In: Computer Vision - ICCV, vol. 2, pp. 1458–1465 (2005)

4. Fei-Fei, L., Fergus, R., Perona, P.: Learning generative visual models from few training examples: An incremental bayesian approach tested on 101 object categories. In: IEEE CVPR Workshop of Generative Model Based Vision (2004)
5. Zhang, H., Berg, A.C., Maire, M., Malik, J.: SVM-KNN: Discriminative Nearest Neighbor Classification for Visual Category Recognition. IEEE-CVPR 2, 2126–2136 (2006)
6. Mutch, J., Lowe, D.G.: Multiclass Object Recognition with Sparse, Localized Features. In: IEEE - CVPR, vol. 1, pp. 11–18 (2006)
7. Boiman, O., Shechtman, E., Irani, M.: In Defense of Nearest-Neighbor Based Image Classification. In: IEEE - CVPR, pp. 1–8 (2008)
8. Xiong, Z., Ramchandran, K., Orchard, M.T., Zhang, Y.-Q.: A Comparative Study of DCT- and Wavelet-Based Image Coding. IEEE Transactions on Circuits and Systems for Video Technology 9(5), 692–695 (1999)
9. Obdržálek, Š., Matas, J.: Image Retrieval Using Local Compact DCT-based Representation. In: Michaelis, B., Krell, G. (eds.) DAGM 2003. LNCS, vol. 2781, pp. 490–497. Springer, Heidelberg (2003)
10. Telagarapu, P., Jagan Naveen, V., Lakshmi Prasanthi, A., Vijaya Santhi, G.: Image Compression Using DCT and Wavelet Transformations. International Journal of Signal Processing, Image Processing and Pattern Recognition 4(3), 61–74 (2011)
11. Mahantesh, K., Aradhya, V.N.M., Niranjan, S.K.: An impact of complex hybrid color space in image segmentation. In: Thampi, S.M., Abraham, A., Pal, S.K., Rodriguez, J.M.C. (eds.) Recent Advances in Intelligent Informatics. AISC, vol. 235, pp. 73–82. Springer, Heidelberg (2014)
12. Telagarapu, P., Jagan Naveen, V., Lakshmi Prasanthi, A., Vijaya Santhi, G.: Image Compression Using DCT and Wavelet Transformations. International Journal of Signal Processing, Image Processing and Pattern Recognition 4(3), 61–74 (2011)
13. Zhu, J., Vai, M., Mak, P.: Face recognition using 2D DCT with PCA. In: The 4th Chinese Conference on Biometrics Recognition (Sinbiometrics 2003), pp. 150–155 (2003)
14. Malik, F., Baharudin, B.: Analysis of distance metrics in content-based image retrieval using statistical quantized histogram texture features in the DCT domain. Journal of King Saud University - Computer and Information Sciences 25(2), 207–218 (2012)
15. Chien, J.-T., Wu, C.-C.: Discriminant wavelet faces and nearest feature classifiers for face recognition. IEEE Transactions on Pattern Analysis and Machine Intelligence 24(12), 1644–1649 (2002)
16. Turk, M., Pentland, A.: Eigenfaces for recognition. Journal of Cognitive Neuroscience 3, 71–86 (1991)
17. Griffin, G., Holub, A., Perona, P.: Caltech 256 object category dataset. Technical Report UCB/CSD-04-1366, California Institute of Technology (2007)
18. Serre, T., Wolf, L., Poggio, T.: Object recognition with features inspired by visual cortex. In: CVPR, San Diego (2005)
19. Holub, A., Welling, M., Perona, P.: Exploiting unlabelled data for hybrid object classification. In: NIPS Workshop on Inter-Class Transfer, Whistler, BC (2005)
20. Lazebnik, S., Schmid, C., Ponce, J.: Beyond bags of features: Spatial pyramid matching for recognizing natural scene categories. In: CVPR, vol. 2, pp. 2169–2178 (2006)
21. van Gemert, J.C., Geusebroek, J.-M., Veenman, C.J., Smeulders, A.W.M.: Kernel codebooks for scene categorization. In: Forsyth, D., Torr, P., Zisserman, A. (eds.) ECCV 2008, Part III. LNCS, vol. 5304, pp. 696–709. Springer, Heidelberg (2008)
22. Cann, S.M., Lowe, D.G.: Local Naive Bayes Nearest Neighbor for Image Classification. In: IEEE CVPR, pp. 3650–3656 (2012)
23. Jianchao, K.Y., Gongz, Y., Huang, T.: Linear Spatial Pyramid Matching Using Sparse Coding for Image Classification. In: IEEE CVPR, pp. 1794–1801 (2009)

AI Based Automated Identification and Estimation of Noise in Digital Images

K.G. Karibasappa and K. Karibasappa

Abstract. Noise identification, estimation and denoising are the important and essential stages in image processing techniques. In this paper, we proposed an automated system for noise identification and estimation technique by adopting the Artificial intelligence techniques such as Probabilistic Neural Network (PNN) and Fuzzy logic concepts. PNN concepts are used for identifying and classifying the images, which are affected by the different type of noises by extracting the statistical features of noises, PNN performance is evaluated for classification accuracies. Fuzzy logic concepts such as Fuzzy C-Means clustering techniques have been employed for estimating the noise affected to the image are compared with the other existing estimation techniques.

Keywords: Fuzzy cluster, Noise Estimation, Noise identification, Neural Network Homogeneous blocks, variance, standard deviation.

1 Introduction

Image noise is unwanted information in an image and noise can occur at any moment of time such as during image capture, transmission, or processing and it may or may not depend on image content. In order to remove the noise from the noisy image, prior knowledge about the nature of noise must be known else noise removal causes the image blurring. Identifying nature of noise and estimation of noise is a challenging problem. The different categories of noise present are film grain noise, thermal noise, photoelectron noise, Gaussian noise and salt-pepper

K.G. Karibasappa
B.V.B College of Engineering and Technology, Hubli
e-mail: karibasappa_kg@bvb.edu

K. Karibasappa
Dayananda Sagar College of Engineering, Bangalore
e-mail: k_karibasappa@hotmail.com

El-Sayed M. El-Alfy et al. (eds.), *Advances in Intelligent Informatics*,
Advances in Intelligent Systems and Computing 320, DOI: 10.1007/978-3-319-11218-3_6

noise etc. Noise identification techniques are broadly classified into different categories based on statistical parameters [3], soft computing approach [4], graphical methods [5,6] and gradient function methods [8] etc. Similarly noise estimation techniques are classified into different categories based on block based method [11,12,14,15,16], filter based method[17,7], hybrid method[13,18] and texture based[18] noise estimation method. These techniques are based on certain assumptions with some advantages and limitations. Although these techniques are very useful for applications where manual image de-noising is acceptable, they fall short of their goals in many other applications that call for automated image identification and estimation techniques. In view of above, automated techniques for identification and estimation of image noise are of considerable interest, because once the type of noise and amount of effect on the image is identified, an appropriate algorithm can then be used to de-noise it. In this paper, we proposed a new methodology based on Artificial Intelligence (neural network and fuzzy logic) for identifying and estimating the noise.

This paper is organized as follows; Section 2 describes the general architecture of the Probabilistic neural network, which is the background for the proposed methodology discussed in Section 3. Implementation details and results have discussed in section 4. Paper concluded with the section-5.

2 Probabilistic Neural Network (PNN)[2]

Neural network is an Artificial intelligence tool widely used for pattern classification [20]. There are different kinds of neural network architectures such as multilayer perceptron, radial basis function, self organizing maps and probabilistic neural network which are proposed by various researchers. Probabilistic neural network model is most effective and efficient in pattern classification because of ease of training and strong statistical foundation in Bayesian model. Even though with these advantages it lacks in addressing other issues like determining the network structure such as size, number of pattern layer neurons and smoothing parameter.

Probabilistic neural network was proposed by Donald Specht. The general architecture of the PNN is shown in the Fig 1. The PNN is multi layered feed forward network with four different layers: the input layer, pattern layer, summation layer and decision layer called as output layer.

Pattern layer of the PNN computes the output based on the input distributed by the input layer using the equation 1

$$\phi_{i,j} = \frac{1}{(2\pi)^{\frac{d}{2}}\sigma^{d}} \exp\left[-\frac{\left(x - x_{i.j}\right)^{T}\left(x - x_{i.j}\right)}{2\sigma^{2}} \right] \quad (1)$$

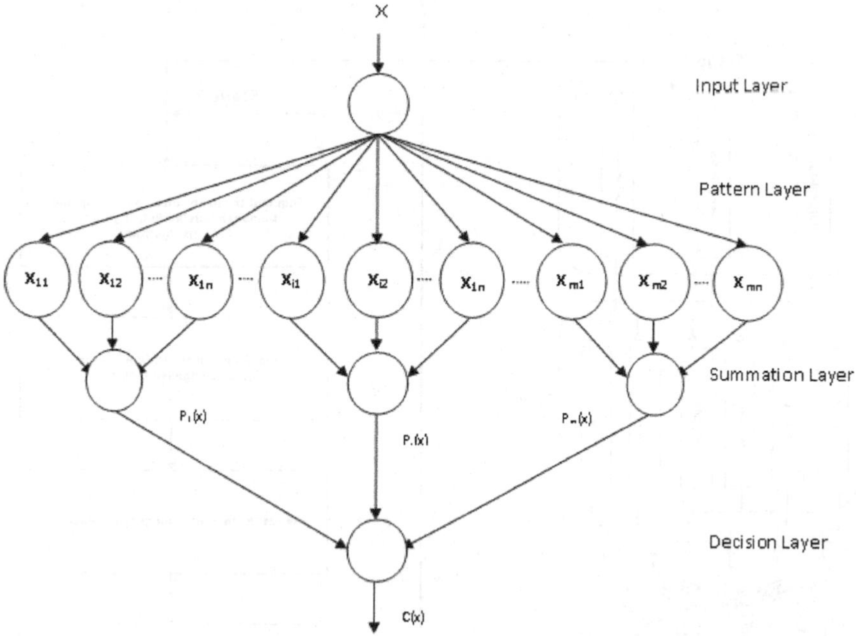

Fig. 1 Architecture of Probabilistic Neural Network

Where d is the pattern vector dimension, σ is the smoothing parameter, x i.j is the neuron. Summation layer computes the pattern xi is classified into class ci by summing and averaging the neurons of the same class as given in equation 2

$$p_{i,j}(x) = \frac{1}{(2\pi)^{\frac{d}{2}}\sigma^{d}} \frac{1}{N_i} \exp\left[-\frac{(x-x_{i.j})^T (x-x_{i,j})}{2\sigma^2} \right] \tag{2}$$

Where N_i is the number of samples in the class c_i. Based on output of neurons of summation layer, decision layer classifies the pattern x as per the Bayesian decision as follows.

$$C(x) = \arg\max \{ p_i(x) \} \tag{3}$$

3 Methodology

Principle architecture of the proposed methodology is shown in the Fig.2 and is classified into two main stages called identification [22] and estimation stage.

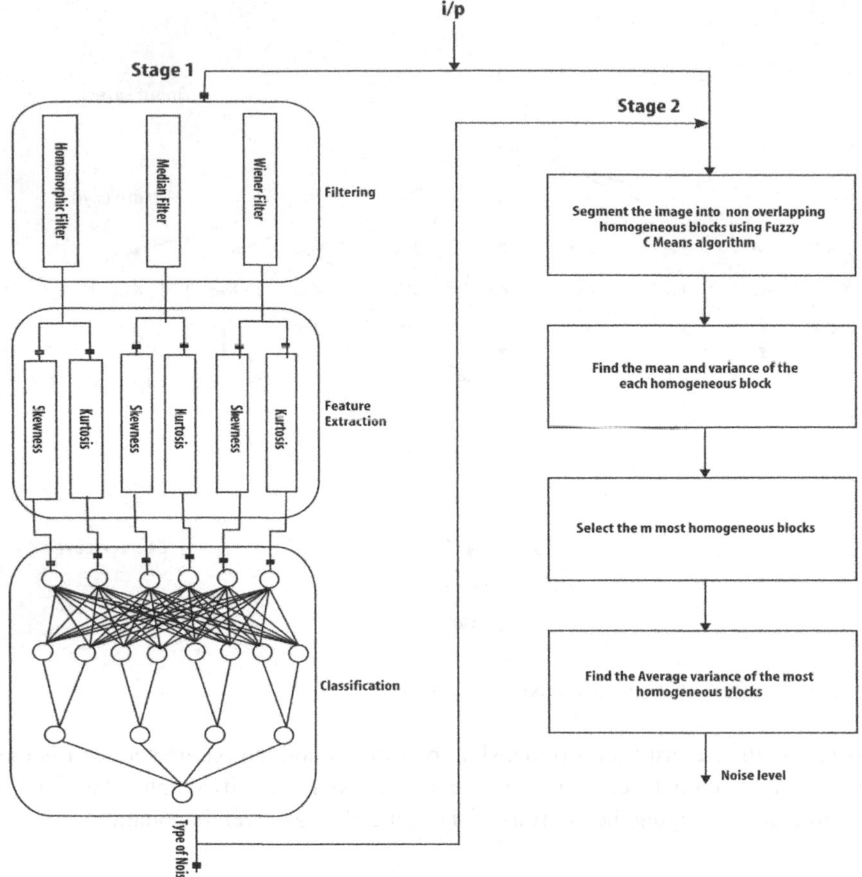

Fig. 2 Architecture of Proposed Methodology

3.1 Noise Identification Stage

Proposed noise identification method consists of the following steps

Step1. Extract some representative noise samples from the given noisy image, Three filters called wiener, median and homomorphic filters are used to the noisy image $f(i, j)$ get the three estimates $y\hat{}_{wiener}(i, j)$, $y\hat{}_{median}(i, j)$ and $y\hat{}_{homo}(i, j)$ as given below.

$$\hat{Y}_{Wiener}(i, j) = f(i, j) * H_{Wiener}(i, j) \tag{4}$$

$$\hat{Y}_{Median}(i, j) = f(i, j) * H_{Median}(i, j) \tag{5}$$

$$\hat{Y}_{Homo}(i, j) = \exp\left[\log\left(f(i, j) * H_{Homo}(i, j)\right)\right] \tag{6}$$

Where, * denotes the spatial filtering operations. Estimate the three noise estimates based on (4), (5) and (6) as given

$$\omega_{Wiener} = f(i,j) - \hat{Y}_{Wiener}(i,j) \tag{7}$$

$$\omega_{Median} = f(i,j) - \hat{Y}_{Median}(i,j) \tag{8}$$

$$\omega_{Homo} = f(i,j) - \hat{Y}_{Homo}(i,j) \tag{9}$$

Noise type is identified based on the output of the three estimators in equations (7) to (9)

Step2. Estimate some of their statistical features [1, 19]
Two features called Skewness and Kurtosis extracted for each of the ω_{Wiener}, ω_{Median} and ω_{Median} for Gaussian or uniform, salt and pepper and speckle noise respectively. These values are compared with reference obtained from by filtering the appropriate noise sequence and Skewness and Kurtosis of the filtered outputs.

Step3. Apply PNN for identifying the type of noise with 6 input nodes, 8 pattern layer nodes, 4 nodes in summation layer and a node in decision layer.

3.2 *Noise Level Estimation Stage*

The noise level estimatiion technique consistes of the stps

Step1. Identify the homogeneous blocks using fuzzy clustering method
Fuzzy C-means clustering technique [21] is the best technique for image homogeneous region classification. Algorithm classifies the input set of data D into *m* homogeneous regions denoted as fuzzy sets *S1, S2, ..., Sm.* to obtain the fuzzy *m*-partitions *S = {S1, S2, .., Sm}* for data set *D = {d1, d2 ..., dn}* and the clusters *m* by minimizing the function *J*

$$J(X,Y,D) = \sum_{i=1}^{m}\sum_{j=1}^{n} \mu_{ik}(d_j)^e |d_j - c_i|^2 \tag{10}$$

Where μ_{ik} is the membership value of the kth data d_k Y={C_1, C_2, ..C_m } are the fuzzy cluster centers.

$X = \mu_{ik}(d_k)$ is m x n matrix i.e., membership function of the kth input data in the ith cluster with the following conditions

$$0 \le \mu_{ik}(d_j) \le 1$$

$$\sum_{i=1}^{n} \mu_{ik}(d_j) = 1$$

$$0 < \sum_{k=1}^{n} x_k < n$$

l is called weighting factor. Its value is varies from 0 to ∞. However common choice is 2.

The Cluster centers are iteratively updated as follows

$$Y_i = \frac{\sum_{k=1}^{n} \mu_{ik}^{l} d_k}{\sum_{k=1}^{n} \mu_{ik}^{l}} \quad \text{where i= 1,2,...k} \tag{11}$$

The membership value of the kth sample data d_k to the ith cluster is given in the equation 3.

$$\mu_{ik} = \frac{\left[\frac{1}{|d_k - c_i|^2}\right]^{\frac{1}{l-1}}}{\sum_{j=1}^{m}\left[\frac{1}{|d_k - c_i|^2}\right]^{\frac{1}{l-1}}} \tag{12}$$

Step2. Identify the Mean and Variance of the each homogeneous blocks
Step3. Select the most homogeneous blocks
Step4. Find the Average variance of the selected homogeneous blocks

4 Implementation and Results

All experiments were carried out using MATLAB. MATLAB function "imnoise" is used to generate Gaussian white noise, speckle noise and salt-and-pepper noise pre determined variance to an input image. Input images are classified into images with more homogeneous regions and less homogeneous. Noise identifier had to identify the type of noise and the estimator estimates the amount of noise.

We used basic MATLAB PNN model for identifying the nature of noise with 6, 8, 4 and 1 nodes at input, pattern, summation and decision layers respectively. Gradient descent back propagation with adaptive learning rate is used. We have conducted simulations on many different image sequences. Fig (3a)-(3d) represent one of the noisy image sequences as an example with four different noise inputs, the calculated kurtosis, skewness and identified noise types are shown in Table 1.

Fig. 3 (a) Gaussian White Noise, (b) Speckle Noise, (c) Salt and Pepper Noise, (d) Non-Gaussian White Noise

Table 1 An Example of Kurtosis, Skewness and Identified Noise Types

. Noise input	Standard Kurtosis	Standard Skewness	Calculated Kurtosis	Calculated Skewness	Noise type
Gaussian	2.8741	0.0320	2.7058	0.0487	1
Speckle	2.6953	0.2066	2.8958	0.1322	2
Salt-and-pepper	30.240	1.2400	29.9806	1.8384	3
Non-Gaussian	2.4184	-0.0191	2.6187	-0.0187	4

Table 2(a) Identification Accuracy of Salt and pepper Noise in Images with more homogeneous and less homogeneous regions

% of Noise	Images Tested	**Salt and Pepper Noise**					
		More homogeneous			**Less homogeneous**		
		Correct	Wrong	Accuracy	Correct	Wrong	Accuracy
10	36	35	01	97.2	32	04	88.9
20	36	30	06	83.8	28	08	77.9
30	36	24	12	66.6	23	13	63.8
40	36	16	20	44.4	16	20	44.4
50	36	09	27	25	16	20	44.4

Table 2(b) Identification Accuracy of Speckle Noise in Images with more homogeneous and less homogeneous

% of Noise	Images Tested	Speckle Noise					
		More homogeneous			Less homogeneous		
		Correct	Wrong	Accuracy	Correct	Wrong	Accuracy
10	36	30	06	83.4	30	06	83.4
20	36	26	10	72.2	25	11	69.5
30	36	18	18	50	20	16	55.6
40	36	06	30	16.6	08	28	22.3
50	36	02	34	5.50	04	22	11.2

Table 2(c) Identification Accuracy of Gaussian Noise in Images with more homogeneous and less homogeneous regions

% of Noise	Images Tested	Gaussian Noise					
		More homogeneous			Less homogeneous		
		Correct	Wrong	Accuracy	Correct	Wrong	Accuracy
10	36	32	04	88.9	30	06	83.4
20	36	32	04	88.9	32	04	88.9
30	36	28	06	77.8	25	11	69.5
40	36	24	12	66.7	16	20	44.4
50	36	24	12	66.7	17	19	47.2

Table 2(d) Identification Accuracy of Non Gaussian Noise Identification more homogeneous and less homogeneous regions

% of Noise	Images Tested	Non-Gaussian Noise					
		More homogeneous			Less homogeneous		
		Correct	Wrong	Accuracy	Correct	Wrong	Accuracy
10	36	31	05	86.1	30	06	83.4
20	36	31	05	86.1	30	06	83.4
30	36	30	06	83.8	30	06	83.4
40	36	28	08	77.7	24	12	66.7
50	36	27	09	75	25	11	69.4

We also conducted testing, for the same training set to different image by adding varies percentage of noise for all the four types. The obtained results are shown in table 2(a) to (d) and few screen shots of the results are also shown in Fig. 4.

From Table2 (a)-(d), we can notice that the Identification accuracy for classifying different types of noise is efficient for less percentage of noise. But, since the values of kurtosis and skewness will vary by adding more noise, the performance at higher rate of noise decreases.

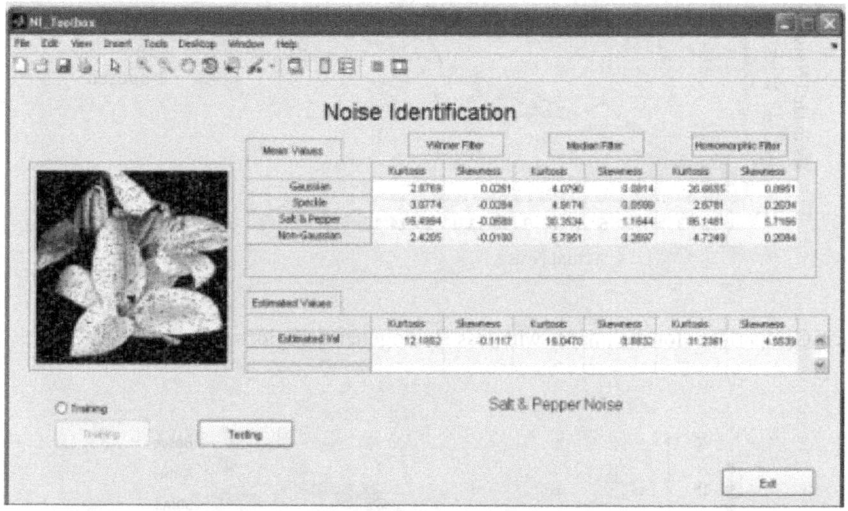

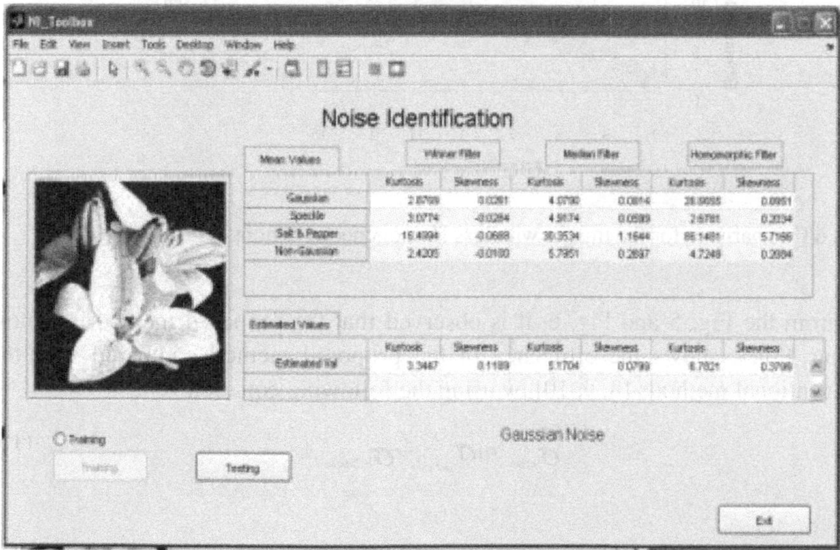

Fig. 4 Snapshots of Results for Noise identification

Noise Level Estimation

Second stage of the proposed method estimates the noise level of the input image affected with Gaussian noise. The performance is mainly depends on the detection of homogeneous regions. Results are also compared with the conventional methods [8, 9, 10] and shows good performance for the images with less and more homogeneous regions as shown in Fig.5 and Fig. 6 respectively.

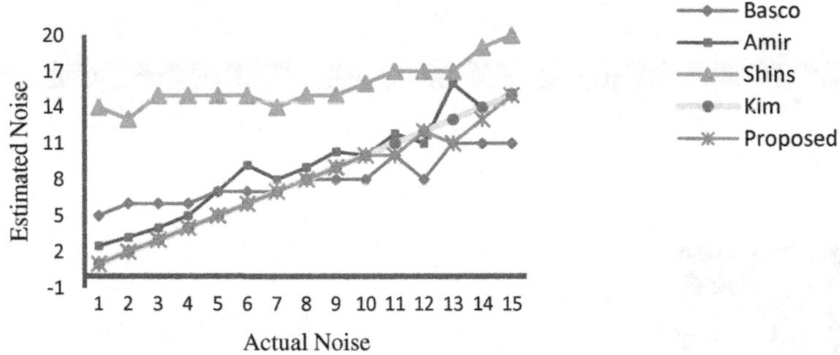

Fig. 5 Comparison for the images with less Homogeneous regions

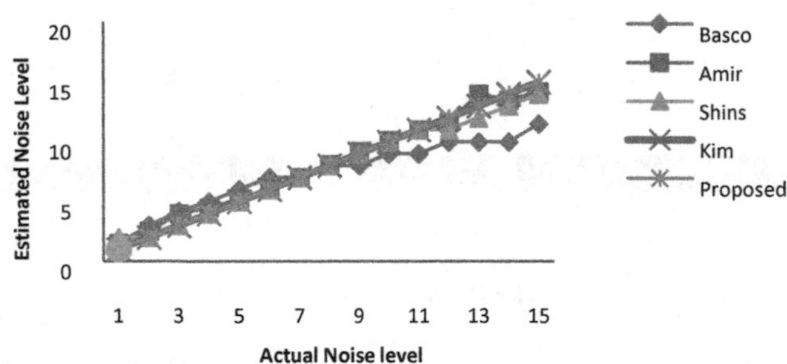

Fig. 6 Comparison for the images with less Homogeneous regions

From the Fig. 5 and Fig. 6, it is observed that the proposed method is almost inline. Fig.7. shows the efficiency of the proposed method compared with the conventional methods [8, 9, 10] by using the following equation

$$\sigma_{error} = |\sigma_{actual} - \sigma_{estimated}| \tag{13}$$

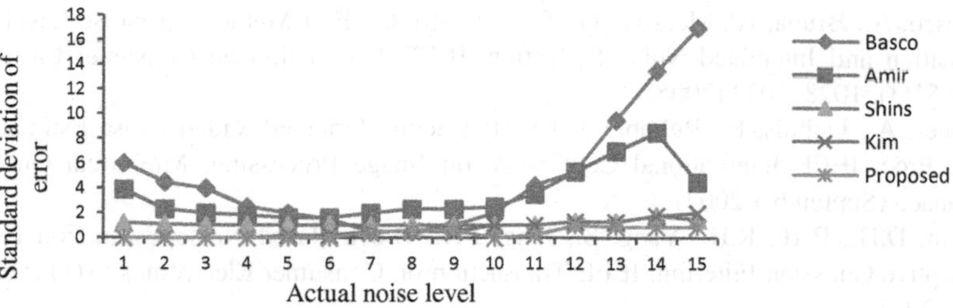

Fig. 7 Performance Comparison of homogeneous and Less homogeneous images

5 Conclusion

From the results it is observed that, performance of the proposed methodology goes on decreases as the noise levels increases i.e., performance is inversely proportional to noise levels for all four types of noises affected to the images which are having more and less homogeneous regions and estimate the noise level of the images which are affected with Gaussian noise only.

References

1. Chen, Y., Das, M.: An Automated Technique for Image Noise Identification Using a Simple Pattern Classification Approach, pp. 819–822. IEEE (2007)
2. Patra, P.K., Nayak, M., Nayak, S.K., Gobbak, N.K.: Probabilistic Neural Network for Pattern Classification. IEEE (2002)
3. Chen, T., Ma, K.K., Chen, L.H.: Tri-state median filter for image denoising. IEEE Transactions on Image Processing 8(12), 1834–1838 (1999)
4. Vozel, B., Chehdi, K., Klaine, L., Lukin, V.V., Abramov, S.K.: Noise Identification and Estimation of its Statistical Parameters by Using Unsupervised Variational Classification. In: IEEE International Conference on Acoustics, Speech and Signal Processing, Proc. 2006, vol. 2, pp. 841–844 (2006)
5. Beaurepaire, L., Chehdi, K., Vozel, B.: Identification of the nature of noise and estimation of its statistical parameters by analysis of local histograms. In: Proceedings of ICASSP 1997, Munich, April 21-24 (1997)
6. Karibasappa, K.G., Karibasappa, K.: Identification and Removal of Impulsive noise using Hypergraph Model. International Journal on Computer Science and Engineering 02(08), 2666–2669 (2010)
7. Liu, X., Chen, Z.: Research on Noise Detection Based on Improved Gradient Function. In: International Symposium on Computer Science and Computational Technology (2008)
8. Olsen, S.I.: Estimation of noise in images An evaluation. Computer Vision Graphics Image Processing, Graphics Models 55(4), 319–323 (1993)

9. Bosco, A., Bruna, A., Messina, G., Spampinato, G.: Fast Method for Noise Level Estimation and Integrated Noise Reduction. IEEE Transaction on Consumer Electronics 51(3), 1028–1033 (2005)
10. Amer, A., Dubois, E.: Reliable and Fast structure-Oriented Video Noise Estimation. In: Proc. IEEE International Conference on Image Processing, Mountreal Quebec, Canada (September 2002)
11. Shin, D.H., Park, R.H., Yang, S., Jung, J.H.: Block Based Noise Estimation Using adaptive Gaussian Filtering. IEEE Transaction on Consumer Electronics 51(1) (February 2005)
12. Sijbers, J., den Dekker, A.J., Van Audekerke, J., Verhoye, M., Van Dyck, D.: Estimation of the noise in Magnitude MR images. Magnetic Resonance Imaging 16(1), 87–90 (1998)
13. Pyatykh, S., Hesser, J., Zheng, L.: Image Noise Level Estimation by Principal Component Analysis. IEEE Transactions on Image Processing 22(2), 687–699 (2012)
14. Muresan, D., Parks, T.: Adaptive principal components and image denoising. In: Proceedings of the International Conference on Image Processing, ICIP 2003, vol. 1, pp. I–101. IEEE (2003)
15. Foi, A.: Noise estimation and removal in mr imaging: The Variance Stabilization Approach. In: 2011 IEEE International Symposium on Biomedical Imaging: From Nano to Macro. IEEE (2011)
16. Konstantiniders, K., Natarajan, B., Yovanof, G.S.: Noise Estimation and Filtering Using Block-Based Singular Value Decomposition. IEEE Transactions on Image Processing 6(3), 479–483 (1997)
17. Nguyen, T.-A., Hong, M.-C.: Filtering-Based Noise Estimation for Denoising the Image Degraded by Gaussian Noise. In: Ho, Y.-S. (ed.) PSIVT 2011, Part II. LNCS, vol. 7088, pp. 157–167. Springer, Heidelberg (2011)
18. Uss, M., Vozel, B., Lukin, V., Abramov, S., Baryshev, I., Chehdi, K.: Image Informative Maps for Estimating Noise Standard Deviation and Texture Parameters. Journal on Advances in Signal Processing 2011, Article ID 806516, 12 pages (2011)
19. Alberola-López, C.: Automatic noise estimation in images using local Statistics. Additive and multiplicative cases. Image and Vision Computing 27, 756–770 (2009)
20. Bezdek, J.: Pattern recognition with Fuzzy Objective Function Algorithms. Plenum Press, New York (1981)
21. Pal, N.R., Bezdek, J.C.: On cluster Validity for the fuzzy C-means Model. IEEE Transaction Fuzzy Systems 3(3), 370–379 (1995)
22. Karibasappa, K.G., Hiremath, S., Karibasappa, K.: Neural Network Based Noise Identification in Digital Images. International Journal of Network Security (2152-5064) 2(3), 28 (2011)

SV-M/D: Support Vector Machine-Singular Value Decomposition Based Face Recognition

Mukundhan Srinivasan

Abstract. This paper presents a novel method for Face Recognition (FR) by applying Support Vector Machine (SVM) in addressing this Computer Vision (CV) problem. The SVM is a capable learning classifier capable of training polynomials, neural networks and RBFs. Singular Value Decomposition (SVD) are used for feature extraction and while SVM for classification. The proposed algorithm is tested on four databases, viz., FERET, FRGC Ver. 2.0, CMU-PIE and Indian Face Database. The singular values are filtered, sampled and classified using Gaussian RBF kernal for SVM. The results are compared with other known methods to establish the advantage of SV-M/D. The recall rate for the proposed system is about 90%. The colossal augmentation is due to the simple but efficient feature extraction and the ability to learn in a high dimensional space.

1 Introduction

Humans have the perpetual ability to recognize fellow beings through their faces. In Computer Vision (CV), the challenge is to make a machine (non-human) recognize a face and identify the person. Research in this field dates back to the early 70's. Through the course of time, Face Recognition has gained predominance by addressing this issue through many perspectives like Pattern Analysis, Machine Learning and Statistics. Recognition of face from static and dynamic sequences is an active area of research with numerous applications. A survey of such applications and methods are given in [5].

As shown in fig.1; two important processes that are key for a good recognition system, viz., feature extraction and classification. On the other hand, a good classifier should be able to differentiate between various face image with high accu-

Mukundhan Srinivasan
Indian Institute of Science (IISc), C.V. Raman Road, Bangalore 560 012 KA, India
e-mail: mukundhan@ieee.org

El-Sayed M. El-Alfy et al. (eds.), *Advances in Intelligent Informatics*,
Advances in Intelligent Systems and Computing 320, DOI: 10.1007/978-3-319-11218-3_7

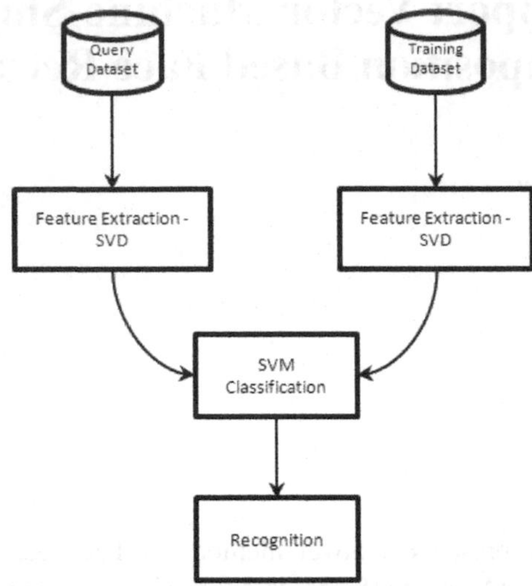

Fig. 1 An outline of the proposed system

racy and computational simplicity. In this paper we present a method using Support Vector machines (SVM) [1, 2, 3, 4] and Singular Value Decomposition (SVD) to efficiently recognize human face images. SVM is an avant garde method to train different kernels like polynomials, neural networks and Radial Basis Function (RBF) classifiers. There are several FR systems and methods developed since the inception pattern analysis. Many systems relied on the geometry like Geometrical Feature Matching. Kanade [6] in the early 70's showed that this method used the concept of matching the geometrical features based on the extraction and achieved close to 70% accuracy. Another developed method known as Eigen faces method [7, 8] used Principal Component Analysis (PCA) to project faces. These methods reported approximately 90% recognition rate on the ORL database. In summary, Eigen faces method is a fast, simple and practical method. With this came the learning perspective to FR systems. The Neural Networks (NN) have been used multiple times in tackling pattern recognition problems. This can be attributed to their non-linear mapping. The methods described in [9, 10] use Probabilistic Decision - Based Neural Network (PDBNN) for recognizing faces with a recall rate of 96%.

The recent methods are the Support Vector Machines (SVM) and Hidden Markov Model (HMM) applied to FR. The SVM is new method for training polynomial as described in [11, 12]. References [13, 14, 15, 16, 17] showed that a probabilistic Markov method would yield much better result and perform the computation in lesser amount of time. The current and contemporary researches are in Bayesian models. A comparison of SVM and HMM is presented in [18]. In [19], a new technique using a probabilistic measure of similarity, based primarily on a Bayesian (MAP) analysis was presented which resulted in a significant increase in computational speed for implementation of very large data sets.

2 Methodology

In this section we describe, in detail the modelling of the SVM and the feature vector extraction procedure.

2.1 Support Vector Machine (SVM)

A framework has been established through statistical learning models by applying methods of machine learning in small sample interval [19, 20]. With the foundation on these lines, the Support Vector Machine (SVM) classifier is developed which is a fast and efficient learning technique to address the classification of faces in a FR system. The underlying principle of SVM is the implementation of structural risk minimization whose goal is to minimize the upper bound on the generalization error.

Definition: A two-class linearly seperate classifier with N observations is defined as

$$\{(x_i, y_i)|x_i \in R_d; y_i = (-1,1); i = 0,1,2...N\} \tag{1}$$

This maximizes the error margin which is equal to the distance from the hyperplane $w : x \cdot w + b = 0$ from the immediate positive observation plus the immediate negative observation distance. A hyperplane with maximized margin yields better generalization. To achieve this, $||w||^2$ is minimized with the following constraints

$$y_i(x \cdot w + b) - 1 \geq 0, i = 1, \cdots, l \tag{2}$$

i.e. every observation is classified based on a specific distance metric from the seperating hyperplane with least distance equal to 1.

We now reformulate this problem in terms of Lagrange multiplier α_i.

$$\mathscr{L}_P - \frac{1}{2}||w||^2 - \sum_{i=1}^{l} \alpha_i (x \cdot w + b) + \sum_{i=1}^{l} \alpha_i \tag{3}$$

This Lagrange has to be minimized subjected to the above constraint in (2). The minimization of primal problem $\mathscr{L}_p$ can also be expressed as the maximization of the dual problem $\mathscr{L}_d$[1].

$$\mathscr{L}_d = \sum_{i=1}^{l} \alpha_i - \frac{1}{2} \sum_{i=1}^{l} \sum_{j=1}^{l} \alpha_i \alpha_j y_i y_j (x_i \cdot x_j) \tag{4}$$

with constraints $\sum \alpha_i y_i = 0$ and $\alpha_i > 0$. $\forall\ \alpha_i$ corresponds to a data point that are included into the solution. These data points shall be the 'support vector' lying on the margin.

When the data points are not linearly seperable, the mapping of input vector space $x \in \mathbb{R}^l$ to higher dimensions $\Phi(x) \in \mathbf{R}^h$ is required. Instead of an explicit feature

[1] Refer Appendix I.

space, a kernal function is defined as $k(a,b)$ such that it is a positive symmentric function of the feature space.

$$\Phi^T(x^1)\cdot\Phi(x^2) = \sum \beta_i \phi(x_i^1)\phi(x_i^2) = K(x^1 \cdot x^2) \tag{5}$$

where the equality is derived from the Mercer-Hilbert-Schmidt theorem[2] for $K(x,y) > 0$ [25]. Replacing $(x_i \cdot x_j)$ with the kernal function in (4), we have

$$\mathscr{L}_d = \sum_{i=1}^{l} \alpha_i - \frac{1}{2}\sum_{i=1}^{l}\sum_{j=1}^{l} \alpha_i \alpha_j y_i y_j K(x_i, x_j) \tag{6}$$

This quadratic problem has to be solved and as a result, the SVM is defined as:

$$f(x) = \left\{ sign\left(\sum_{i=1}^{N} \alpha_i y_i K(s_i, x) + b \right) \right\} \tag{7}$$

In (7), s_i is the number of support vector data point with non-zero Lagrange multiplier.

The contribution of this work is that, the Singular Value Decomposition (SVD) is used for feature extraction and is an input to the SVM. This compliments in finding the best classifying hyper-plane for various inputs. Using SVD with some initial processing gives a better margin between the classifying classes. This linear classifier with an acceptable margin and a classifying hyper-plane is shown in the below fig. 2.

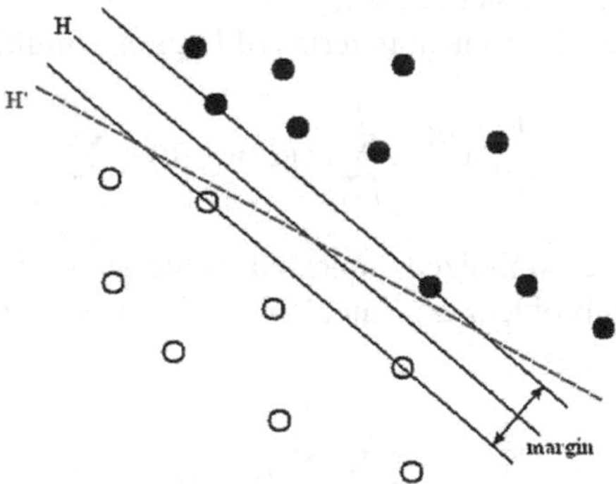

Fig. 2 Classifying plane in a linear SVM

[2] Refer Appendix II.

2.2 *Singular Value Decomposition (SVD)*

In signal processing and statistical data analysis, SVD is considered to be an important tool. The SVD matrix will contain energy levels, noise, rank of the matrix and other useful details. Since the singular value matrices are orthonormal, few characteristic patterns embedded in the signal can be extracted. The singular values of the image vectors are used here because of the following point of views of SVD:

- Transforming correlated data points into uncorrelated ones which better bring out the relationships between the orginal variables.
- The variations of the data points can be identified and ordered.
- Can be used to find the best approximation of the data point without the curse of dimensionality.

Definition: Consider a matrix A to be $m \times n$. The singular values of this matrix are the square root of the eigen values of the $n \times n$ matrix $A^T A$. Let the singular values be denoted by $\sigma_1, \sigma_2, ..., \sigma_n$ which implies $\sigma_1 \geq \sigma_2 \geq ... \geq \sigma_n$.

Theorem 1: A $m \times n$ matrix can be expressed as below

$$A = U \Sigma V^T \tag{8}$$

where $U(m \times n)$ and $V(m \times n)$ are orthogonal matrices and $\Sigma(m \times n)$ is a diagonal matrix of singular values with components $\sigma_{ij} = 0, i \neq j$ and $\sigma_{ij} > 0$.

$$\Sigma_{(m,n)} = \begin{pmatrix} \sigma_{(1,1)} & 0 & \cdots & 0 \\ 0 & \sigma_{(2,2)} & \cdots & 0 \\ \vdots & \vdots & \ddots & \vdots \\ 0 & 0 & \cdots & \sigma_{(m,n)} \end{pmatrix} \tag{9}$$

The stability of SVD, that is, being invarient to small perturbations in face images plays a vital role in selecting this method for extraction. This is shown in the below theorem.

Theorem 2: Consider $A_{(m \times n)}$ and $X_{(m \times n)} \in R_{m \times n}$ with singular values $\sigma_1, \sigma_2, ..., \sigma_n$ and $\tau_1, \tau_2, ..., \tau_n$ respectively. Then,

$$|\sigma_i - \tau_i| \leq \|A - X\|_2 \tag{10}$$

As described above these properties are used to extract features for images with noise, pose, illumination and other variations.

2.2.1 Extraction of Features

The singular value can be used to describe feature vectors as the following: If

$$A = \sum_{i=1}^{k} \sigma_i u_i v_i^T \; ; \; B = \sum_{j=1}^{k} \tau_j u_j v_j^T \tag{11}$$

then,

$$\|A\| = \sqrt[2]{\sigma_1^2 + \sigma_2^2 + \sigma_3^2 + \ldots + \sigma_k^2}\ ;\ \|B\| = \sqrt[2]{\tau_1^2 + \tau_2^2 + \tau_3^2 + \ldots + \tau_k^2} \tag{12}$$

From Theorem 1 and 2,

$$\|A\| + \|B\| \geq \|A - B\| \geq \|A\| - \|B\| \tag{13}$$

$$\begin{aligned} \sqrt[2]{\sigma_1^2 + \sigma_2^2 + \ldots + \sigma_k^2} + \sqrt[2]{\tau_1^2 + \tau_2^2 + \ldots + \tau_k^2} \geq \|A - B\| \geq \\ \left| \sqrt[2]{\sigma_1^2 + \sigma_2^2 + \ldots + \sigma_k^2} - \sqrt[2]{\tau_1^2 + \tau_2^2 + \ldots + \tau_k^2} \right| \end{aligned} \tag{14}$$

Once the face images are pre-processed these vectors can be used as features to distinguish between classes. Many FR systems commonly use pre-processing to improve the performance. In this proposed system we implement a order static filter which directly affects the speed and recognition rate.

2.2.2 Order Static Filtering

The order-static filters are non-linear filter that work on the spatial domain. The OSF is used to reduce the salt noise in face images. A sliding window is moved side ways, up and down with a pixel increment. At each iteration, the center pixel value is replaced by the pixel value of the adjacent cell from the sliding window. To apply the OSF, one typically uses 3 × 3 window. The 2D-OSF can be represented as the following equation

$$f(x,y) = MIN_{(s,t)\in S_{xy}}\{g(s,t)\} \tag{15}$$

Figure 3 shows a simple example demonstrating how minimum order-static filter works. Here a *min* filter using a 3×3 window operates on a 3×3 region of an image. It is evident that this filter has smoothening effect and reduces the information of the image.

78	90	91	87	88	90
92	94	71	69	88	82
84	85	90	99	97	93
80	90	95	92	98	90
89	91	94	85	79	81
84	79	70	74	81	73

→

0	0	0	0	0	0
0	78	71	69	69	0
0	71	71	71	69	0
0	80	85	79	79	0
0	70	70	70	73	0
0	0	0	0	0	0

Fig. 3 Filtering an image using OSF

2.2.3 Quantization

SVD coefficients are continuous values. These coefficients built the observation vector. As the proposed method implements a SVM classifier, these continuous values need to be quantized. This is carried out by the process of rounding off, truncation, or by other nonlinear process.

Consider a vector $X = \{x_1, x_2, x_3, , x_N\}$ as the continuous component. Suppose X needs to be quantized into D levels. Thus the difference between any two successive values will be:

$$\Delta_i = \frac{x_{max} - x_{min}}{D} \quad (16)$$

Once the values of Δ_i are known, it can be replaced by the quantized values

$$x_{quantized} = \frac{x_{max} - x_{min}}{\Delta_i} \quad (17)$$

Hence, all the components of X will be quantized. These quantized values are the input feature vectors for the SVM.

3 Results and Discussion

The proposed system is tested against four standard databases viz., FERET [28], FRGC [31], Indian Face Database [30] and CMU-PIE [29] database. We do this experiment to validate that this proposed method yields satisfactory results on variable environments and standard databases. Major part of this work was carried out in MATLAB with few patches in Octave and DLLs in C as some of the previous work was available only in different formats.

The results of this proposed method are evaluated against three other methods. The performance is reported in terms of two important quantities, viz. Verification Rate (VR)[3] at False Error Rate (FER)[4] = 0.001 and Equal Error Rate (EER)[5]. Tables I and II describe these in detail. Table III and IV deal with the results of CMU-PIE and Indian Face Database.

To conduct the experiments on the proposed system, 40 face image containing 20 male and 20 female were picked at random from FERET, FRGC, CMU-PIE and Indian Face Database. The testing dataset contained 20 face images also picked at random. The feature vector space is obtained by SVD and classified using their data point. The kernal function used is the Gaussian RBF.

---

[3] A statistic metric used to measure the performance whilst at verification.

[4] A statistic metric used to measure the performance of a FR system whilst operating in the verification process.

[5] It is the point on the ROC curve when False Acceptance Rate (FAR) and False Rejection Rate (FRR) are equal. As a FR system isn't set to operate at EER, it is used to benchmark with other existing systems.

Figure 6(a) graphically states the relation between the number of samples required and the number of Support Vector Machine required for classification of face images from their respective classes. In Fig. 6(b), we compare the time required to train the proposed sysem with other contemporary methods. We have compared this SVM+SVD method with HMM+SVD [26] and the HMM+Gabor [14] technique. Evidently, the time required to train the learning algorithm for this proposed method is much lesser than the corresponding time.

It is clear from the above state results that, SVM is supreme in classification using SVD when compared to Hidden Markov Model as in [26]. The choice of using a rather simple and basic feature extraction method is validated to yield satisfactory results due the excellent classification mechanism of SVM.

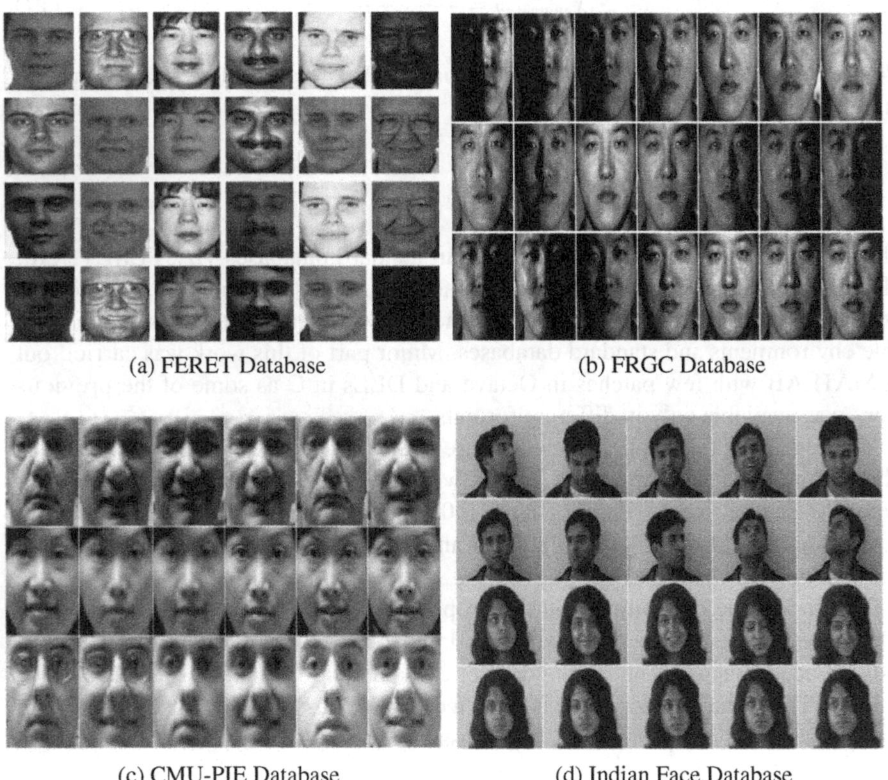

(a) FERET Database (b) FRGC Database

(c) CMU-PIE Database (d) Indian Face Database

Fig. 4 Databases used for testing

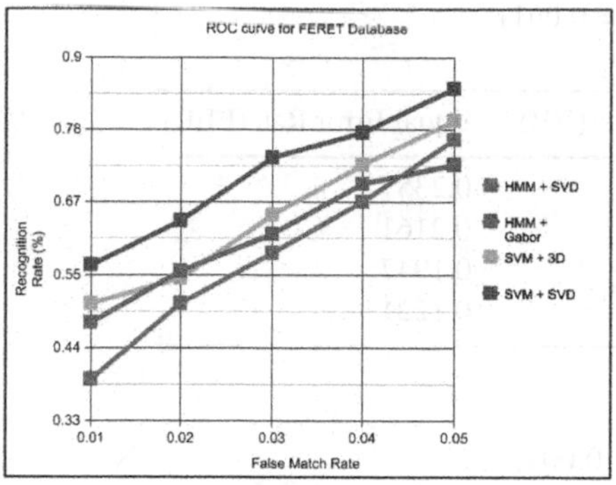

(a) FERET Database

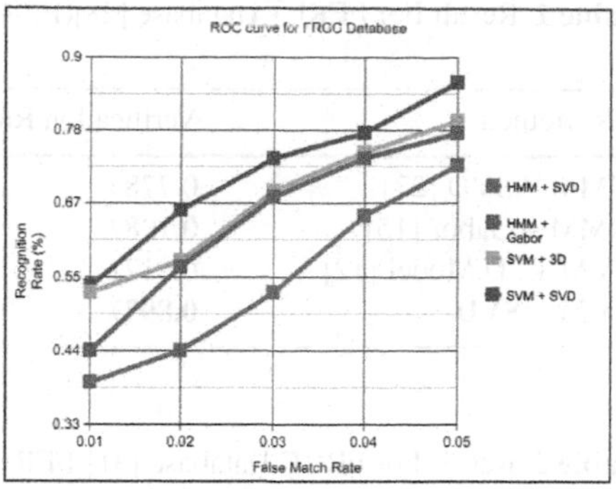

(b) FRGC Database

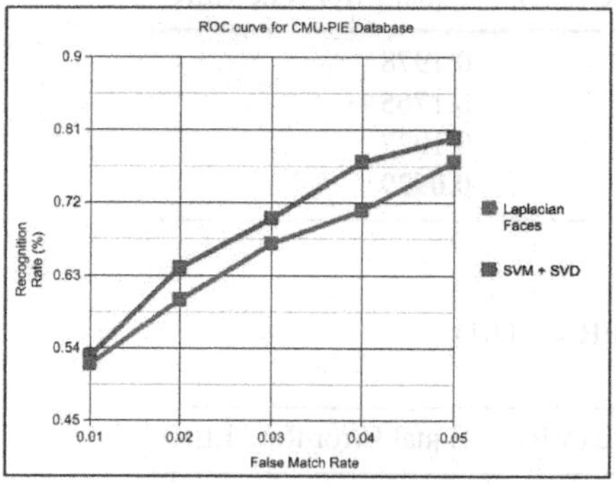

(c) CMU-PIE Database

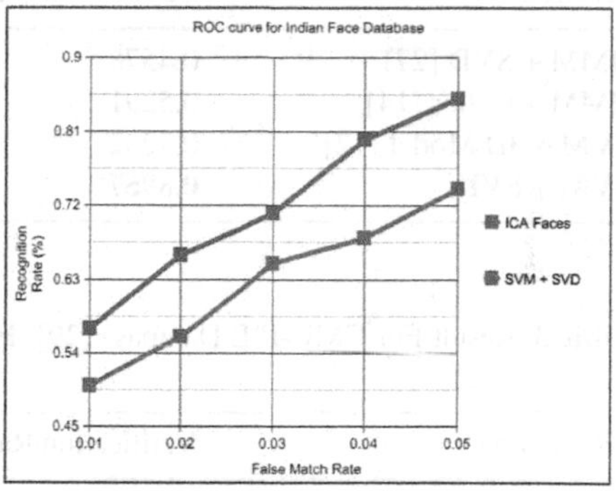

(d) Indian Face Database

Fig. 5 Receiver Operator Characteristics

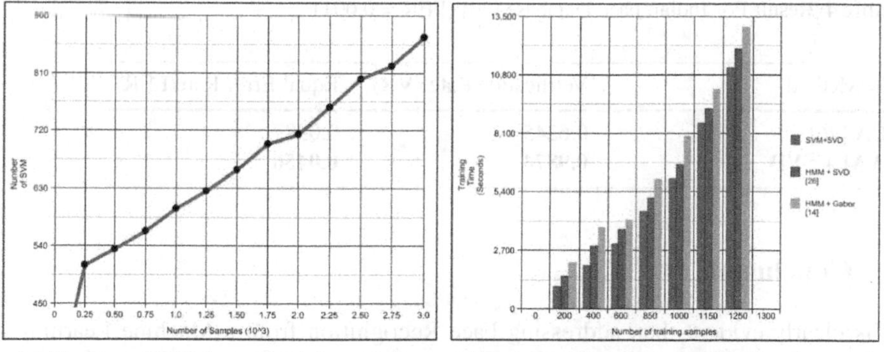

(a) Number of SVMs required for training.

(b) Training Time compared with other methods.

Fig. 6 SVM Results

Table 1 Result For FERET Database [28](FER = 0.001)

FR Method	Verification Rate (VR)	Equal Error Rate(EER)
HMM + SVD [27]	0.7784	0.2357
HMM + Gabor [15]	0.7982	0.2161
SVM + 3D Model [12]	0.8121	0.1947
SVM + SVD	**0.8977**	**0.1221**

Table 2 Result For FRGC Database [31](FER = 0.001)

FR Method	Verification Rate (VR)	Equal Error Rate(EER)
HMM + SVD [27]	0.4578	0.1978
HMM + Gabor [14]	0.5291	0.1765
SVM + 3D Model [12]	0.7232	0.1027
SVM + SVD	**0.8957**	**0.0929**

Table 3 Result For CMU-PIE Database [29] (FER = 0.001)

FR Method	Verification Rate (VR)	Equal Error Rate(EER)
2D Gabor + HMM [14]	0.8457	0.0758
SVM + SVD	**0.9741**	**0.0664**

Table 4 Result For Indian Face Database [30] (FER = 0.001)

FR Method	Verification Rate (VR)	Equal Error Rate(EER)
ICA [26]	0.6245	0.0987
SVM + SVD	**0.9874**	**0.0456**

4 Conclusion

It is clearly evident that, addressing Face Recognition from a Machine Learning perspective yields promising results. As proposed, the SVD is more beneficial to extract features as it is easy to recognize algebraic patterns in the vectors matrices.

This also proves that, SVM is a very stable and efficient classifier for Face Recognition among many other statistical learning models. When compared with other methods, the combination of SVD-SVD stands out by giving appreciable results as shown in Fig. 5 and Fig. 6. It is also noted that, the recognition rate increases with number of samples from each class. A class containing large number of feature vectores can be condensed into much smaller number of SVMs, hence saving memory and improving accuracy. This proposed method yields 90% recognition rate. The novelty of the SV-M/D method is the high recall rate and fast computational speed.

Appendix I - Solving for Dual Optimization Problem[6]

Let us consider the problem defined as:

$$\begin{aligned} &min_w \; f(w) \\ &s.t. \; h_i(w) = 0, i = 1, \cdots, l \end{aligned} \tag{18}$$

Thus the Lagrangian for this problem would be

$$\mathscr{L}(w, \beta) = f(w) + \sum_{i=1}^{l} \beta_i h_i(w) \tag{19}$$

where β_i is the Lagrange multiplier. We then find the partial derivatives of $\mathscr{L}$ and equate them to 0:

$$\frac{\partial \mathscr{L}}{\partial w_i} = 0 \; ; \; \frac{\partial \mathscr{L}}{\partial \beta_i} = 0 \tag{20}$$

We solve for w and β as discussed in [47].

Consider the **primal** optimization problem as:

$$\begin{aligned} &min_w \; f(w) \\ &s.t. \; g_i(w) \leq 0, i = 1, \cdots, k \\ &\quad\quad h_i(w) = 0, i = 1, \cdots, l \end{aligned} \tag{21}$$

Generalizing the Lagrangian, we obtain

$$\mathscr{L}(w, \alpha, \beta) = f(w) + \sum_{i=1}^{k} \alpha_i g_i(w) + \sum_{i=1}^{l} \beta_i h_i(w) \tag{22}$$

α_i and β_i are Lagrange multipliers. Let us consider the quantity $\theta_p(w)$ defined as:

$$\theta_p(w) = \max_{\alpha, \beta; \alpha \geq 0} \mathscr{L}(w, \alpha, \beta) \tag{23}$$

Assume some w be given subjected to the constraints $g_i(w) \geq 0$ or $h_i(w) \neq 0 \; \forall \; i$, then:

$$\begin{aligned} \theta_p(w) &= \max_{\alpha, \beta; \alpha \geq 0} f(w) + \sum_{i=1}^{k} \alpha_i g_i(w) + \sum_{i=1}^{l} \beta_i h_i(w) \\ &= \infty \end{aligned} \tag{24}$$

[6] Readers interested may refer [46] for an elaborate and detailed proof.

Converseley, if the constraints are satisfied for a specific value of w then $\theta_p(w) = f(w)$ Hence,

$$\theta_p(w) = \begin{cases} f(w) & \text{if } w \text{ satisfies primal constraints} \\ \infty & \text{otherwise} \end{cases}$$

Suppose f and g_i are convex[7] and h_i is affine[8]. Considering the minimization problem:

$$\min_w \theta_p(w) = \min_w \max_{\alpha,\beta;\alpha\geq 0} \mathscr{L}(w,\alpha,\beta) \tag{25}$$

Now, when we minimize w.r.t w, we pose a dual optimization problem as below:

$$\max_{\alpha,\beta;\alpha\geq 0} \theta_d(\alpha,\beta) = \max_{\alpha,\beta;\alpha\geq 0} \min_w \mathscr{L}(w,\alpha,\beta) \tag{26}$$

Constructing the Lagrangian for the optimization problem in hand:

$$\mathscr{L}(w,\alpha,b) = \frac{1}{2}\|w\|^2 + \sum_{i=1}^{l} \alpha_i[y^i(w^T x^i + b) - 1] \tag{27}$$

To find the dual form of the optimization problem, we need to minimize $\mathscr{L}(w,b,\alpha)$ w.r.t. w and b to obtain θ_d. Following standard procedure, we take the derivatives of $\mathscr{L}$ w.r.t. w and b to 0. Then we obtain,

$$\nabla_w \mathscr{L}(w,b,\alpha) = w - \sum_{i=1}^{l} \alpha_i y^i x^i = 0 \tag{28}$$

$$\begin{aligned} &\Rightarrow w = \sum_{i=1}^{l} \alpha_i y^i x^i \\ &\frac{\partial}{\partial b}\mathscr{L}(w,b,\alpha) = \sum_{i=1}^{l} \alpha_i y^i x^i = 0 \end{aligned} \tag{29}$$

Substituting (29) in (27) and simplyfing, we get

$$\mathscr{L}(w,b,\alpha) = \sum_{i=1}^{l} \alpha_i - \frac{1}{2}\sum_{i,j=1}^{l} y^i y^j \alpha^i \alpha^j (x^i)^T x^j - b\sum_{i=1}^{l} y^i \alpha^i \tag{30}$$

But from (29), w.k.t. derivative w.r.t. b is zero. Then,

$$\mathscr{L}(w,b,\alpha) = \sum_{i=1}^{l} \alpha_i - \frac{1}{2}\sum_{i,j=1}^{l} y^i y^j \alpha^i \alpha^j (x^i x^j) \tag{31}$$

We obtain the following dual optimization problem as

[7] While f has a Hessian (SOPD square matrix [44]), then it is convex iff the Hessian is positively semi-definite.

[8] $h_i(w) = a_i^T w + b_i$, i.e. linear with intercept term b_i.

$$\max_{\alpha} W(\alpha) = \sum_{i=1}^{l} \alpha_i - \frac{1}{2} \sum_{i,j=1}^{l} y^i y^j \alpha^i \alpha^j \langle x^i x^j \rangle$$
$$s.t.\ \alpha_i \geq 0, i = 1, \cdots, l \tag{32}$$
$$\sum_{i=1}^{l} y^i \alpha^i = 0$$

Appendix II - Mercer-Hilbert-Schmidt Theorem

- **Definition:** If $\int_{\mathscr{X}} \int_{\mathscr{X}} K^2(s,t)\ ds\ dt < \infty$ for a continuous symmentric non-negative K, the there shall exist an orthonormal sequence of eigen function $\Phi_1, \Phi_2, \cdots$ in $L_2[\mathscr{X}]$ and eigen values $\lambda_1 \geq \lambda_2 \geq \cdots \geq 0$ with $\sum_{i=1}^{\infty} \lambda^2 < \infty$ s.t.

$$K(s,t) = \sum_{i=1}^{\infty} \lambda_i \Phi_i(s) \Phi_i(t) \tag{33}$$

- The inner product of $\mathscr{H}$ in function f with $\sum_i (\frac{f_i}{\lambda_i}) < \infty$ then

$$(f,g) = \sum_{i=1}^{\infty} (\frac{f_i g_i}{\lambda_i}) \tag{34}$$

where $f_i = \int_{\mathscr{X}} f(t) \Phi_i(t) dt$.
- Then, feature mapping may be defined as:

$$\Phi(x) = (\sqrt{\lambda_1}\Phi_1(x), \sqrt{\lambda_2}\Phi_2(x), \sqrt{\lambda_3}\Phi_3(x) \cdots \infty) \tag{35}$$

References

1. Boser, B.E., Guyon, I.M., Vapnik, V.N.: A training algorithmfor optimal margin classifier. In: Proc. 5th ACM Workshop on Computational Learning Theory, Pittsburgh, PA, pp. 144–152 (July 1992)
2. Burges, C.J.C.: Simplified support vector decision rules. In: International Conference on Machine Learning, pp. 71–77 (1996)
3. Cortes, C., Vapnik, V.: Support vector networks. Machine Learning 20, 1–25 (1995)
4. Vapnik, V.: The Nature of Statistical Learning Theory. Springer, New York (1995)
5. Chellappa, R., Wilson, C.L., Sirohey, S.: Human and machine recognition of faces: a survey. Proceedings of the IEEE 83(5), 705 (1995)
6. Kanade, T.: Picture Processing by Computer Complex and Recognition of Human Faces. Technical report, Dept. Information Science, Kyoto Univ. (1973)
7. Turk, M., Pentland, A.: Eigenfaces for Recognition. J. Cognitive Neuroscience 3(1), 71–86 (1991)
8. Zhang, J., Yan, Y., Lades, M.: Face recognition: eigenface. Proceedings of the IEEE 85(9) (September 1997), elastic matching and neural nets
9. Lawrence, S., Giles, C.L., Tsoi, A.C., Back, A.D.: Face Recognition: A Convolutional Neural-Network Approach. IEEE Trans. Neural Networks 8, 98–113 (1997)

10. Lin, S., Kung, S., Lin, L.: Face Recognition/Detection by Probabilistic Decision- Based Neural Network. IEEE Trans. Neural Networks 8(1), 114–131 (1997)
11. Osuna, E., Freund, R., Girosi, F.: Training support vector machines: an application to face detection. In: Proceedings of the 1997 IEEE Computer Society Conference on Computer Vision and Pattern Recognition, June 17-19, pp. 130–136 (1997)
12. Srinivasan, M., Ravichandran, N.: Support Vector Machine Components-Based Face RecognitionTechnique using 3D Morphable Modelling Methods. International Journal of Future Computer and Communications 2(5), 520–523 (2013)
13. Samaria, F.S., Harter, A.C.: Parameterisation of a stochastic model for human face identification. In: Proceedings of the Second IEEE Workshop on Applications of Computer Vision 1994, December 5-7, pp. 138–142 (1994)
14. Srinivasan, M., Ravichandran, N.: A 2D Discrete Wavelet Transform Based 7-State Hidden Markov Model for Efficient Face Recognition. In: 2013 4th International Conference on Intelligent Systems Modelling & Simulation (ISMS), January 29-31, pp. 199–203 (2013)
15. Srinivasan, M., Ravichandran, N.: A new technique for Face Recognition using 2D-Gabor Wavelet Transform with 2D-Hidden Markov Model approach. In: 2013 International Conference on Signal Processing Image Processing & Pattern Recognition (ICSIPR), February 7-8, pp. 151–156 (2013)
16. Nefian, A.V., Hayes III, M.H.: Maximum likelihood training of the embedded HMM for face detection and recognition. In: Proceedings of the 2000 International Conference on Image Processing, vol. 1, pp. 33–36 (2000)
17. Nefian, A.V., Hayes, M.H.: An embedded HMM-based approach for face detection and recognition. In: Proceedings of the 1999 IEEE International Conference on Acoustics, Speech, and Signal Processing, March 15-19, vol. 6, pp. 3553–3556 (1999)
18. Srinivasan, M., Raghu, S.: Comparative Study on Hidden Markov Model Versus Support Vector Machine: A Component-Based Method for Better Face Recognition. In: 2013 UKSim 15th International Conference on Computer Modelling and Simulation (UKSim), April 10-12, pp. 430–436 (2013)
19. Moghaddam, B., Jebara, T., Pentland, A.: Bayesian Face Recognition. Pattern Recognition 33(11), 1771–1782 (2000)
20. Burges, C.J.C.: A Tutorial on Support Vector Machines for Pattern Recognition. Data Mining and Knowledge Discovery 2, 121–167 (1998)
21. Smola, A.J., Schlkopf, B.: A Tutorial on Support Vector Regression, NeuroCOLT 2 TR 1998-03 (1998)
22. Osuna, E., Freund, R., Girosi, F.: Training Support Vector Machines: An Application to Face Detection. In: Proc. Computer Vision and Pattern Recognition, pp. 130–136 (1997)
23. Joachims, T.: Text Categorization with Support Vector Machines. In: Nédellec, C., Rouveirol, C. (eds.) ECML 1998. LNCS, vol. 1398, pp. 137–142. Springer, Heidelberg (1998)
24. Papageorgiou, C.P., Oren, M., Poggio, T.: A General Framework for Object Detection. In: International Conference on Computer Vision (1998)
25. Riesz, F., Sz.-Nagy, B.: Functional Analysis. Ungar, New York (1955)
26. Srinivasan, M., Aravamudhan, V.: Independent Component Analysis of Edge Information for Face Recognition under Variation of Pose and Illumination. In: 2012 Fourth International Conference on Computational Intelligence, Modelling and Simulation (CIMSiM), September 25-27, pp. 226–231 (2012)
27. Srinivasan, M., Vijayakumar, S.: Pseudo 2D Hidden Markov Model Based Face Recognition System Using Singular Values Decomposition Coefficients. In: The 2013 International Conference on Image Processing, Computer Vision, & Pattern Recognition (IPCV 2013) (2013) ISBN: 1-60132-252-6

28. Phillips, P., Moon, H., Rizvi, S., Rauss, P.: The FERET Evaluation Methodology for Face-Recognition Algorithms. PAMI 22(10), 1090–1104 (2000)
29. Sim, T., Baker, S., Bsat, M.: The CMU Pose, Illumination, and Expression (PIE) Database of Human Faces. Tech. report CMU-RI-TR-01-02, Robotics Institute, Carnegie Mellon University (January 2001)
30. Jain, V., Mukherjee, A.: The Indian Face Database (2002), `http://vis-www.cs.umass.edu/~vidit/IndianFaceDatabase/`
31. Phillips, P.J., Flynn, P.J., Scruggs, T., Bowyer, K.W., Chang, J., Hoffman, K., Marques, J., Min, J., Worek, W.: Overview of the face recognition grand challenge. In: IEEE Computer Society Conference on Computer Vision and Pattern Recognition, CVPR 2005, June 20-25, vol. 1, pp. 947–954 (2005)
32. He, X., Yan, S., Hu, Y., Niyogi, P., Zhang, H.-J.: Face recognition using Laplacianfaces. IEEE Transactions on Pattern Analysis and Machine Intelligence 27(3), 328–340 (2005)
33. Russell, B.C., Torralba, A., Murphy, K.P., Freeman, W.T.: Label Me: a database and web-based tool for image annotation. Technical Report MIT-CSAIL-TR-2005-056. MIT (2005)
34. Patel, V.M., Wu, T., Biswas, S., Phillips, P.J., Chellappa, R.: Dictionary-Based Face Recognition Under Variable Lighting and Pose. IEEE Transactions on Information Forensics and Security 7(3), 954–965 (2012)
35. Taheri, S., Turaga, P., Chellappa, R.: Towards view-invariant expression analysis using analytic shape manifolds. In: 2011 IEEE International Conference on Automatic Face & Gesture Recognition and Workshops (FG 2011), March 21-25, pp. 306–313 (2011)
36. Lyons, M., Akamatsu, S., Kamachi, M., Gyoba, J.: Coding facial expressions with Gabor wavelets. In: Proceedings of the Third IEEE International Conference on Automatic Face and Gesture Recognition 1998, April 14-16, pp. 200–205 (1998)
37. Fergus, R., Perona, P., Zisserman, A.: Object class recognition by unsupervised scale-invariant learning. In: Proceedings of the 2003 IEEE Computer Society Conference on Computer Vision and Pattern Recognition 2003, June 18-20, vol. 2, pp. II-264–II-271 (2003)
38. LeCun, Y., Chopra, S., Ranzato, M., Huang, F.-J.: Energy-Based Models in Document Recognition and Computer Vision. In: Ninth International Conference on Document Analysis and Recognition, ICDAR 2007, September 23-26, vol. 1, pp. 337–341 (2007)
39. Liao, S., Chung, A.C.S.: A novel Markov random field based deformable model for face recognition. In: 2010 IEEE Conference on Computer Vision and Pattern Recognition (CVPR), June 13-18, pp. 2675–2682 (2010)
40. Liao, S., Shen, D., Chung, A.: A Markov Random Field Groupwise Registration Framework for Face Recognition. IEEE Transactions on Pattern Analysis and Machine Intelligence PP(99), 1
41. Anguelov, D., Lee, K.-C., Gokturk, S.B., Sumengen, B.: Contextual Identity Recognition in Personal Photo Albums. In: IEEE Conference on Computer Vision and Pattern Recognition, CVPR 2007, June 17-22, pp. 1–7 (2007)
42. Wang, Y., Zhang, L., Liu, Z., Hua, G., Wen, Z., Zhang, Z., Samaras, D.: Face Relighting from a Single Image under Arbitrary Unknown Lighting Conditions. IEEE Transactions on Pattern Analysis and Machine Intelligence 31(11), 1968–1984 (2009)
43. Zhao, M., Chua, T.-S.: Markovian mixture face recognition with discriminative face alignment. In: 8th IEEE International Conference on Automatic Face & Gesture Recognition, FG 2008, September 17-19, pp. 1–6 (2008)
44. Binmore, K., Davies, J.: Calculus Concepts and Methods, p. 190. Cambridge University Press (2007)

45. Wang, X., Tang, X.: A unified framework for subspace face recognition. IEEE Transactions on Pattern Analysis and Machine Intelligence 26(9), 1222–1228 (2004)
46. Ng, A.: CS229 Lecture notes, Topic: Part V: Support Vector Machines, Stanford University, http://cs229.stanford.edu/notes/cs229-notes3.pdf
47. Rockarfeller, R.T.: Convex Analysis. Princeton University Press

A New Single Image Dehazing Approach Using Modified Dark Channel Prior

Harmandeep Kaur Ranota* and Prabhpreet Kaur

Abstract. Dehazing is a challenging issue because the quality of a captured image in bad weather is degraded by the presence of haze in the atmosphere and hazy image has low contrast in general. In this paper we proposed a new method for single image dehazing using modified dark channel prior and adaptive Gaussian filter. In our proposed method, hazy images are first converted in to LAB color space and then Adaptive Histogram Equalization is applied to improve the contrast of hazy images. Then, our proposed method estimates the transmission map using dark channel prior. It produces more refined transmission map than that of old dark channel prior method and then Adaptive Gaussian filter is employed for further refinement. The quantitative and visual results show proposed method can remove haze efficiently and reconstruct fine details in original scene clearly.

Keywords: Dark Channel, Image dehazing, Adaptive Histogram Equalization, LAB color space.

1 Introduction

Images of outdoor scenes are degraded as a result of known phenomena which take account of absorption and scattering of light by the atmospheric particles such as haze, fog etc. Haze removal is a critical issue because the haze is dependent on the unknown depth information. Dehazing is the process of removing haze in captured images and to reconstruct the original colors of natural scenes. If the input is only a single haze image, then problem is taken under constraint. Many methods have been proposed by using multiple images. In [1, 2] scene depths can be estimated from two or more images of the same scene that are captured in

Harmandeep Kaur Ranota · Prabhpreet Kaur
Dept. of Computer Science and Engineering, Guru Nanak Dev University,
Amritsar, 143001, Punjab, India

* Corresponding author.

El-Sayed M. El-Alfy et al. (eds.), *Advances in Intelligent Informatics*,
Advances in Intelligent Systems and Computing 320, DOI: 10.1007/978-3-319-11218-3_8

different weather conditions. S. Shwartz [4] and Y.Y. Schechner [5] removed haze using different polarization filters. The major drawback of these methods is that it requires multiple images for dehazing. But in some applications it is not always possible to obtain multiple images.

In order to overcome the drawback of multiple image dehazing methods, single image dehazing methods have been proposed [3, 6, 7]. Tan [3] proposed an automated method that only requires a single input image for dehazing. His proposed method removes haze by maximizing the local contrast of the images. The main drawback of Tan's proposed method generates overstretches contrast.

Fattal [8] proposed a refined image formation model that accounts for surface shading as well as the transmission function under the constraint that the transmission and surface shading are locally not correlated. The drawback of Fattal's method is that this approach cannot well handle heavy haze images.

He et al. [7] proposed a new method based on dark channel prior for single image haze removal and removed the drawbacks of [3,8] methods. In He et al. Proposed method , estimation of the thickness of the haze can be done directly by using dark channel prior and then soft matting algorithm is used to refine transmission value of each pixel to achieve high quality haze-free image. He[7] compared proposed method with various state-of-art methods and showed the superiority of proposed method. The main drawback of these methods is that their results are not much effective.

In our proposed method, dark channel prior [7] is further modified and improved contrast and can handle heavy hazy images. Section 2 reviews the related work. Section 3 presents the proposed methodology. Section 4 discusses result and discussions and section 5 describes the conclusion.

2 Related Work

In this section, we presented Optical model of hazy image [1,2,7] and dark channel prior that are closely to our proposed method.

2.1 Optical Model of Hazy Images

The attenuation of image due to fog can be represented as:

$$I_{att}(x) = J(x)\, t(x) \tag{1}$$

Where J(x) is input image, I(x) is foggy image, t(x) is the transmission of the medium.

The second effect of fog is Airlight effect and it is written as:

$$I_{airtight}(x) = A\,(1 - t(x)) \tag{2}$$

Where A is the Atmospheric light.

As fogy image is degraded by a combination of both attenuation and atmospheric light effect, it is expressed as:

$$I(x) = J(x)\, t(x) + A\,(1- t(x)) \tag{3}$$

When atmosphere is homogeneous, transmission t(x) is represented as:

$$t(x) = e^{-\beta\, d(x)} \tag{4}$$

Where β is the scattering co-efficient of the atmosphere and d(x) is the scene depth of x. This equation describes an exponentially decaying function with depth and its rate is calculated using the scattering coefficient.

2.2 Dark Channel Prior

Statistical data of outdoor images reveal that at least one of color channel values (RGB) is often close to zero in some objects in a haze-free image [4]. The dark channel for an arbitrary image J can be expressed as:

$$J^{dark}(x) = \min_{y\in\Omega(x)} \left(\min_{c\in\{r,g,b\}} J^{c}(y)\right) \tag{5}$$

In this J^{dark} is the dark channel of J, J^{c} is the color channel,(x) is the patch centred at x.

2.3 Transmission Estimation

If J is a haze-free image, the dark channel of J is assumed to be zero.

$$J^{dark}(x) \longrightarrow 0$$

The transmission t(x) is calculated as

$$t(x)=1 - w \min_{y\in\Omega(x)} \left(\min_{c\in\{r,g,b\}} \frac{I^{c}(y)}{A^{c}}\right) \tag{6}$$

The variable w reduces the dark channel and it increases t(x) at a pixel x producing less subtraction in the restored image. Value of w is application dependent and it lies between 0 and 1.

2.4 Restoration of Input Image

After implementing Dark Channel prior and Estimation of transmission map, haze free image can be restored using:

$$J(x) = \frac{I(x)-A}{\max\,(t(x),\, t_o)} + A \tag{7}$$

A typical value of t_o is taken as zero or 0.1as it is prone to noise.

3 Proposed Methodology

In this section, we present our proposed methodology for improve contrast and quality of hazed images. The flow chart of our proposed method is shown in figure 3.The following steps are involved in our proposed strategy:

3.1 Convert RGB to LAB Color Space

In our proposed method, we first convert input colored image to LAB color space. We observed that the fog presence can be better detected by looking at the LAB space of the image than RGB color model. Then split the image into L,A,B channel and Adaptive Histogram Equalization is applied on each channel to improve contrast because hazy images are of low contrast in general.

3.2 Estimation of Refined Transmission Map Using Dark Channel Prior

In section 2, we present the optical model of haze image. Using equations in section 2, we will calculate transmission map and refined transmission map. It is observed that our method will give lower values, resulting into higher values of t(x) because contrast of the images are improved and required lesser correction through equation(7). Figure1 showing our refined transmission map is more clear than old dark channel prior.

(a)Input haze image (b) Image obtained using above steps

(c)Refined Transmission Map [7] (d) Refined transmission map by our proposed method

Fig. 1 Comparisons of transmission Maps

3.3 Apply Histogram Stretching and Adaptive Gaussian Filter

Histogram stretching helps to stretching the range of intensity values of image uniformly and further improves the quality of images. Adaptive Gaussian filter preserve edge information of input hazy images and avoid generation of halo effects produced by dark channel prior and we get reconstruct fine details in original scene clearly as shown in figure 2.

(a)

(b)

Fig. 2 (a) Image after *step B* and Input Image (b) Final Image

4 Experiment Results and Discussions

Our proposed technique has been tested on several different images. Figure 5 shows the input hazy images, dehazed images using classical definition of dark channel prior and images obtained by our proposed method. It is clear that the resultant images using the proposed technique are of good quality and contrast than the old dark channel prior. Not only visual comparison but also quantitative comparisons are confirming the superiority of the proposed method. Peak signal-to-noise ratio (PSNR), mean square error (MSE) and Normalised Absolute Error (NAE) have been implemented in order to obtain some quantitative results for comparison. PSNR is a mathematical measure of image quality based on the pixel difference between two images. Large value of PSNR indicates the high quality of image. PSNR can be calculated by using the following formula [9].

$$\text{PSNR}=10\log_{10}\left(\frac{R^2}{\text{MSE}}\right) \tag{8}$$

Where R represents the maximum variation in the input image and MSE represents the MSE between the given input image I_{in} and the original image I_{org} which can be obtained by the following[9]:

$$\text{MSE} =\sum_{i,j}(I_{in}(i, j)-I_{org}(i, j))^2 \tag{9}$$

Normalised absolute error (NAE) is a measure of how far is the output image from the original image. Low value of NAE means good quality of the image. NAE can be calculated using following formula [9]:

$$\text{NAE} = \frac{\sum_{i=1}^{M}\sum_{j=1}^{N} |[f\,(i,j).f'(i,j)]|}{\sum_{i=1}^{M}\sum_{j=1}^{N} |[f\,(i,j)]} \tag{10}$$

Figure 4 shows the estimation of transmission map and calculated refined transmission map by our method and dehazed output image. The results in Tables 1, Table 2, Table 3 show the comparisons of PSNR, MSE, and NAE values.

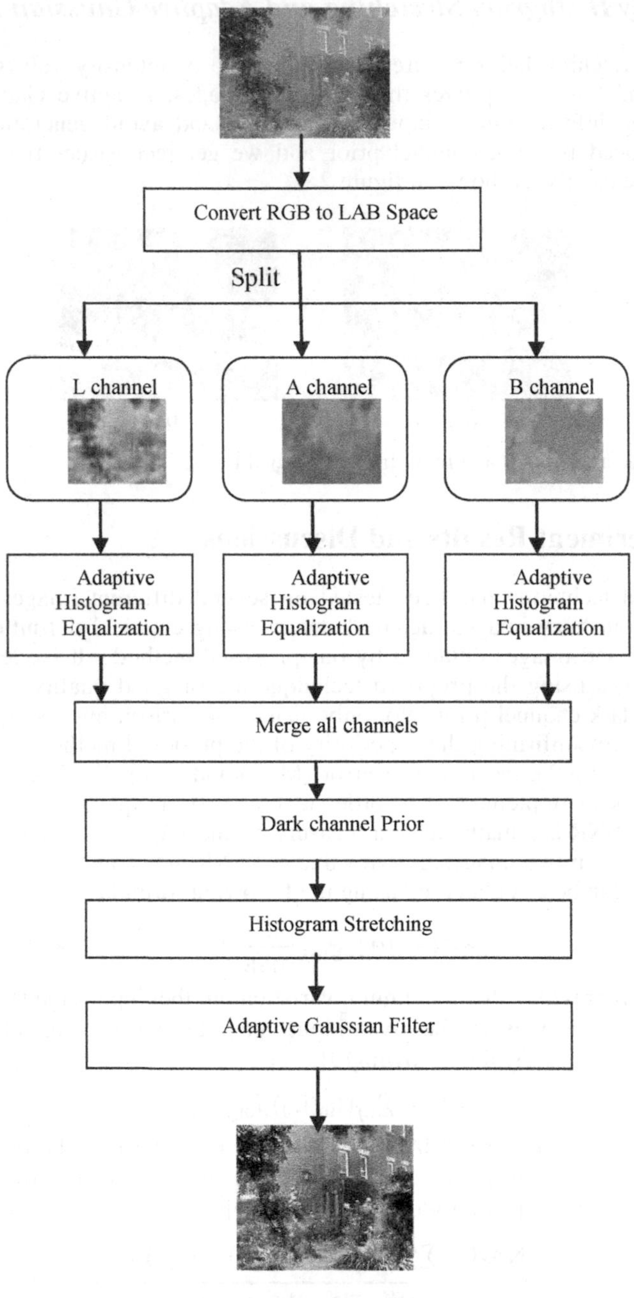

Fig. 3 The Flow Chart of Proposed Method

(a)Input Hazy Image (b) Transmission map (c) refined Transmission (d) Output Image

Fig. 4 Scanorio of Estimation Transmission Map and Refined Transmission Map

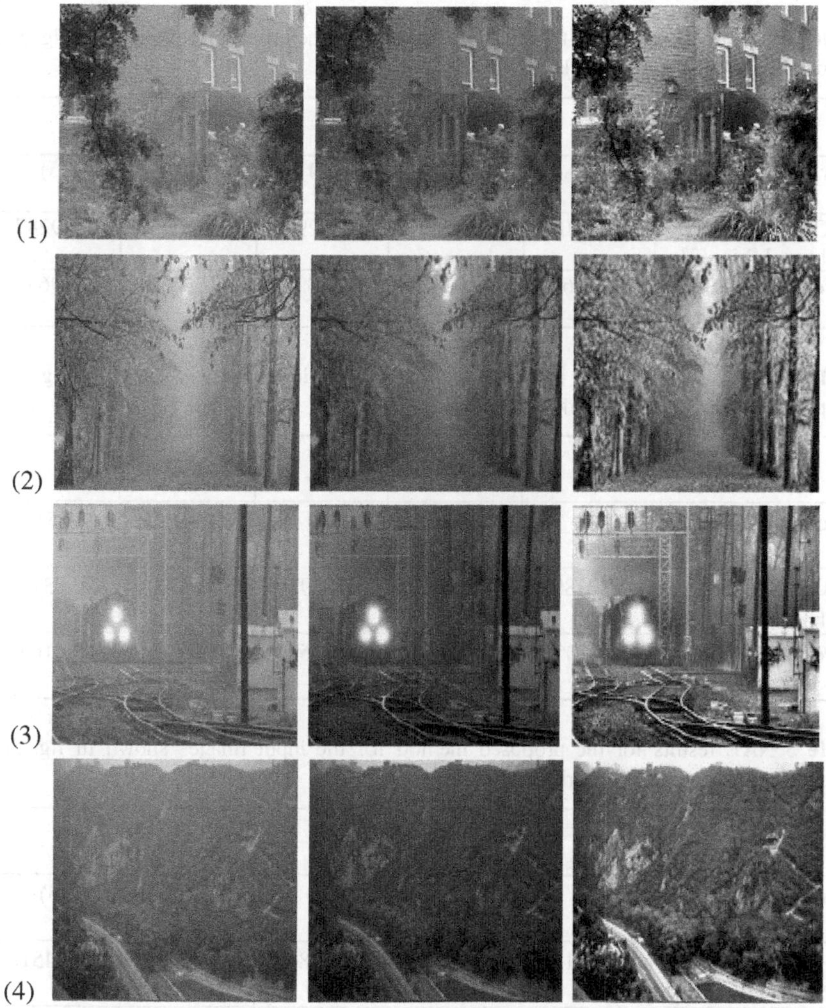

(1)

(2)

(3)

(4)

Fig. 5 (a) Original Images (b) Dark Channel Prior[7] (c) Proposed Method

(5)

Fig. 5 (*continued*)

Table 1 PSNR (dB) results for the proposed method for the input images shown in fig. 5(a) compared with the classical dark channel prior

		PSNR(dB)			
Techniques/images	(1)	(2)	(3)	(4)	(5)
Classical Dark Channel Prior [7]	64.3108	64.5548	59.7293	62.9965	61.9287
Proposed Method	67.1367	66.0319	66.3091	67.5062	65.9630

Table 2 MSE results for the proposed method for the input images shown in fig. 5(a) compared with the classical dark channel prior

		MSE			
Techniques/images	(1)	(2)	(3)	(4)	(5)
Classical Dark Channel Prior [7]	0.0289	0.0291	0.0692	0.0328	0.0314
Proposed Method	0.0105	0.0162	0.0152	0.0115	0.0165

Table 3 NAE results for the proposed method for the input images shown in fig. 5(a) compared with the classical dark channel prior

		NAE			
Techniques/images	(1)	(2)	(3)	(4)	(5)
Classical Dark Channel Prior [7]	0.2115	0.2587	0.4880	0.4420	0.3451
Proposed Method	0.1490	0.1913	0.1997	0.1788	0.2210

5 Conclusion

A new dehazed method is presented using modified dark channel Prior and Gaussian filter for the single haze image. In this paper, we improved contrast of haze images using adaptive Histogram Equalization on LAB color space. Then we calculated refined transmission map of the image and adaptive Gaussian Filter employed for preservation of edge information. Performance of the proposed technique has been compared with the classical dark channel technique. The experimental results show that the proposed technique gives better quality dehazed images. The quantitative measurements such as PSNR, MSE and NAE confirming the superiority of proposed technique over classical dark channel method.

References

1. Nayar, S.K., Narasimhan, S.G.: Vision in bad weather. In: Proc. of CVPR, pp. 820–827 (September 1999)
2. Narasimhan, S.G., Nayar, S.K.: Contrast restoration of weather degraded images. IEEE Trans. Pattern Anal. Mach. Intell. 25(6), 713–724 (2003)
3. Tan, R.T.: Visibility in bad weather from a single image. In: Proc. IEEE CVPR, pp. 1–8 (June 2008)
4. Shwartz, S., Namer, E., Schechner, Y.Y.: Blind haze separation. In: Proc. IEEE CVPR, vol. 2, pp. 1984–1991 (June 2006)
5. Schechner, Y.Y., Narasimhan, S.G., Nayar, S.K.: Instant dehazing of images using polarization 1, 325–332 (December 2001)
6. Tarel, J.P., Hautiere, N.: Fast Visibility Restoration from a Single Color or Gray Level Image. In: IEEE 12th International Conference on Computer Vision (2009)
7. He, K., Sun, J., Tang, X.: Single image haze removal using dark channel prior. In: Proc. IEEE CVPR, pp. 1956–1963 (June 2009)
8. Fattal, R.: Single image dehazing. ACM Trans. Graph 27(3), 1–9 (2008)
9. Gonzalez, R.C., Woods, R.E.: Digital Image Processing, 3rd edn. Pearson Publications

A Fuzzy Regression Analysis Based No Reference Image Quality Metric

Indrajit De and Jaya Sil

Abstract. In the paper quality metric of a test image is designed using fuzzy regression analysis by modeling membership functions of interval type 2 fuzzy set representing quality class labels of the image. The output of fuzzy regression equation is fuzzy number from which crisp outputs are obtained using residual error defined as the difference between observed and estimated output of the image. In order to remove human bias in assigning quality class labels to the training images, crisp outputs of fuzzy numbers are combined using weighted average method. Weights are obtained by exploring the nonlinear relationship between the mean opinion score (MOS) of the image and defuzzified output. The resultant metric has been compared with the existing quality metrics producing satisfactory result.

1 Introduction

Correlation describes the strength of association between two random variables. Whether the variables are positively or negatively correlated is determined by the slope of the line, representing relation between the variables. Regression goes beyond correlation by adding prediction capabilities. Classical regression analysis is used to predict dependent variable when the independent variables are directly measured or observed. However, there are many situations where observations cannot be described accurately and so an approximate description is used to represent the relationship. Moreover, the classical regression analysis deals with precise data while the real world data is often imprecise. In classical regression model, the

Indrajit De
Department of Information Technology, MCKV Institute of Engineering, Liluah, Howrah-711204, West Bengal, India

Jaya Sil
Department of Computer Science and Technology, IIEST (Formerly BESUS), Shibpur, Howrah, West Bengal, India

El-Sayed M. El-Alfy et al. (eds.), *Advances in Intelligent Informatics*,
Advances in Intelligent Systems and Computing 320, DOI: 10.1007/978-3-319-11218-3_9

difference between the measured and expected values of the dependent variable is considered as random error, which can be minimized using different statistical methods. However, the random error may occur due to imprecise observations and in such a situation the uncertainty is due to vagueness, not randomness [1] and so statistical methods fail to provide accurate results.

Uncertainty in image data has been modeled in this paper using fuzzy regression technique where quality of the test image is evaluated as fuzzy number. To quantify the subjective quality class labels of the test image represented by interval type 2 fuzzy sets, corresponding membership functions are modeled by best fitted polynomial equations. The random error is mapped as *residual error*, defined as the difference between observed and estimated output. The residual error is used to calculate spreads of the fuzzy number that defuzzifies the outputs. The crisp outputs are fused by weighted average method to reduce human biasness in assigning quality class labels to the training images. The resultant quality metric has been compared with the existing quality metrics producing satisfactory result.

2 Modeling Image Quality Classes by Fuzzy Regression

Fuzzy regression analysis involves fuzzy numbers and the general form is described by equation (1) [4].

$$\check{Y} = \widetilde{A_0} + \widetilde{A_1}x_1 + \cdots \widetilde{A_n}x_n \tag{1}$$

Where $\check{Y}$ is the fuzzy output, $\widetilde{A_i}$, i = 1,2,….n are fuzzy coefficients and $X = (x_1,\ldots, x_n)$ is non fuzzy input vector. The fuzzy coefficients are assumed triangular fuzzy numbers (TFN) and characterized by membership function $\mu_{A_i}(x_i)$ as defined in figure 1.

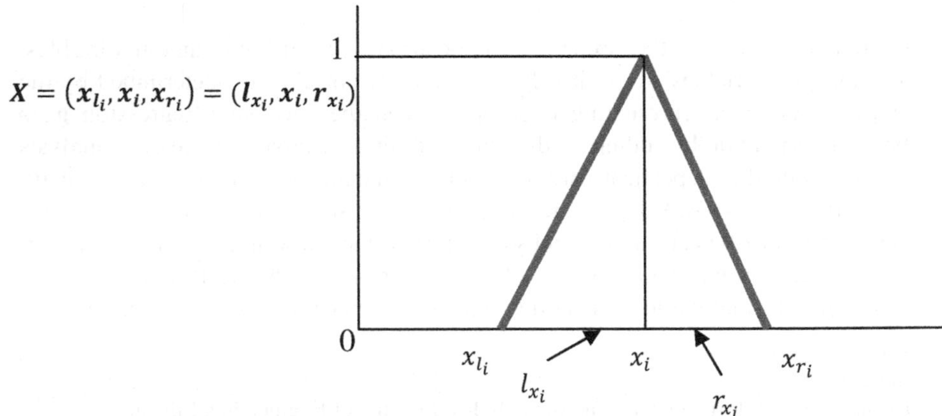

Fig. 1 Membership function Illustrating Triangular Fuzzy Number

When the left spread l_{x_i} and right spread r_{x_i} are equal then the TFN reduces to symmetrical triangular fuzzy number (STFN).

No reference image quality metric designed using fuzzy regression analysis (FRANRIQA) has been described in figure 2.

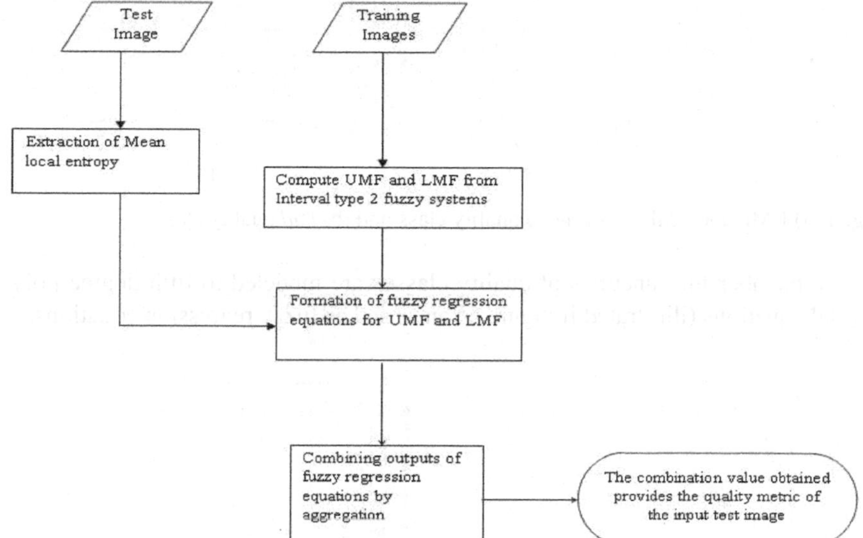

Fig. 2 Flow diagram of Fuzzy Regression Analysis based image quality metric computation

To design the proposed metric for assessing image quality, different training images of BIOIDENTIFICATION Image Database [5] are considered. Upper and lower membership functions (UMF and LMF) of interval type 2 fuzzy sets representing quality classes (*good* and *bad*) of the images (see, figure 3) are obtained using entropy of visually salient regions [10] and kernel density function [11], as shown in figure 4.

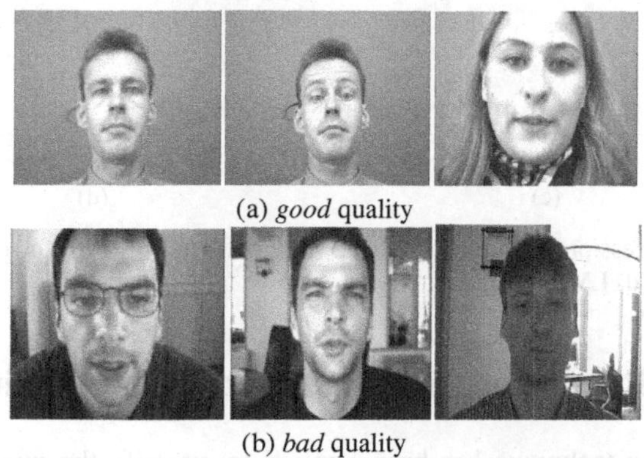

(a) *good* quality

(b) *bad* quality

Fig. 3 Sample training images taken from BIOIDENTIFICATION database

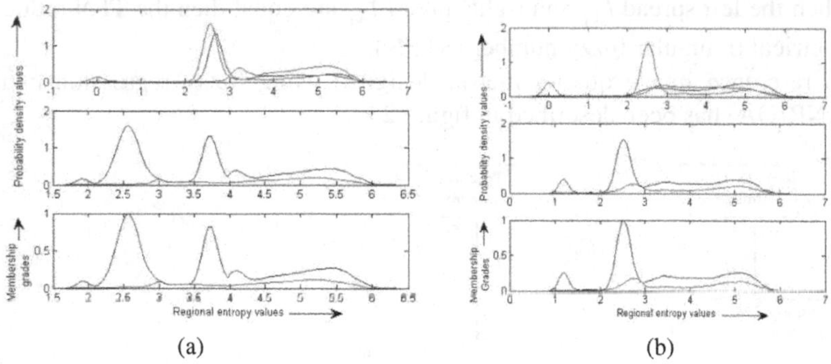

Fig. 4 (a) UMF and LMF for (a) *good* quality class and (b) *bad* quality class

The membership functions of quality classes are modeled to fifth degree polynomial equations (illustrated in figure 5) and used as fuzzy regression equations.

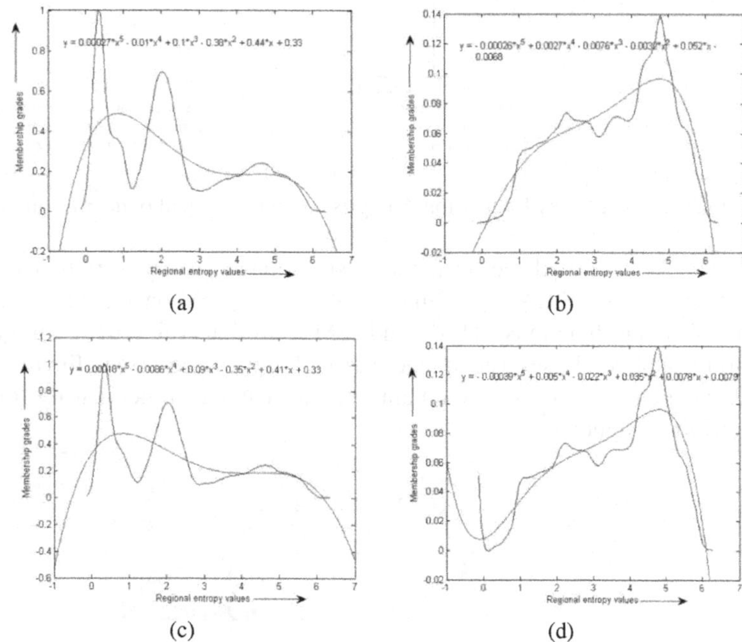

Fig. 5 Fuzzy Regression Equations Modeled (a) UMF and (b) LMF for *good* quality Images; (c) UMF and (d) LMF for *bad* quality Images

3 Fuzzy Regression Analysis Based Image Quality Metric

Fuzzy regression technique has been applied to quantify the quality metric of an image from the quality class label evaluated by interval type 2 fuzzy based no reference image quality assessment method. Average entropy of visually salient regions [6] of the test image is computed, denoted as *Mean_local_ entropy* of the test image. The symmetrical triangular fuzzy number (STFN) is represented by the triangle in figure 6. The value of UMF and LMF corresponding to *Mean_local_ entropy* of the test image is evaluated as illustrated in figure 6 and used as centre of the respective symmetrical triangular fuzzy number (STFN) representing estimated output of the image quality class. The total spread of a symmetrical fuzzy number is twice the *average norm of residual error* calculated considering 100 sample images.

Quality of the test image represented by STFN is defuzzified and for each class four crisp values ($centre_{UMF}$ + $left\_spread_{UMF}$, $centre_{UMF}$ + $right\_spread_{UMF}$, $centre_{LMF}$ + $left\_spread_{LMF}$, $centre_{LMF}$ + $right\_spread_{LMF}$) are obtained. Eight crisp values are fused to assess quality of the test image, which is free from human biasness.

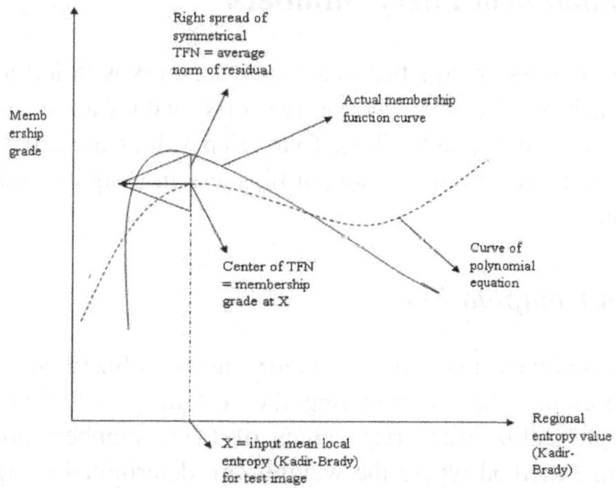

Fig. 6 Symmetrical Triangular Fuzzy Number using Fuzzy Polynomial Regression and Membership function of a particular quality class considering 100 training images

The proposed system is validated using different types of test images taken from TOYAMA image database[7] (Fig.7) having different Peak Signal to Noise Ratio (PSNR), Mean Opinion Score (MOS) and Mean Structural Similarity Index (MSSIM)[8].

(a) (b) (c) (d)

Fig. 7 Sample test images from TOYAMA database- (a) kp0837, (b) kp2320, (c) kp22, (d) kp1315

Example:

Evaluation of quality metric of the test image (Fig. 7(b)) has been explained below.

Say, fifth order polynomial equation which is best fitted with figure 5 (a) given as:

$$y = 0.00027x^5 - 0.01x^4 + 0.1x^3 - 0.38x^2 + 0.44x + 0.33$$

Mean_local_entropy of the test image (kp2320) is used as input *x* for which respective *y* value is calculated representing centre of a STFN. Defuzzified outputs Y_{TFN} = (-0.0174, *y*, 0.0174) are obtained by adding left spread and right spread to *y*, where 0.0174 is the *average residual norm* for the image.

4 Combination of Fuzzy Numbers

Output of fuzzy regression equation is a fuzzy numbers with left and right spread therefore, for each membership function two crisp values are evaluated. Thus for an image in a particular quality class, four crisp values are emerged (UMF and LMF), which are fused to remove human biasness in assigning quality classes to the training images.

4.1 Weight Computation

Evaluation of combined image quality metric using defuzzified outputs of fuzzy numbers are presented here considering the test image of TOYAMA database (Fig. 7(b)). Eight defuzzified crisp values of fuzzy numbers are combined by weighted average method where the weights are determined using a cubic polynomial equation as plotted in figure 8. Nonlinearity between defuzzified values and the normalized MOS of the test image is best fitted with the cubic polynomial equation and therefore, the coefficients of the equation are considered as weights for the particular test image (kp2320). The weighted average value is computed using equation (3).

$$W = \frac{\sum_{i=1}^{N} \prod_{i=1,j=1}^{N/2,M/2} w_{jj} \times F_{ij}}{\sum_{i=1}^{M} w_i} \quad (3)$$

Where W is the weighted average, w_{jj} is respective weight of F_{ij} which is the concerned fuzzy number, *N* is the no. of fuzzy numbers (eight here) and *M* is the number of weights equal to the number of coefficients (four here).

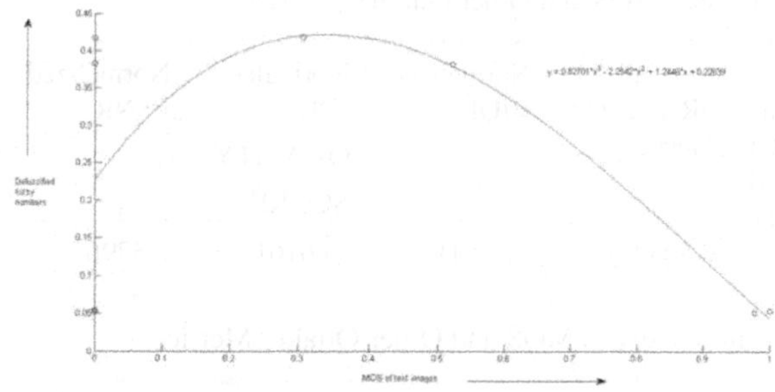

Fig. 8 Polynomial equation for computing weights of image kp2320

Using the coefficients of the equation given in figure 8, which are weights: 0.8270, -2.2542, 1.2446, 0.2263, the quality metric is computed below.

$$\frac{\sum_{i=1}^{8}\prod_{i=1,j=1}^{8/2,4/2} w_{jj}\times F_{ij}}{\sum_{i=1}^{4} w_i} = \frac{\begin{bmatrix} 0.0505 & 0.0516 \\ 0.0535 & 0.0544 \\ 0.3823 & 0.3826 \\ 0.4177 & 0.4174 \end{bmatrix} \times \begin{bmatrix} 0.8270 & 1.2446 \\ -2.2542 & 0.2263 \end{bmatrix}}{0.8270+1.2446-2.2542+0.2263} = 0.8124$$

5 Results of Fuzzy Regression Technique

For sample test images the no reference image quality metrics are evaluated using Fuzzy Regression analysis (FRANRIQA) as shown in table 1. Table 1 provides the normalized values of the FRANRIQA metrics, which is necessary to compare the said metric with others whose scales are different. Table 2 and table 3 provide PPMCC as prediction accuracy [9] and SROCC as prediction monotonicity [9] while comparing with MOS and other quality metrics. From the tables it is evident that the correlation between FRANRIQA metric and MOS of the test images is significantly better compare to other no reference quality metrics and even benchmark objective quality metric, like MSSIM.

Table 1 Comparison between Fuzzy regression based Quality Metric and Other Quality Metrics

Image Name	Normalized FRANRIQA Metric	Normalized MOS	Normalized BIQI	Normalized JPEG QUALITY SCORE	Normalized PSNR	Normalized MSSIM
kp0837	0.9387	0.9778	0.6031	0.8764	0	0.6031
kp22	1.0000	1.0000	0.5067	0.4032	1	0.5067
kp1315	0.8893	0.5222	1.0000	1.0000	0	1.0000
kp 2320	0.8774	0.3079	0.7115	0.8884	0	0.7115

Table 2 PPMCC between MOS and Other Quality Metrics

Pearson product moment correlation coefficient(PPMCC) between Normalized MOS and other quality metrics	Normalized FRANRIQA Metric	Normalized BIQI	Normalized JPEG QUALITY SCORE	Normalized PSNR	Normalized MSSIM
	0.8939	-0.6504	-0.6161	0.5796	-0.6504

Table 3 SROCC values between MOS and Other Quality Metrics

Spearman rank order correlation coefficient(SROCC) values between Normalized MOS and other quality metrics	Normalized FRANRIQA Metric	Normalized BIQI	Normalized JPEG QUALITY SCORE	Normalized PSNR	Normalized MSSIM
	1.0000	-0.8000	-0.8000	0.7746	-0.8000

6 Conclusions

An interval type 2 fuzzy regression analysis based quality metric has been described in this paper to assess quality of distorted images using Shannon entropies of visually salient regions as features. Uncertainty in image features are measured using Shanon's entropy and modelled by Interval type2 fuzzy sets to remove the limitation of selecting type-1 fuzzy membership value. Human perception on visual quality of images are assigned using five different class labels and the proposed method removes human biasness in assigning class labels by combining the crisp outputs of quality metrics. The variation of features are wide enough for capturing important information from the images. The proposed FRANRIQA metric has been compared with the existing quality metrics producing satisfactory result.

References

[1] Chakraborty, C., Chakraborty, D.: Fuzzy Linear and Polynomial Regression Modelling of 'if-Then' Fuzzy Rulebase. International Journal of Uncertainty, Fuzziness and Knowledge-Based Systems 16(2), 219–232 (2008)

[2] Mendel, J.M., John, R.I., Liu, F.: Interval Type-2 Fuzzy Logic Systems Made Simple. IEEE T. Fuzzy Systems 14(6), 808–821 (2006)

[3] Wu, D., Mendel, J.M.: Perceptual reasoning using interval type-2 fuzzy sets: Properties. FUZZ-IEEE, 1219–1226 (2008)

[4] Tanaka, H., Uejima, S., Asai, K.: Linear regression analysis with fuzzy model. IEEE Trans. Sys., Man. Cyber. 12, 903–907 (1982)

[5] Jesorsky, O., Kirchberg, K.J., Frischholz, R.W.: Robust Face Detection Using the Hausdorff Distance. In: Bigun, J., Smeraldi, F. (eds.) AVBPA 2001. LNCS, vol. 2091, pp. 90–95. Springer, Heidelberg (2001)

[6] Kadir, T., Brady, M.: Saliency, Scale and Image Description. International Journal of Computer Vision 45(2), 83–105 (2001)
[7] Tourancheau, S., Autrusseau, F., Sazzad, P.Z.M., Horita, Y.: Impact of subjective dataset on the performance of image quality metrics. In: ICIP, pp. 365–368 (2008)
[8] Wang, Z., Bovik, A.C., Sheikh, H.R., Simoncelli, E.P.: Image quality assessment: from error visibility to structural similarity. IEEE Transactions on Image Processing 13(4), 600–612 (2004)
[9] Recommendation ITU-R BT.500-13: Methodology for the subjective assessment of the quality of television pictures (2012)
[10] Kadir, T., Brady, M.: Scale, Saliency and Image Description. International Journal of Computer Vision 45(2), 83–105 (2001)
[11] Parzen, E.: On Estimation of a Probability Density Function and Mode. The Annals of Mathematical Statistics 33(3), 1065 (1962)

6. Kadir, T., Brady, M.: Saliency, scale and image description. International Journal of Computer Vision 45(2), 83–105 (2001)
7. Tourancheau, S., Autrusseau, F., Sazzad, Z.M.P., Horita, Y.: Impact of subjective dataset on the performance of image quality metrics. In: ICIP, pp. 365–368 (2008)
8. Wang, Z., Bovik, A.C., Sheikh, H.R., Simoncelli, E.P.: Image quality assessment: from error visibility to structural similarity. IEEE Transactions on Image Processing 13(4), 600–612 (2004)
9. Recommendation ITU-R BT.500-13: Methodology for the subjective assessment of the quality of television pictures (2012)
10. Kadir, T., Brady, M.: Scale, Saliency and Image Description. International Journal of Computer Vision 45(2), 83–105 (2001)
11. Parzen, E.: On Estimation of a Probability Density Function and Mode. The Annals of Mathematical Statistics 33(3), 1065–1076 (1962)

Application of Fusion Technique in Satellite Images for Change Detection

Namrata Agrawal, Dharmendra Singh, and Sandeep Kumar

Abstract. The identification of land cover transitions and changes occurred on a given region is required to understand the environmental monitoring, agricultural surveys etc. Many supervised and unsupervised change detection methods have been developed. Unsupervised method is the analysis of difference image by automatic thresholding. In this paper, an approach is proposed for automatic change detection that exploits the change information present in multiple difference images. Change detection is performed by automatically thresholding the difference image thereby classifying it into change and unchanged class. Various techniques are available to create difference image but the results are greatly inconsistent and one technique is not applicable in all situations. In this work, expectation maximization (EM) algorithm is used to determine the threshold to create the change map and intersection method is selected to fuse the change map information from multiple difference images. MODIS 250-m images are used for identifying the land cover changes.

Keywords: change detection, unsupervised techniques, expectation maximization (EM), information fusion, MODIS data.

1 Introduction

Requirement and advancement of monitoring the dynamics of land cover has led to the development of change detection techniques in remote sensing field. Change

Namrata Agrawal · Sandeep Kumar
Department of Computer Science and Engineering,
Indian Institute of Technology Roorkee, Roorkee, Uttarakhand-247667, India
e-mail: agrawal.namrata1@gmail.com, sgargfec@iitr.ac.in

Dharmendra Singh
Department of Electronics and Communication Engineering,
Indian Institute of Technology Roorkee, Roorkee, Uttarakhand-247667, India
e-mail: dharmfec@iitr.ac.in

El-Sayed M. El-Alfy et al. (eds.), *Advances in Intelligent Informatics*,
Advances in Intelligent Systems and Computing 320, DOI: 10.1007/978-3-319-11218-3_10

detection has so many applications that include forest and environment monitoring, land use and land cover change, damage assessment, disaster monitoring and other environmental changes [1]. It is the process of analyzing the multi-temporal images of same geographical area to identify the land cover changes on the earth surface. It is basically of two types: supervised (requires the ground truth information as training data for the learning process of the classifier) and unsupervised (It performs change detection by making a direct comparison of two multi-temporal images considered without using any additional information) [2]. Unsupervised approach is adopted in the paper because it eliminates the expensive and difficult task of collecting the ground truth data.

Typically used automatic change detection techniques adhere to these three phase procedure: (i) data pre-processing, (ii) pixel-by-pixel comparison of multi-temporal images, (iii) image analysis. Pre-processing is performed to reduce noise and improve the visibility of objects (water, urban, grassland, etc.) on the earth surface. After this for pixel-by-pixel comparison difference image is created. Image analysis is then performed to automatically select a threshold to classify the difference image into change and unchanged pixels. Several unsupervised change detection techniques are present in the literature which follows these three steps.

Main challenges of unsupervised change detection technique that are main focus of the paper are 1) it should be fully automatic, unlike some methods [3][4] that require manual parameter tuning, 2) selection of optimal image differencing method is difficult task, 3) context-insensitive techniques [2] are prone to isolated noise pixels, 4) use of MODIS data for change detection is difficult because of low resolution but preferred as it is free and easily available, 5) it should be robust against noise. The proposed technique overcomes all the above problems.

Various combinations of difference image method and thresholding methods exist but there is no existing optimal approach for all cases [5]. Difference image is rich in information and the selection of one of the difference image is a difficult task. Hence the information fusion is performed on the results to get the advantage of all the difference images. Main advantage behind this step is that it gives the common pixels that are assigned changed class in all of the four outputs. "sure change" pixels will be the output after the information fusion.

2 Study Area and Data Used

Study site on which the experiment is performed and the satellite database is discussed.

2.1 Study Area

Study site is the Roorkee region of Uttarakhand and Muzaffarnagar region of western Uttar Pradesh and its nearby area located at 29°35'7.5" N, 77°44'6.17"E. This study area is considered due to its richness in varied landscapes. Two major rivers (Ganga and Yamuna) are present in the study area which is prone to the abrupt changes.

2.2 Data Used

The approach is applied on two sets of MODIS/Terra Surface Reflectance 8-day L3 global 250m data of same geographical area acquired in February 2010 and February 2011. Data of February 2010 and 2011 is selected because river width is changed between this time period. It consists of spectral band 1 with bandwidth of red light (620-670nm) and band 2 with bandwidth of near infra-red light (842-876 nm). Red light is absorbed by the vegetation region and near infra red light is absorbed by the water bodies. So these two bands can distinguish water and vegetation area easily. Figure 1. shows footprint of the data set location on the Google earth LANDSAT image. MODIS image of the region bounded by the black solid boundary is used in the experiment.

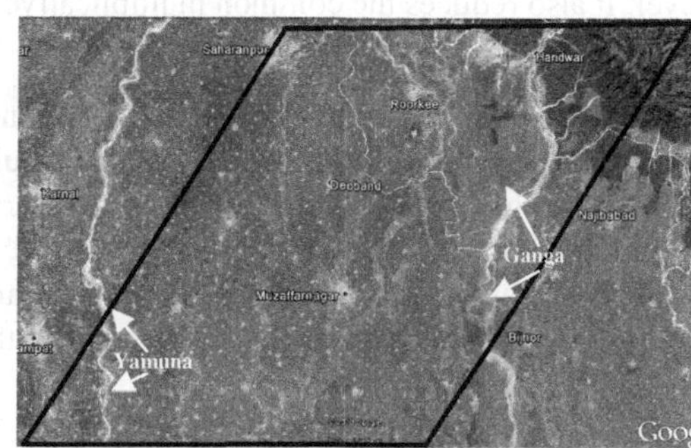

Fig. 1 Google earth image of data set used in the experiment

3 Methodology

For unsupervised change detection, a hybrid approach is proposed in which the change map outputs obtained after the analysis of four main difference image creation techniques are fused for reliable change pixels. Experiment is performed on MODIS data which has low visibility due to low resolution of images. Consequently, to extract the information about the land cover from the images, combination of spectral band is performed which is called the vegetation index calculation. Following four vegetation indexes are used as given in literature [6]: Normalized Difference VI (NDVI), Modified Soil Adjusted VI (MSAVI), Global Environment Monitoring Index (GEMI) and Purified Adjusted Vegetation Index (PAVI). These are helpful in clear identification of different objects on earth surface.

Unsupervised change detection techniques are based on difference image. Pixel by pixel comparison of two images which gives an image is called "difference image". Simple mathematical operations are performed to create difference image. Four techniques used in paper to create difference image are discussed below. Consider two images of same geographical area, $X_1=\{ x_1(i,j) \}$ and $X_2=\{ x_2(i,j) \}$ such that $1 \le i \le H, 1 \le j \le W$, with a size of H×W, acquired at two different time instances, t1 and t2, respectively. Let the output difference image is represented by X_d.

1. Image Differencing:

$$X_d=|X_2-X_1| \tag{5}$$

Simple absolute valued pixel based intensity values subtraction is performed. If there is no change then it gives a zero value and larger values for higher change [7].

2. Image Ratioing:

$$X_d=|X_1/X_2| \tag{6}$$

Pixel-by pixel division is performed to obtain the difference image. This helps in enhancing low intensity pixels and it does not depend on a reference intensity level. It also reduces the common multiplicative error [7].

3. Change Vector Analysis:
This method can process multiple spectral bands of image by creating a feature vector for each pixel. Difference image is calculated by calculating the magnitude of spectral change vector that is created by performing feature vector subtraction [8][9]. In experiment vegetation index is used as band1 and near infra-red spectrum is used as band2, representing greenness and brightness in image, respectively. This method also gives information about the direction of change as given in eq.8

$$X_d=\sqrt{((X_{1(band1)}-X_{2(band1)})^2-(X_{1(band2)}-X_{2(band2)})^2-.....(X_{1(bandN)}-X_{2(bandN)})^2)} \tag{7}$$

$$(\Theta)\ Angle=\tan^{-1}((X_{2(band1)}-X_{1(band1)})/(X_{2(band2)}-X_{1(band2)})) \tag{8}$$

4. Image Regression:
It is assumed that X_2 is a linear function of X_1 image in image regression. X_2 is hence can be estimated using least-squares regression:

$$\tilde{X}_2=aX_1+b \tag{9}$$

Parameters a and b are estimated using squared error of measured and predicted data. Difference image is then calculated as:

$$X_d=|X_2-\tilde{X}_2| \tag{10}$$

Image analysis of difference image is performed to automatically detect the threshold that classifies the pixels into change and unchanged class. Expectation maximization algorithm is used for image analysis.

3.1 Expectation Maximization Based Change Detection

Expectation maximization algorithm is basically used to estimate incomplete data. It is used as the parameter estimation of probability distribution function. When the data required for the estimation is missing or impossible to use then EM is used. As given in [2] EM follows the same steps on all four difference images. Figure 2. shows the steps followed by image analysis technique to create change map.

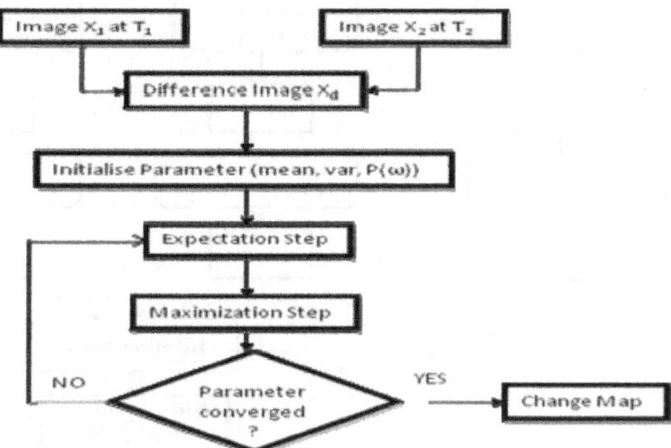

Fig. 2 General scheme of image analysis using EM algorithm

Change map gives the change and no-change pixels. Positive and negative changes are also determined based on the increase and decrease in intensity values in later date image. EM algorithm is fully automatic technique and does not require manual parameter tuning.

EM algorithm is applied on all four types of difference images resulting into different outputs except some common changed areas pixels which are present in every difference image. All differencing techniques have their some disadvantages. For example: Image differencing technique gives the absolute difference but same value may have different meaning depending on the starting class. In image ratioing, scale changes according to a single date so same change on ground may have different score. CVA technique may result in the computation of high dimensional data and it is complex to extract information from such data. Image regression is better for modeling conversion type of change (e.g. Vegetation to non-vegetation change modeling) [9]. Hence experiment is performed to fuse the output of all the four methods to minimize the limitations of the techniques.

3.2 Selection of Common Change

Fusion is performed on the outputs of EM algorithm of all four differencing method to select the common change. Intersection operator is applied for the fusion process. This will give those pixels that are changed in all the outputs. These pixels are sure change pixels. Figure 3 shows all the steps followed in the proposed approach.

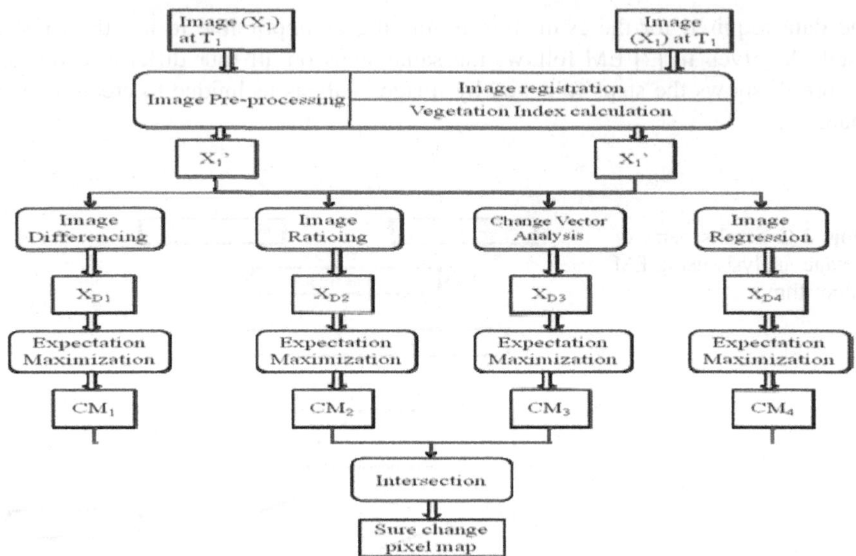

Fig. 3 Architecture of the proposed framework

4 Experiment Results and Discussion

Two MODIS 250m images of Roorkee region are used for change detection. Images of size 400x400 pixels are selected and co-registered to correctly align the images. NDVI, MSAVI, GEMI and PAVI images are obtained by applying these vegetation indexes on the pair of input image using ENVI software. Change detection algorithm is applied in MATLAB environment.

Then four difference images are computed for each type of vegetation indexed input images. After the application of EM algorithm on all four types of difference images, four change maps are obtained for each type of VI images. Figure 4. shows the input image of February 2010 and February 2011. Result of EM algorithm is shown in figure 5. In Figure 5(a)-(l) Red pixels are showing the positive change means intensity of pixel is increased in February 2011 compared to February 2010. Green pixels are showing the negative change i.e. intensity of pixel is decreased in February 2011 compared to February2010. In Figure 5(m)-(p) blue pixels are showing decreased greenness and brightness, red pixels are increased greenness and brightness; white pixels are decreased greenness and increased brightness.

It is seen in the output images that the changes are mostly present near the Ganga River and Yamuna River. There is decrement of vegetation near the river in February 2011.

Fig. 4 Band 1 of MODIS input Image of (a) Feb-2010 and Feb-2011

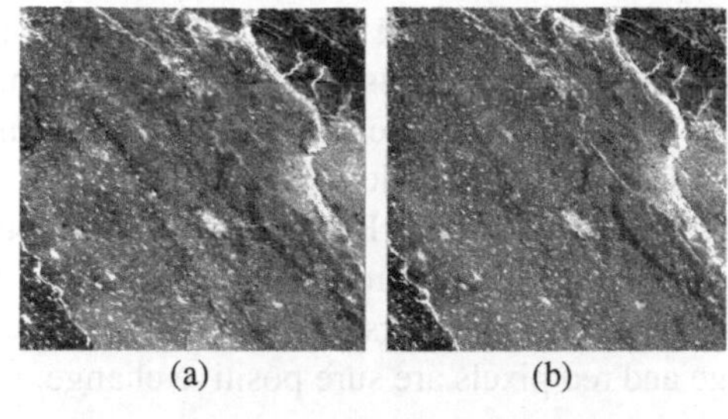

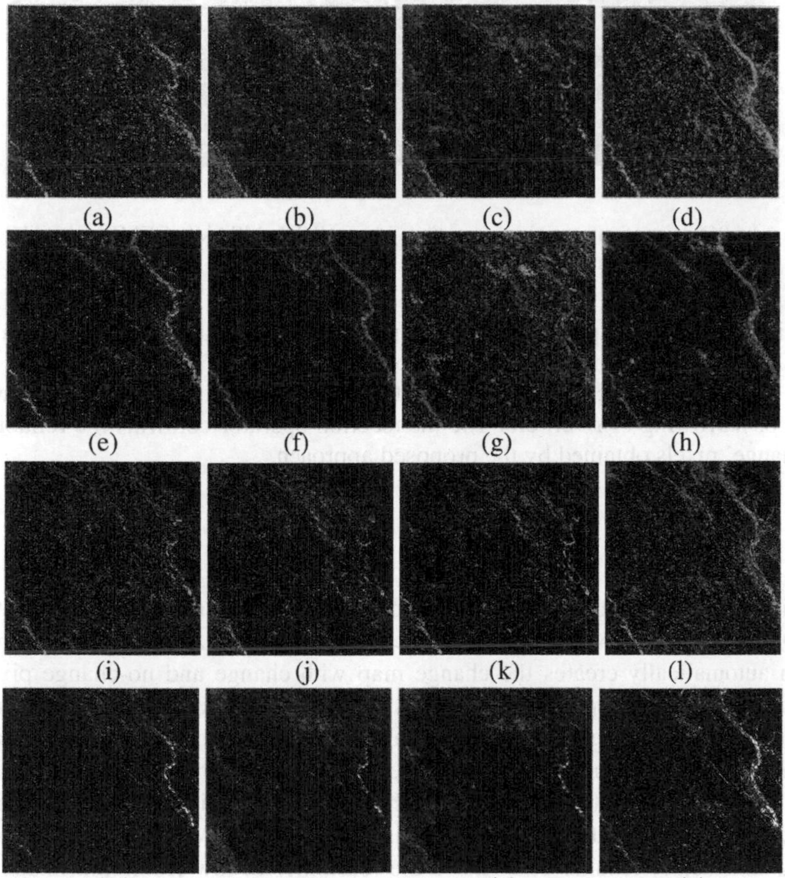

Fig. 5 EM change detection result applied on difference image of (a)-(d) Image Differencing, (e)-(h) Image Ratioing, (i)-(l) Image Regression, (m)-(p) CVA, where input image is Column1: NDVI, column2: MSAVI, column3: GEMI and column4: PAVI

It is clear from the EM outputs that all the outputs are giving different results. So, to reduce this inconsistency, common change pixels selection from all these images is performed in next step. Intersection is performed to get "sure change" pixels. Pixel that is labeled as "change" pixel in all four outputs is marked as "change" in final output. But the pixel that is labeled as "change" in one change map and "no-change" in another change map is marked as "no-change". Figure 6, shows the output of intersection operation where green pixels are sure negative change and red pixels are sure positive change.

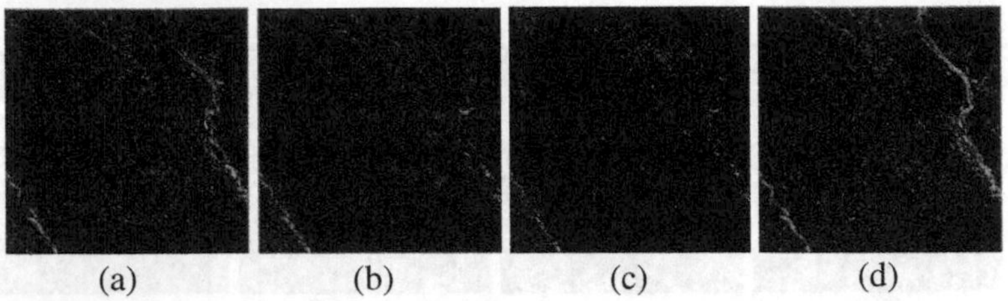

(a) (b) (c) (d)

Fig. 6 (a)-(d)Output of intersection operation when NDVI, MSAVI, GEMI and PAVI image is given as initial input, respectively

This final output helps in inferring the pixel of image that has changed with the period of time. Most of the change is around water area. The reasons behind the water region showing changes are the shifting of river from its original path, widening or narrowing of river, etc. The intersection operator confirms the reliability of 'change' pixels obtained by the proposed approach.

5 Conclusions

In this work for change detection purpose February 2010 and February 2011 MODIS data was used. Unsupervised change detection approach is proposed which automatically creates the change map with change and no-change pixels. Fusion approach is proposed in the thesis which helps in removing the difficulty of selection of one type of difference image. Also this approach does not require any apriori information. Intersection operation helps in removing the isolated noise change pixels that might be present because EM algorithm does not consider the contextual information. Experiment is performed on MODIS data and it is giving satisfactory results. This shows the feasibility of low resolution image for change detection operations. MODIS data is free, easily and very regularly available, hence is encouraged to use. Four difference image algorithms are used namely, image differencing, image ratioing, change vector analysis and image regression. Four difference images are created using these algorithms. The automatic threshold technique is applied on all four difference images separately which results in four change maps. All these change maps are similar with slight differences. So the aim was to obtain a fused change map having "sure-change" pixels. Expectation maximization technique is used for automatic thresholding and Intersection is used for the data fusion in the thesis. Final fused change map tells the change and no-change pixels.

Acknowledgments. Authors are thankful to RailTel for providing the funds to carry out this work.

References

[1] Singh, A.: Review article digital change detection techniques using remotely sensed data. International Journal of Remote Sensing 10(6), 989–1003 (1989)

[2] Bruzzone, L., Prieto, D.F.: Automatic analysis of the difference image for unsupervised change detection. IEEE Transaction on Geoscience and Remote Sensing 38(3), 1170–1182 (2000)

[3] Celik, T.: Unsupervised change detection in satellite images using principal component analysis and k-means clustering. IEEE Geoscience and Remote Sensing Letters 6(4), 772–776 (2009)

[4] Yetgin, Z.: Unsupervised change detection of satellite images using local gradual descent. IEEE Transaction on Geoscience and Remote Sensing 50(5), 1919–1929 (2012)

[5] Liu, S., Du, P., Gamba, P., Xia, J.: Fusion of difference images for change detection in urban areas. In: 2011 Joint Urban Remote Sensing Event (JURSE), pp. 165–168. IEEE Xplore, Munich (2011)

[6] Singh, D., Meirelles, M.S.P., Costa, G.A., et al.: Environmental degradation analysis using NOAA/AVHRR data. Advances in Space Research 37(4), 720–727 (2006)

[7] İlsever, M., Ünsalan, C.: Pixel-based change detection methods. In: Two-Dimensional Change Detection Methods, pp. 7–21. Springer, London (2012)

[8] Roemer, H., Kaiser, G., Sterr, H., et al.: Using remote sensing to assess tsunami-induced impacts on coastal forest ecosystems at the Andaman Sea coast of Thailand. Natural Hazards and Earth System Sciences 10(4) (2010)

[9] Chen, H.D., Cheng, A., Wei, H., et al.: Change detection from remotely sensed images: From pixel-based to object-based approaches. Masroor ISPRS Journal of Photogrammetry and Remote Sensing 80, 91–106 (2013)

A Survey on Spiking Neural Networks in Image Processing

Julia Tressa Jose, J. Amudha, and G. Sanjay

Abstract. Spiking Neural Networks are the third generation of Artificial Neural Networks and is fast gaining interest among researchers in image processing applications. The paper attempts to provide a state-of-the-art of SNNs in image processing. Several existing works have been surveyed and the probable research gap has been exposed.

Keywords: Spiking Neural Networks, Computer Vision, Image Processing.

1 Introduction

The mind boggling performance of the biological neural structure has always been a topic of extensive research. Artificial Neural Networks (ANNs) were developed to mimic the behaviour of biological neurons; the first ideas and models are over fifty years old. Even a simple mathematical model, such as the McCulloch-Pitts threshold model was found to be computationally powerful and has been used to solve a wide variety of engineering problems.

This paper focuses on the third generation of Artificial Neural Networks namely, Spiking Neural Networks (SNNs), which has been recently gaining popularity in Image Processing and Computer Vision applications. A survey of the existing works on Spiking Neural Networks in the field of Image Processing is presented to the reader. An attempt is also made to expose the research gap.

Section 2 introduces the concept of Spiking Neural Networks and its advantages. The various mathematical models of SNNs, which are computationally feasible, are presented in section 3. Section 4 highlights the major works done in

Julia Tressa Jose · J. Amudha · G. Sanjay
Department of Computer Science and Engineering,
Amrita School of Engineering, Amrita Vishwa Vidyapeetham, Bangalore, India
e-mail: {julia.jose.t,sanjaynair1989}@gmail.com,
j_amudha@blr.amrita.edu

El-Sayed M. El-Alfy et al. (eds.), *Advances in Intelligent Informatics*,
Advances in Intelligent Systems and Computing 320, DOI: 10.1007/978-3-319-11218-3_11

the field of Image Processing using these networks and finally Section 5 concludes the survey.

2 Spiking Neural Networks – The Third Generation of ANNs

Biological neurons communicate through action potentials or spikes which are transmitted from one neuron to the other via their chemical junction called synapse. A spike is generated only when the membrane potential of a neuron exceeds a certain threshold. The shapc of the action potential is always the same and hence does not carry any relevant information. Thus a main question is how neurons encode information in the sequence of action potentials that thcy cmit. Several theories have evolved as an answer to this question.

The evolution of ANNs can be classified into three generations [1]. The first generation of McCulloch-Pitts threshold neurons, assumes that the mere occurrence of a spike carries relevant information. Thus neurons are considered as a binary device, where a 'high' denotes spike occurrence and low denotes absence. They have been successfully applied in powerful artificial neural networks like multi-layer perceptron and Hopfield nets. The second generation neuron, called sigmoidal gate, considers that the number of spikes per second (known as firing rate) encodes relevant information. They are more powerful than their predecessors and can approximate any analog function arbitrarily well.

Experiments reveal that the human nervous system is able to perform complex visual tasks in time intervals as short as 150ms [2]. Furthermore, the firing rates of the areas involved in such complex computations are well below 100 spikes per second (see Figure 1). These results indicate that biological neural systems use the exact timing of individual spikes and that the firing rate alone does not carry all the relevant information.

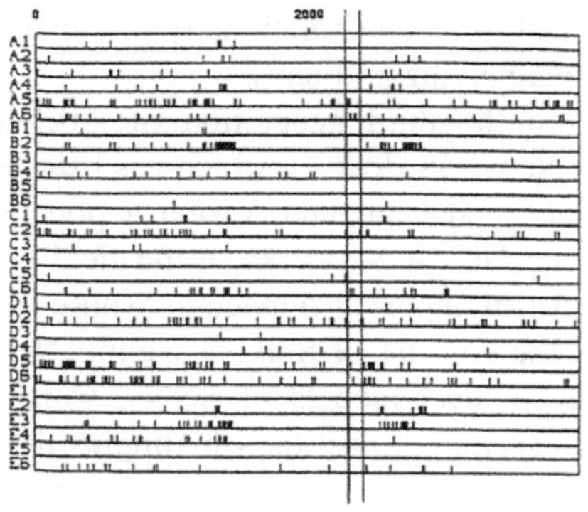

Fig. 1 Simultaneous recordings of 30 neurons in the visual cortex of a monkey. The bar marks a time interval of 150–ms. This short time interval suffices to perform complex computations. However, one sees that there are only a few spikes within this time interval[1].

The third generation, called the Spiking Neuron model, incorporates the timing of individual spikes and hence is more biologically plausible. It can compute the same functions as the sigmoidal gate, which in principle means any function, using a smaller network compared to the former. The computational time required is ~ 100ms. It is also said to be more flexible for computer vision applications [3].

3 Spiking Neuron Models

There are many different schemes for incorporating spike timing information in neural computation. Alan Lloyd Hodgkin and Andrew Huxley received the Nobel Prize in 1963, for their Hodgkin-Huxley (HH) Model which gives the best approximation of a biological neuron. However, the model is too complicated to be used in artificial neural networks. Hence, simplified implementable models, which are close to the HH model, began to be developed. The Integrate and Fire (IF) model [5], and the Spike Response (SRM) model [6] are the most commonly used SNNs for Image Processing.

The IF model and SRM are classified under threshold based SNN models [6], which are based on the fact that, a neuron fires when its membrane potential exceeds a threshold. The IF model is the basic model and is most well-known. It is approximately 1000 times faster than the HH model. However it disregards the refractory capability of neurons.

The SRM model, on the other hand, includes refractoriness. It approximates the very detailed Hodgkin-Huxley model very well and captures generic properties of neural activity. In a public competition of spike-time prediction under random conductance injection, the Spike Response Model was one of the award winning models[7] whereas a standard leaky integrate-and-fire model performed significantly worse. A simplified variant, SRM0 needs less data and is easier to fit to experimental data than the full Spike Response Model.

A variation of the IF model is the conductance based Integrate and Fire model [8,9]. It simulates the behavior of ionic channels in neurons. The conductance based IF is also said to approximate the HH model.

Most of the works presented in this paper are based either on the conductance based IF or the SRM (SRM0) model. However, no literature has been found on a direct comparison of these two models to indicate the better model.

4 SNNs in Image Processing

It has been observed that two main SNN architectures have been repeatedly used in computer vision applications. We name them informally as – Structure I and Structure II. There is a small section on other networks as well; however they have not been explored in detail.

4.1 Structure I

This architecture, based on conductance based IF neurons, was first introduced in [10] for edge detection in images, after which it was extended to include several other applications such as colour segmentation [11], corner detection [13] and visual attention [14].

Structure I is inspired by the nature of the primate visual cortex. It consists of various receptive fields (RFs) for orientation or colour selectivity. For example, figure 2 shows the architecture used for edge detection [10]. The different weight matrices are used for filtering different orientations –up, down, left and right.

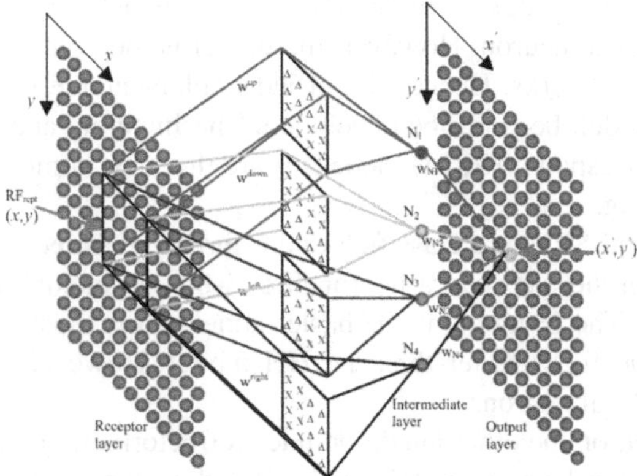

Fig. 2 Structure I: Spiking Neuron Model for edge detection [10]

The neuron firing rate map obtained using this architecture provides better results in comparison with Sobel and Canny edges, shown in fig. 3.

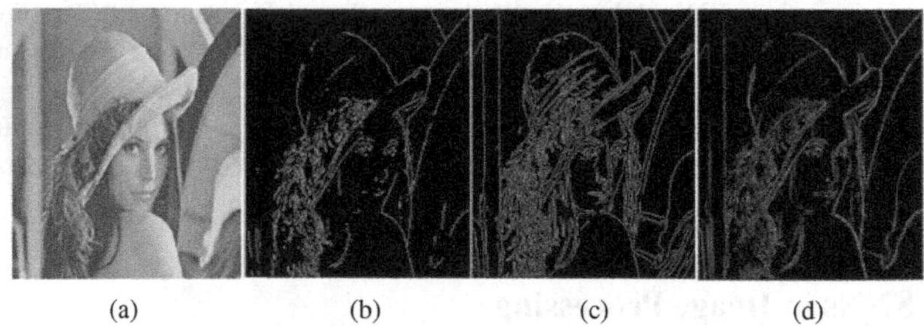

Fig. 3 Comparison of SNN with Canny and Sobel edges [10]. (a) Original image (b) Sobel edge (c) Canny edge (d) Neuron firing rate map.

In [12], hexagonal shaped receptive fields (HSNN) are used instead of the rectangular RFs shown in fig. 2. The processing time was faster compared to the previous model and edge detection was found to be better as the RF size increased, see fig. 4.

In [11], structure I was used to filter RGB images to different colour ON/OFF pathways. In conjunction with a back propagation network, the model could successfully segment images. Corner detection [13] was achieved using an IF model implementation of the structure. It consisted of 4 processing layers: receptor layer, edge detection, corner detection & output layer.

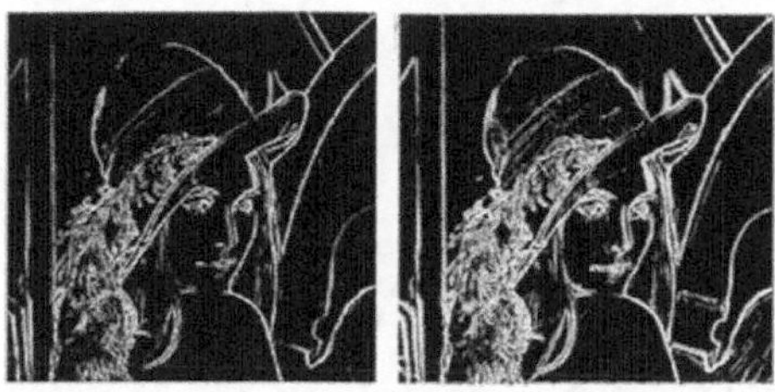

Fig. 4 (a)SNN [10] (b) 37- point HSNN [12]

The authors of [14] propose a hierarchical spiking neural model which uses spike-rate maps from the retina-inspired models of [10, 11] and combines it with a new top–down volition-controlled model to obtain visual attention areas. The saliency index was compared to the Multi-scale model and Itti's model and was found to produce desirable results (see fig.5).

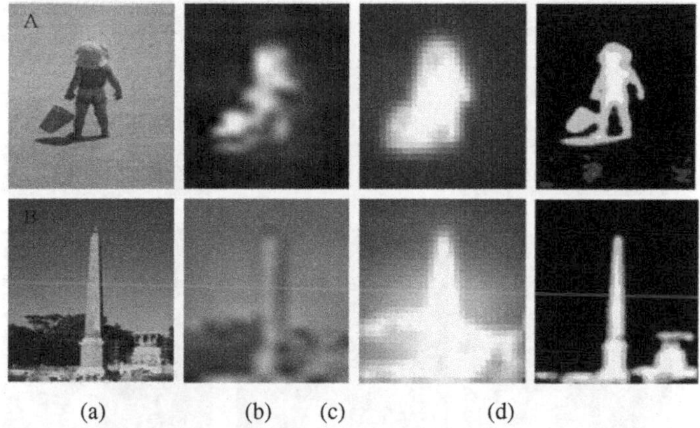

Fig. 5 Column a: Sample Image, Column b: Itti's model, Column c: Multi-scale model, Column d: SNN model for visual attention [14]

4.2 Structure II

The network architecture consists of a feed-forward network of spiking neurons with multiple delayed synaptic terminals as in fig. 6. It was introduced in [6], where spiking neuron behaviour is considered to be similar to a Radial Basis Function (RBF). A neuron fired earlier if the incoming spikes coincided more. Synapses act as a spike pattern storing unit which makes the neuron fix to a certain pattern by adjusting time delays so that all incoming spikes reach the soma simultaneously. The structure is implemented using SRM0 model.

The hidden layer consists of Gaussian receptive field units responsible for converting real values to temporal values (spikes). These temporally encoded values are then transmitted through a multi-synapse network to the output layer. Such synapses are used in brain areas like the neocortex. Each connection from the hidden unit to output neuron consists of multiple connections, each with its own delay and weight, as shown in fig. 6(b). Such a structure enables an adequate delay selection for learning purposes [6].

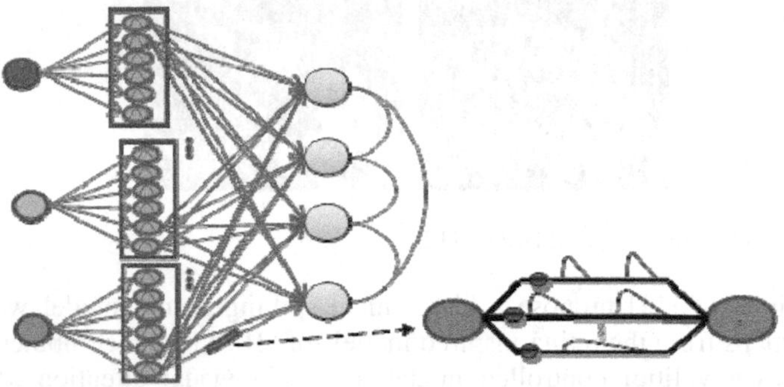

Fig. 6 (a) Structure II (b) Multiple synapse architecture [15]

Fig. 7 Image processing using self-learning Fuzzy spiking neural networks. a) Original image; b) Training set (30% of the original image); c)1st epoch of self-learning fuzzy spiking neural network learning; d) 3rd epoch of self-learning fuzzy spiking neural network learning; e) 3rd epoch of fuzzy cmeans learning; f) 30th epoch of fuzzy c-means learning [16]

In [15], structure II is implemented to perform edge detection and colour segmentation. The colour segmentation results are compared with those using Kmeans and SOM network. Edge detection was compared with Prewitt, Black top hat and Canny edge detection. The spiking neural network was found to perform better in both cases. In [16], a self-learning fuzzy spiking neural network was proposed that combined Structure II with fuzzy probabilistic and fuzzy possibilistic clustering algorithms. It was shown that (see fig.7) this system could reduce the number of learning epochs as compared to conventional fuzzy clustering algorithms.

The author of [6] also talks about replacing the multiple synapse architecture with a single synapse. However this structure has not been experimented much in image processing applications. Table 1 summarizes the features of both the structures and the image processing applications in which they have been used.

Table 1 Comparison of Structure I and Structure II

Model	SNN model	Biological Basis	Applications in Image Processing
Structure I	Conductance based IF	Filtering using Receptive fields	• Edge Detection • Colour segmentation • Corner Detection • Visual Attention
Structure II	SRM0	Learning using Multi synapse	• Edge Detection • Colour segmentation

4.3 Other Structures

In [17] a distributed network of IF neurons was proposed for salient region extraction considering only the luminance feature. The model performed a bit slower and had issues with synchronization. [18] integrates a network of IF neurons with K means clustering for colour segmentation. Several other structures [19-20] based on IF neuron model have also been developed for image segmentation. Table 2 summarizes the major works in image processing using SNNs, in a chronological manner.

Table 2 Summary of major works

Application	**Year**	**Model**	**Other features**
Edge Detection[10]	2007	Structure I	-
Colour segmentation[16]	2009	Structure II using IF	Combination with Fuzzy system
Saliency extraction[17]	2009	Other structures, IF	Distributed network, Only considers bottom up saliency.
Colour segmentation[11]	2010	Structure I	Combined with BP network
Edge Detection and colour segmentation[15]	2010	Structure II	-
Edge detection[12]	2011	Structure I	Hexagonal receptive field (mimics arrangement in retina)
Corner Detection[13]	2011	Structure I using IF	4 processing layers: receptor layer, edge detection, corner detection, output layer.
Colour segmentation[18]	2012	Other structure, IF	Integrated with Kmeans clustering
Visual Attention[14]	2013	Structure I	Top down approach, Hierarchical architecture

5 Conclusion

The paper presented several works on Image processing using Spiking Neural Networks. The following research gaps have been identified:

- Most of the works swirl around edge detection and colour segmentation. Other areas like object recognition have been less explored.
- No literature was found providing a direct comparison of the performances of Structure I and Structure II.
- A detailed performance evaluation of spiking neural models in comparison with other existing models for image processing has to be done, especially in terms computational time. This would help in identifying the disadvantages of the models, if any.
- The models have not been compared with each other. For example, there are four different models for colour segmentation which have not been evaluated against each other.

As a future enhancement, the survey can be extended to include literature on models using various learning algorithms and coding schemes of SNNs. A survey of the works containing hardware implementation of these networks can also be done.

References

[1] Vreeken, J.: Spiking neural networks, an introduction. Technical Report, Institute for Information and Computing Sciences, Utrecht University, pp. 1–5
[2] Natschläger, T.: Networks of Spiking Neurons: A New Generation of Neural Network Models (December 1998)
[3] Thorpe, S.J., Delorme, A., VanRullen, R.: Spike-based strategies for rapid processing. Neural Networks 14(6-7), 715–726 (2001)
[4] Feng, J., Brown, D.: Integrate-and-fire Models with Nonlinear Leakage. Bulletin of Mathematical Biology 62, 467–481 (2000)
[5] Gerstner, W., Kistler, W.: Spiking Neuron Models: Single Neurons, Populations, Plasticity. Cambridge University Press (2002)
[6] De Berredo, R.C.: A review of spiking neuron models and applications. M. Sc. Dissertation, University of Minas Gerais (2005)
[7] Jolivet, R., Kobayashi, R., Rauch, A., Naud, R., Shinomoto, S., Gerstner, W.: A benchmark test for a quantitative assessment of simple neuron models. Journal of Neuroscience Methods 169, 417–424 (2008)
[8] Müller, E.: Simulation of High-Conductance States in Cortical Neural Networks. Master's thesis, University of Heidelberg, HD-KIP-03-22 (2003)
[9] Dayan, P., Abbott, L.F.: Theoretical Neuroscience: Computational and Mathematical Modeling of Neural Systems. The MIT Press, Cambridge (2001)
[10] Wu, Q., McGinnity, M., Maguire, L.P., Belatreche, A., Glackin, B.: Edge Detection Based on Spiking Neural Network Model. In: Huang, D.-S., Heutte, L., Loog, M. (eds.) ICIC 2007. LNCS (LNAI), vol. 4682, pp. 26–34. Springer, Heidelberg (2007)

[11] Wu, Q., McGinnity, T.M., Maguire, L., Valderrama-Gonzalez, G.D., Dempster, P.: Colour Image Segmentation Based on a Spiking Neural Network Model Inspired by the Visual System. In: Huang, D.-S., Zhao, Z., Bevilacqua, V., Figueroa, J.C. (eds.) ICIC 2010. LNCS, vol. 6215, pp. 49–57. Springer, Heidelberg (2010)

[12] Kerr, D., Coleman, S., McGinnity, M., Wu, Q.X.: Biologically inspired edge detection. In: International Conference on Intelligent Systems Design and Applications, pp. 802–807 (2011)

[13] Kerr, D., McGinnity, M., Coleman, S., Wu, Q., Clogenson, M.: Spiking hierarchical neural network for corner detection. In: International Conference on Neural Computation Theory and Applications, pp. 230–235 (2011)

[14] Wu, Q.X., McGinnity, T.M., Maguire, L.P., Cai, R., Chen, M.: A Visual Attention model using hierarchical spiking neural networks. In: Advanced Theory and Methodology in Intelligent Computing, vol. 116, pp. 3–12 (September 2013)

[15] Meftah, B., Lezoray, O., Benyettou, A.: Segmentation and Edge Detection Based on Spiking Neural Network Model. Neural Processing Letters 32(2), 131–146 (2010)

[16] Bodyanskiy, Y., Dolotov, A.: Analog-digital self-learning fuzzy spiking neural network in image processing problems. In: Chen, Y.-S. (ed.) Image Processing, pp. 357–380. In-Teh, Vukovar (2009)

[17] Chevallier, S., Tarroux, P., Paugam-Moisy, H.: Saliency extraction with a distributed spiking neuron network. In: Advances in Computational Intelligence and Learning, pp. 209–214 (2006)

[18] Chaturvedi, S., Meftah, B., Khurshid, A.A.: Image Segmentation using Leaky Integrate and Fire Model of Spiking Neural Network. International Journal of Wisdom Based Computing 2(1), 21–28 (2012)

[19] Buhmann, J., Lange, T., Ramacher, U.: Image segmentation by networks of spiking neurons. Neural Computing 17(5), 1010–1031 (2005)

[20] Rowcliffe, P., Feng, J., Buxton, H.: Clustering within integrate-and-fire neurons for image segmentation. In: Dorronsoro, J.R. (ed.) ICANN 2002. LNCS, vol. 2415, pp. 69–74. Springer, Heidelberg (2002)

Moving Human Detection in Video Using Dynamic Visual Attention Model

G. Sanjay, J. Amudha, and Julia Tressa Jose

Abstract. Visual Attention algorithms have been extensively used for object detection in images. However, the use of these algorithms for video analysis has been less explored. Many of the techniques proposed, though accurate and robust, still require a huge amount of time for processing large sized video data. Thus this paper introduces a fast and computationally inexpensive technique for detecting regions corresponding to moving humans in surveillance videos. It is based on the dynamic saliency model and is robust to noise and illumination variation. Results indicate successful extraction of moving human regions with minimum noise, and faster performance in comparison to other models. The model works best in sparsely crowded scenarios.

Keywords: Moving Region Extraction, Visual Attention, Video Surveillance.

1 Introduction

Surveillance cameras are inexpensive and everywhere these days. However, manually monitoring these surveillance videos is a tiresome task and requires undivided attention. The goal of an automated visual surveillance system is to develop intelligent visual surveillance which can obtain a description of what is happening in a monitored area automatically, with minimum support from an operator, and then take appropriate actions based on that interpretation [1]. Automatic visual surveillance in dynamic scenes, especially for monitoring human activity, is one of the most active research topics in computer vision and artificial intelligence.

G. Sanjay · J. Amudha · Julia Tressa Jose
Department of Computer Science and Engineering,
Amrita School of Engineering, Amrita Vishwa Vidyapeetham, Bangalore, India
e-mail: {sanjaynair1989,julia.jose.t}@gmail.com,
j_amudha@blr.amrita.edu

El-Sayed M. El-Alfy et al. (eds.), *Advances in Intelligent Informatics*,
Advances in Intelligent Systems and Computing 320, DOI: 10.1007/978-3-319-11218-3_12

Extracting regions corresponding to moving humans is one of the most crucial initial steps in visual surveillance. It involves two steps – detection of moving regions and then extracting regions corresponding to humans. A typical approach first models the complex background, and then subtracts the background from each input frame to obtain foreground objects. Although the existing methods have achieved good detection results, most of them are computationally expensive. This factor is of utmost importance and should be kept to the minimum when a system has to be deployed in real time.

In the past few years, Visual Attention algorithms [2,3,4] have been extensively used in image and video processing as they are regarded to be computationally cost effective. These algorithms identify salient regions as foreground, allowing unimportant background regions to be largely ignored. By doing this, they enable processing to concentrate on regions of interest analogous to our human visual system.

Visual Attention has been extensively researched in images but there are relatively few works on video processing. Many models for images can be extended to include video data, however, not all approaches are fast, especially when the video content to be analysed has a large number of frames. Moreover, motion plays a crucial role in video and is more salient compared to other features such as colour and intensity.

This paper proposes a simplified technique for detecting moving humans in a video footage using dynamic visual attention model. The model processes more than a few hundred frames in less than thirty seconds and is robust to environmental noise and illumination.

Section 2 presents a literature survey on the existing visual attention models for moving object detection in video. The proposed model is presented in section 3, followed by results and conclusion in sections 4 and 5 respectively.

2 Visual Attention Models for Object Detection in Video

There are relatively few works which focuses on object detection in video using visual attention models. Salient objects in a video are identified using either static or dynamic attention cues. In a static saliency based model [5], the object stands out from its neighbours based on colour, intensity, orientation etc. No motion is considered. On the other hand, dynamic saliency based models [9, 10] give prominence to objects whose motion is salient to its background. Some models [6, 8], deploy a combination of static and dynamic cues to identify salient objects.

Some models [5, 6], rely on keyframe extraction as the initial step for video processing and object identification. This method summarizes the video content producing only the frames where relevant activity is found. In [5], this process is performed using the histogram keyframe extraction technique. A saliency map is then generated for the key frame using static attention cues based on the extended Itti Koch model [2]. This map indicates regions of relevance in a frame. The model selects human entities in the map through aspect ratio analysis. The average aspect ratio of all salient objects forms the threshold value T_h. Regions having aspect ratios that exceed T_h are identified as most salient by the system. The model has some difficulty in distinguishing objects having aspect ratio similar to humans.

There are models [6], which extract key frames through a combination of static and dynamic visual attention models (VAM), and thereby produce dynamic and static saliency maps. A VAI (Visual attention index) curve is obtained with the help of these maps. The peaks of the curve become the key frame candidate. For motion extraction from the frame, hierarchical gradient based optical flow method is used [7]. Finding two attention maps for each of the video frames becomes computationally expensive.

In contrast to the two methods discussed above, [8] does not use the concept of keyframe extraction at all. Instead, a motion attention map is generated by taking the continuous symmetry difference of consecutive frames. In addition to it, a static feature attention map and a Karhunen-Loeve transform (KLT) distribution map is computed. The final spatiotemporal saliency map is calculated as the weighted sum of these three maps. Each frame requires three attention maps to be obtained and summed which becomes a complex process when the number of frames is large.

In [9], a background reference frame is first created by averaging a series of frames in an unchanging video sequence. Motion vectors corresponding to moving regions are then found by calculating the intensity difference between current and reference frame. A second stage then applies a region growing and matching technique to these motion vectors to obtain motion segmentation. The method relies on an accurate background reference frame which may not be feasible in many cases. In the Mancas Model [10], motion features are extracted by making use of Farneback's optical flow algorithm. However, most optical flow methods are regarded as computationally complex and sensitive to noise.

Though the approaches discussed so far are based on visual attention, they still require extensive computations and a lot of time for processing large number of frames. Moreover only [5] discusses a method to extract humans from a video footage.

Therefore the aim of this paper is to propose a fast, simple yet robust technique to detect moving objects from a video and then segment regions corresponding to humans. The first step uses a modified version of the motion attention map proposed in [8]. For segmenting humans, aspect ratio analysis of [5] is used.

3 Moving Human Detection Using Visual Attention

This paper uses a modified version of the motion attention map, generated by using the continuous symmetry difference method [8]. It considers only motion features; no static cues about colours, grey levels or orientations are included in the model. The block diagram of the proposed approach is depicted in Figure 1.

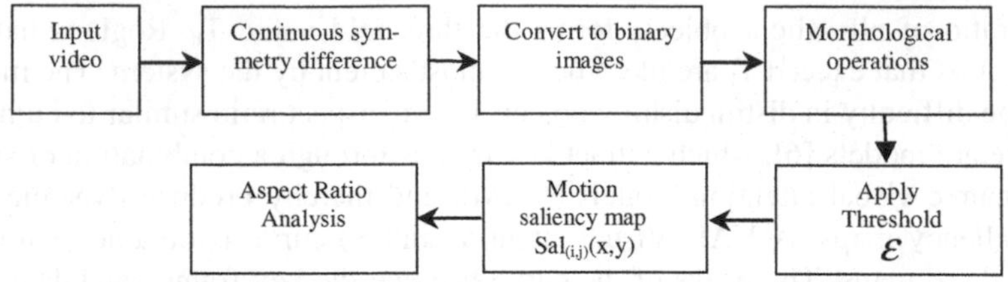

Fig. 1 Block Diagram of Proposed method

The difference of adjacent images is calculated using the following formula:

$$dif_{(i,j)}(x,y) = |w * I_i(x,y) - w * I_j(x,y)| \quad (1)$$

where j = i+n, n, the frame difference, is chosen from 7-9 and w is a Gaussian filter function. Choosing such a high frame difference, does not lead to loss in information as the number of frames to be analyzed is more than a few hundred. This also speeds up the processing. The choice of σ value of the Gaussian filter is also significant. A lower value reduces the overall noise; however there might be more holes in the moving region detected. A higher value reduces the holes but leads to increase in overall noise. The typical value of σ ranges from 0.5-4.

The differenced image sequence is converted from RGB to binary. A few morphological operations (such as the imfill and imclose functions of MATLAB) are performed to fill the holes generated in the moving regions. To reduce output noise, values of difference less than a threshold ε are set to zero. For noise removal and to avoid missing moving object, ε is set to 1-2,

$$Sal_{(i,j)}(x,y) = \begin{cases} 1, \text{ otherwise} \\ 0, \text{ if } dif_{(i,j)}(x,y) < \varepsilon \end{cases} \quad (2)$$

The next important step is to separate moving regions corresponding to humans. Blobs having area greater than 100-150 are preselected to perform aspect ratio analysis. This value may vary depending on how far the camera is positioned from the actual scene. Aspect ratio of each detected blob can be calculated from its bounding box parameters.

$$Aspect\_Ratio(R_i) = \frac{\Delta y}{\Delta x} \quad (3)$$

where, Δy = Difference between two y extremes for the i_{th} blob in a frame,
Δx = Difference between two x extremes for the i_{th} blob in a frame.

This method is adapted from [5] where it is observed that human aspect ratio falls in the range 1 – 1.5. Blobs having lower aspect ratios are masked and thus eliminated.

In some cases an additional threshold θ (eqn. 2) can also be applied to the difference before conversion to binary, for further noise reduction. θ can be in the range 7-9. This alternate method is depicted in figure 2.

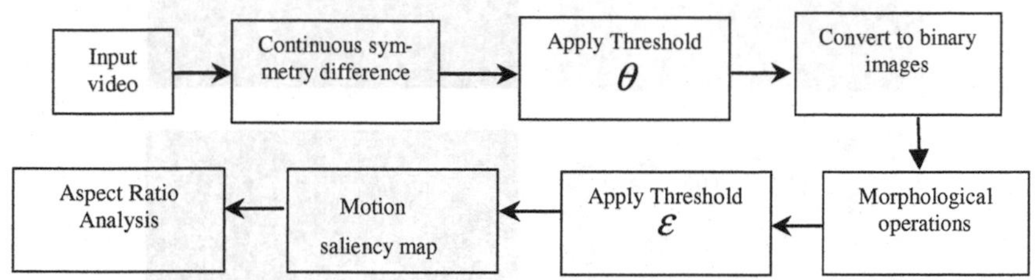

Fig. 2 Alternate Block diagram (for improved noise reduction)

4 Results and Analysis

Four video sequences are chosen for evaluating the proposed method – video1, a video sequence of a public park [13]; video2, a part of PETS2001 dataset [12], and video3 with low lighting [10]; video4, a crowded scenario from the UCSD anomaly detection dataset [14]. All the experiments were conducted using MATLAB R2013a on an Intel Core i5-3210M CPU, running at 2.50 GHz.

Figure 3 shows the results of the proposed method. Regions corresponding to moving humans were successfully extracted from the first three videos in minimum time. The method provides good results even in dim lighting, as indicated by the second last row in fig. 3. The use of frame difference method contributes to the increased computational speed and better performance in low lighting. Setting a threshold value and filtering of detected blobs based on area and aspect ratio are the main factors which lead to decreased noise. However, as the crowd density increases, the model is not able to detect all the regions corresponding to moving humans, depicted in the last row of fig. 3. It works best in sparsely crowded scenarios.

Figure 4 shows a comparison of the execution times of the proposed method with [5], [6] and [10]. The proposed method took approx. 40 seconds for processing 700 frames of video1, whereas even a partial implementation of the other methods took more than a few minutes. This reduced execution time makes our method more suitable for real time applications.

Some models [5,6] identify relevant objects in a scenario by computing saliency map of the video frames. This approach was tested using two static attention models - Itti Koch[2]and Global rarity attention map[11]. The results in figure 5 indicate that the saliency maps contain other objects in the frames, in addition to moving objects. Extracting relevant moving objects from them proves to be a difficult task.

Fig. 3 Column (a)-Input: Moving Regions in video (indicated by yellow boxes); Column (b)-Output: Regions detected by proposed method. Each row indicates one of the four video datasets.

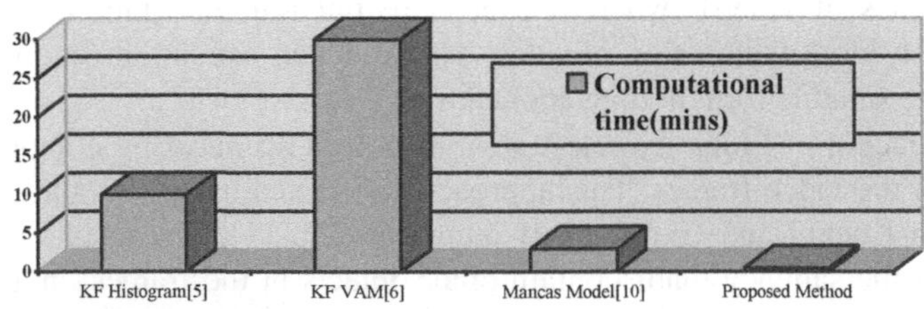

Fig. 4 Comparing execution times of various models

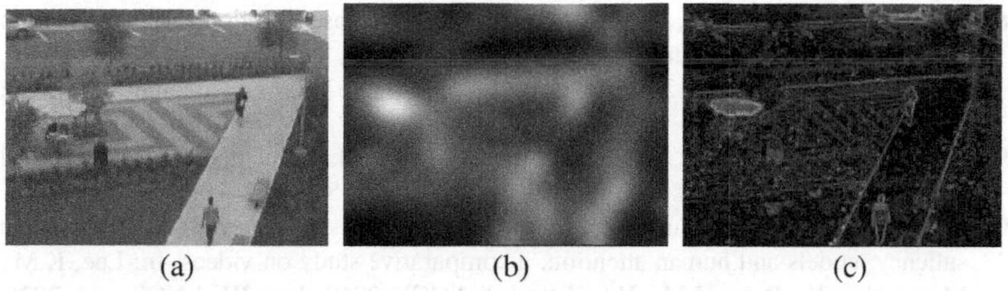

Fig. 5 (a) Original Frame, (b) Itti Koch map, (c) Global Rarity Map

5 Conclusion

A visual attention system for segmenting regions corresponding to moving humans is proposed in this paper. It is based on the dynamic saliency model where attention is mainly based on motion. The results obtained indicate that the model requires less time for computation compared to other models. It is quite robust to illumination and environmental noise. The model is more suited for simple and sparsely crowded scenarios. The values of the parameters: σ of Gaussian filter and n, the frame difference may have to be adjusted for different scenarios. As a future enhancement the system can be implemented in hardware, for example in robotic vision applications and tested for computational efficiency.

References

[1] Dick, A., Brooks, M.: Issues in Automated Visual Surveillance. In: Proceedings of International Conference on Digital Image Computing: Techniques and Application, pp. 195–204 (2003)

[2] Itti, L., Koch, C., Niebur, E.: A model of saliency-based visual attention for rapid scene analysis (1998)

[3] Frintrop, S.: VOCUS: A Visual Attention System for Object Detection and Goal-Directed Search. LNCS (LNAI), vol. 3899. Springer, Heidelberg (2006)

[4] Amudha, J., Soman, K.P., Padmakar Reddy, S.: A Knowledge Driven Computational Visual Attention Model. International Journal of Computer Science Issues 8(3(1)) (2011)

[5] Radha, D., Amudha, J., Ramyasree, P., Ravindran, R., Shalini, S.: Detection of Unauthorized Human Entity in Surveillance Video. International Journal of Engineering and Technology 5(3) (2013)

[6] Amudha, J., Mathur, P.: Keyframe Identification using Visual Attention Model. In: International Conference on Recent Trends in Computer Science and Engineering, Chennai, pp. 55–55 (2012)

[7] Brox, T., Bruhn, A., Papenberg, N., Weickert, J.: High accuracy optical flow estimation based on theory for warping. In: Pajdla, T., Matas, J(G.) (eds.) ECCV 2004. LNCS, vol. 3024, pp. 25–36. Springer, Heidelberg (2004)

[8] Guo, W., Xu, C., Ma, S., Xu, M.: Visual Attention Based Motion Object Detection and Trajectory Tracking. In: Qiu, G., Lam, K.M., Kiya, H., Xue, X.-Y., Kuo, C.-C.J., Lew, M.S. (eds.) PCM 2010, Part II. LNCS, vol. 6298, pp. 462–470. Springer, Heidelberg (2010)
[9] Zhang, S., Stentiford, F.: A saliency based object tracking method. In: International Workshop on Content-Based Multimedia Indexing, pp. 512–517 (2008)
[10] Riche, N., Mancas, M., Culibrk, D., Crnojevic, V., Gosselin, B., Dutoit, T.: Dynamic saliency models and human attention: a comparative study on videos. In: Lee, K.M., Matsushita, Y., Rehg, J.M., Hu, Z. (eds.) ACCV 2012, Part III. LNCS, vol. 7726, pp. 586–598. Springer, Heidelberg (2013)
[11] Mancas, M., Mancas-Thillou, C., Gosselin, B., Macq, B.: A Rarity-Based Visual Attention Map - Application to Texture Description. In: IEEE International Conference on Image Processing, pp. 445–448 (2006)
[12] Performance Evaluation and Tracking and Surveillance, PETS (2001), `http://ftp.pets.rdg.ac.uk/PETS2001/DATASET1/`
[13] Basharat, A., Gritai, A., Shah, M.: Learning object motion patterns for anomaly detection and improved object detection. In: IEEE Conference on Computer Vision and Pattern Recognition, pp. 1–8 (2008)
[14] Mahadevan, V., Li, W., Bhalodia, V., Vasconcelos, N.: Anomaly Detection in Crowded Scenes. In: Proc. IEEE Conference on Computer Vision and Pattern Recognition (CVPR), San Francisco, CA (2010)

Comparative Analysis of Radial Basis Functions with SAR Images in Artificial Neural Network

Abhisek Paul*, Paritosh Bhattacharya, and Santi Prasad Maity

Abstract. Radial Basis Functions (RBFs) is used to optimize many mathematical computations. In this paper we have used Gaussian RBF (GRBF), Multi-Quadratic RBF (MQ-RBF), Inverse-Multi-Quadratic RBF (IMQRBF) and q-Gaussian RBF (q-GRBF) to approximate singular values of SAR (Synthetic Aperture Radar) color images. Simulations, mathematical comparisons show that q-Gaussian RBF gives better approximation with respect to the other RBF methods in Artificial Neural Network.

Keywords: Neural network, Gaussian, Inverse-Multi-Quadratic, Gaussian, q-Gaussian, Radial Basis Function.

1 Introduction

Different Radial Basis Functions (RBFs) of Artificial Neural Network (ANN) are utilized in various areas such as pattern recognition, optimization, approximation process etc. We have chosen few Radial Basis Functions such as Gaussian, Multi-Quadratic, Inverse-Multi-Quadratic and q-Gaussian RBFs for calculation. [2,3,4]. We have take image as input trained with various Radial Basis Functions to approximate the maximum singular value of the matrix of the given image. In Singular value decomposition the singular values of a matrix are arranged in decreasing order gradually. Even most of the time we found the difference of first singular value and second singular value is large. Normally rest of the singular values are very small in size. So, we took maximum singular value for approximation. Synthetic Aperture Radar (SAR) image are generally very large in

Abhisek Paul · Paritosh Bhattacharya
Department of Computer Science and Engineering,
National Institute of Technology, Agartala, India
e-mail: abhisekpaul13@gmail.com

Santi Prasad Maity
Department of Information Technology,
Bengal Engineering and Science University, Shibpur, India

* Corresponding author.

El-Sayed M. El-Alfy et al. (eds.), *Advances in Intelligent Informatics*,
Advances in Intelligent Systems and Computing 320, DOI: 10.1007/978-3-319-11218-3_13

memory size and pixels. We can ignore some smaller singular value of those images for better storage issue. In section 2 architecture of RBF neural network is described. Section 3 shows the analysis. Simulations are given in section 4 and finally conclusion is given in Section 5.

2 Architecture of Radial Basis Function Neural Network

As shown in Fig. 1 Radial Basis Function (RBF) network architecture consist of input layer, hidden layer and output layer respectively. In Fig. 1 inputs are $X = \{x_1, \ldots x_d\}$ which enter into input layer. Radial centres and width are $C = \{c_1 \ldots\ldots, c_n\}^T$ and σ_i respectively. In hidden layer $\Phi = \{ \Phi_1, \ldots\ldots, \Phi n \}$are the radial basis functions. Centres are of $n \times 1$ dimension when the number of input is n. The desired output is given by y which is calculated by proper selection of weights. $w = \{w_{11}, \ldots, w_{1n}, \ldots, w_{m1}, \ldots w_{mn}\}$ is the weight. Here, $w_{j.}$ is the weight of ith centre. [2,3].

$$y = \sum_{i=1}^{m} w_i \phi_i \tag{1}$$

Radial basis functions like Linear, Cubic, Thin plane spline and Gaussian are given in the in the Eqn.2, Eqn.3, Eqn4, and Eqn.5 respectively.

- Multi Qudratic: $$\phi(r) = (r^2 + c^2)^{1/2} \tag{2}$$
- Inverse Multi Qudratic: $$\phi(r) = (r^2 + c^2)^{-1/2} \tag{3}$$
- Gaussian: $$\phi(r) = \exp(-\frac{r^2}{2c^2}) \tag{4}$$
- q-Gaussian: $$\phi(x) = \exp(-\frac{r^2}{(3-q)c^2}) \tag{5}$$

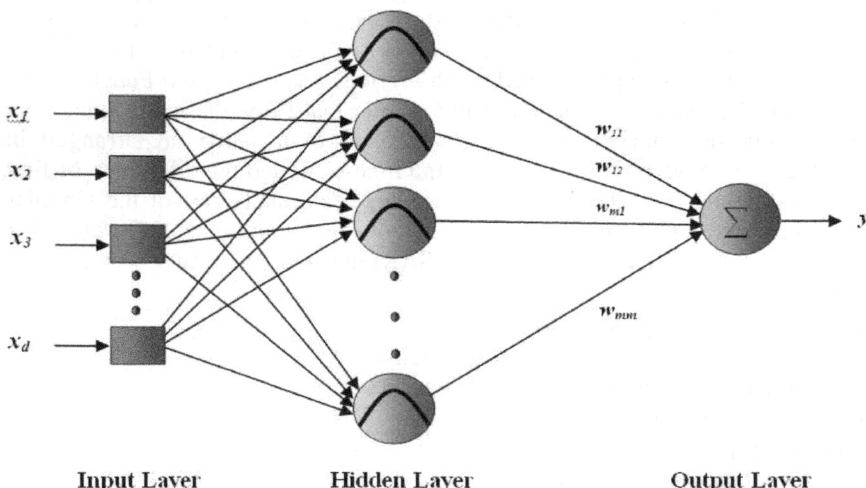

Fig. 1 Radial basis function neural network architecture

3 Analysis

We have taken some colour images. After that we extracted the red, green and blue colour components from the colour images. After that we calculated Y colour component [1] from the Eqn.7. Relation between Y, components and R, G and B colour component is given in Eqn.6. As we have applied Gaussian, Multi-Quadratic, Inverse-Multi-Quadratic and q-Gaussian RBF in artificial neural network, RBF needs optimal selections weight and centres. We have calculated all the method with pseudo inverse technique [5]. As in put we have chosen Flower and Butterfly images. Matrix sizes of all the input images are of 256×256 pixels for every analysis and experiment.

$$\begin{bmatrix} Y \\ I \\ Q \end{bmatrix} = \begin{bmatrix} 0.299 & 0.587 & 0.144 \\ 0.596 & -0.274 & -0.322 \\ 0.211 & -0.523 & 0.312 \end{bmatrix} \begin{bmatrix} R \\ G \\ B \end{bmatrix} \tag{6}$$

$$Y = 0.299R + 0.587G + 0.114B \tag{7}$$

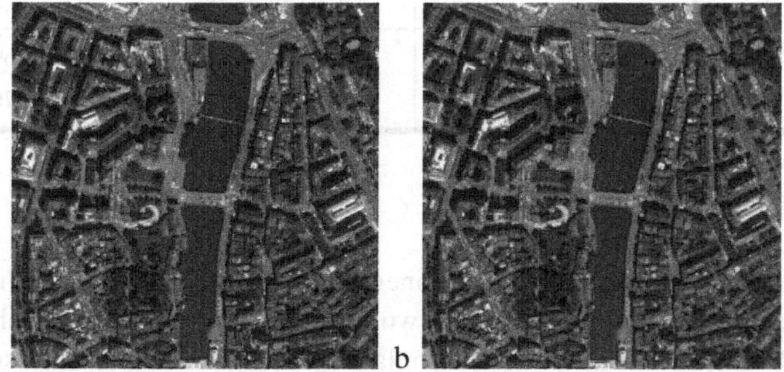

Fig. 2 SAR1 image (a) Original Color Image, (b) Y Color Component

Table 1 RMSE, MAPE and SD of singular values of SAR2 image compared with different Radial Basis Functions when number of hidden layer is 50

RBF function	RMSE	MAPE	SD
GRBF	4.2325169e-006	1.5186503e-008	72.309781650489356
MQRBF	4.1046928e-004	2.8562641e-005	72.309784747336607
I MQRBF	2.7684453e-005	1.5431996e-006	72.309777267616923
qGRBF	1.3381840e-007	6.1880884e-010	72.309784937495991

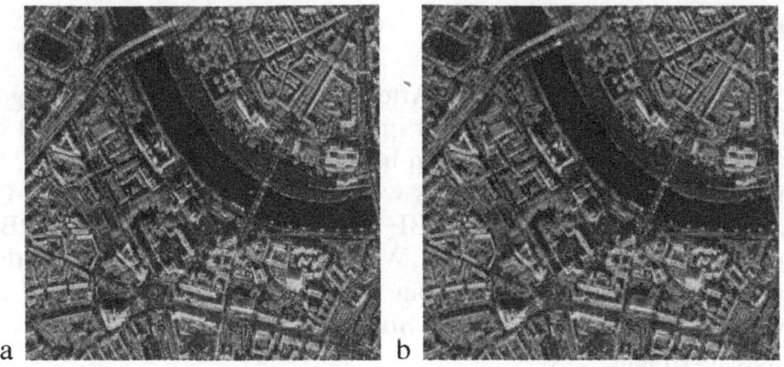

Fig. 3 SAR2 image (a) Original Color Image, (b) Y Color Component

Table 2 RMSE, MAPE and SD of singular values of SAR2 image compared with different Radial Basis Functions when number of hidden layer is 50

RBF function	RMSE	MAPE	SD
GRBF	1.8744234e-006	8.9032985e-009	84.902307224873539
MQRBF	4.0761931e-004	2.1262857e-005	84.902306530010890
I MQRBF	2.5833886e-005	1.2128748e-006	84.902306930706260
qGRBF	2.1812254e-019	3.6899619e-022	84.902307072489975

4 Simulation

Colour images with Y colour components are taken for the simulation. In this paper we have experimented through two different colour images which are SAR1 and SAR2 respectively. We have simulated Singular value of matrixes of these images with normal method, Gaussian, Multi-Quadratic, Inverse-Multi-Quadratic and q-Gaussian. We have used MATLAB 7.6.0 software [6] for the analysis and simulation process.

In Fig.2. SAR1 image and its corresponding Y colour component images are shown. In Fig.3. SAR2 image and its corresponding Y colour components are given. In Fig. 4 image SAR2 with hidden layer 30 is p with all these RBF methods .Singular values of all these images are compared and Root Mean Square Error (RMSE) Mean Absolute Percentage Error (MAPE) of singular values of matrixes of all these images with Gaussian, Multi-Quadratic, Inverse-Multi-Quadratic and q-Gaussian are calculated, simulated and compared. In Table.1 and in Table.2 comparative result of maximum singular values is shown.

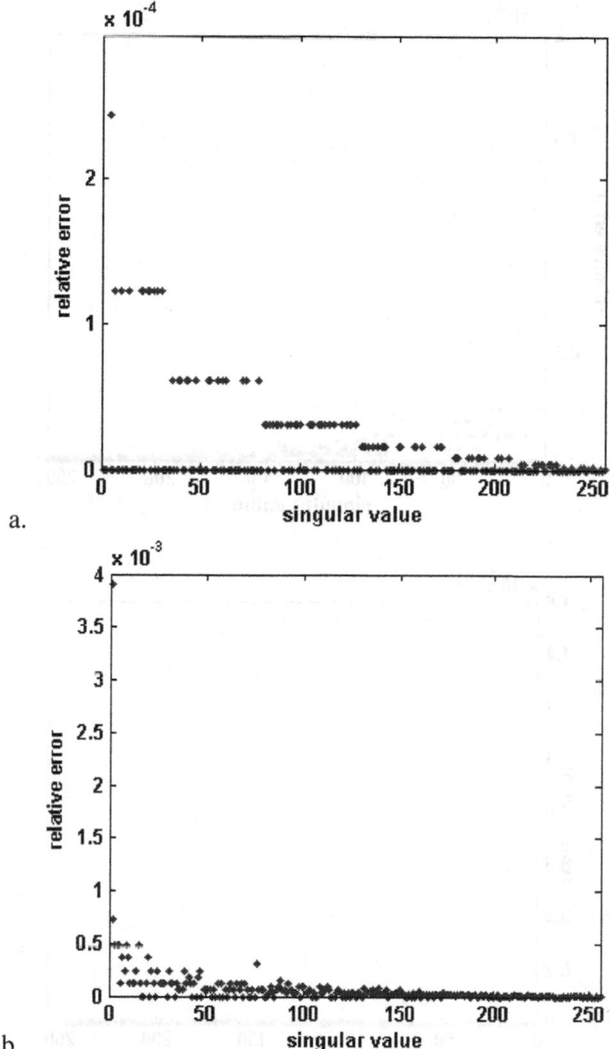

Fig. 4 Singular values of SAR2 image when number of hidden layer is 30. (a) GRBF, (b) MQRBF (c) IMQRBF (d) q-GRBF.

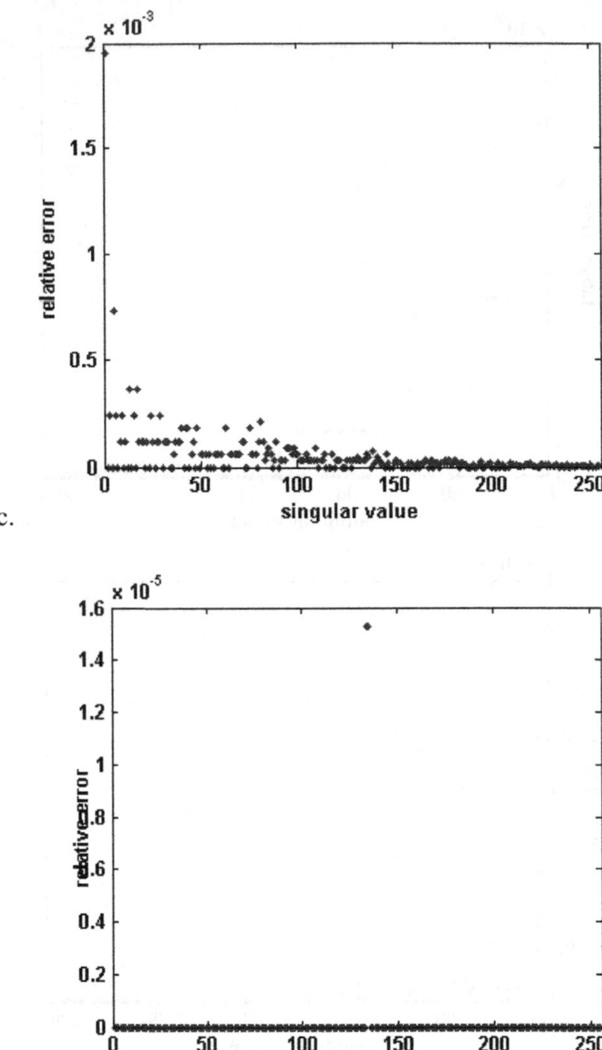

Fig. 4 *(continued)*

5 Conclusion

In this paper various RBF methods like Gaussian, Multi-Quadratic, Inverse-Multi-Quadratic and q-Gaussian RBF methods are utilized for the computation of singular values of various colour images and their corresponding Y colour component matrixes. Simulation results give better result and lesser RMSE, MAPE for q-Gaussian RBF method. So, it can be conclude that q-Gaussian RBF method could be used for calculation compared to the other relative methods in Artificial Neural Network.

Acknowledgements. The authors are so grateful to the anonymous referee for a careful checking of the details and for helpful comments and suggestions that improve this paper.

References

1. Liu, Z., Liu, C.: Fusion of the complementary Discrete Cosine Features in the YIQ color space for face recognition. Computer Vision and Image Understanding 111(3), 249–262 (2008)
2. Schölkopf, B., Sung, K.-K., Burges, C.J.C., Girosi, F., Niyogi, P., Poggio, T., Vapnik, V.: Comparing support vector machines with Gaussian kernels to radial basis function classifiers. IEEE Trans. Signal Process. 45, 2758–2765 (1997)
3. Mao, K.Z., Huang, G.-B.: Neuron selection for RBF neural network classifier based on data structure preserving criterion. IEEE Trans. Neural Networks 16(6), 1531–1540 (2005)
4. Luo, F.L., Li, Y.D.: Real-time computation of the eigenvector corresponding to the smallest eigen value of a positive definite matrix. IEEE Trans. on Circuits and Systems 141, 550–553 (1994)
5. Klein, C.A., Huang, C.H.: Review of pseudo-inverse control for use with kinematically redundant manipulators. IEEE Trans. Syst. Man Cybern. 13(3), 245–250 (1983)
6. Paul, A., Bhattacharya, P., Maity, S.P.: Eigen value and it's comparison with different RBF methods by using MATLAB. In: Sengupta, S., Das, K., Khan, G. (eds.) Emerging Trends in Computing and Communication (ETCC 2014). LNEE, vol. 298, pp. 219–224. Springer, Heidelberg (2014)
7. Paul, A., Bhattacharya, P., Maity, S.P.: Comparative study of Radial Basis Function neural network with estimation of Eigenvalue in image using MATLAB. In: Biswas, G.P., Mukhopadhyay, S. (eds.) Recent Advances in Information Technology (RAIT 2014). AISC, vol. 266, pp. 141–146. Springer, Heidelberg (2014)
8. Math Works. MATLAB 7.6.0 (R2008a) (2008)

Blending Concept Maps with Online Labs for STEM Learning

Raghu Raman, Mithun Haridas, and Prema Nedungadi

Abstract. In this paper we describe the architecture of an e-learning environment that blends concept maps with Online Labs (OLabs) to enhance student performance in biology. In the Indian context, a secondary school student's conceptual understanding of hard topics in biology is at risk because of a lack of qualified teachers and necessary equipments in labs to conduct experiments. Concept map provides a visual framework which allows students to get an overview of a concept, its various sub concepts and their relationships and linkages. OLabs with its animations, videos and simulations is an interactive, immersive approach for practicing science experiments. The blended e-learning environment was tested by systematically developing a concept map for the concept "Photosynthesis" and by successfully integrating it into the OLabs environment. Our blended approach to concept understanding has interesting implications for the teacher who is engaged in training programs.

Keywords: Concept Map, OLabs, Biology, Photosynthesis, simulations, animations, online labs, virtual labs.

1 Introduction

Concept maps are now popularly used as a tool in K 12 education where they are used as linkages in science standards and concept growth dynamics. It is seen that often students get confused between the relationship of various subjects to each other [1] and one of the major educational challenges of modern times is to provide a framework where subjects can be linked to each other. Educators use concept maps as an assisted tool for providing accurate instructions, learning objectives and evaluator criteria. Because of their multi –usability, concept maps have evolved to become an effective productivity booster and a quick accomplisher [2]. Concept maps are actually founded on the principle that it's

Raghu Raman · Mithun Haridas · Prema Nedungadi
Amrita Vishwa Vidyapeetham, Coimbatore, India

El-Sayed M. El-Alfy et al. (eds.), *Advances in Intelligent Informatics*,
Advances in Intelligent Systems and Computing 320, DOI: 10.1007/978-3-319-11218-3_14

easier to make sense of new information when it's presented in a visual way. Novak first categorized concept maps in his cutting edge article and gave them a purpose and nomenclature [3]. These maps are great for summarizing information and presenting them as a holistic linked unit. When a student uses concept maps to derive relationships between apparently unlinked subjects, the higher order thinking skills are being utilized. A concept map is an educational tool that follows the idea of meaningful learning. Concept maps are graphical tools for organizing and representing knowledge and they include concepts, usually enclosed in circles or boxes of some type, and relationships between concepts indicated by a connecting line linking two concepts [3]. Blooms taxonomy actually classifies these skills as synthesis and evaluation [5]. Concepts are defined as perceived regularity in events, objects and the sequential flow is represented by symbolic aids like + or % [6]. When multiple concepts are linked together they form a proposition or a semantic map. Stakeholders in education have begun to realize the importance of addressing student difficulties in comprehending the linkages between different subjects [7]. Interestingly concept maps are now integrated into teacher training programs which help incorporate the principles of modern educational culture. Many educational practitioners believe that K 12 education concept maps can be used to strengthen the academic understanding of students and teachers at least in the field of biology [8].

Biology is a core science subject which deals with all the biological process in the biosphere. Considering its fundamental characteristics and importance, biology is today a standard subject of instruction at all levels of our educational systems. From our literature survey we believe that most of the students show poor performance in biological science because of the lack of meaningful learning, and lack of equipment in labs to do experiments. The research done by Tekkaya [9] on high school students found that the students have misconceptions in various areas of biology like osmosis, diffusion etc. The findings of Ajaja [10] suggest that concept mapping when efficiently used as a study skill could enhance post achievement test scores and retention of biology knowledge well over other study skills used by students. In our traditional method we are teaching students in a linguistic way. Research suggests that information is processed and stored in memory in both linguistic and visual forms. Research in both educational theory and cognitive psychology tells us that visual learning is among the very best methods for teaching and learning. If students start learning using visual ways then they are forced to draw the ideas that they have learnt like how the different concepts are connected to each other, relationships between the concepts and organize their knowledge in a graphical way.

We have online labs (OLabs) for science experiments to perform biological lab experiments that are available to students who have no access to physical labs or where equipment is not available. OLabs comes with detailed theory, interactive simulations, animations and lab videos. We enhance OLabs to improve the students learning skills by incorporating tools to create concept maps and the process of designing concept maps to create a more effective virtual learning environment.

2 Online Labs (OLabs) for Science Experiments

With all the advances in communicative technology that came into prominence during the last few decades, there evolved a simulation based eLearning called online labs [11]. Students may conduct virtual experiments that mimic the features of real life experiments. There are several advantages of using virtual environment such as OLabs. There is a relaxation on time constraints as each student has sufficient time in the lab and the geographical constraints are eliminated. There is an increased economy of scale as the cost of setting up a sophisticated learning environment is reduced. Enhanced safety and security is another big advantage of the simulated environment as per [12].

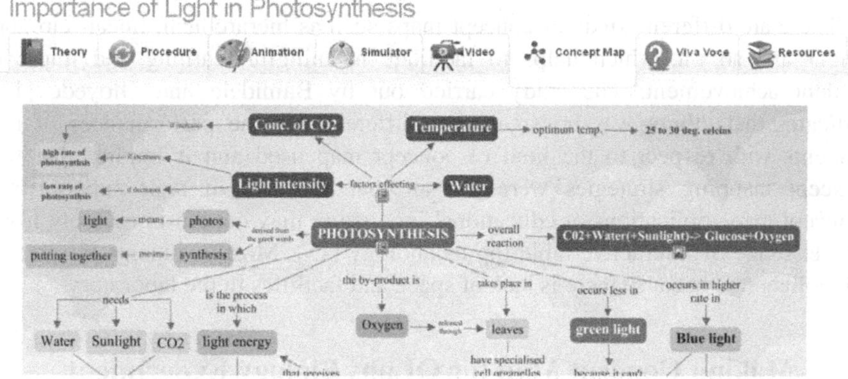

Fig. 1 OLabs Concept Map for Photosynthesis

A concept map typically contains concepts which are represented using circles or boxes. Each related concept can be connected by using lines, and labels on these lines define the relationship between the two concepts. If two or more concepts are connected using linking words and forms a meaningful statement then it known as a preposition. If there is any relationship between concepts in one domain to a concept in another domain then we can connect these concepts using an arrow with linking words. This type of relationship is known as crosslink's. Fig. 2 is an example of a concept map that shows the above characteristics.

In Fig. 2, "Biology", "Animals", "Plants", "living organism" etc. are examples of concepts, and "can be", "made up of", "are capable" etc. are examples of linking words. The relationship between concepts "Plants", "sunlight" and "photosynthesis" is expressed through the linking words "use the energy from" and "to carry out" for forming the **proposition** "Plants use energy from sunlight to carry out photosynthesis". In Fig. 2, the concept "Animals" and "Plants" are two separate domains and are connected to each other to forms a **crosslink**.

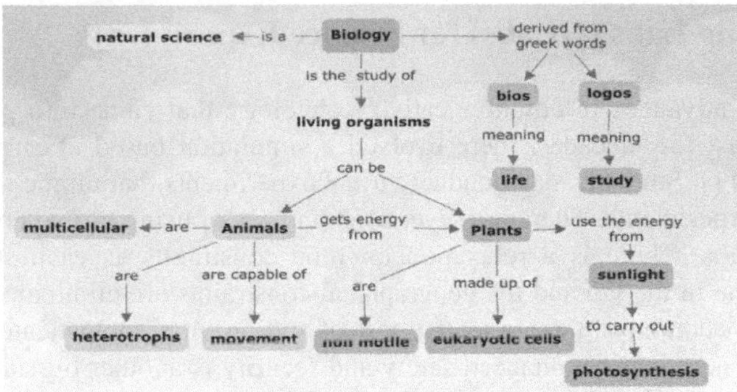

Fig. 2 A concept map about Biology

There are different kinds of concept maps such as hierarchical, linear, circular spider and so on, which helps to facilitate meaningful learning and improve student achievement. The study carried out by Bamidele and Oloyede [13] indicates that "there was no significant difference in the performances of the students with respect to the kind of concept map used and it implies that the concept mapping strategies were not all that different in their superiority". Concept map applications or educational innovation may benefit immensely from proper usage of interactive table top technology [15]. Mostly students in science lab collaborate because there is lack of space and facilities in the laboratory.

3 Making Concept Map for OLabs Biology Experiment

Constructing concept maps is a structured process which is based on a specific topic, revision and re-positioning of the concepts, thus helping to construct a good concept map having a pictorial view of the topic. It helps the students to think efficiently and to reduce complexity of the study. Teachers can take feedback from the students after they finish each step of making the concept map. Concept mapping is also a valuable theory of learning that teachers can use to evaluate a student's level of understanding. This will help to reduce the mistakes of students and to understand each part of the concept map and their reasoning behind the concepts and connections they made.

A. Constructing a good focus question

Novak and Cañas [3] suggested that a helpful way to determine the context of your concept map is to choose a focus question, that is a question that clearly specifies the problem or issue the concept map should help to resolve. Fig. 5 was constructed from the focus question "What is Photosynthesis?" and described all the main concepts related to photosynthesis. Derbenseva et.al [16] found that the structure of the map influences not only the focus question but also the type of relationships that are likely to be constructed in a proposition that links two concepts together. So for constructing good concept map we need specific focus questions and proper concepts.

B. Choose relevant concepts

The next step is to identify the relevant concepts that relate to our main idea or the focus question. Make a list of all the relevant concepts, usually 10 to 25 concepts is sufficient. In Fig. 3(A) it shows that all the main concepts that are related to the topic photosynthesis. From the list, give rank for each concept and place each important concept at the top of the page and arrange all other concepts in a hierarchy underneath that one. Fig. 3(B) shows the ranked list of concepts.

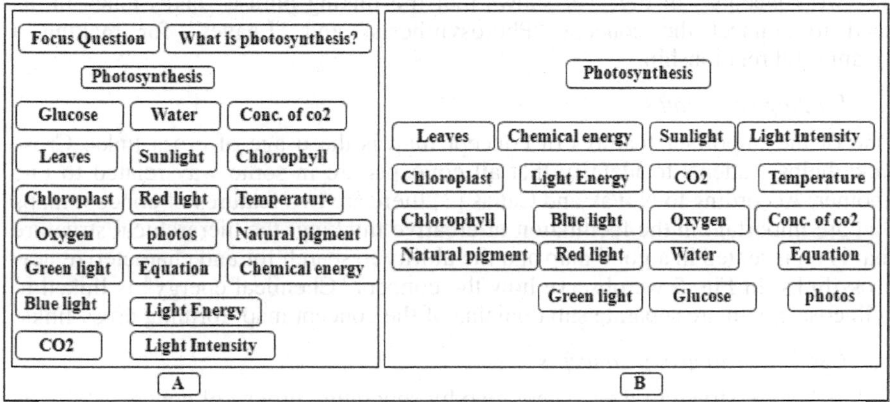

Fig. 3 (A). Focus question and relevant concepts. (B) An example image for arranging important concept at the top and less important below that concept.

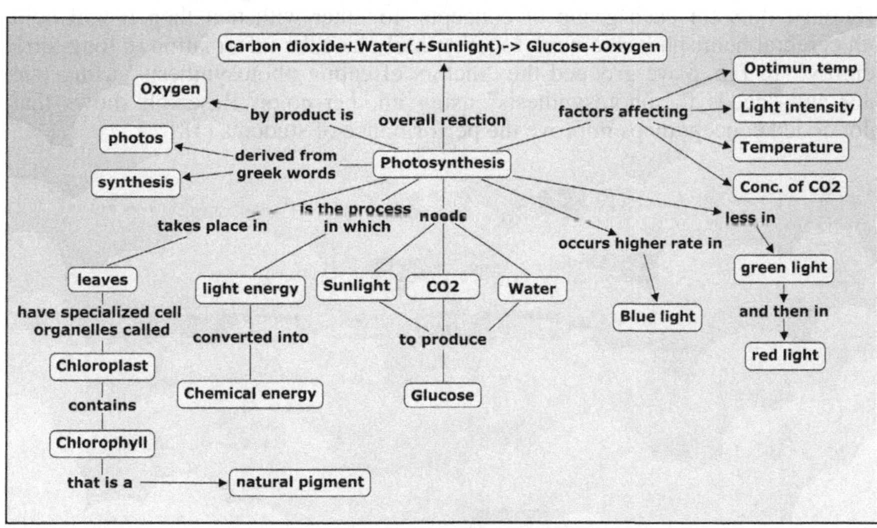

Fig. 4 A preliminary concept map about "Photosynthesis"

C. Construct a preliminary concept map

After choosing relevant concepts we can start making a draft copy of concept map. This will give an overall idea of how the concept map looks. After making a preliminary concept map, a review is essential to improve the map. We can also add, rearrange or remove the concept from the preliminary concept map. In Fig. 4 we can see a preliminary concept map.

D. Choose linking words to connect concepts

Using the appropriate linking words to clearly express the relationship between two concepts is possibly the most difficult task during the construction of concept maps. Linking words usually contains 1-3 words appropriate for relationship of the two concepts. In Fig. 5 it shows that the linking phrase "takes place in" is used to connect the concept "Photosynthesis" and "Leaves" for making a meaningful relationship.

E. Finding cross-links

One of the important features of concept map is the usage of cross links. Cross links helps students to identify that all concepts are in some way related to one another. According to Novak and Cañas [3] there are two features of concept maps that are important in the facilitation of creative thinking: the hierarchical structure that is represented in a good map and the ability to search for and characterize new cross-links. In Fig. 5 we observe how the concept "Chemical energy" is linked to "Glucose"; both are separate sub domains of the concept map, forming cross-links.

F. Use colors to group concepts

Colors help to structure the concept map by separating groups of concepts thereby helping to recall the content of the map after a long time. The left cerebral hemisphere is specialized for linguistic and cognitive processes, whereas the right cerebral hemisphere is specialized for visuospatial processing [17]. If we are using different colors for each group of concepts, together with text then it will help both cerebral hemispheres to work together and place the information in long term memory. In Fig. 5 we grouped the "factors effecting photosynthesis" using one color and "needs for photosynthesis" using another color. Research shows that color coded concept maps improve the performance of students [18].

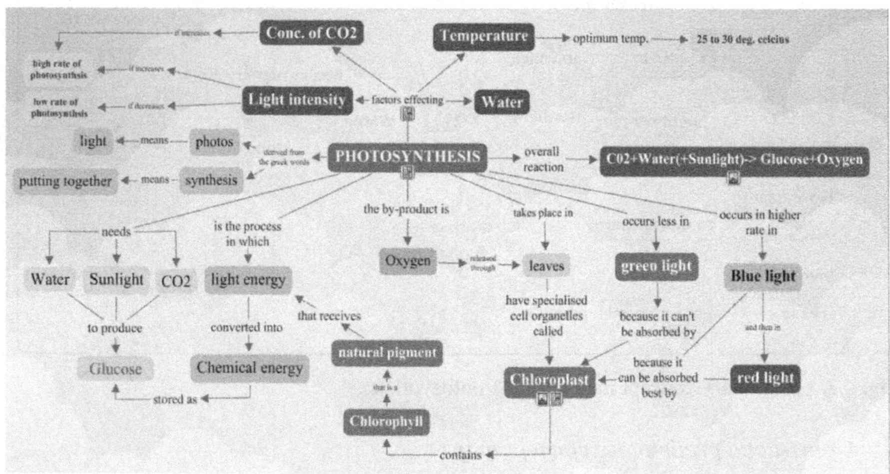

Fig. 5 A Completed concept map about "Photosynthesis"

G. Usage of images, website links, other concept and explanatory notes

Adding images or examples to the map can clarify the meaning of concepts and help to remember the information. We can click on the icon below the concepts to open the images. In Fig. 6 it shows the image of chloroplast inside the concept map. For getting more information about each main concept we also added website links for important concepts. We gave two sample resources from "Amrita" (OLabs) and "Arizona University" for the main concept "Photosynthesis" shown in the Fig. 6.

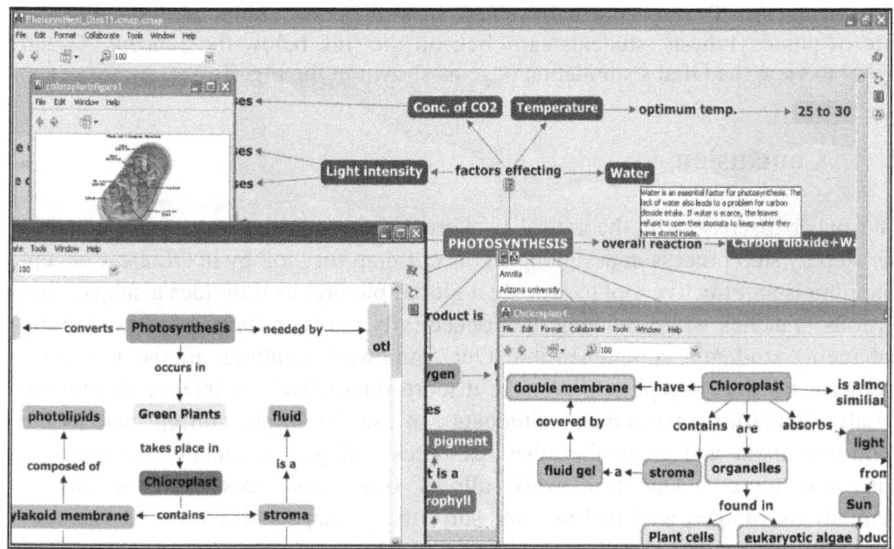

Fig. 6 A Concept map representing "Photosynthesis" with some of the resources

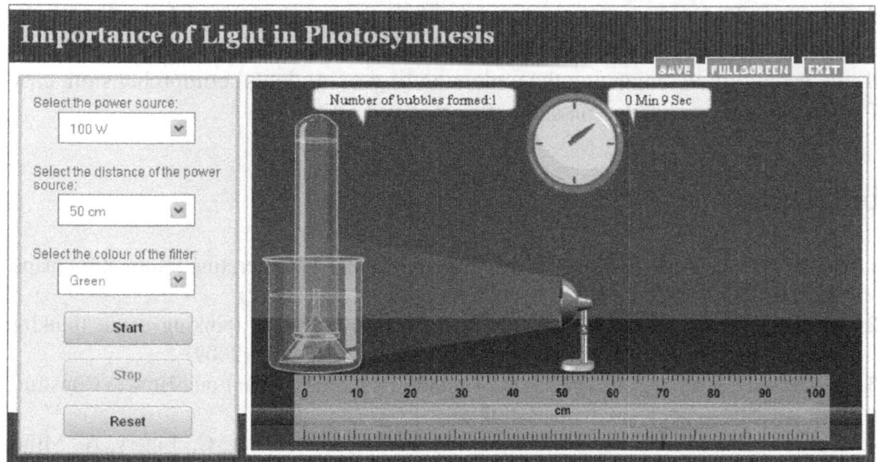

Fig. 7 OLabs simulation for "Importance of Light in Photosynthesis"

We also added sub concept maps for important concepts like "Photosynthesis" and "Chloroplast" as shown in Fig. 6 thus providing further details about the topic. Explanatory notes are used to explain some important concepts in the map. The explanatory notes for the concept "**water**" is shown in Fig. 6. To integrate concept maps and OLabs simulation we have some of the links in concepts that will directly lead to the OLabs simulation environment. Hagemans et.al [18] suggesting that concept map-based learning is relevant for improving student learning in simulation based environments. The completed concept map in Fig. 5 gives overall idea of how different colors like green, blue and red affect the rate of photosynthesis. To experimentally check how different colors are affecting the rate of photosynthesis, students can click on the link below the concept "Green light" to view the OLabs simulation page as shown in the Fig. 7.

4 Conclusion

The paper discusses the theoretical background for concept maps and describes the step by step process in designing a concept map for biology in OLabs. Concept mapping is an effective tool to provide a global picture; as it divides a subject into various branches while showing the connectivity between their subparts, thereby enhancing students' comprehension. Our framework captures all the necessary elements for the concept map to make it more interactive and integrated with the OLabs simulation environment. Students can use the OLabs concept map as the central element and go to the other resources and get information related to a particular topic. Our framework allows other resources such as videos, simulations, images, website links, and sub concept maps and so on to be elements of the OLabs concept map thus providing a complete reference of the subject. The color coded concept map helps students to remember information about a particular subject for longer terms. We thus provide an easy to use system for educators to design concept maps and publish them using the rich media OLabs resources. We are conducting controlled surveys using the concept enriched OLabs pedagogy in enhancing the understanding of students' comprehension, ease of learning, and knowledge on subject matter and attitudes.

References

1. Shahaf, D., Guestrin, C., Horvitz, E.: Trains of thought: Generating information maps. In: Proc. 21st Int. Conf. World Wide Web, pp. 899–908 (2012)
2. Ritchhart, R., Turner, T., Hadar, L.: Uncovering students' thinking about thinking using concept maps. Metacognition and Learning 4, 145–159 (2009)
3. Novak, J.D., Cañas, A.J.: The Theory Underlying Concept Maps and How to Construct and Use Them, pp. 1–36. IHMCC (2008)
4. Crampes, M., Ranwez, S., Villerd, J., Velickovski, F., Mooney, C., Emery, A., Mille, N.: Concept maps for designing adaptive knowledge maps. Inf. Vis. 5, 211–224 (2006)

5. Olney, A.M.: Extraction of Concept Maps from Textbooks for Domain Modeling. In: Aleven, V., Kay, J., Mostow, J. (eds.) ITS 2010, Part II. LNCS, vol. 6095, pp. 390–392. Springer, Heidelberg (2010)
6. Borda, E.J., Burgess, D.J., Plog, C.J., Luce, M.M.: Concept Maps as Tools for Assessing Students' Epistemologies of Science. Electron. J. Sci. Educ. 13, 160–184 (2009)
7. Hao, J.X., Kwok, R.C.W., Lau, R.Y.K., Yu, A.Y.: Predicting problem-solving performance with concept maps: An information-theoretic approach. Decision Support Syst. 48, 613–621 (2010)
8. Gould, P., White, R.: Mental Maps. Penguin Books, New York 197
9. Tekkaya, C.: Remediating high schools' misconception concerning diffusion and osmosis through mapping and conceptual change text. Res. Sci. Tech. Edu., 21, 5–16
10. Ajaja, O.P.: Concept Mapping As a Study skill: Effect on Students Achievement in Biology. Int. J. Sci. 3(1), 49–57 (2011)
11. Kowata, J.H., Cury, D., Silva Boeres, M.C.: Concept maps core elements candidates recognition from text. In: Proc. of Fourth Int. Conference on Concept Mapping, pp. 120–127 (2010)
12. Safayeni, F., Derbentseva, N., Cañas, A.J.: Concept Maps: A Theoretical Note on Concepts and the Need for Cyclic Concept Maps. Sci. Teach. 42, 741–766 (2005)
13. Bamidele, E.F., Oloyede, E.O.: Comparative Effectiveness of Hierarchical, Flowchart and Spider Concept Mapping Strategies on Students' Performance in Chemistry. World J. Edu. 3, 66 (2013)
14. Starr, R.R., Parente de Oliveira, J.M.: Concept maps as the first step in an ontology construction method. Information Systems (2012)
15. Beel, J., Gipp, B., Stiller, J.-O.: Information retrieval on mind maps - what could it be good for? In: 2009 5th Int. Conf. Collab. Comput. Networking, Appl. Work (2009)
16. Derbentseva, N., Safayeni, F., Cañas, A.J.: Experiments on the effect of map structure and concept quantification during concept map construction. In: Proc. of First Int. Conference on Concept Mapping, pp. 209–216 (2004)
17. Corballis, M.: Visuospatial processing and the right-hemisphere interpreter. Brain and Cognition 53, 171–176 (2003)
18. Hagemans, M.G., Meij, H.D., Jong, T.D.: The Effects of a Concept Map-Based Support Tool on Simulation-Based Inquiry Learning. Jrnl. Edu. Psy. 105(1), 1–24 (2013)

Cognitive Load Management in Multimedia Enhanced Interactive Virtual Laboratories

Krishnashree Achuthan, Sayoojyam Brahmanandan, and Lakshmi S. Bose

Abstract. Learning in multimedia enhanced interactive environments has distinctly impacted the cognitive processing of information. Theoretical learning requires conceptual understanding while experimental learning requires cognition of underlying phenomena in addition to a firm grasp of procedures and protocols. Virtual laboratories have been recently introduced to supplement laboratory education. In this paper, an outline of the modes of knowledge representation for virtual laboratories is presented. The results from this work show how the combination of physical and sensory representations in virtual laboratories plays a key role in the overall understanding of the content. Information processing through visual, auditory, pictorial as well as interactive modes offers unique pathways to cognition. An analysis of comprehension for N=60 students showed a significant change in the time taken to answer questions as well as an overall improvement in scores when exposed to multimedia enhanced interactive virtual laboratories (MEIVL). This study also portrayed a reduction in the perception of difficulty in understanding physics experiments. Statistical tests on various modes of assessments were done both online and in classroom quantify the extent of improvement in learning based on the enabling, facilitating and split attention aspects of MEIVL.

Keywords: cognitive load theory, multimedia, enabling, facilitating, split attention.

1 Introduction

A majority of today's youth undergoing higher education continue to learn in traditional settings with little exposure of revolutionary changes predominating the

Krishnashree Achuthan
Amrita Center for Cyber Security Systems and Networks

Krishnashree Achuthan · Sayoojyam Brahmanandan · Lakshmi S. Bose
VALUE Virtual Labs,
Amrita Vishwa Vidyapeetham, Amritapuri, Kollam – 690525
e-mail: krishna@amrita.edu, {sayoojyamb,lakshmisb}@am.amrita.edu

El-Sayed M. El-Alfy et al. (eds.), *Advances in Intelligent Informatics*,
Advances in Intelligent Systems and Computing 320, DOI: 10.1007/978-3-319-11218-3_15

developed countries. One such revolution relates to ICT based learning techniques that has shown prominent impact in the educational system [1]. The integration of ICT into classrooms has improved pedagogic processes significantly [1]. What used to be a difficult and monotonous experience in teaching complex phenomena, has been completed overturned with the introduction of newer tools and techniques. Rote learning which is so prevalent in higher education and traditional classroom teaching has several limitations in delivering the concept to the students. In most cases the students learn only what is exposed to them by the teachers. ICT has facilitated the teachers to tap into the larger expanse of knowledge and empowered them to enhance their teaching methodologies.

Knowledge and comprehension are the fundamental metrics of learning. Application of this knowledge to real problems is the ultimate goal of higher education. Practical education is primarily offered in the form of laboratory sessions in the areas of sciences and engineering. Although the number of hours devoted to practical education is insufficient to expose them to diverse scientific problems, it is expected to provide real world experience. Due to limited lab hours, and inability to repetitively perform the experiments, the cognitive load on the students to learn the underlying phenomena is high within the stipulated time. Additionally, since most laboratory experiments are done in groups, the grasp of critical details by all students within the group can be subpar. One of the practical ways to address the cognition issues have been with enhancing teaching using visual media. Communication through words and pictures helps enrich learning technique [2]. The cognitive loads experienced by students as they are exposed to new information have been studied in detail by several researchers [3-4]. According to the cognitive theory of multimedia learning (CTML) human information systems includes dual channels for visual or pictorial and auditory or verbal processing and each channel has a limited capacity [2] of information adsorption. The use multimedia enhanced interactive virtual laboratory (MEIVL) has resulted in a new dimension to laboratory education [5]. There are various means of reducing the cognitive load during instructional design [3]. MEIVL provides visualization of concepts, and emphasizes enhancing skills by allowing repetition. This work relates to the characterization of MEIVL features and their impact of cognitive load.

2 Related Work

How we learn and process the information is described by the two essential components of human cognitive architecture, i.e., long term memory and working memory. The long term memory relates to information repository that can be exercised without conscious effort and the working memory relates to acquisition of knowledge [6]. The development of working memory includes learning skills that are learnt through human interaction without outstanding effort and those that have to be taught as in a classroom environment. The cognitive load theory (CLT) defines the influence and effective use of resources within the learning technique [7]. Cognitive load is most associated with design and development of instructional materials and their regulation or flow of information into the learner's memory

[3], [8]. Based on the different sources for cognitive load, Sweller et al [9] classified cognitive loads as intrinsic, extraneous or germane. Intrinsic cognitive load arises from the complexity of concepts requiring to be processed. Germane cognitive load, on the other hand, is determined by the degree of effort involved in processing the knowledge. Extraneous cognitive load is derived from the difficulties resulting from the instructional design. Both intrinsic and extraneous loads contribute to the total cognitive loads. Historically, education was imparted first orally which was followed by print media or text books. Pictures improved the cognition and became a regular feature of all learning material. It is impossible to imagine teaching science without the use of pictures and words. On the other hand, complex concepts which may be hard to explain through pictures and text can be taught with the integration of multimedia [6]. Thus visual presentation of any content forms the foundation of multimedia learning.

As mentioned by Mayer [2] dual code learning through sensory mode includes presenting information in multiple states i.e. static and dynamic content with pictures and words. Pictures can be static as well as dynamic. Static graphics includes illustrations and photos, while dynamic graphics includes animations or video. The CLT indicates that the sources of extraneous cognitive load such as split attention, redundancy and transiency should be reduced in multimedia based learning. The work on designing of online learning based on the cognitive load theory has been reported in the paper [10]. When Schar and Zimmerman [11] investigated means to reduce the cognitive load, they showed the presentation of content affected learning of dynamic concepts. They suggest presenting animated content in small chunks as an effective way to reduce the cognitive load. Finkelstein et al [12] in their comparative study between the students showed distinct differences in the influence of multimedia between groups that underwent educational training multimedia and far outnumbered in performance compared to those that did not.

This paper examines relatively unexplored educational innovation i.e. MEIVL from the cognitive load perspective. Some of key advantages of MEIVL are that it enables a learner to conceive invisible phenomena [13] and concepts. With interactivity integrated on individualized learning platform, MEIVL provides a plethora of possibilities in reducing the cognitive exertion required of students over time.

3 Multimedia Enhanced Interactive Virtual Labs (MEIVL)

The significant limitations of traditional laboratories for its inability to impart meaningful learning that allows substantial cognitive processing led to the development of MEIVLs. This was developed as part of a consortium where in over 1500 experiments in nine disciplines of engineering and sciences were designed for students pursuing higher education [5]. MEIVLs have a number of components such as description of the experiment, procedural listing of steps, videos exemplying the overall objectives and methods, a simulator with experimental parameters and an interactive animation. Although the comprehensive nature of MEIVLs make it an attractive supplement to theory and laboratory education, using multimedia for instruction can affect the cognitive load [1], [5]. Challenges in design of MEIVLs include understating if the cognitive processing required exceeds the learner's cognitive capacity. Fig 1 displays the modes of knowledge representation where in the relationship between various sensory elements to the MEIVL components are portrayed. The theory component of MEIVL includes words as well as pictures. Here pictorial representation of the concept behind the experiment given is in congruence with the text. Pictorial representations captivate the learners and grab their visual sensations. Simplicity of style in terms of presentation of the text assists with coherency.

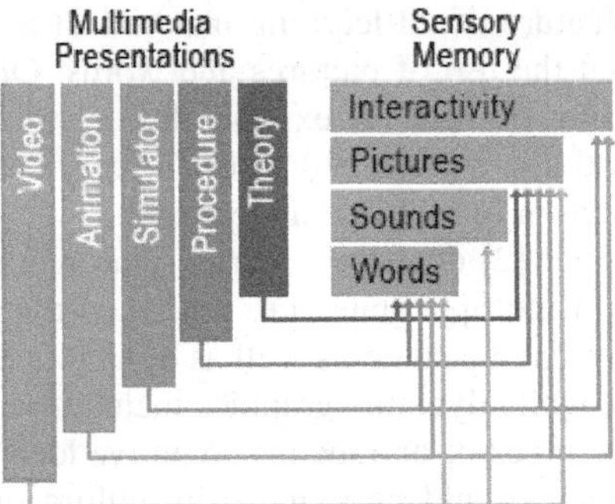

Fig. 1 Modes of Knowledge Representation in MEIVL

In the procedural component, each step for doing the experiment is systematically organized with static pictures. The simulator utilizes four memory factors, mainly focused on the appearance of the whole apparatus mimicking the real lab equipment. This is not only a visual representation but also interactive as the user can handle the apparatus virtually (e.g., adjusting knob). Instructions are given as words besides visual representations. The animations play instrumental role in expanding the imaginative thinking to perceive what may be invisible, and difficult to understand. Video aspects of MEIVL on the other hand does not interactivity, yet does utilize most sensory memory channels. Although all of these components for MEIVLs can be offered in a scalable fashion, the capacity to contain the information in the working memory can be very challenging. The next sections describe the characterization of the impact of components on cognitive load.

4 Methodology

In this study a sample of 60 undergraduate and graduate students pursuing their second year of engineering education were selected. Three experiments that were targeted for the study included measurement of refractive index by spectrometer, characterization of material property from Kundt's tube apparatus and gauging the thermal conductivity from Lee's disc experiment. These experiments were chosen based on the prior knowledge of the difficulty faced by most students in understanding the concept, procedure and significance of experiments. MEIVL components for the former two experiments included animations, theory, procedure, results and applications and the latter had them all except that it had a video instead of animation. The undergraduate students were divided into separate groups of 15 and were exposed to 1) traditional labs wherein the students performed the three experiments in groups or 2) MEIVLs that allowed individualized learning of the experiment on the computer. This was followed with different cognitive tests for perception and conceptual understanding.

4.1 *Coefficient of Thermal Conductivity by Lee's Disc Experiment*

Lee's Disc experiment computes the coefficient of thermal conductivity of a poor conductor such as glass, cardboard etc. The procedure involves placing the poor conductor, with certain dimension i.e. radius r and thickness x, between a steam chamber and two highly conductive identical metal discs. Once in equilibrium, the heat lost by the lower disc to convection is measured and equated to that flowing through the poor conductor. The upper disc temperature T_2 and the lower disc temperature T_1 are recorded. The poor conductor is removed and the lower metal disc is allowed to heat up to the upper disc temperature T_2. Finally, the steam chamber and upper disc are removed and replaced by a disc made of a good insulator. The metal disc is then allowed to cool through $T_1 < T_2$ and toward room temperature T_0. The temperature of the metal disc is recorded as it cools, so a cooling curve can be plotted. Then the slope s1 $=\Delta T/\Delta t$ of the cooling curve at temperature T_1 is recorded. This description, however still lacks step-wise procedure to perform experiment. But by looking at the simulator and video in MEIVL version of Lee's Disc experimental as shown in Fig 2, the learning outcomes can be significantly influenced.

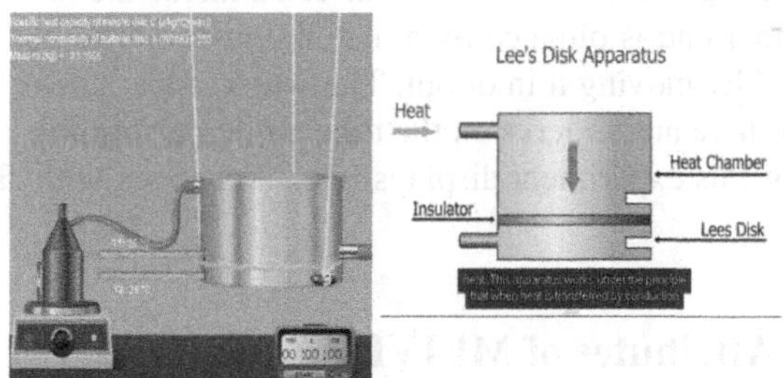

Fig. 2 Simulation and Video of Lees Disc Experiment

4.2 *Refractive Index of a Prism*

The aim of the experiment was to determine the refractive index and angle of a given prism. A detailed explanation of this experiment is given elsewhere [14]. In the MEIVL version, the visual representations like animation and simulation to explain how the light rays strikes on one surface and how the ray is transmitted and reflected through prism (Fig. 3).

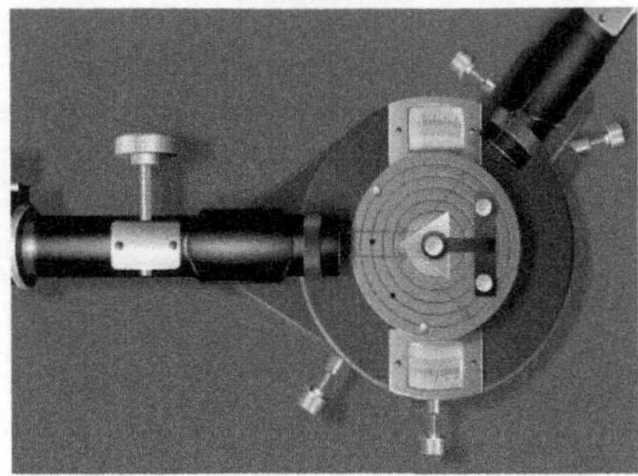

Fig. 3 Top View of MEIVL Spectrometer

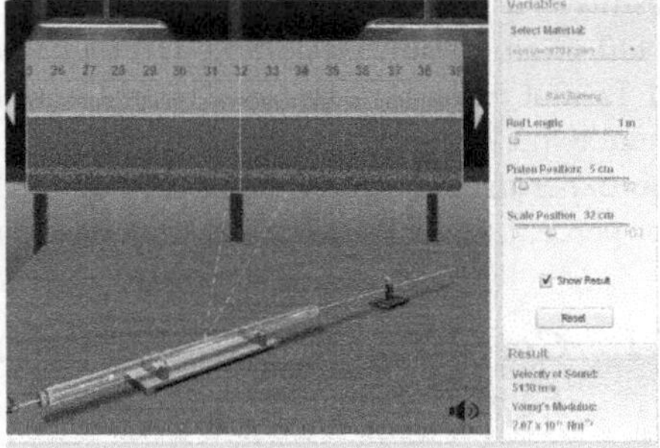

Fig. 4 Kundt's tube Apparatus Simulation

4.3 Sound Velocity and Young's Modulus of Materials

The objective of this experiment is to find the velocity of sound waves and the Young's modulus of the material of the rod. Knowing the speed of sound in air, the speed of sound v in a solid rod can be calculated based on the measurement of sound wavelength, λ. If the frequency of the sound wave, f is known, then the speed of sound can be calculated from v=fλ. The apparatus consists of a long transparent horizontal pipe, which contains a fine powder such as cork dust or talc. At the Simulator Screenshot Video Screenshot ends of the tube, there are metal fittings. At one end of the tube, a metallic rod, of uniform radius having one or two meter length is introduced. This rod is clamped at the middle and carries a circular disc, rigidly fixed at one end. The radius of the disc is slightly smaller than the radius of the glass tube. The rod is inserted inside the tube, without touching it, while its other end is plugged by a metallic piston. The position of the piston can be adjusted by moving it in or out. The whole apparatus is tightly clamped on a table, so that there are no jerks on the tube while performing experiment. The MEIVL version of this experiment displays instant changes with change in length of the rod.

5 Distinct Attributes of MEIVLs

This section describes a few of the distinct attributes of MEIVLs and their direct impact on the cognitive load.

5.1 Enabling Effect

The definition of an enabling effect of MEIVLs is that its features are fundamentally responsible in allowing students to understand specific concepts with clarity. As an example, Fig 5 shows a part of spectrometer called the telescope which is used to view image. For viewing the image we need to adjust the telescope for getting the clear image. So we tried to check whether the student will be able to understand the concept of how light rays travel through the telescope on moving the eye piece of telescope. In the traditional lab, it would be impossible to visualize these ray diagrams. By exposing students to MEIVLs, students were assessed on their understanding of correlation between the eyepiece movement and the flow of light

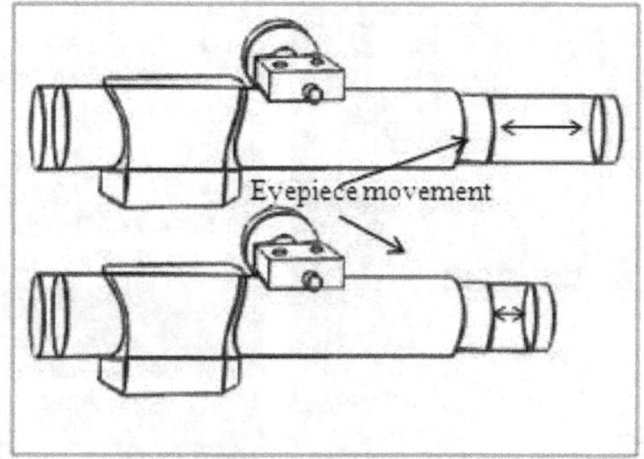

Fig. 5 Telescopic Positions and Eyepiece Movements

5.2 *Facilitating Effect*

Facilitating effect of MEIVLs demonstrate as how a task which was previously difficult to perform became far more effortless with the help of MEIVL. To exemplify, the procedural aspects of laboratory experiments are extremely critical to the successful performance of an experiment. Viewing and interacting with the animation and simulation respectively, the understanding of spectrometer adjustments that is done to determine the angle of prism was done by asking students to sort randomizing pictures (Fig 6) in the right order.

5.3 *Split Attention Effect*

Split attention effect enhances cognitive load by causing learner distraction from the exposed sources of information. This could result from lack of coherent organization of pictorial and textual content. This work investigated if split attention effect was induced by MEIVL. One of the examples that were used to measure this effect, based on Kundt's tube experiment is elaborated here. One set of students completed this experiment in traditional lab, while the other set using Eyepiece movement MEIVLs. Both sets of students were then asked to label the pictures shown in Fig 7 with a title based on their understanding.

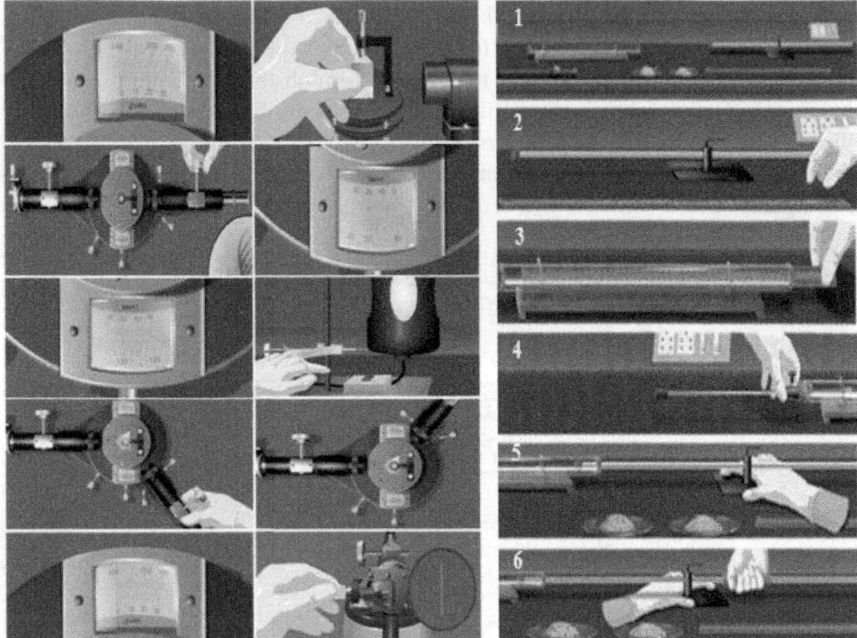

Fig. 6 Random Arrangement of Spectrometer Adjustments

Fig. 7 Sequential Steps in Kundt's Tube Experiment

6 Results and Discussions

The students that were subjected to both traditional physical lab (PL) and MEIVLs were assessed for 1) their cognition of governing principles of three experiments detailed above 2) their understanding of procedural aspects of the experiments and 3) the speed of recollection. These assessments reflected the cognitive demand on the students from both approaches.

6.1 Difficulty Rating

Prior to analyzing the differences in the performance between the labs, the perception of difficulty amongst the students was first ascertained. Ten questions were given as part of a questionnaire. The questions given were of three types: 1) Theory - questions pertaining to theoretical aspects of physics phenomena that governed the experiments 2) Procedure – questions that only related the sequence of practical steps required to perform the experiments and 3) Application – questions that required understanding the concepts and applying them to solve problems. Fig 8a and 8b show the distribution of difficulty rating in these questions between the two sets of students that went through PL or MEIVL. The students that went through MEIVL found most questions in all three categories easier than those that went through PL. Another implication is that application oriented problems were twice as hard for the PL students than the MEIVL.

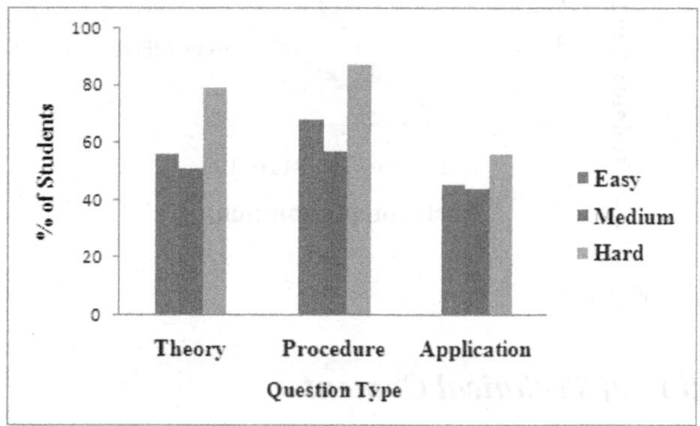

Fig. 8a Difficulty rating of Questions by MEIVL Students

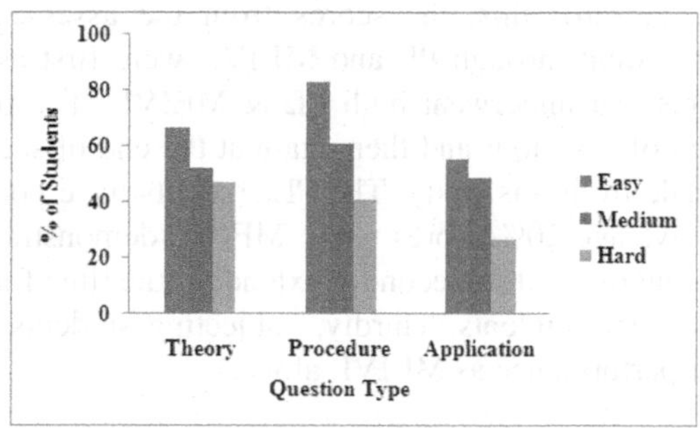

Fig. 8b Difficulty rating of Questions by PL Students

6.2 Speed of Recollection

An indirect measure of conceptual understanding is the speed of factual recollection. In this section, an online instrument that contained twenty questions that included multiple choice questions and pictorial ones were included. Students were divided into two groups those that underwent only PL and those that were exposed to MEIVL. A maximum time of three hours were given to answer the 20 questions or tasks. This being an online test, the time taken by individual students to complete the assessment was individually monitored. Fig 9 shows a plot of the percentage of students completing the task and the time taken to do the same. A significant advantage in terms of completion time is seen with MEIVL.

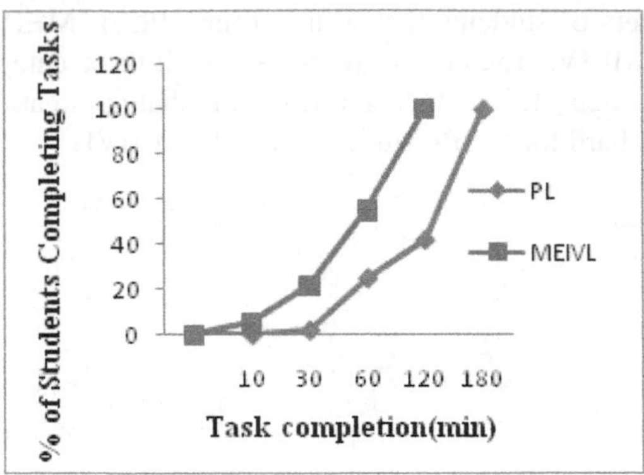

Fig. 9 Speed of Recollection

6.3 Cognition of Technical Content

Although there is significant improvement in the time taken by students to answer questions as seen in the previous section, it is more important to gauge their level of comprehension. Towards this, the scores from the assessed are plotted in Fig 10. Students that went through PL and MEIVL were first assessed and then compared to students that underwent both PL & MEIVL. The performance was monitored at the end of first hour and then again at the end of second hour. Three observations are made from this study. The PL students have poor understanding of the experiments average 20% scores while MEIVL demonstrate a much better performance averaging over 70%. Secondly extending the time for completion did not significantly help the students. Thirdly, subjecting students to both PL and MEIVL had similar performance as MEIVL alone.

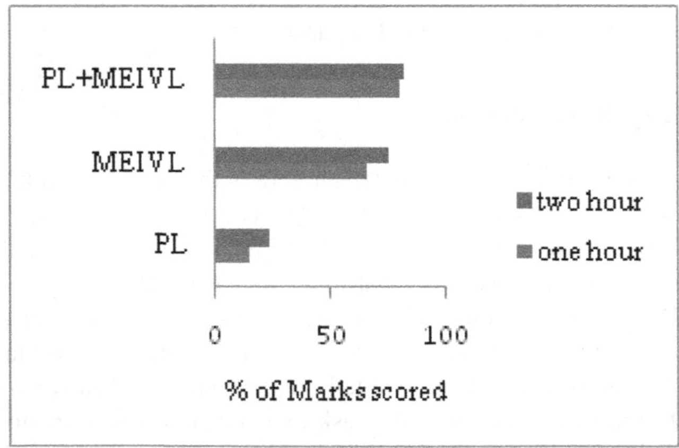

Fig. 10 Cognition of Technical Content

An independent t-test was conducted to examine the effect of MEIVL's attributes described in section 5. The results are tabulated in Table 1. To evaluate the enabling effect, the groups of students that went through MEIVL were able to easily identify the path of light rays through the telescope easily compared to those that did not. This is because unlike in MEIVL, the lab instructor in PL would not be able to show the movement of light rays. So it will be hard for a learner to conceive the idea of diagram without the help of MEIVL. When the facilitating effect of MEIVL was assessed on students by randomizing a sequence of procedural images, they had no issues in ordering them correctly. The challenge remained with students that were exposed to PL. The cognitive load in processing the information presented in the PL was too high and students performed poorly in spite of visual and textual information given to them. The split attention effects from MEIVL were also found to be low based on the high scores from this group. The PL group that was devoid of multimedia enhanced interactive features has to mentally integrate the picture with the text, a process that is cognitively demanding. In spite of the solution for the picture provided to them, the students found it hard to find a suitable answer by analysing the picture. So viewing the picture and there after reading the text or vice versa can imposes a load on the students. Taking an average of all students, the MEIVL group had a much higher mean score (N=15, M=18.52, SD=3.16) compared to the students that went through PL. (N=15, M=11.19, SD=3.09).

Table 1 Comparison of groups with and without animation in three different cases

Case	Group	N	Mean	SD	t Test
Enabling	MEIVL	15	4.86	1.03	t=2.048407
	PL	15	2.2	0.76	df=28, p=0.00000001106
Facilitating	MEIVL	15	5.13	0.83	t=2.048407
	PL	15	3.46	0.89	df=28, p=0.000013
Split Attention	MEIVL	15	8.53	1.30	t=2.048407
	PL	15	5.53	1.44	df=28, p=0.00000196

7 Conclusions

Laboratory education in most institutions is not given sufficient prominence and this in turn has resulted in scientists with poor laboratory skills. MEIVLs have brought in a much needed disruptive intervention in laboratory education practices. Based on the cognitive load theory and its influence on learning and adaptations to MEIVL, this work finds strong correlations to the quality of learning experiences and the cognitive load associated with it. From the various studies conducted MEIVLs have strong influences on the grasp of content due to the integrated visual, auditory, pictorial as well as interactive modes of engaging the students. Over 60% of students found the presentation, organization and delivery of MEIVL so effective in that the average mean score changed by over 50% on learning from MEIVL. The time to learn and present the learning in a coherent fashion is indicative of the enormous impact MEIVLs can have in the landscape of ICT based innovations. Irrespective of the type of assessment i.e. whether online or in the classroom, with limited or extended times, with type of questions asked in descriptive, definitional or multiple choice format, MEIVL group of students always did better by scoring 30% over the PL group of students. One of the reasons for such a large change in the performance is due to the reduction in the extent of difficulty that student faced in comprehending the phenomena with MEIVL. The cognitive loads can be greatly reduced with introduction of MEIVLs in mainstream laboratory education.

Acknowledgements. Our work derives direction and ideas from the Chancellor of Amrita University, Sri Mata Amritanandamayi Devi. The authors would like to acknowledge the contributions of faculty and staff at Amrita University whose feedback and guidance were invaluable.

References

1. Sheorey, T., Gupta, V.: Effective virtual laboratory content generation and accessibility for enhanced skill development through ICT. In: Proceedings of Computer Science and Information, vol. 12, pp. 33–39 (2011)
2. Mayer, R.: Cognitive theory of multimedia learning. In: Mayer, R.E. (ed.) The Cambridge Handbook of Multimedia Learning, pp. 31–48. Cambridge University Press (2005)
3. Hatsidimitris, G.: Learner Control of Instructional Animations: A Cognitive Load Perspective (2012)
4. Sweller, J.: Cognitive load theory and e-learning. In: Biswas, G., Bull, S., Kay, J., Mitrovic, A. (eds.) AIED 2011. LNCS, vol. 6738, pp. 5–6. Springer, Heidelberg (2011)
5. Diwakar, S., Achuthan, K., Nedungadi, P., Nair, B.: Enhanced Facilitation of Biotechnology Education in Developing Nations via Virtual Labs: Analysis, Implementation and Case-studies. International Journal of Computer Theory and Engineering 3, 1–8 (2011)
6. Mayer, R., Moreno, R.: Nine ways to reduce cognitive load in multimedia learning. Educational Psychologist 38, 43–52 (2003)
7. Chandler, P., Sweller, J.: Cognitive load theory and the format of instruction. Cognition and Instruction 8, 293–332 (1991)
8. Sweller, J., et al.: Cognitive Load Theory. In: Explorations in the Learning Sciences, Instructional Systems and Performance Technologies, vol. 1, p. 290 (2011)
9. Sweller, J.: Instructional Implications of David C. Geary's Evolutionary Educational Psychology. Educational Psychologist 43, 214–216 (2008)

10. Burkes, K.: Applying cognitive load theory to the design of online learning (2007)
11. Schär, S., Zimmermann, P.: Investigating Means to Reduce Cognitive Load from Animations: Applying Differentiated Measures of Knowledge Representation. Journal of Research on ... 40, 64–78 (2007)
12. Finkelstein, N., Adams, W., Keller, C., et al.: When learning about the real world is better done virtually: A study of substituting computer simulations for laboratory equipment. Physical Review Special Topics - Physics Education Research (2005)
13. Achuthan, K., Bose, L.: Improving Perception of Invisible Phenomena in Undergraduate Physics Education using ICT. In: ICoICT (accepted, 2014)
14. Raman, R., Achuthan, K., Nedungadi, P., Ramesh, M.: Modeling Diffusion of Blended Labs for Science Experiments among Undergraduate Engineering Student. In: Bissyandé, T.F., van Stam, G. (eds.) AFRICOMM. LNICST, vol. 135, pp. 234–247. Springer, Heidelberg (2014)

The Exploitation of Unused Spectrum for Different Signal's Technologies

Ammar Abdul-Hamed Khader, Mainuddin Mainuddin, and Mirza Tariq Beg

Abstract. Technological advances and market developments in the wireless communication area have been astonishing during the last decade and the mobile communication sector will continue to be one of the most dynamic technological drivers within comparative industries. This paper extend our previous work for detection and discrimination signals, and deals with a cognitive radio system (CR) to improve spectral efficiency for three signals (WiMAX, Frequency Hopping and CDMA2000) by sensing the environment and then filling the discovered gaps of unused licensed spectrum with their own transmissions. We mainly focused on energy detector spectrum sensing algorithm. The simulation shows that the CR systems can work efficiently by sensing and adapting the environment, and showing its ability to fill in the spectrum holes then serve its users without causing harmful interference to the licensed user.

Keywords: Cognitive Radio (CR), Primary User (PU), Secondary User (SU), Cognitive User (CU), Spectrum sensing, Energy Detector.

1 Introduction

Wireless technology has become part and parcel of our daily life. Wireless spectrum is a precious and scarce resource in communication technology. Hence efficient utilization of spectrum is essential. It has been observed by several surveys that a significant portion of licensed spectrum remains unutilized. Conventional fixed

Ammar Abdul-Hamed Khader
Department of Computer Engineering, University of Mosul, Iraq
e-mail: ammar_hameed_eng@yahoo.com

Ammar Abdul-Hamed Khader · Mainuddin Mainuddin · Mirza Tariq Beg
Department of Electronics & Communication Engineering,
Jamia Millia Islamia, New Delhi, India
e-mail: moin_s1@rediffmail.com, mtbeg@jmi.ac.in

El-Sayed M. El-Alfy et al. (eds.), *Advances in Intelligent Informatics*,
Advances in Intelligent Systems and Computing 320, DOI: 10.1007/978-3-319-11218-3_16

assignment of spectrum results in severe under utilization (as low as 10%) of limited licensed/regulated spectrum. So there is a need for innovative techniques that can exploit the wireless spectrum in a more intelligent and flexible manner [1, 2].

Dynamic spectrum access (DSA) techniques have been developed for efficient utilization of regulated spectrum. In DSA, secondary users (who have no spectrum license) can be allowed to use un-utilized licensed spectrum temporarily in opportunistic manner. However primary users (PUs) retain priority in using the spectrum. Cognitive Radio (CR) is a technology that makes use of DSA to utilize the spectrum more efficiently in an opportunistic fashion without interfering with PU [3,4,5].

In CR technology, secondary user's (SU) transmitter changes its parameters based on active monitoring of environment. CR enables SU to sense the empty portion of spectrum, select the best available channel within this portion, coordinate spectrum access with other users and vacate the channel when the primary user enters the environment [6, 7]. Since SU are allowed to transmit data simultaneously through a spectrum meant for PU, it should be aware of the presence or reappearance of it. Thus detection of already existing or entry of new PUs is the first task of CR device. There are various detection techniques such as energy detection, feature detection and matched filter detection [2, 8]. Each detection method has its pros and cons. Signal to noise ratio (SNR) can be maximized using matched filtering but it is difficult to detect signal without signal information. Feature detection approach is based on cyclostationarity; it must also have information about received signal. But in the CR scenario, the devices do not know about primary signal structure and information. Energy detection method does not require any information about the signal to be detected, but it is prone to false detection as it depends on the signal power. Due to simplicity and independence of prior knowledge of signals in CR environment, energy detection approach has been selected here.

In this paper we extend our previous work [9] that deals with the detection and discrimination of the three different signals (CDMA2000, WiMAX and FH signals) only. On this work, we opportunistically utilize the unused spectrums of these signals for unlicensed user.

2 Related Work

An algorithm [10] for detecting mobile WiMAX signals in cognitive radio systems is based on preamble-induced second-order cyclostationarity. They derive closed form expressions for the cyclic autocorrelation function and cyclic frequencies due to the preamble, and use these results in the proposed algorithm for signal detection. In paper [11] of the same authors they proposed a cyclostationarity based algorithm for joint detection and classification of the mobile WiMAX and LTE OFDM signals. The algorithm exploits a priori information on the OFDM useful symbol duration for both signals. Four different approaches for spectrum pooling at the instance of spectrum crunch in the designated block are considered in [12]. In first approach, channel occupancy through random search in complete pooled frequency spectrum is simulated. In second, the channel occupancy through existing regulations based on fixed spectrum allocation (FSA) is simulated. In third, FSA random i.e., allocation of resources to different technologies in the designated slots only through randomized search is simulated. Lastly the channel occupancy in designated slots through Genetic Algorithm (GA) based optimized mechanism is simulated to achieve the desired grade of service (GoS). Paper [13] propose an adaptive spectrum sharing schemes for code division multiple access (CDMA) based cognitive medium access control for uplink communications over the CR networks. The proposed schemes address the joint problems of channel sensing, data transmission, and power and rate allocations. Under this scheme, the SUs can adaptively select between the intrusive spectrum sharing and the non-intrusive spectrum sharing operations to transmit data based on the channel utilization, traffic load, and interference constraints.

3 Spectrum Sensing Schemes

The spectrum sensing feature of CRs introduces techniques for detecting communication opportunities in wireless spectrum by the employment of SUs with low-priority access to the spectrum. Three signal processing methods used for spectrum sensing are discussed below [13,14,15]:

3.1 Matched Filter Detection

Matched filter (MF) provides optimal detection by maximizing SNR but requires demodulation parameters. An optimal detector based on MF is not an option since it would require the knowledge of the data for coherent processing. This means that CR has a priori knowledge of PU signal at both PHY and MAC layers, e.g. modulation type and order, pulse shaping, packet format. The main advantage of MF is that, due to coherency it requires less time to achieve high processing gain. The MF correlates the known signal *s(n)* with the unknown received signal *x(n)*, and the decision is made through:

$$T(x) \triangleq \sum_{n=1}^{N} x(n) s^*(n) -\gtrless_{HO}^{H1} \gamma \tag{1}$$

The test statistic T(x) is normally distributed under both hypotheses:

$$T(x) \sim \begin{cases} N(0, Np_s \sigma_v^2) & under\ H0 \\ N(Np_s, Np_s \sigma_v^2) & under\ H1 \end{cases} \tag{2}$$

Where N is the number of samples and σ_v^2 is the noise variance and p_s is the average primary signal power [16].

3.2 Cyclostationary Feature Extractions

In general, modulated signals are characterized by built-in periodicity or cyclostationarity such as pulse trains, repeating spreading, hoping sequences, or cyclic prefixes. This feature can be detected by analyzing a spectral correlation function. The main advantage of feature detection is robustness to uncertainty in noise power. However, it's computationally complex and requires significantly long observation times [8]. A random process $x(t)$ can be classified as wide sense cyclostationary if its mean and autocorrelation are periodic in time with some period $T0$. Mathematically they are given by [17],

$$Ex(t) = \mu(t + mT0) \tag{3}$$

$$Rx(t, \tau) = \pi(t + mT0, \tau) \tag{4}$$

where, t is the time index, τ is the lag associated with the autocorrelation function and m is an integer.

3.3 Energy Detection

Energy detection (ED) technique is applied by setting a threshold for detecting the existence of the signal in the spectrum. If the receiver cannot gather sufficient information about the PU signal, the optimal detector is an ED. However, the performance of the ED is susceptible to uncertainty in noise power [2,8,13,14,15,16]. Conventional ED consists of a low pass filter to reject out of band noise and adjacent signals, Nyquist sampling A/D converter, square-law device and integrator. The detection is the test of the following two hypotheses:

$$H0\text{: } Y[n] = W[n] \text{ signal absent} \tag{5}$$

$$H1\text{: } Y[n] = X[n] + W[n] \text{ signal present } (n = 1,\ldots, N) \tag{6}$$

Where N is observation interval. The noise is assumed to be additive white and Gaussian (AWGN) with zero mean and variance $\sigma_w^{\ 2}$.

A decision statistic for energy detector is:

$$T = \sum_N (Y[n])^2 \tag{7}$$

Note that for a given signal bandwidth B, a pre-filter matched to the bandwidth of the signal needs to be applied. This implementation is quite inflexible, particularly in the case of narrowband signals and sine waves.

An alternative approach could be devised by using a **P**eriodogram **E**nergy **D**etector (PED) to estimate the spectrum via squared magnitude of the Fast Fourier Transform (FFT). It can be implemented similar to a spectrum analyzer by averaging frequency bins of FFT. Processing gain is proportional to FFT size N and observation/averaging time T. Increasing in value of N improves frequency resolution which helps narrowband signal detection. This architecture also provides the flexibility to process wider bandwidths and sense multiple signals simultaneously [14].

4 The Energy Detection Algorithms

The input signal $x(t)$ is a periodic signal if there exist a constant $T0 > 0$ such that:

$$x(t) = x(t + T0) for \ \ -\infty < t < \infty \tag{8}$$

Where t denotes time, the smallest value of $T0$ that satisfies this condition is the period of $x\ (t)$. And the period of $T0$ defines the duration of one complete cycles of $x\ (t)$.

In analyzing communication signals, it is often desirable to deals with the waveform energy. We classify $x\ (t)$ is an energy signal if and only if, it has non zero but finite energy and ($0 < Ex < \infty$) for all time, where:

$$E_x = \int_{-\infty}^{\infty} x^2\ (t) \tag{9}$$

The Energy Spectral Density (ESD) is:

$$E_x = \int_{-\infty}^{\infty} |X(f)|^2\ df \tag{10}$$

And:

$$F\{x(t)\} = X(f) = \int_{-\infty}^{\infty} x(t)e^{-j2\pi ft}\ dt \tag{11}$$

If $\{x\ (n)\}$ is a sampled data sequence which is available for only a finite time window over $n = 1, 2, 3, \ldots\ldots, N,$ then the discrete Fourier transform (DFT) of $x(n)$ is:

$$\mathrm{DFT}\{x(n)\} = X(k) = \sum_{n=0}^{N-1} x(n)e^{-(j2\pi nk/N)} \tag{12}$$

From Parsevel theorem:

$$\sum_{n=0}^{N-1} |x(n)|^2 \cdot \Delta t = \sum_{n=0}^{N-1} |X(k)|^2 \cdot \Delta f \tag{13}$$

Also the power spectral density (PSD) estimation $S_N(k)$ for a random signal x_N is the *DFT* of the autocorrelation function estimate $R_N(k)$ i.e:

$$S_N(k) = \sum_{-\infty}^{\infty} R_N(k)e^{-(j2\pi nk/N)} \tag{14}$$

This definition relates the two estimates, it is motivated by the fact that the true PSD and autocorrelation function obeys the similar *DFT* relation:

$$S\ (k) = \sum_{-\infty}^{\infty} R\ (k)e^{-(j2\pi nk/N)} \tag{15}$$

The spectral density estimation of (14), can be defined in terms of the sample autocorrelation function, and can be related directly to the observed data. It can be shown that $S_N(k)$ has the following simple relationship to the *DFT* $X_N(k)$ of the data samples:

$$S_N(k) = \frac{1}{N}\left|\sum_{n=0}^{N-1} x_n e^{-(j2\pi nk/N)}\right|^2 = \frac{1}{N}|X_N(k)|^2 \tag{16}$$

Equation (16) can be derived by substituting R_N into (14) where:

$$R_N(k) = \frac{1}{N}\sum_{n=-\infty}^{\infty} x_n \cdot x_{n+k} \quad (17)$$

then $S_N(k)$ become:

$$S_N(k) = \frac{1}{N}\sum_{k=-\infty}^{\infty} \cdot \sum_{n=-\infty}^{\infty} x_n \cdot x_{n+k} e^{-(j2\pi nk/N)} \quad (18)$$

When a factor of $1 = e^{(-j2\pi nk/N)} . e^{(j2\pi nk/N)}$ is introduced into (18), the following can be obtained:

$$S_N(k) = \frac{1}{N}\sum_{n=-\infty}^{\infty} x_n e^{-(j2\pi nk/N)} \cdot \sum_{k=-\infty}^{\infty} x_{n+k} e^{-(j2\pi(n+k)/N)} \quad (19)$$

Changing the summation variable to $m = n + k$ in the second sum, it can be seen that:

$$S_N(k) = \frac{1}{N} X_N(k) \cdot X_N^*(k) = \frac{1}{N}\left|X_N(k)\right|^2 \quad (20)$$

This spectral density estimate is called a periodogram. The periodogram is seen to be the magnitude squared of the *DFT* of the data divided by *N* [18].

5 Simulation Model

Fig.1 shows a simulation model system for IEEE 802.16 (WiMAX), frequency hopping (FH) and CDMA2000-1xRTT systems. The WiMAX (802.16-2004) consist of: data source, forward error correction (FEC) and modulator, IFFT, OFDM transmitter and nonlinear amplifier in the transmitter. While the receiver's blocks are: OFDM receiver, gain and phase compensator, FFT, demodulator and FEC. The signal parameters are: Frequency Band-2GHz, OFDM carriers-256, Adaptive Modulation QPSK, 16QAM, 64QAM, Duplexing TDD and Channel Bandwidth 3.5MHz.

The transmitter in the FH system consists of two important parts, the PN code generator and the frequency synthesizer. The length of m-sequence is $(2^7 - 1)$. The frequency synthesizer generates frequencies that are integer multiple of reference frequency (f_{ref}) corresponding to binary coded decimal (BCD) of the contents of the shift register of PN code generator. The generated frequency will hop over a bandwidth of (0.7 MHz) with the channel spacing (f_{ref}) of (0.1 MHz). The overall hopping signal's bandwidth (BW_{ss}) is approximately:

$$BW_{ss} = m * f_{ref} + BWFSK \quad (21)$$

The data stream from Bernoulli binary generators applied to binary MFSK modulator with carrier frequencies (1MHz). Here the non spread signal has a bandwidth (BWFSK) is approximately

$$fl + rate = 1\ MHz\ + 0.1MHz\ =\ 1.1\ MHz \quad (22)$$

and

$BWss = 7 * 0.1\ MHz\ k + 1.1\ MHz = 1.8\ MHz$

CDMA2000-1xRTT has: radio frequency (RF) bandwidth of (1.25 MHz) radio channels. Adding 64 more traffic channels to the forward link, orthogonal to (in quadrature with) the original set of 64. It supports packet data speeds of up to 153 Kbit/s with real world data transmission averaging 80–100 Kbit/s.

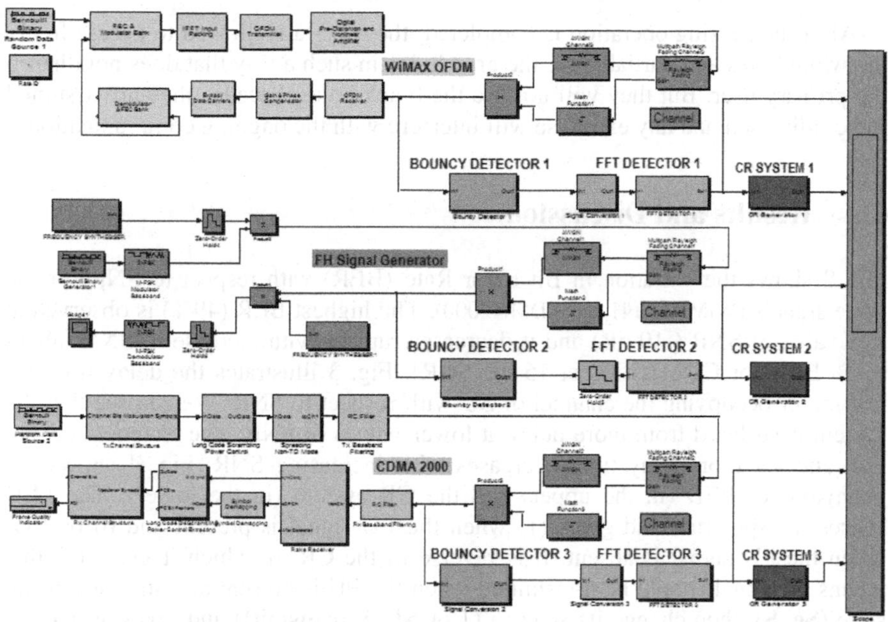

Fig. 1 Matlab Block Diagram for the Entire System

CDMA 2000 1xRTT systems, uses 42-bit PN (Pseudo-Random Noise) Sequence called "Long Code" to scramble voice and data. On the forward link (network to mobile), data is scrambled at a rate of 19.2 Kilo symbols per second (Ksps) and on the reverse link, data is scrambled at a rate of 1.2288 Mega chips per second (Mcps) with Data Capability 9.6/14.4 kbps. Then separate EDs are used after these signals crossing noisy channel (Multipath & Rayleigh Fading Channel with AWGN) and three Bouncy Detectors (details in [9]). EDs is popular for signal detection due to its simple design and small sensing time. The total signal power (energy per unit time) is proportional to the average magnitude squared. A power spectrum describes an energy distribution of a time series in the frequency domain. The results of energy detectors categorize which channel(s) is busy and which is idle by displaying the sorted amount (1) or (0) respectively. Where (1) is used for licences user's signal presence (the amount of energy measured above the threshold level) or (0) is for absent signal (the amount of energy measured below threshold level). The SUs frequently sense the channel. And this information is used in CR system in order to be able to exploit the idle channel case and transmit a signal (unlicensed user) within this period of time. If the information is changing and the PU appeared in any channel, the CR leaves this channel to license user and stop transmitting on it.

Depends on the current information all channels can be classified into three types of spectrum holes [16]:

1. Black spectrum holes, which are fully used,
2. Gray spectrum holes, which are partially used and
3. White spectrum holes, which are not used.

After the sensing operation is completed, the users are allowed to access freely the white holes and partially use the gray holes in such a way that does not disturb the primary user. But they will not use the black holes, because they are assumed to be fully used and any extra use will interfere with the ongoing communication.

6 Results and Discussion

Fig. 2 shows the variation in Bit Error Rate (BER) with respect to SNR for the three signals (WiMAX, FH and CDMA2000). The highest BER (49%) is observed at low value of SNR (-10 dB) and it decreases rapidly with increase in SNR (about 0.1% BER for CDMA2000 at 15 dB SNR). Fig. 3 illustrates the delay time for leaving or occupying the channel of SU with respect to SNR. We can see that the system is suffered from more delay at lower values of SNR. The channel occupation and vacation delay time decreases with increasing SNR. Fig. 4 shows the behaviour of PED (in the upper) and the CR system (in the lower). The PED senses the spectrum and gives (1) when the PU signal is present and (0 or NO) when the PU signal is absent. It is reverse in the CR, in which it gives (0) that means no signal should be transmitted when the PU is present and stay in sensing state (**Se. S**), then change its state to (1 or **SU T(transmit)**) and gives an indication to the SU to opportunistically use the spectrum when the PU is absent.

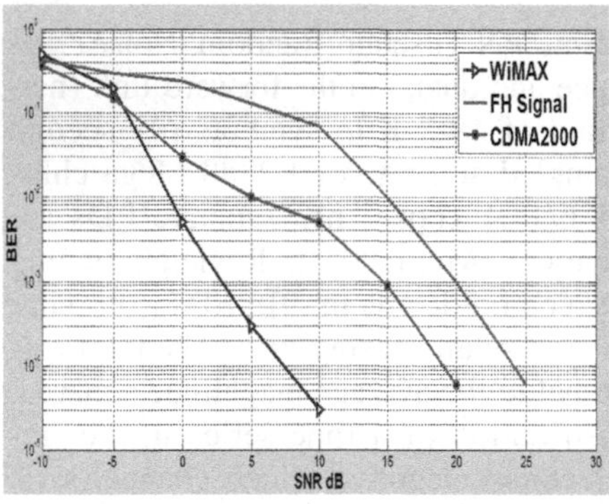

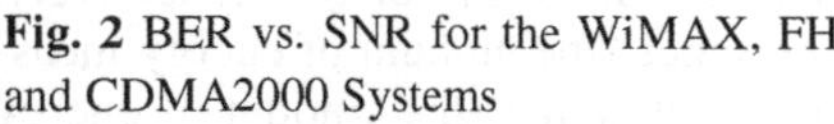
Fig. 2 BER vs. SNR for the WiMAX, FH and CDMA2000 Systems

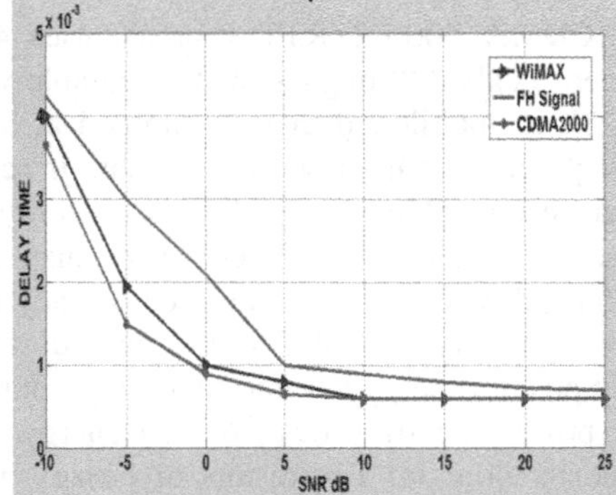

Fig. 3 Delay time vs. SNR for the WiMAX, FH and CDMA2000 Systems

Fig. 4 to Fig. 7 are divided into three time slots (in x axis), where each slot takes about 30 sec. (0-30, 31- 60 and 61- end sec.). That means PU change its state every 30 sec. In Fig. 4, first slot is for the case when WiMAX PU signal is present and has 20 dB SNR. The second slot indicates (NO) transmitted signal, where the processed signal contains noise only. The final slot starts after 60 sec and contains WiMAX PU signal with 10 dB SNR. We can notice that there is no error occurs in the detection of PU because of high SNR values.

Fig. 5 and Fig. 6 also show the behaviour of PED and CR but for FH signals. In Fig. 5, different SNR values for the three time slots are (-1 dB) signal, noise only, (-2 dB) signal, whereas in Fig. 6, these values for three time slots are (-4 dB) signal, noise only, (-8 dB) signal respectively. Also same SNR values for CDMA2000 signal and noise have been used in Fig. 7. It is clear from these Figures that the error in detecting PU (probability of detection P_d) and false alarm error (P_f) is observed when the SNR is too low (**under** (-2dB)) and in same time an error occur in a lower side (CR) and labelled as (**Se. S Er and SU T Er**) respectively because the PED threshold gets confused between two cases signal presents or absents.

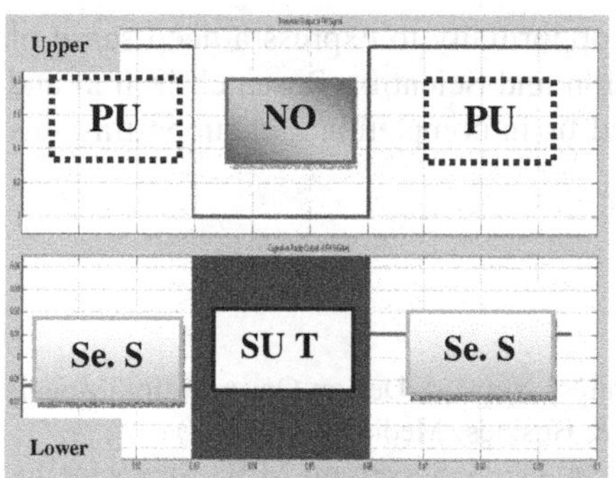

Fig. 4 PED (in the upper) and CR (in the lower) when the WiMAX PU present, absent and present

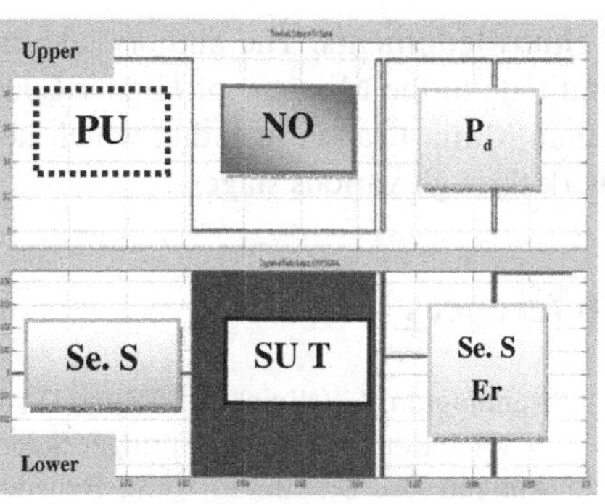

Fig. 5 PED (in the upper) and CR (in the lower) when the FH signal with -1 dB, Noise only and FH with -2 dB

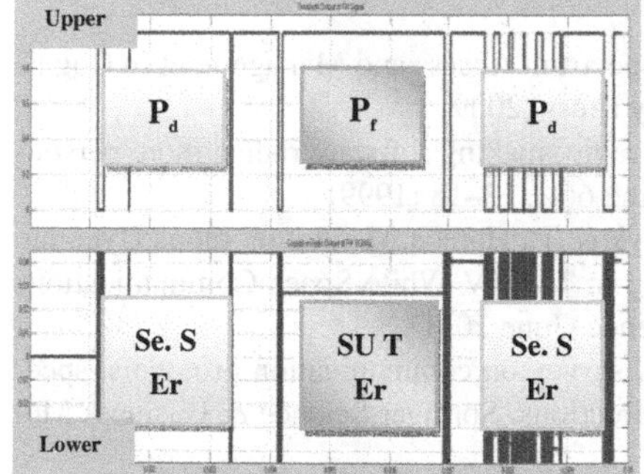

Fig. 6 FH signal with -4 dB, Noise only and FH signal with -8 dB

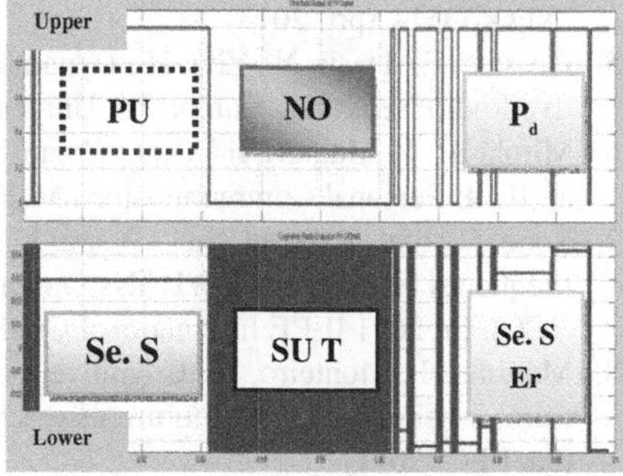

Fig. 7 CDMA2000 signal with -1 dB, Noise only and CDMA2000 signal with -2 dB

7 Conclusions

This paper discusses the problem of detecting the presence and absence of PUs in IEEE 802.16 OFDM, FH signal and CDMA2000 and exploits the idle spectrum(s) for CU(s). It extend our previous work for detection and discrimination signals, and deals with a cognitive radio system (CR) to improve spectral efficiency for three signals. As the knowledge of the authors it is the first time happened in the literature that the behaviour of these three systems are examined together in one system. The Periodogram Energy Detectors have been used to detect the presence of the PUs after these signals pass through bouncy detector. The SUs stay in sensing period and try to exploit the spectrum when the channel is idle. Then vacate the channel as fast as possible when the detectors detect the presence of PU. The simulation results of detection and the exploitation of an idle channel have been done without error at high value of SNR. Simulation results show that an error occurs only with low values of SNR and negligible amount of delay time has been observed between PU and SU changing state.

Acknowledgments. The authors take this opportunity to express a deep sense of gratitude to the Ministry of Higher Education and Scientific Research in Iraq and Jamia Millia Islamia in India, which helped us in completing and supporting this work through various stages.

References

1. Kennington, J., Olinick, E., Rajan, D.: Wireless Network Design Optimization Models and Solution Procedures. Springer Science & Business Media, LLC (2011)
2. Antonio De, D., Emilio, C.S., Benedetto, M.-G.: A Survey on MAC Strategies for Cognitive Radio Networks. IEEE Communications Surveys & Tutorials (2010)
3. Liu, Y., Wan, Q.: Enhanced compressive wideband frequency spectrum sensing for dynamic spectrum access. EURASIP Journal on Advances in Signal Processing (2012)
4. Yoon, S., Li, L.E., Liew, S., Choudhury, R.R., Tan, K., Rhee, I.: QuickSense: Fast and Energy-Efficient Channel Sensing for Dynamic Spectrum Access Networks. In: IEEE INFOCOM (April 2013)
5. Hossain, E., Dusit, N., Zhu, H.: Dynamic Spectrum Access and Management in Cognitive Radio Networks. Cambridge University Press (2009)
6. Mitola III, J., Maguire Jr., G.Q.: Cognitive radio; making software radios more personal. IEEE Personal Communications Magazine 6(4), 13–18 (1999)
7. Song, C., Lan, Z., Sean, S.C., Wang, J., Baykas, T., Harada, H.: Autonomous Dynamic Frequency Selection for WLANs Operating in The TV White Space Communications (ICC). In: 2011 IEEE International Conference (June 2011)
8. Marinho, J., Monteiro, E.: Cognitive radio, survey on communication protocols, spectrum decision issues, and future research directions. Springer Science & Business Media, LLC (2011)
9. Khader, A.A.H., Shabani, A.M.H., Beg, M.T.: Joint Detection and Discrimination of CDMA 2000, WiMAX and Frequency Hopping Signals. American Journal of Scientific Research (89) (July 2013)

10. Al-Habashna, A., Dobre, O.A., Venkatesan, R., Popescu, D.C.: WiMAX Signal Detection Algorithm based on Preamble-Induced Second-Order Cyclostationarity. In: IEEE Globecom (2010)
11. Al-Habashna, A., Dobre, O.A., Venkatesan, R., Popescu, D.C.: Joint Signal Detection and Classification of Mobile WiMAX and LTE OFDM Signals for Cognitive Radio. In: Signals, Systems and Computers (ASILOMAR), Conference Record of the Forty Fourth Asilomar Conference (2010)
12. Sridhara, K., Nayak, A., Singh, V., Dalela, P.K.: Enhanced Spectrum Utilization for Existing Cellular Technologies Based on Genetic Algorithm in Preview of Cognitive Radio. Int. J. Communications, Network and System Sciences 2 (2009)
13. Gorcin, A., Qaraqe, K.A., Celebi, H., Arslan, H.: An Adaptive Thresh-old Method for Spectrum Sensing in Multi-Channel Cognitive Radio Networks. In: 17th International Conference on Telecommunications (2010)
14. Cabric, D., Tkachenko, A., Brodersen, R.W.: Experimental Study of Spectrum Sensing based on Energy Detection and Network Cooperation. In: TAPAS 2006 Proceedings of the First International Workshop on Technology and Policy for Accessing Spectrum, Article No. 12. ACM, New York (2006)
15. Tabassam, A.A., Ali, F.A., Kalsait, S., Suleman, M.U.: Building Cognitive Radios in MATLAB Simulink– A Step Towards Future Wireless Technology. In: IEEE Conference UKSIM International Conference on Computer Modelling and Simulation (2011)
16. Satyanarayana Eerla, V.V.: Performance Analysis of Energy Detection Algorithm in Cognitive Radio. M.Sc. thesis, NIT Rourkela, India (2011)
17. Baldini, G., Giuliani, R., Capriglione, D., Sithamparanathan, K.: A Practical Demonstration of Spectrum Sensing For Wimax Based on Cyclostationary Features. Chapter of book. INTECH (2012) (online March 16)
18. Khuder, A.A.H.: Spectrum Estimation of Frequency Hopping Signal. AL-Mansour Journal, No.14 Special Issue, Part two (2010)

A Sphere Decoding Algorithm for Underdetermined OFDM/SDMA Uplink System with an Effective Radius Selection

K.V. Shahnaz and C.K. Ali

Abstract. Multiuser Detection (MUD) Techniques for orthogonal frequency division multiplexing/space division multiple access (OFDM/SDMA) system remain a challenging area especially when the number of transmitters exceed receivers. Maximum Likelihood (ML) detection is the optimal one, but infeasible due to the high complexity when large number of antennas are used together with high order modulation scheme. Sphere Decoding (SD) algorithm with less complexity but performance near to ML has been explored widely for determined and overdetermined MIMO channels. Very few papers that efficiently deal with an underdetermined OFDM/SDMA channel have been published so far. In this paper a simple pseudo-antenna augmentation scheme has been employed to utilize SD in a rank-deficient case. An effective radius selection method is also included.

1 Introduction

Combination of orthogonal frequency division multiplexing (OFDM) with smart antenna designs have emerged in recent years [1]. They are applied with the main objective of combating the effects of multipath fading on the desired signals, there by increasing both the performance and capacity of wireless systems. An application of smart antennas is space division multiple access (SDMA). Here the users are identified with the help of their spatial signature. Channel distortion due to multipath propagation is easily mitigated with OFDM, while bandwidth efficiency can be increased with use of SDMA. The performance improvement of OFDM/SDMA has

K.V. Shahnaz
Research Scholar, National Institute of Technology, Calicut
e-mail: shahnaznitc@gmail.com

C.K. Ali
Associate Professor, National Institute of Technology, Calicut
e-mail: cka@nitc.ac.in

El-Sayed M. El-Alfy et al. (eds.), *Advances in Intelligent Informatics*,
Advances in Intelligent Systems and Computing 320, DOI: 10.1007/978-3-319-11218-3_17

become an attractive research topic for emerging future wireless systems. Channel estimation and multiuser detection are two major tasks to be handled effectively. In this paper, focus is on multiuser detection (MUD), assuming perfect channel knowledge at the receiver. To solve the problem of multiuser detection in OFDM/SDMA systems, various classical solutions are available, having varying complexity, like zero forcing (ZF), minimum mean square error (MMSE), successive interference cancellation (SIC), parallel interference cancellation (PIC) [2].

OFDM/SDMA systems confront with three channels. Underloaded (number of BS antennas (N_r) larger than transmitters (N_t)), fullyloaded (N_r same as N_t) and overloaded (N_r less than N_t) which is also called a rank-deficient or underdetermined channel. In rank-deficient systems, the $N_r \times N_t$ dimensional OFDM/SDMA channel matrix becomes singular and, hence, noninvertible, thus making the degree of freedom of the detector insufficiently high for detecting the signals of all the transmitters. Above detectors perform well for the first two cases but exhibit very poor performance in the third case which we know is the practical scenario.

We use an efficient Sphere Decoding (SD) algorithm to tackle this problem. SD algorith was introduced by Finke and Phost [3] in 1985. Since then it has been widely used in various communication applications. It searches a lattice point in a hypershere centred at a given vector, there by reducing the search space and hence the tedious computations required for ML detection. There are a few papers in literature that treat underdetermined MIMO channels. But all require some detailed procedures. The pseudo-antenna augmentation scheme in [5] gives a simple and powerful approach to treat rank-deficient MIMO channels and has been extended to an OFDM/SDMA system. The linearly reducing radius selection based on noise variance is also simple and effective.

2 System Model

The multiuser OFDM/SDMA system considered supports N_t mobile stations (MSs) simultaneously transmitting in the Up Link (UL) to the Base Station (BS). Each of the users is equipped with a single transmit antenna, whereas the BS employs an array of N_r antennas. It is assumed that a time division multiple access protocol organizes the division of the available time domain resources into OFDM/SDMA time slots. Instead of one, N_t MSs are assigned to each slot that is allowed to simultaneously transmit their streams of OFDM-modulated symbols to the BS [1].

At the kth subcarrier of the nth OFDM symbol received by the N_r - element BS antenna array we have the received complex signal vector $y[n,k]$, which is constituted by the superposition of the independently faded signals associated with the N_t mobile users and contaminated by AWGN, expressed as:

$$y = Hs + n \tag{1}$$

where the $N_r \times 1$ dimensional vector y, the $N_t \times 1$ dimensional vector s and the $N_r \times 1$ dimensional vector n are the received, transmitted and noise signals, respectively. The indices $[n,k]$ have been omitted for notational convenience. Here,

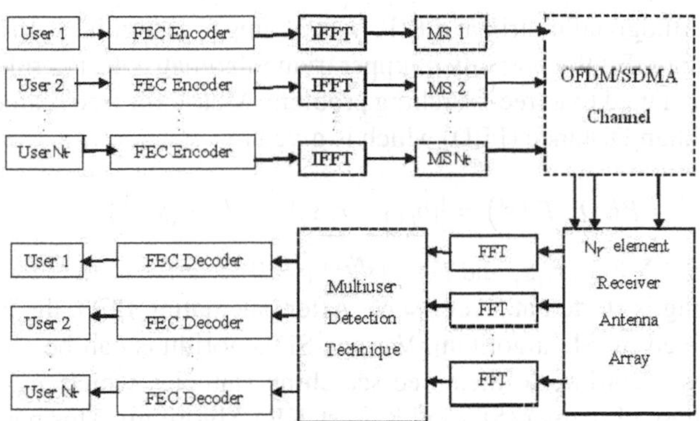

Fig. 1 Schematic of the OFDM/SDMA Uplink System

$y = (y_1, y_2, y_3 ... y_{N_r})^T$, $s = (s^{(1)}, s^{(2)}, s^{(3)} ... s^{N_t)})^T$ and $n = (n_1, n_2, n_3 ... n_{N_r})^T$. $N_r \times N_t$ dimensional matrix H contains the frequency domain channel transfer functions (FD-CHTF) of N_t users.

3 Sphere Decoding Algorithm

Any multiuser detection technique for single carrier MIMO can be used for the latest proposed OFDM/SDMA also, as it is done on per-carrier basis. It is intuitive that the ML Detector gives the optimum result as it searches through all possibility of transmitted symbols. Exhaustive search by optimum ML detector requires 2^{mN_t} evaluations of the decision metric for finding the most likely transmitted N_t user symbol vector $\widehat{s}$.

$$\widehat{s} = arg\left\{min_{s \in M^{N_t}} \|y - Hs\|^2\right\} \tag{2}$$

The set of M^{N_t} number of trial vectors are elements of M_c that denotes the set containing the $M = 2^m$ number of legitimate complex constellation points associated with the specific modulation scheme employed, while m denotes the number of bits per symbol.

To reduce the complexity of ML, SD limits the searching space by radius r. Then the decision criterion of ML detection will be

$$\|y - Hs\|^2 < r^2 \tag{3}$$

By QR decomposition of channel matrix H, Eqn. 3 can be decomposed as follows.

$$\|y - Hs\|^2 = \|y - QRs\|^2 < r^2 \tag{4}$$

which again can be written as

$$\left\|Q^H y - Rs\right\|^2 = \|z - Rs\|^2 < r^2 \tag{5}$$

Q is $Nr \times Nr$ orthogonal matrix and R is $N_t \times N_t$ upper triangular matrix. z is $Q^H y$ and a $N_t \times 1$ vector. By the property of upper triangular matrix R, the rule of ML detection can be mapped to a tree-searching problem. Metric for tree search is defined as Partial Euclidian Distance (PED) which is given by

$$PED : T_i\left(s^i\right) = \left\|b_{i+1} - r_{ii}s_j\right\|^2 + T_{i+1}\left(s^{i+1}\right) \tag{6}$$

where $b_{i+1} = z_i - \sum_{j=i+1}^{N_t} r_{ij}s_j$ and $T_{N_{t+1}}\left(s^{N_{t+1}}\right) = 0$

Tree searching is started at level $i = N_t$. After calculating PED, the next node to visit is determined by SD algorithm. Various SD algorithms can be classified into three categories according to their tree searching stategies which are Fincke and Phost (FP), Schnorr-Euchner (SE) and K-Best(KB) Algorithm. This paper consider SE-SD since it can guarantee fixed throughput without severe performance degradation [4].

The complexity of SD also depends on the radius r of the hypersphere. If r is too large, the hypersphere contains too many lattice points, if r is too small, the hypersphere may contain no lattice point at all. There are no general guidelines for selecting r and it depends on the particular application. Next section discusses a simple method to choose an appropriate r.

3.1 Radius Selection

A deterministic method for selecting an initial hypershere radius has been proposed by Qiao in [8]. This method is designed for communication application.

The steps involved are:

1. Solve $z = Rs$.
2. Round the entries of s to their nearest integer, $\tilde{s} = \lceil s \rfloor$
3. Set $r = \|R\tilde{s} - z\|_2$

This $\tilde{s}$ is known as Babai estimate. The Sphere $\|Rs - z\|_2 = r$ contains atleast one lattice point, namely $\tilde{s}$, the integer vector closest to the real least squares solution s. In communication application, this procedure is nothing but ZF equalization.
Let us examine the size of r.
Take $d = \tilde{s} - s$,

then $R\tilde{s} - z = R(s+d) - z = Rd$, as $z = Rs$

Since $d = \lceil s \rfloor - s$, $\|d\|_2 \leq \sqrt{N_t}/2$.

Thus, $r = \|Rd\|_2 \leq \sqrt{N_t}/2\,\|R\|_2 = \sqrt{N_t}/2\,\|H\|_2 = \|H\|_2$ as we take $N_t = 4$ for simulation.

In this paper, the channel considered is a multipath wi-fi indoor channel modelled by Rayleigh fading. Due to channel power constraint, $\|H\|_2$ is not large. Consequently the radius of the search sphere is not large. The SD algorithm using $r = \|R\tilde{s} - z\|_2$ will find the integer least square solution in this sphere, possibly on its surface. Due to imperfection in channel estimation and rounding errors in calculation of r, the SD algorithm fails to find a lattice point in the computed hypersphere at low SNR. So a value greater than the computed r which does not compromise the complexity and yet give a good performance has to be selected for the simulation.

3.2 Pseudo-antenna Augmentation Scheme

The sphere decoders designed for the case where $N_r \geq N_t$ fail when $N_r < N_t$ since the channel matrix H does not have full column rank and therefore cannot be QR factorized. In paper [5], a modification has been done to the channel matrix estimated, as

$$\widetilde{H} = \begin{bmatrix} eI_{(N_t-N_r)} & 0_{(N_t-N_r)\times N_r} \\ & H \end{bmatrix} \tag{7}$$

to make it a matrix with full column rank.

Here the bottom Nr rows comprise the original channel matrix. I is the identity matrix. e is either a small real or complex number depending on the modulation scheme.

Zeros are augmented to first $N_t - N_r$ rows of final received vector

$$\widetilde{y}_{N_t\times 1} = \begin{bmatrix} 0_{(N_t-N_r)\times N_r} \\ y_{N_r\times 1} \end{bmatrix} \tag{8}$$

Then the psudo received vector may be defined as

$$\begin{bmatrix} es_1 \\ \vdots \\ es_{N_t-N_r} \\ \sum_{l=1}^{N_t} h_{1i}s_i + n_1 \\ \vdots \\ \sum_{i=1}^{N_t} h_{N_r,i}s_i + n_{N_r} \end{bmatrix} \tag{9}$$

and the noise vector as

$$\begin{bmatrix} -es_1 \\ \vdots \\ -es_{N_t-N_r} \\ 0_{(Nr\times 1)} \end{bmatrix} \tag{10}$$

$\widetilde{H}$ has full column rank by this augmentation and can be decomposed using standard QR factorizing algorithms.

The effect of value taken by e has been analyzed in [5]. If e is small, lower bound on the radius with which correct symbol vector included is essentially independent of e. But if e is large, radius needs to be large. With very small e, complexity of SD is independent of e and same as that of usual SD algorithm which is roughly $O(N_t^3)$. So value of e was selected as 0.01.

4 Complexity Comparison

Complcxity of ZF, MMSE, ML and SD are given in terms of number of complex operations involved, in Table 1. For easiness of presentation we take $N_r = N_t$. N is the number of sub carriers used in OFDM. The computational complexity of calculating $||y - Hs||^2$ is determined in [7] to be

$$C_o = 2N_t^2 + 2N_t - 1 \tag{11}$$

Since ML detection searches entire symbol alphabet, its complexity is $2^{mN_t}C_o$ for single ofdm carrier, where 2^m is the number of legitimate complex constellation points associated with the specific modulation scheme as mentioned. The paper [6] examine complexity of SD analytically. We know that the complexity is mainly due to number of lattice points in hypershere with selected radius, and dimensions $k = 1, \ldots N_t$. Let the number of search points searched by the algorithm in the hypershere of a given radius considering all the influencing factors be r_o. Then the complexity per carrier will be proportional to r_oC_o. The worst case value of complexity is exponential, but we always select radius in such a way to restrict the complexity proportional to an integer multiple of of N_t. It is true especially when radius is calculated using Babai estimate. In such case, complexity is almost same as ZF. So let us assume that the complexity will not go beyond $2N_tC_o$. By using Gauss-Jordan Elimination algorithm to compute matrix inversions, complexity of ZF and MMSE per ofdm carrier is given in [7] and this has been used to compute the number of complex operations required.

Table 1 Computational Complexity Comparison of ZF, MMSE, ML and SD

OFDM/SDMA detector	Numer of Complex Operations
ZF	$N((14/3)N_t^3 + 5N_t^2 - (8/3)N_t)$
MMSE	$N((14/3)N_t^3 + 5N_t^2 - (2/3)N_t + 1)$
ML	$N(2^{mN_t}C_o)$
SD	$N(2N_tC_o)$

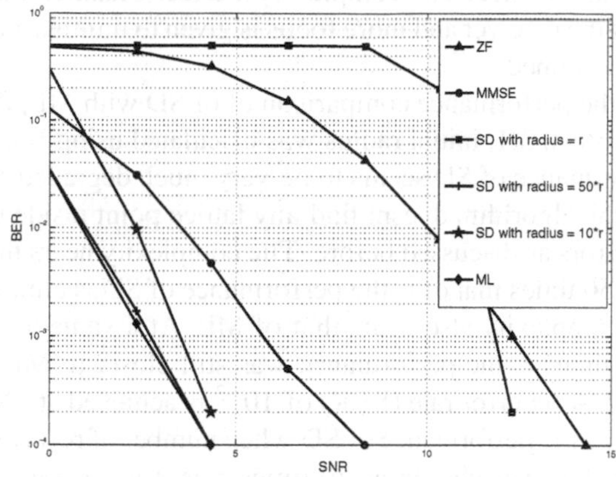

Fig. 2 BER vs SNR performance comparison of SD aided OFDM/SDMA with ML, MMSE, and ZF when Nt = Nr = 4

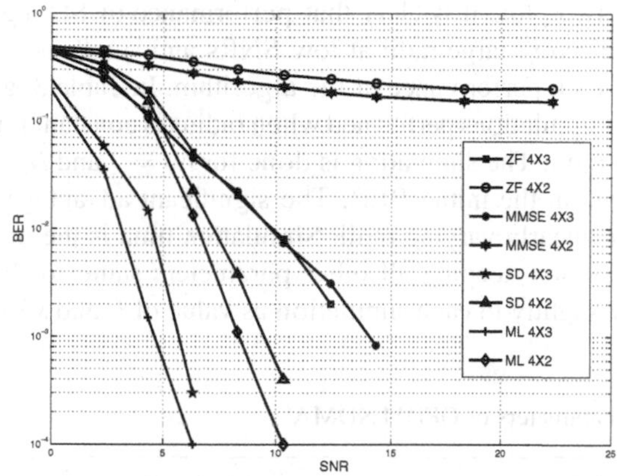

Fig. 3 BER vs SNR performance comparison of SD aided OFDM/SDMA with ML, MMSE and ZF when Nt = 4, Nr = 3, 2

5 Simulation Results

A schematic of multiuser OFDM/SDMA uplink system has been shown in Fig. 1, based on which simulation was performed in MATLAB. IEEE 802.11n high speed WLAN standards are used for the simulation. Bandwidth is 20 MHz. Table 1 give the parameters used for the simulation. MATLAB inbuilt commands ('poly2trellis', and 'convenc' with [133 171] as generator polynomial) was used for half-rate error control coding. Input data was scrambled before convolutional encoding and interleaved after encoding. All these schemes improved the overall perfomance but with some increase in data overhead and complexity. Perfect Channel State Information (CSI) is assumed at the receiver and more focus is given to multiuser detection when channel is underdetermined.

Figure 2 shows the performance comparison of of SD with ML, ZF and MMSE. As explained in Subsect. 3.1, initial radius was calculated using Babai estimate. At low SNRs the performance of SD seems to be very much degraded compared to all other detectors as the algorithm doesnt find any lattice point inside this sphere due to some practical errors as discussed before. The complexity nears that of ZF at this radius. At a radius 50 times that of r, the performance of SD is almost same as ML. It is intuitive that complexity also nears that of ML. At a value less than this, say when radius is 10 times r, the performance was still closer to ML but at a lower complexity for sure. A bit error rate (BER) of 10^{-3} is achieved at SNR $3.5dB$.

Next, Fig. 3 shows the performance of SD when number of receivers are less than transmitters. The radius was chosen as 20 times r as it gives a good performance at lower complexity than ML. Performance of ZF and MMSE keep on degrading severely as the channel become more underdetermined. But performance of SD was found closer to ML, which is really a notable advantage for implementing a practical system.

From the above two plots it is clear that performance of SD algorithm is poor at low SNRs. So selecting large radii at low SNRs and small radii at High SNRs will improve the overall performance of the algorithm. In Table 3, a comparison is shown between two simulations performed when radius is constant for all SNRs, and a linearly reducing radii. The simulation is done for $N_r = 3$ and $N_t = 4$. There was a slight improvement in the initial BER. The significant advantage is the reduced simulation time for linearly varying radii. Simulation time is the time taken to run the MATLAB simulation keeping all other parameters same for both cases. The values of radii vary slightly in each simulation as value of R and $\widetilde{s}$ vary.

Table 2 Simulation Parameters of OFDM/SDMA

Parameters	Specifications
FFT size	$N = 64$
cyclic Prefix length	$L = 16$
modulation	BPSK
channel model	Multipath Rayleigh fading
number of multipaths selected	10
channel State Information	Perfect
number of BS Antennas	$N_r = 4, 3, 2$
number of simultaneous users	$N_t = 4$

Table 3 Performance Comparison of SD algorithm for constant and varying radii

Radius	Simulation Time	BER at SNR 0dB
18 (on an average) for all SNR	211 seconds	0.2375
[82.7532 18.9945 17.1404 15.6659 12.2056 9.6533 7.2722 6.2088 5.3234 5.0695 3.2076 1.2024]	188 seconds	0.1803

6 Conclusion

At full-load, complexity of ZF and MMSE are far less than ML and give acceptable performance. But we can see that SD works far better than them at a complexity less than ML, and nears the performance of ML. Performance of MMSE and ZF are completely degraded when channel is underdetermined. A simple pseudo-antenna augmentation scheme incorporated in SD algorithm makes it an efficient detection scheme for an over-loaded OFDM/SDMA system where the conventional detectors fail severely. The linrarly varying radii selection adopted ensures better BER at an acceptable complexity.

References

1. Vandenameele, P.: A Combined OFDM/SDMA Approach. IEEE Journal on Selected Areas in Communications 18(11), 795–825 (2000)
2. Hanzo, L., Munster, M., Choi, B.J., Keller, T.: OFDM and MC-CDMA for Broadband Multi-User Communications, WLANs and Broadcasting. IEEE Press/Wiley, Piscataway (2003)
3. Fincke, U., Pohst, M.: Improved methods for calculating vectors of short length in lattice, including a complexity analysis. Mathematics of Computation 44(170), 463–471 (1985)
4. Burg, A., Borgmann, M., Wenk, M., Studer, C., Bölcskei, C.: Advanced receiver algorithms for MIMO wireless communications. In: Proceedings of the Conference on Design, Automation and Test in Europe, pp. 502–508, 593–598 (2006)
5. Hung, C.-Y., Sang, T.-H.: A Simple Sphere Decoding Algorithm for MIMO Channels. In: Prceedings of IEEE International Sumposium on Signal Processing and Information Technology (2006)
6. Hassibi, B., Vikalo, H.: On the sphere decoding algorithm.i.expected complexity. IEEE Transactions on Signal Processing 53(8), 2806–2818 (2005)
7. Chang, R.Y., Chung, W.-H., Hung, C.-Y.: Efficient MIMO Detection Based on Eigenspace Search with Complexity Analysis. In: Proceedings of IEEE International Conference on Communications, ICC 2011, Kyoto, Japan (2011)
8. Qiao, S.: Integer least squares: Sphere decoding and the LLL algorithm. In: Proceedings of C3S2E 2008, ACM International Conference Proceedings Series, pp. 23–28 (2008)

Design of Multiband Metamaterial Microwave BPFs for Microwave Wireless Applications

Ahmed Hameed Reja and Syed Naseem Ahmad

Abstract. This work proposes an end-coupled half wavelength resonator dual bandpass filter (BPF). The filter is designed to have a 10.9% fractional bandwidth (FBW) at center frequency of 5.5GHz. Dual-band BPF with more drastic size reduction can be obtained by using metallic vias act as shunt-connected inductors to get negative permittivity (-ε). The process of etching rectangular split ring resonators (SRRs) instead of open-end microstrip transmission lines (TLs) for planar BPF to provide negative permeability (-μ) is presented. The primary goals of these ideas are to get reduction in size, dual- and tri- bands frequency responses. These metamaterial transmission lines are suitable for microwave filter applications where miniaturization, dual- and tri-narrow passbands are achieved. Numerical results for the end-coupled microwave BPFs design are obtained and filters are simulated using software package HFSS. All presented designs are implemented on the Roger RO3210 substrate material that has; dielectric constant ε_r=10.8, and substrate height h =1.27 mm.

Keywords: Bandpass filter, end coupled, metamaterial, microwave, multiband, split ring resonator.

1 Introduction

Microwave bandpass filters are vital components in a wide range of microwave systems, including satellite communications and radar. Transmission line resonator structures consist of a combination of multi-lines of at least one quarter guide

Ahmed Hameed Reja
Department of Electromechanical Engineering,
University of Technology, Baghdad, Iraq

Ahmed Hameed Reja · Syed Naseem Ahmad
Department of Electronics & Communication Engineering
Jamia Millia Islamia, New Delhi-110025, India
e-mail: ahmad8171@yahoo.com, snahmad@jmi.ac.in

El-Sayed M. El-Alfy et al. (eds.), *Advances in Intelligent Informatics*,
Advances in Intelligent Systems and Computing 320, DOI: 10.1007/978-3-319-11218-3_18

wavelength ($\lambda_g/4$), these structures are used as components in different types of filters with different filtering characteristics.

The most popular of various design structures for RF/microwave filters include end-coupled, edge-coupled, hairpin and interdigital filters. Fig. 1 shows a third order end-coupled microstrip filter [1].

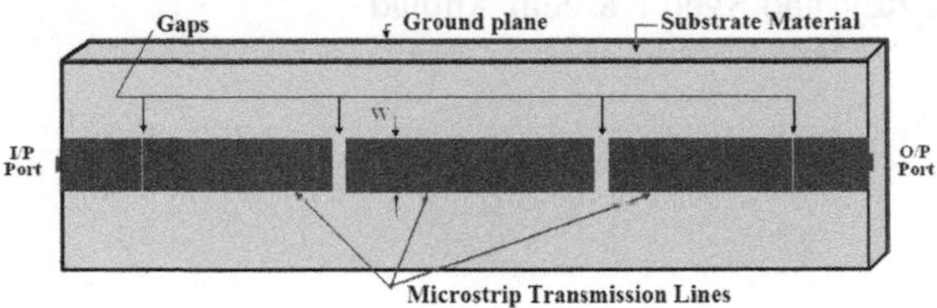

Fig. 1 Three poles end-coupled bandpass filter structure

End-coupled line filters consist of open-end microstrip resonators of approximately half wavelength. The resonators are coupled between them through a gap-spacing (S) for designing of bandpass filters. But it has the disadvantage of consumes more space compared to other microwave filters. In planar technology, filters with small fractional bandwidth (FBW) could be implemented in case of end-coupled resonators, since the coupling between resonators have to be weak. So, this topology cannot be used for the design of wider band filter [1]. There are many advantages of this type of filter such as; low cost, fairly high tolerance, and ease of fabrication. To miniaturize the filter size, the metamaterial idea is presented in this paper. Metamaterials (MTMs) are defined as artificial, effectively homogeneous (average cell size *p* is much smaller than the guided wavelength, λ_g) and exhibiting highly unusual properties (negative values of effective permittivity ε_r, and effective permeability μ_r) not readily available in nature [2, 3].

The electromagnetic properties of metamaterials were already predicted by Veselago in 1968 [2]. Split-ring resonators (SRRs) were one of the first particles for metamaterial structure which were proposed by Pendry in 1999 [4]. The metallic metamaterials comprises double SRRs are the main artificial structures to realize magnetic responses above gigahertz frequencies [5, 6, 7, 8]. The double SRRs topologies depicted in Fig. 2 are used to obtain a negative value of effective

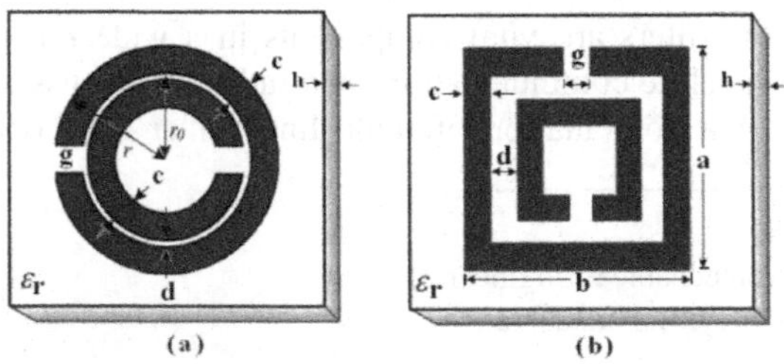

Fig. 2 SRR topologies (a) circle shape, (b) square shape. The relevant dimensions are indicated. (Metal regions are depicted in brown color)

permeability over a desired frequency range. In planar technology, one-dimensional left handed metamaterials and negative permeability transmission lines based on SRRs have been proposed [9, 10].

A dual-band, tri-band and quad-band as a multiband of frequencies are necessary in telecommunication applications. All communication devices which have many channels use multiband frequencies. The need for the design of compact, low-cost, and robust radio frequency (RF) devices operating at multiband frequency to facilitate and develop the next generation wireless systems. These devices are used rather than; complexity, high cost and power demand of multiband parallel transmit and receive signal path circuits. In literature, there are various approaches reported for dual- and tri- bands of BPFs. All these approaches depend on the change of substrate material parameters such as thickness and dielectric constant or change of geometrical parameters of resonators such as unit cell dimensions, strip and split widths, gap widths and separation between the rings. The design of planar dual-band BPFs using modified SRRs as a metamaterial component is presented. The resonance frequencies are analytically derived, and the field distribution on each SRR is studied [11]. A dual-band microstrip filter depending on double resonant property of the complementary split ring resonators (CSRRs) as a resonant unit and T-shaped stub as feeding structure is designed. This filter has two passbands of (2.4 - 2.6) GHz and (3.8 - 4.3) GHz with low insertion loss inside band and high rejection out-of-band [12]. A compact dual-band microstrip filter with equal-length SRRs and zero degrees-feed structure is proposed. A zero degrees-feed structure is applied to realize additional transmission zeros at finite frequencies [13]. A miniaturized dual-band BPF module using double-split (DS)-CSRR and SRR is presented. Two passbands are individually printed on two sides of a substrate material (Rogers 3010) as a compact integration [14].

A compact dual-band BPF using side-coupled octagonal SRRs is proposed. The filter employs two sets of SRRs operating at different frequencies to generate two passbands and provide facilities to control the bandwidth of the two passbands [15]. A compact dual-band substrate integrated waveguide (SIW) filters based on two different types of CSRRs loaded on the waveguide surface is presented. Compact size, good selectivity, stopband rejection and easy fabrication are achieved [16]. A dual-band BPF using SRRs and defected ground structure (DGS) with constant absolute bandwidth is proposed. The inner and outer SRRs are operated for respective passband. DGS can control the coupling coefficient of the filter. This kind of filter with compact size and high selectivity is designed and fabricated [17]. Three SRRs used to obtain a triple-band response is proposed. Two topologies to design triple-band filters with controllable responses and a systematic filter design approach based on a filter coupling model are presented [18]. Miniaturized multiband filters using CSRRs is proposed. A multiple passbands can be generated by loading different types of CSRRs on the waveguide surface. Two types of dual-band filters, a triple-band and a quadruple-band filter are designed and fabricated. The advantages of these filters are compact size, good selectivity and high stopband rejection [19]. A technique based on combination of two parallel multimode resonators and single CSRR to design multi-notch bands ultra-wideband (UWB)-BPF is discussed. The mechanism of realizing the notch-bands is mathematically presented and a triple notch-bands UWB-BPF is designed, simulated and fabricated. The size reduction of around 35% is demonstrated compared to the conventional filter [20].

In this paper, an end-coupled dual- and tri- bands microwave BPF design has been presented. Dual-band filter with size reduction can be obtained by using vias. To enhance the magnetic coupling and obtain multiband, the preferred structure used is etching rectangular SRRs instead of traditional open ended transmission lincs at same external dimensions. It will be shown that the designed metamaterial structures can found applications to the design of dual- and tri-bands microwave filters. These metamaterial structures interest in compact microwave filters design used in wireless applications.

2 End-Coupled, Half-Wavelength Resonator Filter Design

2.1 Traditional End-Coupled Filter Design

Fig. 3 shows the general configuration of three poles (n = 3) end-coupled microstrip BPF, where each open-end microstrip resonator is approximately a half guided wavelength ($\lambda_g/2$) long at the resonant frequency (f_o) of the bandpass filter. A microstrip gap (S) can be represented by an equivalent circuit, as illustrated in Fig. 4 [21].

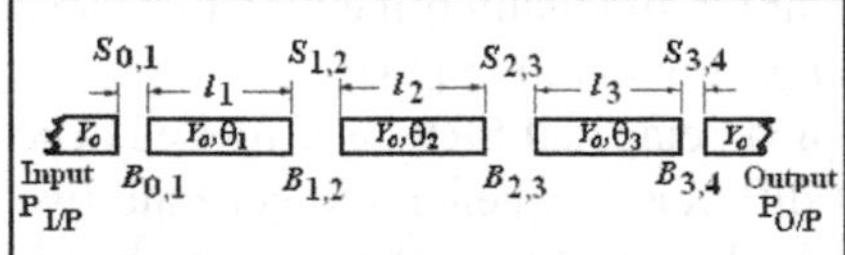

Fig. 3 Three poles configuration of end-coupled microstrip BPF

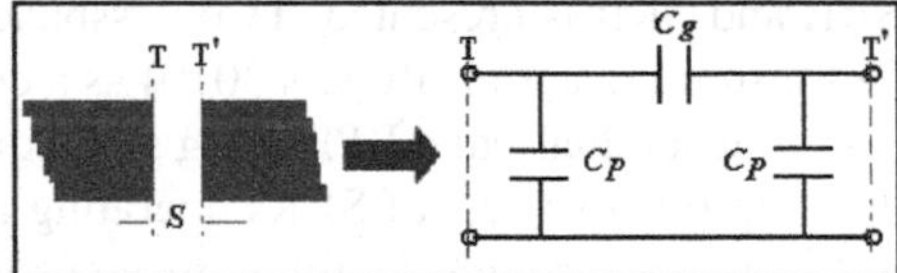

Fig. 4 Microstrip gap and its equivalent circuit

The shunt (C_P) - and series (C_g) - capacitances may be determined by using equations in [22]. These capacitances depend on the width of microstrip (W), dielectric constant of substrate material (ε_r), gap between two adjacent resonators (S), and the thickness of substrate material (h).

$$C_P^{j,j+1}=0.5\ C_e^{j,j+1} \tag{1}$$

$$C_g^{j,j+1}=0.5C_o^{j,j+1}-0.25C_e^{j,j+1} \tag{2}$$

The components C_o and C_e are expressed as

$$\frac{C_o^{j,j+1}}{W}=\left(\frac{\varepsilon_r}{9.6}\right)^{0.8}.\left(\frac{S_{j,j+1}}{W}\right)^{m_o}.e^{k_o} \qquad (PF/m) \tag{3}$$

$$\frac{C_e^{j,j+1}}{W}=12\left(\frac{\varepsilon_r}{9.6}\right)^{0.9}.\left(\frac{S_{j,j+1}}{W}\right)^{m_e}.e^{k_e} \qquad (PF/m) \tag{4}$$

where

$$m_o=\frac{W}{h}[0.619.\log(W/h)-0.3853]$$

$$\text{for } 0.1\leq s/W\leq 1 \tag{5}$$

$$k_o=4.26-1.453.\log(W/h)$$

$$m_e=0.8675$$

$$\text{for } 0.1\leq s/W\leq 0.3 \tag{6}$$

$$k_e=2.043(W/h)^{0.12}$$

The capacitive coupling in open-ends from one resonator structure to the other is through the gap between them, this gap can be represented by J- inverters, which are of the form in Fig. 5 [21]. These J-inverters go about to reflect high impedance levels to the ends of each resonator, and this causes the resonators to exhibit a shunt-type resonance [23]. A third order (n = 3) Chebyshev low pass filter (LPF) prototype with a normalized cutoff Ω_c = 1 and 0.1dB passband ripple is chosen. Whose order of the filter and element values (g_n) are listed in Table 1.

$$J = Y_o \tan\left|\frac{\emptyset}{2}\right|$$

$$\emptyset = -\tan^{-1}\frac{2B}{Y_o}$$

$$\left|\frac{B}{Y_o}\right| = \frac{J/Y_o}{1-(J/Y_o)^2}$$

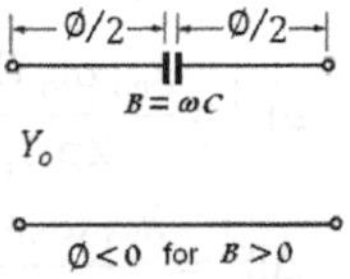

Fig. 5 J-inverters with respect to lumped and transmission line elements

Table 1 The order of filter with element values

n	g_o	g_1	g_2	g_3	g_4
3	1	1.0316	1.1474	1.0316	1

A microstrip end-coupled BPF is designed to have a fractional bandwidth (FBW) of 10.9% at 5.5 GHz central frequency (f_o) related to upper (ω_2) and lower (ω_1) cutoff frequencies and expressed as [21].

$$\text{FBW}=\frac{\omega_2-\omega_1}{\sqrt{\omega_1.\omega_2}} \tag{7}$$

Thus, the filter under consideration operates as a shunt-resonator. The $J_{j,j+1}$ are the characteristic admittances of J-inverters and Y_0 is the characteristic admittance of the microstrip line. Whose general design equations are given as follows [21].

$$\frac{J_{01}}{Y_0}=\sqrt{\frac{\pi.\text{FBW}}{2.g_o.g_1}} \tag{8}$$

$$\frac{J_{j,j+1}}{Y_0}=\frac{\pi.FBW}{2.\sqrt{g_j.g_{j+1}}}, \qquad j=1 \text{ to } n-1 \tag{9}$$

$$\frac{J_{n,n+1}}{Y_0}=\sqrt{\frac{\pi.FBW}{2.g_n.g_{n+1}}} \tag{10}$$

The susceptance $B_{j,j+1}$ under perfect series capacitance condition for capacitive gaps act as in Fig. 5 and expressed as

$$\frac{B_{j,j+1}}{Y_o}=\frac{\frac{J_{j,j+1}}{Y_o}}{1-(\frac{J_{j,j+1}}{Y_o})^2} \tag{11}$$

The electrical lengths of half wavelength resonators are expressed as

$$\theta_j=\pi-\frac{1}{2}\left[\tan^{-1}\left(\frac{2B_{j-1,j}}{Y_o}\right)+\tan^{-1}\left(\frac{2B_{j,j+1}}{Y_o}\right)\right], \quad \text{radians} \tag{12}$$

The effective lengths of the shunt capacitances on the both ends of resonators with guided wavelength (λ_{go}) and angular frequency ($\omega_o = 2\pi.f_o$) can be found by using the following expressions.

$$\Delta l_j^{e1}=\frac{\lambda_{go}.\omega_o.C_P^{j-1,j}}{2\pi.Y_o} \tag{13}$$

$$\Delta l_j^{e2}=\frac{\lambda_{go}.\omega_o.C_P^{j,j+1}}{2\pi.Y_o} \tag{14}$$

The physical lengths of resonators are given by [21]

$$l_j=\frac{\lambda_{go}.\theta_j}{2\pi}-\Delta l_j^{e1}-\Delta l_j^{e2} \tag{15}$$

2.2 Filters Simulation

For microstrip filters simulation, a substrate material with a relative dielectric constant ε_r = 10.8, substrate height h = 1.27mm and conductor thickness t = 0.035mm. The line width for microstrip half-wavelength resonators is also chosen W = 1.85mm. The simulation was conducted using high frequency structure simulator (HFSS) software package. The procedure design of end-coupled filter has been completed, and the final results are listed in Table 2 depending on Fig. 3. The layout dimensions of this design are (10.8mm x 32.411mm).

Table 2 The dimensions of filter components

W (mm)	$S_{0,1} = S_{3,4}$ (mm)	$S_{1,2} = S_{2,3}$ (mm)	$l_1 = l_3$ (mm)	l_2 (mm)	l ($P_{I/P}=P_{O/P}$) (mm)
1.85	0.057	0.52	8.15	8.7	3.123

Several width values of strip lines were taken so as to know which is better to get suitable results of filter design at the desired frequency. It is noted that the preferable value is 1.8mm or 1.85mm. Fig. 6 and Fig. 7 show the return (S_{11}) and insertion (S_{21}) losses frequency response, respectively and related to width of strip line. The return losses are 30dB in the first resonance at 5.5GHz and 20dB in the second resonance at 10.5GHz. The insertion losses at same points of resonances approach to 0.1dB and 2dB, respectively. One of the important parameters in filter design is the group delay, which is expressed as [21].

$$\tau_d = \frac{d\emptyset_{21}}{d\omega} \quad \text{Seconds} \tag{16}$$

Where, $\emptyset_{21}$is the phase angle of S_{21} in radians, and ω is an angular frequency in radians per second. Fig. 8 shows the group delay for different microstrip width values, and it is observed that the group delay remains almost constant out of the passband interval and approximate to 0.12*ns*, whereas in the passband interval there is a sharp rise in group delay.

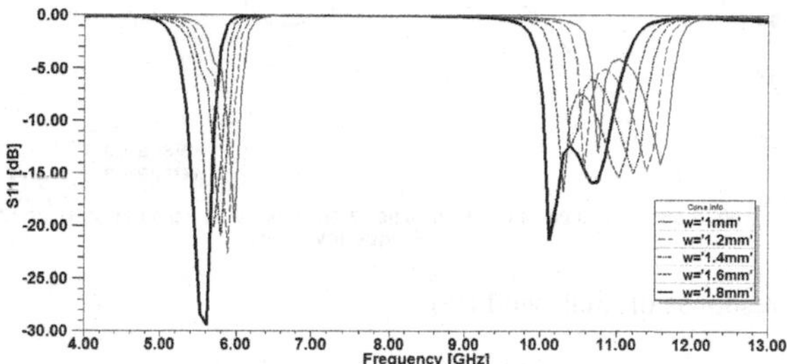

Fig. 6 Return loss responses (S_{11}) of BPF

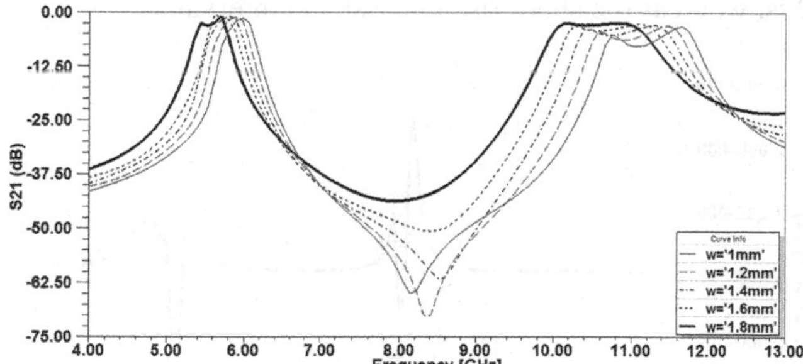

Fig. 7 Insertion loss responses (S_{21}) of BPF

Finally the simulated (through HFSS software, including losses) frequency response of the BPF with better results by taking 1.85mm width of strip line is shown in Fig. 9. The central frequency (f_o) of the filter is 5.5GHz and fractional bandwidth (FWB) equal to 10.9%, and the other resonance at 10.5GHz has FWB of 11.45%.

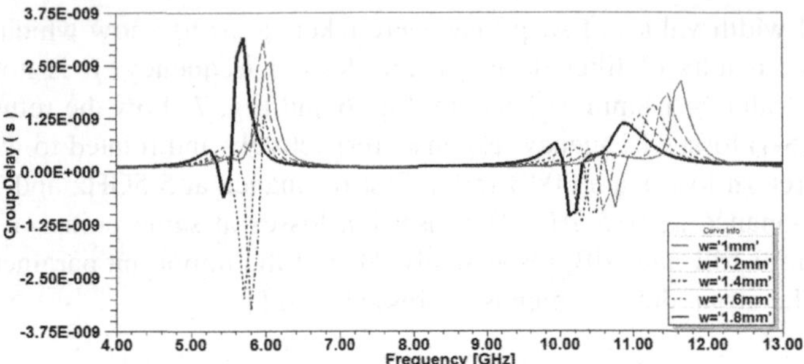

Fig. 8 Group delay variation in different microstrip width values of BPF

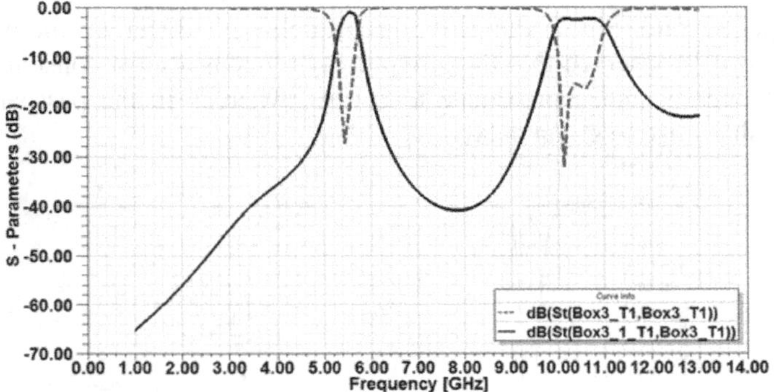

Fig. 9 Frequency response of dual-band BPF

Fig. 10 shows the group delay for dual-band BPF design and it is observed that the group delay remains almost constant out of the passband interval and approximate to 0.12*ns*, whereas the sharp rise in passband interval.

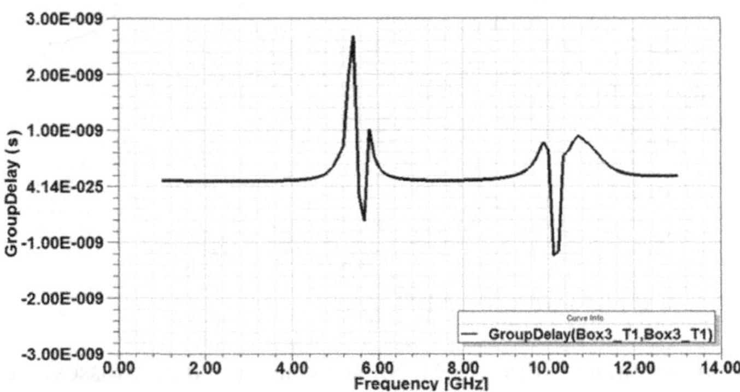

Fig. 10 Group delay variation for conventional dual-band BPF

3 Metamaterial Structures

3.1 Via Structure

In filter design more size reduction and dual-band frequencies can be obtained by using metallic vias at periodic positions. This metallic vias act as shunt-connected inductors, whereas metallic vias provide the required negative permittivity ($-\varepsilon_r$). Fig. 11 shows the geometric design of via intersects the upper conductor and the ground plane. The radius (r) of via in this design is equal to $0.005\lambda_g$.

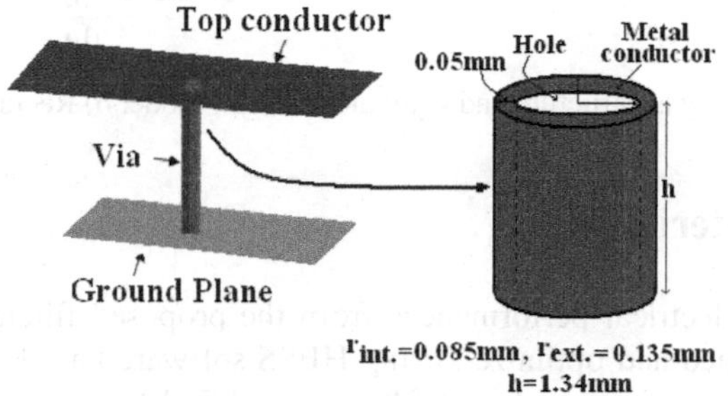

Fig. 11 Geometric design of via structure

3.2 Rectangular SRR Structure

The double SRR structure and its equivalent circuit are shown in Fig. 12. This structure is etching instead of an open-ended transmission line which acts as a shunt capacitance to get miniaturization and multiband resonances. In the double ring configuration, capacitive coupling (C'_1, C'_2) and inductive coupling (L'_1, L'_2) between the external and internal rings were connected by a coupling capacitance (C_m) and a transformer that has transformed ratio (T) [3].

The electromagnetic properties of SRRs structure were studied [4]. The SRRs structure has a negative permeability at specific resonant frequency (f_o) and can be calculated in case of circle shape SRRs as follows

$$\mu_{eff}=\frac{\pi r^2/A^2}{1+\frac{2l.\sigma}{w.r.\mu_o}i-\frac{3l.c_o}{\pi w^2 \ln\frac{2c.r^3}{d}}} \tag{18}$$

Where σ is the resistance per unit length, A, l, are the periods of z- and y-directions, respectively, and c_o is the light velocity. Thereafter changing from circle type parameters to rectangular type parameters with many approximations to get SRRs used in our design as shown in Fig. 12.

The external dimensions of rectangular SRRs have the same external dimensions of open-ended transmission lines. The rectangular type parameters for all resonators are (g = 0.15mm, d = 0.145mm, and c = 0.15mm). The lengths of internal resonators (r_1 and r_2) can be calculated easily from upper parameters.

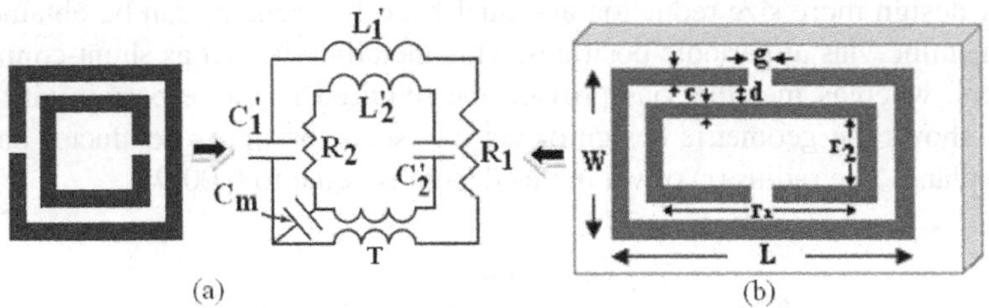

Fig. 12 SRRs topology a) Structure and equivalent circuit model b) Rectangular SRR

4 Metamaterial BPF

To obtain good electrical performances from the proposed filters, the design parameters were tuned and optimized using HFSS software based on finite element method. The layout of conventional filter is modified to get metamaterial filters with new results.

4.1 BPF Using Vias

The end-coupled microstrip poles are connected to the ground plane through metallic vias at periodic positions as shown in Fig. 13. This metallic vias act as shunt-connected inductors, whereas the distance between two adjacent vias is 8.95mm, and the internal radius of each via equal to 0.085mm.

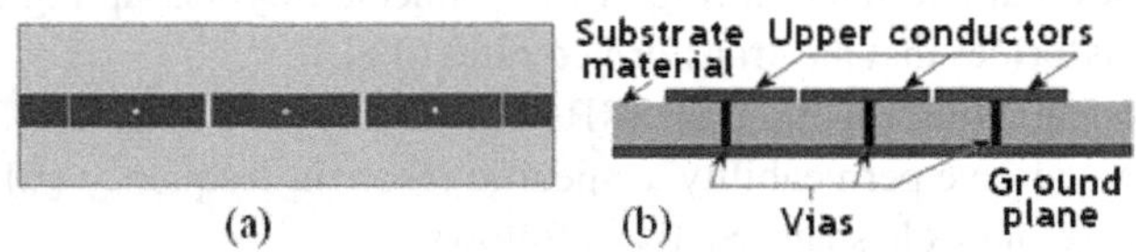

Fig. 13 End-coupled filter design contains vias (a) front view, (b) side view

The frequency response of this filter is shown in Fig. 14. We see that the new resonant frequencies are created in 3.58GHz and 5.58GHz respectively, and rejected at 10.5GHz. The return loss at new resonance frequencies are 22dB and 18dB, respectively. Fig. 15 shows the group delay and it is observed that the group delay remains almost constant approximate to 0.1*ns* along the stop-band interval, whereas beyond the passband there is a sharp rise in group delay.

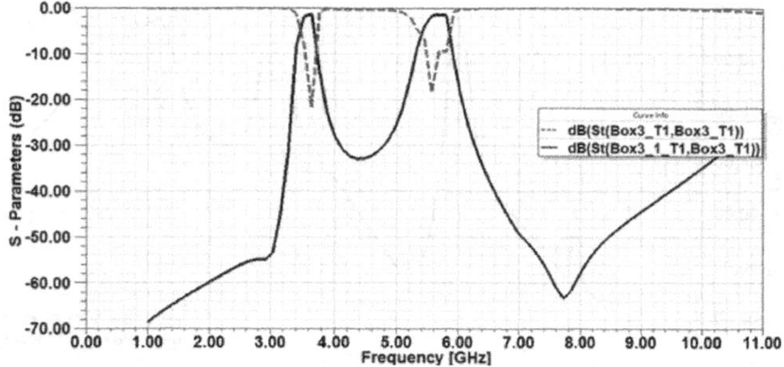

Fig. 14 Frequency response of dual-band BPF contains metallic vias

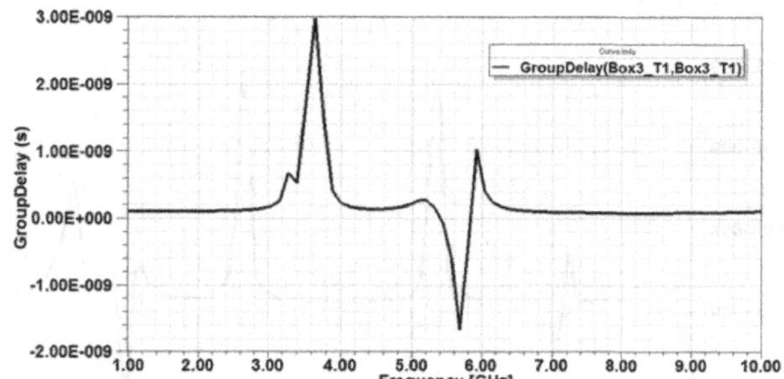

Fig. 15 Group delay of dual-band BPF contains metallic vias

4.2 BPF Using SRRs

The modification of end-coupled filter based on SRRs concept is shown in Fig.16. This idea is achieved by etching rectangular SRRs instead of open-ended transmission lines (l_1, l_2, l_3) in a conventional filter design to get reduction in size up to 20%, and tri-band frequencies at 4.4GHz, 7.9GHz, and 10.5GHz. Compact, low-cost and robust multiband microwave components are obtained when many communication standards are collected in one microwave wireless device. The disadvantage of this design is low return losses and less FBW compared to conventional counterpart. Fig.17 shows the frequency response of modified end-coupled filter. Fig.18 shows the group delay variation of the tri-band filter, and it is observed that the group delay remains almost constant out of the passband interval and approximate to 0.12*ns*.

Fig. 16 Layout of three pole microstrip end-coupled split ring resonators filter

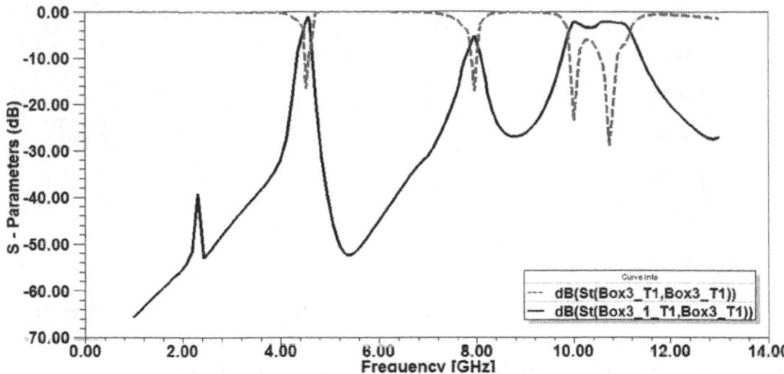

Fig. 17 Frequency response of tri-band filter contains SRRs

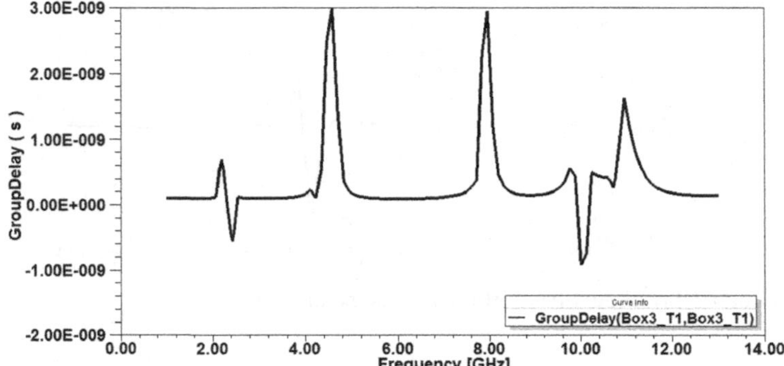

Fig. 18 Group delay variation of tri-band filter Contains SRRs

The comparison between traditional and metamaterial filters with respect to resonance frequencies, fractional bandwidths and return losses are listed in Table 3.

Table 3 Comparison between conventional and metamaterial filters

Filter type	Resonance frequency (GHz)			FBW %	RL (dB)
	f_{o1}	f_{o2}	f_{o3}	at f_{o1}, f_{o2}, f_{o3} respectively	
Conventional	5.5	-	10.5	10.9, -, 11.45	28, -, 18
Metamaterial -Vias	3.58	5.58	-	7, 8, -	22, 18, -
Metamaterial-SRRs	4.4	7.9	10.5	8.7,3.9,12.35	16, 17, 16

5 Conclusions

The end-coupled, half wavelength resonator, bandpass filter is presented for the first time in this paper. A new design achieved by using vias act as shunt inductors between microstrip lines and the ground plane. This idea gives a reduction in size and dual-band frequencies at lower and upper resonance frequency in conventional one. The MTM microstrip lines as periodic SRRs with deliberately manufactured gaps between them have been designed. The main contribution of this work is that proposed structure can be used as a narrow bandpass filter with reasonably good performance. The group delay of BPFs remains almost constant out of the passband interval and is approximately $0.12ns$. The advantages of the end-coupled resonator type of filters that they are fairly high tolerance, ease of fabrication and low cost. The disadvantages are that they consume more space and cannot be used for the design of wideband filters. After modification, more advantages are obtained, such as reduction in size, dual- and tri-bands frequency responses, but it exhibits low return losses at resonance frequencies and less value of fractional bandwidth (FBW) in the first resonance frequencies.

Acknowledgments. The authors would like to express their appreciation to the Ministry of Higher Education and Scientific Research in Iraq and Jamia Millia Islamia in India for supporting this work.

References

1. Lembrikov, B.: Ultra Wideband. Sciyo, Croatia (2010) ISBN 978-953-307-139-8
2. Veselago, V.G.: The electrodynamics of substances with simultaneously negative values of ε and μ. Sov. Phys. Usp. 10, 509–514 (1968)
3. Caloz, C., Itoh, T.: Electromagnetic Metamaterials: Transmission Line Theory and Microwave Applications. Wiley, Hoboken (2006)
4. Pendry, J.B., Holden, A.J., Robbins, D.J., Stewart, W.J.: Magnetism from conductors and enhanced nonlinear phenomena. IEEE Trans. Microwave Theory Tech. 47(11), 2075–2084 (1999)
5. Yen, T.J., Padilla, W.J., Fang, N., Vier, D.C., Smith, D.R., Pendry, J.B., Basov, D.N., Zhang, X.: Terahertz magnetic response from artificial materials. Science 303(5663), 1494–1496 (2004)
6. Moser, H.O., Casse, B.D.F., Wilhelmi, O., Saw, B.T.: Terahertz response of a microfabricated rod split ring resonator electromagnetic metamaterial. Phys. Rev. Lett. 94, 063901 (2005)
7. Casse, B.D.F., Moser, H.O., Lee, J.W., Bahou, M., Inglish, S., Jian, L.K.: Towards three-dimensional and multilayer rod- split-ring metamaterial structures by means of deep x-ray lithography. Appl. Phys. Lett. 90 (2007)
8. Marqués, R., Martín, F., Sorolla, M.: Metamaterials with Negative Parameters Theory, Design, and Microwave Applications. John Wiley & Sons, Inc. (2007)
9. Martin, F., Falcone, F., Bonache, J., Marques, R., Sorolla, M.: Split ring resonator based left handed coplanar waveguide. Appl. Phys. Lett. 83, 4652–4654 (2003)
10. Martin, F., Falcone, F., Bonache, J., Lopetegi, T., Marques, R., Sorolla, M.: Miniaturized CPW stop band filters based on multiple tuned split ring resonators. IEEE Microwave and Wireless Components Letters 13, 511–513 (2003)
11. Garcia-Lamperez, A., Salazar-Palma, M.: Dual band filter with split-ring resonators. In: 2006 IEEE MTT-S International Microwave Symposium Digest, pp. 519–522 (2006)
12. Lai, X., Liang, C.H., Su, T., Wu, B.: Novel dual-band microstrip filter using complementary split-ring resonators. Microwave and Optical Technology Letters 50(1), 7–10 (2008)

13. Fan, J.W., Liang, C.H., Wu, B.: Dual-band filter using equal-length split-ring resonators and zero-degree feed structure. Microwave and Optical Technology Letters 50(4), 1098–1101 (2008)
14. Genc, A., Baktur, R.: Miniaturized dual-passband microstrip filter based on double-split complementary split ring and split ring resonators. Microwave and Optical Technology Letters 51(1), 136–139 (2009)
15. Liu, H.W., Shen, L., Guan, X.H., Huang, D.C., Lim, J.S., Ahn, D.: Compact dual-band bandpass filter using octagonal split-ring resonators with side-coupled stubs. Microwave and Optical Technology Letters 53(5), 1169–1171 (2011)
16. Yuandan, D., Itoh, T.: Miniaturized dual-band substrate integrated waveguide filters using complementary split-ring resonators. In: 2011 IEEE MTT-S International Microwave Symposium, pp. 1–4 (2011)
17. Su, T., Zhang, L.J., Wang, S.J., Li, Z.P., Zhang, Y.L.: Design of dual-band bandpass filter with constant absolute bandwidth. Microwave and Optical Technology Letters 56(3), 715–718 (2014)
18. Geschke, R.H., Jokanovic, B., Meyer, P.: Filter parameter extraction for triple-band composite split-ring resonators and filters. IEEE Transactions on Microwave Theory and Techniques 59(6), 1500–1508 (2011)
19. Dong, Y., Wu, C.T.M., Itoh, T.: Miniaturised multi-band substrate integrated waveguide filters using complementary split-ring resonators. Microwaves, Antennas & Propagation, IET 6(6), 611–620 (2012)
20. Borazjani, O., Nosrati, M., Daneshmand, M.: A novel triple notch-bands ultra wideband band-pass filters using parallel multi-mode resonators and CSRRs. International Journal of RF and Microwave Computer-Aided Engineering 24(3), 375–381 (2014)
21. Jia-Sheng, L.M.J.: Microstrip Filters for RF/Microwave Applications. John Wiley & Sons, Inc. (2001)
22. Gupta, K.C., Garg, R., Bahl, I., Bhartis, P.: Microstrip lines and slotlines. Second Edition. Artech House, Boston (1996)
23. Temes, G.C., Mitra, S.K.: Modern filter theory and design. Wiley, New York (1973)

Studying the Effects of Metamaterial Components on Microwave Filters

Ahmed Hameed Reja and Syed Naseem Ahmad

Abstract. This paper presents a compact stopband filters and bandpass filters using microstrip transmission lines coupled with two parallel sides of square split ring resonators (SRRs) and metallic vias holes. SRRs etched on the upper plane of microstrip line to provide negative value of effective permeability ($\mu < 0$) to the medium in a narrow band above their resonance frequency. Narrow bandpass filter with more drastic size reduction can be obtained by using metallic via holes to get negative effective permittivity ($\varepsilon < 0$) with negative effective permeability ($\mu < 0$) which is generated by using SRRs. Backward wave propagation is achieved as a LHM when SRRs and via holes are applied together. These metamaterial components (SRRs and metallic via holes) are useful for compact narrow stopband and narrow bandpass filters applications at 3GHz resonance frequency. The process of adding many numbers of SRRs and taking many widths of strip conductor (W) with gap variations between resonators and strip line are studied and simulated. The length (l) of the structures can be as small as 0.3 times the signal wavelength at resonance frequency.

Keywords: Bandpass, stopband, via hole, microwave filter, split ring resonator.

1 Introduction

In microwave wireless systems, different microwave filters are used such as stopband filters (SBFs) and bandpass filters (BPFs) in different applications. The miniaturization and the device performance are the key issues when metamaterials

Ahmed Hameed Reja
Department of Electromechanical Engineering,
University of Technology, Baghdad, Iraq

Ahmed Hameed Reja · Syed Naseem Ahmad
Department of Electronics & Communication Engineering
Jamia Millia Islamia, New Delhi -110025, India
e-mail: ahmad8171@yahoo.com, snahmad@jmi.ac.in

El-Sayed M. El-Alfy et al. (eds.), *Advances in Intelligent Informatics*,
Advances in Intelligent Systems and Computing 320, DOI: 10.1007/978-3-319-11218-3_19

(MTMs) are used to design microwave devices such as antenna, filter and diplexer. The first theoretical assumption on the existence of double negative media and prediction of its fundamental properties was mentioned by Russian physicist V. Veselago in 1967 [1]. There are several equivalent terms which have been used as MTMs with negative permittivity and permeability, such as; left-handed (LH) media [1-7], media with negative refractive index (NRI) [1-4], [6], backward-wave media (BW) [8], double-negative (DNG) metamaterials [9], Veselago medium, or negative phase velocity medium (NPV) [1]. Left-handed metamaterial (LHM) presents new structures in modern microwave science; by which the design of novel microwave components such as filters with advantageous characteristics and small dimensions are achieved. Many research groups are studying various aspects of metamaterials, and several ideas, concepts and suggestions for future applications of these materials.

Metamaterials (MTMs) are used due to the possibility for the design of new microwave devices such as filters with more compact size, low insertion loss, high selectivity, low cost and better performance. For compact SBFs design, the split ring resonators (SRRs) which was previously proposed by Pendry et al [10] is the first attractive structure that capable to generate negative magnetic permeability (μ). There are many new filters using SRRs and other structures have been introduced [11], [12]. SRRs are formed by two metallic open-rings separated by a pre- determined distance which is excited by an axial time-varying magnetic field in z-direction (see Fig.1), that induces the electric current inside metallic rings. The gaps or splits (g) present in rings are resistive to the current flow between them. So, the current loop will close through the distributed capacitance that appears between the inner and the outer rings as predicted in Fig.1 [13]. The capacitance $0.5C_o$ is related with each two SRR halves and depends on per unit length capacitance (C_p) and circumference with average radius (r) between rings. This relation can be written as in (1). The self inductance (L_s) of resonator between two metallic rings induces flux density (Φ_m) when the current flows inside rings. Therefore, the angular frequency which results from this circuit model can be calculated from (2) [14].

$$C_o=2\pi.r.C_p \tag{1}$$

$$\omega_o=\frac{1}{\sqrt{L_s.C_s}} \tag{2}$$

where, $\omega_o = 2\pi.f_o$, and f_o is the resonance frequency.

The realization of backward wave propagation using SRR and thin wire (TW) to get negative permeability and negative permittivity, respectively, and several other electrically small resonators was presented by Pendry et al [10], [15]. The double SRR structure and its equivalent circuit shown in Fig. 1 are used to obtain a negative variation of effective permeability to prevent wave propagation at the resonant frequency [16], [17]. Composite medium with simultaneously negative permeability and permittivity as LHM is synthesized and demonstrated by Smith et al [2]. The negative permeability transmission lines based on SRRs and left handed materials based on SRRs and metallic vias have been recently proposed [18],

[19]. The possibility of implementing and manufacturing microstrip lines with negative permeability by etching square SRRs in two parallel sides in the upper conductor line is presented by J. Garcia et al [20].

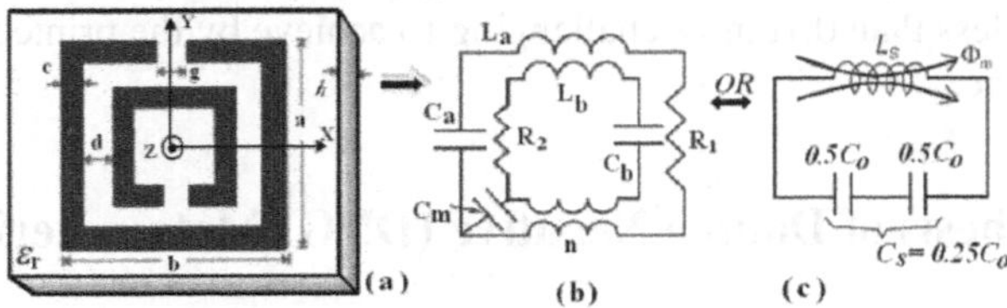

Fig. 1 Unit cell of SRR a) Square shape SRR, b) Equivalent circuit model, c) Simple equivalent circuit model

In this paper two types of filters are proposed depends on SRRs and metallic via holes as metamaterial components. SBFs are presented by adding SRRs in two parallel sides and adjacent to strip line (feed line). This filter transferred into BPFs by drilling holes and adding metallic vias intersect the upper strip line and the ground plane. The process of adding many numbers of SRRs and taking many widths of strip conductor (W) are studied and simulated. To enhance magnetic coupling between line and square SRRs, the distance between strip line and square SRRs must be close as possible. Narrow BPF with more drastic size reduction can be obtained by using metallic vias holes. Backward wave propagation is achieved in presence of two metallic structures (SRRs, vias). The designed filters are applied on Rogers RO3010 substrate with a dielectric constant, ε_r of 10.2 and a substrate height, h of 1.27mm.

2 Synthesis of Negative Permeability Stopband Filter

The narrow stopband filter depicted in Fig. 2 contains microstrip transmission line loaded n-cells of SRRs in two parallel sides with a deliberately simulated gap (S) between resonators and strip line (feed line). The distance (p) between two adjacent cells of SRRs are chosen and it is varies from design to other, but it is constant between cells in one design. The length (l) of the designed filters is the reason of using maximum number of SRRs till 5 cells.

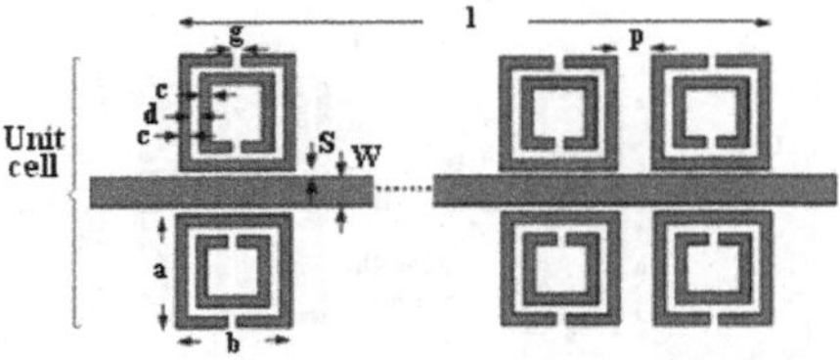

Fig. 2 Geometry of negative permeability metamaterial SBF filter Structure

The better insertion loss performance is achieved when the gap (S) is closed as far as possible to get a strong electromagnetic coupling between the transmission line and adjacent resonators. In this work the optimum value of this gap is 1.25mm. Practically the minimum value of gap is 0.1mm, where the implementation of a gap less than 0.1mm is challenging to achieve by the printed circuit board (PCB) machine.

3 Synthesis of Double Negative (DNG) Metamaterial Filter

In filter design more drastic size reduction can be obtained by using metallic vias at periodic distances. This metallic vias have been represented as shunt connected inductors in equivalent circuit model and provide the required negative permittivity (ε). Fig. 3 shows the geometric design, where a single via intersects the upper (feed line) and lower (ground plane) conductors of microstrip line. The radius (r_{int}) of the metallic via hole in this design is approximately λg/150 and the metallic thickness equal to 0.035mm.

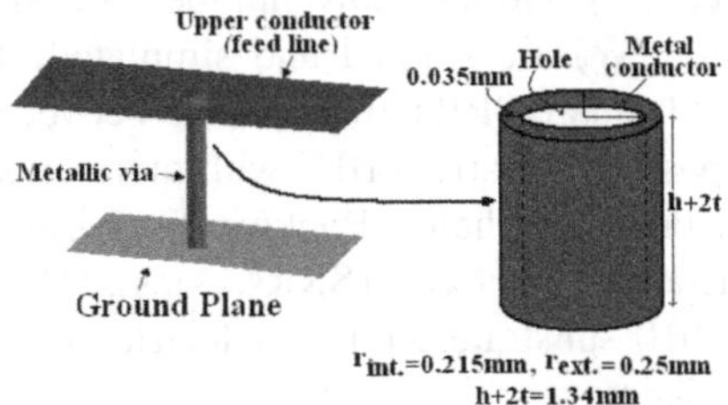

Fig. 3 Geometric design of the metallic via hole structure

Fig. 4 shows the double negative metamaterial narrow band BPF structure. In which the transmission line loaded n-cells of SRRs etched in the upper plane of microstrip line and metallic via holes intersect the feed line and ground plane through the substrate material [18]. Fig. 5 represents an equivalent circuit model of n-cells double negative narrow band filter. The SRRs are modeled by means of inductance (L_s) and capacitance (C_s), which are coupled to the line through a mutual inductance (M). L and C are the per-section inductance and capacitance of the line, respectively, whereas the metallic vias are described as inductance (L_p).

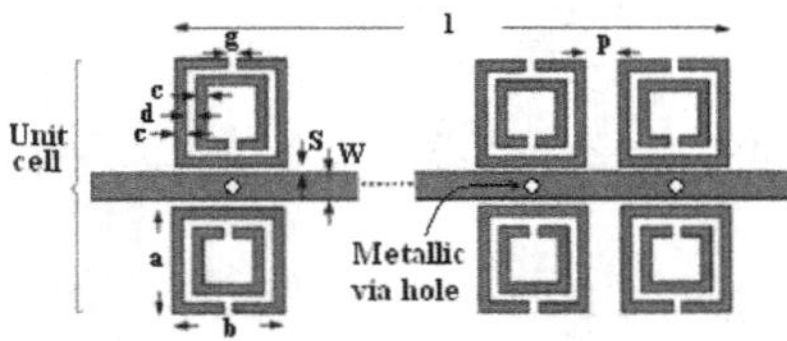

Fig. 4 Double negative metamaterial BPF filter structure

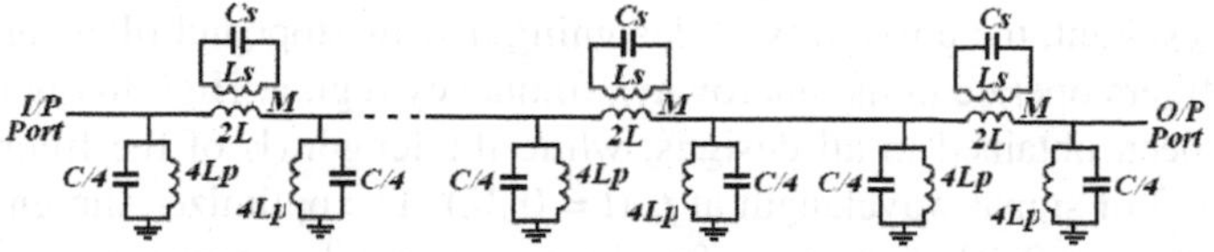

Fig. 5 The double negative metamaterial narrow band BPF equivalent circuit

This periodic structure supports backward waves in a narrow band above the resonant frequency. The lower (ω_L) and upper (ω_H) angular cutoff frequencies are expressed as [18], [21].

$$\omega_L = \sqrt{\frac{1}{C_s'(2L+8L_p)} + \frac{1}{L_s'.C_s'}} \tag{3}$$

$$\omega_H = \sqrt{\frac{1}{2C_s'.L} + \frac{1}{L_s'.C_s'}} \tag{4}$$

where, $C_s' = \frac{L_s}{M^2.\omega_0^2}$, $L_s' = C_s.M^2.\omega_0^2$ and $\omega_0^2 = 1/(L_s'.C_s')$ and $\omega_o = 2\pi.f_o$.

There are many important parameters in filter design such as; selectivity (ξ), fractional bandwidth (δ) and group delay (τ_d). The selectivity or roll-off rate is expressed in (5) [22], [23]. The value of selectivity depends on 3dB attenuation (α_{min}), 20dB attenuation (α_{max}), 20dB stop-band cutoff frequency (f_s) and 3dB cutoff frequency (f_c). The fractional bandwidth (δ) depends on the upper cutoff (f_H)-, lower cutoff (f_L)- and center (f_o) frequencies and expressed as in (6) [24]. The group delay (τ_d) is expressed as in (7) [25].

$$\xi = \frac{\alpha_{min} - \alpha_{max}}{f_s - f_c} \quad dB/GHz \tag{5}$$

$$\delta = \frac{f_H - f_L}{f_o} \tag{6}$$

$$\tau_d = \frac{d\emptyset_{21}}{d\omega} \quad \text{Seconds} \tag{7}$$

Where $\emptyset_{21}$is the phase angle of S_{21} in radians and ω is the angular frequency in radians per second.

4 Simulation Results

In the electromagnetic simulations, all designs are applied on the Rogers R03010 substrate that has dielectric constant, $\varepsilon_r = 10.2$, substrate height, h = 1.27mm. The commercial Ansoft has been used is high frequency structure simulator (HFSS) software as assisted tool for simulation to get results. The important results in this paper are transmission (S_{21})-, reflection (S_{11}) parameters and group delay (τ_d).

From the resultant, the possibility of designing narrow stopband filters and narrow bandpass filters operate in the microwave frequency region. High frequency selectivity has been obtained in all designs, while the length (l) of the filter approximately to 0.3 of signal wavelength at f_o ($l = 0.3\lambda$). The optimized dimensions of a SRRs structure at 3GHz resonance frequency are; a = b = 5mm, g = c = 0.2mm, and d = 0.4mm as described in [14]. In this paper the optimum width of the transmission line (W) is 1.45mm for a characteristic impedance of 50Ω and the optimum distance (S) between SRRs and strip line is 0.125mm.

4.1 Stopband Filters (SBFs) Results

A SBF design is shown in Fig. 6, which has two parallel sides of two SRRs cells and the length of filter (l) is 30mm. The distance (p) between two adjacent cells of SRRs is 20mm. The variation of S-parameters and group delay are shown in Figs. 7 and 8, respectively. Fig. 7 shows the simulated insertion loss (S_{21}) and return loss (S_{11}) of this filter. The maximum attenuation of $|S_{21}|$ is approximately 18dB at 3GHz and the bandwidth of stopband is equal to 200MHz. Fig. 8 shows the group delay variation, which is constant and equal to 0.5*ns* in passband interval.

Fig. 6 SBF contains two cells of SRRs at 1.1mm strip width

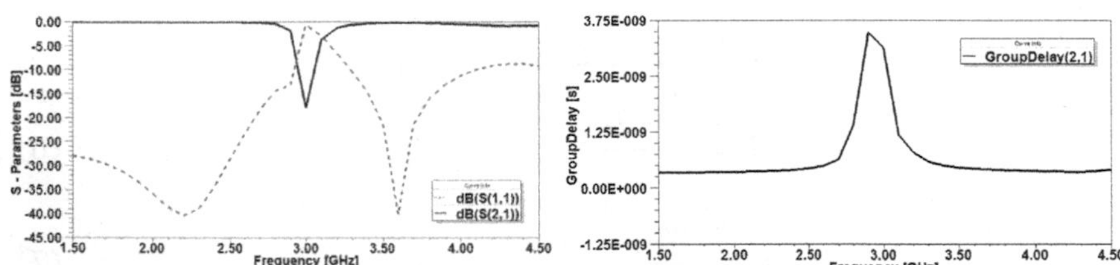

Fig. 7 Frequency response of two cells of SRRs SBF

Fig. 8 Group delay of two cells of SRRs SBF

In case of using two parallel sides of three SRRs cells at 1.3mm strip width, the distance between two adjacent SRRs is equal to 7.5mm and the gap between strip line and SRRs is 0.2mm as shown in Fig 9. The variation in S-parameters and group delay are depicted in Figs. 10 and 11, respectively, where the maximum attenuation of $|S_{21}|$ is approach to 35.2dB at 3GHz and bandwidth of stopband equal to 400MHz. The group delay in passband interval approximate to 0.5*ns*.

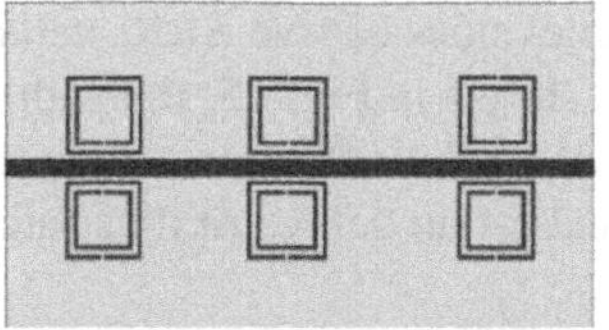

Fig. 9 SBF contains three cells of SRRs at 1.3mm strip width

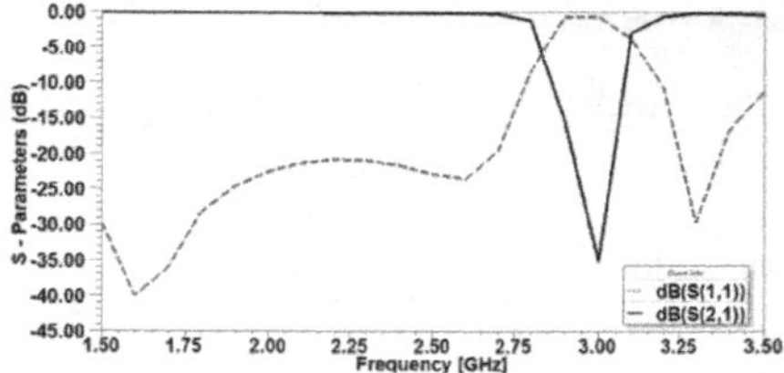

Fig. 10 Frequency response of SBF contains three cells of SRRs

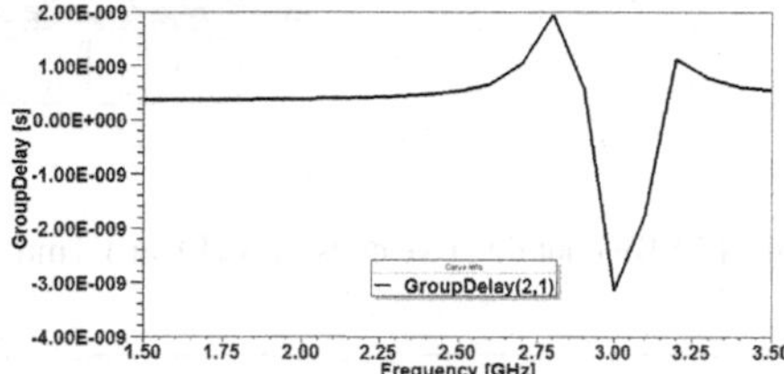

Fig. 11 Group delay of SBF contains three cells of SRRs

In case of using two parallel sides of four SRRs cells with 3.333mm distance between them as shown in Fig.12, the optimum value of maximum attenuation of $|S_{21}|$ at resonance frequency equal to 65dB. The gap (S) equal to 0.125mm and the strip width equal to 1.45mm.

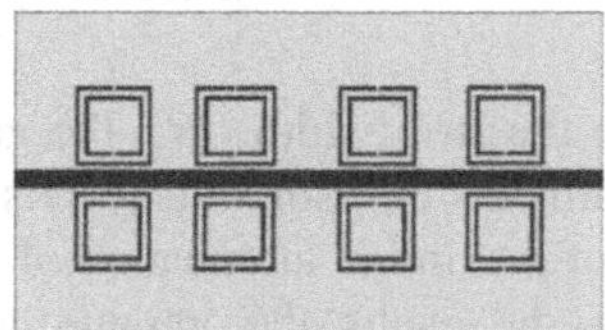

Fig. 12 SBF contains four cells of SRRs at 1.45mm strip width

The transmission S_{21}-dB and reflection S_{11}-dB response are shown in Fig. 13, in which the selectivity is very high and the insertion loss equal to 0.01dB in passband interval. Fig. 14 shows the group delay variation and it is approximately constant and equal to 0.5*ns* in passband interval.

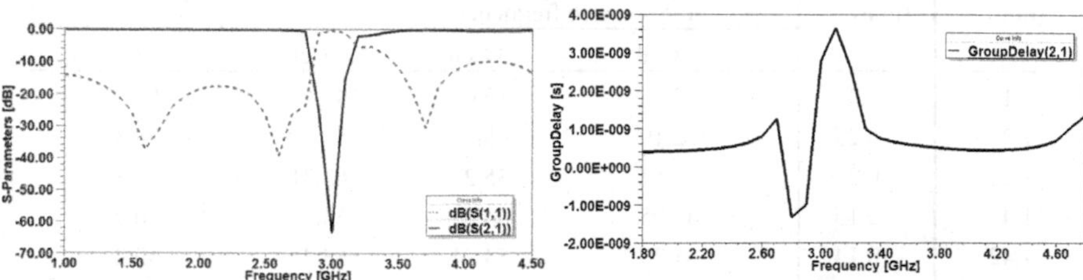

Fig. 13 S-Parameters of SBF contains four cells of SRRs

Fig. 14 Group delay of SBF contains four cells of SRRs

In case of using two parallel sides of five SRRs cells with 1.25mm distance between two adjacent SRRs as shown in Fig. 15, the width of feed line is 1.1mm and the gap between strip line and SRRs is 0.3mm, where the variation with respect to frequency in S-parameters and group delay are illustrated in Figs. 16 and 17.

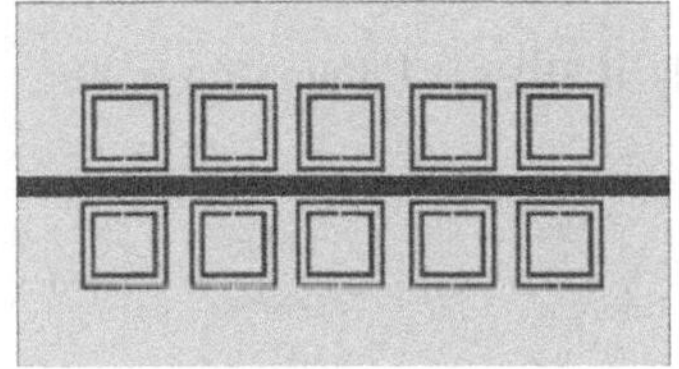

Fig. 15 SBF contains five cells of SRRs at 1.1mm strip width

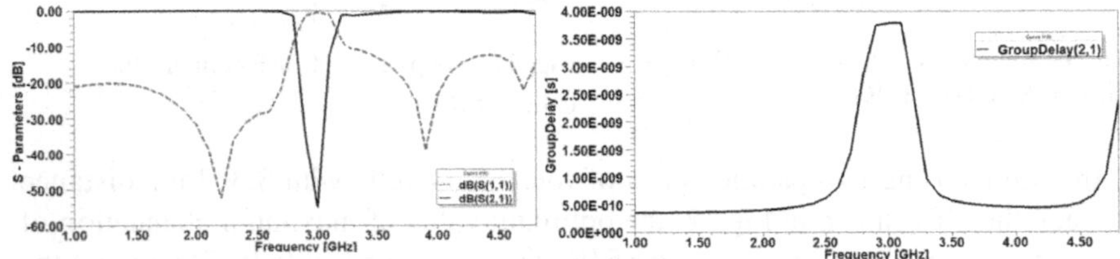

Fig. 16 S-parameters of SBF contains five cells of SRRs

Fig. 17 Group delay of SBF contains five cells of SRRs

The relationship between the strip width (W), the gap (S) between resonators and feed line and different numbers of resonators (SRRs) between two to five cells to get approximated values of maximum attenuation of $|S_{21}|$ at resonance frequency are listed in Table 1. From the results, the optimum values of maximum attenuation of $|S_{21}|$ at resonance frequency are obtained when the strip width equal to 1.45mm and the gap (S) equal to 0.125mm.

Table 1 Maximum value of $|S_{21}|$ at resonance frequency with respect to width of strip line and gap (S) at different numbers of SRRs cells

W (mm)	S (mm)	Maximum attenuation of $|S_{21}|$ (dB) in stopband at resonance frequency			
		2 Cell	3 Cell	4 Cell	5 Cell
1.1	0.3	18	15.6	49	55
1.2	0.25	22.19	25.18	58	58.3
1.3	0.2	27.52	35.2	63.31	58.3
1.4	0.15	42.95	54.23	64.3	60.9
1.42	0.14	51.75	54.24	63.45	59.2
1.45	0.125	57.6	61.6	65	65.3

4.2 Band-Pass Filters (BPFs) Results

In microstrip technology, the SRRs etched as close as possible to the strip line to enhance magnetic coupling. The process of adding metallic via holes with presence of SRRs is appropriate for the design of a narrow band filter obtained above the resonance frequency of the SRRs. The SRRs and metallic vias provide the required negative permittivity and negative effective permeability, respectively, as a DNG media. From the previous design of stopband filters, the transmission line loaded two parallel sides of SRRs with deliberate dimensions for all components in designs. By drilling periodically holes into microstrip line (upper conductor, ground conductor and dielectric) and insert metallic via hole inside it to convert the previous responses into narrow bandpass at resonance frequency. Two cells contain SRRs and metallic vias hole is depicted in Fig. 22. All dimensions of metallic via are indicated in Fig. 3. The metallic cylindrical via has thickness (t) of 0.035mm and height (h+2t) of 1.34mm, whereas the radius of hole inside the metallic cylinder (r_{int}) is of 0.215mm and strip width (W) of 1.1mm. The response obtained from this design is shown in Fig. 23, in which the bandpass is observed at 3GHz and 4.35GHz with 1dB insertion loss and more than 18dB return loss at resonance frequencies.

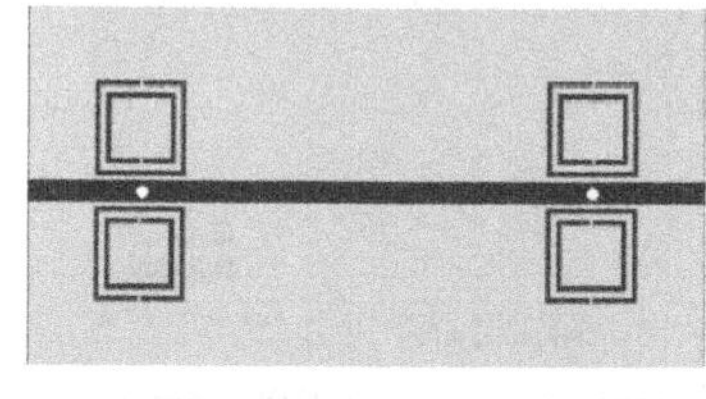

Fig. 22 BPF contains two cells include SRRs and metallic vias at 1.1mm strip width

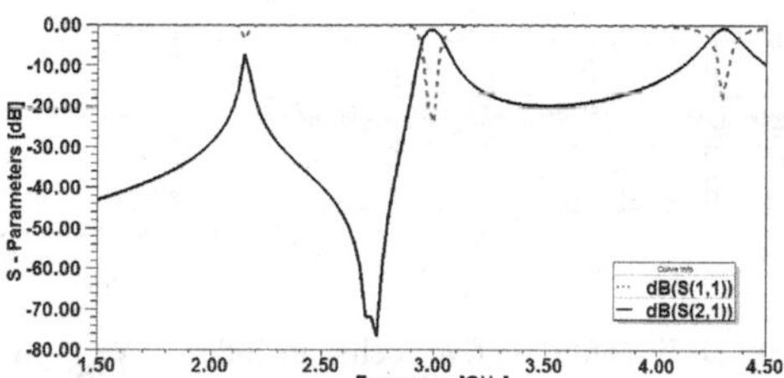

Fig. 23 BPF response contains two cells include SRRs and metallic vias at 1.1mm strip width

In case of BPF, it contains three cells of SRRs and metallic vias as in Fig. 24. The bandpass at 3GHz, 3.8GHz and 4.35GHz resonance frequencies have 1dB insertion loss and more than 17dB return loss as illustrated in Fig. 25.

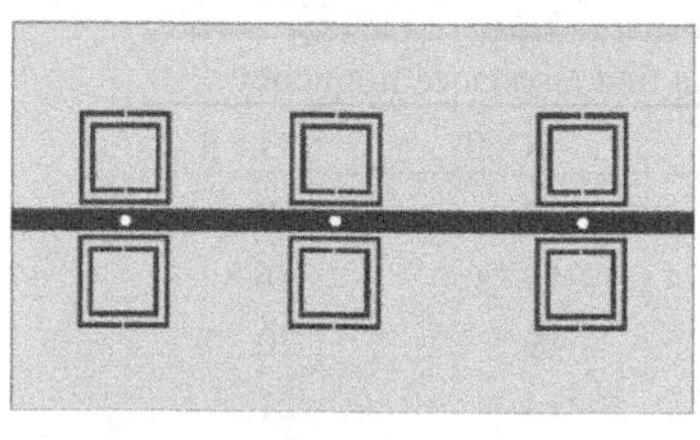

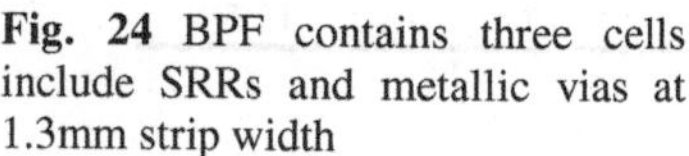

Fig. 24 BPF contains three cells include SRRs and metallic vias at 1.3mm strip width

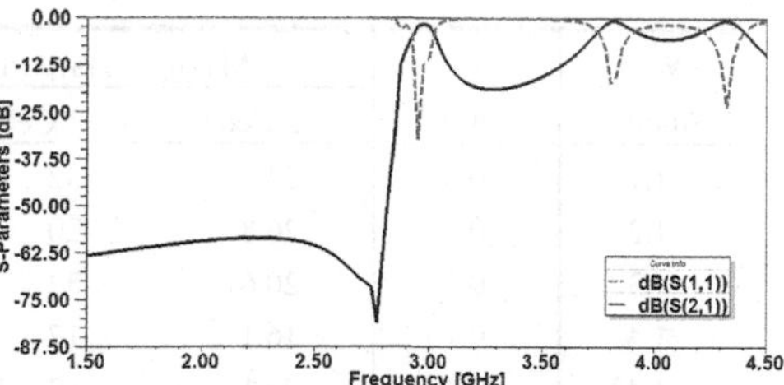

Fig. 25 Frequency response of BPF contains three cells include SRRs and metallic vias

In case of BPF contains four cells of SRRs and metallic vias as in Fig. 26, by which the bandpass at 3GHz and 4.75GHz with 1dB insertion loss and more than 23dB return loss at resonance frequencies is as shown in Fig. 27.

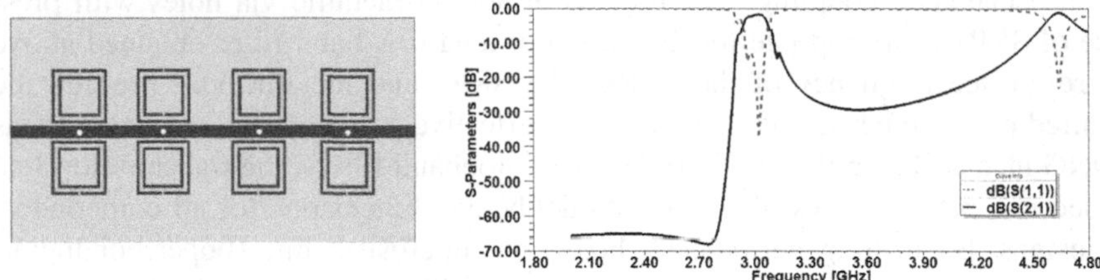

Fig. 26 BPF contains four cells include SRRs and metallic vias at 1.45mm strip width

Fig. 27 Frequency response of BPF contains four cells include SRRs and metallic vias

Finally, in case of BPF contains five cells of SRRs and metallic vias as in Fig. 28, in which the bandpass is at 3GHz and 5.6GHz with 1dB insertion loss and more than 15dB return loss at resonance frequencies is as shown in Fig. 29.

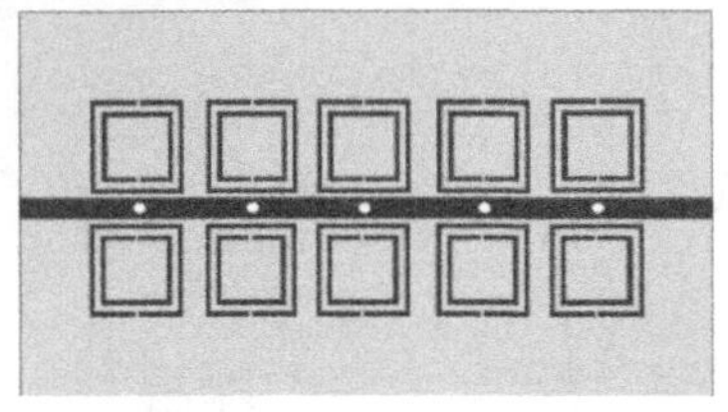

Fig. 28 BPF contains five cells include SRRs and metallic vias at 1.1mm strip width

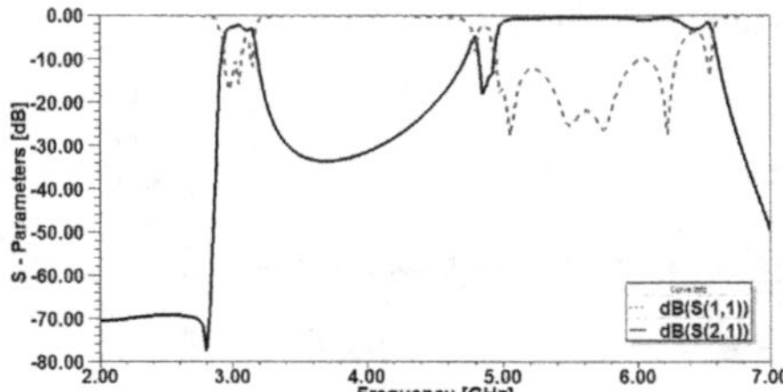

Fig. 29 Frequency response of BPF contains five cells include SRRs and metallic vias

The relationship between the strip width (W), the gap (S) and different numbers of resonators (SRRs) to get maximum approximated values of $|S_{11}|$ at the first resonance frequency are listed in Table 2.

Table 2 Maximum value of $|S_{11}|$ at first resonance frequencies in different strip width (W), gap (S) and different numbers of cells

W	S	Maximum $\|S_{11}\|$ (dB) at first resonance frequency			
(mm)	(mm)	2 Cell	3 Cell	4 Cell	5 Cell
1.1	0.3	24	32	17	18
1.2	0.25	26.8	20	19	16.3
1.3	0.2	20.67	32	19.5	20
1.4	0.15	16.1	17	14	21
1.42	0.14	22.1	13	19	20
1.45	0.125	20	14.5	17	17.5

5 Conclusions

This work is focused on the study of the effects of metamaterial components on the design of microwave filters. The filters are based on two parallel sides of periodically split ring resonators (SRRs) and periodically metallic vias holes as a LHM. These particles are useful for the design of narrow SBFs and narrow BPFs at different results depending on the number of these components, the distance between them and the gap between strip line and SRRs. A left handed transmission line loaded periodic SRRs and metallic vias has been designed and simulated. The main relevant contribution of this work is the fact that this structure can naturally be used as a narrow SBF and narrow BPF with reasonably good performance. The fractional bandwidth is approximately 5%. A very sharp slope at the lower band edge that indicates high frequency selectivity has been achieved. Finally, the dual- band can be obtained by using metallic via holes.

Acknowledgments. The authors would like to express their appreciation to the Ministry of Higher Education and Scientific Research in Iraq and Jamia Millia Islamia in India for supporting this work.

References

1. Veselago, V.G.: The electrodynamics of substances with simultaneously negative values of ε and μ. Usp. Fiz. Nauk. 92, 517–526 (1967)
2. Smith, D.R., Padilla, W.J., Vier, D.C., Nemat Nasser, S.C., Schultz, S.: Composite medium with simultaneously negative permeability and permittivity. Phys. Rev. Lett. 84(18), 4184–4187 (2000)
3. Pendry, J.B.: Negative refraction makes a perfect lens. Phys. Rev. Lett. 85(18), 3966–3969 (2000)
4. Shelby, R.A., Smith, D.R., Schultz, S.: Experimental verification of a negative index of refraction. Science 292(5514), 77–79 (2001)
5. Caloz, C., Chang, C.C., Itoh, T.: Full-wave verification of the fundamental properties of left-handed materials in waveguide configurations. J. Appl. Phys. 90(11), 5483–5486 (2001)
6. Iyer, A.K., Eleftheriades, G.V.: Negative refractive index metamaterials supporting 2-D waves. In: 2002 IEEE MTT International Microwave Symposium (IMS) Digest, Seattle, WA, June 2-7, pp. 1067–1070 (2002)
7. Caloz, C., Okabe, H., Iwai, T., Itoh, T.: Transmission line approach of left-handed materials. In: 2002 IEEE AP-S International Symposium and USNC/URSI National Radio Science Meeting, San Antonio, TX, URSI Digest, June 16-21, p. 39 (2002)
8. Lindell, I.V., Tretyakov, S.A., Nikoskinen, K.I., Ilvonen, S.: BW media-media with negative parameters, capable of supporting backward waves. Microwave Opt. Tech. Lett. 31(2), 129 (2001)
9. Ziolkowski, R.W., Heyman, E.: Wave propagation in media having negative permittivity and permeability. Phys. Rev. E. 64(5), 056625 (2001)

10. Pendry, J.B., Holden, A.J., Robbins, D.J., Stewart, W.J.: Magnetism from conductors and enhanced nonlinear phenomena. IEEE Trans. Microwave Theory and Techniques 47(11), 2075–2084 (1999)
11. Gil, M., Bonache, J., Martín, F.: Metamaterial filters: A review. ScienceDirect. Metamaterials 2, 186–197 (2008)
12. Reja, A.H., Ahmad, S.N., Abdul Raheem, A.K.: A Review of: Metamaterial Based Microwave Filter Design. European Journal of Scientific Research 105(1), 144–165 (2013)
13. Baena, J.D., Bonache, J., Martín, F., Sillero, R.M., Falcone, F., Lopetegi, T., Laso, M.A.G., García-García, J., Gil, I., Portillo, M.F., Sorolla, M.: Equivalent-circuit models for split-ring resonators and complementary split-ring resonators coupled to planar transmission lines. IEEE Transactions on Microwave Theory and Techniques 53, 1451–1461 (2005) ISSN 0018-9480
14. Marque, R., Mesa, F., Martel, J., Medina, F.: Comparative analysis of edge- and broadside-coupled split ring resonators for metamaterial design-theory and experiments. IEEE Trans Antennas Propag. 51, 2572–2581 (2003)
15. Caloz, C., Itoh, T.: Electromagnetic Metamaterials: Transmission Line Theory and Microwave Applications. Wiley, Hoboken (2006)
16. Pendry, J.B., Holden, A.J., Stewart, W.J.: Extremely Low Frequency Plasmons in Metallic Mesostructures. Phys. Rev. Lett. 76(25), 4773–4776 (1996)
17. Lim, H., Lee, J.H., Lim, S.H., Seo, D.W., Shin, D.H., Myung, N.H.: A novel compact coplanar waveguide bandstop filter based on split-ring resonators. In: Proceeding of ISAP 2007. Niigata, Japan (2007)
18. Martin, F., Falcone, F., Bonache, J., Marques, R., Sorolla, M.: Split ring resonator based left handed coplanar waveguide. Appl. Phys. Lett. 83, 4652–4654 (2003)
19. Martin, F., Falcone, F., Bonache, J., Lopetegi, T., Marques, R., Sorolla, M.: Miniaturized CPW stop band filters based on multiple tuned split ring resonators. IEEE Microwave and Wireless Components Letters 13, 511–513 (2003)
20. Garcia-Garcia, J., Bonache, J., Gil, I., Martin, F., Marques, R., Falcone, F., Lopetgi, T., Laso, M.A.G., Sorolla, M.: Comparison of electromagnetic bandgap and split rings resonator microstrip lines as stop band structures. Microwave and Optical Technology Letters (2005), http://dyuthi.cusat.ac.in/purl/1421
21. Gil, I., Bonache, J., Garcia-Garcia, J., Falcone, F., Martin, F.: Metamaterials in microstrip technology for filter applications. In: Antennas and Propagation Society International Symposium, vol. 1A, pp. 668–671 (2005)
22. Sahu, S., Mishra, R., Poddar, D.: Compact metamaterial microstrip low-pass filter. Journal of Electromagnetic Analysis and Applications 3(10), 399–405 (2011)
23. Li, J.L., Qu, S.W., Xue, Q.: Compact microstrip low-pass filter with sharp roll-off and wide stop-band. IEE Electronics Letters 45(2), 110–111 (2009)
24. Edwards, T.C.: Foundations for microstrip circuit design. John Wiley & Sons Ltd. (1988)
25. Jia-Sheng, M., Lancaster, J.: Microstrip filters for RF/Microwave applications. John Wiley & Sons, Inc. (2001)

Incorporating Machine Learning Techniques in MT Evaluation

Nisheeth Joshi, Iti Mathur, Hemant Darbari, and Ajai Kumar

Abstract. From a project manager's perspective, Machine Translation (MT) Evaluation is the most important activity in MT development. Using the results produced through MT Evaluation, one can assess the progress of MT development task. Traditionally, MT Evaluation is done either by human experts who have the knowledge of both source and target languages or it is done by automatic evaluation metrics. These both techniques have their pros and cons. Human evaluation is very time consuming and expensive but at the same time it provides good and accurate status of MT Engines. Automatic evaluation metrics on the other hand provides very fast results but lacks the precision provided by human judges. Thus a need is being felt for a mechanism which can produce fast results along with a good correlation with the results produced by human evaluation. In this paper, we have addressed this issue where we would be showing the implementation of machine learning techniques in MT Evaluation. Further, we would also compare the results of this evaluation with human and automatic evaluation.

1 Introduction

Since the beginning of the time, translation has been the most important activity as the traders, who travelled distant places, needed to communicate with local population in their own language. Moreover while framing strategic alliances, the

Nisheeth Joshi · Iti Mathur
Department of Computer Science, Banasthali University
e-mail: {nisheeth.joshi,mathur_iti}@rediffmail.com

Hemant Darbari
Executive Director, C-DAC, Pune
e-mail: darbari@cdac.in

Ajai Kumar
Applied Artificial Intelligence Group, C-DAC, Pune
e-mail: ajai@cdac.in

El-Sayed M. El-Alfy et al. (eds.), *Advances in Intelligent Informatics*,
Advances in Intelligent Systems and Computing 320, DOI: 10.1007/978-3-319-11218-3_20

rulers of different countries needed translators, so that they may communicate with each other. In modern times this task has become even more important as the world have shrunk down to a global village. Now, not only governments, but also multinational corporations require translations of text from one language to another. Employing human translators for the job is very expensive and time taking. Moreover, skilled human translators are also very hard to find. Since the dawn of computers, computer aided translation or machine translation has been seen as the alternate to human translation. With initial MT engines being just dictionary matchers, current state of the art MT engines have come a long way. Today MT Engines can produce good and comparable results as that of human translators.

For rapid development of an MT Engine, it is required that we get quick and precise evaluation of the outputs of an engine, so that the development process could run smoothly. A manager of an MT system needs this information, so that he man according plan the future course of action. Getting an evaluation result of an MT engine can be achieved by two different approaches: Human Evaluation or Automatic Evaluation. Human evaluation is done by human expert who has an understanding of both source as well as target languages. In this evaluation the human judge is provided with a subjective questionnaire. Based on this questionnaire the judge is required to rate the outputs of an MT engine. Figure 1 shows the working of this approach. Automatic evaluation as the name suggests uses techniques which are independent of human evaluations. In this technique we employ a computer algorithm (popularly known as an evaluation metric) to ascertain the quality of MT output. In this technique, MT output is provided to an evaluation metric which compares the output with a reference translation which has been provided by a human translator. The quality of MT output is assessed by the checking its closeness to the reference translation. Figure 2 shows the working of this approach. This approach although termed as an automatic approach, is not entirely automatic as still it requires human intervention. Until we do not have a

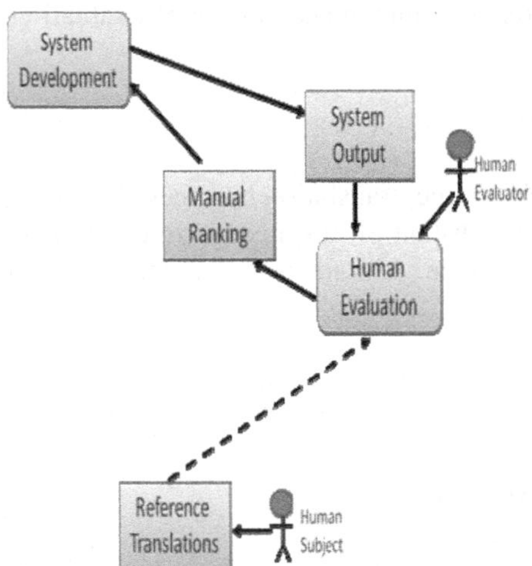

Fig. 1 Human Evaluation Process

translation for a sentence, it cannot be evaluated by an automatic evaluation metric. At times this tends to be a major bottleneck. Thus a need is being felt to have a completely automatic evaluation process. Several researchers are looking into various different techniques where they are trying to develop measures which are completely automatic and could provide evaluation results without any human intervention.

The rest of paper is organized as follows: Section 2 gives a brief review of the evaluation measures that have been used by different researchers. Section 3 shows our experimental setup for a completely automatic MT evaluation. Section 4 shows the evaluation results and its comparison with human and automatic evaluation metrics. Section 5 concludes the paper.

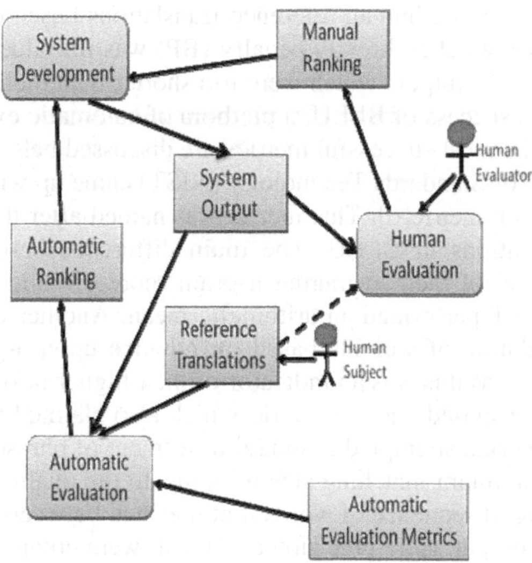

Fig. 2 Automatic Evaluation Process

2 Related Work

During the first formidable years of MT development, evaluation process was completely manual also known as human evaluation process. Miller & Beebe-Center[1] and Pfafflin[2] were the first researchers who proposed methodologies for MT evaluation. Their approach was simple, to provide MT outputs to humans and ask them questions based on the syntax and semantics of the target language. This approach was further modified by Slype[3], who evaluated SYSTRAN MT system. Instead of looking for correctness of the translation, he adjudged SYSTRAN for acceptability. He provided multiple outputs of the system to the evaluators and asked them if translation A is better than translation B. The prime objective of this evaluation was to distinguish between correct translations from incorrect ones. This evaluation, not only gave a measure to check the output of the system, but also found the cost of post editing the incorrect translations. This evaluation changed the view of the people towards MT evaluation. Now, people started looking at the cost of post-editing translations instead of looking for correct translations. In 1980, a more detailed evaluation was carried out for English-French TAUM-Aviation MT system[4]. Here raw MT outputs and post-edited MT outputs were compared with human translations and the results were analyzed in terms of quality and cost of producing such translations.

By the dawn of the current century, automatic evaluation metrics started to emerge, providing an alternate to human evaluation. They gained popularity because they were fast and could provide repeatable results in very less time and in cost effective manner. The very first automatic evaluation metric which caught an eye of research community was BLEU (Bilingual Alignment for Evaluation Undestudy)[5]. The Basic goal of this metric was to calculate similarity between MT output and one or more human reference translations based on n-gram precision. A special measure called brevity penalty (BP) was introduced in this metric which penalized the MT outputs which were too shorter than their human counterparts. Looking at the success of BLEU, a plethora of automatic evaluation metrics followed. Some of the most successful metrics are discussed below.

National Institute of Standards Technology (NIST) came up with their own version of MT Evaluation metric[6]. This metric was named after them and incorporated slight modifications in BLEU. The main difference between BLEU and NIST was the method of their averaging n-gram scores. While BLEU relied on geometric mean, NIST performed an arithmetic mean. Another unique feature of NIST was its calculation of weights based on reliance upon n-grams which occurred less frequently, as this was an indicator of their higher informativeness.

Turian et. al.[7] proposed another metric which also claimed to perform better than BLEU. Here, Turian attempted to model movement of phrases during translation by using the maximum matching size to compute the quality of a translation. It could find the longest sequence of words that matched between human reference translation and MT output. Here precision and recall were computed as the size of the matches divided by the length of lexicons in MT output or human reference translation. Then, the harmonic mean of these two measures was computed for calculation of final GTM score. Snover et.al.[8] proposed another metric which was based on edit distance algorithm proposed by Levenshtein[9]. This metric accounted for no. to shifts along with no. of insertions, deletions and substitutions to compute the results. This metric tried to measure the amount of post editing that a human would have to perform on the machine output so that it exactly matches the reference translations.

Meteor[10] was a major breakthrough in MT evaluation research as this metric not only measured the performance of MT output on lexical level, but also looked at deeper linguistic levels (shallow syntactic and semantic). Meteor was a tunable alignment oriented metric while BLEU was simply a precision oriented metric. This metric used several stages of word matching between the system output and the reference translations.

The major drawback with all these metrics was that they all required one or more human reference translations. The very first completely automatic evaluation was performed by Gammon et.al.[11] where they applied their technique in ascertaining MT quality and fluency. They employed SVM based classifier and used several features from syntactic and semantic levels. Specia et. al.[12] used machine learning techniques in identification of features which can be used for MT evaluation and post editing. They applied a SVM based classifier which was trained in 74 features which were from lexical, shallow syntactic and semantic levels. They concluded that given a trained model, a classifier can work well for a particular language pair and can produce better correlations with human judgments.

3 Experimental Setup

For development of our evaluation system, we used a 3,300 sentence corpus that was built during ACL 2005 workshop on Building and Using Parallel Text: Data Driven Machine Translation and Beyond[13], as the training corpus. The statistics of this corpus is shown in Table 1.

Table 1 Statistics of training corpus used

Corpus	English-Hindi Parallel Corpus	
Sentences	3,300	
	English	Hindi
Words	55,014	67,101
Unique Words	8,956	10,502

We also focused on using supervised machine learning in evaluation of MT engine outputs without using human reference translations. For this we used Decision Tree (DT) based classifier which used J48 algorithm which is a java version of C4.5 decision tree algorithm[14]. We used WEKA toolkit[15] for training this classifier. We also trained a Support Vector Machines (SVM) based classifier which was developed using LIBSVM package developed by Chang and Lin[16]. We used 27 features for training our classifiers. These features were as follows:

1. Length of the source sentence.
2. Length of the target sentence.
3. Average source token length.
4. LM probability of source sentence.
5. LM probability of target sentence.
6. Average no. of occurrences of a target word within a target sentence.
7. Average number of translations per source word in the sentence (as given by IBM 1 table threshold so that prob(t|s) > 0.2).
8. Average number of translations per source word in the sentence (as given by IBM 1 table threshold so that prob(t|s) > 0.01) weighted by the inverse frequency of each word in the source corpus.
9. Percentage of unigrams in quartile 1 of frequency (lower frequency words) in a corpus of the source language (SMT training corpus).

10.Percentage of unigrams in quartile 4 of frequency (higher frequency words) in a corpus of the source language.
11.Percentage of bigrams in quartile 1 of frequency of source words in a corpus of the source language.
12.Percentage of bigrams in quartile 4 of frequency of source words in a corpus of the source language.
13.Percentage of trigrams in quartile 1 of frequency of source words in a corpus of the source language.
14.Percentage of trigrams in quartile 4 of frequency of source words in a corpus of the source language.
15.Percentage of unigrams in the source sentence seen in a corpus.
16.Count of punctuation marks in source sentence Count of punctuation marks in target sentence.
17.Count of punctuation marks in target sentence.
18.Count of mismatch between source and target punctuation marks.
19.Count of content words in the source sentence.
20.Count of context words in the target sentence.
21.Percentage of context words in the source sentence.
22.Percentage of context words in the target sentence.
23.Count of non-content words in the source sentence.
24.Count of non-content words in the target sentence.
25.Percentage of non-content words in the source sentence.
26.Percentage of non-content words in the target sentence.
27.LM probabilities of POS of target sentence.

To test the classifiers, we were also required to have MT Engines. Thus we trained three MT toolkits on tourism domain. These engines were

1. Moses Phrase Based Model[17] (PBM) where phrases of one language are statistically aligned and translated into another language. Moses PBM uses conditional probability with linguistic information to perform the translation.
2. Moses Syntax Based Model[18] (SBM) which implements a Tree to String Model. In this system an English sentence is parsed and its parsed output is matched with the target string and thus transfer grammar is generated which has a parsed output at one end and the string at the other.
3. An Example Based Machine Translation[19] (EBMT) model where examples in the training data which are of higher quality or are more relevant than others are produced as translations.

We registered the outputs of the training corpus against all these three MT engines and asked a human evaluator to judge the outputs. The judging criteria was same as used by Joshi et. al.[20]. All the sentences were judged on ten parameters using a scale between 0-4. Detailed discussion on use of these parameters have been discussed by Joshi et al. [21]. Table 2 shows the interpretation of these scales. The ten parameters used in evaluation were as follows:

1. Translation of Gender and Number of the Noun(s).
2. Identification of the Proper Noun(s).
3. Use of Adjectives and Adverbs corresponding to the Nouns and Verbs.
4. Selection of proper words/synonyms (Lexical Choice).
5. Sequence of phrases and clauses in the translation.
6. Use of Punctuation Marks in the translation.
7. Translation of tense in the sentence.
8. Translation of Voice in the sentence.
9. Maintaining the semantics of the source sentence in the translation.
10. Fluency of translated text and translator's proficiency.

Table 2 Interpretation of HEval on Scale 5

Score	Description
4	Ideal
3	Perfect
2	Acceptable
1	Partially Acceptable
0	Not Acceptable

Once the human evaluation of these outputs was done, we used these results along with the 27 features that were extracted from the English source sentences and Hindi MT outputs. We tested the classifiers using another corpus of 1300 sentences. Table 3 shows the statistics of this corpus.

Table 3 Statistics for Test Corpus

Corpus	English Corpus
Sentences	1,300
Words	26,724
Unique Words	3,515

These 1300 sentences were divided into 13 documents of 100 sentences each. We registered the output of the test corpus on all three MT engines and performed human evaluation on them. Further we also used some of the popular automatic evaluation metrics on the output produced and then we evaluated the results of the systems using the two classifiers. We used BLEU and Meteor metrics for the evaluation. For incorporating automatic evaluation we were also required to have reference translations so we used single human references to be used with automatic evaluation metrics. Since Meteor works on shallow syntactic and semantic levels, we were required to develop tools for this metric. For syntactic matching, we developed a Hindi stemmer based on the light weight stemming algorithm proposed by Rangnathan and Rao [22] and for semantic matching we used Hindi WordNet[23]. For generation of paraphrases we used Moses PBM's phrase table. Moreover, we also compared the results of these evaluations.

4 Results

We correlated the results of the output produced by the MT engines. We used spearman's rank correlation as it produces the unbiased results. Table 4 shows the results of correlation at document level between Human and BLEU & Meteor 1.3 which matches for exact, stem, synonym and paraphrase matches and both the classifiers (DT and SVM). In all the cases EBMT had better correlation with human judgments while Moses PBM showed the poorer results. Correlation of human evaluation with automatic metrics was very low as compared to the results of correlation with classifiers.

Table 4 Document level correlation between human and different automatic evaluation measures

	BLEU	Meteor	DT	SVM
Moses PBM	0.011	0.297	0.089	0.750
Moses SBM	0.181	0.313	0.139	0.773
EBMT	0.490	0.352	0.161	0.781

Table 5 Sentence level correlation between human and different automatic evaluation measures

	BLEU	Meteor	DT	SVM
Moses PBM	0.063	0.045	0.579	0.682
Moses SBM	0.077	0.048	0.635	0.707
EBMT	0.106	0.040	0.590	0.600

Table 6 System level average scores of the all the evaluations

	Moses PBM	Moses SBM	EBMT
Human	0.393300	0.356000	0.464000
BLEU	0.017666	0.012901	0.026553
Meteor	0.069900	0.061205	0.102117
DT	0.538077	0.505577	0.589615
SVM	0.526154	0.492500	0.589038

At sentence level this phenomenon was repeated. In all the cases except for meteor EBMT showed best results and Moses PBM showed poor results. For Meteor, Moses SBM's results were better than EBMT's results. Table 5 shows the results of this study. Here again the correlations of automatic metrics were very low as compared to classifiers. Table 6 shows the average system scores of all the evaluations incorporated. In all the cases EBMT had the best scores and Moses SBM had the poorer scores.

5 Conclusion

In this paper we have discussed the use of machine learning techniques in MT evaluation. We have also compared the results of our study with human and automatic evaluation metrics. By looking at the results we can confidently say that machine learning techniques can prove to be an alternate to human evaluation as it produces consistent results with human evaluations which at times, is not possible with automatic evaluations. Moreover, using machine learning techniques we can have minimum human intervention. In machine learning based MT evaluation, humans are involved only during the training of classifiers while in automatic evaluation human are required every time the evaluation corpus is changed as we would require human reference translations for the new corpus. Thus by looking at these points we can say that machine learning based MT evaluation can be a very good alternative for human and automatic evaluation metrics.

References

[1] Miller, G.A., Beebe-Center, J.G.: Some Psychological Methods for Evaluating the Quality of Translation. Mechanical Translations 3 (1956)

[2] Pfafflin, S.M.: Evaluation of Machine Translations by Reading Comprehension Tests and Subjective Judgments. Mechanical Translation and Computational Linguistics 8, 2–8 (1956)

[3] Slype, G.V.: Systran: evalaution of the 1978 version of systran English-French automatic system of the Commission of the European Communities. The Incorporated Linguist 18, 86–89 (1979)

[4] Falkedal, K.: Evaluation Methods for Machine Translation Systems. An Historical overview and Critical Account. Technical Report, ISSCO, Universite de Geneve (1991)

[5] Papineni, K., Roukos, S., Ward, T., Zhu, W.-J.: Bleu: a method for automatic evaluation of machine translation. RC22176 Technical Report, IBM T.J. Watson Research Center (2001)

[6] Doddington, G.: Automatic Evaluation of Machine Translation Quality Using N-gram Co-Occurrence Statistics. In: Proceedings of the 2nd International Conference on Human Language Technology, pp. 138–145 (2002)

[7] Turian, J.P., Shen, L., Melamed, I.D.: Evaluation of Machine Translation and its Evaluation. In: Proceedings of MT SUMMIT IX (2003)

[8] Snover, M., Dorr, B., Schwartz, R., Micciulla, L., Makhoul, J.: A Study of Translation Edit Rate with Targeted Human Annotation. In: Proceedings of the 7th Conference of the Association for Machine Translation in the Americas (AMTA), pp. 223–231 (2006)

[9] Levenshtein, V.I.: Binary Codes Capable of Correlating Deletions, Insertions and Reversals. Soviet Physics Doklady 8(10) (1966)

[10] Denkowski, D., Lavie, A.: Meteor 1.3: Automatic metric for reliable optimization and evaluation of machine translation systems. In: Proceedings of the Workshop on Statistical machine Translation (2011)

[11] Gamon, M., Aue, A., Smets, M.: Sentence-level MT evaluation without reference translations: Beyond language modeling. In: Proceeding of Annual Conference of European Association of Machine Translation (2005)
[12] Specia, L., Raj, D., Turchi, M.: Machine Translaiton Evaluation and Quality Estimation. Machine Translation 24(1), 24–39 (2010)
[13] Koehn, P., Martin, J., Mihalcea, R., Monz, C., Pedersen, T.: Proceedings of the Workshop on Building and Using Parallel Texts (2005)
[14] Quinlan, J.R.: Improved use of continuous attributes in c4.5. Journal of Artificial Intelligence Research 4, 77–90 (1996)
[15] Hall, M., Frank, E., Holmes, G., Pfahringer, B., Reutemann, P., Witten, I.H.: The WEKA Data Mining Software: An Update. SIGKDD Explorations 11(1) (2009)
[16] Chang, C., Lin, C.: LIBSVM: a library for support vector machines. ACM Transactions on Intelligent Systems and Technology 27, 1–27 (2011)
[17] Koehn, P., Hoang, H., Birch, A., Callison-Burch, C., Federico, M., Bertoldi, N., Cowan, B., Shen, W., Moran, C., Zens, R., Dyer, C., Bojar, O., Constantin, A., Herbst, E.: Moses: Open source toolkit for statistical machine translation. In: Proceedings of ACL: Demonstration Session (2007)
[18] Hoang, H., Koehn, P.: Improved Translation with Source Syntax Labels. In: Proceedings of the Joint 5th Workshop on Statistical Machine Translation and MetricsMATR, Uppsala, Sweden, July 15-16, pp. 409–417 (2010)
[19] Joshi, N., Mathur, I., Mathur, S.: Translation Memory for Indian Languages: An Aid for Human Translators. In: Proceedings of 2nd International Conference and Workshop in Emerging Trends in Technology (2011)
[20] Joshi, N., Darbari, H., Mathur, I.: Human and Automatic Evaluation of English to Hindi Machine Translation Systems. In: Wyld, D.C., Zizka, J., Nagamalai, D. (eds.) Advances in Computer Science, Engineering & Applications. Advances in Soft Computing, vol. 166, pp. 423–432. Springer, Heidelberg (2012)
[21] Joshi, N., Mathur, I., Darbari, H., Kumar, A.: HEval: Yet Another Human Evaluation Metric. International Journal of Natural Language Computing 2(5), 21–36 (2013)
[22] Ramnathan, A., Rao, D.: A Lightweight Stemmer for Hindi, In Proceedings of Workshop on Computational Linguistics for South Asian Languages. In: 10th Conference of the European Chapter of Association of Computational Linguistics, pp. 42–48 (2003)
[23] Narayan, D., Chakrabarti, D., Pande, P., Bhattacharyya, P.: An Experience in Building the Indo WordNet - a WordNet for Hindi. In: Proceedings First International Conference on Global WordNet, Mysore, India (January 2002)

Multi-output On-Line ATC Estimation in Deregulated Power System Using ANN

R. Prathiba, B. Balasingh Moses, Durairaj Devaraj, and M. Karuppasamypandiyan

Abstract. Fast and accurate evaluation of the Available Transfer Capability (ATC) is essential for the efficient use of networks in a deregulated power system. This paper proposes multi output Feed Forward neural network for on line estimation of ATC. Back Propagation Algorithm is used to train the Feed Forward neural network. The data sets for developing Artificial Neural Network (ANN) models are generated using Repeated Power Flow (RPF) algorithm. The effectiveness of the proposed ANN models are tested on IEEE 24 bus Reliability Test System (RTS). The results of ANN model is compared with RPF results. From the results, it is observed that the ANN model developed is suitable for fast on line estimation of ATC.

Keywords: Artificial Neural Network, Available Transfer Capability, Bilateral transaction, Repeated Power Flow.

1 Introduction

Transition of electric industry from its vertically integrated structure to horizontal structure poses many problems to power system engineers and researchers [1].In the environment of open transmission access, US Federal Energy Regulatory Commission (FERC) requires that Available Transfer Capability (ATC) information be made available on a publicly accessible Open Access Same Time Information System (OASIS) [2]. Methods based on DC load flows [3] are faster

R. Prathiba · B. Balasingh Moses
Department of EEE, Anna University, Trichy

Durairaj Devaraj
Department of CSE, Kalasalingam University, Krishnan Koil

M. Karuppasamypandiyan
Department of EEE, Kalasalingam University, Krishnan Koil

El-Sayed M. El-Alfy et al. (eds.), *Advances in Intelligent Informatics*,
Advances in Intelligent Systems and Computing 320, DOI: 10.1007/978-3-319-11218-3_21

than the AC load flow [4], since no iterations are involved. Complexity in computation is also less as the number of data to be used is less. The DC-Power Transfer Distribution Factors (DC-PTDF) [5] are easy to calculate and can give quick estimate of ATC. But the ATC values calculated using them are not very accurate as DC power flow neglect reactive power effects. AC-PTDFs for transfer capability calculation is investigated in [6].AC-PTDFs are based on derivatives around the given operating point and may lead to unacceptable results when used at different operating points to calculate ATC. Also, neither DC nor AC PTDFs based method considers generator limits and bus voltage limits when used to determine ATC. The continuation power flow(CPF)based methods[7] perform full-scale ac load flow solution, and is accurate but due to the complexity involved in the computation it is difficult to be implemented for large systems. The optimal power flow and RPF based methods [8-9] determine ATC formulating an optimization problem in order to maximize the power transmission between specific generator and load subject to satisfying power balance equations and system operating limits. The conventional methods are not applicable for fast estimation of ATC. Neural Network for Estimation of ATC between two areas was developed using ACPTDF [10].MLP Neural network was designed with Quick Prop algorithm to train the network for IEEE 30 bus system and was useful for reliability assessment [11]. A novel MLP with input feature selection and Levenberg-Marquardt algorithm was designed for both bilateral and multilateral transactions in [12] for single and multi-output ATC and the performance is compared with Optimal power flow.

A new model employing multi-output artificial neural networks to calculate transfer capability is developed in this paper. Based on the power flow formulation for calculating real power transfer capability and with the strong generalizing ability of the neural networks, the new model can calculate ATC quickly for a given power system status. This paper is organized as follows. Section 2 provides the methodology for computation of ATC. In section 3 proposed method for ATC estimation and training algorithm to formulate the input-output data set for the ANN is discussed. Section 4 discusses the review of ANN and BPA algoritham. Simulation results are discussed in Section 5. The outcome of the proposed method is concluded in section 6.

2 Computation of ATC

The objective is to estimate the Available Transfer Capability (ATC) for a bilateral contract by increasing the generation at a seller bus/buses and at the same time increasing the same amount of load at the buyer bus/buses, until the power system reaches system limits

Mathematically, each bilateral transaction, between a seller at bus-i and power purchaser at bus-j, satisfies the following power balance relationship.

$$P_{gi} - P_{dj} = 0 \tag{2.1}$$

Where, P_{gi} and P_{dj} are the real power generation at bus-i and real power consumption at bus-j.

ATC for a bilateral contract can be calculated by increasing the generation at a contracted seller bus/buses and at the same time increasing the same amount of load at the contracted buyer bus/buses, until the power system reaches system limits.

$$\text{ATC value is given by} \quad ATC = TTC - ETC \tag{2.2}$$

Where,

TTC-Total Transfer Capability

ETC - base case transfer

Provided TRM and CBM are assumed to be zero for the sake of simplicity.

Subject to the following operating conditions.

$$P_i - \sum_{j=\in N} V_i V_j Y_{ij} \cos(\theta_{ij} + \delta_j - \delta_i) = 0 \tag{2.3}$$

$$Q_i - \sum_{j=\in N} V_i V_j Y_{ij} \sin(\theta_{ij} + \delta_j - \delta_i) = 0 \tag{2.4}$$

$$V_{min} \leq V \leq V_{max} \tag{2.5}$$

$$S_{ijmin} \leq S_{ij} \leq S_{ijmax} \tag{2.6}$$

Where

N- Set of all buses

P_i, Q_i - Real and reactive power at the ith bus

Y_{ij}, δ_i - Bus matrix elements

V_i, δ_i - Magnitude and angle of ith bus

The testing and training data generation for ANN is done using RPF.

3 Proposed Approach for ATC Estimation

The ATC for real time application by ANN approach is proposed. The objective is to estimate the ATC for bilateral transactions under different loading conditions. The real and reactive power for different loading condition is given as the input for NN and the output of NN is the ATC value in MW for different transaction. The schematic diagram of learning stage of neural network is shown in Figure 1. Neural network approach for any application has three stages: Normalization of inputs, training and testing stages. While training the network, the input and output are first normalized between 0 and 1. The input variables after normalization are presented to the neural network for training.

After training, the networks are evaluated through a different set of input– output data. Once the training and testing of the network is over, then the network will be ready for on-line application.

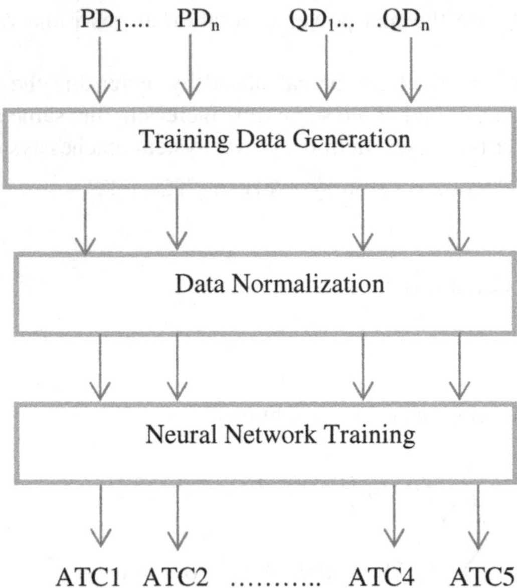

Fig. 1 Schematic diagram of NN learning stage

The various steps involved in the development of multi output ANN – based ATC estimation model are

3.1 *Training Set Generation*

*F*or generating the training set, a large number of load patterns were generated by perturbing the loads randomly (70 % to 130%) at all the buses. The transaction depends on various factors such as load level, generation level, line status, generator status etc., the frequently changing parameter is the load and so it is taken as input to the ANN. The training set generation is done in off-line mode .The dimension of inputs used in this model is 34x250, where both real and reactive powers are taken with 250 varying load conditions. In the proposed model the ANN will estimate the ATC values for different transactions.

3.2 *Normalization of the Data*

Normalization of the data is an important aspect for training of the neural network. Without normalization, the higher valued input variables may tend to suppress the influence of smaller ones. To overcome this problem, neural networks are trained with normalized input data. The value of input variables is scaled between some suitable values (0 and 1 in the present case) for each load pattern. In case of output variables, if the value varies over a wide range, the neural network is not able to map it correctly. The remedy is to scale the output between some suitable range (0 and 1 in the present case).

$$x_n = \frac{(x - x_{\min})}{(x_{\max} - x_{\min})} + startingvalue \tag{3.1}$$

Where x_n is the normalized value, and, x_{min}, x_{max} are the minimum and maximum values of the variable x.

3.3 Training and Testing of Neural Network

The neural network used for ATC estimation consists of three layers. The input layer has neurons equal to the number of inputs selected and output layer has six neuron. The activation functions used in the hidden layer neurons have tangent hyperbolic function and the output neurons have linear activation function. The number of hidden units depends on the input units. Best number of hidden unit is determined by training several networks and estimating the generalization error. Large networks require longer training time. Trial and error procedure is followed to select the suitable number of neurons in the hidden layer. The generated variables after normalization are given as input to the neural network for training. After training, the networks are evaluated through a different set of input–output data. Now the developed multi-output ANN can estimate ATC values for different operating conditions.

4 Review of Artificial Neural Network

An ANN is defined [13,14] as a data processing system consisting of a large number of simple, highly interconnected processing elements (artificial neurons).This architecture is inspired by the structure of the cerebral cortex of the brain.

A structure of multi-layer feed forward network is shown in figure 2.

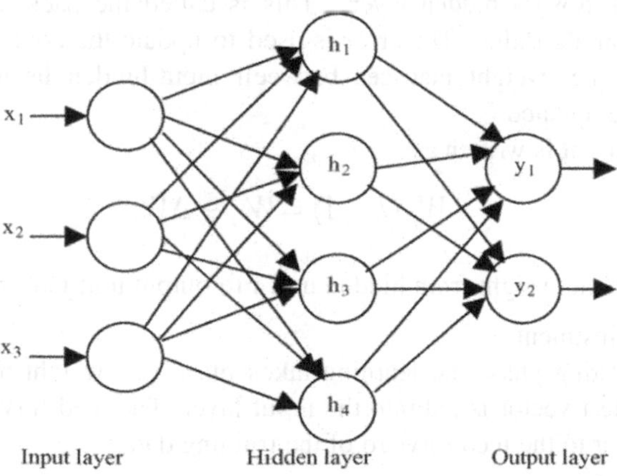

Fig. 2 Artificial Neural Network Structure

The ANN used here consists of three layers, the input vector, hidden layer and output vector. A network is trained so that application of a set of inputs produces the desired (or at least consistent) set of outputs. Each such input (or output) set is referred to as a vector. Training is accomplished by sequentially applying input vectors, while adjusting network weights according to a predetermined procedure. During training, the network weight gradually converts to values such that each input vector produces the desired output vector.

To reduce the computational effort by the conventional method, Back-Propagation Algorithm (BPA) based on Feed forward Neural Network has been utilized to compute the ATC. During the training phase, the training data is fed into the input layer. The data is propagated to the hidden layer and then to the output layer. This is called the forward pass of the Back Propagation Algorithm. In the forward pass, each node in hidden layer gets input from all the nodes of input layer, which are multiplied with appropriate weights and then summed. The output of the hidden node is the non- linear transformation of the resulting sum. Similarly each node in output layer gets input from all the nodes of hidden layer, which are multiplied with appropriate weights and then summed. The output of this node is the non-linear transformation of the resulting sum.

This is mathematically represented as,

$$out_i = f(net_i) = f\left[\sum_{j=1}^{n} w_{ij} out_j + b_i\right] \tag{4.1}$$

Where out_i is the output of the i_{th} neuron in the layer under consideration. out_j is the output of the j_{th} neuron in the preceding layer. w_{ij} are the connection weights between the i_{th} neuron and the j_{th} inputs and b_i is a constant called bias.

The output values of the output layer are compared with the target output values. The target output values are those that we attempt to teach our network. The error between actual output values and target output values are calculated and propagated back toward hidden layer. This is called the backward pass of the back propagation algorithm. The error is used to update the connection strengths between nodes, i.e. weight matrices between input-hidden layers and hidden-output layers are updated.

Mathematically it is written as,

$$W_{ij}(k+1) = W_{ij} + \Delta W_{ij} \tag{4.2}$$

Where W_{ij} is the weight from hidden unit i to output unit j at time k and ΔW_{ij} is the weight adjustment.

During the testing phase, no learning takes place i.e., weight matrices are not changed. Each test vector is fed into the input layer. The feed forward of the testing data is similar to the feed forward of the training data.

5 Simulation Results

This section presents the details of the simulation study carried out on IEEE 24-bus Reliability Test System [15,16] for ATC estimation using the proposed approach. For this system ANN model was developed to estimate the ATC for different bilateral transactions. Neural network toolbox in MATLAB was used to develop the ANN models. The details of the ANN models developed are presented here

5.1 ATC Assessment in IEEE RTS 24 Bus System

IEEE RTS 24 bus system consists of 11 generator buses, 13 load buses and 38 transmission lines. For generating training data for the ANN, the loads at the load buses are varied randomly between 70% to 130% of base load. Based on the algorithm presented in section 3, a total of 250 input-output pairs were generated with 150 for training and 100 for testing. The real and reactive power loads at all load buses are given as input of the neural network and respective ATC values are the output.

The parameters of the network used here are given below:

No. of input – 34
No. of output – 6
No. of hidden neurons – 10
Mean Square Error – 7.730×10^{-5}
The network took 44.38sec to reach the error goal.

Table 1 shows training and testing performance of the network for bilateral transaction between 23-3 under normal operating condition. The ATC value obtained at 115% loading condition is 75.95MW.The mean square error is6.116×10^{5}. The time taken by the network is 1.546s. No. of hidden nodes used are 10 Nos.

Table 1 Training and Testing performance of the Network

Test case	Transaction type	No of Hidden Nodes	Training time(Sec)	Testing Error (mse)
1	23-3	10	1.5756	6.116×10^{-5}

Table 2 shows ATC values for bilateral transaction (23-3), generated by the ANN for different loading condition. In transaction 23-3, bus 23 is source bus and bus 3 is sink bus. ATC obtained from RPF, ANN and the percentage of error in estimating ATC are also given here.

Table 2 ATC values estimated for bilateral transaction (23-3)

% Loading condition	ATC(MW)		% Error
	ANN	RPF	
80.82	159.83	159.00	0.005
86.95	148.39	147.30	0.742
90.95	139.93	139.00	0.006
96.85	126.34	126.00	0.271
100.59	116.92	117.15	0.213
105.2	104.61	105.50	0.84
115.37	75.91	77.20	1.67
119.0	65.31	66.00	0.01
126.07	42.97	42.75	0.51

Table 3 Comparison of RPF and ANN output

Set of Transactions	No of Hidden Nodes	Training time (Sec)	Testing Error (mse)
23-3,21-6,22-5,23-15,18-5	10	44.38	$7.73X10^{-5}$

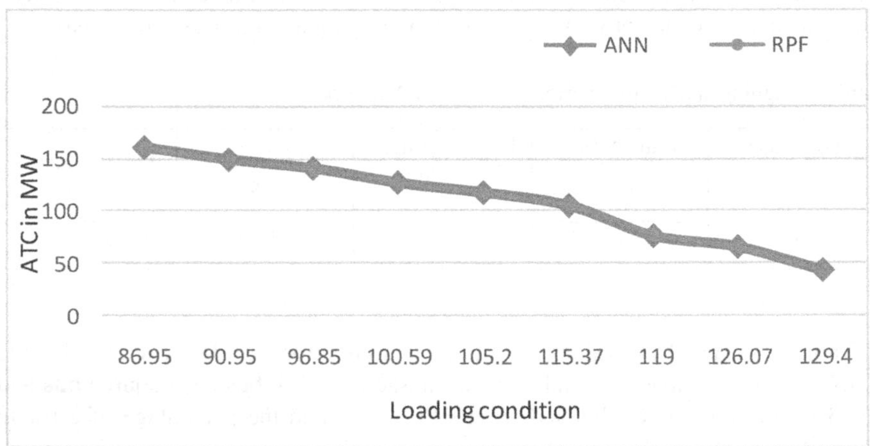

Fig. 3 Comparison of ATC values of RPF with ANN

Table 3 shows training and testing performance of the multi output network for a set of bilateral transaction between 23-3, 21-6, 22-5, 23-15, 18-5 under normal operating condition. The mean square error for the developed model is$7.73x10^{-5}$. The time taken by the RPF is approximately between 30 to 60 minutes depends on different bilateral transaction and for the developed ANN network it is 44.38s thus validating the objective of on line estimation of ATC.

The Fig 3 shows the ANN estimate accurately in comparison with RPF hence, the proposed ANN is computationally efficient and hence it is suitable for on-line estimation of ATC in real time applications for a set of bilateral transaction.

Table 4 Multi-Output ATC Values for set of Transactions

% Loading condition	Transaction	ATC(MW)		%Error
		ANN	RPF	
80.82	23-3	159.1055	159.0000	0.0663
	21-6	159.6060	159.8500	0.1526
	22-5	220.5404	220.2500	0.1318
	23-15	983.2275	978.4000	0.4934
	22-9	217.7627	218.7500	0.4513
	18-5	212.3887	211.9500	0.2069
86.95	23-3	147.2438	147.3000	0.03815
	21-6	145.7208	145.6000	0.0829
	22-5	243.5330	244.1500	0.2527
	23-15	942.2427	943.2000	0.1014
	22-9	271.7608	268.8500	1.0826
	18-5	147.2438	147.3000	0.0381
90.95	23-3	138.8436	139.0000	0.1125
	21-6	136.0366	136.2000	0.1199
	22-5	237.5142	237.2500	0.1113
	23-15	923.8467	919.8500	0.4344
	22-9	296.9590	301.8000	1.6040
	18-5	237.8198	237.1500	0.2824
100.59	23-3	117.2028	117.1500	0.0450
	21-6	113.8265	113.7500	0.0672
	22-5	218.1531	218.2000	0.02149
	23-15	860.3805	863.8000	0.3958
	22-9	281.5755	279.5000	0.7425
	18-5	217.9468	218.1500	0.0931

Table 4 (*continued*)

105.2	23-3	105.5046	105.500	0.0043
	21-6	102.9265	102.900	0.0257
	22-5	207.6826	207.6500	0.0156
	23-15	832.9591	836.4500	0.4173
	22-9	259.5942	259.1000	0.1903
	18-5	207.4827	207.6500	0.0805
115.37	23-3	77.2100	77.2000	0.0129
	21-6	78.8982	78.9500	0.0656
	22-5	181.0146	181.0500	0.0195
	23-15	781.8435	775.4500	0.8244
	22-9	203.4577	204.3500	0.4366
	18-5	181.1349	181.1000	0.0192

The Table 4 shows Multi-output ATC values for six sets of bilateral transaction obtained using ANN and RPF for different loading condition. The percentage error between ANN output and RPF is calculated and presented in Table 4.

6 Conclusion

In a real time operation of deregulated power system, the ISO has to estimate ATC values for many possible proposed transactions. As the ANN can estimate ATC value for more than one proposed transactions simultaneously, the ISO can evaluate many transactions in short time. This enhances the performance of ISO. This paper has presented an ANN-based multi-output ATC estimation method for on-line applications.Simulation was carried out on the IEEE 24-RTS bus system. Results shows that the ANN model with BPA based approach provides estimation of Multi-output ATC for set of transactions under different loading conditions. Hence for large scale practical power systems the proposed ANN model can estimate ATC in lesser computation time with reasonably good accuracy. This makes the ANN model suitable for real time applications.

References

[1] Bhattacharya, K., Bollen, M., Daalder, J.E.: Operation of Restructured Power Systems. Kluwer Academic Publishers (2001)

[2] North American Electricity Reliability Council (NERC). Available transfer capability- Definitions and determinations. NERC Report (June 1996)

[3] Hamoud, G.: Assessment of available transfer capability of transmission systems. IEEE Transactions on Power Systems 15, 27–32 (2000)

[4] Gravened, M.H., Nwankpa, C., Yoho, T.: ATC Computational issues. In: Proc. 32nd Hawaii Int. Conf. System Science, Maui, HI, pp. 1–6 (1999)
[5] Wood, A.J., Woolenberd, B.F.: Power Generation operation and Control, 2nd edn. John Wiley& Sons
[6] Kumar, A., Srivastava, S.C., Singh, S.N.: ATC transmission capability determination in a competitive electricity market using AC distribution factors. Electric Power Components &Systems 32(9), 927–939 (2004)
[7] Ejebe, G.C., Waight, J.G., Frame, J.G., Wang, X., Tinney, W.F.: Available Transfer Capability Calculations. IEEE Transaction on Power Systems 13, 1521–1527 (1998)
[8] Joo, S.K., Liu, C.C.: Optimization techniques for available transfer capability and market calculations. IMA Journal of Management Mathematics 15, 321–337 (2004)
[9] Khaburi, M.A., Haghifam, M.R.: A probabilistic modeling based approach for Total Transfer Capability enhancement using FACTS devices. Electrical Power and Energy Systems 32, 12–16 (2010)
[10] Narasimha Rao, K., Amarnath, J., Kiran Kumar, K., Kamakshiah, S.: Available Transfer Capability Calculations Using Neural Networks in Deregulated Power. In: 2008 International Conference on Condition Monitoring and Diagnosis, Beijing, China, April 21-24 (2008)
[11] Luo, X., Patton, A.D., Singh, C.: Real Power Transfer Capabiltiy Calculations using Multi-Layer Feed –Forward Neural Networks. IEEE Transactions on Power Systems 15(2) (May 2000)
[12] Jain, T., Singh, S.N., Srivastava, S.C.: A Neural Network based method for fast ATC estimation in Electricity Markets. IEEE Transactions on Power Systems 15(2) (May 2007)
[13] Masters, T.: Practical neural network recipes in C++. Academic Press, New York (1993)
[14] Devaraj, D., Preetha Roselyn, J., Uma Rani, R.: Artificial neural network model for voltage security based contingency ranking. Electrical Power Energy System 7, 722–727 (2007)
[15] http://www.ee.washington.edu/research/pstca
[16] Zimmerman, R.D., Murillo-Sanchez, C.E., Thomas, R.J.: MATPOWER: MATLAB power system simulation package

Benchmarking Support Vector Machines Implementation Using Multiple Techniques

M.V. Sukanya, Shiju Sathyadevan, and U.B. Unmesha Sreeveni

Abstract. Data management becomes a complex task when hundreds of petabytes of data are being gathered, stored and processed on a day to day basis. Efficient processing of the exponentially growing data is inevitable in this context. This paper discusses about the processing of a huge amount of data through Support Vector machine (SVM) algorithm using different techniques ranging from single node Linier implementation to parallel processing using the distributed processing frameworks like Hadoop. Map-Reduce component of Hadoop performs the parallelization process which is used to feed information to Support Vector Machines (SVMs), a machine learning algorithm applicable to classification and regression analysis. Paper also does a detailed anatomy of SVM algorithm and sets a roadmap for implementing the same in both linear and Map-Reduce fashion. The main objective is explain in detail the steps involved in developing an SVM algorithm from scratch using standard linear and Map-Reduce techniques and also conduct a performance analysis across linear implementation of SVM, SVM implementation in single node Hadoop, SVM implementation in Hadoop cluster and also against a proven tool like R, gauging them with respect to the accuracy achieved, their processing pace against varying data sizes, capability to handle huge data volume without breaking etc.

Keywords: SVM, Hadoop, Map-Reduce, Classification, Accuracy.

1 Introduction

The amount of data in our world has been exploding, and analyzing large data sets will become a difficult task. Parallel processing of data in a distributed framework

M.V. Sukanya · Shiju Sathyadevan · U.B. Unmesha Sreeveni
Amrita Center for Cyber Security,
Amrita Vishwa Vidyapeetham,
Kollam, India
e-mail: {sukanyaanoop,unmeshabiju}@gmail.com,
Shiju.s@am.amrita.edu

El-Sayed M. El-Alfy et al. (eds.), *Advances in Intelligent Informatics*,
Advances in Intelligent Systems and Computing 320, DOI: 10.1007/978-3-319-11218-3_22

offers efficient data processing in less time. We are familiar with distributed frameworks such as Hadoop, Spark, and Storm etc. Hadoop is a framework for processing parallelizable problems across huge datasets using a large number of commodity hardware, collectively referred to as a cluster. Map-Reduce and Hadoop Distributed File Systems (HDFS) are the key components of Hadoop.

The Map-Reduce paradigm has frequently proven to be a simple, flexible and scalable technique to distribute algorithms across thousands of nodes and petabytes of information. Classic data mining algorithms have been adapted to Hadoop framework for handling the exponential growth in data.

Support Vector Machines (SVMs) is a supervised machine learning algorithm, used to develop a classifier. Here we are using SVM as a tool to classify large data sets into two classes. There are many reasons to prefer SVMs, because they are conceptually easy to understand and solving an SVM is simply solving a convex optimization problem. Parallel SVM loads training data on parallel machines reducing memory requirements through approximate factorization of matrix. It loads only essential data to each machine to perform parallel computation. Training an SVM can easily be broken into parts making it a class of problem well-suited to Hadoop map-reduce.

Apache Hadoop provides the power of scanning and analyzing terabytes of data using a collection of commodity servers working in parallel with the help of machine learning algorithm like Support Vector Machines. The main objective of this paper is the comparison study of the implementation of SVM algorithm in three different leading to the Time and Accuracy findings of three different data sets that are being given into the algorithm.

Among the four different scenarios that is presented in this paper, framework R is a proven data mining analysis tool which has the SVM algorithm implemented in it and is taken as a reference point for the analysis of SVM algorithm against Hadoop cluster, single node Hadoop and Linear SVM. A study of the maximum size of the dataset that can be given as input to SVM algorithm under each of the four different scenarios run on identical configurations was done and the results are presented in this paper. We also present a time complexity analysis of SVM across all the four different test scenarios.

2 Related Works

Distil network is an internet company (Virginia) that specializes in web content protection systems handling large amount of data in their network. They use SVM as a tool to detect botnet's based on their behavior and statistical features [1]. A trained SVM is a set of vectors and much easier to understand than a trained neural network because they are conceptually easy to understand. One big downside to SVMs is their computational complexity with the growing size of the data set used for training. It will take days to train an SVM.

They broke SVM training into subsets and as it suits well to Map-Reduce and train an SVM on each subset [1]. Trained SVM yields Support Vectors which are used in classification. This is again divided into parts and then global SVs are added into it. A new set of SVs is derived from this and it continues. Somewhere along the process SVs provably converge to global optimum. Main advantage of this is that by increasing processing speed via parallelism and divide and conquer makes the problem less intimidating and much more manageable.

Support Vector Machines (SVM) are powerful classification and regression tools. For applying SVM to large scale data mining, parallel SVM methods are proposed. Parallel SVM methods are based on classical MPI model. Map-Reduce is an efficient distribution computing model to process large scale data mining problems. In [2] parallel SVM model based on iterative Map-Reduce is proposed. It is realized with Twister software. Through example analysis it shows that the proposed cascade SVM based on Twister can reduce the computation time greatly.

In the Map task sample data are loaded from local file system according to the partition file. Trained support vectors are sent to the Reduce jobs. And in reduce task all support vectors of Map jobs are collected together and feed back to client. Through iteration, the training process will stop when all sub SVM are combined to one SVM. The efficiency of the method is illustrated in [2] through analyzing practical problems like adult data analysis, forest cover type classification, heart disease classification etc. Parallel SVM based on iterative Map-Reduce is efficient in data intensive problems.

A methodology that could be used to solve management problems involving large data sets using Support Vector Machines [3]. They develop a methodology to help identify the data sets to be deleted when there is a requirement for storage space. Large data is stored using HDFS. HDFS logs give information regarding file access operations. Hadoop Map-Reduce was used to feed information in these logs to SVMs.

Time elapsed in data set classification is dependent on the size of the input HDFS log file. The SVM methodology produces a list of data sets for deletion along with their respective sizes. This methodology was also compared with a heuristic called Retention Cost R_C which was calculated using size of the data set. Accuracies of both were compared by calculating the percentage of data sets predicted for deletion which were accessed at a later instance of time. The methodologies using SVMs proved to be more accurate than using the Retention Cost heuristics.

Clustering-Based SVM (CBSVM) integrates a scalable clustering method with an SVM method and effectively runs SVMs for very large data sets [4]. The existing SVMs are not feasible to run such data sets due to their high complexity on the data size or frequent accesses on the large data sets causing expensive I/O operations.

CB-SVM applies a hierarchical micro-clustering algorithm that scans the entire data set only once to provide an SVM with high quality micro-clusters that carry the statistical summaries of the data such that the summaries maximize the benefit of learning the SVM. In [4] CB-SVM tries to generate the best SVM boundary for very large data sets given limited amount of resource based on the hierarchical clustering. Experiments on synthetic and real data sets show that CB-SVM is very scalable for very large data sets while generating high classification accuracy.

The paper [5] explains about some mathematical properties of support vectors and show that the decision surface can be written as the sum of two orthogonal terms. For almost all values of the parameter, this enables to predict how the decision surface varies for small parameter changes. The important case of feature space of finite dimension m, they also show that there are at most m + 1 margin vectors and observe that m + 1 SVs are usually sufficient to fully determine the decision surface. For relatively small m this latter result leads to a consistent reduction of the SV number.

Support Vector Machines (SVMs) suffer from a widely recognized scalability problem in both memory use and computational time. To improve scalability, the paper [6] proposes a parallel SVM algorithm (PSVM), which reduces memory use through performing a row-based, approximate matrix factorization, and which loads only essential data to each machine to perform parallel computation. Let n denote the number of training instances, p the reduced matrix dimension after factorization (p is significantly smaller than n), and m the number of machines. PSVM reduces the memory requirement from O (n2) to O (np/m), and improves computation time to O (np2/m). From [6] we can show that SVMs can be parallelized to achieve scalable performance. PSVM solves IPM in parallel by cleverly arranging computation order.

3 System Design

3.1 Linear SVM Implementation

As part of the research effort in order to fine tune the performance of the algorithm, SVM was developed from scratch using java in eclipse platform. Accuracy across all four implementations of SVM was also computed by running it against different input data sets of varying sizes. This works for any input dataset given to SVM algorithm, and predict the accuracy of the algorithm by classifying the data. This involves executing a learning algorithm on a set of labelled examples, i.e., a set of entities represented via numerical features along with underlying category labels.

The algorithm returns a trained model that can predict the label for new entities for which the underlying label is unknown. Following section will illustrate the same with the aid of an example.

Input test data set used for classification

x1	x2	Class label
0	0	1
3	4	1
5	9	1
12	1	1
8	7	1
9	8	-1
6	12	-1
10	8	-1
8	5	-1
14	8	-1

→

Intermediate Results obtained from the given input are:

Normalize the given input matrix using Standard Deviation and mean for each columns of the matrix.

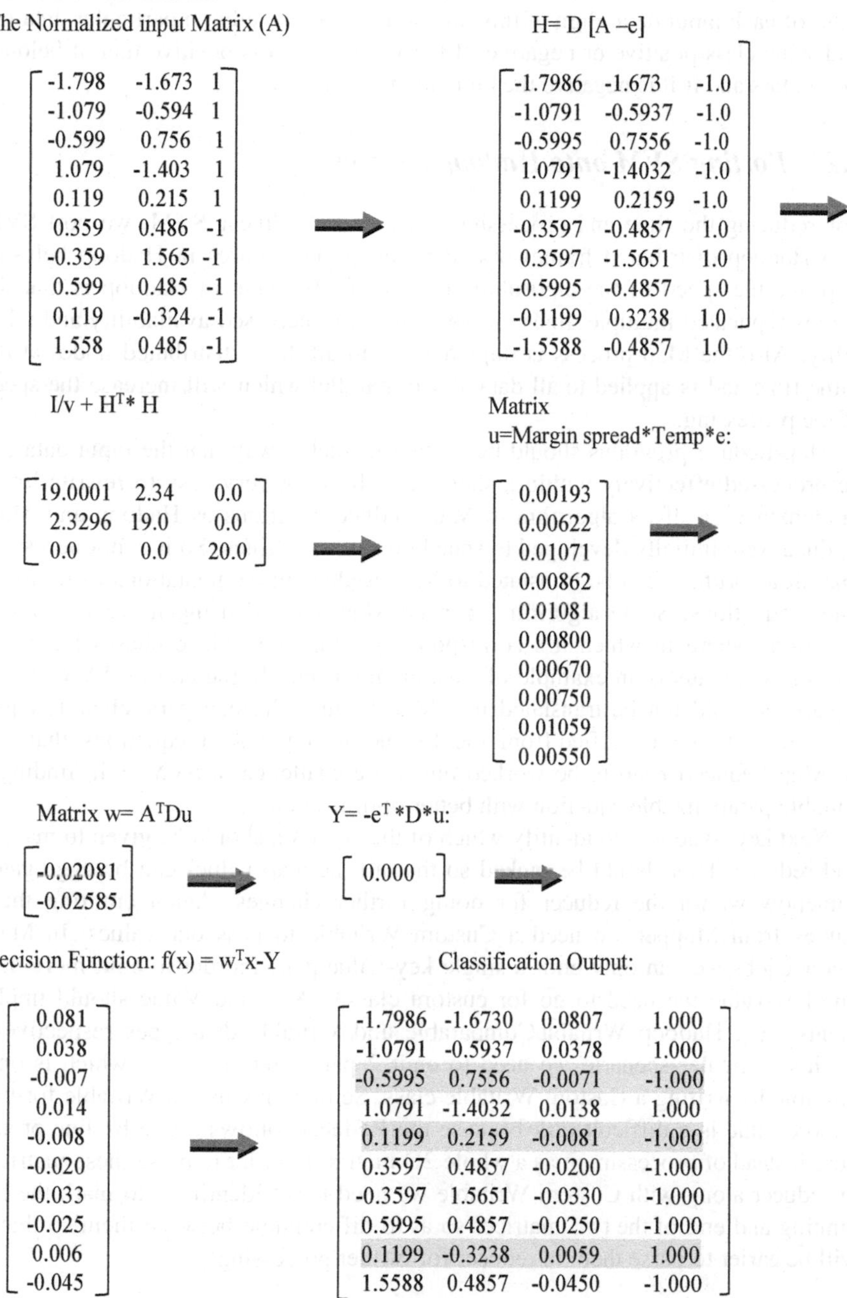

From this output result we can see that 3rd 5th and 9th data points are misclassified. Thus we can say that this dataset has 70% accuracy as there are three misclassifications in the data points. The classification function f(x) determines the class of each input data. Sign of this function decides whether a particular data fall under the class positive or negative. If the value of f(x) is positive then it belongs to +1 class and if it is negative then it belongs to -1 class.

3.2 Porting SVM onto Hadoop Cluster

For reducing the time and processing complexity of linear SVM, we port SVM into Hadoop distributed framework. Parallelization features in Hadoop helps to improve the speed of the algorithm. The main advantage of Hadoop is that the data is replicated multiple times on the system for increased availability and reliability. Also the Map process is implemented in all these distributed nodes at the same time and is applied to all data sets in parallel which will increase the speed of the processing.

Map-Reduce programs should be written in such a way that the input data can be processed effectively within a short time. It is not very easy to rewrite linear implementation of an algorithm to Map-Reduce programs as Hadoop and Map-Reduce were initially developed to speed up search engines. So it is not necessary that all algorithms can be translated to Map-Reduce implementations using its on linear equations. Some algorithms can be Map-reduced using its regular linear equations where in which it's corresponding, map and reduce phases has to be written. K-Means is an example of such an algorithm. In the case of SVM, linear equations could not be translated into Map-Reduce. Instead parallelizable equations need to be identified from the lot and its equivalent equations that can be Map-Reduced need to be worked out. Core challenge in SVM is in finding a suitable parallelizable equation with better optimization.

Next key issue was to identify which of the equation should be given to mapper and reducer. This should be picked so that all the map values can be aggregated somehow within the reducer for doing further changes. Again emitting these values from Mapper we need a Custom Writable to pass our values. In Map-Reduce jobs we can only emit a single key-value pair. Inorder to emit more than one key-value we need to go for custom classes. Key and Value should implements using Hadoop WritableComparable and Writable data types respectively. In this particular scenario we need to emit 2 parameters as value which is only possible by writing a Custom Writable class. Setting up Custom Writable for two matrix value is a difficult task because the Reducer follows a line by line procedure instead of processing it as a whole 2D matrix. In order to parse these matrices in reducer along with Custom Writable we need to set Identifiers to mark the beginning and end of the two matrices so as to differentiate between them so that it will be easier to parse them in reducer for further processing.

3.2.1 Algorithmic Implementation of SVM in Hadoop Cluster

Classification is an everyday task, it is about selecting one out of several outcomes based on their features. We can use Map Reduce to parallelize a linear machine learning classifier algorithm for Hadoop Streaming. Parallel approach encourages computational simplicity in training classification algorithms.

Here we use the parallel processing method for efficient classification of input datasets into two classes. Linear SVM algorithm classifies data based on the classifier function. $\mathbf{f(x) = w^T x\text{-}Y}$

Instead of using this equation, we come up with a parallel equation for Map-Reduce.

For training the classifier assumes numerical training data where each class is either negative or positive and features are represented as vectors of positive floating point numbers.

D -a matrix of training classes
A -a matrix with feature vectors
e -a vector filled with ones,
E = [A -e]
mu= scalar constant used to tune classifier
D -a diagonal matrix with -1.0 or +1.0 values (depending on the class)
Training the classifier using the expression:

$$\mathbf{(Omega, Gamma) = (I/mu + E^T{*}E)^{-1}{*}(E^T{*}D{*}e)}$$

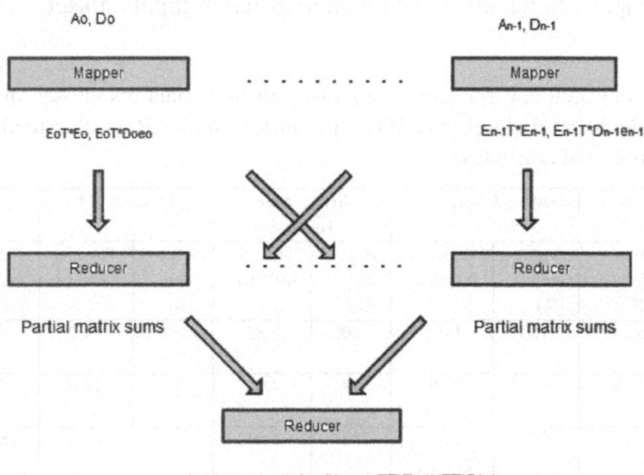

Fig. 1 Functionality of Mapper and Reducer in Hadoop

Classification of an incoming feature vector x can then be done by calculating $\mathbf{x^T {*}Omega - Gamma}$, which returns a number, and the sign of the number corresponds to the class, i.e. positive or negative. The size of the matrices $E^T{*}E$ and $E^T{*}D{*}e$ given as input to the reducer is independent of number of training data, but dependent on the number of classification features. Reducer calculates the partial sums of matrices. Adding more intermediate reducers are useful in calculating the sums of E^T*E and E^T*D*e before sending the sums to the final reducer, means that the final reducer gets fewer matrices to summarize before calculating the final answer. See Fig.2

4 Comparison of SVM Implementation across Linear, Single Node Hadoop, Hadoop Cluster and R

Linear SVM is implemented using the traditional matrix calculation method. In this method, if the training dataset is having high feature dimensions, then the matrix inversion can become a real bottleneck since such algorithms typically has a time complexity of O($n^{3)}$ where n is the number of features. In case of SVM in R, where linear SVM is already implemented, it could handle only a dataset which is half the size of the corresponding system's RAM. Comparing linear SVM implementation and single node Hadoop implementation against its implementation in Hadoop cluster, the later is far too efficient in terms of accuracy, performance and handling large data sets. This is because the Hadoop Cluster implementation can support parallel matrix based inversion and we could make the algorithm more efficient with improved time complexity.

When comparing Map-reduce SVM with linear SVM, we found that Map-Reduce SVM gives higher accuracy with fewer execution times. We have analyzed the comparison results based on three different input datasets.

Table 1 Lists the accuracy and time comparison of three datasets chosen in Linear SVM, SVM in Single node Hadoop, and Hadoop Cluster and in R (with Small dataset size) (* refers to number of attributes)

		Hadoop Cluster		Single node Hadoop		Linear SVM		SVM in R	
Dataset	Dataset size in MB	Accuracy (%)	Time in minutes	Accuracy (%)	Time in minutes	Accuracy (%)	Time in minutes	Accuracy (%)	Time in Minutes
Spam (8)*	52.4	100	1.31	100	3.85	98	14.2	99.5	4
Wine (12)*	55.8	99	3.75	99	4.2	82	14.6	92	4.4
Heart (22)*	629	82	4.83	82	11.32	--	--	80	6.2

From the accuracy and time comparison table, the following things could be inferred. Accuracy and execution time of each of three datasets, namely Spam, Wine and Heart using Linear SVM, R, SVM in Single node and SVM in Hadoop implementations are compared.

Comparison study shows that in linear SVM performs poorly with respect to accuracy and performance. It also failed to process Heart data set successfully because of its data size. Hadoop Map-Reduce performed well in terms of execution time and accuracy across all the three data sets hence can be rated as the best suited solution for this analysis. We did tested the trio against even smaller data sets (KB size) where R and Linear implementations outperformed Hadoop Map-Reduce implementation as expected because of its known processing overheads. This is also due to the fact that the implementation mainly relies on matrix calculations which involves finding inverse matrices which greatly increases the time complexity of the algorithm.

In Table 1, there is a tendency of accuracy reduction with increasing dimension/attributes in the dataset. This is because the matrix calculation method used for implementing the algorithm does not apply any optimization techniques as such. Matrix inversion steps are needed in the implementation procedure of SVM and it really makes the algorithm more time complex and henceforth less efficient. Also in map reduce implementation we are using more than one mapper and a single reducer to perform the classification. We can improve the performance of the algorithm by increasing number of reducers.

Table 2 Accuracy and time comparison of three datasets chosen in Linear SVM, Single node Hadoop and in Hadoop Cluster (with larger dataset sizes) (* refers to number of attributes)

		Hadoop Cluster		Single node Hadoop		Linear SVM		SVM in R	
Dataset	Dataset size in GB	Accuracy (%)	Time in minutes	Accuracy (%)	Time in minutes	Accuracy (%)	Time in minutes	Accuracy (%)	Time in minutes
Spam (8)*	2.5	100	1.31	100	12	--	--	100	30 +
Wine (12)*	2.3	98.8	3.75	98.8	13.4	--	--	97	30 +
Heart (22)*	2.2	82	4.83	82	12.3	--	--	80	30 +

In Table 2, larger datasets are considered. System had 4GB of RAM. Spam data set was 2.5GB and had 5486496 records. The size of wine dataset was 2.3 GB with 40960000 records and Heart data was 2.2 GB with 51200000 records. Linear SVM failed against all the three datasets as it could not handle the dataset size. SVM in single node Hadoop as expected took considerably more time but managed to produce identical accuracy levels against SVM run on Hadoop Cluster. Map-Reduce could handle datasets of any size without any failures using the help of their functional parts; Mapper and Reducer.

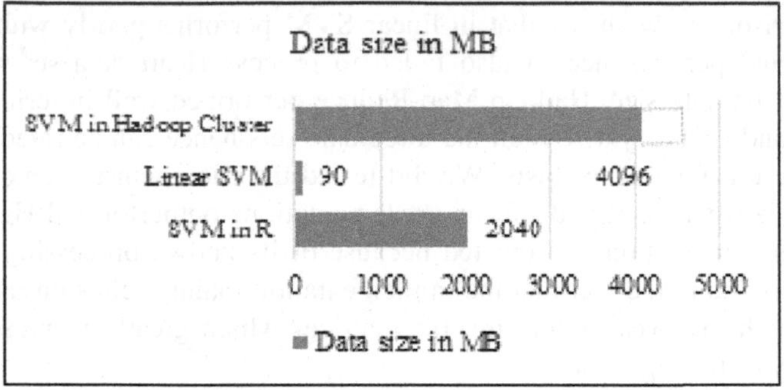

Fig. 2 Comparison of maximum allowable input dataset size for SVM in different frameworks running on a RAM of 4 GB

When comparing the size of the input data with three different frameworks Linear SVM, SVM in R and SVM in Hadoop Cluster, Linear SVM algorithm can run only those datasets within the size of 90 MB. It failed against all datasets larger than 90MB as shown in fig 2. Also in R, the algorithm failed if the size of the dataset is more than half of the corresponding system's RAM size whereas Hadoop implementations could scale itself to handle larger datasets.

Figure 3 plots the average accuracy achieved across all the data sets shown in Table 2. Linear SVM implementation failed to process larger datasets and hence marked in red.

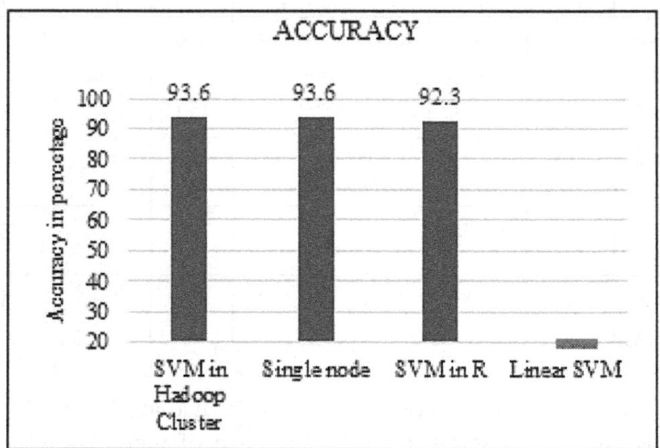

Fig. 3 Average accuracy of SVM in four different frameworks

The time complexity in various SVM implementations is depicted in Figure 4. Off the three implementations that could successfully process the datasets, SVM in single node Hadoop performs considerably well against R with an average execution time complexity clocking at 12.57 minutes as opposed to 30 minutes clocked by R. When the same datasets were presented to the Hadoop cluster implementation (cluster of eight Hadoop nodes), the time complexity sharply dropped bringing the time complexity to almost one eighth of R. Hence it can be concluded that SVM in Hadoop node has a higher efficiency which can be further improved by implementing the same in larger clusters. This validates the fact that parallelizing classification algorithms for larger datasets results in much better performance.

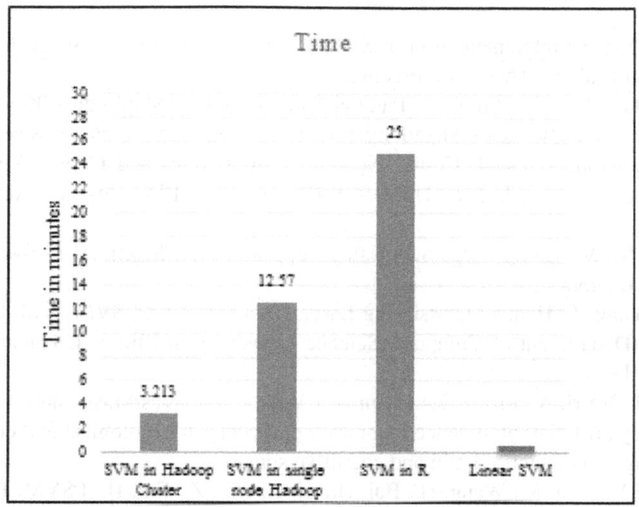

Fig. 4. Average time of execution for SVM implementation in Hadoop cluster, single node Hadoop and SVM in R

5 Conclusion and Future Work

The Map-Reduce programming model has been successfully used for many different purposes. We attribute this success to several reasons. First, the model is easy to use, even for programmers without experience with parallel and distributed systems, since it hides the details of parallelization, fault-tolerance, locality optimization, and load balancing. Second, a large variety of problems are easily expressible in Map-Reduce computation. For example, Map Reduce is used for the generation of data for

Google's production web search service, for sorting, for data mining, for machine learning, and many other systems. Third, the implementation of Map Reduce scales to large clusters of commodity machines. The implementation makes efficient use of the machine resources and therefore is suitable for use on many of the large computational problems encountered.

Map reduce is easy to parallelize and distribute computations and to make such computations fault tolerant. Parallelization techniques are required to process large collection of data. Parallelizing algorithms is a complex task. Using machine learning algorithms like SVM brings simplicity in data processing in distributed environment.

The current implementations can be further improved by using novel optimization techniques such as Sequential Minimal Optimization (SMO), through which time complexity can be greatly improved and higher accuracy can be achieved.

References

1. Freenor, M.: An implementation of SVM for botnet detection in Support Vector Machines and Hadoop: Theory vs. Practice
2. Sun, Z.: Geoffrey Fox Study on Parallel SVM Based on MapReduce Key Laboratory for Computer Network of Shandong Province, Shandong Computer Science Center, Jinan, Shandong, 250014, China 2School of Informatics and Computing, Pervasive Technology Institute, Indiana University Bloomington, Bloomington, Indiana, 47408, USA
3. Srinivas, R.: Managing Large Sets Using Support Vector Machines. University of Nebraska at Lincoln
4. Yu, H., Yang, J., Han, J.: Classifying Large Data Sets Using SVMs with Hierarchical Clusters (Department of Computer Science University of Illinois Urbana-Champaign, IL 61801 USA)
5. Pontil, M., Verri, A.: Properties of Support Vector Machines Massachusetts institute of technology artificial intelligence laboratory and center for biological and computational learning department of brain and cognitive sciences
6. Chang, E.Y., Zhu, K., Wang, H., Bai, H., Li, J., Qiu, Z., Cui, H.: PSVM: Parallelizing Support Vector Machines on Distributed Computers Google Research, Beijing, China
7. Support Vector Machine Tutorial, ung (Ph.D) Dept. of CSIE, CYUT
8. Dean, J., Ghemawat, S.: MapReduce: Simplified Data Processing on Large Clusters Google, Inc.
9. HDFS Under the Hood, Sanjay Radia Sradia Grid Computing, Hadoop
10. Soman, K.P., Loganathan, R., Ajay, V.: Support Vector Machines and Other Kernel Methods by Centre for Excellence in Computational Engineering and Networking. Amrita Vishwa Vidyapeetham, Coimbatore
11. Kiran, M., Kumar, A., Mukherjee, S., Prakash, R.: G Verification and Validation of MapReduce Program Model for Parallel Support Vector Machine Algorithm on Hadoop Cluster
12. Pechyony, D., Shen, L., Jones, R.: Solving Large Scale Linear SVM with DistributedBlock Minimization
13. Bhonde, M., Patil, P.: Efficient Text Classification Model Based on Improved Hypersphere Support Vector Machine with Map Reduce and Hadoop
14. Pechyony, D., Shen, L., Jones, R.: Solving Large Scale Linear SVM with Distributed Block Minimization
15. Yang, H.-C., Dasdan, A., Hsiao, R.-L., Parker, D.S.: Map-Reduce-Merge: Simplified Relational Data Processing on Large Clusters
16. Dean, J., Ghemawat, S.: MapReduce: a flexible data processing tool
17. Chu, C.-T., Kim, S.K., Lin, Y.A., Yu, Y.Y., Bradsky, G., Ng, A.Y., Olukotun, K.: Map-Reduce For Machine Learning on Multicore

Software Analysis Using Cuckoo Search

Praveen Ranjan Srivastava

Abstract. Software analysis includes both Code coverage as well as the Requirements coverage. In code coverage, automatic test sequences are generated from the control flow graph in order to cover all nodes. Over the years, major problem in software testing has been the automation of testing process in order to decrease overall cost of testing. This paper presents a technique for complete software analysis using a metaheuristic optimization technique Cuckoo Search. For this purpose, the concept of Cuckoo search is adopted where search follows quasi-random manner. In requirement coverage, test sequences are generated based on state transition diagram. The optimal solution obtained from the Cuckoo Search shows that it is far efficient than other metaheuristic techniques like Genetic Algorithm and Particle Swarm optimization.

Keywords: Software testing, code coverage, requirements coverage, control flow graph, state transition diagram, xml files.

1 Introduction

Software engineering [1] is a scientific discipline which focuses on cost-effective and high quality development of software systems. Development of a system is controlled by a Software Development Life Cycle (SDLC) model which comprises of Requirement Analysis, Design, Implementation, Testing and Maintenance.

Among all the phases, Software testing [2] [3] is most crucial because it consumes approx. 50% of the total development effort. It also serves as a tool to determine and improve software quality.

In order to implement good software coverage which means covering all the requirements during implementation phase, an optimal set of paths are needed which has to be evaluated based on test data generated automatically [4]. Another

Praveen Ranjan Srivastava
Indian Institute of Management, Rohtak
Haryana, India-124001
e-mail: praveenrsrivastava@gmail.com

El-Sayed M. El-Alfy et al. (eds.), *Advances in Intelligent Informatics*,
Advances in Intelligent Systems and Computing 320, DOI: 10.1007/978-3-319-11218-3_23

factor that lies in the path of software analysis is an optimal generation of test cases because if test data is generated using brute force method, there is a high possibility of generating redundant test cases. Thus, an optimal technique is to be employed. Such techniques which aim at providing optimal or near optimal values are known as metaheuristic techniques [5].

A metaheuristic technique [5] describes a computational method which aims at providing an optimal solution based on some pre-defined quality values. Although many techniques don't guarantee an optimal solution still they implement a stochastically optimal solution. Since test data generation is an optimization problem, metaheuristic techniques can be put to good use. Many metaheuristic techniques [6] [7] [8] [9] [10] have been designed to perform software analysis as well. Typical examples are genetic algorithms [7], particle swarm optimization (PSO) [8], Tabu search [9], Ant colony optimization [10] etc.

In this work, test case generation has been automated by means of a metaheuristic search technique called Cuckoo Search [11], which has not been used for software analysis as yet. The algorithm uses random behaviour of cuckoo bird generally described by Levy's flight and parasitic breeding behaviour of cuckoo bird [12].

The parasitic breeding behaviour of cuckoo bird allows it to lay eggs in the nest of other birds. Some cuckoo birds lay eggs which resemble to great extent eggs of host nest in order to avoid being discovered. If the eggs are discovered by host bird it can destroy its nest with finite probability and lay its eggs on a different location [11] [12] [13].

Division of the paper is as follows: Next section talks about the background of metaheuristic techniques available and flaws associated with them. Third section describes about the Cuckoo Search and the proposed algorithm. Following this proposed methodology is described which is verified in the analysis and conclusion part. Finally the list of references is provided.

2 Background

The test case design holds the central importance for testing, a large number of testing methods designed and developed in last one decade. Existing test case design methods can be differentiated as Search-based method [6], Random test case generation method, Static method [14], and Dynamic method [15].

In software analysis, test data generation too needs to be optimized. This is achieved using one or other metaheuristic techniques. It is common to use Search-Based Software Engineering (SBSE) [6] for test case generation rather than the other existing techniques. It is an application of optimization techniques (OT) in solving software engineering problems. It is most suitable for software coverage because generation of software tests is an indecisive problem because due to the large program's input space, exhaustive enumeration is infeasible [9].

Some Meta-heuristic techniques:

One such technique is proposed [16] using Genetic Algorithm and Tabu Search for Automatic Test Case Generation. Genetic Algorithm is an optimization heuristic technique, uses selection, recombination and mutation operators to generate new candidates (chromosomes) from existing population. The main problem of GA is fitness function and their chromosome length, even increasing the chromosome length has no effect, hence GA based approach is not suitable for strong level software coverage.

Another approach of generating test sequences for software coverage is Particle Swarm Optimization discussed in [17]. It has good exploration abilities, but weak exploitation of local optima. It has weak ability for local search as the convergence rate is slow. This is in contrast with the Cuckoo Search [11] which has quicker convergence rate.

Cuckoo Search in combination with L´evy flights [12] is very efficient for almost all kinds of test problems. It is also a population based algorithm as GA and PSO, but is different in the mode of search as it resembles with the harmony search. It has heavy-tailed step length, which means, it can take steps as large as it wants. Cuckoo search can be applied for one-dimensional as well as for multidimensional problems.

Various optimization problems have been solved using Cuckoo search effectively as it has a quicker convergence rate. The welded beam design problem and spring design optimization is already done by using Cuckoo search optimization [3]. When it employs the L'evy flights, it becomes capable of searching large spaces.

This effective algorithm has been taken for the Software Coverage problem as work has not been done in this aspect yet and results obtained are better than the other metaheuristic techniques discussed above.

3 Cuckoo Search

Cuckoo search is one of the latest nature-based metaheuristic optimization techniques. The detailed description about Cuckoo Search can be get from [11] [12]. Here, a brief description of Cuckoo Search is presented.

A. Overview

The proposed algorithm is based on some species of Cuckoo which lay their eggs in other birds' nest in a quasi-random manner. If the cuckoo's eggs are fit enough then they will survive in the nest by throwing out the host's eggs out of the nest. Otherwise, host mother throw these alien eggs. This process of survival of the fittest is the essence of cuckoo search.

B. L'evy Flight

The efficiency of Cuckoo search is due to the concept called L'evy Flights. L'evy Flights gives scale free search pattern which, in turn, helps to explore a large search space. It uses L'evy distribution in order to find step length. It gives local optimum results for the search mostly but sometimes goes to farther places and hence can explore the search space quicker. So the optimization in cuckoo search is because of the L'evy Flights.

Considering the same large search space for the software coverage, in our case, which has to be covered efficiently, this type of behaviour will be beneficial.

C. **Description of Algorithm**

The basic working of Cuckoo is as follows: Given the fixed no. of nests, the solution is represented by the nests with eggs. It starts with the initially chosen nests by a random walk and lays an egg in it. Then the fitness value of each nest is evaluated. The best nest is chosen and the worst nests are discarded. The best nest will have high quality eggs which will carry over to the next generations. a host can discover an alien egg with a probability pa [0, 1]. In this case, either the eggs are thrown away or the nest is destroyed by the host bird and it can lay eggs on some different location.

Based on this algorithm, the searching of the next best node in a particular iteration can be summarized as:

Next_node = Old_node + α L'evy (λ),* α represents the step-size >0 which is mostly O (1), but can be scaled for mathematical calculations.

To determine the random steps, L'evy distribution is used, mainly for the large steps.

L'evy, $u = t^{-\lambda}$, $(1 < \lambda \leq 3)$, where t represent the iteration number and λ denotes a constant with value ranging between 1 and 3. This power function described the probabilistic nature thereby providing randomness to cuckoo breeding behaviour. Since we have to deal with integer values we cannot use this formula directly. Instead we have to use random function which can be limited to work between 0 and n number of nodes.

4 Proposed Methodology

4.1 Architecture

Above shown architecture (Figure 1) describes the complete flow of events in the system. Application implies the interface where user interacts with the system. Operations start with user providing xml file (denoted by half headed arrow from user to application) which contains CFG for the source code to be tested. Moreover, user is asked to input constant weights through an interface, these weights describe the importance of a node in the CFG according to user. These weights are user driven, and can be justified by him/her only. After taking all the inputs Application first calls PARSER to parse through xml file and prepare the CFG. PARSER is a java code which decodes xml file and converts it into adjacency matrix. XML file format is defined by the application itself. Further, application makes a call to Cuckoo Search mechanism which in turn calculates the in-degree and out-degree of each node. Once it finishes with all the parameters, path calculation procedure is invoked which returns all the possible paths in N number of iterations. N can be given as input by user through an interface or chosen by application to some default value.

Next in the process is fitness value calculation for all the calculated set of paths. The set with maximum fitness is the one shown to the user. Now the test data are generated and fired on the instrumented code. The instrumented code contains information about the CFG mapping. The paths covered by fired test cases and the one obtained from calculations are compared and finally the total coverage percentage is shown.

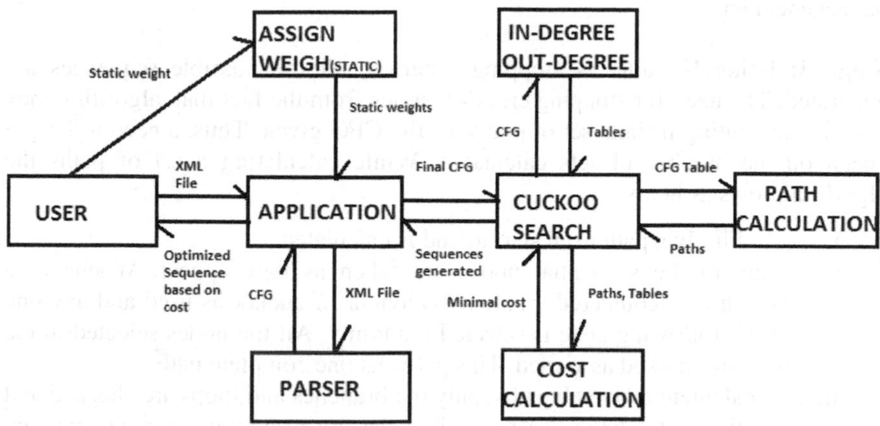

Fig. 1 Basic architecture of software analysis using cuckoo search

4.2 *Algorithm*

First part shows Code Coverage and in next part, Requirement Coverage is discussed.

Part-1

The proposed algorithm is for complete software analysis. It generates a number of test cases and fire them on the paths generated using the maximal fitness criteria. Control Flow Graphs are used as inputs embedded in xml files. These xml files are generated by user according to specified format of the application. Following are the stages involved after taking the inputs:

Stage 1: XML parser parses the graph given as input. From this parsed graph an adjacency matrix is populated which contains all the nodes linked to a particular node thereby giving information about in-degree and out-degree.

Stage 2: In this stage Cyclomatic complexity value is calculated for the CFG using standard formula: **CC = E-N+2**, where E denotes number of edges, N denotes total number of nodes. This formula gives an upper limit on the total number of independent paths available with the given CFG.

Definition of fitness function is given as a product of constant weights provided by user and weights depending on in-degree and out-degree of nodes. Fitness of ith node in a path is given as:

$$\textbf{\textit{Fitness}}_i = -\sum(\textbf{\textit{constant weights}} * (\textbf{\textit{in-degree}}_{i+2} * \textbf{\textit{out-degree}}_i))$$

The value calculated by this formula is an indication for the cost as well as fitness value. More is the value in the positive form more will be the cost which implies less will be the fitness. Hence, negative values of fitness are taken which implies smaller costs. Less will be the cost more will be the fitness of the set of paths generated.

Stage 3: Using N value as stopping criterion, all the possible sequences are generated. The need for stopping criterion arises from the fact that, algorithm may keep on generating infinite set of paths for the CFG given. Thus, a need to keep a check on the number of sets calculated. While calculating a set of paths the algorithm flows as below:

i. Firstly a path from start to end is calculated.
Here all the sequential nodes are taken as they appear. Whenever a branch is encountered, random behaviour of cuckoo is used and any one of the following node is selected randomly. All the nodes selected in the path are marked as visited. This gives us one complete path.

ii. To calculate other sub paths, only the branches and loops are checked and the nodes which aren't visited yet are selection again via random procedure.

iii. Self loops are also handled in the same way with the exception that node containing the self loop can appear at max twice in the same path.

Stage 4: For all the sets generated in N iterations, first the number of paths generated is compared with the Cyclomatic Complexity (CC) measure calculated during the stage 2. One by one total fitness is calculated for each set of paths using the formula in stage 2. This calculation too is done in a random manner as any set can be processed earlier without giving value to its iteration number. Calculations for the invalid sets too are done to check if any of them can give maximum fitness.

Stage 5: Once all the sets have their corresponding fitness value, the one with maximum is selected and taken for further processing.

This completes the path generation from CFG. Following this, sequences for the state diagram are calculated which are included under requirement coverage.

Requirement Coverage: For the requirement coverage part, same algorithm is applied. Only difference lies in place of CFG, state diagrams are used. An adjacency list is created containing all the linked states using the XML parser. Flow of events is same as that in code coverage.

Once the sequence generation is over, next process is test case generation. Algorithm proposed for generation of test cases is Genetic Algorithm [18]. Use of random values is discarded because it will create redundancy by generating same similar set of values thereby triggering same path again and again. Two main aspects of Genetic Algorithm are Mutation and Crossover which are used for generating newer population from older ones.

First, a random set of test cases is generated and fired on the instrumented code which is mapped to the CFG. Test cases which cover maximum number of nodes are taken for new generation and others are discarded based on mutation percentage decided automatically by the code. Based on mutation and crossover criteria new population is generated and fired again. This process is repeated unless all the paths are covered which is shown in Analysis section. All the covered paths are compared with sequences generated from state diagrams to satisfy all the requirements. If all the sequences from requirement part are found in the paths obtained from CFG. The coverage is complete. The nodes which are uncovered after all the test cases are fired are added to uncovered list and consequently percent coverage is calculated. This percentage coverage is shown on the interface. We show the series of steps in the complete process flow in the following section.

5 Analysis and Comparison

To determine the ability and effectiveness of the proposed strategy, the algorithm is implemented in Java and is tested on a widely used benchmark Triangle Classifier Problem taken as case study [23]. The Triangle Classifier estimates the type of a triangle with given edges' length as Scalene, Isosceles, Equilateral, No triangle or invalid input. The automated test data are given as input and results are generated for the specified range of values for edges. A Control Flow Graph (CFG) for the above problem [23] is shown in the Figure 2.

```
var
a,b,c : integer
begin:
if(a<=0|| b<=0|| c<=0)
printf("Side can not be Zero");
else if(a+b<=c|| a+c<=b|| b+c<=a)
printf("Not Triangle/n");
else if(a==b&&b==c)
printf("Equilateral Triangle");
else if(a==b||b==c||c==a)
printf("Isosceles Triangle");
else
printf("Scalene Triangle");
end;
```

Calculation Steps:
The CC value calculated by the code for above shown graph (Figure 3) is 5. Also the table generated for in-degree and out-degree is shown as:

NODE	IN-DEGREE	OUT-DEGREE
0	0	1
1	1	2
2	1	1
3	1	2
4	1	1
5	1	2
6	1	1
7	1	2
8	1	1
9	1	1
10	1	1
11	5	0

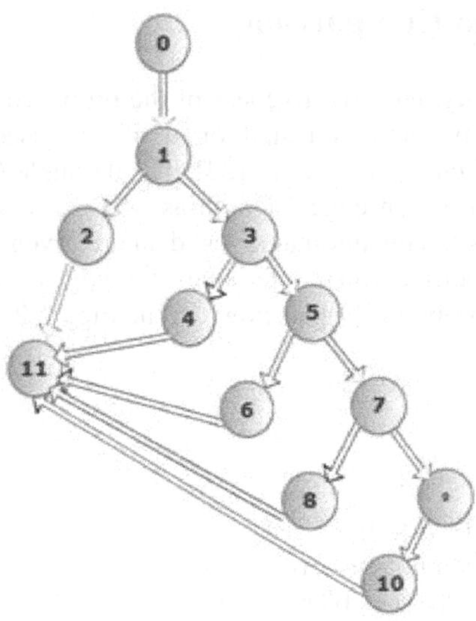

Fig. 2 CFG for triangle classifier problem

The sequences generated are shown as:

Step 1: Find a Path from Start node to end node, name it as main Path
[0, 1, 3, 5, 7, 8, 11]

Step 2: Search for a predicate node in main path or sub path and discover other possible sub paths using constants and varying weights.
[[1, 2, 11], [3, 4, 11], [5, 6, 11], [7, 9, 10, 11]]

Step 3: Form set of all the paths and calculate Cost.
Set 1: [0, 1, 3, 5, 7, 8, 11], [[1, 2, 11], [3, 4, 11], [5, 6, 11], [7, 9, 10, 11]]
Number of Path = 5
Cost = 59

Step 4: Repeat Step 1 to 3 to obtain different set of main path and sub paths.
Set2: [0, 1, 2, 11], [[1, 3, 5, 7, 9, 10, 11], [3, 4, 11], [3, 5, 7, 9, 10, 11], [5, 6, 11], [5, 7, 9, 10, 11], [7, 8, 11], [7, 9, 10, 11]]
Number of paths = 8
Fitness=

Set2: [0, 1, 3, 5, 6, 11], [[1, 2, 11], [3, 4, 11], [5, 7, 9, 10, 11], [7, 8, 11], [7, 9, 10, 11]] :6

Set3: [0, 1, 3, 4, 11], [[1, 2, 11], [3, 5, 6, 11], [5, 6, 11], [5, 7, 8, 11], [7, 8, 11], [7, 9, 10, 11]] :7

Set4: [0, 1, 3, 5, 7, 9, 10, 11], [[1, 2, 11], [3, 4, 11], [5, 6, 11], [7, 8, 11]] :5

Set5: [0, 1, 2, 11], [[1, 3, 5, 7, 8, 11], [3, 4, 11], [3, 5, 6, 11], [5, 6, 11], [5, 7, 8, 11], [7, 8, 11], [7, 9, 10, 11]] :8

Step 5: Select path having maximum fitness whose number of nodes is equal to cyclomatic complexity.

Set 1 and set 5 are selected as they have 5 nodes (Equal to cyclomatic complexity) and having fitness 59.

Here at each branch a decision is taken via the levy flight. Algorithm implements levy flight using random function. Also the number of paths generated for a particular set is given which explains the validation of a set. It can be observed clearly than only Set 4 and Set6 satisfies the CC criteria for independent paths.

Calculations of fitness for a set are shown as:

Assuming user inputs for all edges is 1 for constant weights.

Randomly selected set 3:
$Fitness_3$= -[(1*1*1 + 1*2*1+1*2*5+1*1*1) + (1*2*5+1*1*1) + (1*2*1+1*2*5+1*1*1) + (1*2*5+1*1*1) + (1*2*1+1*2*5+1*1*1) + (1*2*5+1*1*1) + (1*2*1+1*1*5+1*1*1)] = -81

Next random set is 4:
$Fitness_4$= [(1*1*1+1*2*1+1*2*1+1*2*1+1*2*1+1*1*5+1*1*1) +(1*1*5+1*2*1) +(1*2*5+1*2*1) +(1*2*5+1*1*1) +(1*1*5+1*2*1)] = -59

And so on all the calculations are done by selecting sets randomly. It can be observed easily that sets with number of independent paths equal to CC value only give us maximum fitness. Any of these sets can be selected again randomly for further processing with test cases.

Automated test case generation is used here and genetic based approach [15] [16] is applied to make the test case generation process more efficient. This also talks about the fact that generating all the test cases randomly will lead to more redundancy and waste of time. Thus a GA based approach is needed. Initially a number of test cases are generated randomly. Out of the initial population, most efficient test cases are taken and mutated to generate more no. of test cases. At every stage test cases equal to CC are generated and future populations are also generated equal in number to CC.

For the range 0-5:
Randomly generated values of variables are: 1 2 1
Triggered path: 0-1-3-4-11
Randomly generated values: 3 4 5
Triggered path: 0-1-3-5-7-9-10-11
Randomly generated values: 3 3 4
Triggered Path: 0-1-3-5-7-8-11
Randomly generated values: 4 1 0
Triggered Path: 0-1-2-11
Randomly generated values: 0 4 4
Triggered Path: 0-1-2-11

Ideally we can take random generations upto CC only as it defines the independent paths available with the given CFG. Thus, next in the procedure is Genetic mutation and crossover.Assigning probabilities to all the test cases we can generate new population. Here probability assigned to all the test cases is 0.2. They appear on the scale as based on shorted path they cover:

0....0.2...0.4...0.6...0.8...1.0

Next, random numbers are generated in the range [0,1) and values obtained from them are used to decide which all test cases will be a part of new generation and will participate in mutation and crossover that is to be performed.

Random values generated are: 0.1, 0.2, 0.7, 0.4, 0.9.

Bases on these we take (4 1 0), (0 4 4), (3 3 4), (1 2 1), (3 4 5)

Now we use 40% for crossover and rest 60% for mutation.

Crossing over with 4 1 0 and 0 4 4 after 1st index we get (4 4 4) and (0 1 0). While mutation at the same point in rest 3 test cases return (3 4 3) (2 1 1) and (4 5 3)

Firing 4 4 4 gives us a new path i.e. 0-1-3-5-6-11 which is the final path of our CFG. With this all the paths are covered. The number of iterations used is 6 which are shown in Table1.

Table 1 shows the results obtained from testing of the algorithm and compared with the two popular approaches Tabu Search and Genetic Algorithm [16]. It is clear that the proposed strategy is more efficient than the other two as it can cover all paths with less no. of test cases. The Cuckoo Search is giving 100% software coverage while GA is unable to cover node 8 which is for classifying equilateral triangle. Tabu search is also giving 100% coverage but requires more no. of test cases.

Table 1 Number of test cases required to cover all nodes for different set of range

Range	Cuckoo Search	Node Coverage
Range	Tests	Nodes Un-covered
0-5	6	None
0-10	47	None
0-15	62	None
0-50	483	None
0-100	600	None

The results obtained from the Table 1, are plotted on the graph and compared with results obtained in [16]. Graph confirms that the proposed approach is efficient than other meta-heuristics algorithms. It shows the no. of test cases used for finding the complete software coverage within a specified range.

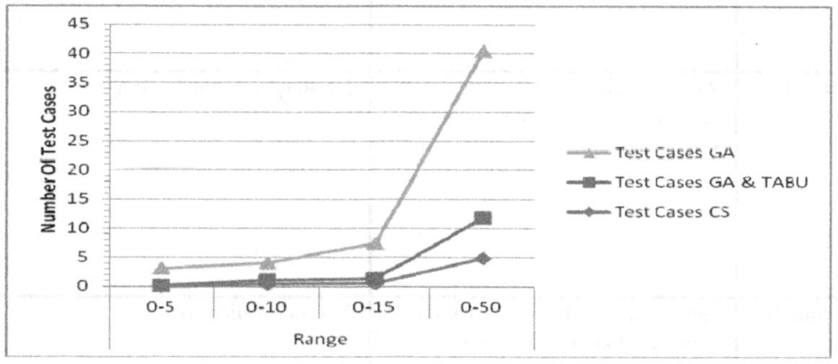

Fig. 3 Comparison graph between CS, TABU and GA

Here Y axis represents number of test cases used to cover all the nodes or maximum number of nodes. X axis represents the Range of input variables.The generation of optimal path given in "Test Sequence Optimization: An Intelligent Approach via Cuckoo Search"[19] is similar to present approach, the differences in both approaches are listed in Table 2.

Comparison study of the performance of CS with GA and Tabu Search leads to the fact that Cuckoo Search working on the concept of randomness [11] is very efficient and proves to be better than other optimization algorithms. The reason behind this is there are very few parameters to be polished in CS than in GA and Tabu search. There is only one parameter pa, denoting the probability of discovering an alien egg by host bird, having a constant value while in Genetic Algorithm population size n is a varying parameter.

Table 2 Comparison of Proposed Approach with Test Sequence Optimization: An Intelligent Approach via Cuckoo Search

Factors	**Proposed Approach**	**Approach Followed in: Test Sequence Optimization: An Intelligent Approach via Cuckoo Search**
Fitness	The strength of each path is measured in terms of fitness.	Only cost is calculated for each path, no fitness involved.
Test Case Generation	Test cases are generated to verify whether all paths covered.	No test case generation.
Requirement Coverage	Along with code coverage, requirement coverage is also done.	No Requirement Coverage.
Path Validation	Generated Paths are validated using Cyclomatic Complexity	No such validation

According to our assumptions, in Cuckoo Search convergence rate [11] is insensitive to the parameters depending on algorithm i.e. we do not have to calibrate these parameters for a specific problem. The use of CC values in the paper further improves the convergence as it restricts the possible number of paths calculated for a particular set. This provides an additional improvement over the test sequence optimization by cuckoo search where only the paths are generated for the CFG and cost of iteration is the only criteria for calculating the best possible set of paths. Instead of cost, fitness is used which is the basic criteria behind implementing the cuckoo search. Finally the test cases are fired based on test data generation discussed earlier.

Moreover, L´evy flights concept [12] is used to represent the search space effectively so that new solutions to be generated are diverse enough. If the search space is large, L´evy flights are unremarkably efficient.

6 Conclusion

This paper presents a technique for complete software coverage using Cuckoo Search. By applying this technique on the problem used, the results are very remarkable. This paper demonstrates the software coverage by generating optimal test sequences for CFG as well as for STT and then comparing it so as to get complete software coverage. The coverage level is verified by applying various test cases generated automatically. By analysing our approach on the case study and statistical data, we conclude that the results incurred using Cuckoo Search is far efficient.

Furthermore, the Cuckoo Search is more generic than any other optimization algorithm. Hence, this potential algorithm can be extended to multiobjective optimization applications. Since a no. of complex problems can be solved by CS efficiently, so lot of work can be done in the future.

References

[1] Sommerville, I.: Software Engineering, 8th edn. Pearson Edition (2009)

[2] Mathur, A.P.: Foundation of Software Testing, 1st edn. Pearson Education (2007)

[3] Beizer, B.: Software Testing Techniques, 2nd edn. Ed. Van Nostrand Reinhold (1990), doi:http://dx.doi.org/10.1002/stvr.4370020406

[4] Korel, B.: Automated software test data generation. IEEE Transactions on Software Engineering 16, 870–879 (1990), doi:10.1109/32.57624, ISSN 8

[5] Yang, X.-S. (ed.): An Introduction with Metaheuristic Applications. John Wiley & Sons (2010)

[6] McMinn, P.: Search-Based Software Test Data Generation: A Survey. Software Testing, Verification and Reliability 14(3), 212–223 (2004)

[7] Lin, J., Yeh, P.: Automatic test data generation for path testing using GAs. Information Sciences, 47–64 (2001) ISSN 131

[8] Kennedy, J., Eberhart, R.: Particle swarm optimization. In: Proceedings of IEEE International Conference on Neural Networks, pp. 1942–1948. IEEE Press, Piscataway (1995)

[9] Geetha Devasena, M.S., Valarmathi, M.L.: Optimized test suite generation using tabu search technique. International Journal of Computational Intelligence Techniques 1(2), 10–14 (2010) ISSN: 0976–0466

[10] Srivastava, P.R., Baby, K.: Automated Software Testing Using Metaheuristic Technique Based on An Ant Colony Optimization. In: International Symposium on Electronic System Design (ISED 2010), pp. 235–240 (2010), doi:10.1109/ISED.2010.52, ISBN: 978-1-4244-8979-4

[11] Yang, X.-S., Deb, S.: Engineering Optimization by Cuckoo Search. Int. J. Mathematical Modeling and Numerical Optimization 1, 330–343 (2010) ISSN 4

[12] Yang, X.-S., Deb, S.: Cuckoo search via lévy flights. In: Proc. World Congress Nature & Biologically Inspired Computing NaBIC, pp. 210–214 (2009)

[13] Kutzelnigg, R.: An Improved Version of Cuckoo Hashing: Average Case Analysis of Construction Cost and Search Operations. In: Proceedings of the 19th International Workshop on Combinatorial Algorithms (IWOCA), pp. 253–266 (2008)

[14] Edvardsson, J.: Proceedings of 2nd Conference on Computer Science and Engineering in Linkoping, pp. 21–28 (1999)

[15] Shen, X., Wang, Q., Wang, P., Zhou, B.: Automatic generation of test case based on GATS algorithm. In: IEEE International Conference on Granular Computing (GRC), pp. 496–500 (2009), doi:10.1109/GRC.2009.5255070, ISBN: 978-1-4244-4830-2

[16] Rathore, A., Bohara, A., Gupta Prashil, R., Lakshmi Prashanth, T.S., Srivastava, P.R.: Application of genetic algorithm and tabu search in software testing. In: COMPUTE 2011 Proceedings of the Fourth Annual ACM Bangalore Conference (2011), doi:10.1145/1980422.1980445, ISBN: 978-1-4503-0750-5

[17] Hassan, R., Cohanim, B., de Wec, O.: A Comparison of particle swarm optimization and the genetic algorithm. Massachusetts Institute of Technology, Cambridge (2004)

[18] Doungsaard, C., Dahal, K., Hossain, A., Suwannasart, T.: An Improved Automatic Test Data Generation from UML State Machine Diagram. In: International Conference on Software Engineering Advances (ICSEA), p. 47 (2007), doi:10.1109/ICSEA.2007.70, ISBN: 0-7695-2937-2

[19] Srivastava, P.R., et al.: Test Sequence Optimization: An intelligent approach via Cuckoo Search. International Journal of Bio-Inspired Computation 4(3), 139–148 (2012)

Memory Based Multiplier Design in Custom and FPGA Implementation

M. Mohamed Asan Basiri and S.K. Noor Mahammad

Abstract. The modern real time applications like signal processing, filtering, etc., demands the high performance multiplier design with fewer look up tables in FPGA implementation. This paper proposes an efficient look up table based multiplier design for ASIC as well as FPGA implementation. In the proposed technique, both the input operands of the multiplier are considered as variables and the proposed *LUT* based multiplier design is compared with other schemes like *LUT* counter, *LUT* of squares and *LUT* of decomposed squares based multiplier designs. The performance results have shown the proposed design achieves better improvement in depth and area compared with existing techniques. The proposed *LUT* based 12×4-bit multiplier achieves an improvement of 34.61% in depth compared to the counter *LUT* based architecture. The 16×16-bit proposed *LUT* based multiplier achieves an improvement factor of 76.84% in the circuit depth over the square *LUT* based multiplication technique using 45 *nm* technology.

1 Introduction

In general, multipliers are the crucial components of cryptography systems like elliptic curve cryptography algorithms [1] and digital signal processing algorithms like fast Fourier transform (FFT) [2]. The multipliers are basic part of multiply accumulate circuit (MAC) [3], which is the heart of digital signal processor, where two input operands are multiplied and the present multiplication result is added to the previous MAC result. Any digital filter application like finite impulse response (FIR) requires a multiplier to perform the operation. The multiplier with lesser depth, lower power requirement and lower area impact the digital system to

M. Mohamed Asan Basiri · S.K. Noor Mahammad
Department of CSE, Indian Institute of Information Technology Design and Manufacturing Kancheepuram
e-mail: {asanbasiri,noorse}@gmail.com

El-Sayed M. El-Alfy et al. (eds.), *Advances in Intelligent Informatics*,
Advances in Intelligent Systems and Computing 320, DOI: 10.1007/978-3-319-11218-3_24

achieve high performance arithmetic. A digital multiplier has two parts, they are partial product generation and addition of partial products. The partial product generation will take $\Theta(1)$ time complexity and depth of the addition of partial products will be varied according to the multiplier structure. In general, carry save multipliers are having two parts in their partial product addition, they are carry save addition tree and final addition. The depth of carry save adder (csa) is $O(1)$. The depth of the carry save addition tree is $O(log_2\, n)$ for Wallace tree [4] multiplier and $O(n)$ for Braun multiplier, where n is the number of bits. The number of carry save stages will be varied in both the above mentioned multipliers. The *sum* and *carry* from the last carry save stage of the multiplier is further added with final adder, where ripple carry adder (RCA) is used with $O(n)$ circuit depth or recursive doubling based carry look ahead adder (CLA) is used with $O(log_2\, n)$ depth.

The n-bit multiplier produces a n number of partial products. If the number of partial products is getting reduced, then the depth of multiplier will be reduced. For this purpose, the higher radix modified Booth algorithm [5] is used. In this case, Booth encoder itself, will cause an increase in multiplier circuit depth. The paper [6] shows the basic complex number multiplier structure, where four multipliers are involved to perform one complex number multiplication. The paper [7] shows the Baugh Wooley based signed multiplication scheme, which is similar to array multiplier with circuit depth $O(n)$. In modern technology, vector processors [8] are playing a major role to achieve data level parallelism (DLP). Here multiple operands are following multiple data paths in the same hardware. So only one instruction is used to perform multiple operations on the vector of data. All the operands are processed in parallel, thus improves the speed of the operation. The twin precision based array multiplier is explained in [9], where the full precision multiplier is used to perform two half precision multiplications with circuit depth of $O(n)$ to achieve DLP. This means that, 8-bit multiplier is used to perform two 4-bit multiplications or one 8-bit multiplication at a time.

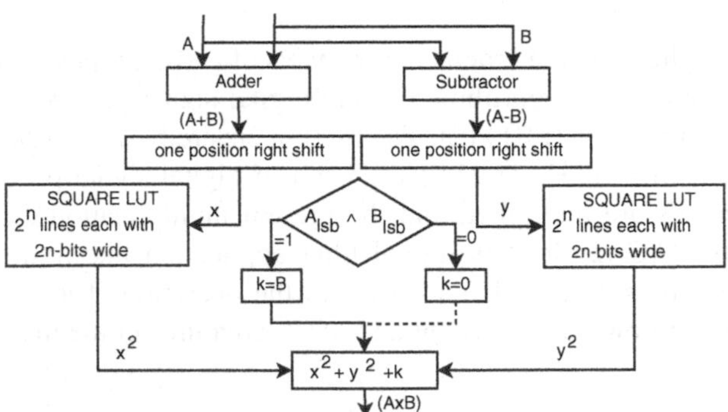

Fig. 1 LUT of squares based multiplier

In the last two decades, look up table (LUT) based multiplier design is becoming an emerging technology in high performance arithmetic circuit design. For example, [10] shows the implementation of LUT based discrete cosine transform, FIR filtering. The papers [11], [12] and [13] explain the LUT based digital signal processing operations. In general LUT based multiplication of two input operands A and B can be found by,

$$x = (A+B)/2 \tag{1}$$

$$y = (A-B)/2 \tag{2}$$

$$AB = x^2 - y^2 + k \tag{3}$$

Here k is equal to 0, if A and B both are even/odd and k is equal to B if any one of A or B is even. The squared values x^2 and y^2 are obtained from the LUT. If the input operands A and B are n-bit wide, then the x and y also will become a n-bit wide. Hence the number of LUT entries will be 2^n and each with $2n$-bits wide. So the size of LUT will be $(2^{n+1}n)$ bits. The major drawback with this approach is the requirement of larger LUT. The Fig. 3 shows the above mentioned LUT based multiplication scheme, where the *xor* value of least significant bit (*lsb*) of A and B (A_{lsb} and B_{lsb}) are used to find the k value. The paper [14] shows that the LUT size reduction techniques for the above mentioned square based multiplication. In the LUT of squares based multiplication, the least n-bits of the squared value ($2n$-bits wide) are periodic with the period $(2^n - 1)$ where the input of the LUT is considered as n-bits wide.

The paper [15] shows the LUT counter based multiplication scheme. If the multiplicand is m-bits wide and multiplier is p-bits wide, then the number of partial products will be p each with m-bits wide, which tends to produce $m+p-1$ columns of partial products. According to [15], t columns of p partial products are sent to the LUT. The number of ones in each column is counted. The counted value is added to the other columns. The outputs from all the $LUTs$ are added to compressor tree. So the size of the LUT will be $2^{pt}q$ bits, where q is the number of bits in the output

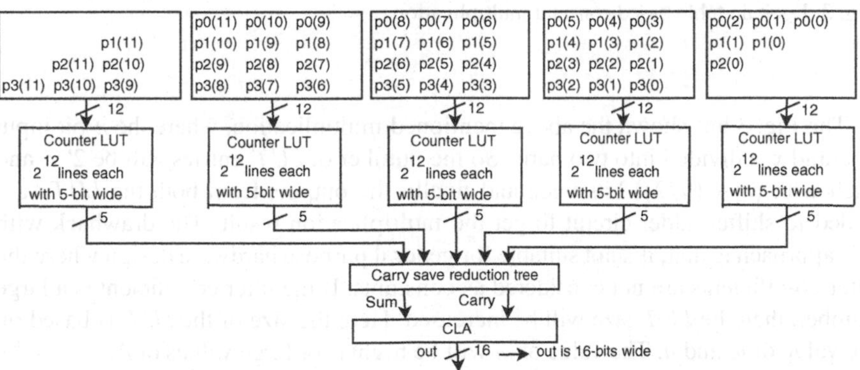

Fig. 2 *LUT* counter based 12x4 multiplier design

from an each LUT. The Fig. 2 shows the LUT counter based 12x4 multiplier design, where 4 partial products each with 12 bits wide are added using five counter $LUTs$. Here 3 columns (12 bits) of partial products are sent to the counter LUT. The maximum number of 1's in each column is 4. So the LUT is having 2^{12} lines each with 5-bits wide. All the three columns should be added in such a way, where the second column (one position left shifted) is added to the third column (two positions left shifted) and the first column. Here the first column is not shifted. If all the three columns are having four 1's, then the shifted addition value three columns will be 11100. So each of the counter LUT is having the 5-bit output line. In the next step, all the output lines of the counter $LUTs$ are added through carry save reduction tree to get the final result. The drawback with this approach is that the LUT size will be increased if the more number of columns are sent to the counter LUT. The number of $LUTs$ will be increased if the less number of columns are sent. This will increase the stages of carry save reduction tree. The paper [16] proposed LUT based FIR filter design, where the filter co-efficients are considered as constant values. If the input signal sample value x is considered as n-bit value, then the number of possible multiplication results between x and the filter co-efficient (A) will be 2^n, where A is assumed as m-bit constant. Hence the LUT will contain 2^n entries, each with $(m+n)$-bits wide.

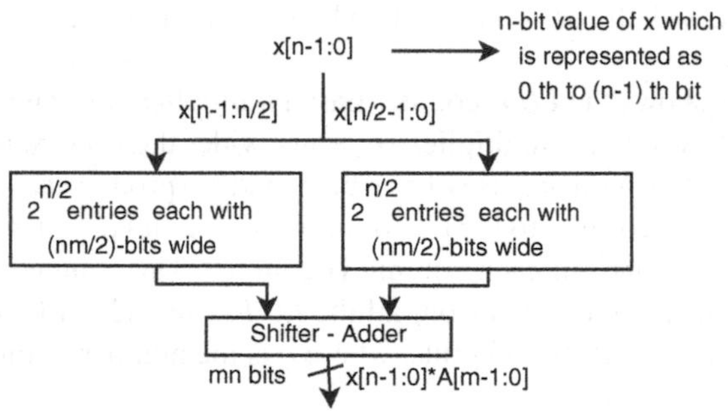

Fig. 3 Look up table based constant multiplication

The Fig. 3 has shown the above mentioned multiplication, where the n-bit input operand x is divided into two parts. So the number of LUT entries will be $2^{n/2}$ and each with $(m+(n/2))$-bits wide and finally the outputs from both the $LUTs$ are added to shifter-adder circuit to get the multiplication result. The drawback with this approach is that, it's not suitable for general purpose hardware design where the filter co-efficients are not considered as constants. If the filter co-efficient is a large number, then the LUT size will be increased. Here the size of the LUT is based on the value of m and n. The value of m will be higher for large values of A.

1.1 Contribution of This Paper

The drawbacks on the existing LUT based multiplication from the above literature gives the motivation towards the proposed LUT based multiplication scheme. Here the size of the LUT and delay, both are considered. The proposed technique uses the smaller $LUTs$, where both the input operands are considered as lesser number of bits. These smaller $LUTs$ are used to build $n \times n$-bit multiplication, where n is considered as larger number bits. This means that, the given input operands are decomposed into smaller multiplications, where each result of the smaller multiplication can be obtained by using the smaller $LUTs$. In the final step, carry save reduction tree is used to add the results from all the smaller LUTs to obtain the desired result. Hence the size of the LUT can be maintained as small. Due to the carry save reduction tree, the depth of the circuit will be in the time bound of $\Theta(log_2 n)$. This proposed LUT based multiplier is used for general purpose multiplication, where both the input operands are considered as n-bit variables. The experimental results show that the proposed LUT based multiplier gives better performance (in terms of delay and area) than the existing LUT based techniques like LUT counter, LUT of squares and LUT of decomposed squares based multipliers. The rest of the paper is organized as, section 2 states the proposed LUT based multiplication scheme. Design modeling, implementation and results are discussed in section 3, followed by a section 4 conclusion.

2 The Proposed LUT Based Multiplication

The proposed 3×3-bit multiplier is shown in Fig. 4, where three $LUTs$ are used to store all the possible results. The 3-bit input operands are A and B. The address line for each LUT will be B. The 8 to 1 multiplexer is used to get all the possible results. The select line for the multiplexer is A. The $LUTs$ (connected to 3, 5 and 7 th input line of multiplexer) are used to store the multiplication result of A by 3, 5 and 7 respectively. Since it is 3-bit multiplication, the line width of each LUT is 6 bits. The number of lines in each LUT will be 8. The data at the 2^i th input of multiplexer can be obtained by left shifting (i times) of A, where $i = 0, 1, 2, ...$ The data at the even number (other than 2^i) input line of the multiplexer can be obtained by left shifting the output of the LUT to the corresponding odd position. Here the left shifting can be implemented by hardwire connection and hence it doesn't require any shifting units. In general, number of $LUTs$ used for $n \times n$ multiplier is $(2^{n-1} - 1)$ and each LUT is consisting of 2^n lines each with $2n$-bits wide. So the size of each LUT will be $(2^{n+1}n)$ bits.

A 2^n to 1 multiplexer is required to design the $n \times n$-bit LUT based multiplier. If n is large, then the requirement of larger multiplexer will cause an overhead of the design. The table 1 shows the comparison between the various LUT based multipliers, which clearly shows about the number of $LUTs$, number of lines per LUT, line width of LUT in bits, the size of each LUT in bits and multiplexer involved in the

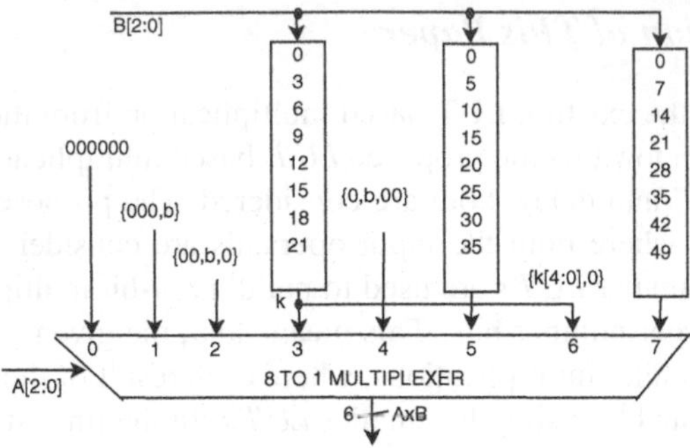

Fig. 4 Proposed *LUT* based 3 × 3-bit multiplier

Table 1 Comparison of *LUT* based multipliers

LUT based multiplier	No. of *LUTs* used	No. of lines per *LUT*	Line width of *LUT* (bits)	Size of each *LUT* (bits)	Multiplexer used
2×2	1	4	4	16	4 to 1
2×3	1	8	5	40	4 to 1
2×4	1	16	6	96	4 to 1
2×5	1	32	7	224	4 to 1
3×3	3	8	6	48	8 to 1
3×4	3	16	7	112	8 to 1
3×5	3	32	8	256	8 to 1
4×4	7	16	8	128	16 to 1
4×5	7	32	9	288	16 to 1
5×5	15	32	10	320	32 to 1

particular multiplier design. These simpler *LUT* based multipliers are used in the larger multiplier design and hence the size of the LUT will be maintained as small.

The Fig. 5 shows the implementation of 16 × 16-bit multiplier using 4 × 4-*LUT* based multipliers. The input operands *A* and *B* both are considered as 16-bits wide. So they can be decomposed into 4 parts, each with 4-bits wide. This is means that, *A* is decomposed into $a0$, $a1$, $a2$ and $a3$. Similarly *B* is decomposed into $b0$, $b1$, $b2$ and $b3$. The multiplication results $a0b0$, $a1b0$, $a2b0$, $a3b0$, $a0b1$, $a1b1$, $a2b1$, $a3b1$, $a0b2$, $a1b2$, $a2b2$, $a3b2$, $a0b3$, $a1b3$, $a2b3$ and $a3b3$ are obtained from 4 × 4-bit *LUTs*. The Fig. 5(a) shows the arrangement of output values obtained from the sixteen 4 × 4 *LUTs*. The Fig. 5(b) shows carry save reduction tree for adding all the partial results $p0$, $p1$, ...$p7$, where *csa* represents the carry save adder with time complexity $O(1)$. So the depth of the carry save reduction tree will be reduced compared to the conventional Wallace structure because the conventional 16 × 16-bit Wallace structure contains 16 partial products. In the proposed design, the number of partial results to the carry save reduction tree is depending on the

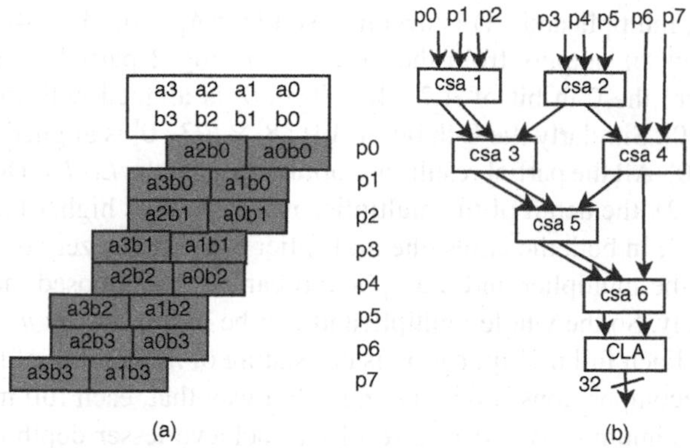

Fig. 5 Proposed *LUT* based 16 × 16-bit multiplier (a) Arrangement of output from sixteen 4 × 4 *LUTs* (b) Carry save reduction tree to add outputs from all the *LUTs*

number of decompositions of the multiplier/multiplicand and the width of the multiplier/multiplicand. In case of 16 × 16-bit multiplier, the number of decompositions of the multiplier/multiplicand is 4 and which is shown in Fig. 5.

In the Fig. 5, both the operands A and B are decomposed by 4. So all the outputs from the *LUTs* are arranged in exactly half of the previous result. In some cases, the decomposition can be done by 2 or 3 or 4 or 5 or any other combination. In this case, the alignment of the output from *LUTs*, tends to give an important role. For example, Fig. 6 shows the various possible decompositions for 10-bit multiplicand (A) which is multiplied by 4-bit multiplier (B). The Fig. 6(a) shows the decomposition of the multiplicand by 3. This means that, multiplicand (A) is decomposed into $A[2:0]$, $A[5:3]$ and $A[9:6]$. The arrangement of outputs from the *LUTs* is having 3 partial results, they are $p0$, $p1$ and $p2$. Here the 0-th bit of the $A[5:3] \times B[3:0]$ should be aligned with 3-rd bit of $A[2:0] \times B[3:0]$. Similarly the 0-th bit of the $A[9:6] \times B[3:0]$ should be aligned with 6-th bit of $A[2:0] \times B[3:0]$.

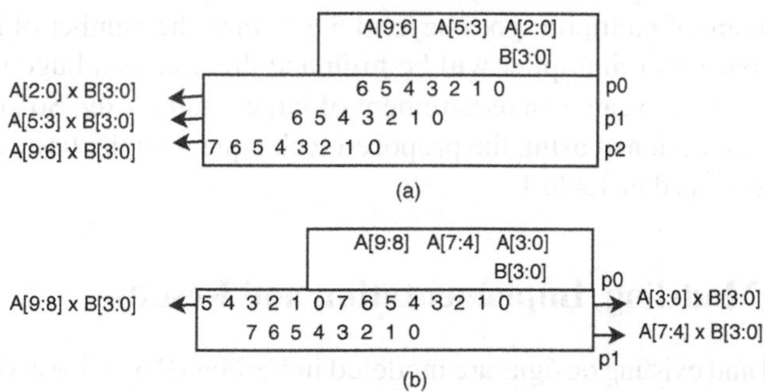

Fig. 6 Proposed *LUT* based 10 × 4-bit multiplication (a) Decomposition with 3 partial results (b) Decomposition with 2 partial results

In Fig. 6(b), multiplicand (A) is decomposed into $A[3:0]$, $A[7:4]$ and $A[9:8]$. The arrangement of outputs from the $LUTs$ is having 2 partial results, they are $p0$ and $p1$. Here the 0-th bit of $A[7:4] \times B[3:0]$ is aligned with the 4-th bit of $A[3:0] \times B[3:0]$. Similarly the 0-th bit of $A[9:8] \times B[3:0]$ is aligned with 4-bit of $A[7:4] \times B[3:0]$. All the partial results are obtained from the $LUTs$. Due to an extra partial result ($p2$), the depth of the multiplier in Fig. $6(a)$ is higher than the multiplier in Fig. $6(b)$. In both the cases, the multiplier is treated as zero decomposition. So in general, the multiplier and multiplicand can be decomposed into $n1$ and $n2$ parts respectively. So the whole multiplicand can be multiplied by $n2$ times of full multiplication. Each full multiplication is consisting of $n1$ number of half multiplications. The decompositions should be in such a way that, each full multiplication should give maximum of two partial results to achieve lesser depth of carry save reduction tree. This will give the way of selecting smaller LUT based multipliers for the decomposition of the required $n \times n$ multiplier.

2.1 *Time Complexity Analysis of Proposed LUT Based Multiplier*

If the number of decompositions of the multiplier is d (where $d > 1$), then the number of partial results to the carry save reduction tree will be kd. Here the proposed logic is considered as the $n \times n$-bit multiplier is made up of basic $LUTs$ mentioned in table 1 only. The value of $k = 2$ if the number of decompositions in multiplicand is more than 1 and $k = 1$ if the number of decompositions in multiplicand is equal to 1. So the depth of the carry save reduction tree will become $O(log_2\ kd)$, which is lesser than the depth of the conventional $n \times n$-bit Wallace tree multiplier ($O(log_2\ n)$). The important thing is to align the multiplication result from the basic $LUTs$ before carry save addition. So the time complexity for the proposed $n \times n$-bit multiplier is $T(n) = T(LUT) + T(mux) + O(log_2\ kd) + O(log_2\ (2n-1))$, where $T(LUT)$ represents the time taken by accessing the LUT, $T(mux)$ represents the depth of the multiplexer used, $O(log_2\ kd)$ shows the depth of the carry save reduction tree and $O(log_2\ (2n-1))$ shows the time complexity for the recursive doubling based carry look ahead adder (CLA) used in the last stage of the multiplier. The basic LUT based multipliers mentioned in that table are varied from 2×2 to 5×5. If the requirement of multiplier goes beyond 5×5, then the number of $LUTs$ used to design the particular multiplier will be high and this causes a huge memory requirement and this also causes a requirement of larger multiplexer. So any $n \times n$-bit multiplier can be designed using the proposed technique with the basic LUT based multipliers mentioned in Table 1.

3 Design Modeling, Implementation and Results

The proposed and existing designs are modeled in Verilog HDL. These Verilog HDL models are simulated and verified using the Xilinx ISE simulator. The timing, area and power analysis of this implementation has been done with Cadence 6.1 ASIC design tool. All the designs are implemented for 45 *nm* technology, where the library *tcbn45gsbwpbc088_ccs.lib* is used for estimating the timing/area/power details.

Table 2 Performance analysis of 12 × 4 multiplier using 45 *nm* technology

	LUT counter based	Proposed *LUT* based
Worst path delay (*ps*)	439.4	287.3
Total area (μm^2)	758.71	537.49
Net power (*nw*)	9520.08	6890.34

The Table. 2 shows the worst path delay, total area and net power comparison between the counter and proposed *LUT* based 12 × 4-bit multiplier using 45 *nm* technology. Here depth of the *LUT* counter based 12 × 4-bit multiplier seems to be higher than the proposed *LUT* based multiplier due to the increase in the depth of the carry save tree which is mentioned in section 1. The proposed *LUT* based 12 × 4-bit multiplier achieves an improvement of 34.61% in depth compared to the counter *LUT* based architecture. The Table 3 shows the details about the *LUT* of decomposed squares for the multipliers 4 × 4, 8 × 8, 12 × 12 and 16 × 16-bit multipliers. Here the conventional square *LUT* is decomposed into two *LUTs* and both are having different address lines as the inputs. The number of lines and the line width (in bits) for both the *LUTs* of decomposed squares are varied according to the each multiplier, these details are mentioned in the Table 3. In Table 3, the input address line is mentioned as *in* and the output from the *LUT* is mentioned as *out*. The proposed *LUT* based 4 × 4, 8 × 8, 12 × 12 and 16 × 16-bit multipliers are compared with the LUT of squares based and decomposed LUT of squares based [14] techniques.

Table 3 Decomposed square *LUT* based multipliers

	First *LUT* of decomposed squares				Second *LUT* of decomposed squares			
	Input address	No. of lines	Line width	Output	Input address	No. of lines	Line width	Output
4 × 4	in[1:0]	4	2	out[1:0]	in[3:0]	16	6	out[7:2]
8 × 8	in[4:0]	32	5	out[4:0]	in[7:0]	256	11	out[15:5]
12 × 12	in[6:0]	128	7	out[6:0]	in[11:0]	4096	17	out[23:7]
16 × 16	in[8:0]	512	9	out[8:0]	in[15:0]	65536	23	out[31:9]

The Fig. 7, 8 and 9 are showing the worst path delay, area and net power comparison between the proposed with an other existing *LUT* based 4 × 4, 8 × 8 and 12 × 12-bit multipliers using 45 *nm* technology respectively. In these cases, the proposed *LUT* based multiplier seems to be better than the other existing *LUT* based multiplication techniques, they are *LUT* of squares based and *LUT* of decomposed squares based multipliers. The Table 4 is showing the worst path delay, total area and net power comparison between the 16×16-bit proposed LUT based multiplier with the other existing LUT based techniques. The 16×16-bit proposed LUT based multiplier achieves an improvement factor of 76.84% in the circuit depth over the square LUT based multiplication technique using 45 *nm* technology and the same achieves an improvement of 95.2% in area reduction over the square LUT based multiplication.

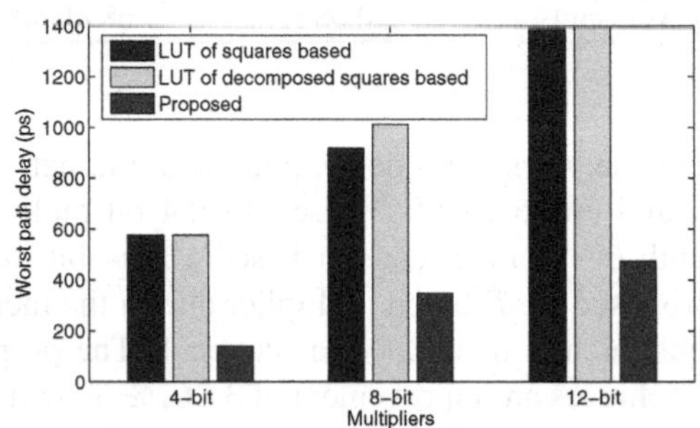

Fig. 7 Worst path delay (*ps*) comparison for 4×4, 8×8 and 12×12-bit multiplier using 45 *nm* lib

The above mentioned whole designs, are implemented with FPGA, where the device *EXC7A100T* from the family of *Artix* 7 with package *CSG*324 is used. The Table 5 shows the comparison between the counter based and proposed LUT based 12×4-bit multiplier in FPGA implementation. The number of $LUTs$ used in LUT counter based and proposed LUT based 12×4-bit multiplier are 294 and 134 respectively. The Table 6 shows the comparison of delay and number of $LUTs$ for 4×4, 8×8, 12×12 and 16×16-bit multipliers using the existing and proposed LUT based multiplier designs. In all the cases, the proposed technique requires a lesser number of $LUTs$ than other techniques.

Table 4 Performance analysis of 16×16 multiplier using 45 *nm* technology

	LUT of squares based	LUT of decomposed squares based	proposed
Worst path delay (*ps*)	2331.8	2228.9	540.9
Total area (μm^2)	95876.4	96668.08	4219.31
Net power (*nw*)	1684823.36	1709193.57	75155.92

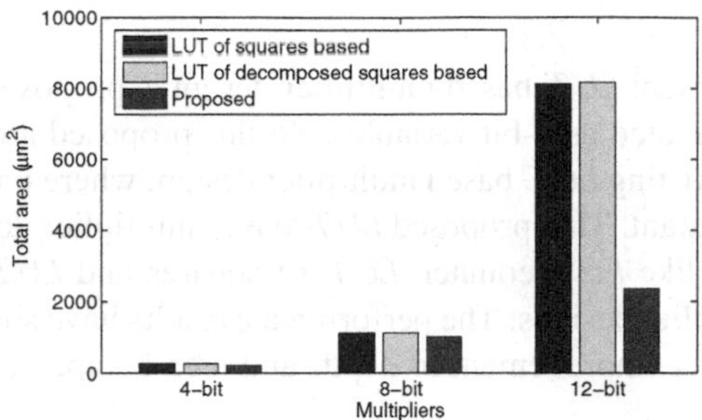

Fig. 8 Total area (μm^2) comparison for 4×4, 8×8 and 12×12-bit multiplier using 45 *nm* lib

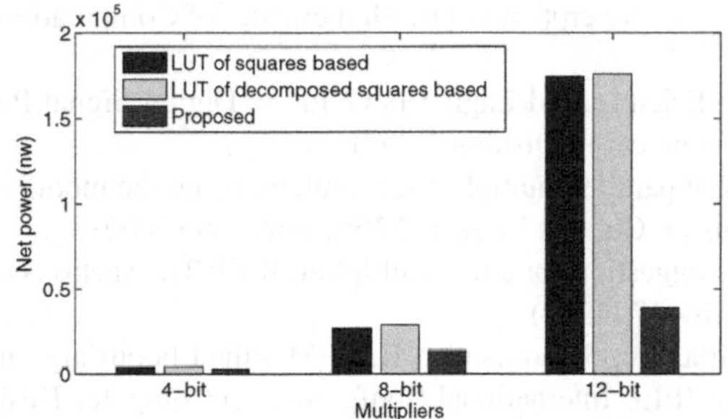

Fig. 9 Net power (*nw*) comparison for 4×4, 8×8 and 12×12-bit multiplier using 45 *nm* lib

Table 5 Comparison of number of $LUTs$ and delay in FPGA implementation for 12×4 multiplier

	Delay (*ns*)	No. of $LUTs$
12×4-bit LUT counter based	6.975	294
12×4-bit proposed LUT based	5.253	134

Table 6 Comparison of number of $LUTs$ and delay (*ns*) in FPGA implementation

	LUT squares		LUT of decomposed squares		Proposed LUT	
	No. of $LUTs$	Delay	No. of $LUTs$	Delay	No. of $LUTs$	Delay
4×4	44	5.047 *ns*	44	5.0447 *ns*	27	2.368 *ns*
8×8	224	10.901 *ns*	229	10.968 *ns*	153	5.581 *ns*
12×12	1670	13.295 *ns*	1670	13.295 *ns*	417	8.160 *ns*
16×16	64483	19.046 *ns*	61679	18.915 *ns*	723	9.378 *ns*

4 Conclusion

In this paper, an efficient *LUT* based multiplier design is proposed, where both the input operands are treated as n-bit variables. So this proposed architecture has the advantage over an existing *LUT* based multiplier design, where one of the operands is considered as constant. This proposed *LUT* based multiplier design is compared with other schemes like *LUT* counter, *LUT* of squares and *LUT* of decomposed squares based multiplier designs. The performance results have shown the proposed design achieves better improvement in depth and area compared with an existing technique.

References

1. Koblitz, N.: Elliptic curve crptosystems. Mathematics of Computation 48(177), 203–209 (1987)
2. Smith, S.W.: The Scientist and Engineers Guide to Digital Signal Processing, pp. 551–566. California Technical Publishing (1997)
3. Elguibaly, F.: A fast parallel multiplier accumulator using the modified Booth algorithm. IEEE Transactions on Circuits Systems 27(9), 902–908 (2000)
4. Wallace, C.S.: A suggestion for a fast multiplier. IEEE Transactions on Electronic Computers EC-13(1), 14–17 (1964)
5. Madrid, P.E., Millar, B., Swartzlander, E.E.: Modified booth algorithm for high radix multiplication. In: IEEE International Conference on Computer Design: VLSI in Computers and Processors, pp. 118–121 (1992)
6. Ismail, R.C., Hussin, R.: High Performance Complex Number Multiplier Using Booth-Wallace Algorithm. In: IEEE International Conference on Semiconductor Electronics, pp. 786–790 (2006)
7. Sjalander, M., Larsson-Edefors, P.: High-Speed and Low-Power Multipliers Using the Baugh-Wooley Algorithm and HPM Reduction Tree. In: IEEE International Conference on Electronics, Circuits and Systems, pp. 33–36 (2008)
8. Kozyrakis, C.E., Patterson, D.A.: Scalable, vector processors for embedded systems, Micro. IEEE Journals and Magazines 23(6), 36–45 (2003)
9. Sjalander, M., Larsson-Edefors, P.: Multiplication acceleration through twin precision. IEEE Transactions on Very Large Scale Integration (VLSI) Systems 17(9), 1233–1246 (2009)
10. Kim, H., Somani, A.K., Tyagi, A.: A Reconfigurable Multifunction Computing Cache Architecture. IEEE Transactions on Very Large Scale Integration (VLSI) Systems 9(4), 509–523 (2001)
11. Guo, J.I., Liu, C.M., Jen, C.W.: The efficient memory-based VLSI array design for DFT and DCT. IEEE Trans. Circuits Syst. II, Analog Digit. Signal Process 39(10), 723–733 (1992)
12. Chiper, D.F.: A systolic array algorithm for an efficient unified memory-based implementation of the inverse discrete cosine transform. In: IEEE International Conf. Image Process, pp. 764–768 (1999)
13. Chiper, D.F., Swamy, M.N.S., Ahmad, M.O., Stouraitis, T.: Systolic algorithms and a memory-based design approach for a unified architecture for the computation of DCT/DST/IDCT/IDST. IEEE Trans. Circuits Syst. I, Reg. Papers 52(6), 1125–1137 (2005)

14. Vinnakota, B.: Implementing Multiplication with Split Read-only Memory. IEEE Transactions on Computers 44(11), 1352–1356 (1995)
15. Mora-Mora, H., Mora-Pascual, J., Sanchez-Romero, J.L., Chamizo, J.M.G.: Partial product reduction by using look-up tables for $M \times N$ multiplier. Integration, the VLSI Journal 41, 557–571 (2008)
16. Meher, P.K.: New Approach to Look-Up-Table Design and Memory-Based Realization of FIR Digital Filter. IEEE Trans. Circuits Syst. II, Regular Papers 57(3), 592–603 (2010)

Application of Simulated Annealing for Inverse Analysis of a Single-Glazed Solar Collector

Ranjan Das

Abstract. This work presents the application of simulated annealing (SA)-based evolutionary optimization algorithm to solve an inverse problem of a single-glazed flat-plate solar collector. For a given configuration, the performance of a solar collector may be expressed by heat loss factor. Four parameters such as air gap spacing, glass cover thickness, thermal conductivity and the emissivity of glass cover have been simultaneously estimated by SA to meet a given heat loss factor distribution. Many possible and nearly unique combinations of the unknowns are observed to satisfy the same requirement, which results in satisfactory reconstruction of the required heat distribution.

Keywords: solar collector, simulated annealing, inverse problem, optimization, heat loss factor.

1 Introduction

The analysis of solar collectors is one of the recent fields of research in the area of renewable energy. For assessing the performance of a solar collector, evaluation of the heat loss factor is one of the important considerations. For analyzing the performance of any system it is found that on many occasions computational approaches are useful in saving money, manpower and time [6]. The available literatures indicate that the heat loss factor of a solar collector depends on many parameters such as air gap spacing, tilt angle, plate and glass emissivity, air properties, etc. [2, 3, 10, 15, 17, 20-22]. The evaluation of thermal field from knowledge of collector dimensions and other parameters is called forward problem [12]. However, when few parameters are unknown and a given thermal distri-

Ranjan Das
School of Mechanical, Materials and Energy Engineering,
Indian Institute of Technology Ropar, Rupnagar, Punjab, India
e-mail: ranjandas@iitrpr.ac.in

El-Sayed M. El-Alfy et al. (eds.), *Advances in Intelligent Informatics*,
Advances in Intelligent Systems and Computing 320, DOI: 10.1007/978-3-319-11218-3_25

bution is to be achieved, then the problem is known as inverse problem [24] and it contributes in the design of any engineering system. For solving inverse problems, an optimization algorithm is necessary for regularization [25].

Details of few good studies can be found which deal with the optimization of solar systems [1, 5, 9, 11, 13, 19 and 23]. It is observed that investigation of inverse problems for solar collectors is very rare and that most studies deal with optimizing few parameters [4]. However, it is well-known that the performance of a solar collector depends upon many factors such as the plate and glass distance, thickness of glass covers, the glass emissivity, environment, etc., and it is important to optimize more factors affecting the heat losses. Therefore, the present work is aimed at estimating important thermo-physical parameters of a single-glazed collector satisfying a given heat loss factor distribution. For optimization, it is well-known that the evolutionary methods perform better than the other methods [18], therefore, in the present work, SA is used. The following section presents description of the problem.

2 Formulation

Let us consider a single-glazed flat-plate solar collector. At steady-state condition and assuming the sky temperature equal to the ambient temperature, the heat flux between the absorber plate to the glass cover and from the glass cover to the ambient air can be respectively expressed as below [17, 20],

$$\ddot{Q}=\left(h_{r,p1}+h_{\text{conv.},p1}\right)\left(T_p-T_1\right)=\left(h_{r,1a}+h_{\text{conv.},1a}\right)\left(T_1-T_a\right) \tag{1}$$

where, the subscripts 1, p and a are used to represent the glass cover, absorber plate and ambient condition, respectively.

For known values of various properties and ambient conditions, the set of equations may be recursively solved to yield the intermediate glass cover temperature (T_1). Then, the total heat loss coefficient (U_t) is calculated. From the view of accuracy and simplicity, the following correlation is used [17, 20],

$$U_t^{-1}=\left[\frac{12.75\left\{\left(T_p-T_1\right)\cos\beta\right\}^{0.264}}{\left(T_p+T_1\right)^{0.46}L^{0.21}}+\frac{\sigma\left(T_p^2+T_1^2\right)\left(T_p+T_1\right)}{1/\varepsilon_p+1/\varepsilon_g-1}\right]^{-1} + \left[h_w+\frac{\sigma\varepsilon_g\left(T_1^4-T_a^4\right)}{\left(T_1-T_a\right)}\right]^{-1}+\frac{t_g}{k_g} \tag{2}$$

where, the glass cover temperature (T_1) can be evaluated by using the following relationship [17, 20],

$$T_1 = T_a + h_w^{-0.38}\left[0.567\varepsilon_p - 0.403 + {T_p}/{429}\right](T_p - T_a) \quad (3)$$

Let us now assume a case where a given heating requirement is needed, which is based upon attainment of a prescribed heat loss factor distribution (say, $\tilde{U}_t$). However, parameters such as air gap spacing, L, glass thickness, t_g, glass emissivity, ε_g and thermal conductivity of glass, k_g are unknown. In order to estimate the feasible values of the unknowns (L, t_g, ε_g and k_g), the following objective function is minimized,

$$F = \left(U_t - \tilde{U}_t\right)^2 \quad (4)$$

where, U_t is calculated using guessed values of the unknowns (L, t_g, ε_g and k_g) which are updated iteratively. Thus, the objective function, F is also updated during the iterative process. For minimizing the objective function, F, SA is used, which is briefly discussed below.

The SA mimics the annealing process of heat treatment. In this, the objective function, F is initially assigned to a higher SA-temperature, T' level and through successive iterations; it is gradually minimized to a lower level. For an initial guess, a random number generator creates another point in the vicinity of the initial guess, and the difference of objective functions between the initial and the new point (e) is computed. If the difference decreases ($e<0$), the new point is accepted. Otherwise (i.e., $e \geq 0$), then it is accepted only if the random number is less than the following probability [7, 8, 14, 16],

$$P = \exp\left(-{e}/{T'}\right) \quad (5)$$

The above operation is known as uphill approach of the SA [7, 8, 14, 16] which eliminates the possibility of converging at local minimum. Further details about SA can be found in [7, 8, 14, 16]. When all random points are checked, the next SA-temperature, T' level is obtained by an exponentially decreasing annealing schedule given below [7, 8, 14, 16],

$$T'_{k+1} = 0.95T'_k \quad (6)$$

where, k is the index for the iteration level. The above cycle is repeated till the termination condition is satisfied. In the present work, the algorithm is terminated either when the objective function is minimized to zero or SA exceeds 150 iterations, which have been found enough for minimizing the objective function, F to a sufficiently minimum value. The following section presents the results and discussions.

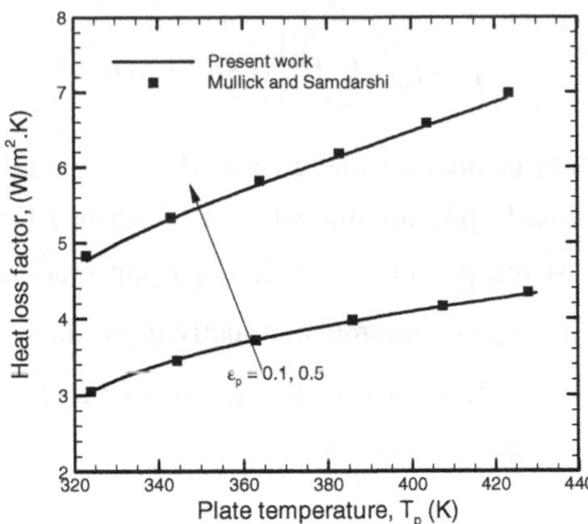

Fig. 1 Validation of the forward method algorithm

3 Results and Discussion

This section presents the results and discussion regarding estimation of unknowns (L, t_g, ε_g, k_g) using SA by inverse analysis. In Fig. 1 validation of the forward method is done with the relevant results available in the literature [17]. The results have been compared for two different values of plate emissivity such as $\varepsilon_p = 0.1, 0.5$. Other relevant parameters are air gap spacing, L = 0.025m, glass thickness, 0.003m, glass emissivity, ε_g = 0.85, thermal conductivity of glass, k_g = 1.02 W/m·K, tilt angle, $\beta = 45^{\circ}$, ambient temperature, T_a = 30°C and wind heat transfer coefficient, h_w= 20W/m^2·K. It is observed from Fig. 1 that the results are in satisfactory agreement with the literature results. Table 1 presents the estimated values of unknowns (L, t_g, ε_g, k_g) for 10 different runs of SA. Since the operation of SA is random in nature, therefore, results may differ from one run to another. The exact heat loss factor field $\tilde{U}_t$ is obtained by solving the forward problem with plate emissivity, $\varepsilon_p = 0.5$. It is observed from Table 1 that the estimated parameters (L, t_g, ε_g, k_g) for all 10 runs remain nearly unique with respect to those considered in the forward method. For run 1 of Table 1, the variation of the objective function, F and estimated parameters (L, t_g, ε_g and k_g) with number of iterations of SA are shown in Figs. 2-6. It is observed from Fig. 2 that the objective function, F gradually reduces with number of iterations, but, it does not become zero even in 150 iterations and minimizes closer to O (10^3). This is found sufficient for yielding satisfactory reconstruction of the heat loss factor field as shown later in Fig. 7. From Figs. 3-6, it is also seen that beyond approximately 75-80 iterations of SA, there is no significant update in the unknowns (L, t_g, ε_g, k_g), thus inferring 75 iterations to be sufficient.

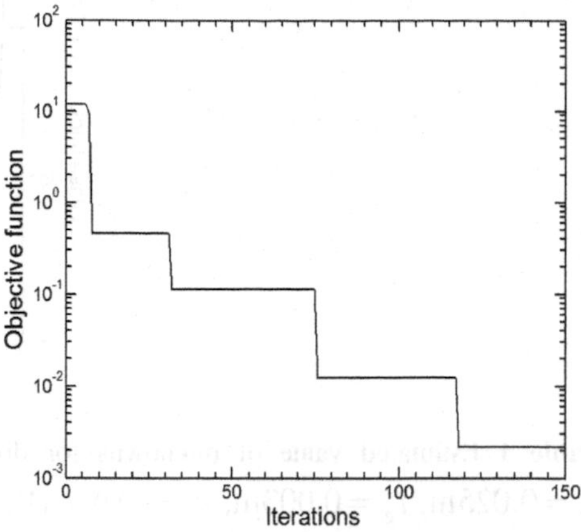

Fig. 2 Iteration variation of the objective function

Figure 7 shows the comparison between exact and reconstructed heat loss factors at different absorber plate temperatures. The exact field is computed using the same set of values mentioned earlier ($L = 2.5\times10^{-2}$ m, $t_g = 3\times10^{-3}$ m, k_g =1.02W/m·K, $\beta = 45^\circ$, ε_p=0.5, T_a= 30°C, h_a =20W/m^2·K and ε_g =0.85). The reconstructed field is computed using estimated values of the unknown parameters (L, t_g, ε_g, k_g). The comparison is only presented for run 1 of Table 1. It is seen that exact and reconstructed fields are in good agreement with each other.

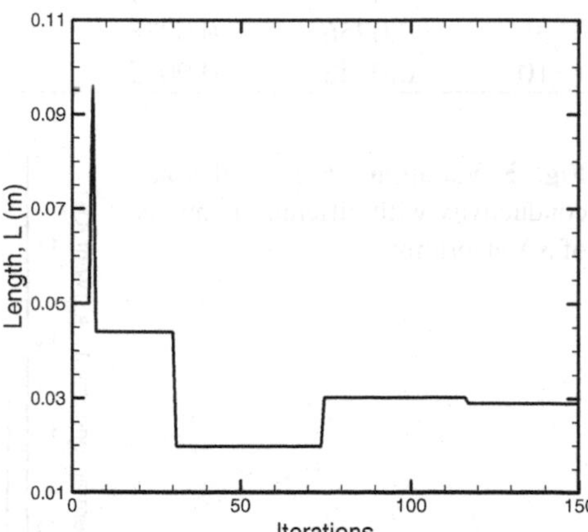

Fig. 3 Variation of the length with different iterations of SA algorithm

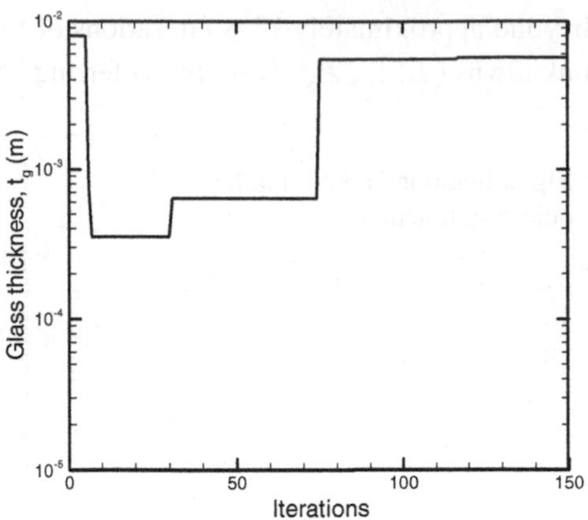

Fig. 4 Variation of the glass thickness with iterations of SA algorithm

Table 1 Estimated value of unknowns for different runs of SA; (Forward method: $L = 0.025$m, $t_g = 0.003$m, k_g =1.02W/m·K, $\beta = 45°$, ε_p = 0.5, T_a = 30°C, h_a =20W/m^2·K and ε_g =0.85)

Run	L (m)	t_g (m)	k_g (W/m·K)	ε_g
1	0.0291	0.0057	1.1280	0.9252
2	0.0245	0.0037	1.2985	0.8447
3	0.0235	0.0021	1.2585	0.8156
4	0.0245	0.0034	1.1737	0.8461
5	0.0249	0.0037	1.2855	0.8492
6	0.0230	0.0021	1.3899	0.8086
7	0.0235	0.0028	1.0043	0.8355
8	0.0254	0.0041	1.1707	0.8646
9	0.0256	0.0038	1.1375	0.8629
10	0.0248	0.0032	1.1348	0.8472

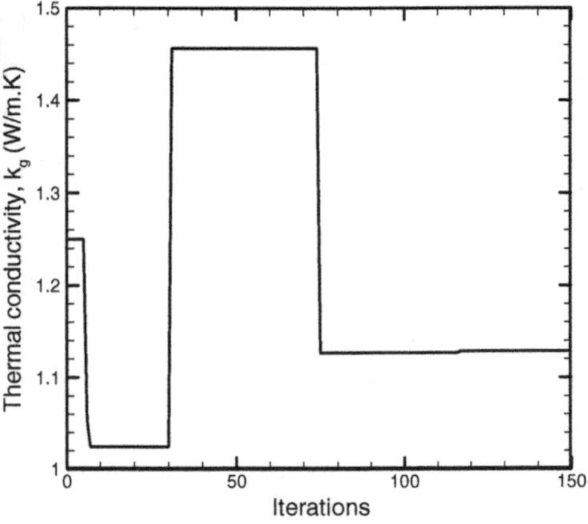

Fig. 5 Variation of glass thermal conductivity with different iterations of SA algorithm

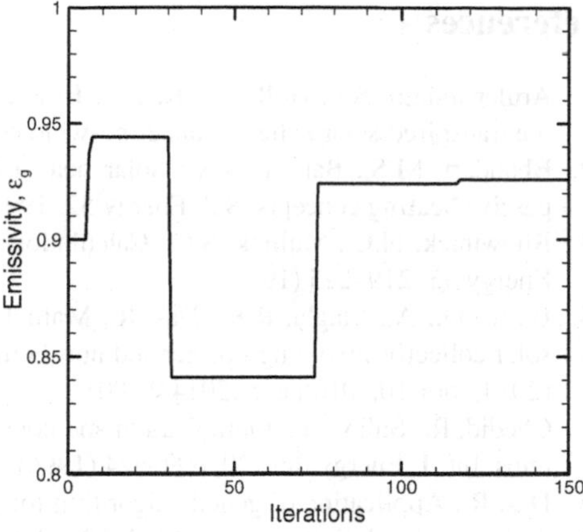

Fig. 6 Variation of glass emissivity with different iterations of SA algorithm

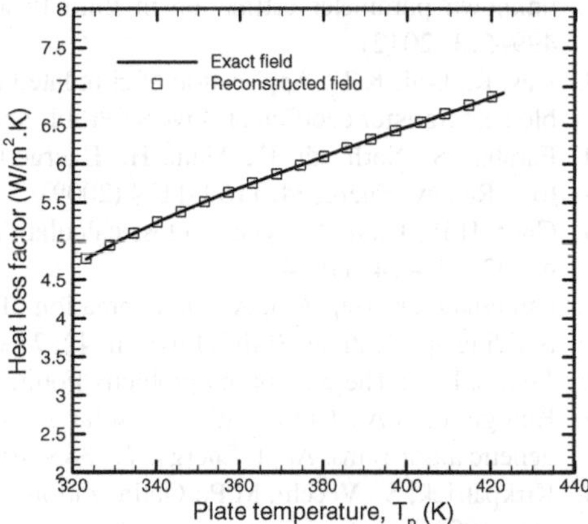

Fig. 7 Comparison of exact and reconstructed fields (Run1 of Table 1)

4 Conclusion

An inverse problem is solved using simulated annealing (SA) algorithm for estimating critical parameters in a single-glazed flat-plate solar collector. Four parameters such as air gap spacing, glass thickness, thermal conductivity and emissivity of glass have been estimated by SA for satisfying a given heat loss factor distribution and variation of the objective function and four parameters have been studied for different iterations. Some results of the forward method have been compared with results available in the relevant literature. The reconstructed heat loss factor field is observed to be in satisfactory agreement with the exact field. It is concluded that SA is suitable in solving the present inverse problem of single-glazed flat-plate solar collector.

References

1. Arulanandam, S.J., Hollands, K.G.T., Brundrett, E.: A CFD heat transfer analysis of the transpired solar collector under no-wind conditions. Sol. Energy 67, 93–100 (1999)
2. Bhandari, M.S., Bansal, N.K.: Solar heat gain factors and heat loss coefficients for passive heating concepts. Sol. Energy 52, 199–208 (1994)
3. Bhowmick, N.C., Mullick, S.C.: Calculation of tubular absorber heat loss factor. Sol. Energy 35, 219–225 (1985)
4. Bhowmik, A., Singla, R.K., Das, R., Mallick, A., Repaka, R.: Inverse modeling of a solar collector involving Fourier and non-Fourier heat conduction. Appl. Math. Model. (2014), doi:10.1016/j.apm.2014.04.001
5. Chedid, R., Saliba, Y.: Optimization and control of autonomous renewable energy systems. Int. J. Energy Res. 20, 609–624 (1996)
6. Das, R.: Application of genetic algorithm for unknown parameter estimations in cylindrical fin. Appl. Soft Comput. 12, 3369–3378 (2012)
7. Das, R.: A simulated annealing-based inverse computational fluid dynamics model for unknown parameter estimation in fluid flow problem. Int. J. Comput. Fluid D. 26, 499–513 (2012)
8. Das, R., Ooi, K.T.: Application of simulated annealing in a rectangular fin with variable heat transfer coefficient. Inverse Probl. Sci. Eng. 21, 1352–1367 (2013)
9. Farahat, S., Sarhaddi, F., Ajam, H.: Exergetic optimization of flat plate solar collectors, Renew. Energ. 34, 1169–1174 (2009)
10. Garg, H.P., Datta, G.: The top loss calculation of flat-plate solar collectors. Sol. Energy. 32, 141–143 (1984)
11. Gunerhan, H., Hepbasli, A.: Determination of optimum tilt angle of solar collectors for building applications. Build. Environ. 42, 779–783 (2007)
12. Jaynes, E.T.: The well-posed problem. Foundations of Physics 3, 477–492 (1973)
13. Kalogirou, S.A.: Optimization of solar systems using artificial neural-networks and genetic algorithms. Appl. Energy. 77, 383–405 (2004)
14. Kirkpatrick, S., Vecchi, M.P.: Optimization by simulated annealing. Science 220, 671–680 (1983)
15. Klein, S.A.: Calculation of flat-plate collector loss coefficients. Sol. Energy 17, 79–80 (1975)
16. Metropolis, N., Rosenbluth, A.W., Rosenbluth, M.N., Teller, A.H.: Equation of state calculations by fast computing machines. J. Chem. Phys. 21, 1087–1092 (1953)
17. Mullick, S.C., Samdarshi, S.K.: An improved technique for computing the top heat loss factor of a flat-plate collector with a single glazing. J. Sol. Energy Eng.-T. ASME 110, 262–267 (1988)
18. Pastorino, M.: Stochastic optimization methods applied to microwave imaging: A review. IEEE T. Antenn. Propag. 55, 538–548 (2007)
19. Shariah, A., Al-Akhras, M.A., Al-Omari, I.A.: Optimizing the tilt angle of solar collectors. Renew. Energ. 26, 587–598 (2002)
20. Samdarshi, S.K., Mullick, S.C.: Analysis of the top heat loss factor of flat plate solar collectors with single and double glazing. Int. J. Energy Res. 14, 975–990 (1990)
21. Samdarshi, S.K., Mullick, S.C.: An analytical equation for top heat loss factor of flat plate solar collector with double glazing. J. Sol. Energy Eng.-T. ASME 113, 117–122 (1991)

22. Samdarshi, S.K., Mullick, S.C.: Generalized analytical equation for the top heat loss factor of a flat-plate solar collector with N glass covers. J. Sol. Energy Eng.-T. ASME 116, 43–46 (1994)
23. Shi, J.H., Zhu, X.J., Cao, G.Y.: Design and techno-economical optimization for stand-alone hybrid power systems with multi-objective evolutionary algorithms. Int. J. Energy Res. 31, 315–328 (2007)
24. Singla, R.K., Das, R.: Application of Adomian decomposition method and inverse solution for a fin with variable thermal conductivity and heat generation. Int. J. Heat Mass Transfer. 66, 496–506 (2013)
25. Tsirakos, D., Baltzopoulos, V., Bartlett, R.: Inverse optimization: functional and physiological considerations related to the force-sharing problem. Crit. Rev. Biomed. Eng. 25, 371–407 (1997)

A Novel Image Encryption and Authentication Scheme Using Chaotic Maps

Amitesh Singh Rajput and Mansi Sharma

Abstract. The paper presents an amalgam approach for image encryption and authentication. An ideal image cipher should be such that any adversary cannot modify the image and if any modifications are made, can be detected. The proposed scheme is novel and presents a unique approach to provide two level security to the image. Hashing and two chaotic maps are used in the algorithm where hash of the plain image is computed and the image is encrypted using key dependent masking and diffusion techniques. Initial key length is 132-bits which is extended to 148-bits. Performance and security analysis show that the proposed scheme is secure against different types of attacks and can be adopted for real time applications.

1 Introduction

Storage of data in open networks and information exchange across the internet has created an environment in which illegal users can obtain important information. Images are the most important utility of our life and are used in many applications. In open environments, focusing on processing and transmission of digital images, there are several security problems associated with them. Hence, reliable, fast and robust image security techniques are required to store and transmit digital images. Due to large size of digital images, traditional data encryption algorithms cannot be directly applied to them. An important property of digital images is that they are less sensitive as compared to text data and if some pixels values of the image are modified, the modification cannot be identified by human eye. Therefore, a

Amitesh Singh Rajput
Department of CSE, Sagar Institute of Science & Technology, Bhopal, India
e-mail: amiteshrajput@gmail.com

Mansi Sharma
School of IT, Rajiv Gandhi Proudyogiki Vishwavidalaya, Bhopal, India
e-mail: mansisharma1245@gmail.com

El-Sayed M. El-Alfy et al. (eds.), *Advances in Intelligent Informatics*,
Advances in Intelligent Systems and Computing 320, DOI: 10.1007/978-3-319-11218-3_26

small degradation in the digital image is acceptable which is not possible with text data but this is also an advantage from the attacker's point of view. If an adversary access the image and make some little modifications, the image will be acceptable.

To prevent image from unauthorized access and modifications, strong image encryption and authentication techniques are essential. Image encryption techniques transform the pixel values of the plain image and convert the plain image to another one that is hard to understand. Most applications require confidentiality as well as integrity. Therefore, it is important to develop techniques that are capable of providing both the services simultaneously. To provide authentication, hash functions should be preferred. Using hash functions, integrity of the image is checked and if any modifications have been made, can be determined. Integrity of the received image is checked by comparing the received hash with the one calculated at the receiving end. The general approach to obtain this is to combine encryption and authentication schemes in some way. During the past some years, some image encryption and authentication schemes have been proposed [1, 2, 6].

Another technique for securing digital images is based on the use of chaotic functions. Chaotic techniques provide a good combination of complexity which results into high security, speed and reasonable computational overheads. In order to overcome image encryption problems, a number of different image encryption schemes have been proposed based on chaotic maps [2-5, 7-10, 13-18]. The chaos-based encryption was first proposed in 1989 [9], since then, many researchers have proposed and analyzed a lot of chaos-based encryption algorithms. Two chaotic logistic maps are used in [10]. Initial conditions of both the chaotic maps are calculated using the secret key. Eight different operations are used and at an instance one of them is used to encrypt the pixels of the image. Hyper chaotic functions are used to encrypt the image in [14]. The scheme is divided into two parts. In the first part, total shuffling of the image pixels takes place whereas the shuffled image is then encrypted using hyper chaotic function in the second part. Cryptanalysis of the scheme is presented by [12]. It has been shown that the reuse of the key stream more than once makes the cryptosystem weak against chosen cipher text, chosen plain text attacks. According to [12], three couples of plain text/cipher text were enough to break the cryptosystem in a chosen cipher text and chosen plain text attacks scenario. Two solutions are also projected for changing the key stream. In [15], chaotic image encryption algorithm where key stream is generated by nonlinear Chebyshev function is proposed. The secret keys in encryption process are dependent on each other and provide good correlation results. Diffusion and substitution based gray image encryption scheme is proposed in [11]. The scheme consists of substitution-diffusion techniques. The algorithm is based on the number of iterations (rounds).

To provide more security to the image, authentication and encryption techniques can be pooled. An authenticated image encryption scheme based on memory cellular automata is proposed in [2]. The image is divided into blocks and hash of each block is generated. The encryption is conceded with the linear memory cellular automata (LMCA) to diffuse pixel values of the image. The hash values are used to verify integrity of the image. The scheme proposed in [1] uses digital signature to verify integrity of the received image. The digital signature of the plain image is computed and added to the encoded version of the original image. Image encoding is done by an appropriate error control code, such as Bose Chaudhuri Hochquenghem (BCH) code. A fast image encryption and authentication scheme is proposed in [6]. Message authentication code (MAC) is used to validate integrity of the image. MAC of the plain image is embedded and the entire image is then encrypted using a simple masking function. A novel scheme for image encryption and authentication is proposed in this paper. The rest of the paper is classified as follows: in Section 2, the detailed algorithm of the proposed image cipher is discussed. In Section 3, security and performance analysis of the proposed algorithm is provided. Finally, section 4 concludes the paper.

2 Proposed Approach

The proposed scheme encrypts the plain image and validates integrity of the image at the receiving side. Along with encryption, authentication is also an important aspect for image security and the proposed scheme provides image encryption as well as authentication. The proposed scheme consists of two phases: image encryption phase and, decryption and integrity validation phase.

Chaotic techniques provide a good combination of complexity which results into high security, speed and reasonable computational overheads. Logistic maps have very complicated dynamic behavior. In the proposed scheme, we keep the value of system parameter of both the logistic maps to be constant (3.9999) which corresponds to highly chaotic behavior and we used them to generate initial key stream and provide randomness to the image being encrypted. The two chaotic maps used in the algorithm are –

$$x_{n+1} = 3.9999x_n(1 - x_n) \tag{1}$$

$$y_{n+1} = 3.9999y_n(1 - y_n) \tag{2}$$

A 148-bit key is used in the algorithm. The initial values of both the chaotic maps are computed using the scheme presented by [10] with some modifications. Here, we use 36-bits of the secret key to generate initial values for the first chaotic map. Initial values for the second chaotic map are generated from 24-bits of the secret key. The chaotic maps are executed multiple times, only their initial values are different depending on the key bits selected. The chaotic maps are executed and chaotic values are generated. Initial values of the chaotic maps are different for different phases. Use of diverse values of the chaotic maps in different phases makes the cryptosystem robust such that any adversary cannot obtain meaningful information from the cipher image. The sub-phases of the proposed scheme are explained below.

2.1 Image Encryption Phase

Input: Plain image with initial key
Output: Cipher image with extended key

In the proposed scheme, 512-bit hash of the original image is generated using a standard hash generation technique and generated 512 hash bits are converted into 64 bytes such that each byte represents a decimal number. In this phase, the plain image and initial key are given as input, and encrypted image and extended key are obtained. The initial key length is 132-bits and 16-bits are appended to the key showing the difference value of the hash values and pixels of the plain image. This can be explained as follows.

1. Generate 512-bit hash of the plain image and convert the generated 512-bits into 64 bytes (such that each byte represents a decimal number).
2. Add all the 64 decimal numbers obtained in the last step and store the sum in S_1.
3. Iterate first and second chaotic maps (64 times) and add all the resulting 64 pixel values. Store the sum in S_2, and obtain the difference between S_1 and S_2. Convert the difference value into binary form and append the bits at the end of the initial key.

Now, the plain image is encrypted using masking and diffusion techniques. The encryption phase consists of three sub-phases: masking, horizontal diffusion and vertical diffusion. The sub-phases of the scheme are explained below:

2.1.1 Masking Process

In this step, each pixel of the input image is replaced with a new pixel obtained by mixing the properties of the current pixel with the previous pixel and the key stream. The current pixel is added to the previous masked pixel and then the entire sum is XORed with the chaotic key stream generated from the first chaotic map using different set of initial values. Each pixel is masked by the following encryption function –

$$E(i) = M(i) \oplus \{[P(i) + E(i-1)] mod\ L\} \quad (3)$$

Where, P(i) is the plain image pixel value, M(i) is the chaotic key stream generated by executing first chaotic map, E(i) is the encrypted pixel output and L is the gray level. The chaotic key stream is derived from the first chaotic map which is dependent on the key. In this way, all the pixels of the image are masked resulting a masked image.

2.1.2 Horizontal Diffusion

Diffusion refers to the process of rearranging or spreading out the pixels in the image so that redundancy in the image is spread out over the complete cipher image. The masked image obtained in the previous phase is given as input in this segment. In horizontal diffusion, we mix the properties of horizontally adjacent pixels of the plain image and XOR them concurrently with the resulting values of the first chaotic map. In an image cryptographic system, the process of diffusion removes the possibility of differential attacks which is done by comparing the pair of plain and cipher images.

2.1.3 Vertical Diffusion

Here, the image obtained in last step is given as input. Like horizontal diffusion, here we mix the properties of vertically adjacent pixels of the input image and XOR them concurrently with the same resulting values of the first chaotic map as of horizontal diffusion respectively. Complete cipher image is obtained as output of this phase.

2.2 Image Decryption and Integrity Validation Phase

Input: Cipher image
Output: Detecting whether the received image is tampered or not.

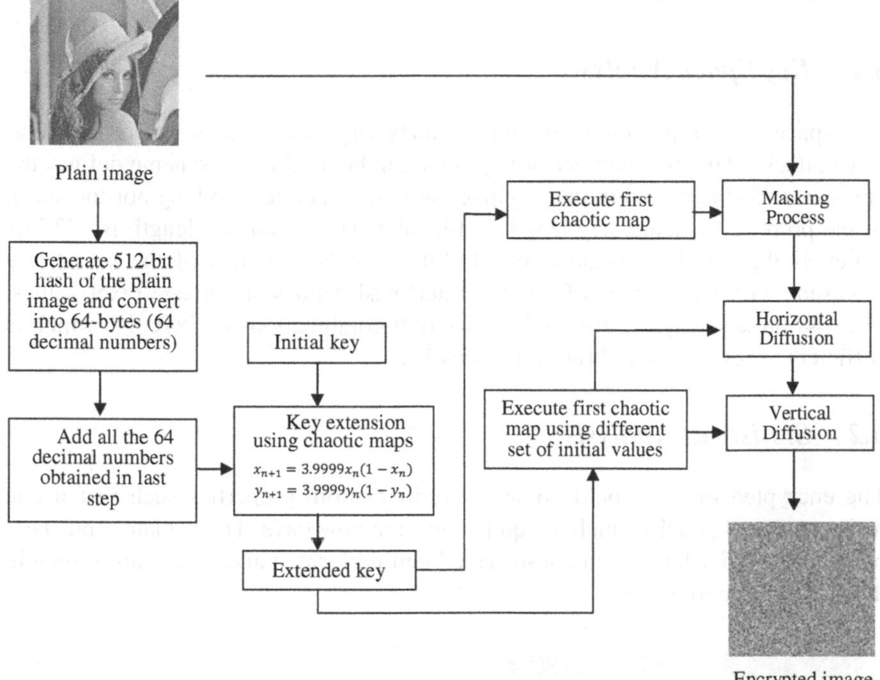

Fig. 1 Block diagram of the proposed scheme

The reverse process of encryption is applied to the cipher image to get the recovered plain image. Integrity of the received image is tested out here. The steps of this phase are explained below.

1. Extract the decimal value from the last 16 bits of the key (say D_1).
2. Generate 512-bit hash of the recovered plain image and convert these 512-bits into 64 bytes such that each byte represents a decimal number and add all the 64 decimal numbers obtained. Store the sum in S_3.

3. Execute first and second chaotic maps (64 times), and add all the resulting pixel values and store the sum in S_4 and obtain the difference value (D_2) between S_3 and S_4.

Compare D_1 and D_2 for integrity. Use the same hash generation algorithm as used at the sender side. If D_1 and D_2 are equal, then integrity of the recovered image is validated.

3 Performance and Security Analysis

An ideal image cryptographic scheme should resist all kinds of attacks. In this section, security and performance of the proposed image cryptographic scheme are discussed.

3.1 Key Space Analysis

Key space of a cryptosystem should be fairly large such that it can resist brute-force attacks. The total number of keys that can be used in the scheme defines the key space. The key used in image cipher should be neither too long nor too short. In the proposed scheme, key length is 148-bits. The initial key length is 132-bits which is then further extended to 148-bits. The last 16 bits of the secret key represents difference value of the generated hash values and image pixel values. The 148-bit key length causes 2^148 different combinations (3.5×10^44) which is sufficient to resist various brute-force attacks.

3.2 Statistical Analysis

The encrypted image should possess certain random properties such that it can resist statistical attacks which are quite common now days. To evaluate robustness of the proposed scheme, some tests have been performed and results are shown in the next sub sections.

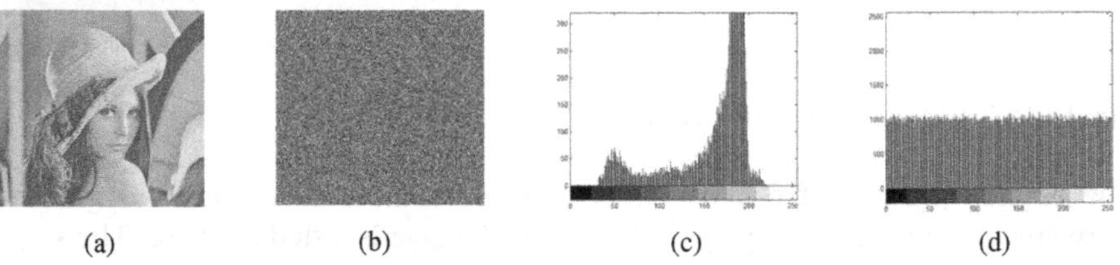

Fig 2. Histogram analysis: Frames (a) and (b) show the plain image 'Lena' and its corresponding cipher image respectively. Frames (c) and (d) show histogram of images shown in (a) and (b).

3.2.1 Histogram

Histogram of the encrypted image should be distributed uniformly showing that the pixels of the encrypted image are fairly distributed. We performed this test on the plain image 'Lena'. The histogram of the plain image and its corresponding encrypted image are shown in figure 2. The histogram of the encrypted image shows fair distribution of pixels which does not reveal any information to the adversary.

3.2.2 Correlation of Adjacent Pixels

Neighboring pixels are highly correlated in images. Correlated pixels can be horizontally, vertically or diagonally adjacent. For an ideal image cipher, the less correlation of two adjacent pixels is the stronger ability of resisting statistical attacks. The proposed scheme exhibits this property well. In this section, the correlations between horizontally, vertically and diagonally adjacent pixels of the plain and encrypted image are discussed. The correlation between adjacent pixels is calculated by the following equations:

$$E(x) = \frac{1}{N}\sum_{i=1}^{N} x_i \tag{4}$$

$$D(x) = \frac{1}{N}\sum_{i=1}^{N}(x_i - E(x))^2 \tag{5}$$

$$cov(x,y) = \frac{1}{N}\sum_{i=1}^{N}(x_i - E(x))(y_i - E(y)) \tag{6}$$

$$r_{xy} = \frac{cov(x,y)}{\sqrt{D(x)}\times\sqrt{D(y)}} \tag{7}$$

Where x and y are gray-scale values of two adjacent pixels of the image, cov(x, y) is the covariance, D(x) is the variance and E(x) is the mean. In order to test correlation between two adjacent pixels, we selected 3600 random pairs of adjacent pixels (horizontal, vertical and diagonal) of the original and encrypted image. Table 1 illustrates the comparison between the proposed scheme and scheme presented by [15] and shows that the proposed scheme exhibits good correlation results than [15].

Table 1 Correlation coefficients of two adjacent pixels of the plain image 'Lena' and corresponding encrypted image

	Plain image (Lena)	Encrypted image by the proposed scheme	Encrypted image by [15]
Horizontal	0.9189	0.0006	-0.09736
Vertical	0.9028	-0.0057	0.04844
Diagonal	0.9266	-0.0077	-0.07068

3.3 Differential Analysis

The changes in cipher image should be significant such that known-plaintext attack and chosen-plaintext attack can be avoided. Two common measures: number of pixels change rate (NPCR) and unified average changing intensity (UACI) are calculated to test how one pixel change in the plain image affects the corresponding cipher image. The percentage of different pixels in two encrypted images is calculated by NPCR. On the other hand, the average of intensity – difference between the pixels of two cipher images is measured by UACI. The NPCR and UACI are calculated by the following equations:

$$NPCR = \frac{\sum_{ij} D(i,j)}{W \times H} \times 100 \quad (8)$$

$$UACI = \frac{1}{W \times H}\left[\sum_{ij} \frac{C_1(i,j) - C_2(i,j)}{255}\right] \times 100 \quad (9)$$

Where C_1 and C_2 are two encrypted images, whose corresponding original images have only one-pixel value difference. D (i, j) is a two dimensional array of the same size as of C_1 or C_2. D (i, j) is derived from C_1 and C_2. If C_1 and C_2 are equal, D (i, j) =0, otherwise D (i, j) =1. W and H are width and height of the image. We applied the proposed scheme over a number of images of the USC-SIPI image database and find that the proposed scheme exhibits NPCR $\geq$ 99%. Hence, the scheme exhibits good properties to resist differential attacks. The NPCR and UACI test results are shown in table 2 and 3 respectively.

Table 2 NPCR

Image name	NPCR (%)
Boat	99.0435
Lake	99.4996
Lena	99.5011
Pepper	99.5941

Table 3 UACI

Image name	UACI (%)
Boat	33.2626
Lake	33.0106
Lena	33.0998
Pepper	33.1628

3.4 Performance Analysis

For an ideal image cipher, encryption/decryption time and computational complexity of the algorithm is an important aspect. Images are of large size as compared to the text, hence chaotic techniques are used. Due to large size of the images, time is an important aspect to be considered when developing an image cipher. In the previous sections, we have already shown that the proposed scheme is good for resisting different kind of attacks. The proposed scheme consists of iterating chaotic maps and the operations involved in the scheme consists of simple addition/subtraction and XORing, yet the scheme is robust against different types of attacks. Hence, the proposed scheme is efficient in terms of computational complexity. Time analysis has been done on 2.30 GHz i3-2350M CPU, with 2 GB RAM computer. The operating system is Windows 7 and programming environment is MATLAB 7.7. The size of the image is 124×124. The average encryption speed of the proposed scheme is 90 ms which shows that the proposed scheme is good in terms of execution time as well.

4 Conclusion

Images are the most important utility of our life. They are used in many applications. Ideal image security techniques should be such that any adversary cannot modify the image and if any modifications are made can be detected. In the paper, we proposed a novel approach to provide confidentiality and integrity to the image. Hash of the plain image is generated and then the entire image is encrypted using masking and diffusion techniques which are key dependent. Initially 132-bit key is used in the algorithm which is further extended to 148-bits. A significant feature of the proposed scheme is that integrity of the image is checked which is not achieved with most image encryption schemes. Experimental, security and performance analysis shows that the scheme is secure against different types of attacks and can be adopted for real time applications. In future, the proposed scheme can be extended to provide more security to the image using some mathematical models and more arbitrary chaotic functions.

References

1. Sinha, A., Singh, K.: A technique for image encryption using digital signature. Optics Communications 1, 229–234 (2003)
2. Bakhshandeh, A., Eslami, Z.: An authenticated image encryption scheme based on chaotic maps and memory cellular automata. Optics and Lasers in Engineering 51(6), 665–673 (2013)
3. Chen, G., Mao, Y., Chui, C.: A symmetric image encryption scheme based on 3D chaotic cat maps. Chaos Solitons Fractals 21, 749–761 (2004)
4. Kwok, H.S., Tang, W.K.S.: A fast image encryption system based on chaotic maps with finite precision representation. Chaos, Solitons and Fractals 32, 1518–1529 (2007)
5. Shateesh Sam, I., Devraj, P., Bhuvaneshwaran, R.S.: A novel image cipher based on transformed logostic maps. Springer Science + Business Media, LLC (2010)
6. Qiu, J., Wang, P.: An image encryption and authentication scheme. In: 2011 Seventh International Conference on Computational Intelligence and Security (CIS), December 3-4, pp. 784–787. IEEE (2011)
7. Sabery, M., Yaghoobi, M.: A New Approach for Image encryption using Chaotic logistic map. 978-0-7695-3489-3/08 ©. IEEE (2008)
8. Mao, Y., Chen, G., Lian, S.: A novel fast image encryption scheme based on 3D chaotic baker maps. Int. J. Bifurcat. Chaos. 14, 3613–3624 (2004)
9. Matthew, R.: On the derivation of a chaotic encryption algorithm. Cryptologia 8(1), 29–42 (1989)

10. Pareek, N.K., Patidar, V., Sud, K.K.: Image encryption using chaotic logistic map. Image Vis. Comput. 24, 926–934 (2006)
11. Pareek, N.K., Patidar, V., Sud, K.K.: Diffusion-Substitution bsed gray image encryption scheme. Digital Signal Processing 23(3), 894–901 (2013)
12. Rhouma, R., Belghith, S.: Cryptanalysis of a new image encryption algorithm based on hyper-chaos. Physics Letters A 372, 5973–5978 (2008)
13. Sun, F., Liu, S., Li, Z., Lu, Z.: A novel image encryption scheme based on spatial chaos map. Chaos Solitons Fractals 38, 631–640 (2008)
14. Gao, T., Chen, Z.: A new image encryption algorithm based on hyper-chaos. Physics Letters A 372, 394–400 (2008)
15. Huang, X.: Image encryption algorithm using chaotic Chebyshev generator. Nonlinear Dyn. 67, 2411–2417 (2012)
16. Wei, X., Guo, L., Zhanga, Q., Zhang, J., Lian, S.: A novel color image encryption algorithm based on DNA sequence operation and hyper-chaotic system. The Journal of Systems and Software 85, 290–299 (2011)
17. Zhang, Q., Guo, L., Wei, X.: Image encryption using DNA addition combining with chaotic maps. Math. Comput. Model. 52, 2028–2035 (2010)
18. Shatheesh Sam, I., Devaraj, P., Bhuvaneswaran, R.S.: An intertwining chaotic maps based image encryption scheme. Nonlinear Dyn. 69, 1995–2007 (2012)

An Imperceptible Digital Image Watermarking Technique by Compressed Watermark Using PCA

Shaik K. Ayesha and V. Masilamani

Abstract. To provide secure communication, a modified digital watermarking scheme using discrete cosine transform (DCT) and principal component analysis (PCA) has been proposed. This scheme uses DCT for watermarking and PCA for compressing the watermark. In this technique, PCA in addition to DCT is used for watermarking the digital image to improve the quality of the watermarked image.

1 Introduction

Due to the rapid growth of networked multimedia systems, the distribution of digital data have become very efficient, which in-turn made the distribution of illegal copies easier without degradation in the data quality. So, there is a need to provide the authentication of the owner as well as data to prevent illegal distribution when the information is transmitted over World Wide Web environment. For this purpose, Digital watermarking came into existence, which is a kind of copy right inserted in a noise tolerant signal such as image or audio or video data, which is used for identifying ownership of the copyright. Digital watermarks are classified as fragile and robust based on the robustness. If a watermark is not detectable after some slight modification in the watermarked image, then that watermark is said to be "fragile". If a watermark resists some kind of unintentional attacks, then it is said to be robust. Based on the imperceptibility, watermarks are classified as perceptible and imperceptible. If the watermark is visible, then that is perceptible watermark. If the watermark is invisible then that watermark is imperceptible.

Digital Image watermarking is a technique of inserting data into an image in such a way that it can be used to make an assertion about the image [7]. Watermark insertion and extraction can be done in spatial domain or transform domain. Widely

Shaik K. Ayesha · V. Masilamani
Dept. of CSE., Indian Institute of Information Technology Design and Manufacturing Kancheepuram, Chennai, 600127, India
e-mail: ayeshanoormd@gmail.com, masila@iiitdm.ac.in

El-Sayed M. El-Alfy et al. (eds.), *Advances in Intelligent Informatics*,
Advances in Intelligent Systems and Computing 320, DOI: 10.1007/978-3-319-11218-3_27

used transform domain techniques are DCT and DWT. Techniques of incorporating watermarks in spatial-domain, transform domain and sub-band filtering approaches have been proposed in [1].

Earlier spatial domain watermarking techniques were based on modification of LSB pixels, but prone to attacks. In transform domain, the information in the original data is spread throughout the spectrum. For an invisible watermark, hiding the data in insignificant regions will be susceptible to attacks. So, invisible watermarks should be inserted in the significant regions where small change to the region of watermark has to result in the more change in image quality. A technique using KL transform or PCA has been proposed [2]. Here, PCA is closely related to the statistical properties of the image, which has the advantage of high energy concentration. This method is versatile in choosing a suitable region for data hiding. It allows a multi-layered key system, providing a high degree of data security. An overview of different information hiding techniques has been briefly discussed in [3]. Stegnography, which is a information hiding technique for secret communication and watermarking is for content authentication and copyright management, also part of information hiding. A detailed survey on existing and newly proposed stegnographic and watermarking techniques and classification based on different domains has been explained in [4]. In this paper, a modified robust invisible watermarking scheme based on PCA has been proposed. This technique will provide the advantages of using both DCT and PCA together. The remainder of this paper is organized as follows: Section 2 provides a survey on detailed related work on digital image watermarking techniques. Section 3 provides detailed description of the proposed technique. Section 4 discusses about the analysis of the technique. Conclusions and Future references are presented in Section 5.

2 Related Work

The watermarks are embedded in the Watermarked image for several purposes as discussed in [10]. Most of the digital image watermarking schemes incorporate watermarks into spatial or transform domain representation [10]. One of the techniques in spatial domain is adding a modified maximal length linear shift register sequence (m-sequence) to the pixel value. Two types of m-sequences are proposed: unipolar and bipolar. The elements of unipolar sequence are $\{0, 1\}$ and bipolar sequence are $\{-1, 1\}$ [5]. Another watermarking technique in spatial domain is by modifying LSB values of the image with the MSB values of the watermark, where the watermark will be invisible [10].

In transform domain DCT and DWT are widely used. In DCT watermarking technique, the watermark is embedded in DCT coefficients of the image. This technique is robust and susceptible to various attacks such as cropping, very low-bit rate JPEG compression and D/A conversion [1]. A modified DCT watermarking technique has been discussed in [10], where the watermark is embedded in largest coefficients being the perceptually significant regions. The authentication is determined by finding correlation between the embedded watermark and the extracted watermark.

The watermark embedding sometimes will degrade the quality of the image due to the loss of the significant information. So, by considering the just noticeable difference (jnd) values, the watermarking is done. If the original image pixel value is greater than jnd value then watermark is embedded, otherwise the pixel value is unmodified [1]. In [5], two two-dimensional watermarking techniques: Constant-W 2D watermark and Variable-W 2D watermark, have been discussed where the second technique offers better security. Millennium watermark system proposed for DVD copy protection purpose, where the encryption is done in addition to watermarking for DVD copy prevention [6]. The relation between the PCA and other transforms like SVD have been detailed in [7].Visible watermarks are placed on top of the image. Television networks place visible watermarks in upper-or-lower right hand corner of the screen [10]. The techniques discussed earlier have performed watermarking in spatial or transform domain, either visible or invisible where every technique is unique for its purpose. The technique proposed in [2] will use the PCA for compressing the Original image and then watermark is embedded into PCA compressed image. The detailed analysis of PCA have been discussed in [11].

In this paper, we present a digital watermarking scheme in DCT transform domain and use PCA for compressing the watermark. Here, Novelty is that the PCA based compressed watermark is embedded into the DCT coefficients of Original image. But in [2], original image itself is compressed and watermarking is done on the compressed image. Here the watermark is compressed by PCA inorder to increase the quality of the watermarked image. If the size of the watermark is more than the image size, by compressing the watermark using PCA most significant information of the watermark can be embedded into the Original image. To reduce degradation in the quality of the watermarked image, compressed watermark is used.

3 Proposed Work

In this paper,an invisible digital image watermarking technique is proposed which uses PCA based compressed watermark instead of watermark. Here the watermark is compressed by the PCA and the compressed watermark is embedded into the DCT coefficients of the image. This technique will ensure that the quality of the watermarked image is maintained as high as the original image (input image) without degradation in the quality of the watermarked image. The detailed watermark embedding scheme has been discussed in Fig. 1. The scheme is as follows:

Watermark Embedding Scheme

Watermark, w of size P×Q is given as input to PCA-based compression. The output of the PCA is the compressed watermark, w' of size K×Q, where K $\leq$ P. The image that we want to watermark say f, of size M×N and compressed watermark, w' is given as input to the watermarking algorithm. The output of the watermark embedding algorithm is the expected watermarked image, $f_{w'}$. The watermark is extracted by using watermark extraction scheme as shown in Fig. 2. The explanation of watermark extracting scheme is as follows:

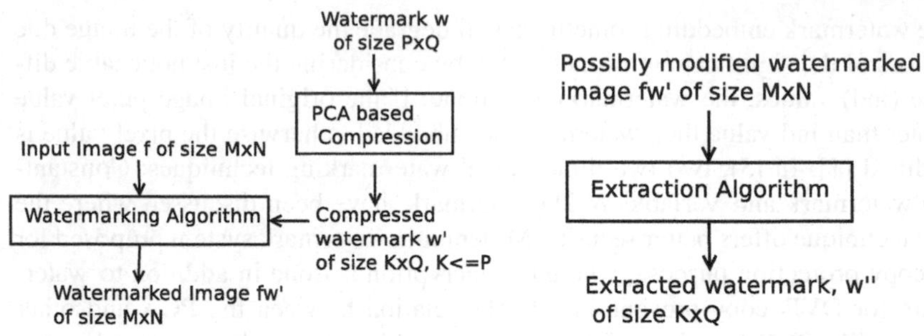

Fig. 1 Watermark Embedding scheme

Fig. 2 Watermark Extracting Scheme

Watermark Extracting Scheme
Assumed to be modified watermarked image, $f_w^{\prime\prime}$ is given as input to the extraction algorithm. The output is the watermark thats been extracted. The procedure for extraction algorithm is as shown in Fig. 4.

Algorithm 1. Watermarking Algorithm

Input: Image f of size M×N.
Output: Watermarked Image $f_{w'}$ of size M×N.
Step 1: Perform DCT on f, result is F of size M×N.
Step 2: Embed the PCA based compressed watermark, w' into the DCT coefficients of f i.e., F by using following equation.

$$F_{w'} = \mathrm{F}(1+\mathrm{a}w')$$

Step 3: Perform inverse DCT on $F_{w'}$, result is $f_{w'}$, which is the watermarked image.

Algorithm 2. :Extraction Algorithm

Input: Possibly modified watermarked image,$f'_{w'}$ and original Image f.
Output: Extracted Watermark.
Step 1: Perform DCT on $f'_{w'}$, result is $F'_{w'}$.
Step 2: Perform DCT on f, result is F
Step 3: Extract the watermark by using following equation

$$w'' = \frac{1}{a}\left(\frac{F_{w'}}{F} - 1\right)$$

Result is w'', which is the extracted watermark

To determine the authenticity, we need to perform correlation between the original watermark, w and the extracted watermark, w''. If the correlation coefficient is greater than some predefined threshold T, then the data is authenticated.

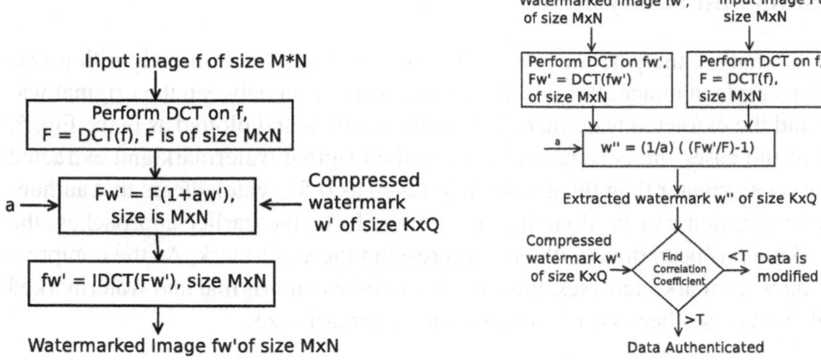

Fig. 3 Watermarking Procedure

Fig. 4 Watermark Extracting Scheme

3.1 Principal Component Analysis (PCA)

An application of PCA has been discussed in [2] is as follows:

Input is the image w of size P×Q.

Consider pixels in each column as a sample denoted as x_i, and find mean vector m_x by equation:

$$m_x = \frac{1}{q}\sum_{i=1}^{q} x_i, 1 \le i \le q \tag{1}$$

And find co-variance matrix C_x by equation:

$$C_x = \frac{1}{q}\sum_{i=1}^{q}(x_i - m_x)(x_i - m_x)^T; 1 \le i \le q \tag{2}$$

Now find eigen values λ_i and eigen vectors V_i for this covariance matrix, then arrange the matrix in such a order that the vector corresponding highest eigen values should be placed in the first row of transformation matrix A of size P×P. From A, by selecting K number of rows, say A_k. Then find y_i as follows:

$$y_i = A_k(x_i - m_x); 0 \le i \le N-1 \tag{3}$$

Now the input image has been compressed to size K×Q. The transformation matrix A will have high energy concentration in the first row and less energy concentration in the second row compared to the first row and so on. When we transform input image by K selected columns of A, the output image will also contain high energy concentration in the first k rows, which are compressed. Now the regions which have high energy concentration are significant and suitable for data hiding, which can be used for inserting watermark.

4 Results and Analysis

In order to analyze the performance of the proposed approach, the algorithm was tested by using lena image. For this image, the correlation between the original watermark and the extracted watermark has been measured and plotted as in the Fig. 5. In most of the cases the correlation between the original watermark and extracted watermark was greater than the threshold T(taken as 0.5), which clears that authenticity determination can be done by our approach. In the earlier approaches, the watermarking has been done without compressing the watermark. As the compression on the watermark increases, the correlation between original and watermarked image also increases because of reduction in watermark size.

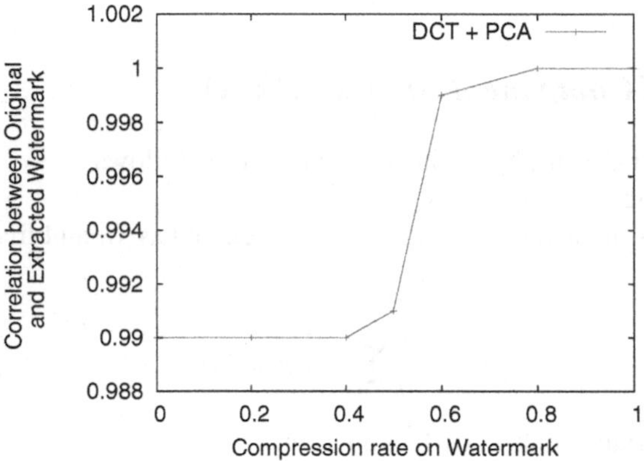

Fig. 5 Correlation between original and extracted watermark

In this approach, only K largest DCT coefficients are being watermarked, while performing watermarking and extracting, the correlation between extracted watermark and original watermark and correlation between original and watermarked images for different compressions on watermark is as shown in Fig. 6. To evaluate our approach in a better way, the correlation between the original image and watermarked image, and also correlation between extracted and original watermark for various DCT watermarked coefficients are shown in Fig. 7. The correlation between watermarks for various DCT watermarked coefficients and for various compressions on the watermark are plotted in Fig. 8. Different watermarked images by different compression values have been shown in Fig. 11 Fig 9 and Fig 10 shows the original and watermark images respectively. PSNR of watermarked image for watermarking various DCT coefficients of the Original image are shown in Fig. 12. PSNR of watermarked image for compression on watermark and without compression on watermark are plotted in Fig. 13. For the experimental study, a standard data set given in [8, 9] has been used.

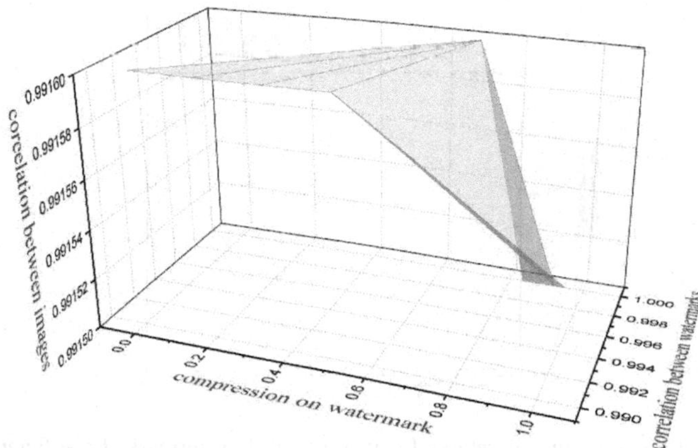

Fig. 6 Correlation between the images and correlation between watermarks for various compressions on the watermark

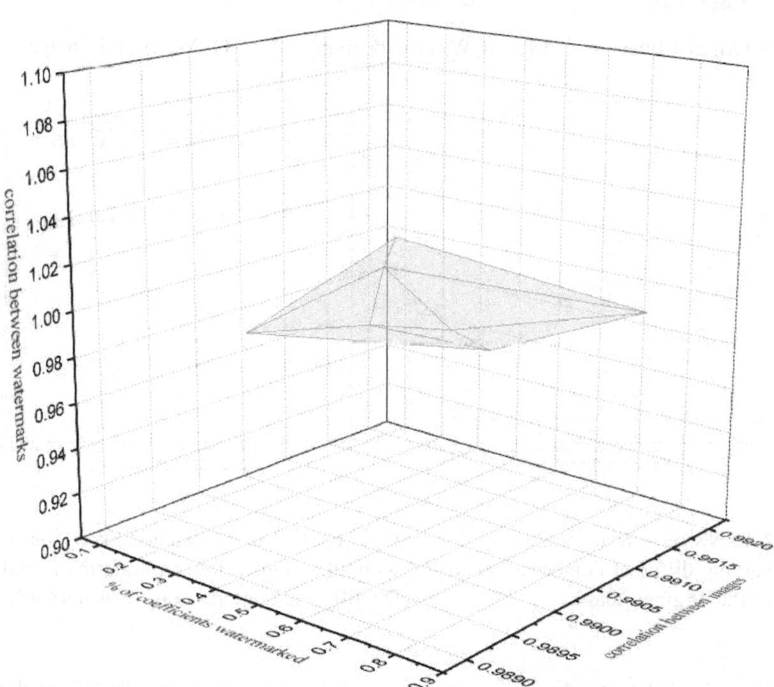

Fig. 7 Correlation between the images and correlation between watermarks for various DCT watermarked coefficients

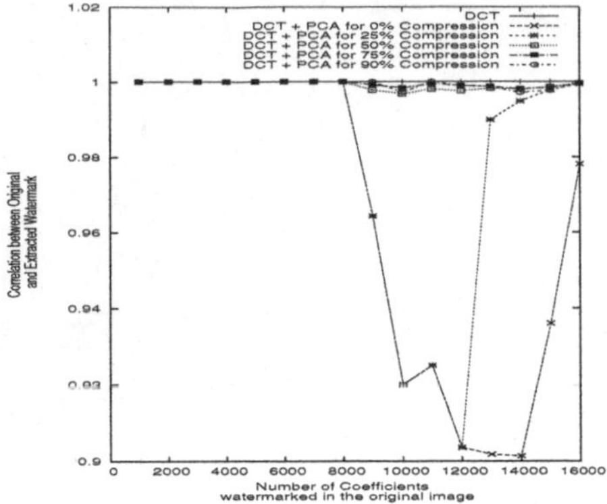

Fig. 8 Correlation between watermarks for various DCT watermarked coefficients and for various compressions on watermark

Fig. 9 Original Image **Fig. 10** Watermark Image **Fig. 11** Watermark Image

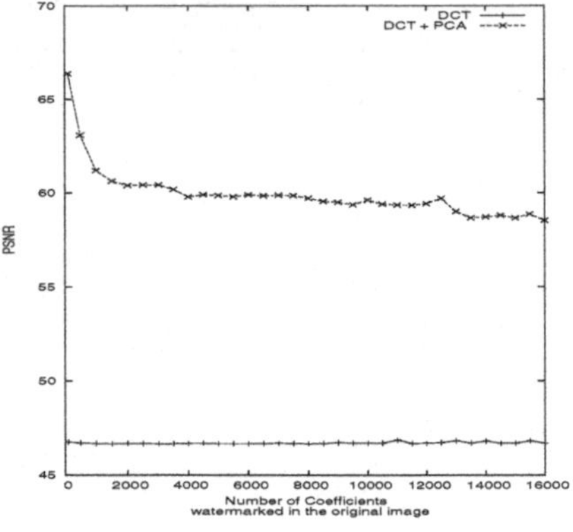

Fig. 12 PSNR of watermarked image for watermarking different number of coefficients of the original image

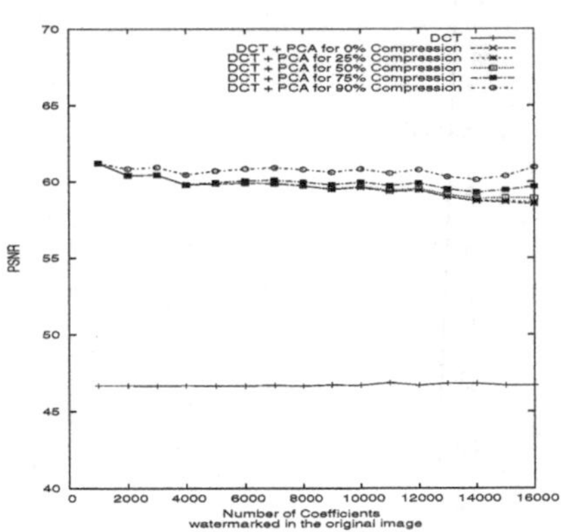

Fig. 13 PSNR of watermarked image without compression on watermark and for different compressions on watermark

In Fig. 13. if the compression on watermark is increasing , the PSNR of the watermarked image is also increasing. From Fig. 12 it's clear that the PSNR of the proposed technique is high compared to the DCT coefficient modulation technique [10]. In proposed technique compressed watermark will be containing high energy concentration, if the selected DCT coefficients of Original image is watermarked with compressed data having highly significant information about the watermark only those coefficients will be watermarked resulting high PSNR of the watermarked image.

5 Conclusion and Future Scope

A Digital image watermarking technique using DCT in addition to PCA has been presented in this paper which utilizes the advantages of both PCA and DCT. PCA is used for compressing the watermark before watermarking is implemented. The aim of this paper is to achieve the following objectives: First, to propose a watermarking technique, which will provide high quality watermarked image. Second, to ensure that this technique will offer high content security. The proposed technique can be implemented by other transform techniques such as SVD and can be applied to real time video applications.

References

1. Wolfgang, R.B., Delp III, E.J.: Overview of image security techniques with applications in multimedia systems. In: Proc. SPIE 3228, Multimedia Networks: Security, Displays, Terminals and Gateways, pp. 297–308 (February 1998)
2. Wang, S.-Z.: Watermarking based on principal component analysis. Journal of Shanghai University (English Edition) 4(1), 22–26 (2000)
3. Petitcolas, F.A.P., Anderson, R.J., Kuhn, M.G.: Information hiding a survey. Proceedings of the IEEE 87(7), 1062–1078 (1999)
4. Potdar, V.M., Han, S., Chang, E.: A survey of digital image watermarking techniques. In: Proceedings of 3rd IEEE International Conference on Industrial Informatics, INDIN 2005, pp. 709–716 (2005)
5. Wolfgang, R.B., Delp, E.J.: A watermarking technique for digital imagery: further studies. In: International Conference on Imaging, Systems and Technology, pp. 279–287 (1997)
6. Maes, M., Kalker, T., Linnartz, J.-P.M.G., Talstra, J., Depovere, F.G., Haitsma, J.: Digital watermarking for DVD video copy protection. IEEE Signal Processing Magazine 17(5), 47–57 (2000)
7. Gerbrands, J.J.: On the relationships between SVD, KLT and PCA. Pattern Recognition 14(1), 375–381 (1981)
8. `http://lear.inrialpes.fr/jegou/data.php#copydays`
9. `http://www.petitcolas.net/fabien/watermarking/image_database`
10. Woods, R.E., Gonzalez, R.C.: Digital Image Processing, 3rd edn. Pearson Education Publications (2008)
11. Choras, R.S.: Image Processing and Communications Challenges, vol. 3. Springer Fachmedien (2011)

Spread Spectrum Audio Watermarking Using Vector Space Projections

Adamu I. Abubakar, Akram M. Zeki, Haruna Chiroma, Sanah Abdullahi Muaz, Eka Novita Sari, and Tutut Herawan

Abstract. Efficient watermarking techniques guarantee inaudibility and robustness against signal degradation. Spread spectrum watermarking technique makes it harder for unauthorized adversary to detect the position of the embedded watermark in the carrier file, because the watermark bits are spread in the carrier medium. Unfortunately, there is a high possibility that synchronization of the watermark bits and carrier bits will go out of phase. This will lead to watermark detection problem in the carrier bit sequence. In this paper, we propose a vector space projections approach on spread spectrum audio watermarking technique, in

Adamu I. Abubakar · Akram M. Zeki
Department of Information Systems,
International Islamic University Malaysia, Malaysia
e-mail: 100adamu@gmail.com, akramzeki@iium.edu.my

Haruna Chiroma
Department of Artificial Intelligence,
University of Malaya, Malaysia

Haruna Chiroma
Federal Collage of Education, Gombe, Nigeria
e-mail: freedonchi@yahoo.com

Sanah Abdullahi Muaz
University of Malaya, Malaysia
e-mail: samaaz.csc@buk.edu.ng

Eka Novita Sari
AMCS Research Center, Indonesia
e-mail: eka@amcs.co

Tutut Herawan
Universiti Malaysia Pahang, Malaysia
e-mail: tutut@um.edu.my

El-Sayed M. El-Alfy et al. (eds.), *Advances in Intelligent Informatics*,
Advances in Intelligent Systems and Computing 320, DOI: 10.1007/978-3-319-11218-3_28

order to presents both the watermark bits and carrier bits as vectors. Similarities of watermark vector to a carrier vector are resolve by the normalized dot product of the cosine of angle between them for embedding. After embedding and extraction by the technique, signal processing methods in the form of attacks were applied. Our approach proved robust when compared with other audio watermarking techniques. This technique gives good results and was found to be robust on performance test.

Keywords: Audio watermarking; Spread spectrum; Vector space projections.

1 Introduction

The process of embedding a watermark on audio streams is describes as audio watermarking [1]. It is one of the complex watermarking techniques compared to image, text and video watermarking. This is because audio watermarking technique cannot be visible [2]. Spread spectrum (SS) watermarking technique is a borrowed model from communication engineering, which refers to the communication signals generated at a specific bandwidth and intentionally spread in a communication system [3]. In the process, a narrow band signal is transmitted over a large bandwidth signal [4], making it undetectable as it is overlapped by the larger signal. SS is used in watermarking because the watermark to be embedded in the carrier file consists of a low band signal that is transmitted through the frequency domain of the carrier file in the form of a large band signal. The watermark is thus spread over multiple frequency bands so that the energy of one bin is very low and undetectable [3]. It becomes more difficult for an unauthorized party to detect and remove the watermark from the host signal.

There are two types of spread spectrum: the frequency-hopping spread spectrum (FHSS) and the direct-sequence spread spectrum (DSSS). In DSSS, a watermark signal is embedded directly by introducing a pseudorandom noise sequence (PN), which implies that a key is needed to encode and decode the bits [4]. In FHSS, carrier frequency 'hops' across the spectrum at set intervals of time, and in a pattern determined by the PN sequence [3]. Previous research on spread spectrum audio watermarking technique utilizes many approaches. Crucial to that is orthogonal frequency division multiplexing (OFDM) approach [1] and field-programmable gate array (FPGA) implementation of by using chirp spread spectrum [2]. In Cox *et al.* [3] spread spectrum's approach dwells on multimedia in general. Frequency hopping approach spread spectrum is seen in Cvejic and Seppanen [4]. Machine learning approach was also use in audio watermarking mostly to make detection easier through a training based approach [5]. Recently an adaptive audio watermarking approach by wavelet-based entropy is propose in [6], this method is related to the work of Liang *et al.* [7] which uses chaotic and wavelet transform, and also in Wang *et al.* [8] that utilized discrete wavelet transform. The technique is intended for blind audio watermarking; unfortunately the embedding of the watermark is in the vector norm of the segmented approximation components, which depends on the size of the watermark image. In our approach vector space projections is utilized, in order to presents both the watermark bits and carrier bits as vectors. Similarities of watermark vector to the carrier vector are resolve by the normalized dot product of the cosine of angle between them for embedding. This will circumvent the weakness of depending on the size of watermark image.

Following this section is section 2, which presents an overview on audio watermarking techniques. Section 3 presents an introduction to vector space projection and section 4 discusses research methodology with an experiment procedure in digital watermarking. Section 5 presents experimental analysis and results, with the comparison of this research finding with previous research findings. Finally, section 6 is the conclusion the research work.

2 Related Work

Generally, to embed watermark data into an audio stream requires the transformation of the watermark into a binary format and, to some extent, to an encoded state. Different methods of transformation involve discrete Fourier transform (DFT), discrete cosine transforms (DCT) and discrete wavelet transforms (DWT) [9]. If $g(x)$ represents binary watermark data and $g(x) \in \{-1, +1\}$ this can then be directly embedded into the audio signal, say $f(x)$ in its time/spatial or transform domain. The reverse of the embedding leads to extraction. There are various models for watermarking techniques [22].

Least significant bit (LSB) involves direct substitution of LSB at each audio data sampling point by a watermark coded binary string; the substitution operation hides the bits to substitute the original [9]. The weakness of this model is that a resampling of the content of the audio signal or compression will lead to the loss or destruction of the watermark [3]. An echo hiding algorithm embeds a watermark into a host audio signal by inserting a small watermark as an echo that is perceptible by humans [3]. Unfortunately, this technique is suitable for only small size watermark [9]. Phase Coding Audio Watermarking technique relies on the phase difference of the audio segment, that is, the short-term phase of the signal over a small time interval [10]. Unfortunately, if a single segment within a phase is removed, compressed or reassembled, it will lead to damage or loss of part of the watermark, which will eventually affect the entire watermark [3]. Prior research on spread spectrum audio watermarking technique proves it efficiencies [3]. Spread spectrum model for audio watermarking is utilized in many different ways, like such as OFDM approach [1], FPGA [2]. FHSS [4], Machine learning approach [5], wavelet-based entropy [6], chaotic and wavelet transform [8]. Peng and Wang [11] utilized genetic optimization with variable-length mechanism in order to search for position of watermark in the carrier and many other approaches. Generally these techniques have shown a good performance, because watermarking performance is measure mostly by robustness performance metrics [12].

3 Vector Space Projection

A projection for any dimension follows a linear transformation with an inner product for which it presents an orthogonal projection [13]. Consider a v as a t-dimensional subspace of a vector space w with inner product, when v is in the x-plane where $Z(x,y)=(x,0)$, then by orthogonal projection v could be projected to w where v has an orthonormal basis $\{v_1,...,v_t\}$ such that

$$proj_v(w)=\sum_{i=1}^{t}\langle w,v_1\rangle v_1 \tag{1}$$

This becomes the sum is the orthogonal projection onto v. Thus by definition the projection of vector w on v is $w_1 = w_1\vec{v} = (|a|\cos\theta)\vec{v}$. An analogy of application of vector space projection is in vector space model, where is predominantly used in information retrieval applications. It is one of the models that can simply relate term weighting and relevance feedback in ranking [14]. In order to consider using vector space model in audio watermarking applications just as vector space projection, the watermark w and the carrier c should be transformed into vectors $v(w,c)$, so that the vectors of equal length can be considered as a pair of (watermark/carrier) bits for embedding.

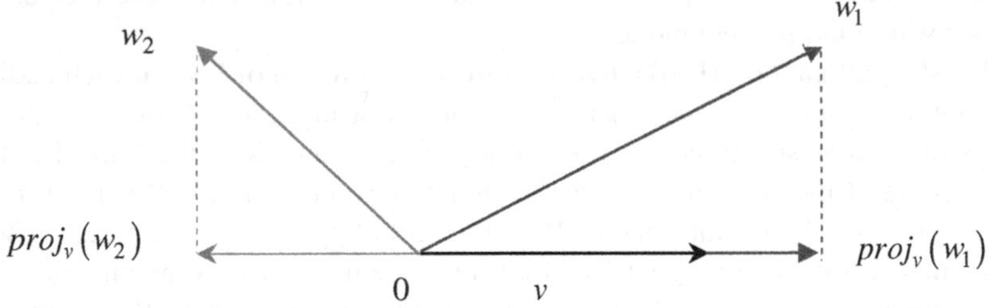

The cosine of angle between a pair of normalized two identical vectors will be 1, if the angle is 0^0 otherwise 0 if the angle is 90^0 hence the vector will be resolve by $\cos\theta$ $(x.y=|x||y|\cos\theta)$ and the normalized cosine measured will be derived by equation 2 as follows

$$v(w,c)=\frac{w_i.c}{|w_i||c|}=\frac{\sum_j w_{i.j}\times w_{c.j}}{\sqrt{\sum_j w^2_{i.j}}\sqrt{\sum_j w^2_{c.j}}}. \tag{2}$$

The vector space model will generate constraints set that will be generalized with vector space projection. This method ensures that a specific location which lies within intersection of a number of sets of watermark bits and carrier bits that satisfies both are derived by vector space projection. The technique selects an initial point in the vector space and projects the subsequent iteration. The sequence of projection finds the closet point at which both the watermark bits sequence and the carrier bits sequence cosine angle is at 0^0 and 90^0 using equation 1.

4 Proposed Method

The steps involves for this study (see Figure 1) are as follows:

- Collection of dataset
- Preprocessing and building a vector space model
- Developing synchronization code
- Embedding of watermark and application of selected audio watermarking attacks
- Extraction of watermark and measuring the state of the watermark before and after extraction
- Repeat the step 3 to 6 for until an optimum result is obtained by step 5 and 6.

Figure 1 shows the phases of this research. The original watermark is first converted to binary, and the audio file is decomposed into segments. In the same phase a synchronization code is generated. Both DWT and DCT were performed in the preliminary step. The binary contents of the watermark image and the audio segments where used for the vector space projection. Thereafter embedding and extraction processes were carried out as shown in Figure 1.

4.1 Dataset

Data for experimental analysis in audio watermarking are generally the same, since the primary requirement is the audio file. The watermark file could be different depending on the goal and objectives of research. This research uses an audio file as the carrier file and image file was chosen as the watermark file. The audio file sampling rate is 44,100Hz and resolution 16bit, hit rate 705 kbit/sec and mono. The watermark file is binary image with pixel 96× 84.

4.2 Preprocessing

The preprocessing stage prepares both the watermark and carrier files for the watermarking process, similar to study performed in [15-17]. The carrier file is first decomposed into non-overlapping frames of 1024 samples each, this are transformed to the vector of equal length where the total length of the audio signal containing is a set of vector that are independently and equally distributed. The original audio carrier signal (c) it is divided into segments i such that $c=\{c(i),1\le i\le L\}$ is the carrier with L samples, where $c(i)$ the non-overlapping

Original Audio (host)
Image Watermark file
Generate Synchronization code
Decomposing into segments
Convert to binary
Perform DWT on each segment
Decomposing into sets of specific lengths
1 0 1 0 1 1 0 1 0 1 → Host
0 1 0 1 1 0 1 0 1 1 → Watermark
Building the cosine model for each pair
Model not fit
Host and Watermark aggregates
Embed Synchronization code
Synchronization code
Model fit
Watermark Watermark
Embed the watermark to host
Watermarked file
Watermarked audio
Segmenting
Perform DWT
Check the synchronization code
Synchronization code found?
NO
YES
Watermarked audio
Next Segment
Watermarked segment
Reconstruct segments
Watermarked Audio
Exhausted all the segments?
NO
YES
Segments Reconstruction

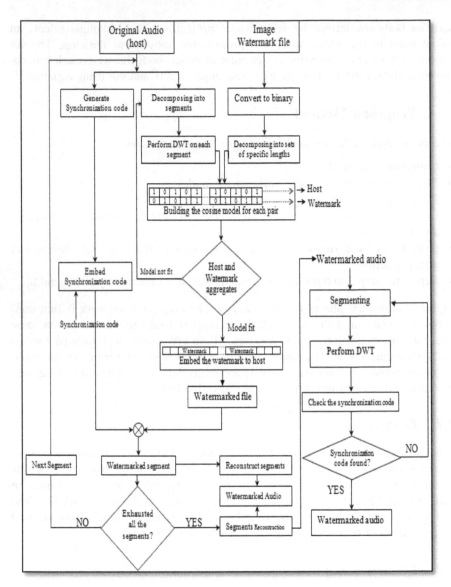

Fig. 1 The Embedding and Extraction Technique

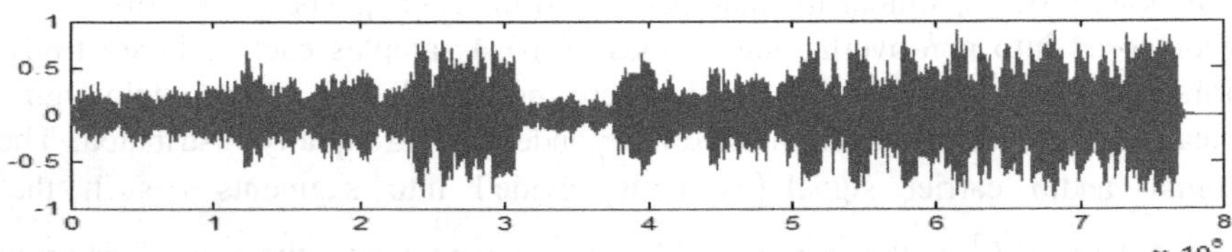

Fig. 2 Original audio un-watermarked

segments and transformed into vector $c(c_1, c_2, \ldots, c_i)$. The watermark file (w) is a binary image $w = \{w(i,j), 1 \le i \le P, 1 \le j \le P\}$ where $w(i,j) \in \{0,1\}$ are the pixel value at (i,j) for which $f(w(i,j))$ mapped the carrier $c(i)$. Synchronization code is used for the detection of watermark. The correlation corresponding to $c = \{c(i), 1 \le i \le L\}$ of the watermarked and un-watermarked carrier values is prepared as the Synchronization code.

4.3 Embedding the Watermark

The proposed method embeds the watermark bits within the sample of an audio signal. For each carrier vector $c(i)$, a corresponding watermark vector was mapped into its transformed into discrete cosine transformation (DCT) giving rise to the DCT coefficients as well as maintaining the quantizing state, the watermark sequence of data $w_1(i,j), w_2(i,j), \ldots, w_n(i,j)$ are then embedded into the carrier, in the following phase:

$$c' = \begin{cases} \dfrac{w_n}{w_n + x} \ge U & if \;\; c(i) = 1 \\[2ex] \dfrac{w_n + x}{w_n} \ge U & if \;\; c(i) = 0 \end{cases}, \tag{3}$$

where x is the selected sequence of the watermark segment and U is the length of each segment of the audio carrier file.

4.4 Detecting and Extraction of the Watermark

The watermarked audio signal c' is segmented into $c(i)$ where $i = 1, 2, \ldots, L$ of the length of the each segment. M×M is the number of bits in the watermark image. Thus, we first establish the start position by representing $x \cdot y$ as the normalized inner product of vectors x and y, i.e. $x \cdot y \equiv N^{-1} \sum_i x_i y_i$ with $x^2 \equiv x \cdot x$. Inverse DCT is performed on each segment $c^*(i)$ and the proceedings are as follows

$$c^*(i)^1, c^*(i)^2, \ldots, c^*(i)^{n-1}.$$

Thus, a watermark (w') is detected by correlating a given signals and subsequently extracted from each of the preceding segments in the following

$$w'_n(i) = \begin{cases} 1 & if \;\; c^*(i)^n > 0, \\ 0 & if \;\; c^*(i)^n \le 0 \end{cases} \quad (n = 0, 1, \ldots, n-1). \tag{4}$$

5 Experiment Results

The watermark is embedded on the carrier using equation 3 and extracted by using equation 4. Watermarked file was formed as a result of this embedding process. Thereafter the embedded watermark was extracted. In order to measure the performance of the technique, bit error rate (BER) is used after applying some attacks. This is the measure that is usually used to calculate the robustness of the watermarking algorithm [14]. It is presented in equation 5 as follows:

$$BER = \frac{\sum_{i=1}^{m}\sum_{j=1}^{m} w(i,j) \oplus w'(i,j)}{M \times M}, \tag{5}$$

where $w(i,j)$ and $w'(i,j)$ are the watermark before embedding and the watermark after extraction. The BER values will indicate the degree of noise caused by embedding the watermark. The higher the distortion caused by the watermark the more the bit synchronization errors and eventually renders the technique used inadequate. Normalized correlation (NC) evaluates the correlation between the extracted and the original watermark. It is define by

$$NC(w,w') = \frac{\sum_{i=1}^{M}\sum_{j=1}^{M} w(i,j)w'(i,j)}{\sqrt{\sum_{i=1}^{M}\sum_{j=1}^{M} w^2(i,j)}\sqrt{\sum_{i=1}^{M}\sum_{j=1}^{M} w'^2(i,j)}}, \tag{6}$$

where, w and w' are the original and the extracted watermarks, respectively, and i, j are indices in the binary watermark image. If $NC\ (w,\ w')$ is close to 1, then the correlation between w and w' is very high, while close to zero indicate a very low correlation. The following attacks used:

1) Additive noise: White noise with 10% of the power of the audio signal is added, until the resulting signal has an SNR of 20 dB
2) Low-pass filtering: The low-pass filter with cut-off frequency of 11,025 Hz is applied to the watermarked audio signal
3) Pitch shifting: This is applied by shifting one degree higher and one degree lower
4) Re-quantization: This process involves re-quantization of a 16-bit watermarked audio signal to 8-bit and back to 16-bit.

Table 1 shows the result of BER and NC obtained for our technique. The four attacks are applied to the same watermarked filed. This attacked were selected based on the fact that there were used by previous studies to measure the performance of their watermarking technique [18-21]. Our results obtained based on the attacks indicates a good performance reasonable percentages of BER and the NC values are all 1, reflecting high correlation between the original watermark and the extracted watermark.

Table 1 The extracted watermark with the NC and BER

Attacks	NC	BER(%)	Extracted watermark
Re-assembling	1.000	0	
Pitch shifting	1.000	0.432	
Additive noise	1.000	0.212	
Low-pass filtering	1.000	0.342	

The results of BER obtained by the proposed technique are compared to previous research findings. It indicates that our technique performs better under some selected attacks.

Table 2 Comparisons of previous findings with different techniques in terms of BER (%)

Attacks	Propose	[18]	[19]	[20]	[21]
Re-assembling	0	2.3437	4.3945	0	2.54
Pitch shifting	0.432	-	0.5156	-	-
Additive noise	0.212	-	0.0195	-	1.32
Low-pass filtering	0.342	3.5156	49.707	5.664	0.54

6 Conclusion

This paper presents the use of vector space projection approach to spread spectrum audio watermarking techniques in order to ensure inaudibility and robustness against signal degradation. Spread spectrum watermarking technique has been identified to be a technique that makes it harder to detect the position of the embedded watermark in the carrier file by an unauthorized user. The model provide for hiding watermark bits in a low band signal over a large bin of the carrier medium. However, in order to avoid any possibility that will lead to the lost of watermark when spread in the carrier file, we make the watermark and carrier bits vectors. Where for each segment of the carrier bits, a corresponding vector of the watermark bit which is also a vector undergoes the embedding process. This will ensure a synchronization of the watermark bits with the carrier bits throughout the audio stream. As a result it will be easier to detect a watermark in the carrier bits sequence. The model gave good projections, and the experimental result after applying some attacks, showed a good performance. We have compared the performance of our algorithm with other, recent audio watermarking algorithms. Overall, our technique has high embedding capacity and achieves low BER against attacks reassembling, pitch shifting, additive noise, and low-pass filtering. The most unique characteristic of the method proposed in this study lies in its utilization of the projections by vector space.

Acknowledgement. This research is supported by High Impact Research Grant, University of Malaya Vote no UM.C/628/HIR/MOHE/SC/13/2 from Ministry of Higher Education Malaysia.

References

1. Garcia-Hernandez, J., Parra-Michel, J., Feregrino-Uribe, R., Cumplido, C., High, R.: payload data-hiding in audio signals based on a modified OFDM approach. J. Expert Syst. Appl. 40(8), 3055–3064 (2013)
2. Karthigaikumar, P., Kirubavathy, K.J., Baskaran, K.: FPGA based audio watermarking—Covert communication. J. Microelectron 42(5), 778–784 (2011)
3. Cox, I.J., Joe, K., Leighton, F.T., Shamoon, T.: Secure spread spectrum watermarking for multimedia. IEEE T. Image Process 6(12), 1673–1687 (1997)
4. Cvejic, N., Seppanen, T.: Spread spectrum audio watermarking using frequency hopping and attack characterization. Signal Process 84(1), 207–213 (2004)
5. Peng, H., Li, B., Luo, X., Wang, J., Zhang, Z.: A learning-based audio watermarking scheme using kernel Fisher discriminant analysis. J. Digit. Signal Process 23(1), 382–389 (2013)
6. Chen, S., Huang, H., Chen, C., Seng, K.T., Tu, S.: Adaptive audio watermarking via the optimization point of view on the wavelet-based entropy. J. Digit. Signal Process 23(3), 971–980 (2013)
7. Liang, T., Bo, W., Zhen, L., Mingtian, Z.: An Audio Information Hiding Algorithm with High Capacity Which Based on Chaotic and Wavelet Transform. ACTA Electronica Sinica 38(8), 1812–1824 (2010)
8. Wang, X., Wang, P., Zhang, P., Xu, S., Yang, H.: A norm-space, adaptive, and blind audio watermarking algorithm by discrete wavelet transform. Signal Process 93(8), 913–922 (2013)
9. Bender, W., Gruhl, D., Morimoto, N., Lu, A.: Techniques for data hiding. IBM Systems Journal 35(3.4), 313–336 (1996)
10. Hussain, I.: A novel approach of audio watermarking based on image-box transformation. Math. Comput. Model. 57(3), 963–969 (2013)
11. Peng, H., Wang, J.: Optimal audio watermarking scheme using genetic optimization. Ann. Telecommun. 66(5-6), 307–318 (2011)

12. Petitcolas, F.A., Ross, J., Anderson, G., Markus, K.: Information Hiding-A Survey. Proceedings of the IEEE Special Issue on Protection of Multimedia Content 87, 1062–1078 (1999)
13. Abardia, J., Bernig, A.: Projection bodies in complex vector spaces. Advances in Mathematics 227(2), 830–846 (2011)
14. Croft, B., Metzler, D., Strohman, T.: Search Engines: Information Retrieval in Practice. Addison Wesley, Pearson Education Inc., Uppper Saddly River, New Jersy (2010)
15. Zeki, A.M., Abubakar, A.I., Chiroma, H., Gital, A.Y.: Investigating the dynamics of watermark features in audio streams. In: 2013 IEEE Symposium on Industrial Electronics and Applications (ISIEA), pp. 61–65. IEEE (2013)
16. Zeki, A.M., Abubakar, A.I., Chiroma, H.: Investigating Digital Watermark Dynamics on Carrier File by Feed-Forward Neural Network. In: 2013 International Conference on Advanced Computer Science Applications and Technologies (ACSAT), pp. 162–165. IEEE (2013)
17. Zeki, A.M., Azizah, A.M., Abubakar, A.I., Zamani, M.: A robust watermark embedding in smooth areas. Research Journal of Information Technology 3(2), 123–131 (2011)
18. Megías, D., Serra-Ruiz, J., Fallahpour, M.: Efficient self-synchronised blind audio watermarking system based on time domain and FFT amplitude modification. Signal Process 90(2), 3078–3092 (2010)
19. Baritha-Begum, M., Venkataramani, Y.: LSB based audio steganography based on text compression. Procedia Engineering 30, 702–710 (2012)
20. Lei, B., Soon, I.Y., Zhou, F., Li, Z., Lei, H.: A robust audio watermarking scheme based on lifting wavelet transform and singular value decomposition. Signal Process 92(9), 1985–2001 (2012)
21. Hu, H., Chen, W.: A dual cepstrum-based watermarking scheme with self-synchronization. Signal Process 92(4), 1109–1116 (2012)
22. Zeki, A.M., Abubakar, A.I., Azizah, A.M.: Steganographic Software: Analysis and Implementation. International Journal of Computers and Communications 6(1), 35–42 (2012)

RBDT: The Cascading of Machine Learning Classifiers for Anomaly Detection with Case Study of Two Datasets

Goverdhan Reddy Jidiga and Porika Sammulal

Abstract. The inhuman cause of behavior in computer users, lack of coding skills pursue a malfunctioning of applications creating security breaches and vulnerable to every use of online transaction today. The anomaly detection is in-sighted into security of information in early stage of 1980, but still we have potential abnormalities in real time critical applications and unable to model online, real world behavior. The anomalies are pinpointed by conventional algorithms was very poor and false positive rate (FPR) is increased. So, in this context better use the adorned machine learning techniques to improve the performance of an anomaly detection system (ADS). In this paper we have given a new classifier called rule based decision tree (RBDT), it is a cascading of C4.5 and Naïve Bayes use the conjunction of C4.5 and Naïve Bayes rules towards a new machine learning classifier to ensure that to improve in results. Here two case studies used in experimental work, one taken from UCI machine learning repository and other one is real bank dataset, finally comparison analysis is given by applying datasets to the decision trees (ID3, CHAID, C4.5, Improved C4.5, C4.5 Rule), Neural Networks, Naïve Bayes and RBDT.

Keywords: Anomaly detection, C4.5, Decision tree, Naïve bayes, RBDT.

1 Introduction

Anomaly detection [2] is a kind of intrusion detection to model the behavioral patterns in image and medical applications, novelties in industrial machine malfunctions, fraud accounting actions in financial banking sectors and also the

Goverdhan Reddy Jidiga
Department of Technical Education, Govt. of Andhrapradesh, Hyderabad, India
e-mail: jgreddymtech@gmail.com
Porika Sammulal
JNTUH College of Engineering, Karimnager, JNTU University, Hyderabad, India
e-mail: sammulalporika@gmail.com

El-Sayed M. El-Alfy et al. (eds.), *Advances in Intelligent Informatics*,
Advances in Intelligent Systems and Computing 320, DOI: 10.1007/978-3-319-11218-3_29

anomalies in all kinds of network applications. Today the intrusion detection systems (IDS) [1] are modeled by the various conventional, adorned machine learning approaches (or classifiers) and Meta classifiers. The anomalies have various dimensions and detecting them as false is depending on the ratio of dynamic, online issues to be considered in model. As on today the world is moving in competitive directions and no one is follow the ethical values. Like in medical diagnosis, anomaly boundaries need to model into 3-layer security [8], it is help to focus on the awareness and required to reduce the design cost of algorithms in information security. The multilevel classification in adorned machine learning is more decent method to identify and model the zero day attacks in latest critical infrastructure applications.

1.1 Machine Learning in Anomaly Detection

The Anomaly based intrusion detection system [2, 3] is a system for detecting computer intrusions and anomalous behavior by monitoring system activity, incoming and outgoing internet traffic based on applications and classifying it as either normal or anomalous. In this paper the Anomaly detection system (ADS) is a module to detect the abnormal samples (records) or anomalies by monitoring decision tree transition and categorizes them as either regular behavior or anomalous through observing class label. In this security field many IDSs [2,3,5,7] techniques have been developed for detecting and modeling anomalies in structured and unstructured multi dimensional data in traditional, real time and critical infrastructure applications, but still impossible to catch all latest anomalies. So machine learning is useful in this area to achieve satisfactory results in ADS. Machine learning techniques (predict known) [6,7] are generally different from data mining (predict unknown)[4] and facilitate users to extract new features from datasets and construct a new system to solve predictive issues like novelties, anomalies, outliers. The detection of all done by anomaly detection by imposing classification, ranking of features, learning control and outcomes decision analysis. Hence the machine learning in anomaly detection [5] always enables to automate the system massively from big database sources by constructing novel rules using mathematical hardness in real time critical infrastructure applications and improve the results. In this paper the outline of the work is as follows, in section-2 we have given huge literature review, section-3 gives the framework and algorithms of proposed work, sectin-4 briefly explains about case studies, section-5 confer experiments with results and finally discussions, future work given.

2 Related Work

The decision tree (DT) [35] is powerful learning classifier used in anomaly detection [7] and today it is completely designed in terms of dynamic rule sets extracted from learning. Initially the popular classification algorithm ID3 [10] designed based on heuristics and it is a top down, divide and conquer, greedy based and entropy-gain technique. The C4.5 [11-13] is a posterior technique for ID3 and used still now in many applications successfully giving optimal solutions by change of some parameters. There are many classification techniques available in machine learning like normal rule based classifiers , Naïve Bayes(NB)[34] , Support vector machines (SVM), Neural networks (NN)[31] and DT based algorithms[35] CLS, GUIDE, QUEST, CHAID[9], CART[16], C5.0 and multivariate decision trees. All these algorithms have advantages and disadvantages when use in ADS given in this paper as follows.

The QUEST [24] is based on univariate attribute domain using ANOVA F-test, but it uses 10-fold cross validation in training is only positive here. The CHAID [9] designed to handle nominal attributes based on p-value, chi-square test and likelihood ratio. The CHAID tree is an attractive DT over its precedents like AID, THAID and its construction begins with complete data space and continues based on repeated and homogeneous subspaces into two or more child nodes. But this algorithm treats the missing instances as same unit of single and also lack in pruning. CART [16, 23] is a recursive method used in regression and also good for classification, but the split is limited to only two. It uses cost complexity pruning takes additional cost.

The popular C4.5 [11-14] introduced with a many new features and it constructs a tree by considering all attributes get equal priority assume most significant and optimize the decision rule by well pruning. C4.5 also suffering with null instances, overfitting and insignificant attributes in some cases. This is still good due to efficiency of algorithm and adjusts with other bagging and boosting techniques. In [18] they used 'one against all approach' with C4.5 on three sets of UCI datasets, got good estimation and in [19] uses this for outlier detection by hybrid process, but they have not given correct results with their data, Later in [22] uses hierarchical clustering with same data shown well. In [21] uses this C4.5 for remote sensing data along rough sets also got good results. Finally C4.5's successor C5.0 shown some improvements to previous, but not well due to heard to learn compare to learning of C4.5 generally fast [14, 20]. C4.5 Rule [14, 15, 30] is rule based induction tree and it is uses pruning heuristics to improve the accuracy by remedial strategy of derive, generalize, group and ordering the rules. But the problem with continuous value attribute is rules are needed to update continuously. Other rule based CN2 presented in [25], evaluates possible conjunction of attribute tests conditions of a rule based on I_g (entropy) measure. Later in [26], the measure is changed to get accuracy more, but frequent change in rules then training is a complicated in learning take many iterations for rule search. The STAGGER, FLORA3, AQ-PM also kind of rule learning systems uses numerical data. Some uses decision rules by overlapping and in all above adapting new rules and remove old is drift in accuracy also. Improved C4.5 [27, 29] is a successor of C4.5 uses generalized entropy with new parameter β and improved gain ratio used instead of I_g (standard). Later cost sensitive C4.5 [28] (version of C4.5) uses misclassification costs matrix in training process, in [30] Enhanced C4.5 algorithm for intrusion detection in networks by extracting a set of classification rules from KDD dataset.

Neural networks [31] used in ADS with different approaches and like multi layer perceptrons (MLP) is popular to classify the data with threshold, activation function and error adjustment to improve the accuracy. Naïve bayes (NB) [34] is simple and better predictive accuracy over ID3 and also refining NB classifier [42] is good. C4.5 in some well organized data without any discrepancies; it is extremely efficient like ensemble classifiers (bagging and boosting) to combine classifier predictions under the Gaussian naive bayes (GNB) assumptions. AdaBoost [6] algorithm is developed with incremental refinement over large datasets to classify the data instances and they are finalized the boosting with k-NN is better than bagging with C4.5 in terms of accuracy. Random Forest [6, 16] is again shows the performance is better than AdaBoost. Finally the latest work in this field is structure learner [41], uses the features of DT and mapped into markov network structures to improve the learning very fast and even for complex data instances. Other extensions of DTs are like oblivious decision tree, RBDT-1[39], neurotree [40] (light weight IDS) and fuzzy decision tree also useful in ADS, but in case large datasets parallel ID3, SLIQ, SPRINT like useful to get good results. In [38, 43], uses decision semantics to remove the rules based on irrelevant conditions in tree over the process of converting the rules.

3 Proposed Work: RBDT

Decision tree (DT) [35] is a kind of machine learning [5] algorithm to solve the classification problems and make simplify the solution with genuine performance parameters like predictive accuracy (maximize) and false positive rate (minimize).

The RBDT (Rule Based Decision Tree) is multi-level classification model. It is a combination or a kind of cascading of Naïve bayes and C4.5 rules used to classify the on-the-fly data by framing dynamic rules from existing classifier including with new behavior. In RBDT the rules are not predictable, but dynamic while learning. The decision tree is addressing the frequent anomalies by calculating performance parameters associated with ADS at every node which is label the class C as leaf. The RBDT algorithm also support multi-way split which has more than two out comes at each node. The RBDT is constructed for the given datasets is shown in Fig.1 (bottom). Here the attribute PAN is selected as a root due to high information gain (I_g) value compared to the candidate attribute set has {PAN, P-MODE, DATE}. The RBDT is constructed in this paper is shown partially in Fig.1, but full tree is very complex and invalid records treated as error data in this dataset-2.

3.1 Frame Work of RBDT

The frame work of RBDT is given in Fig.1, it is a process begin with applying the Gaussian naïve bayes (GNB) on selected datasets as input and generation of Gaussian segments $S = \{S_0, S_1, S_2 \ldots S_n\}$like criteria used ten-fold cross validation in training of sample data. The rules are generated from each segment separately taken into one set (α-rules) and collectively from all segments (original dataset) create another set (β-rules). In this the rules are generate from applying C4.5 rules algorithm and form a pair of rule set $\{C_\alpha, C_\beta\}$ and same as with naïve bayes (NB) form a rule set $\{N_\alpha , N_\beta\}$ shown in Fig.2 .The GNB is based on naïve bayes algorithm used to classify the data or samples, but in this case we have used it as pre-classification to observe the analysis of training help to improve the performance in large datasets.

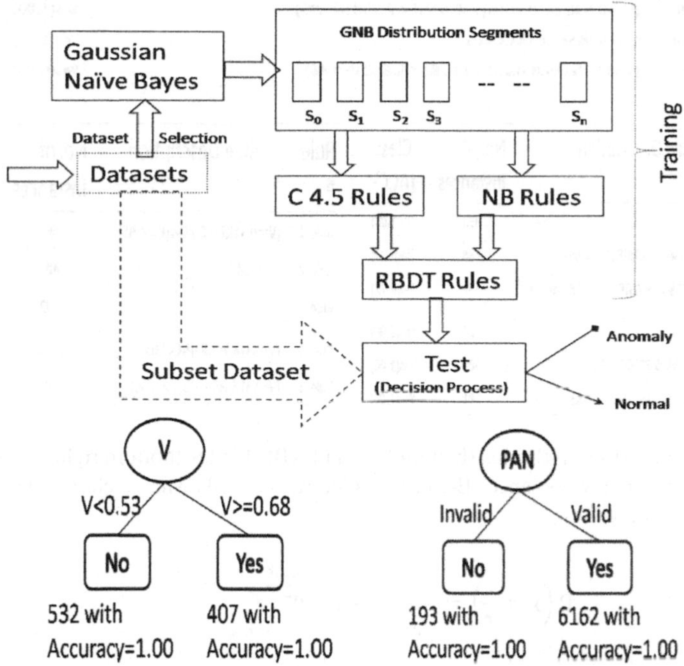

Fig. 1 The frame work of RBDT has RBDT rules $\{ R_\alpha , R_\beta \}$conjunction of NB$\{N_\alpha , N_\beta\}$, C4.5$\{C_\alpha, C_\beta\}$ Rules (top) and construct the decision tree based on RBDT rules and continue the test process to determine the outcome of new data as input. In Fig (bottom left) shows the simple RBDT for dataset-1(Banknote authentication) and for training set of dataset-2 (real-time bank dataset), the tree constructed on attribute {PAN} based on high I_g by PAN (bottom right).

We have selected GNB only, because it supports for continuous data type well compared to Gaussian based Bernoulli and multinomial models used in discrete data as input. In this the dataset (X) has instances$\{X_0, X_1, \ldots X_n\}$ of continuous type segmenting into S_i by computing a mean, variance; co-variance is given by following GNB instead of only two segments of class-yes and class-no.

Rule #	Rule Description	Class (ACC)
Rule-1	V>=0.26&V<0.53&S>=5.85	No(0.98)
Rule-2	(V>=0.53&E>=0.28)&(V>=0.63&C>=-4.95)	No(0.99)
Rule-3	(V>=0.53&E>=0.28)&(V<0.63&C<-2.25&S>=5.95)	No(0.99)
Rule-4	(V>=0.53&E>=0.28)&(V<0.63&C>=-2.25&E<0.79)	No(0.99)
Rule-5	(V>=0.53&E>=0.28)&(V<0.63&C>=-2.25)&(E>=0.79&C>=1.85)	No(1.00)
Rule-6	(V>=0.53&E>=0.28)&(V<0.63&C>=-2.25)&(E>=0.79&C<1.85&S>=3.85)	No(1.00)
Rule-7	(V<0.53&S<5.85)&(C>=3.5&S>=-1.85&V>=0.35)	No(1.00)
Rule-8	V<0.53&(S>=5.85&V<0.26)	Yes(0.96)
Rule-9	(V>=0.53&E>=0.28)&(V<0.63&C<-2.25&S<5.95)	Yes(1.00)
Rule-10	V<0.53&S<5.85&C<3.05	Yes(1.00)
Rule-11	(V<0.53&S<5.85)&(C>=3.05&S<-1.85&V<0.46)	Yes(0.99)

Rule #	Rule Description	No. of Instances	Class (ACC)
Rule-1	V>=0.68	393	No(1.00)
Rule-2	(V>=0.53&E>=0.28)&(C>=-4.95)	484	No(0.99)
Rule-3	(S>=5.15&S<=9.65)&(C<-4.45&C<8.85)	192	No(0.72)
Rule-4	V<0.53	305	Yes(1.00)
Rule-5	(V<0.53&S<5.85)&C<3.05	305	Yes(1.00)
Rule-6	(V<0.53&S<5.85)&(C>=3.05&S<-1.85)	181	Yes(1.00)

Rule #	Rule Description	No. of Instances	Class (ACC)
Rule-1	(V>=0.31&E>=0.28)&(C>=4.95)	484	No(0.99)
Rule-2	V>=0.68	407	No(1.00)
Rule-3	V<0.53	532	Yes(1.00)
Rule-4	(V<0.31&S<5.85)&C<3.05	305	Yes(1.00)
Rule-5	(V<0.31&S<5.85)&(C>=3.05)	181	Yes(1.00)

Fig. 2 C4.5 rules (top), NB rules (bottom left) and RBDT rules (bottom right) for Banknote authentication, here V-variance, E-entropy, C-kurtosis, S-skewness, class labels(Yes/No) and ACC- accuracy.

$$P\left(x=\frac{v}{c}\right)=\frac{1}{\sqrt{2\pi\sigma_c^2}}.-e^{\frac{(v-\mu_c)^2}{2\sigma_c^2}} \quad (1)$$

Where x-continuous attribute, μ_c -mean value of x of class c for each segment, σ - is variance of x of class, P is a probability density of some value in segment for a given class C. v- a instance and new instance (in test) to determine its class.

The rules collected from GNB segments shown in Fig.2. Here the rules are given for collective segments maximum and very less at individual segment. The rules which give accuracy high only shown in Fig.2 and remaining rules having less support and less accuracy.

3.2 *Issues in Decision Trees*

Overfitting: Overfitting is problem of decision tree create training set error and reduce the accuracy by poor criteria and selection of rules. Generally this can avoided by general pruning techniques like cross validation (10-fold), cost complexity pruning[16], reduced error pruning [11,12], minimum error pruning, pessimistic pruning, error based pruning [14], optimal pruning. All these are maximum use the bottom up approach to get good performance. Optimal pruning with Δ is good for all.

Pruning: The Tree pruning [17] is a property of decision trees to avoid the overfitting problems and if any sub tree is unlikely to make less accuracy then possibility to prune the tree at any level to increase accuracy until satisfactory performance. In these case many techniques available based on pre and post pruning. The pruning cost is depending on pruning criteria and technique. If pruning is done at probability of node(v) , which has a set of positives (p) and negatives(n) with expected count of irrelevant P_k and N_k in each subsets P_s and N_s. If v=n+p, then

$$P_k = p * \frac{P_k+N_k}{p+n}, \quad N_k = n * \frac{P_k+N_k}{p+n} \tag{2}$$

$$\Delta = \sum_{k=1}^{d} \frac{(P_s-P_k)^2}{P_k} + \frac{(N_s-N_k)^2}{N_k} \tag{3}$$

Δ is distributed according to χ2 and take decision to accept or not. It will determine the cost of pruning if tree grows in unordered by describing error or noise.

3.3 Algorithms

In this paper, our work flow is divided into two algorithms one is pre-classification and other is generating decision tree. In algorithm-1, how the RBDT framework to be carried out in terms of pseudo code statements.

```
Algorithm-1: RBDT Pre-Classification
Input: Samples Set (S), Gaussian parameters,
Output: Rule_set (R)
(1). Prepare dataset;
(2). Do Normalization of data;
(3). Apply GNB classifier and segmenting the dataset;
(4). Apply C4.5 rule algorithm and Naïve Bayes algorithm;
(5). Prepare RBDT rules Rule_set(R);
(6). Call generate_RBDT;
```

In algorithm-2, the RBDT construction is explained for only two childs and shown one of RBDT for dataset-2 in Fig.1 (right). O_i is depends upon creation of no. of childs at L_i for A_i. For this the dataset (sample-set) has proper attribute domain (AD) represented as AD (A_n U C_n), here A denote attribute set contains 'n' attributes then A= {A_1, A_2 A_3 …A_n} and C is a class label {yes, no}. In training the set of instances (samples or records) are described by T(s) = {(x_1, y_1), (x_2, y_2) … (x_n, y_n)}, here x_i€X and X is data value (except class) is determined by attribute domain and y is a class instance of domain Y. The generalized classification is simply done by y=f(x), f is a function determined by rules.

```
Algorithm-2: Generate_RBDT
Input: Sample_Set (S), attribute_list (A), Key_Attribute,
          Rule_set(R) and Candidate_attribute_set (K);
Output: RBDT
(1).Create node N;
      If samples of dataset is same class;
      Then return N as a leaf node and labeled with Class C;
(2).If Ai ∈ Ki
    Then calculate information-gain (I) for each attribute Ai;
(3).If I(Ai)>I(Aj)then Ai is a test attribute as Root; Else Aj;
(4).Create Li and Ri to Root by applying rule Ri ∈ R[R1… Rn];
(5).Divide the samples into sub-roots Li and Ri ;
    If Li or Ri not generated then apply Ri+1 and goto step (4);
(6).For each sub-root compute optimality cost Oi;
      If Oi (Ai) at Level Li ≠ Oi (Ai) at Level Li+1;
     Then goto step (7); else PRUNE (T, ti) at Level Li+1  ;
(7).For each sub-root, if not Labeled, A ≠ NULL, S ≠ NULL
     Goto step (5) else Labeled as class C;
(8)Generate_RBDT(S, A);
```

The information gain is calculated by following steps and the Entropy (H) is provide the splitting criteria variable used to find the purity (or impurity) of attribute.

$$H\left(\frac{p}{p+n},\frac{n}{p+n}\right) = -\frac{p}{p+n}\log_2\frac{p}{p+n} - \frac{n}{p+n}\log_2\frac{n}{p+n} \tag{4}$$

Here p-the probability of positives (yes or true samples), q-the probability of negatives (no or false samples).Generally the training set has combination of p positives and n negatives and which are input to root to train. In this the attribute variance (in dataset-1) and PAN (in dataset-2) contains different values so the expected entropy is required, so the expected entropy (EH) (Average Entropy of children's) is calculated for attribute A as:

$$EH(A) = \sum_{i=1}^{k}\frac{p_i+n_i}{p+n}\,.\,H\left(\frac{p_i}{p_i+n_i},\frac{n_i}{p_i+n_i}\right) \tag{5}$$

Where the k-distinct value or partitions, i- particular child and $p_i + n_i$ is assume as parent then information-gain (A) is:

$$I(A) = H\left(\frac{p}{p+n},\frac{n}{p+n}\right) - EH(A) \tag{6}$$

For dataset-2, the decision tree constructed based on the attribute candidate set C has three attributes PAN, P-MODE, DATE and for all these then I_g is calculated, but attribute PAN has I_g and variance attribute has highest information-gain for dataset-1.

4 Case Studies

4.1 Datasets

For our experiment, we have chosen two datasets 1) Banknote Authentication from UCI machine learning repository and 2) Real time bank dataset collected from private CA (Chartered Accountant).

Banknote authentication: This dataset provided at UCI machine learning repository [36] and it is collection of feature data extracted from image, transformed into 4 attributes of multivariate data. This dataset is actually extracted to authenticate the bank specimens and the data was digitized by 400x400 pixel ratio with 600 dpi resolution. This dataset has 1372 observations (instances) without any missing values with 4 continuous and 1 class attribute. For our experiment the classification task is trained with 500 observations and tested on remaining samples with yes or no (1 or 0) basis.

Real time bank dataset: The dataset is a bank database consists of 51095 records about transactions made during 2006 and 2013, but in that maximum transactions are held in 2008 to 2010. Out of these 6356 unique records are identified and consider as training set with learning rules .The original data set has 9 attributes, but we have concentrated on attributes which has highest I_g (E). We have given the experimental statistics on this dataset and 3 core attributes was considered as key attribute set to elevate results in work.

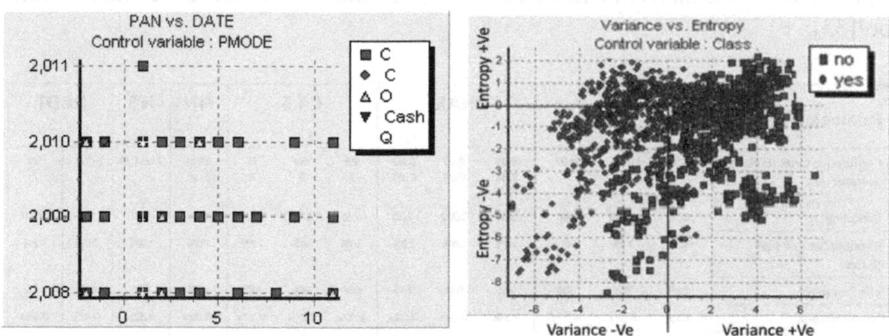

Fig. 3 Scatter plots for two datasets. Dataset-2 is depending on control variable as PMODE (left) shows how many records are grouping into bins in each year, here {Cash, C,C} are same kind, Q-cheque, O-other. In Fig (right) shows on dataset-1 control variable as class and the regions determine negative instances of entropy and variance.

4.2 Normalization of Datasets

Banknote authentication: This dataset has two main key attributes which are Variance and Entropy has high I_g, but as per statistical ethics the variance and entropy are a non-negative [37] and also mutually inversely proportional to each other and plots shown in Fig.3 (right). In Gaussian (normal distribution) instance, the mutual relation and dependency is exist between variance and entropy: it is like maximizes entropy (E) for a given variance (V) and at the same time minimizes the variance for given entropy. In our experiment we need to fit the data in specific range and classify the samples effectively by different normalization criteria. We have chosen the function mat2grey () to give the range of values between 0 and 1. The mat2grey function is used to convert matrix in scaling of image values between 0 and 1

Real time bank dataset: The dataset has different combination of attributes and we know that the attribute PAN has highest I_g , but it is continuous and alphanumeric based value. So it is not possible to classify the records as per original set, for this we have uses groups of PAN status and substituted with decimal value and we have given some substitutions for PAN status given in Fig.3 (left). The PAN is split attribute and the next level attribute is PAN-Status is normalized by decimal scaling ($V/10^2$).

5 Experiments and Results

In this paper, for our experimental work, the datasets related to same application was taken and consider as benchmark. The both case studies (Banknote authentication from UCI [36] and Real time bank dataset) are simulated on Intel Pentium CPU 3GHz speed, 2GB RAM, Windows-XP OS , Matlab -32 bit version[32] for Rules and performance benchmark for Case studies is SIPINA data mining tool[33].

Parameters \ Alg'm	ID3			CHAID			C4.5			NN	NB	RBDT	
Confidence Level (P-Value) -parameters	0.05	0.05	0.07	0.05 0.01	0.07 0.01	0.07 0.01	25 2	50 2	75 2	100 I 5 n/l	U.C.C.D	25 2	50 2
Sampling	500	1000	500	500	1000	500	500	500	1000	500	500	500	1000
Random Sample size (Filter)	50%	72%	50%	50%	50%	72%	50%	50%	72%	50%	50%	50%	72%
Idle Samples	686	385	686	686	686	385	686	686	385	676	676	676	385
Learning Time	0.32s	0.34s	0.33s	0.34s	0.34s	0.34s	0.37s	0.36s	0.37s	0.32s	0.12s	0.62s	0.69s
Testing Time	0.31s	3.44s	0.32s	0.37s	0.31s	0.32s	0.34s	0.32s	0.34s	0.38s	0.25s	0.52s	0.64s

Fig. 4 Describe the learning time and testing time with different kind of training parameters taken for different algorithms on dataset-1 (Banknote authentication). Here NB takes very less time compare to all, ID3 also take less time due to simplicity in algorithm, where as others CHAID, C4.5, NN takes more time due to additional parameters and rules play major role. In RBDT, take much more time.

In Fig.4 and Table.1, the algorithms like ID3, CHAID, C4.5 and RBDT are based confidence level (or p-value). For NN, multi layer perceptrons (MLP) considered with parameters are no. of iterations, nodes per layer, one hidden layer and maximum error =0.05. All parameters are taken at default initially and gradually changed. In naïve bayes (NB) the default prior consider as unconditional class distribution (UCCD) and same as all classes (All C). In Table.1, we have shown all possible results with different combinations and highlighted some values showing good in performance.

Table 1 Shows the performance of all algorithms tested on dataset-1. Here FPR-false positive rate, DR-detection rate, ACC-accuracy, ER-error rate. All results were taken at training set of 500 samples.

Samples (Observations) Tested:	50%				72%				100%			
Method	FPR	DR	ACC	E.R	FPR	DR	ACC	E.R	FPR	DR	ACC	E.R
ID3 p = 0.05	0.05	0.90	0.93	0.07	0.05	0.89	0.92	0.08	0.07	0.92	0.92	0.08
ID3 p = 0.07	0.10	0.90	0.91	0.09	0.07	0.89	0.93	0.07	--	--	--	--
CHAID p = 0.05, 0.01	0.14	0.94	0.89	0.11	0.07	0.91	0.92	0.08	0.08	0.92	0.92	0.08
CHAID p = 0.07, 0.01	0.05	0.90	0.93	0.07	0.09	0.90	0.91	0.09	--	--	--	--
C4.5 25, 2	0.08	0.93	0.93	0.07	0.09	**0.96**	0.93	0.07	0.09	**0.95**	**0.93**	0.07
C4.5 15, 2	**0.01**	0.90	0.94	0.06	0.11	0.91	0.90	0.10	--	--	--	--
C4.5 50, 2	0.08	0.89	0.91	0.09	0.07	0.90	0.91	0.09	0.10	0.93	0.91	0.09
C4.5 75, 2	0.09	0.90	0.91	0.09	0.06	0.88	0.92	0.08	--	--	--	--
C4.5 75, 3	0.08	0.89	0.91	0.09	--	--	--	--	0.10	0.93	0.91	0.09
I-C4.5 25, 2	0.12	0.86	0.87	0.13	0.06	0.90	0.92	0.08	--	--	--	--
I-C4.5 15, 2	0.04	0.95	0.93	0.07	0.07	0.07	0.84	0.16	0.10	0.95	0.92	0.08
I-C4.550, 2	**0.02**	0.82	0.87	0.13	0.12	0.82	0.86	0.14	--	--	--	--
I-C4.5 75, 2	0.13	0.86	0.87	0.13	--	--	--	--	0.09	0.88	0.85	0.15
I-C4.5 75, 3	0.11	0.87	0.90	0.10	--	--	--	--	0.10	0.81	0.86	0.14
Rule C4.5 25	0.05	**0.96**	**0.94**	0.06	0.06	0.95	0.92	0.08	0.09	0.92	0.91	0.09
Rule C4.5 50	--	--	--	--	0.09	0.94	0.92	0.08	0.09	**0.99**	**0.99**	**0.01**
NN (MLP) 100, 5	**0.01**	0.91	0.96	0.04	**0.01**	0.62	0.72	0.28	**0.01**	0.92	0.96	0.04
NN (MLP) 200, 5	**0.01**	0.91	0.96	0.04	--	--	--	--	**0.01**	0.92	0.96	0.04
NB Prior=0.5 UCCI	0.06	0.92	0.93	0.07	0.08	0.95	0.93	0.07	0.07	**0.95**	**0.94**	0.06
NB Prior=0.5 All C	0.06	0.93	0.94	0.06	0.06	0.92	0.93	0.07	0.07	0.95	0.94	0.06
RBDT 25, 2	0.03	**0.98**	**0.97**	0.03	0.05	**0.96**	**0.96**	0.04	0.07	**0.96**	**0.98**	0.02
RBDT 50, 2	0.03	**0.97**	**0.98**	0.02	0.05	0.95	**0.96**	0.04	0.04	**0.97**	**0.97**	0.03

Parameters \ Alg'm	ID3			CHAID			C4.5			RBDT	
Confidence Level (P-Value) -parameters	0.05	0.05	0.1	0.05 0.01	0.05 0.02	0.1 0.01	25 2	50 2	25 2	25 2	50 >=2
Sampling	500	1000	5000	500	1000	5000	500	1000	6355	1000	5000
Random Sample size (Filter)	50%	75%	97%	50%	75%	97%	50%	75%	100%	75%	100%
Idle Samples	3177	1589	191	3178	1589	191	3178	1589	0	1589	0
Learning Time	0.94s	0.78s	0.63s	0.34s	0.78s	0.63s	0.62s	0.62s	0.62	0.82s	0.96s
Input Ratio	97% 0.03% 2.87%	97% 0.9% 2.3%	97% 0.03% 2.87%	97% 0.03% 2.87%	97% 1% 2%	97% 0% 3%	96% 1% 3%	97% 0% 3%	97% 0% 3%	96% 1% 3%	97% 0.03% 2.87%

Fig. 5 Describe the learning time and testing time with different kind of training parameters taken for different algorithms on dataset-2 (Real time bank dataset). Here we also shown input ratio of participating samples in training.

In Fig.5 and Table.2, the dataset-2 is simulated and results were given on default parameters and after little changes. For all the training parameters were consider same as in dataset-1, but compare to first the second case study showing poor performance. The accuracy is almost equal in both cases, but not in FPR. The some of the best results are highlighted for all algorithms applied on dataset. In this section we have shown the possible experimental work on both datasets with visual plots in Fig.6 and Fig.7. The plots are used here to present comparison of performance parameters. In Fig-6, we observe that our algorithm RBDT is well in predictive accuracy to classify the instances correctly, but for dataset-2 it is little bit low compares to C4.5 and ID3. The RBDT algorithm is also good for the indication of low FPR compare to all, in dataset-1 NN also give 0.1 only.

Table 2 Shows the performance of all algorithms tested on dataset-2. Here FPR-false positive rate, DR-detection rate, ACC-accuracy, ER-error rate. All the results were noted on training of 6356 records (data instances) and test on remaining.

Records Tested	**50%**				**100%**			
Method	**FPR**	**DR**	**ACC**	**E.R**	**FPR**	**DR**	**ACC**	**E.R**
ID3 p = 0.05	0.77	0.98	0.97	0.03	0.73	**0.99**	0.97	0.03
ID3 p = 0.07	**0.47**	0.99	**0.98**	0.02	**0.42**	**0.99**	**0.99**	0.01
CHAID p = 0.05, 0.01	0.80	**0.99**	0.97	0.03	0.73	0.99	0.97	0.03
CHAID p = 0.03, 0.02	0.80	0.98	0.97	0.03	--	--	--	--
C4.5 25, 2	0.70	0.98	0.97	0.03	0.42	**0.99**	**0.99**	0.01
C4.5 15, 2	0.80	**0.99**	0.97	0.03	--	--	--	--
C4.5 75, 2	0.61	**0.99**	0.97	0.03	0.71	**0.99**	**0.99**	0.01
I C4.5 25, 2	0.80	**0.99**	0.97	0.03	0.69	0.98	0.97	0.03
I C4.5 15, 2	0.61	**0.99**	**0.98**	0.02	--	--	--	--
Rule C4.5 25	0.65	**0.99**	**0.98**	0.02	0.72	0.99	0.97	0.03
Rule C4.5 15	0.73	0.99	0.97	0.03	--	--	--	--
NN (MLP) 100, 5	0.74	0.99	0.97	0.03	0.85	0.98	0.97	0.03
NN (MLP) 200, 5	0.84	0.99	0. 97	0.03	--	--	--	--
NB Prior=0.5 UCCI	0.84	0.99	**0.98**	0.02	0.85	0.98	0.97	0.03
NB Prior=0.5 All C	0.80	0.98	0.97	0.03	0.84	0.99	0.97	0.03
RBDT 25, 2	**0.38**	0.98	**0.98**	0.02	**0.31**	0.96	0.97	0.03

5.1 Discussions

From above Fig.6 and Fig.7, the proposed learning algorithm is shown some good results compare to ID3, CHIAD, C4.5, NB. So the conjunction rules of both C4.5 and Naïve Bayes always show affective performance. The rule C4.5 is good for dataset-1 when confidence level=25, 0.94 accuracy is given and also FPR is low, but while testing 72% of data instances its performance is not well compare to total instances. In above all classifiers used in dataset-1, the FPR is very less by NN (MLP) only, but same NN will gill give poor results in detecting true positives (DR). The DR is recorded less for RBDT in both datasets, 0.98 (average) for dataset-1 with all kind of combinations and testing 50%, 72% and 100% samples(observations). The NB also good in accuracy and DR noted 0.95 after testing

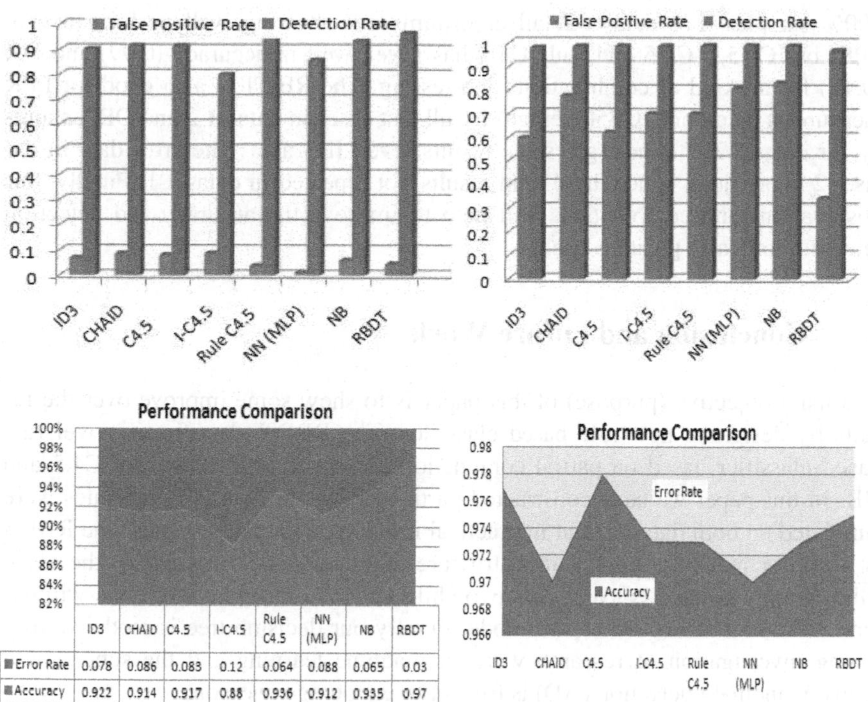

	ID3	CHAID	C4.5	I-C4.5	Rule C4.5	NN (MLP)	NB	RBDT
Error Rate	0.078	0.086	0.083	0.12	0.064	0.088	0.065	0.03
Accuracy	0.922	0.914	0.917	0.88	0.936	0.912	0.935	0.97

Fig. 6 The performance parameters and comparison for both datasets (1- Left, 2-Right) and classification algorithm on X-axis and rate of performance on Y-axis to be taken: The Fig (top left) shows the FPR and DR for dataset-1 and top right for dataset-2. Bottom left shows the performance comparison of all classifiers used in this paper (minimum 82%) for dataset-1and bottom right shows for dataset-2.

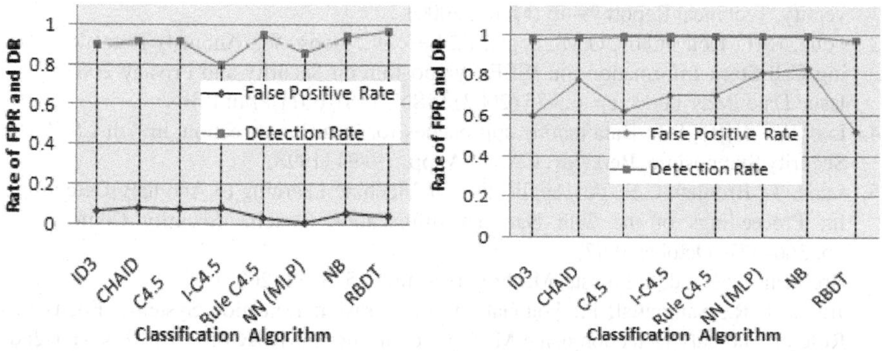

Fig. 7 The ROC is plotted for both datasets-1(left) and dataset-2(right) between FPR and DR with respect to classifiers

100% samples. For dataset-2, all algorithms not showing well performance in FPR, but C4.5, I-C4.5 and Rule C4.5 has given well in accuracy (0.99) and DR (0.98) for all kind of combinations and testing. The RBDT is also good for FPR measure, it noted only 0.35(average) for all test cases in dataset-2 and DR, accuracy for dataset-2 is almost got same results, even ID3 also. The error data in dataset-2 is problem, hence these kind results not repeated in dataset-1. Finally, our observations and combinations will help to improve the accuracy and detection rate with low false positive rate.

6 Conclusion and Future Work

The main objective (purpose) of this paper is to show some improve over the results by designing new rule based classifier. The RBDT classifier is novel rule based classifier based on partial conjunction and cascading rules from C4.5 and NB. In this paper we have constructed partial RBDT for dataset-2 and rules were simulated on both datasets, but in practical it is long and complex one. The RBDT is good for real time critical infrastructure applications. In this paper, the algorithm is only pointed conventional two child split. So in future work, we will extend it for more than two at each node. Finally our decision tree is modeled after strong investigation of research work in machine learning and the subject will focus in anomaly detection (AD) is fruitful to encourage everyone.

References

1. Denning, D.E.: An intrusion detection model. IEEE Transactions on Software Engineering (1987)
2. Axelsson, S.: Intrusion Detection Systems: A Survey and Taxonomy, Chalmers University. Technical Report 99-15 (March 2000)
3. Feng, H.H., Kolesnikov, O.M., Fogla, P., Lee, W., Gong, W.: Anomaly Detection Using Call Stack Information. In: IEEE Symposium on Security and Privacy 2003, CA, Issue Date: May 11-14, pp. 62–75 (2003) ISSN: 1081-6011 Print ISBN
4. Lee, W., Stolfo, S.J.: Data mining approaches for intrusion detection. In: 7th USENIX Security Symposium, Berkeley, CA, USA, pp. 79–94 (1998)
5. Lane, T., Brodley, C.E.: An Application of Machine Learning to Anomaly Detection. In: Proceedings of the 20th National Information Systems Security Conference, pp. 366–377 (October 1997)
6. Breiman, L.: Random Forests. Machine Learning 45, 5–32 (2001)
7. Jidiga, G.R., Sammulal, P.: Foundations of Intrusion Detection Systems: Focus on Role of Anomaly Detection using Machine Learning. In: ICACM - 2013 Elsevier 2nd International Conference (August 2013) ISBN No: 9789351071495
8. Jidiga, G.R., Sammulal, P.: The Need of Awareness in Cyber Security with a Case Study. In: Proceedings of the 4th IEEE Conference (ICCCNT), Thiruchengode, TN, India, July 4-6 (2013)
9. Kass, G.V.: An Exploratory Technique for Investigating Large Quantities of Categorical Data. Applied Statistics 29(2), 119–127 (1980)

10. Quinlan, J.R.: Induction of decision trees, Machine Learning 1, pp. 81–106. Kluwer Publishers (1986)
11. Quinlan, J.R.: Simplifying decision trees. International Journal of Man Machine Studies 27, 221–234 (1987)
12. Quinlan, J.R.: Decision Trees and Multivalued Attributes. In: Richards, J. (ed.) Machine Intelligence, vol. 11, pp. 305–318. Oxford Univ. Press, Oxford (1988)
13. Quinlan, J.R.: Unknown attribute values in induction. In: Proceedings of the Sixth International Machine Learning Workshop Cornell. Morgan Kaufmann, New York (1989)
14. Quinlan, J.R.: C4.5: Programs for Machine Learning. Morgan Kaufmann, Los Altos (1993)
15. Quinlan, J.R., Rivest, R.L.: Inferring Decision Trees Using The Minimum Description Length Principle. Information and Computation 80, 227–248 (1989)
16. Breiman, L., Friedman, J., Olshen, R., Stone, C.: Classification and Regression Trees. Wadsworth Int. Group (1984)
17. Russell, S., Norvig, P.: Artificial Intelligence: A Modern Approach, 3rd edn. Prentice-Hall (2009)
18. Polat, K., Güne, S.: A novel hybrid intelligent method based on C4. 5 decision tree classifier and one against all approach for multi-class classification problems. Expert Systems with Applications 36, 1587–1592 (2009)
19. Jiang, S., Yu, W.: A Combination Classification Algorithm Based on Outlier Detection and C4. 5. Springer Publications (2009)
20. Cohen, W.W.: Fast effective rule induction. In: Proceedings of the Twelfth International Conference on Machine Learning Chambery, France, pp. 115–123 (1993)
21. Yu, M., Ai, T.H.: Study of RS data classification based on rough sets and C4. 5 algorithms. In: Proceedings of the Society of Photo-Optical Instrumentation Engineers (SPIE) Conference Series (2009)
22. Yang, X.Y.: Decision tree induction with constrained number of leaf node. Master's Thesis, National Central University (NCU-T), Taiwan (2009)
23. Michael, J.A., Gordon, S.L.: Data mining technique for marketing, sales and customer support. Wiley, New York (1997)
24. Loh, W.Y., Shih, Y.S.: Split selection methods for classification trees. Statistica Sinica 7, 815–840 (1997)
25. Clark, P., Niblett, T.: The CN2 induction algorithm. Machine Learning 3, 261–283 (1989)
26. Clark, P., Boswell, R.: Rule induction with CN2: Some recent improvements. In: Kodratoff, Y. (ed.) EWSL 1991. LNCS, vol. 482, Springer, Heidelberg (1991)
27. Rakotomalala, R., Lallich, S.: Handling noise with generalized entropy of type beta in induction graphs algorithm. In: Proceedings of International Conference on Computer Science and Informatics, pp. 25–27 (1998)
28. Chauchat, J.H., Rakotomalala, R., Carloz, M., Pelletier, C.: Targeting customer groups using gain and cost matrix: a marketing application. In: Proceedings of Data Mining for Marketing Applications Workshop (PKDD), pp. 1–13 (2001)
29. Rakotomalala, R., Lallich, S., Di Palma, S.: Studying the behavior of generalized entropy in induction trees using a m-of-n concept. In: Żytkow, J.M., Rauch, J. (eds.) PKDD 1999. LNCS (LNAI), vol. 1704, pp. 510–517. Springer, Heidelberg (1999)
30. Rajeswari, P., Kannan, A.: An active rule approach for network intrusion detection with enhanced C4.5 Algorithm. International Journal of Communications Network and System Sciences, 285–385 (2008)

31. Ghosh, A., Schwartzbard, A.: A study using NN for anomaly detection and misuse detection. Reliable Software Technologies, `http://www.docshow.net/ids/useni`
32. `http://www.mathworks.in/products/matlab/`
33. `http://eric.univ-lyon2.fr/~ricco/sipina.html`
34. Benferhat, A.S., Elouedi, Z.: Naive Bayes vs Decision Trees in Intrusion Detection Systems. In: Proc. ACM Symp. Applied Computing (SAC 2004), pp. 420–424 (2004)
35. Rokach, L., Maimon, O.: Decision Trees
36. Bache, K., Lichman, M.: UCI Machine Learning Repository. University of California, School of Information and Computer Science, CA (2013), `http://archive.ics.uci.edu/ml`
37. Usta, I., Kantar, Y.M.: Mean-Variance-Skewness-Entropy Measures: A Multi-Objective Approach for Portfolio Selection. Entropy 13, 117–133 (2011), doi:10.3390/e13010117
38. Abdelhalim, A., Traore, I.: Converting Declarative Rules into Decision Trees. In: Proceedings of the World Congress on Engineering and Computer Science, Vol-I WCECS 2009, San Francisco, USA, October 20-22 (2009)
39. Abdelhalim, A.: Issa Traore, The RBDT-1 method for rule-based decision tree generation. Technical report (ECE-09-1), University of Victoria, STN CSC, Victoria, BC, Canada (July 2009)
40. Siva, S., Sindhu, S., Geetha, S., Kannan, A.: Decision tree based light weight intrusion detection using a wrapper approach. Elsevier-Expert Systems with Applications 39, 129–141 (2011), doi:10.1016/j.eswa.2011.06.013
41. Lowd, D., Davis, J.: Improving Markov Network Structure Learning Using Decision Trees. Journal of Machine Learning Research 15, 501–532 (2014)
42. Zaidi, N.A., Cerquides, J., Carman, M.J.: Alleviating Naive Bayes Attribute Independence Assumption by Attribute Weighting. Journal of Machine Learning Research 14 (2013)
43. Anchiang, D., Chen, W., Fanwang, Y., Jinnhwang, A.: Rules Generation from the Decision Tree. Journal of Information Science and Engineering 17, 325–339 (2001)

Analysis and Evaluation of Discriminant Analysis Techniques for Multiclass Classification of Human Vocal Emotions

Swarna Kuchibhotla*, Hima Deepthi Vankayalapati, BhanuSree Yalamanchili, and Koteswara Rao Anne

Abstract. Many of the classification problems in human computer interaction applications involve multi class classification. Support Vector Machines excel at binary classification problems and cannot be easily extended to multi class classification. The use of Discriminant analysis how ever is not experimented widely in the area of Speech emotion recognition. In this paper Linear Discriminant Analysis and Regularized Discriminant Analysis are implemented over Berlin and Spanish emotional speech databases. Prosody and spectral features are extracted from the speech database and are applied individually and also with feature fusion. Based on the results obtained, LDA classification performance is poor than RDA due to the singularity problem. The results are analysed using ROC Curves.

Keywords: Mel Frequency Cepstral Coefficients (MFCC), Pitch, Energy, Linear Discriminant Analysis (LDA), Regularized Discriminant Analysis(RDA).

1 Introduction

The Automatic recognition of emotions has received much attention for building more intuitive human computer interfaces. Vocal information provides two types of information that are relevant for emotions: acoustic properties and linguistic

Swarna Kuchibhotla
Research Scholar, Acharya Nagarjuna University, Guntur, A.P., India

Hima Deepthi Vankayalapati
Department of Computer Science and Engineering, VRSEC, Vijayawada, India

BhanuSree Yalamanchili · Koteswara Rao Anne
Department of Information Technology, VRSEC, Vijayawada, A.P., India

* Corresponding author.

El-Sayed M. El-Alfy et al. (eds.), *Advances in Intelligent Informatics*,
Advances in Intelligent Systems and Computing 320, DOI: 10.1007/978-3-319-11218-3_30

content. In this paper acoustic features are considered. In the design of speech emotion recognition system the extraction of suitable features place a major role since pattern recognition techniques are rarely independent of problem domain, it is believed that a proper selection of features significantly affects the classification performance[1]. Usually speech features can be grouped into four categories: prosody, qualitative, spectral and TEO based. Based on many researchers prosody and spectral features convey much of the emotional content for an utterance[2].

Classification tasks aim to assign a pre defined class to each instance. It can help to understand existing data and be used to predict how the new instances will behave. The typical examples of classification includes optical character recognition, part of speech tagging, text categorization, speech emotion recognition etc.,[3]

This paper is organized as follows. Section 2 describes the feature extraction techniques, section 3 describes the classification algorithms, Section 4 describes the emotional databases, Section 5 describes the experimental results and Section 6 describes the conclusion.

2 Feature Extraction

The speech signal is divided into number of frames with 256 samples per each frame, a hamming window is applied for each frame and an overlapping of 100 samples is used. Features are evaluated for each frame.

Prosody Statistics: we refer pitch and energy curves as prosody features and these values are estimated over simple 6 statistics (mean, variance, minimum, range, skew and kurtosis) so we call it as prosody statistics. The pitch and energy rate provides new useful information so their first and second derivatives are also calculated. [5]. Energy and pitch were estimated for each frame together with their first and second derivatives, providing 6 features per frame and the same 6 statistics are applied totally we get 36 features. The speech signal and its energy, pitch curves are as shown in Fig.1.

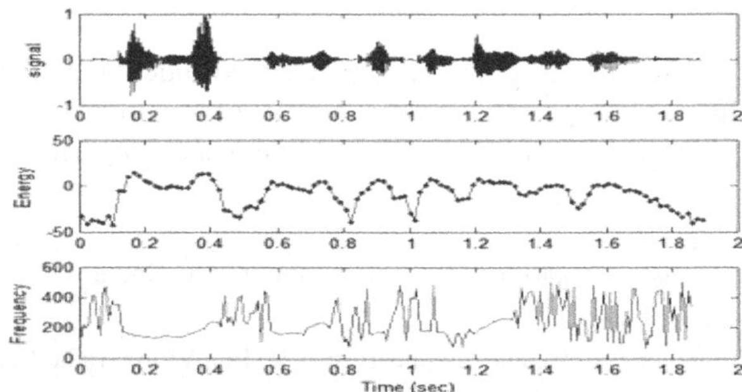

Fig. 1 Energy and Pitch plots of a speech signal

Spectral Statistics: Mel frequency cepstral coefficients (MFCC), the spectral representation of speech are most efficient in order to extract the correct emotional state. Reference [4] shows spectral features outperform prosody. The Mel Frequency Cepstral Coefficients are shown in Fig. 2.

Eighteen MFCC Coefficients and their first and second derivatives are estimated for each frame giving a total of 54 spectral features .The same six statistics applied to these 54 values we get 54*6=324 coefficients.

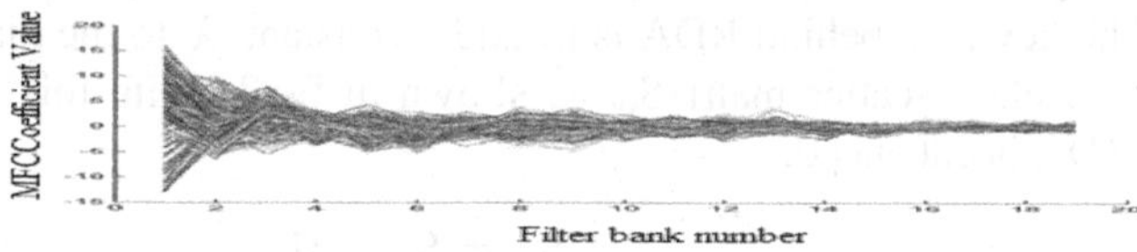

Fig. 2 Obatained MFCCoefficents

3 Classification of Speech Samples

The Purpose of classifier is to assign a test speech sample to one of the pre determined classes. Here two types of classifiers viz Linear Discriminant Analysis and Regularized Discriminant Analysis are implemented and a detailed analysis of their comparative study of the results are discussed in section 5.

3.1 Linear Discriminant Analysis (LDA)

LDA is a statistical method which minimizes the within class covariance and maximizes the between class covariance in the given data base of speech signals [6]. The within class scatter matrix S_w is the sum of class dependent scatter matrix, which is the sum of covariance matrices of individual classes and is given by Eq. 1.

$$S_w = \frac{1}{N}\sum_{i=1}^{n}\sum_{x\epsilon c_i}(x-\bar{\mu}_i)(x-\bar{\mu}_i)' \quad (1)$$

The between class scatter matrix S_b gives the sum of variances between different classes and is given by the Eq. 2. where N_i is the no of samples in class i.

$$S_b = \frac{1}{N}\sum_{i=1}^{n} N_i(\bar{\mu}_i-\mu)(\bar{\mu}_i-\mu)' \quad (2)$$

The transformation matrix W is computed using the following Eq.3.

$$S_w{}^{-1}S_b W = W\lambda \quad (3)$$

The train and test samples are to be projected on to this transformation matrix. To classify the test speech signal we consider the difference between the projected train and test samples. This algorithm is optimal only whenhe scatter matrix S_w is non-singular. If the matrix is singular, we get a warning that Matrix is close to singular or badly scaled. We call it as a singularity problem. The problem occurs because the dimensionality of the data is more when compared with number of classes .This is solved by using Regularized Discriminant Analysis which is discussed in section 3.2

3.2 Regularized Discriminant Analysis(RDA)

In this paper Regularized Discriminant Analysis is proposed for speech emotion recognition. The key idea behind RDA is to add a constant λ to the diagonal elements of with in class scatter matrixS_w as shown in Eq.3. With this the performance of the RDA is enhanced.

$$S_w = S_w + \lambda I \tag{4}$$

where λ is regularized parameter which is relatively small such that S_w is positive definite. In our paper the value of λ is 0.001. It is somehow difficult in estimation of regularization parameter value in RDA as higher values of λ will disturb the information in the within class scatter matrix and lower values of λ will not solve the singularity problem in LDA [7].

Similar to the LDA, once the transformation matrix W is given, the speech samples are projected on to this W. After projection, the Euclidian distance between each train sample and the test sample are calculated, the minimum value among them will classify the test speech sample.

4 Speech Databases

In this paper we used Berlin and Spanish emotional speech data bases. The reason for choosing these databases is both are multi speaker databases, so it is possible to perform speaker independent tests. The databases contain the following emotions: anger, boredom, disgust, fear, happiness, sadness and neutral. The Berlin database [8] contains 500 speech samples and are simulated by ten professional native German actors,5 male and 5 female [9].The number of speech files are as shown in Table 1.

Table 1 Number of emotional speech files in Berlin Database

Emotion	Anger	Boredom	Happy	Fear	Sad	Disgust	Neutral	Total
No.of Files	127	81	71	69	62	46	79	535

Spanish database contains 184 sentences for each emotion which include numbers, words, sentences etc. as shown in Table 2.The corpus comprises of recordings from two professional actors, one male and one female.

Table 2 Item identfier for Spanish Database

Item identifier	Spanish Corpus contents
1 to 100	Affirmative sentences including short and longer ones
101 to 134	Interrogative and (5) stressed sentences.
135 to 150	Paragraphs
151 to 160	Digits
161 to 184	Isolated Words

5 Experimental Result

In this section, we evaluate the performance of LDA and RDA classifiers for each emotion based on feature extraction approaches using Berlin and Spanish databases.

A) Performance evaluation of Linear Discriminant Analysis (LDA)
The emotion recognition performance of LDA is evaluated as follows

Table 3(a) Confusion Matrix for LDA with feature fusion

LDA	Berlin				Spanish			
Emotion	Happy	Neutral	Anger	Sad	Happy	Neutral	Anger	Sad
Happy	25	2	18	5	24	5	15	6
Neutral	6	29	-	15	4	28	7	11
Anger	11	5	34	-	9	-	32	9
Sad	-	8	6	36	-	12	3	35

Table 3(b) Accuracy percentage of each emotion on different parameters using LDA

LDA	Berlin			Spanish		
Emotion	Prosody	Spectral	Feature fusion	Prosody	Spectral	feature fusion
Happy	38	43	50	35	43	48
Neutral	37	40	58	40	45	56
Anger	50	62	68	45	56	64
Sad	43	60	72	39	52	70

Table 3(a) & 3(b). Summarizes the results with confusion matrices and performance accuracy of each feature along with their feature fusion. The recognition accuracy is very low with Prosody features and high with spectral features in both the databases. This shows that recognition rate is improved with spectral features and is increased further by concatenating these two features represented with column feature fusion in Table3(b).

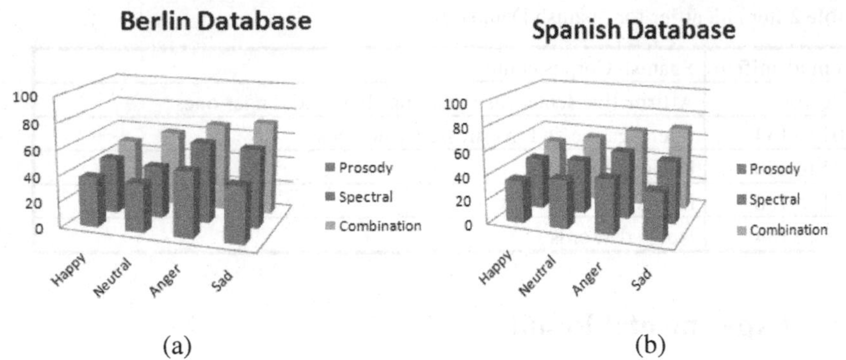

(a) (b)

Fig. 3(a)(b) Graphical representation of Accuracy for each emotion using LDA

The Blue, Red and Green lines in Fig. 3 shows the graphical representation of efficiencies of prosodic, spectral and their fusion. The recognition rate of Anger and sad are good when compared with recognition rate of happy and neutral for both the databases. The recognition rate is slightlymore for Berlin when compared with spanish.

B) Performance evaluation of Regularized Discriminant Analysis (RDA)

The Overall performance of LDA is less because of the singularity problem. To improve the accuracy we used Regularized discriminant analysis which solves the singularity problem.

Table 4(a) Confusion Matrix for RDA with feature fusion

RDA	Berlin				Spanish			
Emotion	Happy	Neutral	Anger	Sad	Happy	Neutral	Anger	Sad
Happy	36	2	10	2	35	2	9	4
Neutral	6	35	-	10	-	30	7	13
Anger	7	-	41	2	11	-	34	5
Sad	-	1	-	49	1	-	-	49

Table 4(b) Accuracy Percentage of each emotion on different parameters using RDA

RDA	Berlin			Spanish		
Emotion	Prosody	Spectral	feature fusion	Prosody	Spectral	feature fusion
Happy	47	70	72	42	54	70
Neutral	46	68	70	37	47	60
Anger	77	80	82	40	53	68
Sad	97	97	98	57	90	98

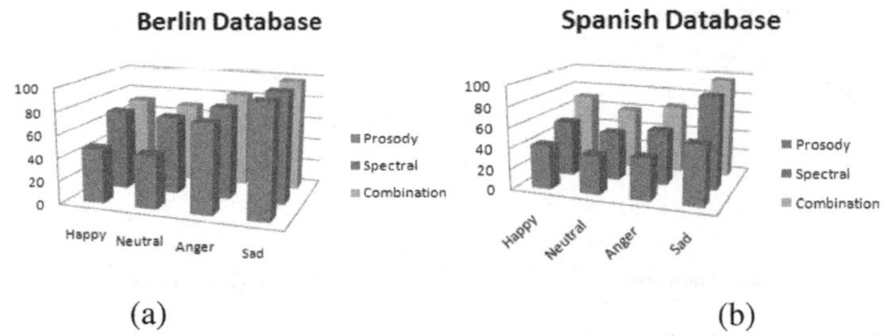

(a) (b)

Fig. 4 Graphical representation of Accuracy for each emotion using RDA

The performance of RDA is further enhanced with each feature and also with feature fusion. The results are shown with confusion matrices along with their efficiencies for each feature are shown in Table 4(a) & 4(b).

C) Performance Comparision of LDA and RDA

The overall performance of the classifiers is as shown in Table 5. With feature fusion, the performance is improved in RDA with 81% for Berlin and 74% for Spanish. The efficiency is more for RDA when compared with LDA in both databases. In terms of features, efficiency is more for spectral features when compared with prosody features.

Table 5 Performance comparison of LDA and RDA

Database	Berlin(%)		Spanish(%)	
Algorithm	LDA	RDA	LDA	RDA
Prosody	40	67	42	44
Spectral	49	79	51	61
Feature fusion	62	81	60	74

D) Comparison of classifiers using ROC Curves

ROC curve is a graphical representation of the relationship between both sensitivity and specificity. Online plotting of ROC mechanism is used to draw these curves. The Fig. 5 shows the ROC plot for LDA and RDA with all the features. From this figure we extract accuracy, sensitivity and specificity as well as area under curve as shown in the Table 6. If the Area Under Curve (AUC) is more, classifier efficiency is more, and vice versa.

The Area Under Curve is greater than 0.8 for RDA, then we can say that it is a very good classifier and AUC of LDA is greater than 0.6 then we say that this classifier is of sufficient one.

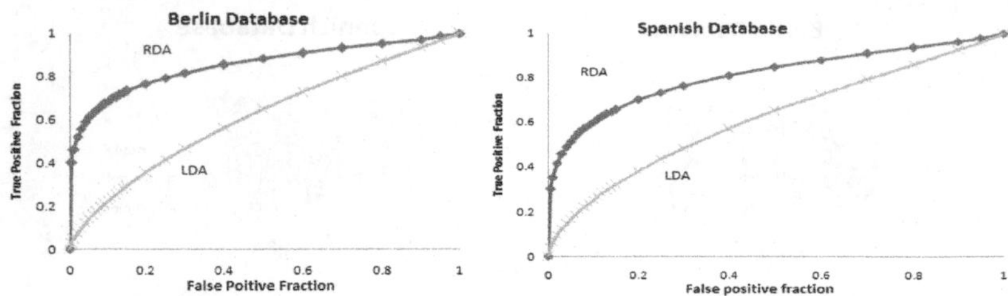

Fig. 5 (a) ROC plot for Berlin database (b) ROC plot for Spanish Database

Table 6 Values extracted from Berlin ROC Plot,AUC is Area Under Curve

Database	Berlin(%)		Spanish(%)	
Algorithm	LDA	RDA	LDA	RDA
Accuracy	62.00	80.80	59.00	74.00
Sensitivity	60.30	75.70	57.70	70.70
Specificity	64.30	88.20	60.80	78.40
AUC	0.619	0.854	0.601	0.814

6 Conclusions

Speech emotion recognition has been systematically evaluated by using Berlin and Spanish databases. To improve the performance of speech emotion recognition Regularization technique has been proposed. The results shows that the emotion recognition performance is considerably increased with RDA. The feature fusion and spectral features outperforms the prosody features and yields better results. The ROC Curves plotted against the classifiers shows the performance deviation very efficiently.

Future work should focus on the extraction of other features and detection of voiced and unvoiced frames to improve the recognition accuracy.

References

1. El Ayadi, M., Kamel, M.S., Karray, F.: Survey on speech emotion recognition: Features, classification schemes, and databases. Pattern Recognition 44(3), 572–587 (2011)
2. Cowie, R., et al.: Emotion recognition in human-computer interaction. IEEE Signal Processing Magazine 18(1), 32–80 (2001)
3. Ververidis, D., Kotropoulos, C.: Emotional speech recognition: Resources, features, and methods. Speech Communication 48, 1162–1181 (2006)
4. Luengo, I., Navas, E., Hernáez, I.: Feature Analysis and Evoluation for Automatic Emotion Identification in Speech. IEEE Transctions on Multimedia 12(6) (October 2010)

5. Luengo, I., et al.: Automatic emotion recognition using prosodic parameters. INTERSPEECH (2005)
6. Vankayalapati, H.D., Kyamakya, K.: Nonlinear feature extraction approaches for scalable face recognition applications. ISAST Transactions on Computers and Intelligent Systems 2 (2009)
7. Ye, J., et al.: Efficient model selection for regularized linear discriminant analysis. In: Proceedings of the 15th ACM International Conference on Information and Knowledge Management. ACM (2006)
8. Berlin emotional speech database, `http://www.expressive-speech.net/` (last accessed on October 25, 2012)
9. Milton, A., Sharmy Roy, S., Tamil Selvi, S.: SVM Scheme for Speech Emotion Recognition using MFCC Feature. International Journal of Computer Applications (0975 – 8887) 69(9) (2013)
10. Murray, I., Arnott, J.: Toward the simulation of emotion in syn-thetic speech: A review of the literature on human vocal emotion. Journal of the Acoustical Society of America 93(2), 1097–1108 (1993)

An Extended Chameleon Algorithm for Document Clustering

G. Veena and N.K. Lekha

Abstract. A lot of research work has been done in the area of concept mining and document similarity in past few years. But all these works were based on the statistical analysis of keywords. The major challenge in this area involves the preservation of semantics of the terms or phrases. Our paper proposes a graph model to represent the concept in the sentence level. The concept follows a triplet representation. A modified DB scan algorithm is used to cluster the extracted concepts. This cluster forms a belief network or probabilistic network. We use this network for extracting the most probable concepts in the document. In this paper we also proposes a new algorithm for document similarity. For the belief network comparison an extended chameleon Algorithm is also proposed here.

1 Introduction

Artificial Intelligence have a successful history in the area of concept mining. The introduction of semantic web followed by the ontology to this area have increased its efficiency and application to a wide range of problems. The earlier works related to the concept mining were based on the statistical approach ie, statistical analysis of the term frequency were considered as the basis for concept mining. This strategy of concept mining provided only the keyword based analysis and did not give any importance to the semantic of the term or phrase ie, the semantic of the concept retrieved was not preserved. This paper provides a novel approach for concept mining. This work also extend to the area of concept comparison and concept clustering. For the purpose of concept representation a new model called as the semantic net is constructed. A new conceptual based comparison model is also proposed here.

In a nutshell, the goal of our paper is to propose a new concept representation based on which comparison and clustering is done. We achieve our goal by

G. Veena · N.K. Lekha
AmritaVishwaVidyapeetham , Dept. of Computer Science and Application
e-mail: veenag@am.amrita.edu, lekhak37@gmail.com

El-Sayed M. El-Alfy et al. (eds.), *Advances in Intelligent Informatics*,
Advances in Intelligent Systems and Computing 320, DOI: 10.1007/978-3-319-11218-3_31

using different phases such as the preprocessing phase, concept extraction phase and a comparison and clustering phase which are explained in the following sections.In this paper, Section 2 describes the related works, section 3 describes the proposed solution followed by the experimental results in section 4 and conclusion in section 5.

2 Related Works

A great deal of work was done in the area of concept mining for past few years. The paper introduced in [1], gave an introduction to a concept based mining model, which included two modules a concept based mining model and concept based similarity measure. The concept based mining model retrieved concepts precisely but did not preserve the semantic of the concept which is more essential. In the next paper [2], emphasis was given to the concept mining which was further extended to the area of information retrieval. A Conceptual ontological graph(COG) was introduced as a part of the paper. The comparison method described in the paper was based on the concept word length. Considering the concept word length, the similarity was done on the basis of word length of the concept which did not preserve the semantic of the concept. In paper [3], a conceptual ontological graph (COG) was included and along with it a new module called as concept based weighting analysis was also introduced. The concept based weighting analysis assigns weight to the concepts and the highest assigned weighted concept was taken as the main concept of that sentence.

Most widely used text similarity measure was based on Vector Space Model (VSM) in [5,6,7]. Here the similarity was measured based on feature vector. In paper [8], the semantic matching of the concepts was done based on a match operator. The work did not focused on the overlaying relations. Paper [9], described semantic matching based on the comparison of labels of the node ie, on the bases of relation. Paper [10], proposed a Fuzzy Similarity based Concept Mining model (FSCMM) which was based on three measures Sentence level, Document level and Integrated corpus level. A Fuzzy Feature Category Similarity Analysis was also done for the similarity analysis. Paper [11], proposed a similarity measure based on the information distance and Kolmorgov Complexity. In this paper the similarity of concepts was analyzed and the similarity is measured between every pair of object according to the most dominant shared feature.

3 Proposed Solutions

An efficient concept based mining model is proposed here and finally we extend this model for the purpose of similarity analysis and clustering. The similarity analysis is done both on sentence and document level. The core part of our paper is an introduction of an efficient concept mining model. This model is designed based on a set of algorithms and graphs. The proposed solution approach gives answers to the following questions:

1)How does our model lead to an efficient concept extraction?

2)How does concept extraction lead to an efficient calculation of the similarity analysis and clustering?

The answers for the above mentioned questions can be obtained from three modules i)Preprocessing module ii)Concept extraction module iii) Comparison and Clustering module which is described in the subsection 3.1, 3.2 and 3.3 respectively.

3.1 Preprocessing

This module describes an efficient way for the generation of a verb-argument structure which finally leads to the concept extraction, Fig.1.

The pre-processing step includes document cleaning, Part of speech Tagging(POS Tagging) and Phrase structure tree generation. Document cleaning includes removal of unwanted characters such as numbers, punctuation etc. and tokenizing (ie,breaking the sentence into tokens with the help of delimiters). After this, Stanford Parser is used for the POS Tagging. Stanford Parser is a statistical parser which can parse input data written in several languages. The properly tagged sentences are then given to the Shallow semantic Parsing for the generation of verb-argument structures.

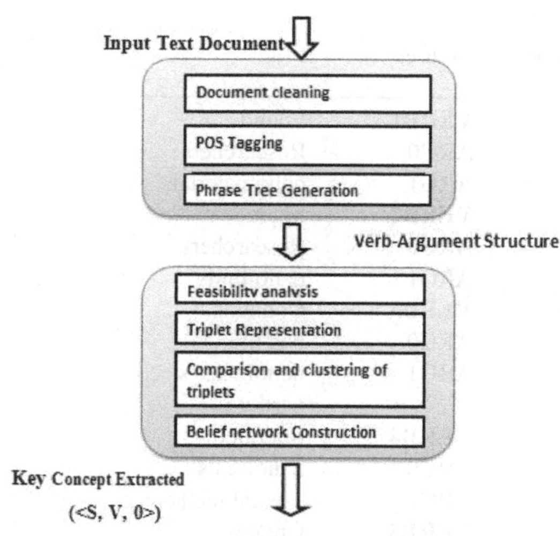

Fig. 1 Key Concept Extraction

For the formation of verb argument structure a phrase tree is generated from the tagged sentences and then a semantic role labeling is done based on the prop bank notation. Prop bank uses predicate independent labels such as ARG0, ARG1 etc as the labels. A sentence can contain more than one verb argument structure according to the number of verbs contained in it. The formation of the verb-argument structure after the labeling is done based on the path analysis.

Example 1: ***Researchers found nanomaterials and made nanofluids. Nanofluids contains many nanometer-sized particle. These fluids are supplied by several methods. Nanofluids and nanomaterials creates a greater evolution in the research area.***

In the above example there are three sentences and their corresponding verbs are found, made, contains, supplied and creates and their corresponding arguments (Subjects, objects) are Researchers, nanomaterials, nanofluids, nanometer-sized particles, Several methods and greater evolution in the research area ie, the verb-argument structures generated according this example is listed out in Table1.

3.2 Concept Extraction

From the verb-argument structure the root verb is extracted and finally its corresponding arguments ie, its left and right phrases known as the noun phrases are taken. For each <subject, verb, object> weight analysis is done. A TF-IDF measure is used for the calculation of weight of each verb, subject and object. We use a term frequency(tf) and an inverse document frequency(idf) for this calculation. The

Table 1 Verb-Argument Structures

VERB1	Found
ARG0	Researchers
ARG1	nanomaterials
VERB2	Made
ARG0	Researchers
ARG1	nanofluids
VERB3	Contains
ARG0	nanofluids
ARG1	nanometer-sized particles
VERB4	Supplied
ARG0	nanofluids
ARG1	several methods
VERB5	Creates
ARG0	nanomaterials
ARG1	greater evolution in the research area
VERB6	Creates
ARG0	nanofluids
ARG1	greater evolution in the research area

weight of a term t_i or a phrase p_i in jth document, $x(pij)$, is calculated as,

$$W(p_{i,j}) = tf(p_{i,j}) \times idf(p_i) \quad (1)$$

$$idf(p_i) = log \frac{(|S|)}{(|d_j: p_j \Sigma d_j|)} \quad (2)$$

where $tf(pi,j)$ is the number of times p_i occurs in jth document. S is the total number of sentences in the document.

A feasibility analysis is done for retrieving the most important concept. The feasibility analysis is done based on the term frequency of the concept in document and corpus level. The term frequency is calculated for each term or phrase. After the term frequency analysis of each concepts, the concept with higher term frequency is taken as the most feasible concept and their corresponding triplet representation is generated. The formation of the root verb along with its left and right entity gives us the triplet form based on that particular root verb. The triplet form is represented as the <S,V,O>, where V is the root verb, S is the subject and O is the object. Likewise for all verb argument structures in a document and their corresponding triplet forms are generated. A triplet generator algorithm is described in Algorithm 1.

Algorithm 1. Triplet Generation Algorithm

```
S is a new sentence
Declare Lv an empty list of verb
Declare Sub an empty list of subjects
Declare Obj an empty list of objects
for each verb-argument structures do
    Add verb to Lv
    Add Subject to Sub
    Add Object to Obj
end for
for each Subj_i and Obj_i do
    Calculate the feasibility
    if feasibility ≤ threshold(0.1) then
      remove it from Sub and Obj list
    end if
end for
for each each feasible term do
    retrieve their corresponding arguments (from the list of Sub,Obj) and create a link
between them.
end for
```

The triplets generated are associated with their corresponding weight which is the calculated TF-IDF value. The triplet generated for the the VERB0,ARG0 and ARG1 from Table1 and Algorithm1 will be <Researchers,Found,nanomaterials>.

3.3 Comparison and Clustering

The novel approach of this paper is the introduction of comparison and clustering of concepts. A similarity analysis is done on the concept. The similarity analysis is calculated on the bases of cosine similarity measures.

3.3.1 Cosine Similarity(CS)

Here the concepts in a document is compared according to a single term similarity measure, a cosine correlation similarity measure is adopted along with the Term/Inverse document frequency (TF-IDF) term weighting[17]. The cosine measure calculates the cosine of the two angles between the concept vectors. The similarity measure (sim_s) is:

$$sim_s(c_1, c_2) = cos(x, y) = \frac{(c_1.c_2)}{(\| c_1 \| . \| c_2 \|)} \tag{3}$$

The vectors c_1 and c_2 are represented as single-term weights calculated by using the TF-IDF weighting scheme.

The clustering is done using a proposed concept based DB scan algorithm which preserve the semantic of the document. This algorithm is found to outperforms all the other methods. A concept based clustering algorithm is proposed based on the DB scan:

Algorithm 2. Extented DB scan Algorithm

```
M is a set of Triplet T
Q is an empty queue
Pick T from M such that
if T is not yet classified then
    if T is having heightest weight(TF-IDF) then
        Compare with all other concepts in M based on CS(Cosine Similarity) value
        if match found then
            Add the matched concepts to Q and assign it to new cluster clstr_i
            for each C ∈ Q do
                Repeat steps (6) to (9)
            end for
        end if
    end if
end if
```

The proper comparison and clustering will lead to the semantic net construction. A semantic net construction is done based a similarity comparison between the subject-subject and subject-object from the constructed clusters. A semantic net is a knowledge representation model which is a directed graph consisting of node and their corresponding links. The nodes in the semantic graph represents the concepts and the links represent their relation. The basic idea of semantic nets is that it provides a graph theoretic structure for the concepts. The semantic net construction can be considered as a mathematical model called as a belief network.

Belief Network

A belief network is a probabilistic graphical model, that shows a set of random variables and their conditional dependencies via a directed acyclic graph (DAG). The belief network here represents the nodes as the subject and object and their conditional dependency to the other nodes. The conditional dependency explains the relationship between the subjects and objects.

For the construction of the belief network a noun phrase comparison is done inside each cluster. The comparison is based on the subject-subject and subject-object. The similarity analysis is also based on an efficient database which gives the synonym for each word called as the WordNet. The set of synonyms for a word is called synset. If there is no exact word similarity, then the corresponding synset will be checked for matching. So that original semantics are preserved.

The belief network generated based on Example1 is given in Fig.2.

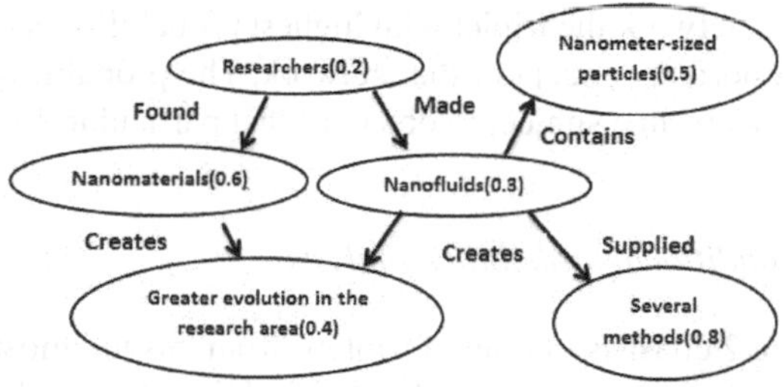

Fig. 2 Belief Network

Algorithm 3 explains the Belief network constructor algorithm:

Algorithm 3. Belief Network Constructor Algorithm

Clstr is a set of clusters
for each cluster $clstr_i$ **do**
  Compare the concepts in $clrstr_i$ (Noun Phrase Comparison)
  if matching is found **then**
    Add a link between the similar nodes
  else
    Remove the dissimilar concepts from $clstr_i$
    Create a Belief network(BN)corresponding to each $clstr_i$
  end if
end for

A conditional probability measure is calculated upon this belief network. The conditional probability calculation gives us the most feasible concept from the belief network ie, when the conditional probability is applied the most probable concept of that document can be retrieved which can be furthur used for the purpose of indexing, information retrieval etc. The calculation of the conditional probability on a belief network to find the most feasible concept is novel approach proposed in this paper. Likewise for a set of phrases and terms the belief network generated is shown in Fig.3.

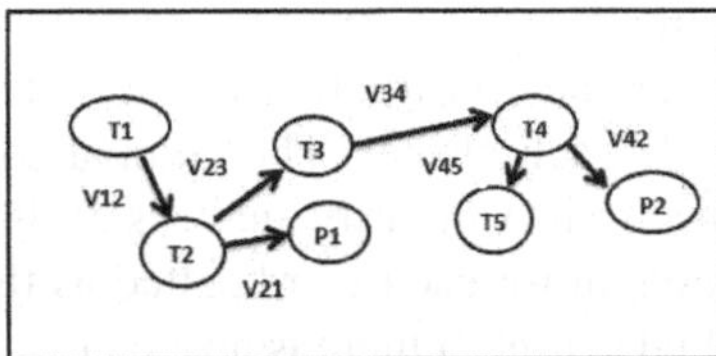

Fig. 3 Belief Network for a set of terms and phrases

For each subject and object in the belief network after the calculation of the probability of each subject and object with respect to all other subjects-objects in the network the triplet with highest probability is taken as the most feasible or more important concept of that network. The probability is calculated on the basis of the links from a subject to object of that particular document.

Conditional probability analysis

Fig.2 consists of a set of triplets, from this the most probable triplet in the document level can be calculated by the applying the conditional probability and it is calculated as:

P(Researchers | nanomaterials)=P(nanomaterials | Researchers) P(Researchers)/ P(nanomaterials)

P(Researchers | nanofluids)=P(nanofluids | Researchers) P(Researchers)/ P(nanofluids)

P(Fluids | Several methods)=P(Several methods | Fluids) P(Fluids)/ P(Several methods)

P(Nanofluids | nano-sized particles)=P(nano-sized particles | Nanofluids) P(Nanofluids)/ P(nano-sized particles)

Chameleon:A two phase Clustering Algorithm for Belief Network Comparison

This is a graph partition based algorithm which operates on graph in which nodes represents the data items and edges represents relation between them [15]. This algorithm is basically used for the inter comparison between the belief networks for more efficient clustering and to find the more accurate concept. After the application of this algorithm again the condition probability is applied which is used to pick the most imporatant concept. This algorithm leads to a more efficient inter documents clustering.This method is based on two measures i) Relative interconnectivity ii) Relative Closeness.

i) *Relative interconnectivity*

The relative interconnectivity between the belief networks of different documents can be found out by:

Consider bel_i and $bel_i + 1$ be two belief networks ,then the relative interconnectivity can be measured using eq(4):

$$RI(bel_i, bel_i + 1) = \frac{(\mid EC(bel_i, bel_i + 1) \mid) * 2}{(\mid EC(bel_i) \mid\mid EC(bel_i + 1) \mid)} \tag{4}$$

where EC(bel_i,$bel_i + 1$)= sum of weights of edges that connect Ci with Cj. EC(bel_i) = weighted sum of edges that partition the cluster into roughly equal parts.

ii)*Relative Closeness*

The relative Closeness between the belief networks of different documents can be found out by:

Consider bel_i and $bel_i + 1$ be two belief networks ,then the relative Closeness can be measured using eq(5):

$$RC(bel_i, bel_i + 1) = \frac{(S(bel_i, bel_i + 1))}{(\mid bel_i \mid S(bel_i) + \mid bel_i + 1 \mid S(bel_i + 1))} \tag{5}$$

A Chameleon algorithm is explained below:

Algorithm 4. Concept based Chameleon Algorithm

---

belf is a set of belief networks
for each belief network bel_i **do**
  Compare bel_i with the adjacent belief network $bel_i + 1$
  if bel_i= $bel_i + 1$ **then**
   add an edge between the equal nodes of bel_i and $bel_i + 1$ Calculate RI and RC:
  end if
  if RI(bel_i)* RC($bel_i + 1$) $> T_R$ (T_R= Threshold) **then** Combine bel_i and $bel_i + 1$
  end if
end for

Example: Consider three belief network bel_i, bel_j and bel_k which is named as A, B and C respectively. The edge weight between the clusters is given by $W_c li$ and it is given by the equation:

$$W_c li = \frac{(Weight of R1)}{(Sum of total number of RI in B)} \tag{6}$$

where RI is the concept in cluster A The inner edge weight is assigned by the equation:

$$I_c li = \frac{(Weight of RI)}{(Total number of ougoing edges from RI)} \tag{7}$$

where RI is the concept in cluster A From the above equation we calculate EC from which RI and RC between the three belief networks is calculated, the pair having value more than thresold is selected and combined together to form a single cluster. Here in the above example A and B have highest RC and RI value than A, C and B, C ie, from the Fig.4 and Fig.5 we can conclude that A and B form a cluster, but A and C does not form a cluster.

Cluster A
Cluster B
1
R1 (0.2)
R1 (0.1)
0.5
R2 (0.3)
R3 (0.4)
R3 (0.1)
P1 (0.2)
R4 (0.6)
More RI and RC

Fig. 4 Graph Representing more RI and RC

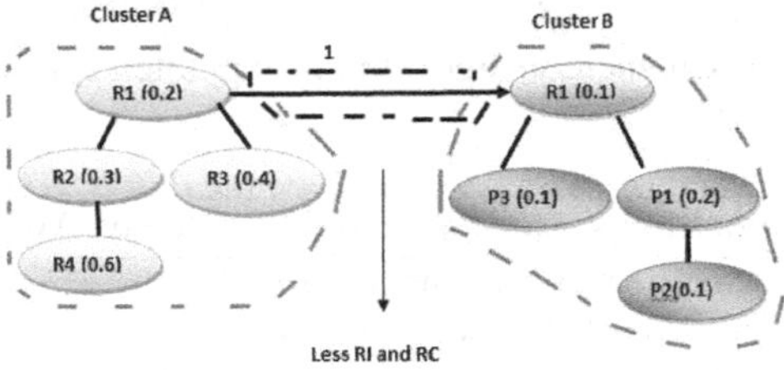

Fig. 5 Graph Representing less RI and RC

4 Experiments and Evaluation

The experimental setup consisted of two data sets. The first data set contains 200 ACM abstract articles collected from the ACM digital library. The ACM articles are classified according to the ACM computing classification system into five main categories: general literature, hardware, computer system organization, software and data. The second data contains 50 PubMed abstract articles collected from the site http://www.ncbi.nlm.nih.gov/pubmed.

PubMed is a online database which is a bibliographic database of life sciences and biomedical information. It include many articles based on bibliographic information from academic journals covering medicines, nursing etc. It also covers even more fields which includes literature in biology and biochemistry ,as well as molecular evolution.

- The Sample data set:

Example1: ***Researchers found nanomaterials and made nanofluids. Nanofluids contains many nanometer-sized particle. These fluids are supplied by several methods. Nanofluids and nanomaterials creates a greater evolution in the research area.***

The runtime of concept based DB scan algorithm was analyzed and it was obtained as O(nlogn) where n is the number of concepts.

The result analysis is done based on two measures:

1) Concept based Clustering Efficiency

The concept based Clustering efficiency is done based on two measures F-measure and Entropy. The F-measure consists both the Precision P and Recall R. The measure of cluster j with respect to class i given by:

$$P = \frac{(M_{i,j})}{(M_j)} \tag{8}$$

$$R = \frac{(M_{i,j})}{(M_i)} \tag{9}$$

where,

$M_{i,j}$- Number of members of class i in cluster j
M_j - Number of members of cluster j
M_i - Number of members of i

The F-measure is calculated as:

$$F(i) = \frac{(2PR)}{(P+R)} \tag{10}$$

When evaluating class i, the cluster having the main F-measure is measured and finally taken as the cluster that draws to class i.The entropy is measured for evaluation of consistancy of the clusters. Higher the consistancy the entropy will be low and vise versa.The entropy is calculated as:

$$E_c = \sum_{j=1}^{n} \frac{(M_j)}{(M)} * E_j \tag{11}$$

M_j- Amount of cluster j
M - Sum of concepts

Table2 gives the cluster efficiency according to the calculated F-measure and Entropy value.

Table 2 Clustering efficiency

	Concept based DB scan clustering	K-Nearest Neighbor(KNN)
F-measure	1	0.75
Entropy	0.30	0.89

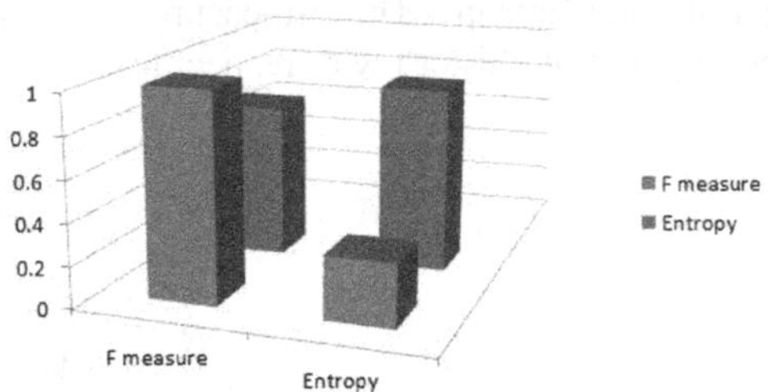

Fig. 6 Clustering efficiency

2)Accuracy Measure

An accuracy measure is calculated for the comparison of concept mining by several methods. Here comparison is done between thwo concept based mining model i)Keyword based concept mining mode (Vector Space Model) ii)Proposed concept based minig model.

The accuracy measure is calculated as:

Accuracy Measure = $\sum_{i=1}^{n} C_i$

Table 3 Analysis Results

Mining Types	Percentage Accuracy
Keyword based mining	0.75
Concept based mining	0.89

Table3 shows the calculated percentage accuracy of the concepts retrieved from a document using two concept mining model. Fig 7 provides the result analysis of the comparison.

The result analysis of accuracy measure plotted in a graph which represents a correct comparison between the keyword based concept mining model and proposed concept mining model. The accuracy rate is higher for the proposed concept based mining when compared to the keyword based concept mining model.

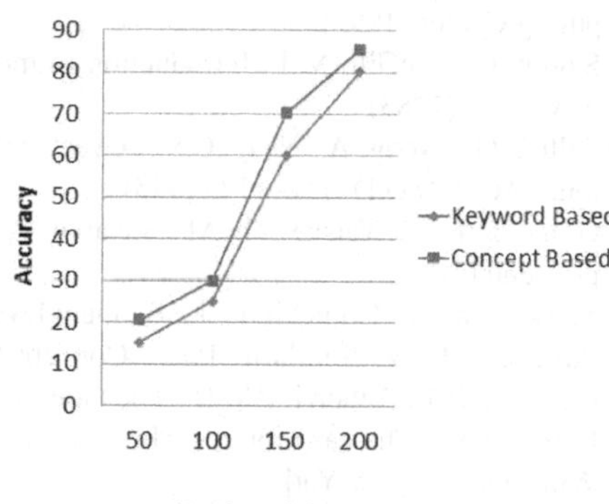

Fig. 7 Analysis Graph

5 Conclusion

A new concept-based mining model, composed of a triplet representation and a belief network construction of the concepts is introduced in this paper, which is followed by a concept based comparison and clustering model. The triplet representation of the concepts followed by a belief network construction of that concept presents a new method for concept mining in an efficient way when compared to other methods.This representation captures the structure of the sentence semantics in an efficient way. The new concept based similarity measure analyzes the similarity between the concept based on the belief network structure. A concept based clustering is also applied , which is done in the sentence level which increases the accuracy of the clusters.

The future work in this area includes the introduction to a concept based indexing which in a far way improve the clustered concepts usage to the application of information retrieval.Many concept based information retrieval systems already existed but applying this model for the purpose of such an application makes a challenge in the area of information retrieval.

Acknowledgements. We are thankful to Dr M R Kaimal, Chairman Department of Computer Science, Amrita University for his valuable feedback and suggestions.

References

[1] Shehata, S., Karray, F., Kamel, M.S.: Enhancing Text Clustering Using Concept-based Mining Model. In: ICDM 2006, pp. 1043–1048 (2006)
[2] Shehata, S., Karray, F., Kamel, M.S.: Enhancing Text Retrieval Performance using Conceptual Ontological Graph. In: ICDM Workshops 2006, pp. 39–44 (2006)
[3] Shehata, S., Karray, F., Kamel, M.S.: An efficient concept-based retrieval model for enhancing text retrieval quality. Knowl. Inf. Syst. 35(2), 411–434 (2013)
[4] Aas, K., Eikvil, L.: Text categorisation: a survey. Technical report 941, Norwegian Computing Center (1999)
[5] Salton, G., McGill, M.J.: Introduction to modern information retrieval. McGraw-Hill, New York (1983)
[6] Salton, G., Wong, A., Yang, C.S.: A vector space model for automatic indexing. Commun. ACM 18(11), 112–117 (1975)
[7] Giunchiglia, F., Yatskevich, M., Shvaiko, P.: Semantic Matching: Algorithms and Implementation
[8] Yatskevich, M., Giunchiglia, F.: Element level semantic matching using WordNet
[9] Puri, S.: A Fuzzy Similarity Based Concept Mining Model for Text Classification
[10] Cilibrasi, R.L., Vitanyi, P.M.B.: The Google Similarity Distance
[11] Fillmore, C.: The case for case. In: Universals in linguistic theory. Holt, Rinehart and Winston, Inc.,New York
[12] Jurafsky, D., Martin, J.H.: Speech and language processing. Prentice Hall Inc., Upper Saddle River (2000)
[13] Kingsbury, P., Palmer, M.: : the next level of treebank. In: Proceedings of treebanks and lexical theories (2003)
[14] Ramos, J.: Using TF-IDF to Determine Word Relevance in Document Queries
[15] Han, J., Han, J., Kamber, M., Pei, J.: Data Mining: Concepts and Techniques

User Feedback Based Evaluation of a Product Recommendation System Using Rank Aggregation Method

Shahab Saquib Sohail, Jamshed Siddiqui, and Rashid Ali

Abstract. The proliferation of the Internet has changed the daily life of a common man. There is a diverse effect of rapid growth of Internet in the daily life. The influence of Internet has changed the way we live and even the way we think. The use of the Internet for purchasing different products of the daily needs has increased exponentially in recent years. Now customers prefer online shopping for the acquisition of the various products. But the huge e-business portals and increasing online shopping sites make it difficult for the customers to go for a particular product. It is very common practice that a customer wishes to know the opinion of other consumers who already have acquired the same product. Therefore we tried to involve the human judgment in recommending the products to the users using implicit user feedback and applied a rank aggregation algorithm on these recommendations. In this paper we chose few products and their respective ranks arbitrarily taken from previous work. For obtaining user's purchase activities a vector feedback is taken from the user and on the basis of their feedback, products are scored; hence they are again ranked which gives each user's ranking. We propose a rank aggregation algorithm and apply it on individuals ranking to get an aggregated final users' ranking. Finally we evaluate the system performance using false negative rates, false positive rates, and precision. These measures show the effectiveness of the proposed method.

Keywords: Customer review; opinion mining; recommendation technique; rank aggregation method; vector feedback.

Shahab Saquib Sohail · Jamshed Siddiqui
Department of Computer Science, Faculty of Science,
Aligarh Muslim University, Aligarh 202002, India

Rashid Ali
Department of Computer Engineering, ZH College of Engineering and Technology,
AMU, Aligarh 202002, India

El-Sayed M. El-Alfy et al. (eds.), *Advances in Intelligent Informatics*,
Advances in Intelligent Systems and Computing 320, DOI: 10.1007/978-3-319-11218-3_32

1 Introduction

The proliferation of the Internet has changed the daily life of a common man. There is a diverse effect of rapid growth of Internet in the daily life. The influence of Internet has changed the way we live and even the way we think. The use of the Internet for purchasing different products of the daily needs has increased exponentially in recent years. Now customers prefer online shopping for the acquisition of the various products. But the huge e-business portals and increasing online shopping sites make it difficult for the customers to go for a particular product. It is very common practice that a customer wishes to know the opinion of other consumers who already have acquired the same product. A good number of researches have been made to propose product recommendation techniques for making online shopping easy and reliable [1,2,3].

Opinion mining is one of the emerging and efficient methods amongst other known and frequent used techniques for recommendation of products being used worldwide.

Customers' reviews are the basis for opinion mining technique. In [4] the author proposed to use a human judge to decide the overall opinion about a product on the basis of the available opinion data. The advantage of this user feedback based approach is that it exploits human intelligence in deciding the overall opinion about a product. Therefore, one can expect a real opinion about a product and can rank different products correctly on the basis of these extracted opinions. For this, we need feedback from the users/ human judges [5].

We arbitrarily chosen ranked products from [5] and presented these products along with their respective reviews to the users. We choose 5 different items each consist of 10 products, all the products are ranked. The users are asked to give their feedback in terms of five vector values namely V,T,P,S and E which gives the importance of a particular product in the eye of a consumer, these terms are defined in the subsequent section, these values convey the log information of the user i.e. their activities are stored. Now we score the products according to the weights given to these vector values and the feedback we obtained from the users. Sorting these products in descending order will give the individual user's ranking for every product. We propose a rank aggregation algorithm; we apply this algorithm to get a final ranking. These rankings are user's ranking as performed on the basis of their feedback, the ranking we choose arbitrarily is the system ranking of the method proposed in [5]. Further to check the performance of the system, the parameters False Positive Rate (FPR), False Negative Rate (FNR) and precision are tested for top 5 positions.

2 Back Ground

2.1 User Feedback Based Opinion Mining

Usually user's feedbacks are taken by asking them to fill in a feedback form for providing his opinion about a particular product for which summarized reviews (opinion data) are presented before him. The form-based approach faces several issucs, this approach demands a lot from the users, there are a lot of difficulties for a careless user who might either fill it casually or not fill it at all. And on the basis of this feedback, we assign different weights for scoring the feedback. Then we may rank different products on the basis of these scores. These types of feedbacks are termed as explicit user feedback. There is another way to get feedback from the users, instead of asking them to fill the forms; we watch their actions while they browse the summarized reviews of a specific product presented before them [5]. We call these feedbacks as implicit user feedback. Here, we present the summarized reviews of different arbitrarily chosen ranked products before the users. Then, we observe the actions of the users on these and infer user feedback implicitly from their actions.

2.2 *Rank Aggregation Method*

Given a set of n customers say $C=(C_1,C_2,C_3,\ldots,C_n)$, a set of m products say $P=(P_1,P_2,P_3,\ldots,P_m)$, and a ranked list l_i on C for each customer i. Then, l i(j) < l i(k) indicates that the customer i prefers the product j to k. The Rank Aggregation Problem is to combine the m ranked lists l_1, l_2, l_3,..., l_m into a single list of candidates, say l that represents the collective choice of the customers. Several good rank aggregation methods are available; the classical Borda's method [6] is one of them. Markov chain based methods [7] and soft computing based methods [8] are also better options, but the problem with the methods proposed in [6, 8] is that they work well for full lists only. If we want to use them for partial lists, we need to convert each of the partial lists into full lists. The methods proposed in [9] and [10] can work well for partial lists but, there we need a decision attribute (overall ranking) in training phase to train the system [5]. Since, while combining the ranking from different users, we do not have any overall ranking and the different user's rankings are also partial lists, we propose a new rank aggregation method that can be used to obtain the aggregated ranking. The algorithm is illustrated in section 3.4.

3 Proposed Recommendation Scheme

In this section, first we give architecture for our recommendation method followed by detailed explanation then the procedures are elaborated.

The architecture of the recommendation technique is depicted in fig 1. As discussed in the previous sections, in this paper we intend to evaluate the ranking system presented in [5] by providing a new users ranking and evaluating the system ranking with the new user ranking based on implicit user feedback using vector feedback technique and applying rank aggregation method. We provide ranking of 5 different items each consist of 10 feedback in terms of five vectors, these five vectors convey the user log information and hence their activities are interpreted to score the products and finally ranking is done by applying rank aggregation method on different users' ranking obtained.

3.1 Vector Feedback

We use five components as user feedback each having their significance in the recommendation process. Here, we characterize the feedback of the user by a five component vector (V, T, P, S, and E) which consists of the following:

(a)The sequence V in which the user visits the product reviews, V = (V1, V2, ….VN). If product i is the kth product visited by the user, then we set Vi = k.
(b) To know whether the user prints the reviews for product i, this is denoted by the Boolean Pi. We denote the vector (P1, P2, …, PN) by P.
(c) Whether or not the user saves the reviews for product i. we denote it by the Boolean Si. We denote the vector (S1, S2, …, SN) by S.
(d) The time duration 't' for which the user remain on the page to browse the contents of the review. We denote the vector (T1, T2, …., TN) by T.

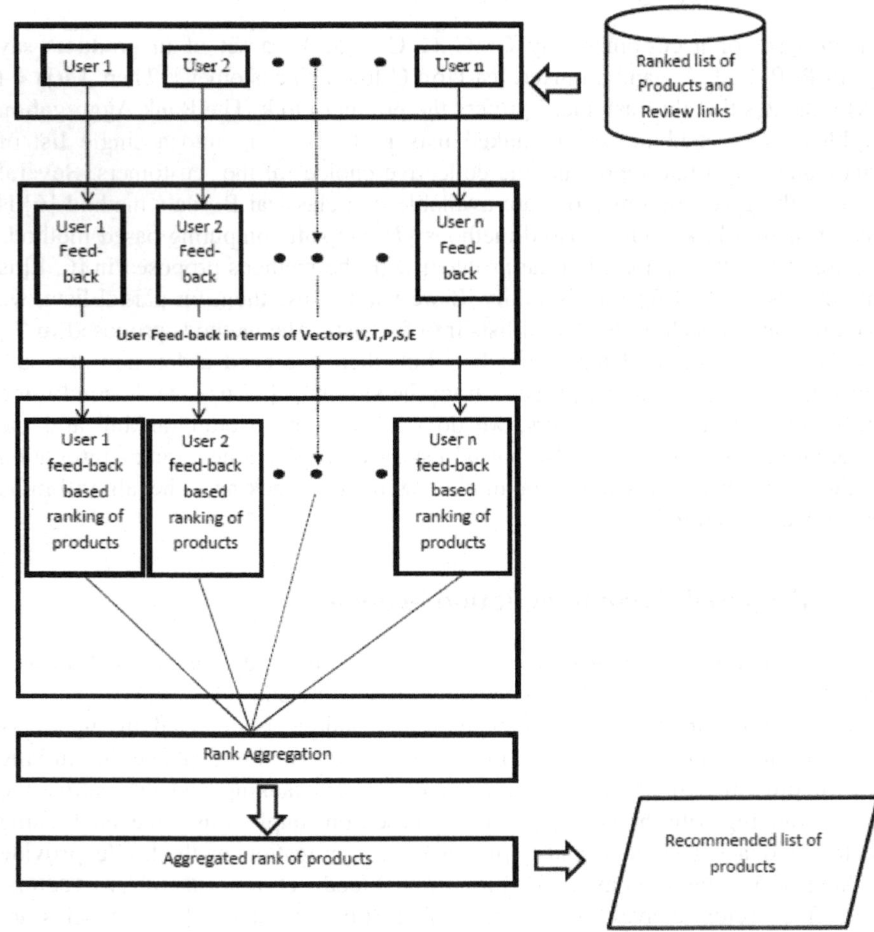

Fig. 1 Architecture of the recommendation scheme

(e) Whether or not the user e-mailed the reviews for product i to someone. This is denoted by the Boolean Ei. We denote the vector (e1, e2, …, eN) by E.

Each of these five components depicts the importance of a particular product in eyes of the user.

The value of V ranges from 1 to 10 only if the user visits the review page of the products concerned. Value of V shows which product is given priority by the user, if a user visits a particular product in very first browsing; v is assigned value '1'. If V=0, it means user does not visit the review page of the product, that implies user does not like the product. T is the time duration for which user remain visiting the review of the products. If a user does not visit the page then T also becomes '0'. S, P and E are Boolean values, either yes or no, 1 is assigned for yes and '0' for no. If the review is not printed, saved and e-mailed then P, S and E are assigned 0 respectively.

3.2 Score Computation

We calculate,

$$z = \sum\left\{Ej + Pj + Sj + Tj + \left(\frac{1}{Vj}\right)\right\} \text{------- (I), for } Vj \neq 0,$$

If Vj=0 we assign all vector to zero, that means z=0, j is showing the user number ranges from 1 to 10 i.e. V_5 means the value of V for fifth user. And Tj is calculated as follows;

Tj =0 if Vj=0

Tj =1 if T = 1 sec to 119 sec,

Tj =2 if T = 120 sec to 179 sec,

Tj =3 if T = 180 sec to 239 sec,

Tj =4 if T = 240-299 sec,

Tj =5 if T >= 300 sec

Finally we compute Tj as Tj= Tj/Tmax;

Here Tmax=5; If the third user browse a review site for 180 seconds, we assign T3=3,

Finally T3=T3/Tmax;

T3=3/5;

T3=0.6;

So in calculating z we use T3 as 0.6. E, P and S are set either 1 or 0 according to the user performance as depicted earlier.

Finally we compute, $Z = z/5$ ------- (II)

Below table 1 is showing above calculation for user 1.

Table 1 Score computation for user 1

	User 1					
Products	**V**	**T**	**P**	**S**	**E**	**Z**
L1	1	3	1	1	1	**0.92**
L2	4	2	1	1	0	**0.53**
L3	2	2	1	1	0	**0.58**
L4	3	2	0	1	0	**0.346**
L5	5	3	0	0	1	**0.36**
L6	9	2	0	1	0	**0.302**
L7	6	1	1	0	0	**0.272**
L8	7	2	0	0	1	**0.308**
L9	0	0	0	0	0	**0**
L10	8	2	1	1	0	**0.509**

3.3 Individual User's Ranking

We calculate the value of 'Z' as discussed in the above section. We sort the value of 'Z' for all users, and for each item's products in descending order that gives the ranking of each user in ascending order. Thus we have 10 different ranking for each item. We call it as individual user's ranking. We propose a rank aggregation algorithm to get a final ranking. The method is discussed below.

3.4 Product Ranking

Let us assume that 'U' represents the union of all the ranked lists of the users concerned, 'n' is the cardinality of the union, and we have 'm' different ranking from m different users, we have a matrix, say R (n × m) with n rows and m columns.

We give a rank aggregation algorithm to compute the final rank.

```
1:  find the ranked position of the i^th product for each user 'j', in R(i,j) ;
2:  If (product's ranking is present for user 'j')
    {
3:   K ← ranked place of the product
    k=1,2,.. n, 0<k<n
4:  R(i,j) ← [(m+k) – {(2*k)- 1}] ;
    }
5:  else (If product is missing i.e. k=0)
    {
    R(i,j) ← 0
    }
6:   repeat the step 1-5 for all positions of R (i,j)
    where i=1,2,…,m and j=1,2,…n
7:  find the sum of scores all entries of R(i,j) of each product 'i' for
    every user 'j'.
8:  sort the products in descending order of scores
9:  Rank in ascending order
```

Fig. 2 Rank Aggregation Algorithm

4 Experiments and Results

The experimental results and different tables used are illustrated below.

4.1 Scoring Table

Since we are evaluating a recommendation system, therefore first we proposed a recommendation technique that gives user's ranking on the basis of their feedback and then we use evaluation parameters FPR@5 and FNR@5 to evaluate the system's ranking with respect to customer's choice. We performed our task for 10 different ranked products of 5 different items. The scoring table for user 1 is depicted in table 1, in the same way we compute for every user for each product.

4.2 User's Ranking

The way we calculate scores in table 1, similarly we calculate it for every user for all products of each item, for illustration, we give individual user's ranking for all the products of item 'L' in table 2. Here symbol 'Li' denotes product Li of item L.

Table 2 Different users ranking

User / Ranking	System Ranking	User 1 Ranking	User 2 Ranking	User 3 Ranking	User 4 Ranking	User 5 Ranking	User 6 Ranking	User 7 Ranking	User 8 Ranking	User 9 Ranking	User 10 Ranking
1	L1	L1	L2	L4	L1	L10	L2	L1	L1	L1	L2
2	L2	L3	L1	L2	L2	L9	L1	L5	L2	L2	L1
3	L3	L2	L4	L6	L3	L8	L4	L2	L3	L3	L4
4	L4	L10	L6	L3	L5	L5	L10	L3	L4	L4	L5
5	L5	L5	L5	L5	L4	L1	L6	L10	L5	L5	L3
6	L6	L4	L7	L8	L6	L7	L3	L9	L6	L6	L8
7	L7	L8	L3	L9	L7	L2	L9	L4	L7	L7	L7
8	L8	L6	L8	L10	L8	L3	L7	L8	L8	L8	L6
9	L9	L7	L9	L1	L9	L4	L8	L6	-	L10	L10
10	L10	-	L10	L7	L10	L6	L5	L7	-	-	L9

Table 3 Final aggregated user's ranking for all the products of each item

Rank / Products	1	2	3	4	5	6	7	8	9	10
Laptop	L1	L2	L3	L4	L5	L6	L8	L7	L10	L9
Headphone	H1	H2	H3	H5	H4	H6	H7	H10	H8	H9
Tablet	T1	T2	T3	T6	T4	T5	T7	T10	T8	T9
Smartphone	S1	S2	S3	S5	S4	S6	S9	S7	S10	S8
Printer	P1	P2	P3	P4	P5	P6	P9	P7	P10	P8

4.3 Final Users' Ranking

We apply algorithm proposed in fig 2, the user's ranking depicted in table 2 yields a new final aggregated ranking shown in table 3.

4.4 Performance Evaluation

There are usually two common characteristic error discussed in a recommendation system [5], false negative and false positive. The false negative error refers to a situation when the items which are preferred by the customers, found missing in the recommendation. The "false positive" refers to a situation in which products that are recommended are the ones which customers do not like, and this is the worst condition as it causes the irritation for the customers and discourages them for any further buying. Also we test the precision, denote it by P@5.

We use three parameters, FPR@5, FNR@5 and P@5 defined as;

- FNR@5 - False Negative Rate at top-5 position.

FNR@5 = (Number of products missing in recommendation but preferred by customer in top-5 position) / 5;

- FPR@5 - False Positive Rate at top-5 position

FPR@5 = (Number of products recommended in top-5 position but not preferred by customer) / 5;

- P@5- we define the precision at top-5 positions

P@5= (Number of products recommended in top-5 positions that are also preferred by customer) / 5;

The table 4 gives the values for FNR@5, FPR@5 and P@5 for final aggregated users' ranking of each item. The high precision and very low error values indicates the an exceptional good performance of the system.

Table 2 FPR@5, FNR@5 and P@5 for all the items

Parameters / Items	FPR@5	FNR@5	P@5
Laptop	0	0	1
Headphone	0	0	1
Tablet	0.2	0.2	0.8
Smartphone	0	0	1
Printer	0	0	1
Average	**0.04**	**0.04**	**0.96**

5 Conclusion

We intend to evaluate a recommendation system using user feedback scores. We presented a recommendation technique using user feedback score, taken the implicit feedback from users to score their activities and propose a rank aggregation algorithm to get a better recommendation that may satisfy the customer.

On the basis of the scores given to the products, and applying proposed rank aggregation method, we ranked the products and this user ranking serves as a basis for the evaluation of the recommendation system we arbitrarily chosen.

The two characteristic errors false negative and false positive, and precision for top 5 ranked products were tested. The high precision 0.96 and low error rate 0.04 for both FPR@5 and FNR@5 shows that the proposed system would be very helpful for customers to get a good recommendation for their products of choice.

In future, one can compare the algorithm with existing one to check the relative performance; also to we can enlarge the customer size to perform the job on a big data set. The different parameters are used for top 5 positions; it can be tested for top 10 positions for the entire product as well.

References

1. Andreevskaia, A., Bergler, S.: Mining WordNet for Fuzzy Sentiment: Sentiment Tag Extraction from WordNet Glosses. In: EACL 2006, pp. 209–216 (2006)
2. Carenini, G., Ng, R.T., Pauls, A.: Interactive Multimedia Summaries of Evaluative Text. In: IUI 2006 (2006)
3. Hu, M., Liu, B.: Mining and summarizing customer reviews. In: KDD 2004 (2004)
4. Ali, R.: Development of a product recommendation system using web based opinion mining: Pro-Mining. project report, College of Computers and Information Technology, Taif University
5. Ali, R.: Pro-Mining: Product recommendation using web-based opinion mining. IJCET 4(6), 299–313 (2013)
6. Borda, J.C.: Memoire sur les election au scrutiny. Histoire de l'Academie Royale des Sciences (1781)
7. Dwork, C., Kumar, R., Naor, M., Sivakumar, D.: Rank aggregation methods for the web. In: Proceedings of the Tenth International Conference on World Wide Web, pp. 613–622 (2001)
8. Beg, M.M.S., Ahmad, N.: Soft Computing Techniques for Rank Aggregation on the World Wide Web. World Wide Web – An International Journal 6(1), 5–22 (2003)
9. Ali, R., Beg, M.M.S.: Modified Rough Set Based Aggregation for Effective Evaluation of Web Search Systems. In: Proceedings of the 28th North American Fuzzy Information Processing Society Annual Conference (NAFIPS 2009). IEEE Press, Cincinnati (2009)
10. Ali, R., Beg, M.M.S.: A Learning Algorithm for Meta searching using Rough Set Theory. In: Proceedings of the 10th International Conference on Computer and Information Technology (ICCIT 2007), pp. 361–366. IEEE Press, Dhaka (2007)

Misalignment Fault Prediction of Motor-Shaft Using Multiscale Entropy and Support Vector Machine

Alok Kumar Verma*, Somnath Sarangi, and Mahesh Kolekar

Abstract. Rotating machines constitutes the major portion of the industrial sector. In case of rotating machines, misalignment has been observed to be one of the most common faults which can be regarded as a cause for decrease in efficiency and can also for the failure at a time. Till date the researchers have dealt only with the vibration samples for misalignment fault detection, whereas in the present work both stator current samples and vibration samples has been used as a diagnostic media for fault detection. Multiscale entropy (MSE) based statistical approach for feature extraction and support vector machine (SVM) classification makes the proposed algorithm more robust. Thus, any non-linear behavior in the diagnostic media is easily handled. The proposed work has depicted an approach to analyze features that distinguishes the vibration as well as current samples of a normal induction motor from that of a misaligned one. The result shows that the proposed novel approach is very effective to predict the misalignment fault for the induction motor.

Keywords: Misalignment, Wavelet denoising, Multiscale entropy, Support vector machine, Fault diagnosis.

1 Introduction

Till date, online condition monitoring and fault detection of rotating machinery technique have been of great significant attentions among researchers. Recently it

Alok Kumar Verma · Mahesh Kolekar
Department of Electrical Engineering, Indian Institute of Technology, Patna, India
e-mail: {alokverma,mahesh}@iitp.ac.in

Somnath Sarangi
Department of Mechanical Engineering, Indian Institute of Technology, Patna, India
e-mail: somsara@iitp.ac.in

* Corresponding author.

El-Sayed M. El-Alfy et al. (eds.), *Advances in Intelligent Informatics*,
Advances in Intelligent Systems and Computing 320, DOI: 10.1007/978-3-319-11218-3_33

has observed that misalignment of motor shaft is one of the most important and easily encountered faults in the vast majority of rotating machinery. Misalignment condition in rotating machine is a condition where the centerlines of coupled shafts do not coincide with each other [1]. Loads increases due to misalignment on bearings and couplings. This increased load may lead to decrease in motor efficiency or damage of machine. Major factors that give rise to this respective misalignment conditions are asymmetry in applied loads, unequal settlement of foundation, improper assembly of machines etc, hence called the keys of misalignment [1, 2]. Thus good knowledge about the rotor shaft vibration and motor current signature analysis can be considered as the key elements for diagnosis and analysis of the misalignment in the rotating machines.

The non-linear behaviors encountered in mechanical system due to loading condition or damping, vibration in the friction, may change the normal vibration and current signals to the complex and non-linear [3]. It has been observed that commonly used signal processing techniques including time and frequency domain techniques as well as advanced signal processing techniques, like wavelet transform and time-frequency domain may all have limitations. [3]. Therefore, it resulted for need of techniques for non-linear dynamic parameter estimation which could provide a good alternative to extract the defect-related features hidden in the complex as well as non-linear vibration and current samples [3, 4]. An experiment [4] has been already done on non-linear dynamic parameters used for feature extraction and fault diagnosis and approximate entropy (ApEn). Approximate entropy was well illustrated and was selected as a working tool for rolling bearing fault detection. ApEn had also found its ways in the fields of physiological signal as well as vibration signal processing of rotating machine [3, 4], However ApEn reported more similarity in the time series and self-matching property of ApEn makes it to be heavily dependent on the length of time series [5].To overcome the limitations of ApEn, Richman and Moorman [6] introduced a new kind of entropy technique called sample entropy (SampEn) which has gained a lot of attention which excludes self-matching property [6, 7].

In this recent paper [8], a new entropy measure known as multi-scale entropy (MSE) has been introduced .The traditional entropy measure was measuring entropy on the single scale; there was no correspondence between the regularity and the complexity of the time-series. The researchers used this newly developed MSE technique to distinguish between young healthy hearts and congestive heart failure of a person. Considering a rotating machine as a combination of bearings, shafts and other mechanical components [9] and having a sufficient numbers of machine complexity that implies non-linear dynamic parameters applied on single scale (ApEn and SampEn of original time series) may be insufficient for characterizing machine vibration signals. Due to this reason, the multi-scale method was introduced and tried in the presented study with the idea of improving performances of machine fault diagnosis. With the best effort from author's literature survey in the field of fault diagnosis, very few work was done in which MSE has been applied and that was only with vibration signals. Long Zhang et al. [10] discussed about multi-scale entropy (MSE), taking into account multiple time scales, was introduced for feature extraction from fault of vibration signal. MSE with the support of vector machines constitutes the proposed intelligent fault diagnosis method. Jun-Lin Lin et al. [11] deals an approach to discover methods that distinguish the vibration signals of aligned motor from those of a misaligned one. Experimental results shows that classifiers based on these features obtain better and more accurate rates than those based on frequency-related features. Verma et al. [12] used MSE and grey-fuzzy algorithm to predict stator winding fault. They used MSE as a feature to deal with the nonlinearity exists in the stator winding vibration. Long Zhang et al. [13] introduced a bearing fault diagnosis method based on MSE and adaptive neuro-fuzzy inference system (ANFIS), in order to tackle the nonlinearity existing in bearing vibration as well as the uncertainty inherent in the diagnostic information.

Therefore, in this work authors investigate the misalignment fault by using diagnostic media like stator current as well as rotor vibration with the MSE, and that will be a novel approach, in order to tackle the nonlinearity existing in misalignment vibration and current. Presented work intuitively, a motor in a rotating machine is analogous to a heart in a person. This analogy motivates the use of MSE on vibration and current signals of motor in the presented work. Presented work proposes a method for detecting motor shaft misalignment by measuring the rotor vibration as well as stator current of an induction motor. Experimental result shows that the proposed algorithm can be used to classify effectively between aligned and misaligned motors using both rotor vibration as well as stator current of an induction motor.

2 Experimental Analysis

2.1 Experimental Setup

The entire experimental setup is shown in fig. 1 which comprises of four major sub-systems. It constitutes a three phase induction motor (Marathon Electric, 0.75HP) along with its accessories, constituting a data acquisition system, various sensors and a computer storage and a display, as shown in fig. 1. The induction motor with a wiring enclosure are adjoined on the left side of the system arrangement and that wiring enclosure of the motor support permits access to the 3-phase power supply and motor supply leads. The wiring enclosure attached with current probe is shown in Fig. 1 (b)

The roller bearings acts as the support factor to the rotor shaft of the given induction motor. The length of the shaft between two roller bearings is 0.72390 metre and the shaft diameter is of 0.0127 metre. Triaxial industrial accelerometer and current probe are two important sensors of this experiment. To collect vibration signals, an accelerometer was set up above the roller bearing on the right hand side which is shown in fig. 1 (d). Precision laser alignment kit has been used

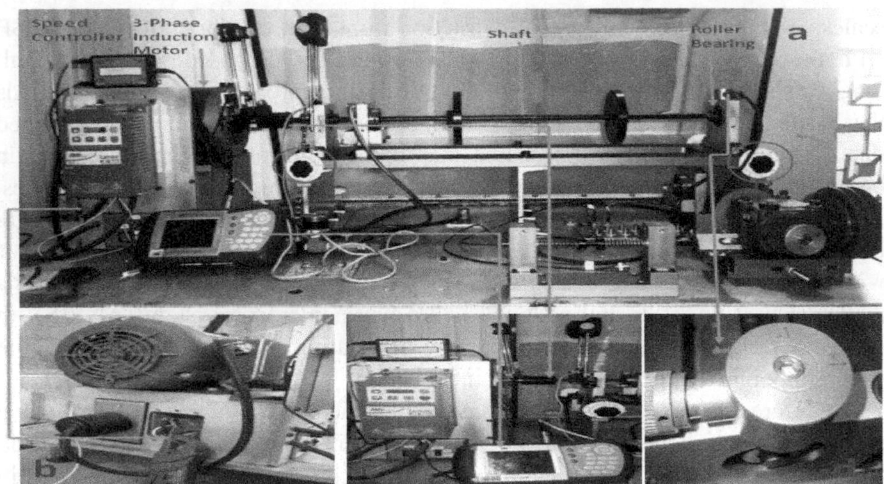

Fig. 1 Experimental set-up used: (a) Full view of the system arrangement, (b) Current probe, (c) Precision laser alignment kit and flexible coupling, (d) Triaxial industrial accelerometer with roller bearing

for investigating the alignment of shaft. The respective instrument constitutes one transmitter, a receiver and a controlling section as shown in fig. 1 (e). The complete specification of all the components used is mentioned in the following Table 1.

Table 1 Specifications of the components used in experimental setup

Sensors/ Machine	Manufacturer	Model/Serial No.	Sensitivity/ Specification
Accelerometer	IMI Sensors	604B31	Sensitivity=10.2 mV/(m/s2)
Current Probe	Fluke	I200s	100 mV/A
Laser Alignment Kit	Optalign smart	ALI 12.200	-------------
Induction motor	Marathon Electric	HVN 56T334F53033	0.75 HP, 50 Hz, 2850 RPM

2.2 *Experimental Details*

The present work enunciates the entire experimental procedure to differentiate the misaligned motor from that of the properly aligned one. The experimental setup along with its accessories in the aligned position is shown in fig 2. The misalignment condition is generated in the same setup by moving the base plate leftwards on horizontal plane which resulted in the movement of the support structure of the base plate as shown in fig 3 (with solid line). The corresponding schematics for the aligned and misaligned condition are illustrated using fig. 2 (a) and 2 (b). As shown in fig. 2 (b), a misalignment of 30 mils or 0.03 inches or 0.000762 meters between the driving and the driven shaft is provided to achieve the misalignment condition.

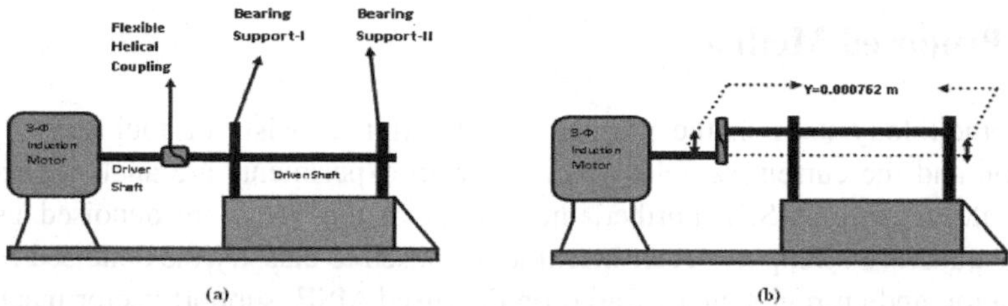

Fig. 2 (a) Aligned Setup, (b) Experimental Setup with misalignment of 0.000762 metre

Fig. 3 Misalignment generation by moving support structure

Collectively, four sets of vibration as well as current signals were collected through this experiment. The first set of aligned vibration and current data (denoted by A_1) was collected maintaining the motor running at aligned condition. The motor was then misaligned with 0.000762 metre (on both side) by the process described above, and then the second set which is a misaligned data (denoted by M_1) was collected. Later the motor was adjusted back to its form of aligned condition, and the third set of aligned data (denoted by A_2) was obtained and stored. Lastly, the fourth set of data (denoted by M_2) was procured by misaligning the motor with 0.000762 meters (on both sides as process described above). Each set of data contained 26 records, recording the vibration and current signals at 26 different speeds ranging from 760 rpm to 1510 rpm with an increment of 30 rpm. Each record was a time series and containing 15364 signal values. The vibrational signals were obtained using the accelerometer and the current signals were collected by current probe at 5.12 s/s. Parameters used and their levels are shown in Table 2.

Table 2 Factors and levels considered for experiments

	Control Factor	Level
A	Rotational Speed (rpm)	760 to 1510
B	Alignment Cond. (mils)	0 & 2×0.000762 m

3 Proposed Method

The methodology used in the present investigation consists of four steps. The vibration and the current samples obtained from experiments are analyzed using Multiscale entropy (MSE). Further, the results of the MSE are denoised using wavelet transform. Support vector machine was used to classify the denoised MSE of vibration and current signals. Based on denoised MSE, support vector machine is employed to model the entire system and to give the performance. The result obtained from the above analysis is validated by performing the confirmation experiments. The step by step details of the analysis is mentioned in the sub sections 3.1 and 3.2.

3.1 Multiscale Entropy

It is the most prominent task to find out the regularity in a time series for both classifying and predicting future values in a time series. In 1991, Pincus [5], proposed a statistical measure for noisy time series, called approximate entropy (ApEn), to quantify the regularity of time series. In this proposed work, the MSE algorithm is used which is based on the SampEn for different scales of the same process instead of previously used regularity measure ApEn statistics. SampEn is used as replacement of ApEn and it measures the regularity in serial data. It measures the two sequences of m consecutive data within a tolerance r remain similar when one consecutive point is included.

ApEn (m, r, n) may be calculated for the time series of length N, according to the equation

$$\text{ApEn}\,(m, r, N) = \frac{1}{N-m}\sum_{i=1}^{N-m}\left(-\ln\frac{n_i^{m+1}}{n_i^{m}}\right), \quad (1)$$

where, r is tolerance of time series, m is pattern length and n is number of matching (including self match). ApEn is reported with certain disadvantages, such as heavy dependence on the length of the time series, smaller values for the shorter time series and lack of consistent results for different values of m and r. To overcome the shortcomings of ApEn, sample entropy (SampEn) as a new kind of entropy, this excludes the disadvantages of ApEn. As a refinement of ApEn, SampEn is used and it measures the regularity in series data. The SampEn is defined as

$$\text{SampEn}\,(m, r, N) = -\ln\left(\frac{\sum_{i=1}^{N-m} n_i^{m+1}}{\sum_{i=1}^{N-m} n_i^{m}}\right). \quad (2)$$

The value of r is taken as 0.15 times the standard deviation and m is taken as 2 for present analysis to avoid distortion. Costa et al. [13] proposed an algorithm called MSE, which considers SampEn at multiscale. MSE has been already successful to analyze physiological signals. Consider a time series, {X1, . . . , XN}, which is a coarse-grained time series in MSE is given by

$$y_j^{(\tau)} = \frac{1}{\tau}\sum_{i=(j-1)\tau+1}^{j\tau} \mathrm{X_i} \tag{3}$$

Where τ is known as scale factor, and for $\tau = 1$, that coarse grained time series is the original series. It is find that as increases, the length of the coarse-grained time series decreases. SampEn of a white noise falls quickly as the scale factor rises as shown in fig. 4. However, the SampEn of a pink or noise remains approximately stable as the scale factor rises. Therefore, white noise is more regular than pink noise.

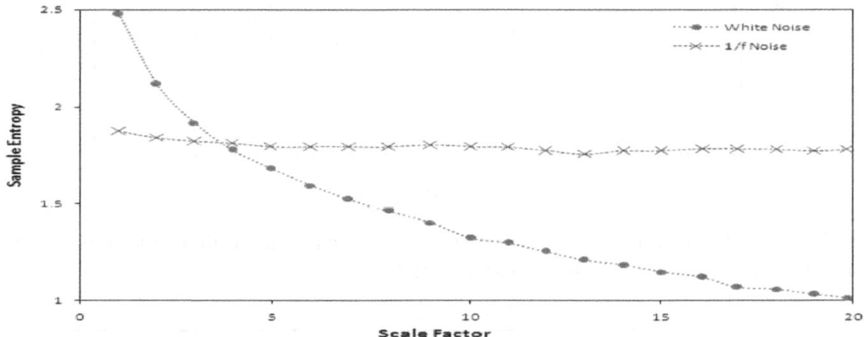

Fig. 4 Sample entropy with respect to the scale factor for coarse-grained time series of white and 1/f noises

Wavelet transform is one of the most effective and preferred method of denoising signals, although the effect of wavelet denoising depends on the signal types. Sample Entropy at different scale factor of normal and misaligned motors at 1100 rpm before and after denoising for vibration and current signals are enunciated using Fig. 5 and 6 respectively. It can be observed through fig. 5, that there exists increment of SampEn when the scale factor is small and reduction in SampEn when the scale factor is large. Similarly, the nature of periodicity of the current signals can also be observed from fig. 6. Thus, the information or the distinguishable characteristic present in both vibration and current signals are extracted or inferred from its respective sample entropies and can be consequently deployed to distinguish misaligned motors from that of the healthy ones.

Diagnosis or detection of motor shaft misalignment is carried out based on the vibration and current signals. The sample entropy of de-noised vibration and current signals is calculated for a scale factor ranging from 1 to 20. These calculated values are used as input to a SVM based classification algorithm to detect misaligned fault. Based on MSE the performance of the classifiers is discussed in result section, the denoising process is carried out using wavelet transform for both current and vibration samples. Daubechies wavelet transform was chosen for both vibration and current, which was implemented using Matlab and the parameter settings used for vibration samples are (tptr = ''rigrsure''; n = ''2''; wav = ''db4'') and for the current samples, the parameter settings were (tptr = ''rigrsure''; n = ''6''; wav = ''db2'').

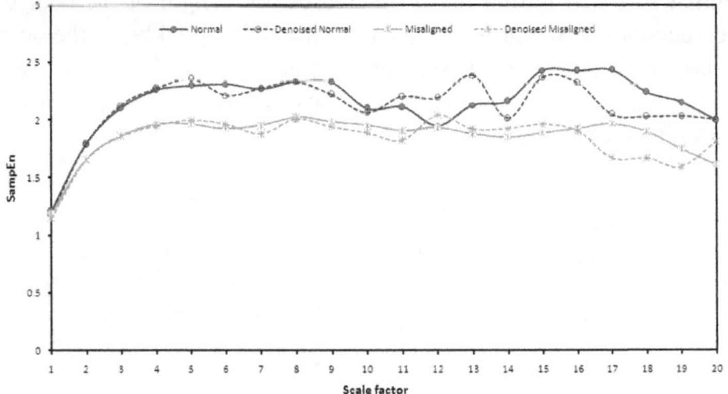

Fig. 5 Sample Entropy at different Scale factor of normal and misaligned motors at 1100 rpm before and after denoising for vibration signals

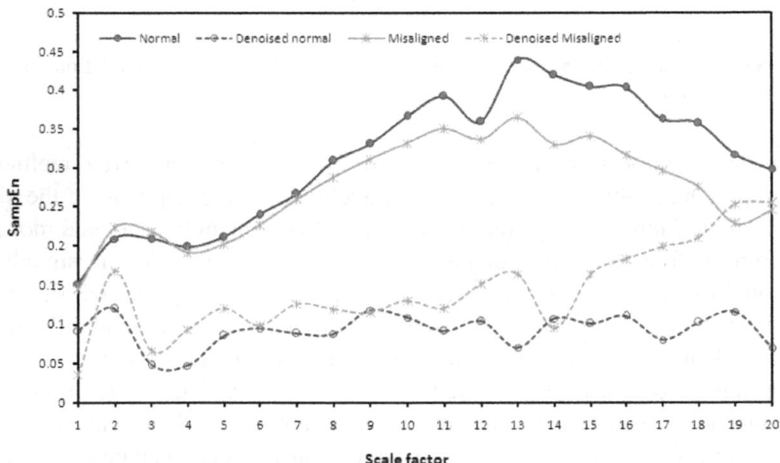

Fig. 6 Sample Entropy at different Scale factor of normal and misaligned motors at 1100 rpm before and after denoising for current signals

3.2 *Support Vector Machines*

The denoised MSE were used as input to the support vector machines (SVM) for detection of misalignment fault. SVM are relatively new method used for binary classification. The present work also requires a binary classification of the given samples to distinguish a healthy motor from that of a misaligned one. The basic idea is to find a hyperplane which separates the n-dimensional data perfectly into its two classes. First the input vectors are mapped into feature space (possible with higher dimension), either linearly or non-linearly, which is relevant with the selection of the kernel function. Then within the feature space, a hyperplane is constructed which separated the two classes (this can be extended to multiclass). As shown in fig. 7, the two hyperplanes are constructed on each side of the hyperplane that separates the data. The two classes are then separated by an optimum hyperplane, minimizing the distance between closest misaligned class points (+1 class) and properly aligned data points (-1 class), which are known as support vectors. The right side of the separating hyperplane represents the +1 class and the left-hand side represents the -1 class.

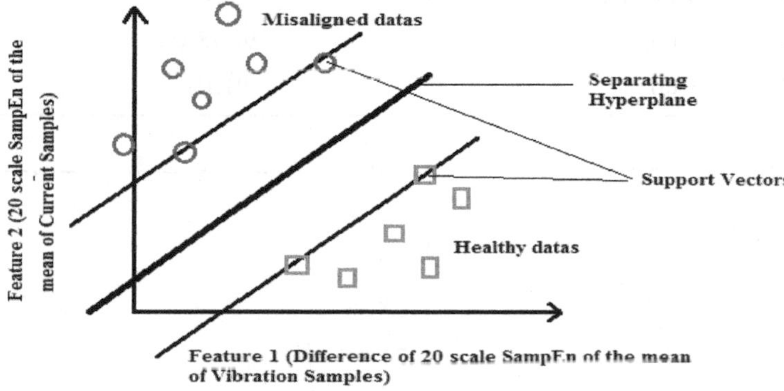

Fig. 7 SVM for misaligned and properly aligned motor classification

4 Results and Discussions

MSE technique is the only underlying basis of the present method of distinguishing or classifying the misaligned motors from that of healthy ones. The respective statistical based approach (MSE) easily tackles any existing non-linear behavior in the misalignment vibration or current samples, thus describing the regularity in the diagnostic information The allocated method examines the features that distinguishes the rotor vibration as well as stator current samples of aligned induction motor from that of a misaligned one with the help of an SVM

Table 3 Prediction accuracy on the training set for Test-1 and 2 of vibration and current (i.e., Test-1 = A_1 and M_1, Test-2 = A_2 U M_2 , total 52 samples, training set= 26 samples and testing set= 26 samples

Test	Media	Training Data	Cross Validation Accuracy (%)
1	Vibration	26	91.8%
2	Vibration	26	94.4%
1	Current	26	92.1%
2	Current	26	90.11%

Table 4 Prediction accuracy on the testing set for Test 1 and of vibration and current (i.e., Test-1 = A_1 and M_1, Test-2 = A_2 U M_2, total 52 samples, training set= 26 samples and testing set= 26 samples

Test	Media	Testing Data	Accuracy (%)	Time(Sec.)
1	Vibration	26	90.68%	2
2	Vibration	26	94.1%	2
1	Current	26	92.18%	2
2	Current	26	91.1%	2

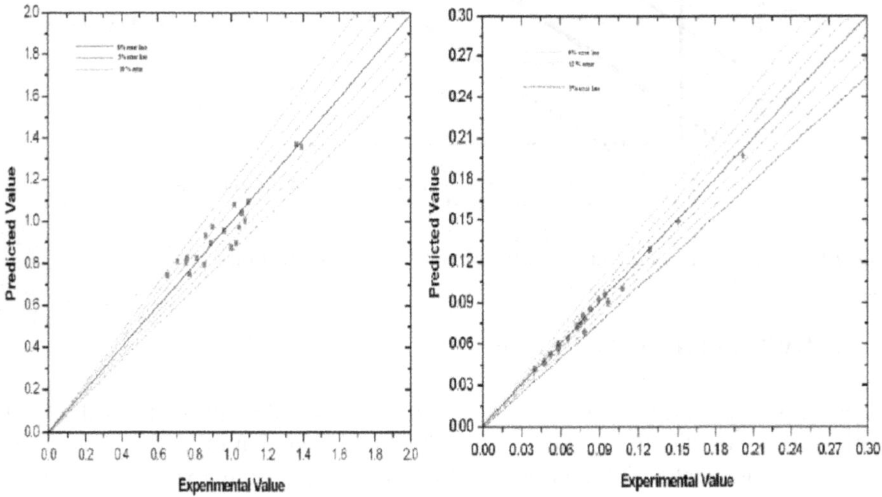

Fig. 8 Variation of predicted value with experimental value used for vibration and current of test-1

classifier. Test 1 was performed for both vibration and current samples used the sets A_1 and M_1, having 26 samples each. From the total of 52 numbers of samples, 26 samples were is randomly chosen as training set data and the remaining 26 samples were chosen as testing data set. Test 2 was used the sets A_2 and M_2, having 26 samples each and total 52 samples. As Test 1 26 samples were chosen as training set and the remaining 26 samples were chosen as testing data. Further, the SVM based classification was carried out using the following data: (a) stator current samples and (b) Rotor vibration samples. In this presented work, the number of input to the classifier is two (speed and normal or fault) and output is one (fault detection using vibration or current). In case of test-1, gives the maximum accuracy and is considered best for all three tests. The above classifier was then trained and tested for various values. The optimal value for the fault detection using vibration is shown in Table 3 and 4 and fig 8 shows the Variation of predicted value with experimental value used for vibration and current of test-1.

5 Conclusions

The main aim of the proposed approach was to investigate the robustness and effectiveness of distinguishing capability of the algorithm. The results infer that the misaligned motors can be easily distinguished from that of the healthy ones using stator current and rotor vibrational signals as the diagnostic media that were not done before. The developed system will lead to alter the long winded task of modeling and analysis of the vibration and current during misalignment fault with the help of MSE based statistical approach. From the given results, it can be observed that the best vibration accuracy is 94.68% and best current accuracy is 92.18% for all cases of the experimental values. The Present study clearly shows that the SVM model can be trained to predict the misalignment fault using vibration and current with reasonable accuracy. This approach can be used to detect different faults as well.

References

1. Piotrowski, J.: Shaft alignment handbook. CRC Press, New York (2006)
2. Verma, A.K., Sarangi, S., Kolekar, M.: Shaft Misalignment Detection using Stator Current Monitoring. International Journal of Advanced Computer Research 3, 305–309 (2013)
3. Yan, R.Q., Gao, R.X.: Approximate Entropy as a Diagnostic Tool for Machine Health Monitoring. Mechanical Systems and Signal Processing 21, 824–839 (2007)
4. Yan, R.Q., Gao, R.X.: Machine Health Diagnosis Based on Approximate Entropy. In: ICMT Instrumentation and Measurement Technology Conference, pp. 2054–2059. IEEE Press, Italy (2004)
5. Pincus, S.M.: Approximate entropy as a measure of system complexity. PNAS 88, 2297–2301 (1991)

6. Richman, J.S., Moorman, J.R.: Physiological Time-Series Analysis Using Approximate Entropy and Sample Entropy. Am. J. Physiol. H. 278, 2039–2049 (2000)
7. Haitham, M.A., Alan, V.S.: Use of Sample Entropy Approach to Study Heart Rate Variability in Obstructive Sleep Apnea Syndrome. IEEE Trans. Bio. Eng. 50, 1900–1904 (2007)
8. Costa, M., Goldberger, A.L., Peng, C.K.: Multiscale Entropy Analysis of Complex Physiologic Time Series. Phys. Res. Lett. 89, 68–102 (2002)
9. Fan, X.F., Zuo, M.J.: Machine Fault Feature Extraction Based on Intrinsic Mode Functions. Meas. Sci. Technol. 19, 245105, 12 (2008)
10. Zhang, L., Xiong, G., Liu, H., Zou, H., Guo, W.: An Intelligent Fault Diagnosis Method Based on Multiscale Entropy and SVMs. In: Yu, W., He, H., Zhang, N. (eds.) ISNN 2009, Part III. LNCS, vol. 5553, pp. 724–732. Springer, Heidelberg (2009)
11. Lin, J.L., Liu, J.Y., Li, C., Tsai, L., Chung, H.: Motor shaft misalignment detection using multiscale entropy with wavelet denoising. Expert Systems with Applications 37, 7200–7204 (2010)
12. Verma, A., Sarangi, S., Kolekar, M.: Stator winding fault prediction of induction motors using multiscale entropy and grey fuzzy optimization methods. Comput. Electr. Eng. (2014), `http://dx.doi.org/10.1016/j.compeleceng.2014.05.013`
13. Zhang, L., Xiong, G., Liu, H., Zou, H., Guo, W.: Bearing fault diagnosis using multi-scale entropy and adaptive neuro-fuzzy inference. Expert Systems with Applications 37, 6077–6085 (2010)

A Learning Based Emotion Classifier with Semantic Text Processing

Vajrapu Anusha and Banda Sandhya

Abstract. In this modern era, we depend more and more on machines for day to day activities. However, there is a huge gap between computer and human in emotional thinking, which is the central factor in human communication. This gap can be bridged by implementing several computational approaches, which induce emotional intelligence into a machine. Emotion detection from text is one such method to make the computers emotionally intelligent because text is one of the major media for communication among humans and with the computers. In this paper, we propose an approach which adds natural language processing techniques to improve the performance of learning based emotion classifier by considering the syntactic and semantic features of text. We also present a comprehensive overview of emerging field of emotion detection from text.

Keywords: Emotion Recognition, Affective Computing, Machine Learning, Natural Language Processing.

1 Introduction

With the rapid growth of computer technology and its applications, there is growing need for the computers which can work together with humans. Making the computer emotionally intelligent improves the effectiveness of human-machine communication by the transfiguration of a computer into a human-like partner.

The role of emotions in human-machine communication was stated by Picard by introducing the concept of Affective computing [20]. It is an interdisciplinary field spanning the psychology, cognitive science and computer science, and it aims to develop the systems that can recognize, interpret, process and simulate human affects.

Vajrapu Anusha · Banda Sandhya
MVSR Engineering college, Hyderabad, India
e-mail: anu.v503@gmail.com, sandhya_cse@mvsrec.edu.in

El-Sayed M. El-Alfy et al. (eds.), *Advances in Intelligent Informatics*,
Advances in Intelligent Systems and Computing 320, DOI: 10.1007/978-3-319-11218-3_34

Human beings express emotions through various media such as text, speech, facial expressions and gestures. Text-based emotion detection is important because text is a main medium in computer-mediated communication in the form of emails, chat rooms, product reviews and web blogs.

Applications of text-based emotion detection can be found in business, education, psychology and in any other fields where there is a paramount need to understand and interpret emotions.

This paper describes our system implemented for text-based emotion detection, which uses a combination of machine learning and natural language processing techniques to recognize the affect in the form of six basic emotions proposed by Ekman [6]. Section 2 provides the brief background of emotion detection from text and detailed study of various text-based emotion detection methods and related works. Section 3 describes our system architecture and its functionality. Section 4 explains the experimental setup and results. Section 5 discusses the conclusion, difficulties and areas to be improved.

2 Background and Related Work

Emotional Intelligence [22] refers to the ability to perceive, control and evaluate emotions. To induce emotional intelligence into a computer, the first aspect is emotion perception also known as emotion recognition or emotion detection.

The basic necessity of an emotion recognition system is emotion representation. The popular models for emotion representation are dimensional model and categorical model.

Dimensional model represents emotions in a dimensional form which relates each other by a common set of dimensions and are generally defined in two dimensional (Valence, Arousal) or three dimensional (pleasure, Arousal, Dominance) space.

Categorical model presents emotions as discrete and fundamentally different constructs. A very popular categorical model is Ekman [6] emotion model, which specifies six basic emotions: anger, disgust, fear, happiness, sadness, surprise. Other popular models are Plutchik's emotions wheel [21] and OCC (Ortony/Clore/Collins) model [17] which is mainly designed to model human emotions in general. Emotion can also be viewed in terms of more generalized categories: positive vs negative, and detection in the form of these categories is called sentiment analysis or opinion mining.

The methods which exist in the literature of content based emotion detection from text can be categorized based on three viewpoints: process of detection, data sources used for the process and user perspective as shown in Figure 1.

2.1 Based on the Process of Detection

This subsection explains the types of emotion detection methods based on how the detection is made.

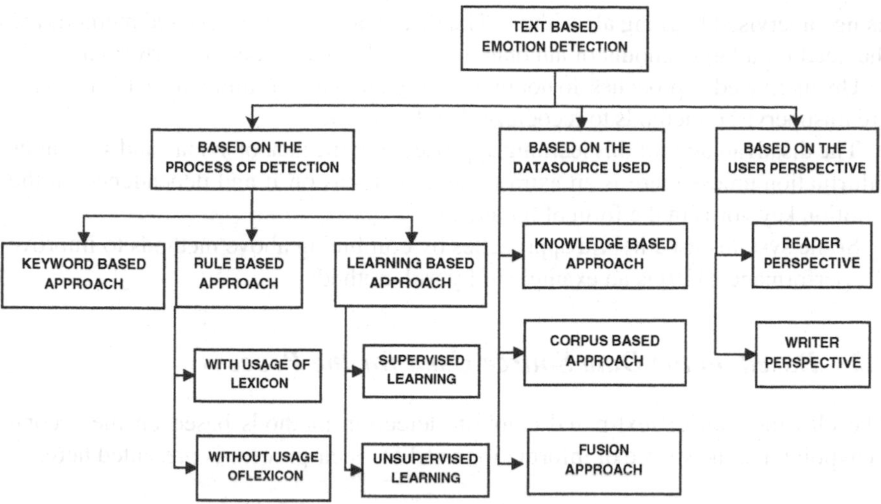

Fig. 1 Taxonomy of Text based Emotion Detection methods

Keyword based approach: Keyword based methods are applied at the basic word level. They need an Emotional keyword dictionary which has keywords with corresponding emotion labels. These are based on the assumption of keyword independence, and they ignore the possibilities of using different types of keywords simultaneously to express complicated emotions. [11], [10], [13] and [19] are the approaches to detect emotion based on keywords.

The disadvantages of this approach are: the meanings of keywords could be multiple and changes according to the usages and contexts, inability to recognize emotions of sentences which do not contain any keywords and the lack of syntactic and semantic information.

Rule based approach: This approach is used to develop various rules to define a language structure which are useful for identifying emotion. To define the rules some of the systems use an affective lexicon which contains a list of lexemes annotated with their affect and some others don't take any help of affective lexicons. The approaches [5], [16] detect emotions by defining rules with the help of affective lexicon and [9] detects without the usage of a lexicon respectively.

The disadvantages of this approach are: Complexity in designing and modifying rules, the inflexibility of catering to the emotions other than those already listed and the rules are specific to the representation of the source from which knowledge is extracted.

Learning based approach: This approach formulates the problem of detecting emotion from text as classification of the input texts with emotions as classes. Supervised approach uses manually annotated datasets to train a classifier which is used for further classifications. Methods in [1], [4], [12] and [3] detect emotions using supervised learning algorithms. The disadvantages of supervised methods are the need for a large amount of annotated data and domain specific performance.

Unsupervised approaches do not require any annotated training data. [26] and [2] are unsupervised methods to recognize the affect of text.

The disadvantages of this learning approach are: the lack of syntax and semantics information unless there is an extra module to focus on it and dependence on the emotion keywords in the form of features.

Some systems used hybrid approaches by combining above methods to improve the performance. [29] is an example of hybrid method.

2.2 Based on the Data Source Used for the Process

The classification of text-based emotion detection methods based on the second viewpoint, i.e. the source of information used for their process is presented here.

Knowledge based approach: Thesaurus (knowledge) contains a list of words with synonyms and related concepts. These approaches use synonyms or glosses of lexical resources in order to determine the emotion or polarity of words, sentences and documents.

There are many knowledge bases available for assistance in affect detection. Wordnet-Affect [27] is an affective lexical hierarchical resource which is developed and extended from Wordnet with additional hierarchy of affective domain labels and it is used to find the affect of words. SentiWordnet [7], GI [24], LIWC [18] are some resources to find the sentiment of text.

[11], [5], [26], [10], [16] and [19] are the methods to recognize the affect of text using different knowledge bases.

Corpus based approach: A corpus is a collection of large and structured set of texts especially related to particular subject or author. Researchers take a variety of corpora pertinent to their experiments such as texts from web blogs, news articles, fairy tales, stories, message boards and some survey data.

SemEval2007 Task14(Affective Task) [25] presented two standard corpora of news headlines collected from Google News and CNN, whose main theme is to classify headlines into positive/negative emotion category or distinct emotion categories like anger, disgust, fear, happiness, sadness and surprise. ISEAR(International Survey on Emotion Antecedents and Reactions) [23] is a collection of 7666 manually annotated sentences, which are surveyed from the people of different fields in 37 countries across five continents. Each sentence is annotated with one of seven basic emotions: anger, disgust, fear, guilt, joy, sadness, shame along with some other information.

The approaches [9], [29], [4] and [3] use different corpora to detect affect from text according to their experiments.

Fusion approach: The fusion approach is a kind of hybrid method that makes use of both the corpus and knowledge-based approaches to have the advantages of both. [1] and [12] are such methods that use both types of background knowledge.

2.3 Based on the Perspectives

Based on the third viewpoint emotion can be detected from the two different perspectives: Reader and Writer perspective.

Reader perspective: From reader perspective one particular text segment can evoke multiple emotions. The procedures in [11], [26], [12], [13] and [3] are focused to recognize multiple emotions from a single text document.

Writer perspective: Most of the previous works performed emotion analysis in the perspective of the writer where one text segment portrays only one emotion. The approaches [11], [9], [1], [19], [29], [10], [16], [4] and [2] are aimed to detect final single emotion of a text portion.

Some of the related works with their functionality and their categorization from the above three viewpoints are presented in table 1.

Even though there has been a lot work done in this area, emotion detection from text is still immature and needs a lot of improvements.

Table 1 Previous related works in the literature of emotion detection from text

Article	**Task Description**	**Detection process**	**Data sources for background knowledge**	**persp-ective**
Al Masum et al [16]	Classify news sentences to OCC categories using semantic parsing, valences of linguistic components and user account preferences	Rule based approach with affect lexicon	Knowledge based approach which uses SenseNet to find valences	Writer
Binali et al [4]	Identify sentiment of blog sentences using some linguistic processing and SVM classifier	Supervised learning approach	corpus based approach which uses annotated blog dataset	Writer
Phil Katz et al [12]	Finds emotion of news sentences using synonym expansion using thesauri and manually annotated news corpus	Supervised learning approach	Fusion approach which uses Rogets thesaurus and news corpus	Reader
Chunling et al [11]	Detects affect of messages in chat system using Ekman model and displays corresponding avatars	keyword based approach	knowledge based approach which uses Wordnet-Affect, Wordnet	Writer
Chaumartin [5]	Recognizes emotion of words using factors like Wordnet category membership, negation etc. and detects final emotion of news headlines by exploiting dependency graphs and using patterns of words.	Rule based approach with usage of affective lexicon	knowledge based approach which uses Sentiwordnet, Wordnet, Wordnet-Affect	Reader

Table 1 (*Continued*)

Liu et al [9]	Generated four common sense affect models by considering the semantics of language and combined them to classify the emotion of text	Rule based approach without affective lexicons	corpus based approach which uses OMCS(Open Mind Common Sense)	writer
Aman and Szpakowicz [1]	Detects the emotion of text in terms of Ekman model categories using SVM classifier	Supervised learning approach	Fusion approach uses OMCS corpus, lexicon built from Rogets thesaurus, Wordnet-affect	Writer
Strapparava and Mihalcea [26]	compared supervised and unsupervised methods and Implemented different systems to detect the affect of text in unsupervised way which of them three are variations of LSA and WA-presence is another method	Unsupervised learning approach	knowledge based approach which uses WordNet-Affect	Reader
Perikos and Hatzilygeroudis [19]	Detects emotion at sentence level using dependency graph and patterns of words where affect of each word is recognized by using lexicons	Keyword based approach	knowledge based approach which uses WordNet-Affect, WordNet	Writer
Yang et al [29]	Performs sentiment analysis to recognize affect of suicide notes by developing a model that incorporates several NLP techniques to handle the complicated features of text	Hybrid approach comprises of keyword, learning methods	Corpus based approach which uses Cincinatti Medical NLP challenge dataset	Writer
P.Kumar Bhowmick et al [3]	Detects multiple emotions of each news headline by applying RAKEL algorithm using features such as words, polarity of patterns, semantic frames	Supervised learning approach	Corpus based approach which uses an archive from Times of India	Reader
A.Agrawal and Aijun An [2]	A semantic approach to detect emotion of sentence using features like NAVA words, PMI scores, syntactic dependencies	Unsupervised learning approach	does not use any detailed affective lexicon, annotated dataset	Writer
Zornitsa Kozareva et al [13]	categorize the news sentences to Ekman categories using statistics from three web search engines and PMI scores calculated from news corpus	keyword based approach	does not need any annotated data or affective lexicon	Reader
Hancock et al [10]	A online chat experiment conducted to report the strategies in expression of positive, negative emotions	keyword based approach	knowledge based method which uses LIWC for linguistic analysis	Writer

3 Natural Language Processing Approach to Improve the Traditional Learning Based Emotion Classifier

Supervised learning methods are significantly used for inducing intelligence into machines as their approach is analogous to human learning which builds knowledge from past experiences and uses it for performing in new unseen situations. But as discussed in section 2 the major drawback of supervised learning approach for emotion classification is its incapability to consider syntactic and semantic information. To overcome this drawback, we added an extra module to traditional learning system architecture. This extra module (NLP module) focuses on analyzing syntactic and semantic information using NLP techniques.

System architecture for our approach is shown in Figure 2.

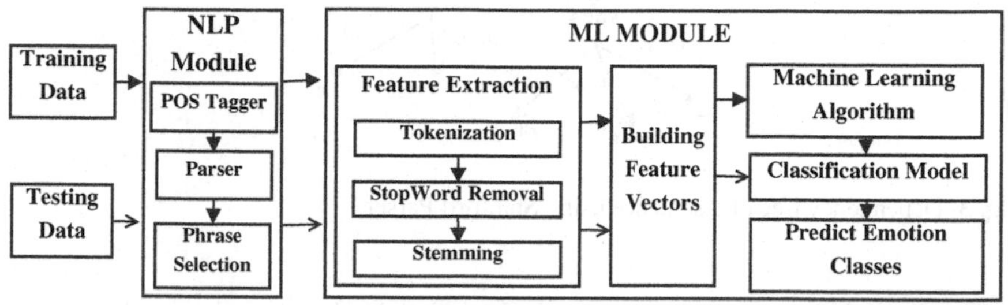

Fig. 2 Architecture of Our System for Learning Based Emotion Classification With Natural Language Processing Techniques

The NLP module uses POS (part-of-speech) tagger to specify each word's grammatical role and creates the dependency tree based on the words relationships, using the Stanford CoreNLP toolkit [14]. Then phrase selection is done using the rules on dependency relationships, which leads to give priority to the semantically important information for the classification of sentence's emotion.

The ML module receives the phrases selected from NLP module and performs the emotion classification using the standard approach similar to any type of text classification, which is clearly explained in [8]. This module uses 'Porter Stemmer' algorithm for stemming and 'english.stop' stopwords file for stopwords removal. The selected stopwords file was mainly designed for information extraction, so before using it we removed some negation words (Ex: never, not, didn't etc.) from the file which are important for classifying emotion of text. Then TF-IDF (Term Frequency-Inverse Document Frequency) statistic is used to build the feature vectors. We considered Naive Bayes Multinomial (NBM) and Support Vector Machines (SVM) as classification algorithms. We used WEKA [28] for building our ML module.

So the procedure followed by our system in classifying a sentence is as follows:

1. Use Stanford CoreNLP pipeline (comprises tokenizer, POS tagger and parser [15]) to get the dependencies and dependency tree of a sentence
2. pass the dependency tree to phrase selection sub module which outputs the parts of sentence which are semantically important for classification
3. Then employs learning approach to classify the sentence's emotion
   a. extract the features of sentence
   b. construct the sentence's feature vector
   c. if it is training data use it for building the model, else classify emotion of sentence using the model already built

As an example, consider the sentence “I thought I would pass exam but I failed it.”. In figure 3 the sentence’s dependency tree and its dependencies are specified.

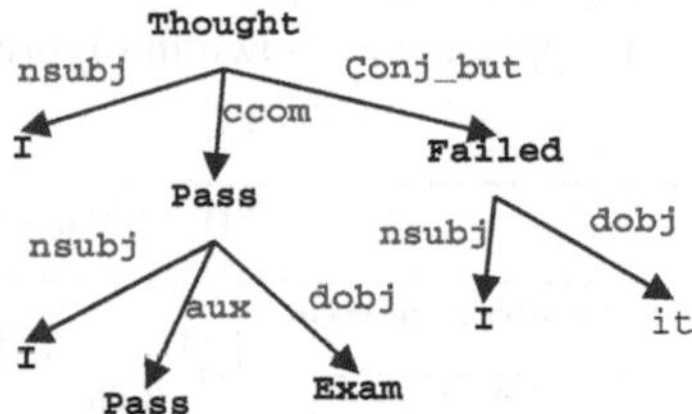

Fig. 3 Dependency tree of a sentence using Stanford Parser

The dependency represents grammatical relationship between the words in a sentence, which is a triplet: Name of relation, governor, dependent. For example, nsubj (i,thought) is nominal subject relationship between two words, showing that ‘i’ is the subject of ‘thought’.

In phrase selection module, dependency tree of sentence is received and it selects phrases by processing the relationships of root with other words. For the example in figure 3, the root ‘thought’ and ‘failed’ are connected by using conjunction ‘but’. In general, ‘but’ specifies the phrases which are opposite in sense, whenever ‘but’ is encountered, the phrase following ‘but’ is the main contribution to final emotion of sentence. So the phrase with head ‘failed’ is selected by the module.

By passing this selected phrase to the ML module, it will be classified to corresponding emotion which will be assigned as final emotion to the complete sentence.

We added some more rules to NLP module such as rules on ‘yet’, ‘not’, ‘though’, ‘after’ etc. according to their behavior in representing sentences.

4 Experiments and Results

For supervised learning method the main problem is with the selection of qualified dataset because a large scale dataset is needed to cover most of the words in given language and it should have emotional content independent from different cultures to eliminate cultural affects of emotion. These requirements lead us to use ISEAR [23] corpus as training dataset. As discussed in Section 2.2, ISEAR contains seven emotion categories from which we selected five categories (Anger, Disgust, Fear, Joy, Sadness) which are specified in Ekman emotion theory. Out of the five considered emotions, we considered ‘joy’ as positive emotion category and remaining into negative emotion class to perform the Sentiment Analysis experiment on ISEAR. To evaluate the performance of the system F1-Measure and Kappa statistics are used.

First, we evaluated the performance of standard machine learning approach on ISEAR dataset by applying 10-fold cross validation using ML module and Table2

Table 2 Performances of classifiers using 10-fold cross validation on ISEAR dataset for Ekman's model classification

Classifier	Stem	performances (Unigrams)		performances (N-Grams)	
		AvgF1	kappa	AvgF1	kappa
NBM	no	63.6	0.54	65.3	0.56
	yes	64.1	0.55	66.7	0.58
SVM	no	66.0	0.57	64.4	0.55
	yes	67.1	0.59	66.5	0.58

Table 3 Performances of classifiers using 10-fold cross validation on ISEAR dataset for Sentiment Analysis

Classifier	Stem	performances (Unigrams)		performances (N-Grams)	
		AvgF1	kappa	AvgF1	kappa
NBM	no	81.8	0.50	88.2	0.64
	yes	85.3	0.56	88.6	0.65
SVM	no	88.3	0.61	88.3	0.62
	yes	86.7	0.56	88.7	0.63

and 3 shows the corresponding results of Ekman's model classification and Sentiment analysis respectively.

Apart from performances, it leads to some major error classifications using unigram features. For example, "I am not happy" and "I thought I would pass the exam, but I failed it" are misclassified to 'joy' instead of 'sadness'. This is partly handled by N-gram approach, but it makes the process computationally intensive by increasing dictionary size and also it looks for the exact match of word sequences, which is not suitable for real-world data classification.

To correct these errors, we used our approach of applying NLP module to sentences before passing them to ML module. With this approach, we limited our process to recognize emotion from simple short sentences as we focused on analyzing syntactic and semantic information at the level of sentences. The average sentence length of ISEAR dataset is 2.6, which makes it impossible to apply our approach directly on ISEAR. So, we evaluated it on a dataset of simple short sentences collected with help of ISEAR. The sample of our dataset for 'sadness' emotion is shown in Figure 4.

Table 4 shows the performances of 10-fold cross validation using ML module alone and NLP module in combination with ML module for Ekman's model classification. And results for sentiment analysis are presented in Table 5. These results

I am not happy.
I was happy to receive a letter from home but it was telling me that my mother was very ill.
Though i had worked throughout the year, i did not get good percentage.
I thought i would pass exam but i failed it.
Love is not a happy feeling.
I purchased a new dress but was upset to find it torn.
I did not succeed to enter at the University.
I had not understand anything after a lecture.
My results were not satisfactory though I worked at my best.

Fig. 4 sample of dataset used for NLP approach

Table 4 Performances with 10-fold cross validation on sample dataset for Ekman's model classification

Classifier	Performances AvgF1	kappa	Approach
NBM	43.1	0.29	without
SVM	43.0	0.31	NLPmodule
NBM	52.9	0.43	with
SVM	63.1	0.49	NLPmodule

Table 5 Performances with 10-fold cross validation on sample dataset for Sentiment Analysis

Classifier	Performances AvgF1	kappa	Approach
NBM	42.3	-0.15	without
SVM	52.7	0.15	NLPmodule
NBM	62.2	0.27	with
SVM	58.6	0.14	NLPmodule

proved that our semantic approach with the help of NLP techniques attains significant improvements on plain supervised learning method.

So, by selecting the phrases which are important for classification will consider the semantically important information, reduce the confusion for classifier, and improves the performance.

As the future work, we will implement this system by adding a complete set of rules using deep analysis of linguistic features based on dependency trees of data, and we will try to make the classifier to learn these rules from text automatically.

5 Conclusion

In this paper, we presented an extensive survey of content based emotion detection from text and we hope this will be helpful in directing the future research in related fields.

We also explained our experiments of emotion recognition from text using supervised machine learning approach alone and NLP techniques in combination with machine learning to improve the performance. But this method is limited to single short sentences. This can be applied to long sentences by applying some sort of text compression/summarization using dependency trees before sending to the process of learning.

Emotion detection from text needs to be improved from multiple directions. Recognizing emotion from implicit cues is difficult compared to detection from explicit cues, but can be improved by constructing a knowledge base of emotion events and its usage in the process of classifying emotion. There is also need of personalized emotion model to make process of recognition more realistic because some situations lead to different emotions based on different people, which can be handled efficiently by it.

As emotions are entities with fuzzy boundaries and each emotion occurs alone rarely, emotion recognition in terms of a small number of discrete categories with sharp boundaries is difficult and impractical. The multi-labeled and fuzzy classifications may improve the performance of recognition and make more practical recognition of emotion.

The work in this area may also focus on intensities and duration of emotions, the time when the situation happened and the person's mood/mental state before the situation happened because of the fact that a person who is already frustrated may get angry very easily than others.

References

1. Aman, S., Szpakowicz, S.: Using Roget's thesaurus for fine-grained emotion recognition. In: Proceedings of the Third International Joint Conference on Natural Language Processing, IJCNLP 2008, pp. 296–302 (2008)
2. Ameeta, A., Aijun, A.: Unsupervised emotion detection from text using semantic and syntactic relations. In: Proceedings of the 2012 IEEE/WIC/ACM International Joint Conferences on Web Intelligence and Intelligent Agent Technology, WI-IAT 2012, vol. 01, pp. 346–353. IEEE Computer Society, Washington, DC (2012), http://dl.acm.org/citation.cfm?id=2457524.2457613
3. Bhowmick, P.K., Basu, A., Mitra, P.: Reader Perspective Emotion Analysis in Text through Ensemble based Multi-Label Classification Framework. Computer and Information Science 2(4), 64–74 (2009)
4. Binali, H., Wu, C., Potdar, V.: Computational approaches for emotion detection in text. In: 2010 4th IEEE International Conference on Digital Ecosystems and Technologies (DEST), pp. 172–177 (2010), doi:10.1109/DEST.2010.5610650
5. Chaumartin, F.R.: UPAR7: A knowledge-based system for headline sentiment tagging. In: Proceedings of the Fourth International Workshop on Semantic Evaluations (SemEval-2007), Prague, Czech Republic, pp. 422–425 (2007), http://hal.archives-ouvertes.fr/hal-00611242
6. Ekman, P.: Basic Emotions. In: Handbook of Cognition and Emotion (1999)
7. Esuli, A., Sebastiani, F.: SENTIWORDNET: A Publicly Available Lexical Resource for Opinion Mining. In: 5th Conference on Language Resources and Evaluation, pp. 417–422 (2006), http://citeseerx.ist.psu.edu/viewdoc/summary?doi=10.1.1.61.7217
8. Ikonomakis, M., Kotsiantis, S., Tampakas, V.: Text classification using machine learning techniques. WSEAS Transactions on Computers 4(8), 966–974 (2005)
9. Liu, H., Lieberman, H., Selker, T.: A model of textual affect sensing using real-world knowledge. In: Proceedings of the 8th International Conference on Intelligent User Interfaces, IUI 2003, pp. 125–132. ACM, New York (2003), http://doi.acm.org/10.1145/604045.604067, doi:10.1145/604045.604067
10. Hancock, J.T., Landrigan, C., Silver, C.: Expressing emotion in text-based communication. In: Proceedings of the SIGCHI Conference on Human Factors in Computing Systems, pp. 929–932. ACM (2007)
11. Ma, C., Prendinger, H., Ishizuka, M.: Emotion estimation and reasoning based on affective textual interaction. In: Tao, J., Tan, T., Picard, R.W. (eds.) ACII 2005. LNCS, vol. 3784, pp. 622–628. Springer, Heidelberg (2005), doi:10.1007/11573548_80
12. Katz, P., Singleton, M., Wicentowski, R.: Swat-mp: The semeval-2007 systems for task 5 and task 14. In: Proceedings of the 4th International Workshop on Semantic Evaluations, SemEval 2007, pp. 308–313. Association for Computational Linguistics, Stroudsburg (2007), http://dl.acm.org/citation.cfm?id=1621474.1621541

13. Kozareva, Z., Navarro, B., Vázquez, S., Montoyo, A.: Ua-zbsa: A headline emotion classification through web information. In: Proceedings of the 4th International Workshop on Semantic Evaluations, SemEval 2007, pp. 334–337. Association for Computational Linguistics, Stroudsburg (2007), http://dl.acm.org/citation.cfm?id=1621474.1621546
14. Manning, C.D., Surdeanu, M., Bauer, J., Finkel, J., Bethard, S.J., McClosky, D.: The Stanford CoreNLP natural language processing toolkit. In: Proceedings of 52nd Annual Meeting of the Association for Computational Linguistics: System Demonstrations, pp. 55–60 (2014), http://www.aclweb.org/anthology/P/P14/P14-5010
15. de Marneffe, M.C., MacCartney, B., Manning, C.D.: Generating typed dependency parses from phrase structure trees. In: LREC (2006), http://nlp.stanford.edu/pubs/LREC06_dependencies.pdf
16. Al Masum, S., Prendinger, H., Ishizuka, M.: Emotion sensitive news agent: An approach towards user centric emotion sensing from the news. In: IEEE/WIC/ACM International Conference on Web Intelligence, pp. 614–620 (2007), doi:10.1109/WI.2007.124
17. Ortony, A., LClore, G., Collins, A.: The Cognitive Structure of Emotions. Cambridge University Press (1990)
18. Pennebaker, J., Francis, M., Booth, R.: Linguistic inquiry and word count (computer software). Erlbaum Publishers, Mahwah (2001)
19. Perikos, I., Hatzilygeroudis, I.: Recognizing emotion presence in natural language sentences. In: Iliadis, L., Papadopoulos, H., Jayne, C. (eds.) EANN 2013, Part II. CCIS, vol. 384, pp. 30–39. Springer, Heidelberg (2013)
20. Picard, R.W.: Affective Computing. The MIT Press, Cambridge (1997)
21. Plutchik, R.: The Emotions. University Press of America, Inc. (1991)
22. Salovey, P., Brackett, M.A.P., Mayer, J.: Emotional Intelligence: Key Readings on the Mayer and Salovey Model. Natl Professional Resources Inc. (2004)
23. Scherer, K.R., Wallbott, H.: International survey on emotion antecedents and reactions (isear) (1990), http://www.unige.ch/fapse/emotion/databanks/isear.html
24. Stone, P.J., Dunphy, D.C., Smith, M.S., Ogilvie, D.M.: The General Inquirer: A Computer Approach to Content Analysis. MIT Press, Cambridge (1966)
25. Strapparava, C., Mihalcea, R.: Semeval-2007 task 14: Affective text. In: Proceedings of the 4th International Workshop on Semantic Evaluations, SemEval 2007, pp. 70–74. Association for Computational Linguistics, Stroudsburg (2007), http://dl.acm.org/citation.cfm?id=1621474.1621487
26. Strapparava, C., Mihalcea, R.: Learning to identify emotions in text. In: Proceedings of the 2008 ACM Symposium on Applied Computing, SAC 2008, pp. 1556–1560. ACM, New York (2008), http://doi.acm.org/10.1145/1363686.1364052, doi:10.1145/1363686.1364052
27. Strapparava, C., Valitutti, A.: Wordnet-affect: an affective extension of wordnet. In: Proceedings of the 4th International Conference on Language Resources and Evaluation, pp. 1083–1086 (2004)
28. Witten, I.H., Frank, E.: Data Mining: Practical Machine Learning Tools and Techniques, 2nd edn. Morgan Kaufmann Series in Data Management Systems. Morgan Kaufmann Publishers Inc., San Francisco (2005)
29. Yang, H., Willis, A., de Roeck, A.: BN A hybrid model for automatic emotion recognition in suicide notes. Biomedical Informatics Insights 5, 17–30 (2012), www.la-press.com/a-hybrid-model-for-automatic-emotion-recognition-in-suicide-notes-article-a3017, doi:10.4137/BII.S8948

An Empirical Study of Robustness and Stability of Machine Learning Classifiers in Software Defect Prediction

Arvinder Kaur and Kamaldeep Kaur

Abstract. Software is one of the key drivers of twenty first century business and society. Delivering high quality software systems is a challenging task for software developers. Early software defect prediction, based on software code metrics, has been intensely researched by the software engineering research community. Recent knowledge advancements in machine learning have been intensely explored for development of highly accurate automatic software defect prediction models. This study contributes to the application of machine learning in software defect prediction by investigating the robustness and stability of 17 classifiers on 44 open source software defect prediction data sets obtained from PROMISE repository. The Area under curve (AUC) of Receiver Operating Characteristic Curve (ROC) for each of the 17 classifiers is obtained for 44 defect prediction data sets. Our experiments show that Random Forests, Logistic Regression and Kstar are robust as well as stable classifiers for software defect prediction applications. Further, we demonstrate that Naïve Bayes and Bayes Networks, which have been shown to be robust and comprehensible classifiers in previous on software defect prediction, have poor stability in open source software defect prediction.

1 Introduction

With computers, smart devices and appliances deeply implanted in business and homes, the software systems, operating them, are no doubt, one of the key drivers of twenty first century business and society. Delivering high quality software systems is a tough challenge for software engineers and managers. As a means

Arvinder Kaur · Kamaldeep Kaur
USICT , Guru Gobind Singh Indraprastha University
arvinderkaurtakkar@yahoo.com, kdkaur99@gmail.com

El-Sayed M. El-Alfy et al. (eds.), *Advances in Intelligent Informatics*,
Advances in Intelligent Systems and Computing 320, DOI: 10.1007/978-3-319-11218-3_35

towards the end of delivering high quality software with-in budget and time, metrics based, early software defect prediction models have been intensely explored by a number of software engineering researchers [1-12]. Metrics based early software defect prediction models intend to classify modules of software under development into two types: defect prone and defect free, before actual beginning of formal testing.

The intention is to focus meagre testing time and resources on more risky software components, which are likely to contain defects. Researchers have also highlighted some principal advantages gained from metrics based defect prediction models as under:-

- The maintenance phase is easier for both developers and customers. The overall maintenance costs are lowered and there is an increase in the overall project success rate of complex software systems[2].
- Defect prediction models are particularly useful in today's highly competitive market where companies operate with very tight profit margins [2].
- Defect prediction models are useful not only for closed source proprietary software development but also for open source software as many commercial software products like web browsers and office suites are based on open source software[2][3].
- Defect prediction models enable architectural improvements by suggesting a more rigorous design for high-risk software code segments [4].

Software defect prediction offers a unique opportunity for empirical testing of machine learning concepts and theories. In an attempt to automate defect prediction, software engineering researchers have intensely explored the applicability and relative advantages of machine learning based classifiers, within metrics based software defect prediction. There are many influential studies on metrics based defect prediction that explore numerous statistical and machine learning classifiers like logistic regression[4], linear and quadratic discriminant analysis[4], artificial neural networks[6], support vector machine based classifiers[2,4], fuzzy subtractive clustering[7], decision tree approaches[8], decision table[9], voting feature intervals[2], k-nearest neighbor[4], Naïve Bayes classifiers[1] and Bayesian networks[3,13]. However, there are several reports of inconsistencies of predictive performance of classifiers across defect prediction data sets from different application domains. Recently, the focus has shifted towards benchmarking classification models over large number of software systems [4]. The current empirical study contributes to defect prediction literature by studying robustness and stability dimensions of classifiers in software defect prediction, conjunctively. The concepts of robustness and stability of classifiers are borrowed from machine learning literature[5] and are presented as under for the convenience of readers:-

- "Robustness of a classifier refers to the ranking of the average performance of a classifier, among a set of classifiers[5]". For example, if we assume Area under Curve (AUC) of Receiver Operating Characteristic Curve(ROC) is a performance measure, all classifiers are ranked according to their AUC over all studied defect prediction data-sets, and the ranking order of a classifier is used to capture the robustness of a classifier, with a smaller ranking number denoting a more robust learner. Thus the concept of robustness is similar to the commonly used concept of predictive accuracy.

- Stability of a classifier refers to the ranking of the variance of a classifier in a set of classifiers [5]. For example, if we assume variance of the AUC of ROC Curve, as a performance measure, we rank all classifiers according to AUC variance over all studied defect prediction data-sets. The smallest ranking number denotes the classifier with the lowest variance. "This ranking order is used to capture a classifier's stability, with a smaller ranking number denoting a more stable classifier[5]". Stability also conceptually means the degree to which the performance of a classifier is affected by changes in training data sets.

In this study we conduct experiments for comparing robustness and stability of 17 classifiers over 44 data sets pertaining to Java based software systems. In conformance with recent research trends, our comparisons of multiple classifiers over multiple data sets are based on Area under Curve (AUC) of Receiver Operating Characteristic Curve (ROC). Our experiments can be considered as a sequel to the experiments of Menzies et al.[1] and Lessman et al [4]. The remainder of this paper is organized as follows: In section 2, the stability dimension of classifier assessment is highlighted. In section 3, the details of experimental design and procedure are presented. Section 4 reports experimental results. In section 5, threats to validity of this study are presented. In section 6, conclusions and future research directions are presented.

2 Stability Dimension of Classifier Assessment

Popular machine learning classifiers for software defect prediction include Naïve Bayes[1,4], Bayes Networks[3,13], Random Forests[4,12], Neural Networks[6], Case based reasoning[11], Kstar[4], Artificial immune systems[10] and ensemble learning[2]. Intrigued by the inconsistent results regarding accuracy of classifiers, Lessman et al[4] conducted a benchmarking experiment of 22 classification models for software defect prediction over 10 data sets obtained from NASA MDP repository[14]. They concluded that in software defect prediction, the accuracy or robustness of most of the classifiers does not differ significantly in the statistical framework of classifier comparison established by Demsar[15]. They recommended the development of a multidimensional classifier assessment system based on accuracy, comprehensibility, computational efficiency and ease of usability. However the concept of stability of classifiers, which is closely related to generalization ability [15], has not been studied much in software defect prediction till now. In this study, we propose to add stability dimension to multidimensional classifier assessment system proposed by Lessman et al [4]. A classifier is said to be stable if its predictive performance does not vary with changes in data sets. Software defect prediction would benefit from more stable classifiers if the same set of classifiers can be applied to software systems catering to different application domains. We propose robustness and stability trade-off while considering candidate classifiers for software defect prediction. Thus, if two classifiers are similarly ranked in terms of robustness on a large number of defect prediction data sets, the more stable ones should be preferred. For example, Naïve Bayes and Bayesian Networks have been deeply investigated in software defect prediction [1,3,4,13], however our experiments demonstrate that Naïve Bayes and Bayesian networks are unstable classifiers for software defect prediction.

3 Experiment Design

3.1 Empirical Data

In this study, software code metrics data sets of 13 open source software systems are used to assess the robustness and stability of 17 machine learning based classifiers. A total of 44 data sets relating to these 13 software systems are available in the PROMISE repository. These data sets are used because their source code is freely available in open source repositories and their properties can be studied by other researchers as well [3]. The data sets are freely available in the PROMISE repository [17]. The software systems and the data sets along with their class distribution are summarized in Table 1. All of the 13 open source software systems have been developed on Java programming language platform. Each instance in these data sets represents a single Java class in an object-oriented (OO) software system. The attributes of each of these data sets consist of 20 software metrics. These attributes measure complexity, cohesion, coupling, size and defect proneness properties of a Java class in a software system. The metrics or code attributes are listed and summarized in Table 2. If a software class has number of defects greater than zero, it is labelled as defect-prone, otherwise as not defect prone. The19 software metrics are used as independent variables and 20th metric defect-proneness is taken as the dependent variable.

3.2 Experimental Procedure

In this study, we use 17 machine learning classifiers. These classifiers are selected based on prior literature on software defect prediction as well as machine learning literature [1-12]. The summary of classifiers is presented in Table 3. For conducting experiments pertaining to this study, an Intel Core2 Duo PC with 8 GB RAM and Windows XP operating system was used. Java based machine learning toolkit WEKA [18] (Waikato Environment for Knowledge Analysis) version 3.6.0 provided for all the classifiers. To overcome the sampling bias, M*N –way cross validation was employed, where both *M* and *N* were selected as ten [1]. Ten bins were created. Nine out of these ten bins were used as training set, while the last one was used as the test set. The dataset was randomized *M=10* times and N=10 sets were created in each iteration. The experimental procedure can be best understood by means of Fig 1.

```
DS=List of Data_Sets
    M=10
    N=10
    CLASSIFIERS(C) = List of Classifiers
    for all DS ∈ Data_Sets do
        for i= 1 to M do
        DP' = Randomize the order of DP
        S= Generate N equal sized subsets from DP'
            for j=1 to N do
                Test = S[j], Train=S-S[j]
                for all C ∈ CLASSIFIERS
                Model=Apply C on Train
                Prediction C=Apply Model on Test
                end for
            end for
        end for
    end for
```

Fig. 1. Experimental Procedure

Table 1 Summary of Data Sets [10][12]

Project Name	No. of Classes	No. of Defect-Free Classes	No. of Defective Classes	Project Name	No. of Classes	No. of Defect-Free Classes	No. of Defective Classes	Project Name	No. of Classes	No. of Defect-Free Classes	No. of Defective Classes
ant 1.3	125	105	20	xerces 1.2	440	369	71	jedit 4.3	492	481	11
ant 1.4	178	138	40	xerces 1.3	453	384	69	lucene 2.0	195	104	91
ant 1.5	293	261	32	xerces 1.4	588	151	437	lucene 2.2	247	103	144
ant 1.6	351	259	92	xerces init	162	85	77	lucene 2.4	340	137	203
ant 1.7	745	579	166	ivy-1.1	111	48	63	camel 1.0	339	326	13
synapse 1.0	157	141	16	ivy-1.4	241	225	16	camel 1.2	608	392	216
synapse 1.1	222	162	60	ivy-2.0	352	312	40	camel 1.4	872	727	145
synapse 1.2	256	170	86	poi-1.5	237	96	141	camel 1.6	965	777	188
velocity1.4	196	147	49	poi-2.0	264	227	37	log4j 1.0	135	101	34
velocity1.5	214	142	72	poi-2.5	385	137	248	log4j 1.1	109	72	37
velocity 1.6	229	151	78	poi-3.0	442	161	281	log4j 1.2	205	16	189
xalan-2.4	723	613	110	jedit 3.2	272	182	90	Pbeans1	26	6	20
xalan-2.5	803	416	387	jedit 4.0	306	231	75	Pbeans2	51	41	10
xalan-2.6	885	474	411	jedit 4.1	312	233	79	Tomcat	858	781	77
xalan-2.7	909	898	11	jedit 4.2	367	319	48				

Table 2 Summary of Metrics[12]

Software Metric		Software Metric	Definition
WMC	Weighted Methods per Class	Ca	Afferent Couplings
DIT	Depth of Inheritance Tree	Ce	Efferent Couplings
NOC	Number of Children	LCOM	Lack of Cohesion in methods
CBO	Coupling between object classes	LOC	Lines of Code
RFC	Response for a Class	CAM	Cohesion Among Methods
NPM	Number of Public Methods	CBM	Coupling Between Methods
DAM.	Data Access Metric	AVG_CC	Average Cyclomatic complexity.
MOA	Measure Of Aggregation	AMC	Average Method Complexity
MAX_CC	Maximum Cyclomatic Complexity	Num_Defects	The defect count of a class as observed in logs of source code repository

Table 3 Classifiers Employed in the Study [11]

Naïve Bayes(NB)	Bayes Network(BNET)	Support Vector Machine(SVM)
Logistic Regression(LR)	Kstar	Multi-Layer Perceptron(MLP)
IB1	OneR	Random Forest(RF)
Naïve Bayes Tree (NBTree)	Voting Feature Intervals (VFI)	Decision Stump(DS)
RandomTree(RDTree)	Decision Table (DT)	OneR
REPTree	J48	PART

3.3 Evaluation Criteria

In this study, area under curve (AUC) of ROC curves is used as an evaluation metric to compare the robustness and stability of 17 classifiers over 44 defect prediction data sets. An ROC curve is a two dimensional plot in which the x-axis denotes the false positive rate (FPR) and the y-axis denotes the true positive rate (TPR) of a classifier. In an ROC curve, the upper left point (0,1) is the most desired point, known as "ROC heaven", representing 100% TPR and zero FPR, while the point (1,0) is the least desired point called "ROC hell". As the AUC of ROC curve gets larger, the classifier gets better. The benefit of using ROC curves for classifier comparison is that the AUC metric of ROC curve is not biased against the minority class. The AUC metric enables the evaluation of defect predictors over all possible combinations of misclassification costs and prior probabilities of defect prone and non- defect prone modules. Thus ROC curve is a good way of visualizing the performance of a prediction model. The AUC of ROC curve is a desirable way to obtain a single performance metric for comparing a number of defect prediction models. At the current state of art in machine learning and software defect prediction literature, it is highly recommended to use non-parametric significance tests for statistical comparison of robustness of classifiers [15]. Significance tests allow a rigorous and fair comparison of classifiers by distinguishing random and chance observations from significant observations [15].

According to Demsar[15], the non-parametric Wilcoxon test can be utilized for comparison of two classifiers over multiple data sets. Demsar[15] also advises the usage of Friedman test followed by post-hoc Nemenyi test for comparison of multiple classifiers over multiple data sets. For application of Friedman test, all the classifiers are ranked according to their AUC performance on each data set. The classifier with highest AUC value gets a rank of 1, the second highest gets a rank of 2 and so on. Next, the mean rank of a classifier is obtained over N data sets. After this, the Friedman test compares the mean ranks of classifiers to check whether there are significant differences between mean ranks of classifiers. The test statistic of Friedman test is calculated as under:-

$$\chi_F^2 = \frac{12N}{k(k+1)}\left[\sum_j R_j^2 - \frac{k(k+1)^2}{4}\right] \tag{1}$$

In (1) , k is the number of classifiers, N is the number of data sets and R_j is the average rank of a classifier j over N data sets . $R_j = \frac{1}{N}\sum_i r_i^j$ where r^j is the rank of jth classifier over i^{th} data set. The test statistic χ_F^2 is Chi-square distributed with k-1 degrees of freedom. In this study N=44 and k=17. If the value of test statistic is large enough and significant at $\alpha = 0.05$ then the null hypothesis, that all classifiers have similar performance, is rejected. Subsequently, Nemenyi post hoc test is used to perform pair wise comparison of classifiers. Two classifiers are significantly different from each other if the difference in their ranks is greater than the critical difference given by :-

$$CD = q_\alpha \sqrt{\frac{k(k+1)}{12N}} \quad \text{where } q_\alpha \text{ is studentized range statistic} \tag{2}$$

Thus classifiers with lowest mean rank value on all studied data set is rated as the most robust. The classifiers with lowest rank variance over all studied data sets are rated as most stable.

4 Experimental Results Analysis

Table 4 presents the results of AUC performance of 17 classifiers over 44 defect prediction data sets. The second last row of Table 4 reports the mean rank of each of the 17 classifiers over 44 data sets and is related to robustness and the last row reports the variance of ranking of classifiers over 44 data sets and is related to stability concept. The test statistic of Friedman test in our study is chi=312.86, degrees of freedom=16 and significance=0.000, therefore, there are significant performance differences among 17 classifiers. The critical difference (CD) as per Nemenyi post hoc test is 3.723. Fig. 2 presents the empirical comparison of AUC performance of 17 classifiers, where x-axis denotes the ranking order of the classifiers and y-axis denotes the average rank of the classifiers AUC values. The error bars in Fig. 2 indicates the critical difference of the Nemenyi post hoc test. The AUC performance of two classifiers is considered to significantly different if the corresponding error bars in Fig. 2 do not overlap. Fig 3 plots the ranking order of robustness on x-axis and ranking order of stability on y-axis, for each of the 17 classifiers. The following observations may be made from the results shown in Table 4, Fig. 2 and Fig. 3 .

- From second last row of Table 4, Random Forest classifier has the smallest AUC mean rank value of 2.85. Thus, Random Forest is the top performing classifier in terms of robustness in open source defect prediction. This result is consistent with the results of Lessman et al[4] obtained on 11 NASA MDP data sets[14]. But our experiments are at a much larger scale than that of Lessman et al[4] as we have evaluated 44 open source defect prediction data sets[17].
- The logistic regression classifier is the second most robust classifier with a mean AUC rank value of 5.125. In contrast to results obtained by Lessman et al[4] results, in our experiments Kstar, an analogy based classifier, performs well and is the third most robust classifier with an AUC rank value of 5.147. The Naïve Bayes classifier is fourth most robust classifier with AUC rank value of 5.636. The Bayes Network classifier is fifth most robust classifier with AUC rank value of 6.199.
- From Fig. 2, it is observed that the error bars for top five most robust classifiers- Random Forests, logistic regression, Kstar, Naïve Bayes, and Bayes Network are overlapping, thus the top five classifiers do not have statistically significant accuracy differences and are equally robust in defect prediction. SVM is the least robust classifier with low AUC performance in this study. This result may be because we do not perform any SVM parameter tuning for reasons of simplicity.
- From last row of Table 4, the Random Forest classifier has smallest AUC rank variance of 6.332 and is the most stable classifier as well. The second most stable classifier is Kstar with AUC rank variance of 8.786. The third most stable classifier is Logistic regression with AUC rank variance of 9.594. From last row Table 4, it is observed that Bayes network classifier has the highest AUC rank variance of 29.62 and is the most unstable classifier. Fig. 3 presents the two-dimensional classifier assessment system proposed in this study. The x-axis represents the robustness dimension and y-axis represents the stability dimension. From Fig 3, although the Bayes Network and Naïve Bayes are ranked high in robustness, they are ranked poor in stability. In previous research on

software defect prediction, Bayes Networks[3,13] have been shown to be robust classifiers. Thus, in the two dimensional classifier assessment, Random Forest, Logistic Regression and Kstar have good performance on robustness as well as stability and are thus preferred over Naïve Bayes and Bayes Network classifier. As per Fig. 3, the neural network classifier- MLP is averagely robust and stable.

- Random Forest classifier ranks variables in order of importance along with classification . Logistic Regression classifiers have in built variable selection mechanism. This could be considered as a third dimension of classifier assessment.
- As compared to the study of Lessman et al[4], from Table 4, in our study several models have AUC below 0.7. Therefore, more future research is needed to find out the degree to which static code attribute based defect prediction is useful in open source software development context.

5 Threats to Validity

There can be several sources of threats to the validity of an empirical study. One of them is the errors in defect data collection process, due to the comments in source code version control systems of open source software sometimes not being well written[19]. Another source of threat to validity is that we have not applied any data filtering and normalization techniques and results regarding stability of Naïve Bayes and Bayes Networks may be correct to a first approximation only. As compared to the study of Lessman et al[4], we have not applied model selection procedures to neural networks and support vector machines and we have not investigated various support vector machine formulations such as least square and Lagrangian SVMs , for reasons of simplicity. But we have evaluated much larger number of data sets as compared to Lessman et al.[4] and we have added stability dimension to classifier assessment. The average data set size is medium and is representative of medium-scale open source software systems. Many of the studied data sets have class imbalance problem, which may impact the performance of several classifiers considered in this study. The scope of this study is intra-project validation. The experiments in this study are performed in the context of intra-project validation only, as large number of classifiers is considered over 44 data sets. The large number of classifiers would multiply the computing resource requirements if intra-project and inter-project validation is studied in a single study. A useful extension of this work will be to perform experiments using these 17 classifiers and 44 data sets in the inter-project validation context.

Table 4 AUC Performance of 17 Classifiers over 44 open source defect prediction data sets

Project / Classifier	ant 1.3	ant -1.4	ant -1.5	ant -1.6	ant -1.7	synapse 1.0	synapse1.1	synapse1.2	\velocity1.4	velocity1.5	velocity1.6	xalan 2.4	xalan 2.5	xalan 2.6	xalan-2.7	xerces-init	xerces-1.2	xerces-1.3
LR	0.696	0.659	0.811	0.814	0.814	0.672	0.732	0.743	0.781	0.763	0.747	0.760	0.678	0.804	0.776	0.714	0.637	0.776
NB	0.769	0.598	0.748	0.809	0.806	0.747	0.733	0.757	0.756	0.709	0.712	0.742	0.759	0.786	0.853	0.709	0.588	0.782
BNET	0.828	0.584	0.801	0.834	0.810	0.747	0.729	0.743	0.790	0.698	0.730	0.791	0.638	0.813	0.717	0.805	0.676	0.802
IB1	0.668	0.613	0.632	0.646	0.665	0.641	0.670	0.804	0.803	0.676	0.659	0.618	0.676	0.711	0.770	0.679	0.647	0.690
K*	0.634	0.615	0.753	0.757	0.750	0.699	0.734	0.804	0.813	0.782	0.729	0.724	0.759	0.766	0.889	0.745	0.74	0.789
VFI	0.660	0.644	0.713	0.735	0.745	0.707	0.706	0.719	0.729	0.675	0.696	0.69	0.596	0.730	0.846	0.699	0.687	0.784
DT	0.714	0.479	0.748	0.809	0.791	0.509	0.692	0.698	0.675	0.768	0.679	0.74	0.682	0.787	0.864	0.795	0.657	0.805
OneR	0.531	0.507	0.555	0.739	0.682	0.59	0.587	0.662	0.684	0.669	0.552	0.523	0.598	0.677	0.499	0.773	0.499	0.586
PART	0.515	0.669	0.610	0.736	0.727	0.441	0.719	0.721	0.679	0.712	0.682	0.684	0.689	0.798	0.656	0.718	0.68	0.771
DS	0.708	0.587	0.65	0.742	0.726	0.723	0.611	0.674	0.695	0.652	0.684	0.711	0.586	0.660	0.795	0.639	0.563	0.723
J48	0.518	0.576	0.601	0.689	0.665	0.589	0.689	0.727	0.703	0.627	0.665	0.545	0.667	0.760	0.656	0.699	0.710	0.666
RDTree	0.525	0.654	0.638	0.658	0.525	0.586	0.714	0.656	0.769	0.683	0.618	0.621	0.647	0.691	0.771	0.706	0.665	0.658
NBTree	0.719	0.585	0.756	0.719	0.719	0.665	0.674	0.729	0.707	0.793	0.690	0.728	0.695	0.795	0.747	0.804	0.669	0.762
REPTree	0.525	0.503	0.638	0.765	0.525	0.422	0.659	0.705	0.728	0.691	0.690	0.729	0.688	0.800	0.742	0.773	0.601	0.712
RF	0.675	0.684	0.799	0.818	0.816	0.704	0.752	0.778	0.868	0.794	0.785	0.794	0.761	0.837	0.900	0.817	0.781	0.838
MLP	0.792	0.721	0.679	0.765	0.755	0.740	0.742	0.742	0.740	0.773	0.782	0.721	0.687	0.797	0.820	0.737	0.668	0.769
SVM	0.515	0.500	0.500	0.621	0.624	0.500	0.623	0.672	0.724	0.562	0.589	0.498	0.590	0.707	0.500	0.627	0.500	0.534

Table 4 (*continued*)

Project / Classifier	xerces-1.4	ivy-1.1	ivy-1.4	ivy-2.0	poi-1.5	poi-2.0	poi-2.5	poi-3.0	jedit-3.2	jedit-4.0	jedit-4.1	jedit-4.2	jedit-4.3	lucene 2.0	lucene 2.2	lucene 2.4	camel-1.0	camel-1.2	camel-1.4	camel-1.6
LR	0.992	0.669	0.494	0.774	0.721	0.580	0.800	0.794	0.85	0.776	0.823	0.806	0.641	0.75	0.631	0.766	0.613	0.627	0.71	0.708
NB	0.845	0.667	0.66	0.769	0.718	0.719	0.729	0.805	0.787	0.741	0.773	0.836	0.594	0.755	0.631	0.731	0.743	0.574	0.682	0.675
BNET	0.992	0.697	0.460	0.783	0.751	0.646	0.855	0.873	0.85	0.775	0.784	0.861	0.500	0.712	0.536	0.697	0.429	0.492	0.682	0.638
IB1	0.866	0.641	0.507	0.613	0.609	0.596	0.775	0.758	0.715	0.713	0.666	0.65	0.532	0.68	0.583	0.674	0.557	0.587	0.576	0.593
K*	0.933	0.711	0.544	0.708	0.761	0.71	0.853	0.86	0.798	0.749	0.829	0.77	0.658	0.701	0.598	0.714	0.689	0.686	0.646	0.673
VFI	0.908	0.644	0.479	0.72	0.698	0.709	0.782	0.763	0.753	0.744	0.741	0.798	0.708	0.729	0.641	0.683	0.639	0.53	0.611	0.653
DT	0.929	0.641	0.435	0.704	0.748	0.609	0.836	0.859	0.794	0.739	0.74	0.806	0.458	0.677	0.531	0.665	0.437	0.499	0.608	0.636
OneR	0.895	0.691	0.496	0.559	0.59	0.54	0.745	0.659	0.609	0.661	0.694	0.537	0.499	0.603	0.541	0.527	0.497	0.532	0.513	0.524
PART	0.924	0.64	0.611	0.712	0.778	0.564	0.785	0.718	0.717	0.641	0.72	0.588	0.323	0.643	0.584	0.668	0.641	0.58	0.654	0.657
DS	0.860	0.629	0.640	0.685	0.633	0.550	0.705	0.749	0.722	0.675	0.685	0.772	0.507	0.631	0.547	0.541	0.559	0.487	0.612	0.559
J48	0.913	0.622	0.453	0.558	0.731	0.469	0.777	0.772	0.672	0.649	0.669	0.707	0.500	0.596	0.606	0.687	0.436	0.636	0.615	0.616
RDTree	0.866	0.655	0.556	0.624	0.653	0.587	0.77	0.734	0.687	0.661	0.633	0.608	0.656	0.6	0.546	0.655	0.557	0.573	0.557	0.609
NBTree	0.929	0.697	0.456	0.65	0.768	0.668	0.83	0.834	0.79	0.667	0.726	0.675	0.456	0.671	0.579	0.728	0.497	0.538	0.656	0.641
REPTree	0.886	0.648	0.419	0.585	0.730	0.637	0.842	0.798	0.778	0.695	0.721	0.704	0.458	0.723	0.569	0.743	0.434	0.568	0.631	0.621
RF	0.946	0.711	0.65	0.761	0.742	0.747	0.883	0.871	0.840	0.765	0.803	0.795	0.672	0.729	0.647	0.777	0.546	0.688	0.65	0.717
MLP	0.901	0.661	0.475	0.697	0.771	0.647	0.823	0.767	0.768	0.808	0.729	0.745	0.500	0.740	0.654	0.672	0.583	0.627	0.623	0.705
SVM	0.783	0.678	0.500	0.534	0.654	0.539	0.710	0.726	0.706	0.567	0.654	0.540	0.500	0.709	0.610	0.668	0.500	0.513	0.500	0.504

Table 4 (*continued*)

Project / Classifier	log4j-1.0	log4j-1.1	log4j-1.2	Pbeans1	Pbeans2	Tomcat
LR	0.736	0.815	0.739	0.730	0.655	0.747
NB	0.835	0.824	0.705	0.721	0.542	0.799
BNET	0.828	0.841	0.408	0.350	0.411	0.800
IB1	0.671	0.761	0.609	0.675	0.652	0.603
K*	0.732	0.73	0.75	0.800	0.608	0.751
VFI	0.787	0.793	0.686	0.75	0.454	0.747
DT	0.821	0.808	0.422	0.421	0.488	0.702
OneR	0.71	0.79	0.495	0.425	0.577	0.556
PART	0.703	0.752	0.625	0.779	0.491	0.697
DS	0.676	0.700	0.57	0.575	0.461	0.689
J48	0.701	0.587	0.686	0.725	0.470	0.619
RDTree	0.631	0.748	0.588	0.783	0.465	0.637
NBTree	0.751	0.764	0.55	0.283	0.602	0.759
REPTre	0.653	0.718	0.626	0.546	0.479	0.754
RF	0.784	0.761	0.776	0.813	0.657	0.752
MLP	0.710	0.795	0.775	0.742	0.468	0.762
SVM	0.72	0.742	0.500	0.500	0.526	0.500

Classifier	Rank Var.	Mean Rank
LR	9.594	5.125
NB	13.80	5.636
BNET	29.62	6.199
IB1	12.07	11.08
K*	8.786	5.147
VFI	12.84	8.318
DT	21.34	9.261
OneR	10.98	13.74
PART	18.34	9.863
DS	13.03	12.25
J48	13.13	12.21
RDTree	12.29	12.07
NBTree	14.31	8.238
REPTre	13.67	10.24
RF	6.332	2.85
MLP	10.89	6.482
SVM	10.64	14.24

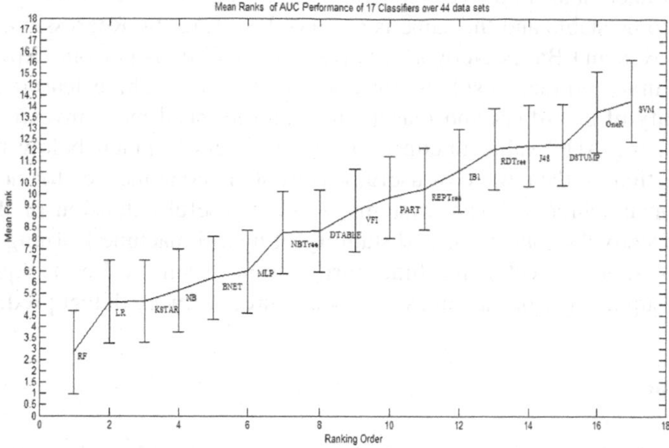

Fig. 2 AUC performance for 17 classifiers : x-axis –ranking order: y-axis Mean rank AUC performance: error bars=critical difference

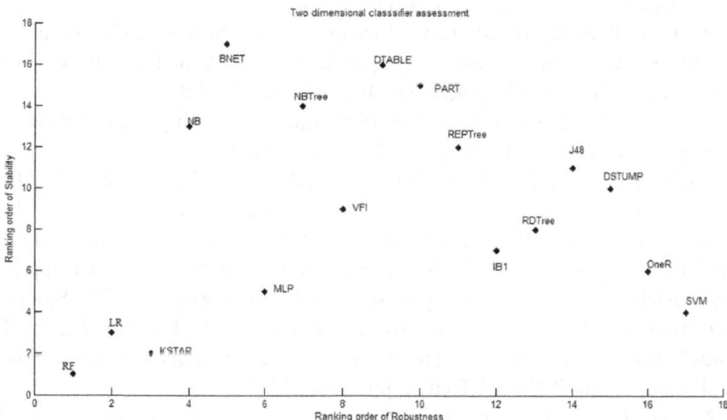

Fig. 3 Two dimensional classifier assessment. Lower ranking orders means more robust or stable

6 Conclusion

In this study we evaluated 17 classifiers over 44 defect prediction data sets pertaining to open source software systems with respect to two dimensional classifier assessment of robustness and stability. Our experimental results indicate that Random Forests, Logistic Regression and Kstar are robust as well as stable classifiers for software defect prediction. These results are consistent with theoretical concepts of machine learning as Random Forest is an ensemble technique which reduces variance and improves generalization. Statistical classifiers are also considered to be stable and the same is observed for Logistic Regression. However, Naïve Bayes and Bayes Network are poor in stability as per our experimental results and this empirical result is not consistent with machine learning theory. Thus, stability of classifiers and reasons for stability need more investigation. In many cases, the AUC value obtained from classifiers is much below 0.7. This raises a question on the degree of usefulness of static code metrics for defect prediction of open source software systems. Another useful extension of this work would be to study the robustness and stability of hybrid machine learning methods and various support vector machine formulations such as Least Square and Lagrangian support vector machines for open source software defect prediction.

References

1. Menzies, T., Greenwald, J., Frank, A.: Data mining static code attributes to learn defect predictors. IEEE Transactions on Software Engineering 3, 2–13 (2007)
2. Bener, A.B., Turhan, B.: An industrial case study of clas-sifier ensembles for locating software defects. Software Quality Journal 19(3), 515–536 (2011)
3. Okutan, A., Yildiz, O.T.: Software defect prediction us-ing Bayesian networks. Empirical Software Engineering 19(1), 154–181 (2014)
4. Lessmann, S., Baesens, B., Mues, C., Pietsch, S.: Benchmarking classification models for software defect prediction: A proposed framework and novel findings. IEEE Transactions on Software Engineering 34(4), 486–496 (2008)
5. Liang, G., Zhu, X., Zhang, C.: An Empirical Study of Bagging Predictors for Different Learning Algorithms. In: AAAI, pp. 1802–1803 (2011)
6. Khoshgoftaar, T.M., Allen, E.B., Hudephol, J.P., Aud, S.J., Allen, E.B., Hudephol, J.P., Aud, S.J.: Application of neural networks toquality modeling of a very large telecommunication system. IEEE Transactions on Neural Networks 8, 902–909 (1997)
7. Yuan, X., Khoshgoftaar, T.M., Allen, E.B., Ganesan(2000), K.: An application of fuzzy clustering to software quality prediction. In: Proc. The 3rd IEEE Symposium on Application-Specific Systems and Software Engineering Technology. IEEE (2000)
8. Khoshgoftaar, T.M., Seliya, N.: Tree-Based Software Quality Estimation Models For Fault Prediction. In: IEEE METRICS, pp. 203–213 (2002)
9. He, Z., Shu, F., Yang, Y., Li, M., Wang, Q.: An investigation on the feasibility of cross-project defect prediction. Autom. Softw. Eng. 19(2), 167–199 (2012)
10. Catal, C., Diri, B.: Software Fault Prediction with Object-Oriented Metrics Based Artificial Immune Recognition System. In: Münch, J., Abrahamsson, P. (eds.) PROFES 2007. LNCS, vol. 4589, pp. 300–314. Springer, Heidelberg (2007)
11. Khoshgoftaar, T.M., Seliya, N., Sundaresh, N.: An empirical study of predicting software faults with case-based reasoning. Software Quality Journal 14(2), 85–111 (2006)
12. Guo, L., Ma, Y., Cukic, B., Singh, H.: Robust Prediction of Fault-Proness by Random Forests. In: Proc. 15th Int'l Symp. Software Reliability Eng. (2004)
13. Dejaeger, K., Verbraken, T., Baesens, B.: Towards comprehensible Fault Prediction Models using Bayes Network Classifiers. IEEE Transactions on Neural Networks 39(2), 237–257 (2013)

14. Chapman, M., Callis, P., Jackson, W.: Metrics Data Program. NASA IV and V Facility (2004), http://mdp.ivv.nasa.gov/
15. Demsar, J.: Statistical Comparisons of Classifiers over Multiple Data Sets. Journal of Machine Learning Research 7(12), 1–3 (2006)
16. Bousquet, O., Elisseeff, A.: Stability and Generalization. Journal of Machine Learning Research 2, 499–526 (2002)
17. http://promisedata.org/
18. Hall, M., Frank, E., Holmes, G., Pfahringer, B., Reutemann, P.: The WEKA Data Mining Software: An Update. SIGKDD Explorations 11(1), 10–18 (2009)
19. Jureczko, M., Madeyski, L.: Towards identifying software project clusters with regard to defect prediction. In: PROMISE (2010)

A Two-Stage Genetic *K*-harmonic Means Method for Data Clustering

Anuradha D. Thakare and Chandrashkehar A. Dhote

Abstract. Clustering techniques are aimed to partition the entire input space into disconnected sets where the members of each set are highly connected. K-harmonic means (KHM) is a well-known data clustering technique, but it runs into local optima. A two stage genetic clustering method using KHM (TSGKHM) is proposed in this research, which can automatically cluster the input data points into an appropriate number of clusters. With the best features of both the algorithm, and TSGKHM the first stage overcomes the local optima and results in optimal cluster centers, and in the second stage, results into/in optimal clusters.

The proposed method is executed on globally accepted, four real time data sets. The intermediate results are produced. The performance analysis shows that TSGKHM performs significantly better.

Keywords: Clustering, Genetic Algorithm (GA), K-harmonic means, and the TSGKHM algorithm.

1 Introduction

Cluster analysis is an important and primitive research field, which measures the identical relationship among the objects in the absence of prior knowledge.

Anuradha D. Thakare
Department of Computer Engineering,
Pimpri Chinchwad College of Engineering,
Pune, India
e-mail: adthakare@yahoo.com

Chandrashkehar A. Dhote
Department of Computer sc. and Engineering,
Prof. Ram Meghe Institute of Technology and Research, Badnera,
Amravati, India
e-mail: vikasdhote@rediffmail.com

El-Sayed M. El-Alfy et al. (eds.), *Advances in Intelligent Informatics*,
Advances in Intelligent Systems and Computing 320, DOI: 10.1007/978-3-319-11218-3_36

K-means is a centroid based clustering method [1], which classifies given input objects in to k clusters and for many applications; it was used effectively for producing the clusters [2], and becomes popular for its feasibility. It has the ability to deal with a large amount of data, but its computational complexity becomes high. Also, due to the random choice of initial centroids, it easily runs into local optima. Several modifications are done by the researchers to improve the K-means clustering algorithm. The improved version of K-means was proposed called K-harmonic means [3, 4]. The KHM is not sensitive to the selection of initial seeds as centroids like the k-means algorithm because the clustering objective of KHM is to minimize the harmonic average from all the data points to all the cluster centers [5, 6]. The improvements in the KHM are suggested and various hybrid models are proposed by the researchers. Both K-means and KHM easily run in to local optima.

GA is a stochastic general search method [7], capable of effectively exploring the large search spaces. It is widely used because of its abilities of self-adaptation and self-organization. Mostly, it is used to solve some complicated optimization problems. The global optimization technique called the genetic algorithm is used to produce the optimal clusters. The GA has the capability to automatically divide the input space into many regions and then through a number of generations, produces accurate clusters. The quality of the solutions that GA has showed in the different types of fields and problems it makes perfect sense to try to use it in clustering problems [8]. GA minimizes the square error of the cluster of dispersion. The complex problems such as unsupervised clustering or non-parametric clustering are often dealt with by employing an evolutionary approach.

A method called the genetic K-means algorithm (GKM) was proposed [9], which defines a distance-based mutation. This is a basic mutation operator specific to clustering.

A new hybrid data clustering algorithm based on KHM and improved GSA, called IGSAKHM, was proposed [10], which not only helps the KHM clustering escape from local optima, but also overcomes the slow convergence speed of the IGSA.

A hybrid algorithm for clustering called PSOKHM was proposed [11], which uses Particle Swarm Optimization to help KHM escape from local optima at a certain level, and results in better clustering. IGSAKHM is superior to the KHM algorithm and the PSOKHM algorithm in most cases.

In this paper, we have proposed a two stage genetic *k*-harmonic means method for data clustering, which overcomes the problem of local optima by integrating the clustering objective function of KHM as a fitness function in GA. In the first stage, GA results in globally optimal cluster centers and in the second stage, GA uses the KHM function for cluster formation. The proposed *TSGKHM* method is found to work very satisfactorily.

1.1 K-Harmonic Means Clustering

KHM uses the Harmonic Averages of the distances from each data point to the centers as it is constituent to its performance function [10]. The K-Harmonic Means algorithm overcomes the drawbacks of K-Mean in terms of initialization of centers. KHM uses different objective functions [10] one of, which is as follows:

$$KHM = \sum_{k=0}^{n} \left(\frac{\text{K cluster centers}}{(\sum \textit{average} \text{ of distance from all points to cluster center})^{p}} \right) \tag{1}$$

Where, P is an input parameter.

In KHM, the objective function computes the membership function and weight associated to/with each of the data points. Then it recomputes the centers location from all the data points and assigns the data point xi to cluster Ci, which is having the highest membership value.

2 Related Work

A self-organized genetic algorithm for document clustering was proposed [12] based on semantic similarity measure. The Genetic Algorithm in conjunction with the hybrid strategy, gets the best clustering performance. The self-organized genetic algorithm, considering the influence between the diversity of the population and the selective pressure, efficiently, evolves the clustering of the documents in comparison with the standard k-means algorithm in the same similarity strategy.

Another hybrid approach is a combination of K-harmonic means and the Particle Swarm Optimization (PSOKHM) algorithm. The PSO method is a population based global optimization technique in, which the solution space of the problem is expressed as a search space. Each position in the search space is an interrelated solution of the problem [5]. The KHM algorithm tends to converge faster than the PSO, but the drawback is that it gets stuck in local optima. The hybrid clustering algorithm called PSOKHM [11], maintains the qualities of KHM and PSO. The objective function of KHM is the fitness function of PSOKHM; KHM is applied for four iterations to the particles in the swarm where it is a collection of particles. To improve fitness value, the Particle Swarm Optimization algorithm is applied for eight generations.

Various clustering approaches using a center-based clustering algorithm, KHM have been proposed by the researchers. The main objective of the clustering is to produce the accurate and effective clusters with less computation time. The performance of the KHM [13], is measured based on a function, which calculates Harmonic averages of the distances from each data point to the centers. The objective function in KHM, computes the membership function and weight associated to/with each of the data points. The centers location is recalculated from all the data points having the highest membership value, to assign a data point to the cluster. A two-stage genetic clustering algorithm can automatically, determine the proper number of clusters and the proper partition from a given data set [14]. The two-stage selection and mutation operations are implemented to exploit the search capability of the algorithm.

In the Gravitational Search Algorithm, the objects are considered to be with masses, which attract each other by the gravity force. All the objects move towards the ones with heavier masses due to force. The information is being transformed by the gravitational force among the objects and the objects, which have a higher mass become heavier. The Gravitational Search Approach using the K-Harmonic Means method called GSAKHM [10] was proposed, which is an effective way of clustering the documents and can be achieved by using a combination of the Gravitational Search Method and the K-Harmonic Means algorithm.

The hybrid clustering method using two popular clustering techniques, ant based clustering and the K-Harmonic Means algorithm (ACAKHM) was proposed [15]. ACA works on the principle of an ant's behaviour of collecting or organizing the feedback of the behaviour of the ants. Even though, ACA provides appropriate partitions without defining the initial cluster centers, it takes a long time to get a better result. To overcome this drawback, ACAKHM makes a better use of the advantage of both ACA and KHM. ACA is effectively used to get the initial canters and then KHM is used to avoid local optima.

Candidate group search is based on some selection rules to isolate the candidate groups for each center (CGSKHM). Screening through all the data sets. If it fits in to the candidate group, the center has to be interchanged and a new solution is achieved by using KHM. The Candidate Group Search [16], offers a scheme combining of some haphazardness and deterministic selection rules coming from the data set, CGS outperforms KHM and it requires less computation time. Variable neighbourhood search (VNS) is a meta-heuristic technique intended to resolve combinatorial and global optimization problems. Variable neighbourhood search for harmonic means clustering (VNSKHM) was proposed [17] in, which, the Variable neighbourhood search is experimented to solve the problem of KHM clustering, which is easily trapped in local optima. The elementary idea is to continue to a systematic change of neighbourhood within a local search algorithm.

3 Proposed Method

The KHM algorithm is found to be a good clustering algorithm for many applications. The best feature is that , it requires less function evaluations and hence, converge is faster. But it gets stuck into local optima most of the times and therefore, the technique with global search abilities is required to improve the performance.

The proposed method integrates the KHM and GA to produce optimal clusters, called Two-Stage Genetic KHM. The best features of both the techniques are used for cluster formation. The input space contains the clusters on, which the proposed two-stage GA is applied for cluster optimization in two stages with two objective functions, one for each stage. In the first stage, the clustering objective minimum intracluster distance is used as the fitness function in a genetic process, which results in subclusters formation. In the second stage, the KHM function is used as a fitness function for cluster optimization. The steps carried out are as given in the algorithm.

Algorithm: ***TSGKHM***
Steps:

1. Initial parameter setting with number of generations and population size
2. Generate the population
3. Fitness evaluation with *fitnessfun_1*
4. Apply GA operators & generate a new population
5. Repeat step 3 & 4 till convergence
6. Store Sub clusters with their fitness values
7. Fitness evaluation with *fitnessfun_2*
8. Apply GA operators & generate a new population
9. Repeat step 7 & 8 till convergence
10. Optimal clusters

3.1 Fitness Function_1: Intracluster Distance [18]

In the first stage GA, the fitness function is intracluster distance *E*. The data points with minimum *E* value are the most fitted members for the cluster formation. On this basic theme, the subclusters are formed. Suppose a data point x_i is assigned to the jth cluster based on minimum euclidean distance. The intra-cluster distance [19] is calculated by following formula:

$$E = \sum_{x_i \subseteq c_j} \frac{\|x_i - x_j\|}{N_j} \qquad for\ i = 1, 2, 3, \ldots N \tag{2}$$

Where, x_j is the center of cluster c_j, and *N* is the size of population i.e. number of data points.

3.2 Fitness Function_2: (KHM) K-Harmonic Means [10]

The K-Harmonic Mean of the data point, is calculated by equation 1.

$$KHM(X, C) = \sum_{i=1}^{m} \frac{n}{\sum_{j=1}^{n} \frac{1}{\| x_i - c_j \|^p}} \tag{3}$$

Where, *N* is number of clusters, x_i is number of objects, and c_j is cluster center. In order to evaluate the membership of a data point, the function [10] in equation 2 is used.

$$m\left(\frac{C_j}{X_i}\right) = -i\left(\frac{\| X_i - C_j \|^{-p-2}}{\sum_{j=0}^{k} \| X_i - C_j \|^{-p-2}}\right) \tag{4}$$

The center is recalculated in the KHM algorithm [10] by using the formula given below.

$$C_j = \frac{\sum_{i=1}^{n} m\left(\frac{C_j}{X_i}\right) w(X_i) X_i}{\sum_{i=1}^{n} m\left(\frac{C_j}{X_I}\right) w(X_i)} \tag{5}$$

4 Results and Discussion

4.1 Dataset Used

The experimentation results, initially, four globally accepted real life data sets are taken. These data sets are described in terms of the number of points present, dimensions, and the number of clusters. These four real life data sets are taken from a UCI machine learning repository [19], which represents examples of data with low, medium, and high dimension. The data set with this description are necessary for calculating the objective function values of clusters. Initially we have tested the results of our proposed work on selected four datasets in future we may test it on more datasets.

(a) *Fisher's Iris* dataset *(n=150, d=4, k=3)*: It contains three different species of iris flower and 50 samples were collected from the four features that are: sepal length, sepal width, petal length, and petal width.
(b) *Wine* dataset *(n=178, d=13, k=3)*: It consists of 178 objects characterized by 13 features. These features are obtained by a chemical analysis of wines that are produced in the same region in Italy.
(c) *Cancer* dataset *(n=683, d=9, k=2)*: It contains 683 objects characterized by nine features. All the data is categorized in to two classes, malignant tumors (444 objects) and benign tumors (293 objects).
(d) Glass dataset (*n* = *214*, *d* = *9*, *k* = *6*): It contains 214 objects characterized by nine features. The objects are categorized into six classes.

4.2 Performance Measures

In order to evaluate the performance of a proposed method we have used two measures i.e. KHM and *F*-measure. The runtime is also recorded for all the methods. A measure, which is used to evaluate the performance of a classification model [20, 21] where all the class labels are known is used. *F*-measure is the measure of the test of accuracy with an optimum score as 1 giving good clustering and combines both precision and recall to compute the scores. Precision is the fraction of retrieved instances that are relevant. Recall is the fraction of retrieved instances that are relevant. KHM is the summation of all data points of the harmonic average of the distance from a data point to all the centers, as defined in equation (3). The smaller value of sum indicates higher quality of clustering.

4.3 Discussion on Results

In order to form good quality clusters, the proposed work is divided into two stages. In the first stage, the sub clusters are formed using GA. The objective function, intracluster distance is used as a fitness function for GA. Smaller the values of intracluster distance, higher the quality of clusters. The clusters formed are considered as subclusters so that final clusters can be formed in the second stage, using the KHM as a fitness function for GA. The best results are tabulated for each data set. Table 4.1, shows the results of four datasets in terms of sub-clusters formed in the first stage. These subclusters are nothing, but the actual clusters formed by *fitness function_1*. The input parameter P, *F*-measure, and runtime values are taken from the second stage calculations. *F*-measure values represent the clustering accuracy. These values change with the change in input. The runtime increases with respect to the size of the dataset and input parameter values.

Table 4.1 Results of proposed *TSGKHM*

Performance Criteria	Datasets			
	Iris	Wine	Cancer	Glass
Size of dataset	150	178	683	214
Sub-clusters	9	8	12	5
Inputs (P)	4	2	4	2
F-measure	0.89	0.86	0.65	0.89
Runtime	78	1950	3806	3463

The second stage results for each dataset are recorded in the following tables. Table 4.2 shows the intermediate results for the various datasets. The subclusters resulted from the first stage and the number of data points in each subcluster is tabulated. The KHM values for some subclusters of iris dataset are exceeded even ~600 because these are the intermediate results and the subclusters will be more refined further at the time of final cluster formation. When these subclusters are given as input, in the second stage, we got the results in terms of KHM and *F*-measure. The KHM values are varying for each subcluster because of the number of data points present. Also, we got minimum values of KHM for some subclusters, as the subclusters are already optimized using GA. The reason for using KHM is to optimize the number of clusters by subcluster merging. In order to merge the subclusters, the KHM values are compared. The KHM value reflects the quality of the subcluster based on, which the decision will be taken whether to merge the subcluster with the other or to keep it separate. The exact logic for subcluster merging may be enhanced to improve the results.

Table 4.2 Results of proposed *TSGKHM* in terms of subclusters on the datasets

Subcluster no.	# of Data points	Runtime	KHM	F-measure
Iris Dataset for P=4				
1	01	125	0.0141	1.05
2	01	78	0.0453	0.30
3	04	63	0.1205	0.44
4	45	47	117.87	0.74
5	35	78	117.88	0.89
6	11	78	117.90	1.04
7	03	62	117.92	1.02
8	29	62	147.86	1.63
9	21	62	578.27	1.08
Wine Dataset for P=2				
1	06	2059	0.0108	1.05
2	64	2106	0.0109	0.30
3	02	1872	0.0434	0.44
4	21	1856	0.0456	0.74
5	38	1950	0.1632	0.89
6	06.	2090	0.4085	0.49
7	03	1934	0.4101	0.52
8	38	1669	0.0098	0.59
Glass Dataset for P=2				
1	05	3385	0.00010	0.30
2	14	3463	0.00110	0.89
3	18	2886	0.00100	1.02
4	29	3447	0.00120	1.05
5	17	2885	0.00010	1.08
6	19	3.403	0.00214	0.28
7	22	3385	0.00031	0.27
8	24	3432	0.00045	0.37
9	18	3338	0.00160	0.36
10	20	3385	0.00010	0.32
11	08	3354	0.00821	0.35
12	20	3417	0.00032	0.45
Cancer Dataset for P=4				
1	10	3088	0.5210	0.25
2	198	3027	0.3425	0.56
3	12	2793	0.2489	0.50
4	434	2823	0.2560	0.46
5	29	3806	0.0241	0.65

Table 4.3 Comparison of intermediate results of *TSGKHM* with the existing methods PSOKHM and GSAKHM[10]. Here, KHM is the KHM function values, *F*-m is the *F*-measure, and Rt is the runtime parameter.

Dataset	Criteria	PSOKHM			GSAKHM			TSGKHM		
		KHM	*F*-m	Rt	KHM	*F*-m	Rt	KHM	*F*-m	Rt
Iris	P=4	106.06	0.751	1.967	704.48	0.8471	2.799	0.0141	1.05	125
Wine	P=2	59844	0.829	9.525	7.046	0.9440	4.461	0.0108	1.05	2059
Glass	P=2	1196.7	0.424	17.669	7.0236	0.9897	3.994	0.0001	1.08	2885
Cancer	P=4	NA	NA	NA	701.03	0.6824	2.352	0.0241	0.65	3806

The intermediate results of TSGKHM are tabulated. These will be used for analysis purposes so as to find out the exact criteria for subcluster merging. The results of the proposed work are compared with the existing hybrid KHM methods i.e. PSOKHM and GSAKHM [10] as tabulated in Table 4.3 Here, The best values for each dataset are taken for the comparison. These comparisons are made to analyze the deviations in the values of performance measures. Further, the results would be more refined after final cluster formation.

5 Conclusions

The two-stage genetic algorithm using K-harmonic means for data clustering is proposed, which can automatically cluster the entire data to produce the optimal clusters. We presented the results of the first stage in terms of subclusters using GA and results of the second stage in terms of KHM values and *F*-measure. The GA utilizes two fitness functions, intraclsuter distance to form subclusters and KHM to get the optimal number of clusters. We got the desired results for the first stage and in the second stage; the experimentation is done for calculating KHM values. The minimum values for KHM reflects good clustering hence, this observation will be used further for subclusters merging to get the final clusters.

References

[1] Macqueen, J.B.: Some Methods for Classification and Analysis of Multivariate Observations. In: Proceedings of 5th Berkeley Symposium on Mathematical Statistics and Probability, vol. 1, pp. 281–297. University of California Press, Berkeley (1967)

[2] Jiawei Han, M.K.: Data Mining Concepts and Techniques, Morgan Kaufmann Publishers, An Imprint of Elsevier (2006)

[3] Zhang, B., Hsu, M., Dayal, U.: K-harmonic means – a data clustering algorithm. Technical Report HPL-1999-124. Hewlett-Packard Laboratories (1999)
[4] Zhang, B., Hsu, M., Dayal, U.: K-harmonic means. In: International Workshop on Temporal, Spatial and spatio-Temporal Data Mining, TSDM 2000, Lyon, France (September 12, 2000)
[5] Cui, X., Potok, T.E.: Document clustering using particle swarm optimization. In: IEEE Swarm Intelligence Symposium, Pasadena, California (2005)
[6] Kao, Y.T., Zahara, E., Kao, I.W.: A hybridized approach to data clustering. Expert Systems with Applications 34(3), 1754–1762 (2008)
[7] Karegowda, A.G., Manjunath, A.S., Jayaram, M.A.: Application of Genetic Algorithm Optimized Neural Network Connection Weights For Medical Diagnosis of Pima Indians Diabetes. International Journal of Computer Applications (0975 – 8887) 43(1) (April 2012)
[8] Painho, M., Bação, F.: Using Genetic Algorithms in Clustering Problems. In: GeoComputation. Higher Institute of Statistics and Information Management, New University of Lisbon, Travessa, EstêvãoPinto (Campolide) P-1070-124 Lisboa (2000)
[9] Krishna, K., Murty, M.N.: Genetic k-means algorithm. IEEE Transactions on Systems, Man and Cybernetics B Cybernetics 29, 433–439 (1999)
[10] Yin, M., Hu, Y., Yang, F., Li, X., Gu, W.: A novel hybrid K-harmonic means and gravitational search algorithm approach for clustering. Expert Systems with Applications 38, 9319–9324 (2011)
[11] Yang, F., Sun, T., Zhang, C.: An efficient hybrid data clustering method based on K-harmonic means and Particle Swarm Optimization (2009)
[12] Song, W., Park, S.C.: An Improved GA for Document Clustering with semantic Similarity measure. In: Fourth International IEEE Conference on Natural Computation, pp. 536–540 (2008)
[13] Zhang, B., Hsu, M.: K-Harmonic Means - A Data Clustering Algorithm Umeshwar Dayal Software Technology Laboratory HP Laboratories, Palo Alto HPL-1999-124 (October 1999)
[14] He, H., Tan, Y.: A two-stage genetic algorithm for automatic clustering. Neurocomputing 81, 49–59 (2012)
[15] Jiang, H., Yi, S., Li, J., Yang, F., Hu, X.: Ant clustering algorithm with K-harmonic means clustering (2010)
[16] Hung, C.H., Chiou, H.-M., Yang, W.-N.: Candidate groups search for K-harmonic means data clustering (2013)
[17] Alguwaizani, A., Hansen, P., Mladenovic, N., Ngai, E.: Variable neig
[18] Everitt, B.S., Landau, S., Leese, M., Stahl, D.: Cluster Analysis. Willey series in probability and statistics
[19] `ftp://ftp.ics.uci.edu/pub/machine-learning-databases/`
[20] Sahaa, S., Bandyopadhyay, S.: A generalized automatic clustering algorithm in a multiobjective framework. Applied Soft Computing 13, 89–108 (2013)
[21] Handl, J., Knowles, J.: An evolutionary approach to multiobjective clustering. IEEE Transaction on Evolutionary Computation 11(1), 56–76 (2007)

Time-Efficient Tree-Based Algorithm for Mining High Utility Patterns

Chiranjeevi Manike and Hari Om

Abstract. High utility patterns mining from transaction databases is an important research area in the field of data mining. Due to the unavailability of downward closure property among the utilities of the itemsets it becomes great challenge to the researchers. Even though, efficient pruning strategy called, transaction weighted utility downward closure property used to reduce the number of candidate itemsets, total time to generate and test candidate itemsets is more. In view of this, in this paper we have proposed a time-efficient tree-based algorithm (TTBM) for mining high utility patterns from transaction databases. We construct conditional pattern bases to generate high transaction weighted utility patterns in the second pass of our algorithm. We used an efficient tree structure called, HP-Tree and tracing method to keep high transaction weighted utility patterns and for discovering high utility patterns respectively. We have compared the performance against Two-Phase and HUI-Miner algorithms. The experimental results show that the execution time of our approach is better.

1 Introduction

The problem of mining frequent patterns have been playing an important role in different stages of data mining and knowledge discovery process such as association rule mining [1], classification [4], clustering [7] etc. Big challenge faced by researchers is finding efficient pruning strategy to reduce the large number of candidate patterns. Agrawal et al.[1] was introduced an efficient pruning strategy called, Apriori. This strategy says that the support count of a superset of low support subsets is low. In frequent pattern mining the patterns with support count above the

Chiranjeevi Manike · Hari Om
Indian School of Mines, Dhanbad - 826 004, Jharkhand, India
e-mail: `chiru.research@gmail.com`, `hari.om.cse@ismdhanbad.ac.in`

El-Sayed M. El-Alfy et al. (eds.), *Advances in Intelligent Informatics*,
Advances in Intelligent Systems and Computing 320, DOI: 10.1007/978-3-319-11218-3_37

specified threshold are generated as frequent patterns. In real time scenario all items have different profits or importance, and to answers the few queries like the customers whose contributing more to the profit of the company, the itemsets contributing more to the total profit, only the knowledge of frequent patterns is not sufficient. In view of this utility mining was introduced with the assumption of different utility values of items in transaction database and objective of usefulness [17]. Utility of an item in a transaction is different measures of items like profit, cost, and quantity sold etc. Pruning strategy like Apriori cannot be applied in utility pattern mining because of the absence of downward closure property among the utilities of itemsets. Therefore, finding efficient pruning strategy becomes a great challenge to the researchers. Liu et al. [11] proposed an efficient algorithm called, Two-Phase by incorporating a pruning strategy namely, transaction weighted utilization (TWU) downward closure property. This strategy says that the transaction weighted utilization of itemsets hold downward closure property. Many researchers have been proposed different algorithm to efficiently mine high utility patterns with limited resources. In this paper we have proposed a time-efficient tree based algorithm for mining high utility patterns with the objective of reducing execution time.

The remaining part of this paper is organized as follows. In Sect. 2, we describe the related work. Problem definition is given in Sect. 3. In Sect. 4, we describe our proposed algorithm with example. In Sect. 5, experimental results are presented and analyzed. Conclusions and future works are given in Sect. 6.

2 Related Work

Utility mining theoretical model and fundamental definitions are given by Yao et al. [16]. This model is based on the level-wise candidate generation and test methodology, in each level low utility itemsets are pruned by using two properties called Utility bound property, and support bound property. To overcome the drawbacks, Liu et al. [11] proposed an efficient two-phase algorithm with a pruning strategy called transaction weighted utilization downward closure property. This property says that transaction weighted utility of any superset of a low transaction weighted utility itemset is low. This algorithm suffers from level-wise candidate generation and test problem. Yao et al. [15] proposed two algorithms namely UMining and Umining_H with two efficient pruning strategies by analyzing the utility upper bound property which is introduced in [16]. These algorithms are also suffers from level-wise candidate generation and test problem and loss some patterns. Another algorithm was proposed by Li et al. [9] called isolated itemset discarding strategy (IIDS) based on the level-wise candidate generation and test problem. To overcome the drawbacks with level-wise candidate generation and test problem of above algorithms Erwin et al. [6] was proposed an efficient algorithm called CTU-Mine using pattern growth approach [8]. Even though the runtime efficiency of CTU-Mine is better for low utility values on dense datasets, overall performance is better than the Two-Phase algorithm. Another efficient algorithm was proposed by same authors called CTU-PRO [5] that is performed well on both sparse and dense datasets. Ahmed et al.

[2] was proposed an algorithm called HUC-Prune based on the pattern growth approach. Tseng et al. [14] proposed an efficient two pass algorithm called UP-Growth by using a data structure called UP-Tree. Shi et al. [12] proposed modified version of Two-Phase algorithm to improve the performance. Above algorithms [16, 15, 9] suffers from the problem of level-wise candidate generation and test problem and algorithms which are implemented based on the pattern growth require more memory hence, recently Liu et al. [10] proposed an efficient algorithm called HUI-Miner. Even though it needs no candidate generation process and requires very less amount of memory, takes more time to construct utility-lists and generate high utility patterns. Yen et al. [18] proposed an efficient algorithm for mining high utility patterns with out need of generating candidate itemsets. Ahmed et al. [3] proposed three efficient data structures for interactive and incremental high utility pattern mining. Tseng et al. [13] proposed two algorithms called UP-Growth and UP-Growth$^+$ with more efficient strategies to generate candidate itemsets with two scans of the database. In view of the runtime performance and memory consumed by above algorithms, we proposed a tree based algorithm to improve the runtime performance of mining high utility patterns.

Table 1 Transaction Database

Tid	Item A	Item B	Item C	Item D	Item E
T_1	0	1	2	3	1
T_2	2	1	1	0	0
T_3	1	2	1	1	1
T_4	0	1	0	14	1
T_5	1	0	0	1	0

Table 2 Utility Table

Item Name	A	B	C	D	E
Profit($)	3	5	1	1	10

3 Problem Definition

We have followed the identical definitions conferred in the preceding works [16, 11, 15]. Let $I = \{i_1, i_2, i_3, \ldots, i_m\}$ be a finite set of items and TD be a transaction database $\{T_1, T_2, T_3, \ldots, T_n\}$ in which each transaction $T_i \in TD$ is a subset of I. each item in a transaction associated with purchased quantity that is also called as internal utility of item defined by $iu(i_p, T_q)$, for example, $iu(A, T_2) = 2$, in Table 1. External utility of an item is the unit profit value defined by $eu(i_p)$, for example, $eu(B) = 5$, in Table 2.

Definition 1. The utility of an item i_p in a transaction T_q is the product of its internal utility and external utility and it is defined by $u(i_p) = iu(i_p, T_q) \times eu(i_p)$, for example, $u(C, T_1) = 2 \times 1 = 2$, in Table 1 and Table 2.

Definition 2. The database utility of an item i_p is the sum of all utilities in the database defined by $du(i_p) = \Sigma_{i_p \in T_q \in TD} u(i_p)$, for example, $du(D) = u(D, T_1) + u(D, T_3) + u(D, T_4) + u(D, T_5) = 3 + 1 + 14 + 1 = 19$, in Table 1 and Table 2.

Definition 3. The utility of an itemset X in transaction T_q is defined as $u(X, T_q) = \Sigma_{i_p \in X} u(i_p, T_q)$, for example, $u(AB, T_2) = 11$, in Table 1 and Table 2.

Definition 4. The database utility of an itemset X in a database TD is defined by $u(X, TD) = \Sigma_{i_p \in X \in TD} u(i_p, T_q)$, for example $u(AB, TD) = u(AB, T_2) + u(AB, T_3) = 11 + 13 = 24$, in Table 1 and Table 2.

Definition 5. Transaction utility can be defined as the sum of the individual utilities of all items in that transactionT_q, defined by $tu(T_q) = \Sigma_{i_p \in T_q} u(i_p), tu(T_1) = 20$, in Table 1 and Table 2.

Definition 6. The total database utility is the sum of the all transaction utilities in the database, defined by $tdu(TD) = \Sigma_{T_q \in DB} tu(T_q)$, $tdu(TD) = tu(T_1) + tu(T_2) + tu(T_3) + tu(T_4) + tu(T_5) = 20 + 12 + 25 + 29 + 4 = 90$, in Table 1 and Table 2.

Definition 7. Transaction weighted utility of an item i_p in the transaction database TD is defined as $twu(i_p, TD) = \Sigma_{i_p \in T_p \in TD} tu(T_q)$, $twu(A, TD) = tu(T_1) + tu(T_3) + tu(T_5) = 12 + 25 + 4 = 41$, in Table 1 and Table 2.

Definition 8. Minimum utility threshold is given by the percentage of the total transaction database utility, defined by $minUtil = tdu(TD) \times \delta$.

Definition 9. High utility itemset is an itemset with utility $\geq minUtil$.

Definition 10. High utility pattern mining is the process of finding patterns with utility more than the specified minimum utility threshold.

4 Proposed Algorithm

In this section, we describe our proposed algorithm called TTBM, this algorithm requires three database scans. During the first scan it finds the high transaction weighted utility of each item, in second scan constructs the tree based on the high transaction weighted utility patterns. Next it updates the actual utilities of pattern in tree by scanning database again. Brief description of the three scans of the algorithm for the above transaction database(Table 1) is given below.

4.1 First Scan

During the first scan, algorithm loads each transaction one by one and accumulates the transaction weighted utility of the items. For example, consider the transaction

Algorithm 1. TTBM Algorithm

```
Input: Transaction Database TD; Minimum Utility Threshold δ
Output: High utility Patterns
while input ≠ null do
    for each item do
        claculate TWU value
        if TWU > δ then
            add item to the itemsList
    sort itemsList TWU desc order
while input ≠ null do
    for each transaction do
        if item TWU > δ then
            add item to the oList
            TU = TU + itemUtility
        sort oList items as in itemsList
        tree.addTrans(oList, TU)
HuiList = mfpgrowth(tree)
tree.addHui(HuiList) while input ≠ null do
    for each transaction do
        load items and generate subitemsets
        calculate Utility of each subset
        tree.addUtilities(subitemsets, listutility)
hui_list = tree.tracing(δ)
return hui_list
```

database and utility table in Table 1 and Table 2, after first scan algorithm finds the following transaction weighted utilities of items, A:41, B:86, C:57, D:78, E:74. If any item transaction weighted utility is lower than the minimum utility that will be discarded, and that will not be consider for further computation so number of candidate patterns effectively reduced.

4.2 Second Scan

During second scan, algorithm loads all items of a transaction except the items which are discarded in the first scan. No item will be discarded if minimum utility is set to 30, item A will be discarded for minimum utility 50 (i.e. twu(A) = 41). Transaction utility of each transaction will also be calculated in parallel, if any item is discarded that utility is subtracted from transaction utility. The below table (Table 3) represents the transaction database during second scan, items are ordered in transaction weighted utility ascending. Next by reading each transaction TWU-Tree is constructed that is shown in Fig. 1(a). HP-Tree in Fig. 1(a) is constructed for high transaction weighted utility patterns {CEDB}, {DE}, {BE}, {BD}, and

{BDE}. While loading first transaction root node is null so a new node with item C will be constructed and transaction utility is assigned to the utility of that node. Next we check child list of current node for the next item E, as the item C do not have any child so E will be attached as child to the node C, in this fashion all items will be processed. While processing next transaction first we will check first item exists in the children list of root node, if not it will be created else transaction utility is accumulated to existing. Next modified pattern growth is applied to find high transaction weighted utility patterns. HP-Tree is constructed for all high transaction weighted utility patterns, Fig. 1(b) represent the HP-Tree after adding all high transaction weighted utility patterns.

Table 3 Updated Transaction Database

Tid	Item C	Item E	Item D	Item B	TU
T_1	2	1	3	1	20
T_2	1	0	0	1	6
T_3	1	1	1	2	22
T_4	0	1	14	1	29
T_5	0	0	1	0	1

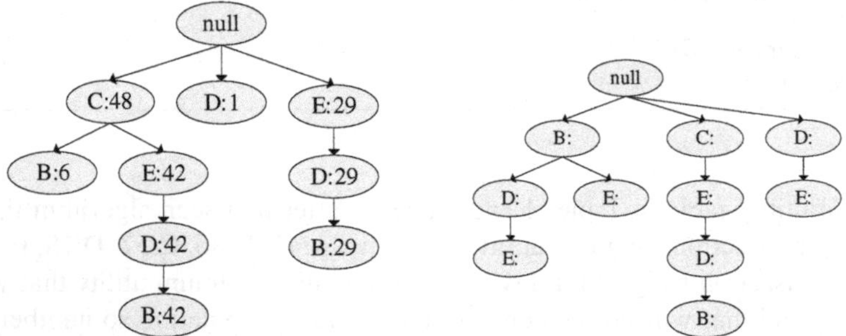

(a). Transaction weighted utilities (b). High Transaction weighted utility itemsets

Fig. 1 HP-Tree

4.3 *Third Scan*

In third scan, candidate patterns are generated and utility also calculated in parallel. These candidate utilities are updated in HP-tree only if candidate is exist in the tree, Fig. 2 represents the HP-Tree after adding all high transaction weighted utility patterns. After adding all candidate utilities by traversing HP-Tree all high utility patterns with utility more than the specified utility are generated. Algorithm generates the high utility patterns {BDE:68} and {BE:50} for minimum utility 50.

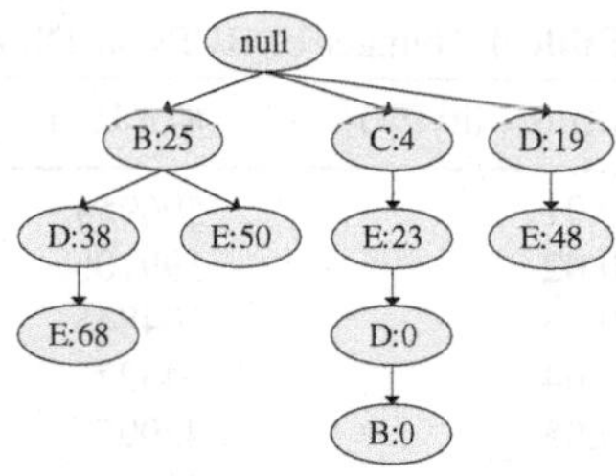

Fig. 2 HP-Tree with utilities

5 Experimental Evaluation

All algorithms are implemented in java and experiments are preformed on a PC with processor Intel $Core^{TM}$ i7 2600 CPU @ 3.40 G_{HZ}, 2GB Memory and the operating system is Microsoft Windows 7 32-bit.

5.1 Synthetic Datasets

IBM synthetic dataset generator was used for generating datasets. The parameter settings for generating datasets are followed from [1]. IBM generator generates the binary database, so fit this into the scenario of utility pattern mining purchased quantities of items in every transaction is generated randomly ranging from 1 to 5. External utility (unit profit) value is also generated randomly ranging from 1 to 20 by following lognormal distribution as shown in Fig. 3.

5.2 Experimental Results

Experiments are done over synthetic datasets to evaluate the performance of proposed approach in different cases. In first case, we have performed experiments with varying minimum utility thresholds (low range 0.01 to 0.09 and high range 0.1 to 0.9). In next and subsequent cases, experiments are done to check the scalability of algorithms with varying number of distinct items, number of transactions, and transaction lengths respectively. In the Fig. 4(a) we have shown only the performance comparison of our approach with HUI-Miner, because Two-Phase algorithm

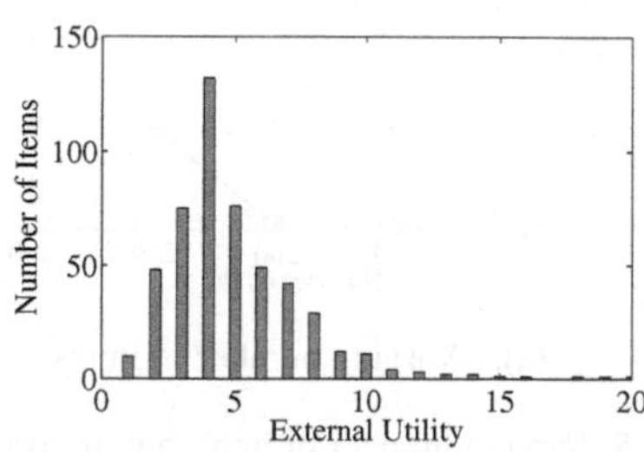

Fig. 3 External utility distribution for 300 items

Table 4 Number of HUPs on T8I4D50K

MinUtility(%)	# of HUP	MinUtility(%)	# of HUP
0.01	596954	0.1	12350
0.02	199106	0.2	3435
0.03	104845	0.3	1574
0.04	64322	0.4	879
0.05	42997	0.5	555
0.06	31073	0.6	376
0.07	23597	0.7	271
0.08	18729	0.8	199
0.09	14983	0.9	149

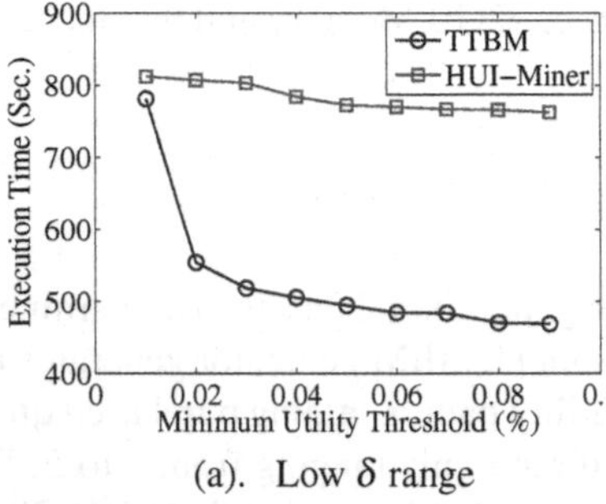

(a). Low δ range

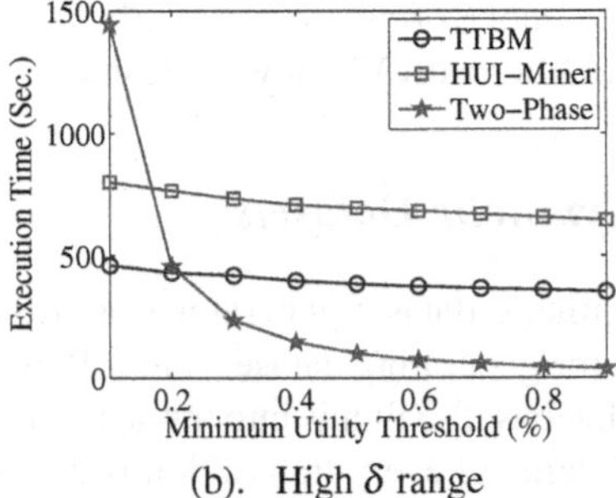

(b). High δ range

Fig. 4 Varying minimum utility thresholds

performance is not good at low minimum utility thresholds that can also be observed from Fig. 4(b) that the execution time is exponentially increasing with the decreasing utility threshold. Fig. 4(a) & (b) shows that performance of our proposed algorithm is far better than the performance of HUI-Miner for different utility ranges. From Fig. 4(b) we can also observe that execution time of Two-Phase algorithm is more efficient than proposed and HUI-Miner.

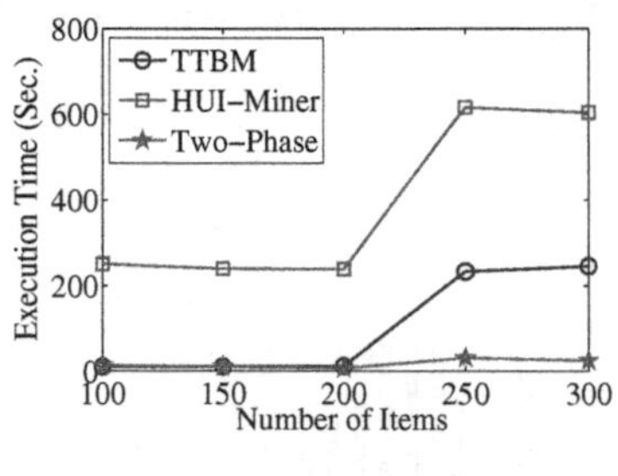

(a). Varying number of items

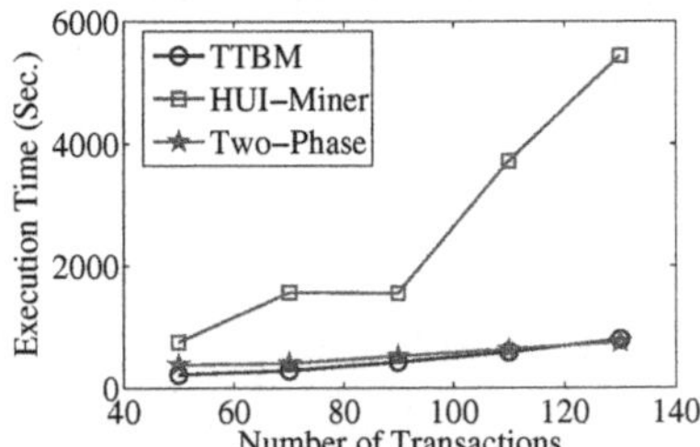

(b). Varying number of transactions

Fig. 5 Varying number of items and transactions

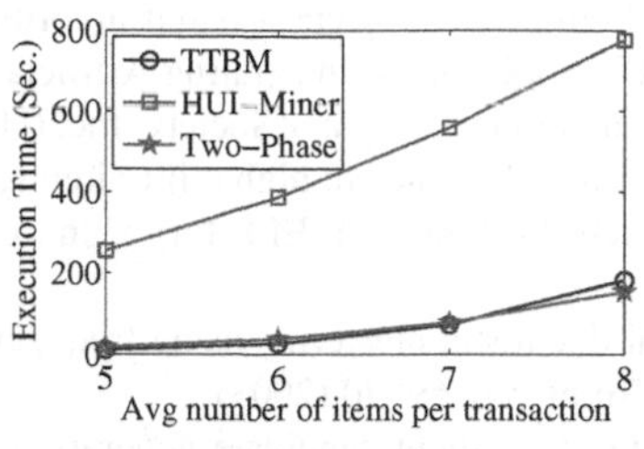

(a). Varying average length of transactions

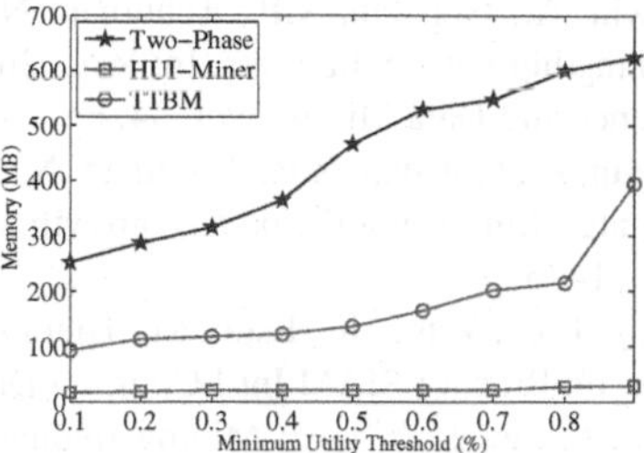

(b). Memory consumption

Fig. 6 Varying average length, minimum utility

From Fig. 5(a) we can observe that effect of increasing number of distinct items on the execution time of the algorithms. From Fig. 5(a) we can also observe that there is slight variation on the performance of proposed and Two-Phase algorithm, but meager effect on the HUI-Miner algorithm. Fig. 5(b) and Fig. 6(a) shows that effect of execution performance on varying number of transactions, and transaction length is similar on proposed and Two-Phase algorithm and smaller than that of HUI-Miner. Fig. 6(b) shows memory consumption of three algorithms.

6 Conclusion

In this paper we have proposed a time-efficient algorithm for effectively mine the high utility patterns from transaction database. Our algorithm only need maximum three database scans compare to the existing few algorithm those require multiple database scans. Experimental performance analysis shows that proposed algorithm execution time is better in all cases. However, the execution time of our algorithm is efficient, memory consumption is more. So in our future work we make our algorithm more efficient in both time and memory in finding the complete set of high utility patterns.

References

1. Agrawal, R., Srikant, R., et al.: Fast algorithms for mining association rules. In: Proc. 20th Int. Conf. VLDB, vol. 1215, pp. 487–499 (1994)
2. Ahmed, C.F., Tanbeer, S.K., Jeong, B.-S., Lee, Y.-K.: An efficient candidate pruning technique for high utility pattern mining. In: Theeramunkong, T., Kijsirikul, B., Cercone, N., Ho, T.-B. (eds.) PAKDD 2009. LNCS (LNAI), vol. 5476, pp. 749–756. Springer, Heidelberg (2009)
3. Ahmed, C.F., Tanbeer, S.K., Jeong, B.S., Lee, Y.K.: Efficient tree structures for high utility pattern mining in incremental databases. IEEE Transactions on Knowledge and Data Engineering 21(12), 1708–1721 (2009)
4. Brin, S., Motwani, R., Silverstein, C.: Beyond market baskets: generalizing association rules to correlations. ACM SIGMOD Record 26, 265–276 (1997)

5. Erwin, A., Gopalan, R.P., Achuthan, N.: A bottom-up projection based algorithm for mining high utility itemsets. In: Proc. 2nd Int'l Workshop on Integrating Artificial Intelligence and Data Mining, vol. 84, pp. 3–11. Australian Computer Society, Inc. (2007)
6. Erwin, A., Gopalan, R.P., Achuthan, N.: Ctu-mine: An efficient high utility itemset mining algorithm using the pattern growth approach. In: Proc. 7th IEEE Int'l Conf. on CIT, pp. 71–76 (2007)
7. Fung, B.C., Wang, K., Ester, M.: Hierarchical document clustering using frequent itemsets. In: Proc. of SIAM Int'l Conf. on Data Mining, pp. 59–70 (2003)
8. Han, J., Pei, J., Yin, Y.: Mining frequent patterns without candidate generation. ACM SIGMOD Record 29, 1–12 (2000)
9. Li, Y.C., Yeh, J.S., Chang, C.C.: Isolated items discarding strategy for discovering high utility itemsets. Data & Knowledge Engineering 64(1), 198–217 (2008)
10. Liu, M., Qu, J.: Mining high utility itemsets without candidate generation. In: Proc. 21st ACM Int'l Conf. on Information and Knowledge Management, pp. 55–64 (2012)
11. Liu, Y., Liao, W.-k., Choudhary, A.K.: A two-phase algorithm for fast discovery of high utility itemsets. In: Ho, T.-B., Cheung, D., Liu, H. (eds.) PAKDD 2005. LNCS (LNAI), vol. 3518, pp. 689–695. Springer, Heidelberg (2005)
12. Shi, Y., Liao, W.K., Choudary, A., Li, J., Liu, Y.: High utility itemsets mining. Int'l Journal of Information Technology & Decision Making 09(06), 905–934 (2010)
13. Tseng, V., Shie, B.E., Wu, C.W., Yu, P.: Efficient algorithms for mining high utility itemsets from transactional databases. IEEE Transactions on Knowledge and Data Engineering 25(8), 1772–1786 (2013), doi:10.1109/TKDE.2012.59
14. Tseng, V.S., Wu, C.W., Shie, B.E., Yu, P.S.: Up-growth: an efficient algorithm for high utility itemset mining. In: Proc. 16th ACM SIGKDD Int'l Conf. on Knowledge Discovery and Data Mining, pp. 253–262 (2010)
15. Yao, H., Hamilton, H.J.: Mining itemset utilities from transaction databases. Data & Knowledge Engineering 59(3), 603–626 (2006)
16. Yao, H., Hamilton, H.J., Butz, C.J.: A foundational approach to mining itemset utilities from databases. In: Proc. 4th SIAM Int'l Conf. on Data Mining, pp. 482–486 (2004)
17. Yao, H., Hamilton, H.J., Geng, L.: A unified framework for utility-based measures for mining itemsets. In: Proc. ACM SIGKDD 2nd Workshop on Utility-Based Data Mining, pp. 28–37 (2006)
18. Yen, S.J., Chen, C.C., Lee, Y.S.: A fast algorithm for mining high utility itemsets. In: Behavior Computing, pp. 229–240 (2012)

A Lexicon Pooled Machine Learning Classifier for Opinion Mining from Course Feedbacks*

Rupika Dalal, Ismail Safhath, Rajesh Piryani,
Divya Rajeswari Kappara, and Vivek Kumar Singh

Abstract. This paper presents our algorithmic design for a lexicon pooled approach for opinion mining from course feedbacks. The proposed method tries to incorporate lexicon knowledge into the machine learning classification process through a multinomial process. The algorithmic formulations have been evaluated on three datasets obtained from ratemyprofessor.com. The results have also been compared with standalone machine learning and lexicon based approaches. The experimental results show that the lexicon pooled approach obtains higher accuracy than both the standalone implementations. The paper, thus proposes and demonstrates how a lexicon pooled hybrid approach may be a preferred technique for opinion mining from course feedbacks and hence suitable for develpment in a practical caurse feedback mining system.

1 Introduction

Opinion Mining or Sentiment Analysis is defined as a quintuple $< O_i, F_{ij}, S_{kijl}, H_k, T_l >$ [1]; where O_i is the focused object, F_{ij} is the particular feature of the object, S_{kijl} is the sentiment polarity (positive, negative or neutral) of the object O_i by opinion holder k on the jth feature expressed at time l. A lot of work has been done, in the past, on opinion mining from different data sources (such as movie reviews, product reviews, cellphone reviews etc). While some researchers tried to mine the opinion for a text document as a whole (document-level), others have tried to take the task to a deeper level of granularity by computing opinion at sentence-level

Rupika Dalal · Ismail Safhath · Rajesh Piryani ·
Divya Rajeswari Kappara · Vivek Kumar Singh
Text Analytics Laboratory, South Asian University, Akbar Bhawan,
Chanakyapuri, New Delhi-110021, India
e-mail: {rupika08,ismaail.s,rajesh.piryani,divyakappara,
vivekks12}@gmail.com

* This work is supported by UGC, India Major Research Project Grant No.-41-624/2012.

El-Sayed M. El-Alfy et al. (eds.), *Advances in Intelligent Informatics*,
Advances in Intelligent Systems and Computing 320, DOI: 10.1007/978-3-319-11218-3_38

or aspect-level. Irrespective of the level of analysis, broadly two approaches have been used for the opinion mining task, the lexicon-based approach and the machine learning classification approach.

The overwhelming trend towards E-learning and Massive Online Open Courses (MOOCs) is an interesting phenomena to observe. Users of these course often have a forum to express their views about the course. Even in traditional classroom teaching, institutions are now increasingly focusing on student course feedbacks. In either of the scenarios, the number of students writing the feedback on one or more courses may be large. It is in this context that we have tried to position our work on opinion mining from course feedbacks. While most of the course feedbacks have closed ended questions with fixed responses (a summary of which can be plotted visually), it is also common to have open ended free-form text boxes in the course feedback forms, which give liberty to students to freely write down his/her overall experiences on a course. However, when reports on feedback are generated, they use only numeric value fields and not the free-form text. It is difficult to read all the texts in feedback responses manually. We have, therefore, designed a opinion mining framework to automatically compute polarity of opinions expressed in course feedback responses by the students.

The rest of the paper is organized as follows: the section 2 briefly describes the existing approaches (machine learning classifier and lexicon based approach)for opinion mining, including our SentiWordNet implementation. We then describe how we designed the machine learning and lexicon hybrid using lexicon pooling, in section 3. The section 4 presents details about the dataset, evaluation measures and results. The paper concludes in section 5, with a discussion of our results and the usefulness of this work.

2 Existing Approaches

Opinion Mining as a task can be approached through a machine learning classification approach or by using a lexicon-based method. Dave et al. in their work reported in 2003 [2], used scoring, smoothing, Naive Bayes (NB), Maximum Entropy (ME) and Support Vector Machine (SVM) for assigning sentiment to documents. Kim and Hovy in a work reported in 2004 [3], used a probabilistic method for assigning sentiment to expressions. Pang and Lee in their work reported in 2002 [4] and 2004 [5], applied NB, SVM and ME classifiers for document-level sentiment analysis of movie reviews. In a later work reported in 2005 [6], they have applied SVM, regression and metric labeling for assigning sentiment of a document using a 3 or 4 point scale. Gamon in a published work in 2004 [7], used SVM to assign sentiment of a document using a 4-point scale. Bikel et al. in their work in 2007 [8], implemented subsequence kernel based voted perceptron and compared its performance with standard SVM. Durant and Smith [9], tried sentiment analysis of political weblogs. Since then many researchers have been trying to use some machine learning classifier for sentiment analysis of unstructured texts. All these classifiers are supervised in nature and require annotated data for training. Since we aimed to do opinion mining from course reviews, we first tried to evaluate the applicability of machine learning classifiers and implemented NB, SVM and ME using the R programing language.

Lexicon-based methods are unsupervised, in nature used for opinion mining and do not require any training or prior annotated data. Instead this approach relies on use of sentiment lexicons. However, initially, these methods were not very accurate primarily due to inappropriate and limited sentiment lexicons being used. This is why, Turney [10], [11], used the World Wide Web itself as a corpus and implemented the unsupervised SO-PMI-IR algorithm on movie and travel reviews. The sentiment lexicon used in this approach are either created manually or automatically [12] and [11]. The research in lexicon-based approach focused on using adjectives as pointers of the semantic orientation [12], [13], [14] and [15]. Taboada et al. [16] also built a dictionary for lexicon based methods. With the extension of WordNet as a sentiment lexicon, Sebastiani and Esuli [17], [18], worked towards gloss analysis. We have implemented the SentiWordNet-based lexicon approach. For this, we used different linguistic features and a suitable score aggregation for the overall opinion class computation. Computational Linguists suggest that adjectives are good pointers of opinions and also Adverbs further modify the opinion expressed in review sentences. For example, the sentence "He is very good lecturer" expresses a more positive opinion about the teacher than the sentence "He is good lecturer". For our linguistic feature selection we mine both 'adjective' and 'verb', along with any 'adverb' prior to them. As adverbs are modifying the scores of succeeding terms, therefore, it needs to be decided in what proportion the sentiment score of an 'adverbs' should modify the succeeding 'adjective' or 'verb' sentiment score. We have used the modifying weightage (scaling factor) of adverb score as 0.35, based on the conclusions reported in [19] and [20]. For aggregating the two pattern (extracted 'adverb+adjective' and 'adverb+verb' combines) scores, we have attempted different weight factors varying from 10% to 100%, and found that 30% weight for verb score produces best accuracy levels. The pseudo-code for our implementation, SentiWordNet (AAAVC) [21], is depicted in Algorithm 1. Here AAAVC refers to Adverb+Adjective and Adverb+Verb Combine.

3 The Lexicon Pooled Hybrid Approach

The machine learning and lexicon-based approach, each have its own merits and demerits. The main problem with machine learning approach is that it requires substantial amount of training (and hence labeled training data), which may not always be available. The lexicon-based approach on the other hand is largely dependent on availability of a good sentiment lexicon. We tried to design a hybrid approach that combines both the approaches with the aim of harnessing their advantages and reducing the drawbacks. It may be very interesting to visualize that a hybrid machine learning classifier can perform better even in presence of very limited training data. The hybrid design allowed us to achieve higher accuracy levels with lesser training data. A different kind of combining (sequential application of multiple methods) have been attempted earlier in the work reported by Prabowo and Thelwall [22]. We have, however, tried to combine the two different kind of approaches through a multinomial pooling of lexicon within the classification process. This is based on the sentiment analysis work done on Blogposts by Melville et al. [23]. We first quickly summarize the NB and then explain how lexicon pooling has been incorporated within its classification process.

3.1 Naive Bayes

It is a supervised probabilistic classifier, in which the opinion mining problem can be visualized as a 2-class text classification task. Every text document is assigned to one of the two opinion classes, 'positive' and 'negative' class. The main idea in this classification is to classify the document based on statistical pattern of occurrence of terms in the documents vis-a-vis the classes. The best class in NB classification is the Maximum A-Posteriori (MAP) class, computed as:

$$c_{map} = \arg\max_{c \in C}[logP(c) + \sum_{1 \leq k \leq n_d} logP(t_k|c)] \quad (1)$$

In equation 1, each $P(t_k/c)$ value refer to the weight which specify how good an indicator the term t_k is for class c, and in similar way the prior $logP(c)$ indicates the relative frequency of class c. In our case the labeled datasets have been fed to the NB algorithm as k-folds (with k=3). If the amount of training data is not sufficient, NB will perform badly as it will be difficult to learn term-class probabilities for majority of terms. We tried to address this issue by incorporating the knowledge from sentiment lexicon.

3.2 The Pooled Multinomial

We aimed to build a composite NB classifier which incorporates the background knowledge from a publicly available sentiment lexicon with the training data. The multinomial NB classifier for opinion mining relies on three assumptions: (a) documents are considered as bag of words; (b) there is one to one correspondence between each bag of word component and its class, and (c) given a document, the words in the document are produced independently of each other. For the background knowledge model we need to estimate a-prior probabilities from the sentiment lexicon. The prior class probability $P(c_j)$ is estimated from the labeled training data as done in normal NB classifier. The probabilities from NB and lexicon-based knowledge are then aggregated by linear pooling [24].

Combining probability distributions form multiple experts is generally known as pooling distributions. In our case the 'experts' are the two approaches to calculate membership of a term to an opinion class. We have used linear pooling approach to combine the probability distributions, which has been found to perform better in a past work [23]. The linear pooling approach dates back to Laplace [24]. In this approach the aggregate probability is calculated as:

Algorithm 1. SWN (AAAVC)

For each sentence, extract adv+adj and adv+verb combines.
For each extracted adv+adj combine do:
if $score(adj) = 0$ **then**
  ignore it.
end if
if $score(adv) > 0$ **then**
  if $score(adj) > 0$ **then**
    $f(adv, adj) = min(1, score(adj) + sf * score(adv))$
  end if
  if $score(adj) < 0$ **then**
    $f(adv, adj) = min(1, score(adj) - sf * score(adv))$
  end if
end if
if $score(adv) < 0$ **then**
  if $score(adj) > 0$ **then**
    $f(adv, adj) = max(-1, score(adj) + sf * score(adv))$
  end if
  if $score(adj) < 0$ **then**
    $f(adv, adj) = max(-1, score(adj) - sf * score(adv))$
  end if
end if
For each extracted adv+verb combine do:
if $score(verb) = 0$ **then**
  ignore it.
end if
if $score(adv) > 0$ **then**
  if $score(verb) > 0$ **then**
    $f(adv, verb) = min(1, score(verb) + sf * score(adv))$
  end if
  if $score(verb) < 0$ **then**
    $f(adv, verb) = min(1, score(verb) - sf * score(adv))$
  end if
end if
if $score(adv) < 0$ **then**
  if $score(verb) > 0$ **then**
    $f(adv, verb) = max(-1, score(verb) + sf * score(adv))$
  end if
  if $score(verb) < 0$ **then**
    $f(adv, verb) = max(-1, score(verb) - sf * score(adv))$
  end if
end if
Add the positive and negative scores to respective pools.

$$P(t_i|c_j) = \alpha_{NB} P_{NB}(t_i|c_j) + \alpha_{LB} P_{LB}(t_i|c_j) \quad (2)$$

Here, NB and LB are the experts (read methods) we used, and $P_e(t_i/c_j)$ represents the probability assigned by those experts on a term t_i occurring in a document of class c_j. The α_{NB} and α_{LB} are the weights associated to the experts NB and LB where:

$$\alpha_{NB} = \log\left(\frac{acc_{NB}}{1 - acc_{NB}}\right); \alpha_{LB} = \log\left(\frac{acc_{LB}}{1 - acc_{LB}}\right)$$

and acc_{NB} and acc_{LB} are the accuracies of experts NB and LB on the training set. All acc_{NB} and acc_{LB}'s are normalized to sum to one. The final probability score for a word is thus a linear weighted combination of probabilities computed from NB and the lexicon-based method. The weights depend on the classification accuracies of the respective individual methods.

4 Dataset and Results

We have used 3 datasets in our experimental evaluation. These datasets are collected from the website ratemyprofessor.com, which contains feedbacks written by students across the world on their courses (including instructors). The total collection comprised of 811 student feedbacks, out of which 556 are positive and 255 are negative reviews. These 'positive' and 'negative' label assignments have been done by a set of three annotators and we obtained very accurate inter-indexer consistency values. The table 1 describes the dataset statistics.

Table 1 Datasets

S.No.	Dataset	#Reviews	Positive	Negative	Avg. Word Length
1	Dataset 1	468	342	126	37
2	Dataset 2	227	136	91	36
3	Dataset 3	116	78	38	23

We have evaluated the machine learning classifiers with six different n-gram feature selection models: (i) Unigram, (ii) Unigram with Specific POS Tag, (iii) Bigram (iv) Bigram with specific POS Tag (v) Trigram, and (vi) Trigram with Specific POS Tag. All text was first tagged by using Stanford POS tagger. With unigram+POS scheme, we extract unigram with POS tags as adjective. For bigram+POS and trigram+POS, we select two and three word patterns corresponding to patterns used by Turney in [11]. We have removed from the feature selection process, those terms which do not occur in at least two documents. For the lexicon-based implementations, features were extracted in a similar manner by using POS tag information. We used Recall, Precision, F-measure, Accuracy and Entropy for computing standard performance measures. The table 2 presents the results for the NB classifier with different feature selection schemes. The table 3 shows the accuracy levels of the three machine learning classifiers, on unigram feature selection (one that proved better than others) and the SWN based implementation. The tables 4, 5 and 6 shows the summary of results of NB, lexicon-based and the lexicon-pooled approach, with three different sentiment lexicon usage, respectively. While the table 4 corresponds to pooling with the Bing Liu Opinion Lexicon (BLOL); the table 5 is for pooling with SWN lexicon (only polarity classes are used and not strengths) and the table 6 shows results for pooling with SWN lexicon polarity values.

Table 2 Naive Bayes Results

Dataset	Scheme	#Features	Accuracy	Precision	Recall	F-Measure	Entropy
Dataset 1	Unigram	2000	**79.91**	0.74	0.77	0.76	0.24
	Unigram+POS	355	73.08	0.67	0.70	0.69	0.27
	Bigram	10958	76.92	0.71	0.73	0.71	0.24
	Bigram+POS	987	64.96	0.53	0.53	0.53	0.20
	Trigram	17558	68.16	0.65	0.68	0.67	0.31
	Trigram+POS	839	62.82	0.58	0.60	0.59	0.32
Dataset 2	Unigram	1456	76.21	0.75	0.75	0.75	0.23
	Unigram+POS	265	**77.53**	0.77	0.76	0.77	0.23
	Bigram	6536	71.81	0.71	0.71	0.71	0.25
	Bigram+POS	547	60.35	0.65	0.64	0.65	0.31
	Trigram	8931	61.67	0.66	0.65	0.66	0.31
	Trigram+POS	394	48.02	0.69	0.56	0.62	0.14
Dataset 3	Unigram	844	**71.55**	0.73	0.76	0.74	0.18
	Unigram+POS	147	53.45	0.59	0.59	0.59	0.19
	Bigram	2706	68.97	0.67	0.70	0.68	0.24
	Bigram+POS	186	70.69	0.71	0.74	0.72	0.20
	Trigram	3189	69.83	0.65	0.61	0.63	0.23
	Trigram+POS	121	70.69	0.72	0.75	0.73	0.18

As observed, the simple unigram feature selection performs best with all the three machine learning classifiers implemented. Further, the NB matches the performance levels of ME and SVM in all these three datasets. Results on dataset 1 is better than others, apparently because it has larger reviews thereby allowing more features to be used in the learning and classification process. As seen in table 3, the results of SWN are quite close to that of machine learning classifiers. The main thing to observe in the results (table 4) is that when we pool the sentiment lexicon in the NB process, the results are better than both the individual approaches (except in dataset 3). The tables 5 and 6 show the results for pooling with SWN and its polarity strength use. While the first one shows improvements over NB and SWN implementations, use of polarity does not have a significant improvement. It may also be observed that the results of pooled multinomial are improved substantially if the results for lexicon-based methods are good. In case, the lexicon-based method does not perform well, results for pooled approach do not improve.

Table 3 Results for NB, Lexicon-based and Lexicon Pooling with BLOL

Datasets	Method	# Feature	Accuracy	Precision	Recall	F-Measure	Entropy
	Naive Bayes	2000	79.91	0.75	0.77	0.76	0.24
Dataset 1	Lexicon Based	2000	76.50	0.70	0.70	0.70	0.23
	Lexicon Pooled	2000	**82.69**	0.78	0.80	0.79	0.22
	Naive Bayes	1456	76.21	0.75	0.75	0.76	0.23
Dataset 2	Lexicon Based	1456	70.93	0.76	0.65	0.70	0.12
	Lexicon Pooled	1456	**77.53**	0.79	0.74	0.76	0.17
	Naive Bayes	843	**71.55**	0.73	0.74	0.74	0.18
Dataset 3	Lexicon Based	843	63.79	0.62	0.63	0.62	0.25
	Lexicon Pooled	843	63.07	0.69	0.70	0.70	0.14

Table 4 Results for NB, Lexicon-based and Lexicon Pooling with SentiWordNet Lexicon

Datasets	Method	# Feature	Accuracy	Precision	Recall	F-Measure	Entropy
	Naive Bayes	2000	79.91	0.75	0.77	0.76	0.24
Dataset 1	Lexicon Based	2000	73.93	0.66	0.55	0.60	0.07
	Lexicon Pooled	2000	**79.99**	0.72	0.71	0.71	0.2
	Naive Bayes	1456	76.21	0.75	0.75	0.76	0.23
Dataset 2	Lexicon Based	1456	67.40	0.72	0.60	0.66	0.095
	Lexicon Pooled	1456	**76.48**	0.71	0.66	0.68	0.17
	Naive Bayes	843	**71.55**	0.73	0.76	0.74	0.18
Dataset 3	Lexicon Based	843	50.86	0.60	0.59	0.60	0.14
	Lexicon Pooled	843	71.03	0.71	0.74	0.73	0.18

Table 5 Results for NB, Lexicon-based and Lexicon Pooling with SentiWordNet+Polarity

Datasets	Method	# Feature	Accuracy	Precision	Recall	F-Measure	Entropy
	Naive Bayes	2000	**79.91**	0.75	0.77	0.76	0.24
Dataset 1	Lexicon Based	2000	48.08	0.50	0.50	0.50	0.38
	Lexicon Pooled	2000	76.50	0.72	0.61	0.66	0.11
	Naive Bayes	1456	**76.21**	0.75	0.75	0.76	0.23
Dataset 2	Lexicon Based	1456	54.63	0.56	0.56	0.56	0.34
	Lexicon Pooled	1456	69.16	0.76	0.63	0.68	0.11
	Naive Bayes	843	**71.55**	0.73	0.76	0.74	0.18
Dataset 3	Lexicon Based	843	51.72	0.47	0.47	0.47	0.26
	Lexicon Pooled	843	62.07	0.72	0.71	0.71	0.11

5 Discussion

We have first implemented and empirically evaluated the performance levels of NB, SVM and ME machine learning classifiers and the SWN implementation of lexicon-based approach. Then, we designed a lexicon-pooled classifier by pooling the sentiment lexicon knowledge into the machine learning classifier process. The results how that pooling the knowledge from sentiment lexicon improves the accuracy of the NB classifier. However, the proportion of change varies with the way pooling is done. The BLOL appears to be more suitable for the pooling process. The SentiWordNet lexicon pooling also obtains better results than standalone NB. However, pooling with sentiment strength information from the lexicon does not obtain the expected improvement in performance. It can be concluded that the sentiment polarity class of a term as obtained from the sentiment lexicon, is a suitable and more appropriate choice for the lexicon pooling than the polarity strengths. In fact for basic NB, the class information of a term is the key input required for classification. It is also important to mention that the improvement in results of lexicon pooling are substantial if results of lexicon based approach are good. This suggests that a good sentiment lexicon is requires for pooling.

It would be relevant to state that the lexicon-pooled design could be very useful in cases where the amount of training data available for the machine learning classifier is not sufficient. The lexicon-based knowledge compensates for the lack of adequate training data by estimating the term-class probabilities from the sentiment lexicon. In the domain of course feedbacks it becomes very useful. Further, this opinion mining formulation allows to obtain accurate analysis from the course feedback texts and can complement the closed-ended question response summaries. The empirical analysis shows that standalone implementations of machine learning classifier approach and the lexicon-based approaches can be made to perform better in a hybrid formulation. The hybrid formulation combines advantages of both. The missing term-class knowledge for the NB is compensated by the information from the sentiment lexicon, thereby improving the result accuracy. It remains to be seen as to how the sentiment lexicon may be combined with other machine learning classification approaches such as SVM and ME. For NB, it is done through a simple linear multinomial pooling but will be a bit complex for other methods. Further, it would be relevant to apply the hybrid on other data types and also for aspect-based opinion mining task.

References

1. Liu, B.: Sentiment analysis and subjectivity. In: Handbook of Natural Language Processing, 2nd edn. Taylor & Francis, Boca (2010)
2. Dave, K., Lawrence, S., Pennock, D.M.: Mining the peanut gallery: Opinion extraction and semantic classification of product reviews. In: Proceedings of the 12th International Conference on World Wide Web, WWW 2003, pp. 519–528. ACM, New York (2003)
3. Kim, S.-M., Hovy, E.: Determining the sentiment of opinions. In: Proceedings of the 20th International Conference on Computational Linguistics, COLING 2004 (2004)
4. Pang, B., Lee, L., Vaithyanathan, S.: Thumbs up? In: Proceedings of the ACL Conference on Empirical Methods in Natural Language Processing, EMNLP 2002 (2002)
5. Pang, B., Lee, L.: A sentimental education. In: Proceedings of the 42nd Annual Meeting on Association for Computational Linguistics, ACL 2004 (2004)
6. Pang, B., Lee, L.: Seeing stars. In: Proceedings of the 43rd Annual Meeting on Association for Computational Linguistics, ACL 2005 (2005)
7. Gamon, M.: Sentiment classification on customer feedback data. In: Proceedings of the 20th International Conference on Computational Linguistics, COLING 2004 (2004)
8. Bikel, D.M., Sorensen, J.: If we want your opinion. In: International Conference on Semantic Computing (ICSC 2007) (2007)

9. Durant, K.T., Smith, M.D.: Mining sentiment classification from political web logs. In: Proceedings of Workshop on Web Mining and Web Usage Analysis of the 12th ACM SIGKDD International Conference on Knowledge Discovery and Data Mining (WebKDD 2006), Philadelphia, PA (2006)
10. Turney, P.D.: Mining the web for synonyms: Pmi-ir versus lsa on toefl. In: Flach, P.A., De Raedt, L. (eds.) ECML 2001. LNCS (LNAI), vol. 2167, pp. 491–502. Springer, Heidelberg (2001)
11. Turney, P.D.: Thumbs up or thumbs down?: semantic orientation applied to unsupervised classification of reviews. In: Proceedings of the 40th Annual Meeting on Association for Computational Linguistics, pp. 417–424. Association for Computational Linguistics (2002)
12. Hatzivassiloglou, V., McKeown, K.R.: Predicting the semantic orientation of adjectives. In: Proceedings of the 35th Annual Meeting of the Association for Computational Linguistics and Eighth Conference of the European Chapter of the Association for Computational Linguistics, pp. 174–181. Association for Computational Linguistics (1997)
13. Wiebe, J.: Learning subjective adjectives from corpora. In: AAAI/IAAI, pp. 735–740 (2000)
14. Hu, M., Liu, B.: Mining and summarizing customer reviews. In: Proceedings of the 2004 ACM SIGKDD International Conference on Knowledge Discovery and Data Mining, KDD 2004 (2004)
15. Taboada, M., Anthony, C., Voll, K.: Methods for creating semantic orientation dictionaries. In: Proceedings of the 5th International Conference on Language Resources and Evaluation (LREC), Genova, Italy (2006)
16. Taboada, M., Brooke, J., Tofiloski, M., Voll, K., Stede, M.: Lexicon-based methods for sentiment analysis. Computational Linguistics 37(2), 267–307 (2011)
17. Sebastiani, F.: Machine learning in automated text categorization. ACM Computing Surveys (CSUR) 34(1), 1–47 (2002)
18. Esuli, A., Sebastiani, F.: Determining the semantic orientation of terms through gloss classification. In: Proceedings of the 14th ACM International Conference on Information and Knowledge Management, pp. 617–624. ACM (2005)
19. Benamara, F., Cesarano, C., Picariello, A., Recupero, D.R., Subrahmanian, V.S.: Sentiment analysis: Adjectives and adverbs are better than adjectives alone. In: ICWSM (2007)
20. Berger, A.L., Pietra, V.D., Pietra, S.D.: A maximum entropy approach to natural language processing. Computational Linguistics 22(1), 39–71 (1996)
21. Singh, V.K., Piryani, R., Uddin, A., Waila, P.: Sentiment analysis of movie reviews: A new feature-based heuristic for aspect-level sentiment classification. In: 2013 International Multi-Conference on Automation, Computing, Communication, Control and Compressed Sensing (iMac4s), pp. 712–717. IEEE (2013)
22. Prabowo, R., Thelwall, M.: Sentiment analysis: A combined approach. Journal of Informetrics 3(2), 143–157 (2009)
23. Melville, P., Gryc, W., Lawrence, R.D.: Sentiment analysis of blogs by combining lexical knowledge with text classification. In: Proceedings of the 15th ACM SIGKDD International Conference on Knowledge Discovery and Data Mining, KDD 2009 (2009)
24. Clemen, R.T., Winkler, R.L.: Combining probability distributions from experts in risk analysis. Risk Analysis 19(2), 187–203 (1999)

Quality Metrics for Data Warehouse Multidimensional Models with Focus on Dimension Hierarchy Sharing

Anjana Gosain and Jaspreeti Singh

Abstract. Data warehouses, based on multidimensional models, have emerged as powerful tool for strategic decision making in the organizations. So it is crucial to assure their information quality, which largely depends on the multidimensional model quality. Few researchers have proposed some useful metrics to assess the quality of the multidimensional models. However, there are certain characteristics of dimension hierarchies (such as relationship between dimension levels; sharing of some hierarchy levels within a dimension, among various dimensions etc.) that have not been considered so far and may contribute significantly to structural complexity of multidimensional data models. The objective of this work is to propose metrics to compute the structural complexity of multidimensional models. The focus is on the sharing of levels among dimension hierarchies, as it may elevate the structural complexity of multidimensional models, thereby affecting understandability and in turn maintainability of these models.

Keywords: Data warehouse, Multidimensional model, Quality metrics, Dimension hierarchies.

1 Introduction

Data warehousing systems play an important role in company's decision making processes. Inmon [15] defined data warehouse as "subject-oriented, integrated, time-varying, non-volatile collections of data that is used primarily in organizational decision making". These systems homogenize and integrate large volumes of organization's data in order to provide a single and comprehensive

Anjana Gosain · Jaspreeti Singh
University School of Information and Communication Technology
Guru Gobind Singh Indraprastha University, New Delhi - 110078, India
e-mail: {anjana_gosain,jaspreeti_singh}@yahoo.com

El-Sayed M. El-Alfy et al. (eds.), *Advances in Intelligent Informatics*,
Advances in Intelligent Systems and Computing 320, DOI: 10.1007/978-3-319-11218-3_39

representation of current as well as historical information. As data warehouses have become the main tool for strategic decision making, it is essential to guarantee their information quality from the early stages of the project [33]. The information quality in a data warehouse includes data presentation quality and data warehouse system quality, where one of the main components of the latter is multidimensional model quality [5].

It is widely accepted that development of data warehouse systems is based on multidimensional modeling [15, 20]. A variety of approaches to multidimensional modeling for data warehouse systems have been proposed in the past. [9] presents a survey on multidimensional modeling approaches available in the literature. Multidimensional modeling is a design technique which represents data as if placed in an n-dimensional space and specify two fundamental notions: the 'fact' and the 'dimension' [15, 18, 20]. The subjects of interest for an analyst are represented as 'facts' which are described in terms of a set of attributes called 'measures' or 'fact attributes' for the business process. A 'dimension' represents different ways in which the 'facts' can be analyzed. The 'dimensions' are in turn organized into 'hierarchies' at various granularity 'levels' that allows the data analyzers to pose meaningful on-line analytical processing (OLAP) queries [15, 16, 18, 20]. A dimension may consist of a single level referred to as 'non-hierarchy dimension' [24]. Besides, there can be shared dimensions as well as dimensions that share some hierarchy levels [2, 21, 22, 24, 29] as discussed in Section 4. Furthermore, there can be different relationships between aggregation levels within a dimension hierarchy [2, 16, 26, 29].

These components of the multidimensional model i.e. facts, dimensions, hierarchies, levels etc. contribute to its structural complexity [12]. As stated in [40], structural complexity is one of the important determinants of quality objectives for software artefacts. Indeed, data warehouse schemas are software artefacts, so it is important to investigate the possible associations that may exist between the structural elements of these schemas and their quality factors [40].

Metrics are widely recognized as an effective means to evaluate the quality factors in a consistent and objective manner [4, 7, 27]. Few works have been done in the past based on defining quality attributes for conceptual model of data warehouse [3] or metrics proposals based on structural complexities to evaluate these quality attributes [5, 6, 10, 28, 34, 39]. The proposed metrics are based on the facts, dimensions, attributes, hierarchy relationships, hierarchy depth, foreign keys, shared dimensions, multiple hierarchies, alternate paths etc. However, there are certain characteristics of dimension hierarchies such as relationship between dimension levels, sharing of some hierarchy levels within a dimension, among various dimensions etc. that have not been considered so far and needs further investigation as they are crucial to understand any multidimensional modeling approach [2, 21, 23, 25, 26, 29, 31, 32]. Notably, a single metric for shared hierarchies [10] is already proposed, but it does not incorporate all the aspects related to sharing of dimensional hierarchies, particularly when sharing of few hierarchy levels occur, situations that often arise in practice. The objective of this work is to propose metrics to compute structural complexity of multidimensional models with emphasis on sharing of levels among dimension hierarchies. The sharing of levels may enhance its structural complexity, thereby affecting understandability and in turn maintainability of multidimensional model.

This paper is organized as follows. Sub-section 1.1 provides an example to be used throughout this paper. Section 2 describes the related work. The categorization of hierarchies in data warehouses is discussed in Section 3. The types of dimension sharing are discussed in Section 4. Further, Section 5 describes the metrics proposed in this paper followed by an example for the same. Finally, conclusion and future work are discussed in the last section.

1.1 Running Example

In this section, we outline the approach used in this paper for representing structural properties of multidimensional model of a data warehouse. We adopt a conceptual perspective and consider the multidimensional model, namely MultiDimER, proposed by Malinowski et al. [23]. This model, based on entity-relationship (ER) approach, defines the schema as a set of dimensions and fact relationships. A dimension represents a concept that groups data with common semantic meaning for the domain being modeled. It can be either a level, or one or more hierarchies. A level represents an entity type and its instances are addressed as members. A level contains key attributes and may also contain descriptive attributes. A hierarchy consists of several related levels that are used for drill-down and roll-up OLAP queries. For the two related levels, lower level is referred to as a child, the higher level as a parent, and the relationship between them is called child-parent relationship. The cardinalities are used to indicate the participation of the two related levels in the child-parent relationship. Furthermore, two or more child–parent relationships can be exclusive which is represented using $\otimes$ symbol. The hierarchies in a dimension may specify different conceptual structures for analysis which is expressed using the analysis criterion as (Criterion). A fact relationship is represented as a n-ary relationship between levels. The measures associated to the fact relationship are represented as its attributes.

Consider the example shown in Fig. 1 [23]. We have used this example in order to explain various concepts related to categorization of dimensional hierarchies (Section 3) and dimension sharing (Section 4). Roughly speaking, it illustrates a sample scenario where three fact relationships, i.e. Sales, Employees' Salaries and Sales Incentive Program, indicate the facts of interest. The Sales fact is structured in a four-dimensional space and allows analysis: "who" bought (dimension Customer), from "where" (dimension Store), "what" (dimension Product), and "when" (dimension Time). Similarly, Employees' Salaries fact relationship can be analyzed along Employee, Store and Time dimensions, and Sales Incentive Program fact can be analyzed along Program Type, Sales Region and Time dimensions. The dimensions are further arranged into hierarchies using desired granularity levels.

2 Related Work

Very little research has been conducted in the field of metrics or objective indicators for data warehouse models, to analyze the quality attributes such as understandability, maintainability, analyzability etc.

The first proposal of metrics for logical data warehouse schemas is given by Calero et al. [5]. The authors have provided three sets of metrics based on their applicability to table, star or schema levels. They formally validated the schema level metrics using Zuse framework [41]. Serrano et al. [35] have presented the empirical validation of the schema level metrics identified in [5] and further replicated their experiments [36, 38] to prove that a subset of metrics proposed in [5] are useful as quality indicators. Furthermore, Serrano et al. [40] carried out the theoretical validation of a subset of metrics identified in [36, 38] using Briand's framework [4] and subsequently carried out the experimental validation to conclude that the metrics are significantly related to understandability of the data warehouse schema. Gosain et al. [11] and Kumar et al. [19] replicated the empirical validation of 'useful' metrics identified in [40].

Berengur et al. [3] have defined quality indicators and their corresponding metrics for conceptual models for data warehouse. The authors categorized the metrics as diagram level metrics and package level metrics.

The paper [37] presents a set of metrics to evaluate quality of conceptual data warehouse models. These metrics are proposed on the basis of unified modeling language (UML) and are categorized into three parts as: class scope, star scope and diagram scope metrics. The metrics are based on classes related to facts, dimensions, their relationships, attributes, dimension hierarchies etc. The authors have also empirically validated the metrics to show their practical utility. They have theoretically validated these metrics using distance framework [30].

Cherfi and Prat [6] proposed metrics to measure simplicity and analyzability of multidimensional schema. However, these metrics are neither theoretically nor empirically validated, so could not prove their utility as practical indicators to measure quality of data warehouse models.

Gosain et al. [10] proposed metrics considering some aspects related to dimension hierarchies such as multiple hierarchies, shared hierarchies, etc. The authors have validated the metrics theoretically (using Briand's framework [4]) as well as empirically [12].

Sushama et al. [28] proposed a complexity metric to consider complexity caused by relationships among various elements present in data warehouse multidimensional models. The authors have done preliminary validation of the metric using Kaner's framework [17].

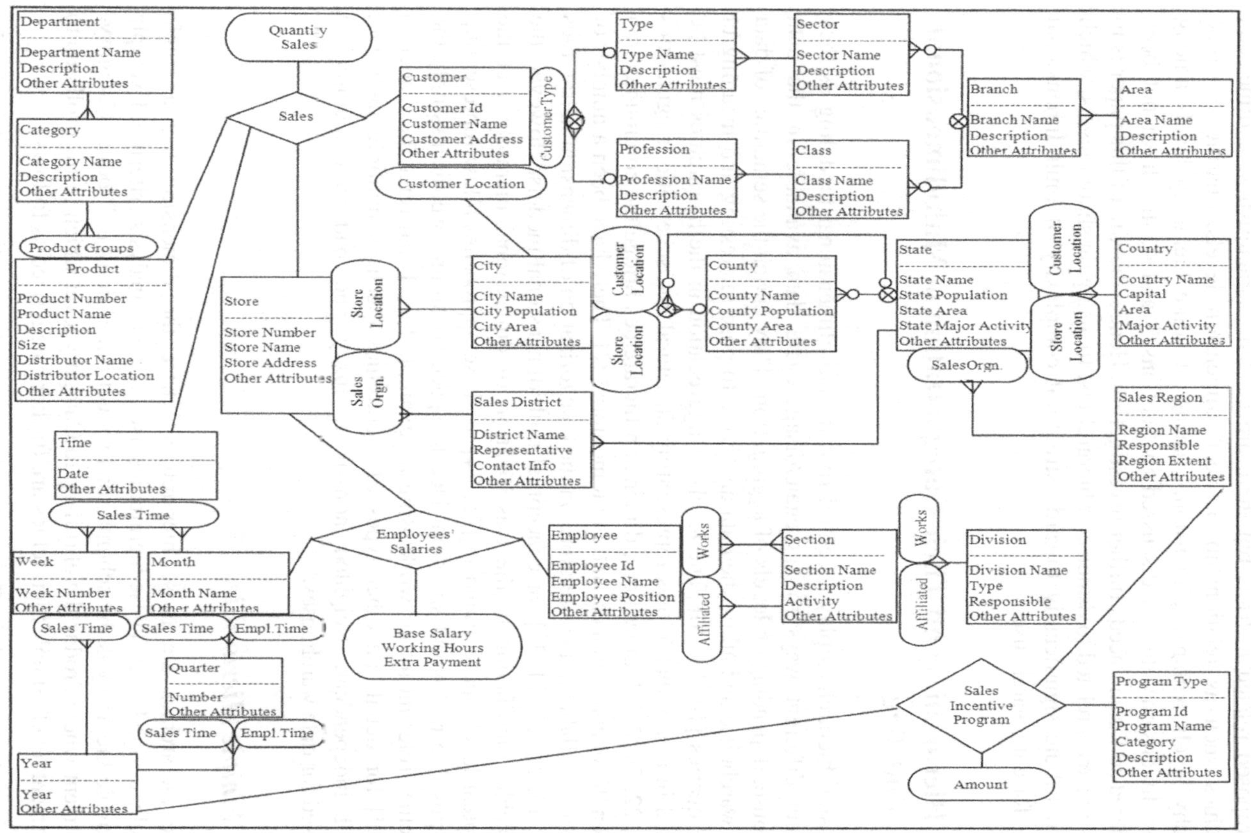

Fig. 1 Sample scenario containing different hierarchies reusing hierarchy levels [23]

Although few researchers have given due importance to dimension hierarchies and proposed related metrics, but there are certain characteristics of dimension hierarchies such as relationship between dimension levels, sharing of some hierarchy levels within a dimension, sharing of some hierarchy levels among various dimensions etc., as discussed in Sections 3 and 4, that have not been considered so far and needs further investigation. Hence, the aim of this paper is to propose metrics related to dimension hierarchies, focusing on shared levels, which may contribute significantly towards structural complexity of multidimensional models for data warehouse.

3 Hierarchies and Their Categorization in Multidimensional Modeling

Dimension hierarchies play a central role in multidimensional modeling as they represent different ways to aggregate/disaggregate data available in the data warehouse, depending on levels of aggregation. They enrich the semantics of data in the warehouse and allow the data analyzers to pose interesting and meaningful OLAP queries [16, 18, 20]. The levels of aggregation in the hierarchies must be modeled in a way that allows representing structural and schematic irregularities [2, 21, 22, 24, 29]. Furthermore, dimension hierarchies bridge the communication between the decision makers and implementers [22]. There have been a number of proposals available in the literature on the categorization of dimension hierarchies [2, 13, 14, 21, 22, 24, 32] for conceptual multidimensional models. However, the remarkable contribution to the classification of dimensional hierarchies at the conceptual level can be found in [22, 24] as these approaches allow representing every possible real-world situation. Due to space constraint, we discuss only the hierarchies relevant to our work using an example Fig. 1 and refer the reader to [22, 24] for details on other types of hierarchies (such as multiple inclusive, parallel independent, weighted non-strict, fuzzy and patterned hierarchies) proposed for data warehouses.

3.1 Simple Hierarchies

Simple hierarchies are those hierarchies where the dimension has a single hierarchy defined on it. These hierarchies use only one analysis criterion. Example of a simple hierarchy is the schema for Product dimension i.e. Product-Category-Department where Product level roll-up to Category level which in-turn rolls-up to Department level. Simple hierarchies are further categorized as follows:

- Symmetric hierarchies: This is a sub-type of simple hierarchy where at the schema level only one path exists and all the levels in the hierarchy are mandatory. For instance, the hierarchy of the schema Product-Category-Department is symmetric.

- Asymmetric hierarchies: This is also a sub-type of simple hierarchy where at the schema level only one path exists, but some lower levels of the hierarchy are not mandatory. To keep it simple, this hierarchy is not shown in Fig. 1.

3.2 *Multiple Alternative Hierarchies*

These represent the scenario where more than one hierarchy is defined in a dimension, i.e. there exists multiple non-exclusive simple hierarchies that may share some levels. Notably, all such hierarchies associated to a dimension are based on the same analysis criterion. An interesting aspect about this type of hierarchy is that it is not semantically correct to use different composing hierarchies in combination as grouping conditions within a query [22, 24]. The entire Time dimension of the Sales fact relationship is a classical example of multiple hierarchies where the pairs of aggregation paths (Date-Week-Year and Date-Month-Quarter-Year) share the two levels i.e. Date and Year and is based on the same analysis criteria, namely Sales Time.

3.3 *Generalized Hierarchies*

A dimension may include subtypes that result from generalization/specialization relationship between some levels [1] i.e. a level is super class for other levels. Such hierarchies are termed as generalized hierarchies. At the schema level, this hierarchy contains multiple exclusive paths sharing some levels [22]. All these exclusive paths represent a single hierarchy and are based on the same analysis criteria. At the instance level, every member of the hierarchy is associated to only one path. These hierarchies include special cases of specialization and generalization as categorized in [24] (namely, generalization, specialization, disjoint specialization, overlapping specialization, mixed and non-covering) on the basis of splitting and joining levels i.e. the levels at which alternative paths split and join respectively, strictness of hierarchy or partial roll-ups. Although, [22] considers only non-covering hierarchies as a special case of generalized hierarchies in which exclusive paths are obtained by partial roll-up relationships. An example of generalized hierarchy is the Customer Type hierarchy of the Customer dimension where Branch and Area are two common hierarchy levels.

3.4 *Parallel Hierarchies*

Parallel hierarchies in a dimension are based on different analysis criteria. On account of shared levels, these hierarchies can be specialized into parallel independent and parallel dependent hierarchies. This work uses the concept of parallel dependent hierarchies that have sharing of some levels. An example of parallel dependent hierarchies can be observed in Store dimension: 1) location hierarchy: to represent geographical location of the store address (Store-City-County-State-Country) and 2) sales organization hierarchy: to represent its organizational division (Store-Sales district-State-Sales region). Here, the composing hierarchies i.e. location and sales organization hierarchies are non-covering and symmetric respectively. In these hierarchies, State being the common level plays different roles based on the specified analysis criteria.

3.5 Strict Hierarchies

If all the cardinalities from parent to child levels in a hierarchy are one-to-many, then such hierarchies are called strict hierarchies, i.e. at the schema level, a child is related to at most one parent member and a parent may have several child members. An example of strict hierarchy is the schema Date-Month-Quarter-Year which has at most one outgoing roll-up relationship.

3.6 Non-Strict Hierarchies

Such hierarchies arise as a result of at least one many-to-many cardinality between their child and parent level, which is a very common scenario in real-life applications. An example of such hierarchy is the relationship that exists in Employee dimension (Employee-Section sub-dimension) of Employees' Salaries fact where an employee is working in several sections leading to a many-to-many cardinality in this relationship, thereby acting as a non-strict hierarchy. Such hierarchies require special care while performing aggregation, to avoid the well-known problem of double counting [2, 21, 23, 25].

An important point about categorization of hierarchies is their orthogonal behavior. Various kinds of hierarchies discussed above can be either strict or non-strict. For instance, it can be observed that the Employee dimension has a hierarchy (Employee-Section-Division) which is non-strict as well as symmetric.

4 Dimension Sharing

A multidimensional data model may allows multiple facts which can share dimensions among each other, leading to a multi-fact scheme generally referred to as a galaxy or fact constellation [18]. The smallest unit that can be shared is a dimension level whereas the highest shareable unit can be a complete dimension [23, 25]. Within a single fact, there can be sharing of levels within a dimension and between dimensions. Furthermore, going beyond a single fact, the sharing of levels as well as sharing of dimensions can occur across fact schemes also [23, 25].

Ref. [25] identified different types of dimension sharing based on whether the shared unit is a complete dimension or a part of the dimension (i.e. few hierarchy levels). In this paper, we are focused on the scheme complexity due to shared hierarchy levels, i.e. the case of partial sharing, which is discussed below.

4.1 Partial Sharing

Partial sharing can arise due to the sharing of dimension level(s) within or across fact schemes. The dimension level being shared is modeled once and referred by different fact-dimensional relationships.

There may be different cases of partial sharing as follows:

- sharing of dimension level(s) within a dimension, as are the cases for generalized hierarchies, parallel dependent hierarchies and multiple alternative hierarchies. For instance, the level State is shared between the two hierarchies of the Store dimension (i.e. the Sales organization, the Store location) in the Sales fact.
- sharing between dimensions in a same fact scheme. For example, the aggregation path comprising four levels namely, City, County, State, Country is shared between the Store location hierarchy in Store dimension and the Customer location hierarchy in Customer dimension of Sales fact.
- sharing between dimensions across different fact schemes. An example of this kind of sharing is the sharing of the aggregation path comprising levels Month, Quarter and Year of the Time dimension among two facts i.e., Sales and Employees' Salaries. Another example is the Sales region level in Store dimension of Sales fact being shared in the Sales Incentive Program fact.

Notably, sharing of dimensions/levels allows reusing of existing data as well as enhances analytic capability of the scheme by opening possibilities to analyze measures associated to different fact relationships [21, 23]. On the contrary, the sharing of dimensions or some hierarchy levels among different dimension hierarchies may significantly increase the structural complexity of the scheme at the conceptual level.

5 Metrics for Multidimensional Models

The complexity of a system largely depends on the number and the variety of elements and the relationships among those elements [8]. Our goal is to define a set of metrics to measure the structural complexity of multidimensional models for data warehouse. Specifically, we have focused on the sharing of levels among dimension hierarchies, within and across fact schemes, at conceptual level. The proposed metrics fulfill the properties desired out of the software metrics in general i.e. they are simple, objective and empirical [7]. The metrics are defined as follows based on the example scenario described in Introduction:

Number of Non-Hierarchy Dimensions (NNHD): This metric count the non-hierarchical dimensions in the multidimensional model for data warehouse.

Number of Hierarchy Dimensions (NHD): This metric count the hierarchical dimensions in the multidimensional model for data warehouse.

Number of Shared Levels Within Dimensions (NSLWD): Various granularity levels can be shared by more than one hierarchy within a dimension. For all the dimensions existing in the multidimensional model, NSLWD metric counts the total number of such shared levels within dimensions. If the dimension itself is shared among different fact schemes, then the shared levels of that particular dimension are counted only once.

Number of Shared Levels between Dimensions within a Fact Scheme (NSLBD): Within the same fact scheme, there can be sharing of granularity level(s) among hierarchies associated to different dimensions. NSLBD counts the total number of such shared levels between dimensions, for all the fact schemes existing in the model.

Number of Shared Levels between Dimensions across Different Fact Schemes (NSLAF): In a multi-fact scheme, the sharing of levels can occur among hierarchies associated to different dimensions, across different fact schemes. The metric NSLAF counts the total number of such shared levels existing in the model. If the same level is shared more than once across different fact schemes, then that shared level is counted only once.

Total Number of Shared Levels (NSL): This metric counts the total number of shared levels in the multidimensional model for data warehouse. So,

$$\mathbf{NSL = NSLWD + NSLBD + NSLAF} \tag{1}$$

Total Number of Attributes in Shared Levels (TASL): If there are shared levels in the multidimensional model, TASL metric counts the total number of attributes in those shared levels. If the same level is shared more than once on account of different types of dimension sharing, then, we include the shared level only once to count its attributes.

Let NASL denote the number of attributes of a shared level and NRSL[1] denote the total number of shared levels, excluding repetition of the shared level that may occur due to different types of dimension sharing (i.e. NRSL = NSL - repeated counting of shared levels), then

$$\mathbf{TASL = \sum_{k=1}^{NRSL} NASLk} \tag{2}$$

Number of Non-Strict Hierarchies (NNSH): This metric counts the total number of non-strict hierarchies for all the fact schemes existing in the model.

Consider the sample scenario discussed in Introduction. The proposed metrics are computed for the multidimensional model shown in Fig. 1.

[1] Here, NRSL is just a term used for computation of TASL and not considered as metric for computing structural complexity of multidimensional model.

NNHD = 1: There is only one dimension (i.e. Program Type dimension of Sales Incentive Program fact scheme) which involves a single level.

NHD = 5: Except the dimension namely Program Type, all the other dimensions (i.e. Customer, Store, Product, Time, and Employee) involve at least one hierarchy, so the total number of hierarchical dimensions is 5.

NSLWD = 4: In the Sales fact, there are four such levels - The level State is shared between the parallel hierarchies of the Store dimension (i.e. the Sales organization, the Store location); Branch and Area are two shared levels in generalized hierarchy of the Customer dimension; and the level Year is shared between multiple hierarchies of the Time dimension (i.e. Date-Week-Year and Date-Month-Quarter-Year). However, there is no such shared level in other fact schemes. Here, the Store dimension being shared among Sales and Employees' Salaries fact is considered only once while counting shared levels.

NSLBD = 4: The four levels namely, City, County, State and Country, are shared among Customer and Store dimensions of the Sales fact. Again, there is no level being shared between dimensions in Employees' Salaries and Sales Incentive Program fact.

NSLAF = 4: Considering the pair of facts namely, Sales Incentive Program and Sales, the Sales Region level and Year level are shared among these two facts; the levels Month, Quarter and Year are shared among the Sales fact and Employees' Salaries fact; and no sharing of levels occur among Employees' Salaries and Sales Incentive Program facts. Notably, the Store dimension being shared among Employees' Salaries and Sales facts is the case of complete dimension sharing, hence not included in the count for this metric. Also, as mentioned above, the level Year being shared twice, needs to be counted only once.

NSL = **4 + 4 + 4 = 12,** which is the total of the metrics namely NSLWD, NSLBD, and NSLAF.

TASL = 24. For the computation of TASL, we first need the values of NRSL and NASL which can be obtained as follows:

Here, NRSL = 10, since, NRSL = NSL - repeated counting of shared levels, where repeated counting of shared levels = 1 + 1 = 2 (1 count for State level and 1 count for Year level, because: the level State is being counted twice: (i) in the Store dimension while counting NSLWD and (ii) between Store and Customer dimensions while counting NSLBD. Similarly, the level Year is counted twice while computing the number of shared levels taking into account different types of dimension sharing) and NSL = 12 as computed above. Therefore, NRSL = 12 - 2 = 10.

Also, while computing NASL for the shared levels using example Fig. 3, we have not taken into consideration 'Other Attributes' field. For instance, City level includes four attributes namely City Name, City Population, City Area and Other Attributes. So NASL for City level i.e. $NASL_{City}$ = 4 - 1 = 3 (i.e. 'Other Attributes' field excluded). Similarly, NASL value for all other shared levels is as follows: $NASL_{Branch}$ = 2, $NASL_{Area}$ = 2, $NASL_{County}$ = 3, $NASL_{State}$ = 4, $NASL_{Country}$ = 4, $NASL_{Sales\ region}$ = 3, $NASL_{Month}$ = 1, $NASL_{Quarter}$ = 1, $NASL_{Year}$ = 1.

So, TASL = $\sum_{K=1}^{10}$ NASLK= 24, adding up all the individual NASL values mentioned above.

NNSH = 1: The only hierarchy involving many-to-many relationship is the simple hierarchy of Employee dimension in the Employees' Salaries fact, thereby acting as a non-strict hierarchy.

The values of the proposed metrics are shown in Table 1.

Table 1 Computed values of proposed metrics

Proposed metrics	Value
NNHD	1
NHD	5
NSLWD	4
NSLBD	4
NSLAF	4
NSL	12
TASL	24
NNSH	1

However, the metrics definition is only one step in the complete process of obtaining a set of valid and useful metrics. It is essential to validate the metrics in order to confirm and understand the implication of the metrics.

An important point to mention here is that we have used an ER-based multidimensional model to explain metrics proposed in this paper, but these metrics are based on general structural properties of the multidimensional models for data warehouse, so these metrics can very well be applied to object oriented multidimensional models also.

6 Conclusion and Future Work

Multidimensional data model quality depends largely on its structural features like facts, dimensions, hierarchies, relationship among various elements etc. Though dimension hierarchies allow analyzing data at different levels of granularity, but they may elevate structural complexity of the data models, thereby affecting their understandability and in turn maintainability. In this paper we have proposed metrics related to dimension hierarchies, with emphasis on sharing that may be possible among those hierarchies, in order to assure the quality of the multidimensional models used in the early stage of a data warehouse design. These metrics will help to measure the structural complexity of multidimensional models for data warehouse. However, it is fundamental to perform theoretical as well as empirical validation of the defined metrics to prove their utility. For this, we are now working on the theoretical validation of these metrics.

References

1. Abello, A., Samos, J., Saltor, F.: Understanding analysis dimensions in a multidimensional object-oriented model. In: Proc. of the 3rd Int. Workshop on Design and Management of Data Warehouses, pp. 1–9 (2001)
2. Abelló, A., Samos, J., Saltor, F.: YAM2 (Yet Another Multidimensional Model): An Extension of UML. In: International Database Engineering and Applications Symposium, pp. 172–172. IEEE Computer Society (2002)
3. Berenguer, G., Romero, R., Trujillo, J., Bilò, V., Piattini, M.: A Set of Quality Indicators and Their Corresponding Metrics for Conceptual Models of Data Warehouses. In: Tjoa, A.M., Trujillo, J. (eds.) DaWaK 2005. LNCS, vol. 3589, pp. 95–104. Springer, Heidelberg (2005)
4. Briand, L.C., Morasca, S., Basili, V.R.: Property based software engineering measurement. IEEE Trans. Softw. Eng. 22, 68–86 (1996)
5. Calero, C., Piattini, M., Pascual, C., Serrano, M.: Towards data warehouse quality metrics. In: Proc. of Third Int. Workshop on Design and Management of Data Warehouse, Interlaken, Switzerland, pp. 1–10 (2001)
6. Si-said Cherfi, S., Prat, N.: Multidimensional schemas quality: assessing and balancing analyzability and simplicity. In: Jeusfeld, M.A., Pastor, Ó. (eds.) ER Workshops 2003. LNCS, vol. 2814, pp. 140–151. Springer, Heidelberg (2003)
7. Fenton, N.: Software measurement: a necessary scientific basis. IEEE Trans. Softw. Eng. 20(3), 199–206 (1994)
8. Flood, R.L., Carson, E.R.: Dealing with Complexity: An Introduction to the Theory and Application of Systems Science. Plenum Press, Springer, New York (1993)
9. Gosain, A., Singh, J.: Conceptual Multidimensional Modeling for Data Warehouses: A Survey. In: Communicated to 3rd International Conference on Frontiers in Intelligent Computing Theory and Applications (FICTA) to be held on 14-15 November, Proceedings to be Published in Springer AISC (2014)
10. Gosain, A., Nagpal, S., Sabharwal, S.: Quality metrics for conceptual models for data warehouse focusing on dimension hierarchies. ACM SIGSOFT Softw. Eng. Notes 36(4), 1–5 (2011)
11. Gosain, A., Sabharwal, S., Nagpal, S.: Assessment of quality of data warehouse multidimensional model. International Journal of Information Quality 2(4), 344–358 (2011)
12. Gosain, A., Nagpal, S., Sabharwal, S.: Validating dimension hierarchy metrics for the understandability of multidimensional models for data warehouse. IET Software 7(2), 93–103 (2013)

13. Hurtado, C.A., Gutiérrez, C., Mendelzon, A.O.: Capturing summarizability with integrity constraints in OLAP. ACM Trans. Database Syst. 30(3), 854–886 (2005)
14. Husemann, B., Lechtenborger, J., Vossen, G.: Conceptual data warehouse design. In: Proc. of the Int. Workshop on Design and Management of Data Warehouses, p. 6 (2000)
15. Inmon, W.H.: Building the Data Warehouse, 4th edn. John Wiley & Sons, Inc., New York (2005)
16. Jagadish, H.V., Lakshmanan, L.V.S., Srivastava, D.: What can hierarchies do for data warehouses? In: Proc. of the 25th International Conference on Very Large Databases, pp. 530–541 (1999)
17. Kaner, C., Bond, P.: Software Engineering Metrics: What Do They Measure and How Do We Know? In: Proc. of 10th International Software Metrics Symposium (Metrics 2004), Chicago, IL (2004)
18. Kimball, R.: The data warehouse toolkit. John Wiley & Sons, Chichester (2006)
19. Kumar, M., Gosain, A., Singh, Y.: Empirical validation of structural metrics for predicting understandability of conceptual schemas for data warehouse. International Journal of System Assurance Engineering and Management, 1–16 (2013)
20. Lenzerini, M., Vassiliou, Y., Vassiliadis, P.: Fundamentals of data warehouses, Jarke, M. (ed.). Springer (2002)
21. Luján-Mora, S., Trujillo, J., Song Il, Y.: A UML profile for multidimensional modeling in data warehouses. Data & Knowledge Engineering 59(3), 725–769 (2006)
22. Malinowski, E., Zimányi, E.: OLAP hierarchies: A conceptual perspective. In: Persson, A., Stirna, J. (eds.) CAiSE 2004. LNCS, vol. 3084, pp. 477–491. Springer, Heidelberg (2004)
23. Malinowski, E., Zimányi, E.: Hierarchies in a multidimensional model: From conceptual modeling to logical representation. Data & Knowledge Engineering 59(2), 348–377 (2006)
24. Mansmann, S., Scholl, M.H.: Empowering the OLAP technology to support complex dimension hierarchies. International Journal of Data Warehousing and Mining (IJDWM) 3(4), 31–50 (2007)
25. Mansmann, S., Scholl, M.H.: Extending the Multidimensional Data Model to Handle Complex Data. Journal of Computing Science and Engineering 1(2), 125–160 (2007)
26. Mazón, J.N., Lechtenbörger, J., Trujillo, J.: A survey on summarizability issues in multidimensional modeling. Data & Knowledge Engineering 68(12), 1452–1469 (2009)
27. Melton, A.: Software Measurement. International Thomson Computer Press, London (1996)
28. Nagpal, S., Gosain, A., Sabharwal, S.: Complexity metric for multidimensional models for data warehouse. In: Proc. of the CUBE International Information Technology Conference, pp. 360–365. ACM (2012)
29. Pedersen, T.B., Jensen, C.S., Dyreson, C.E.: A foundation for capturing and querying complex multidimensional data. Information Systems 26(5), 383–423 (2001)
30. Poels, G., Dedene, G.: Distance: a framework for software measure construction, Research Report DTEW 993 (1999)

31. Prat, N., Akoka, J., Comyn-Wattiau, I.: A UML-based data warehouse design method. Decision Support Systems 42(3), 1449–1473 (2006)
32. Rizzi, S.: Conceptual modeling solutions for the data warehouse. In: Data Warehouses and OLAP: Concepts, Architectures and Solutions, pp. 1–26 (2007)
33. Rizzi, S., Abelló, A., Lechtenbörger, J., Trujillo, J.: Research in data warehouse modeling and design: dead or alive? In: Proc. of the 9th ACM International Workshop on Data Warehousing and OLAP, pp. 3–10 (2006)
34. Serrano, M.: Definition of a Set of Metrics for Assuring Data Warehouse Quality. Univeristy of Castilla, La Mancha (2004)
35. Serrano, M., Calero, C., Piattini, M.: Validating metrics for data warehouse. IEE Proceeding - Software 149(5), 161–166 (2002)
36. Serrano, M., Calero, C., Piattini, M.: Experimental validation of multidimensional data models metrics. In: Proc. of 36th Annual Hawaii Int. Conf. on System Sciences, Hawaii (2003)
37. Bilò, V., Calero, C., Trujillo, J., Luján-Mora, S., Piattini, M.: Empirical validation of metrics for conceptual models for data warehouse. In: Persson, A., Stirna, J. (eds.) CAiSE 2004. LNCS, vol. 3084, pp. 506–520. Springer, Heidelberg (2004)
38. Serrano, M., Calero, C., Piattini, M.: An experimental replication with data warehouse metrics. International Journal of Data Warehousing and Mining 1(4), 1–21 (2005)
39. Serrano, M., Trujillo, J., Calero, C., Piattini, M.: Metrics for data warehouse conceptual models understandability. Journal of Information and Software Technology 49, 851–870 (2007)
40. Serrano, M., Calero, C., Sahraouli, H., Piattini, M.: Empirical studies to assess the understandability of data warehouse schemas using structural metrics. Software Quality Journal 16(1), 79–106 (2008)
41. Zuse, H.: Framework of Software Measurement. Walter de Guyter, Berlin (1998)

A Method to Induce Indicative Functional Dependencies for Relational Data Model

Sandhya Harikumar and R. Reethima

Abstract. Relational model is one of the extensively used database models. However, with the contemporary technologies, high dimensional data which may be structured or unstructured are required to be analyzed for knowledge interpretation. One of the significant aspects of analysis is exploring the relationships existing between the attributes of large dimensional data. In relational model, the integrity constraints in accordance with the relationships are captured by functional dependencies. Processing of high dimensional data to understand all the functional dependencies is computationally expensive. More specifically, functional dependencies of the most prominent attributes will be of significant use and can reduce the search space of functional dependencies to be searched for. In this paper we propose a regression model to find the most prominent attributes of a given relation. Functional dependencies of these prominent attributes are discovered which are indicative and lead to faster results in decreased amount of time.

Keywords: Prominent Attributes,Indicative Functional Dependencies.

1 Introduction

The super change in information technology paved the way for high dimensional data exchange. For the readiness in communication and effective manipulation of high dimensional data, knowledge discovery became inevitable. Knowledge Discovery and Data Mining is a multidisciplinary area which has the main focus on methodologies for extracting useful knowledge from data. If we have a method to extract enough knowledge to represent the high dimensional data, then the effective manipulation of data will be possible. When

Sandhya Harikumar · R. Reethima
Department of Computer Science and Engineering,
Amrita Vishwa Vidyapeetham, Clappana P.O, Kollam
e-mail: sandhyaharikumar@am.amrita.edu, reethu7@gmail.com

El-Sayed M. El-Alfy et al. (eds.), *Advances in Intelligent Informatics*,
Advances in Intelligent Systems and Computing 320, DOI: 10.1007/978-3-319-11218-3_40

we have a set of data pertaining to a particular domain, the relationships existing between the attributes can efficiently be understood by the functional dependencies provided.

A functional dependency is a constraint between two sets of attributes in a relation from a database. An attribute or set of attributes X is said to functionally determine another attribute Y (written X→Y) if and only if each X value is associated with at most one Y value [1]. Customarily we call X as determinant set and Y as dependent set. So if we are given the value of X we can determine the value of Y. Thus we understand that X is enough to represent Y. For example, in an address database, zip code determines city. The discovery of functional dependencies from relations has received considerable interest [2,3]. Formally, a functional dependency over a relation schema R is an expression X→Y, where X,Y⊆R.

The dependency holds or is valid in a given relation r over R if for all pairs of tuples t, u∈r we have: if t[B] = u[B] for all B∈X, then t[A] = u[A] (we also say that t and u agree on X and A). A functional dependency X→A is minimal (in r) if A is not functionally dependent on any proper subset of X, i.e. if Y→A does not hold in r for any Y⊂ X. The dependency X→A is trivial if A∈X [2].

An approximate functional dependency is a functional dependency that almost holds. For example, gender is approximately determined by first name [2]. Such dependencies arise in many databases when there is a natural dependency between attributes, but some tuples contain errors or represent exceptions to the rule. The discovery of unexpected but meaningful approximate dependencies seems to be an interesting and realistic goal in many data mining applications. They also have applications in database design [4].

The main challenge to induce indicative functional dependency is to find the most prominent or relevant attributes. In order to find which attributes are prominent we need to analyze the huge volume of data. Processing each and every data value is inefficient and hence we need an efficient method to analyze the data and understand which attributes are prominent. The analysis may be easy if the data values corresponding to each attribute posses high variation or large dissimilarity to each other. Very rarely this is the case since the schema and data are tightly coupled. The context of the information is also lacking. Hence in this paper our main focus is to find the most prominent attributes with the help of which we seek for functional dependencies that can be approximate but provide some important relationships existing among the attributes of the data. We propose a regression model to find the prominent attributes from high dimensional data. By applying linear regression we are removing the attributes which has an error value less than a particular threshold value and keep rest of the attributes as prominent ones. Further, the indicative functional dependencies associated with the prominent attributes are found by using the FD_ Mine algorithm [2].

For an instance, consider the following Table which contains five attributes describing course details. It is evident from the table that a student can attend a single semester at a time. When we are considering the attributes LECTURE and Teaching Assistant, it can be seen that the same lecture is not taught by more than one Teaching Assistant. These are all some relationships that we can find from the table. If we have some information on the important attributes of the table then it is enough to consider the relationships based on these important attributes. So the first step is to obtain the prominent attributes from the data and then find the functional dependencies associated with them. We can consider these as the indicating functional dependencies since these are good enough to indicate the important relationships among the attributes, within the data.

Table 1 Sample entries in table of course data

St_ ID	SEM	LECTURE	Teaching_ Assistant	CLASSROOM
1001	6	Algorithm	A	107
1017	5	DB	B	103
1109	3	MOBILE	C	106
1001	6	Maths	E	108
1017	5	cs	D	104
1017	5	DB	B	107
1108	3	MOBILE	C	105
1111	3	Maths	E	110

1.1 Motivation

Several algorithms for the discovery of functional dependencies [5, 11, 12] and Approximate functional dependencies [5, 13] have been presented, amongst which TANE algorithm [5] is based on considering partitions of the relation and deriving valid dependencies from the partitions. The algorithm searches for dependencies in a breadth-first or level-wise manner. The paper presents how the search space can be pruned effectively and how the partitions and dependencies can be computed efficiently. The method works well with relations up to hundreds of thousands of tuples. The method is at its best when the dependencies are relatively small. When the size of the (minimal) dependencies is roughly one half of the number of attributes, the number of dependencies is exponential in the number of attributes and the situation is more or less equally bad for any algorithm. When the dependencies are larger, the level-wise method that starts the search from small dependencies is obviously farther from the optimum. The level-wise search can, in principle, be altered to start from the large dependencies. Nevertheless, the partitions could not be computed efficiently.

Problem of discovering multivalued dependencies is dealt by fdep algorithm [6]. They elaborate the connection between the problem of dependency discovery and the problem of inductive learning or learning from examples from the field of machine learning. The results show that the TANE algorithm performs better, when the number of the attributes of the input relation is small. An inductive learning has also been followed in [10] to find the dependencies. As the number of attributes of the input relation increases, the performance of fdep is better than the performance of the TANE algorithm. The main drawback lies in the large number of hypotheses which have to be considered by the proposed algorithm.

A novel search method, for computing minimal FDs using heuristic-driven, depth-first search is described by the algorithm fastFDs [7]. The main drawback is that the performance is dependent on the heuristic we have selected. Discovering the Functional Dependencies when there is more than one relation is described as qualified FDs[8]. They gave propagation rules of application in restructuring operators. [11,12,13] also have shown various methods of finding the functional dependencies from the relational databases.

FD_ Mine[2] is one amongst the most efficient algorithms. In the paper, several properties of relational databases relevant to the search for functional dependencies, are identified. Using these, a theorem has been proved that allows equivalences among attributes to be identified based on nontrivial closures. This algorithm finds all minimal functional dependencies in a database. Like TANE, FD_ Mine is based on partitioning the database and comparing the number of partitions. However, FD_ Mine provides additional pruning rules, based on the analysis of the theoretical properties of functional dependencies. These pruning techniques are guaranteed not to eliminate any valid candidates, and whenever they are relevant, they reduce the size of the dataset or the number of checks required. The results show that the pruning rules in the FD_ Mine algorithm are valuable because they increase the pruning of candidates and reduce the overall amount of checking required to find the same FDs.

2 Linear Regression Model

A regression model is a compact mathematical representation of the relationship between the response variable and the input parameters in a given design space [9,10]. Linear regression models are widely used to obtain estimates of parameter significance as well as predictions of the response variable at arbitrary points in the design space. In Linear Regression, the model specification is that the dependent variable, y_i is a linear combination of the parameters (but need not be linear in the independent variables). For example, in simple linear regression for modeling n data points consisting of one independent variable x_i, and two parameters, β_0 and β_1, the equation of a straight line with varying parameters gives the best fit.

$$y_i = \beta_0 + \beta_1 x_i + \varepsilon_i, \quad i = 1, \ldots, n. \tag{1}$$

In multiple linear Regression, there are several independent variables or functions of independent variables.

Adding a squared term of x_i to the preceding regression gives parabola equation:

$$y_i = \beta_0 + \beta_1 x_i + \beta_2 x_i^2 + \varepsilon_i,\ i = 1, \ldots, n. \tag{2}$$

This is still Linear Regression. Although the expression on the right hand side is quadratic in the independent variable x_i, it is linear in the parameters β_0,β_1 and β_2.

In both cases, ε_i is an error term and the subscript i indexes a particular observation.

Given a random sample from the population, we estimate the population parameters and obtain the sample linear regression model:

$$\widehat{y_i} = \widehat{\beta_0} + \widehat{\beta_1} x_i. \tag{3}$$

The residual, $e_i = y_i$ -$\widehat{y_i}$, is the difference between the value of the dependent variable predicted by the model, $\widehat{y_i}$, and the true value of the dependent variable, y_i. One method of estimation is ordinary least squares. In the more general multiple regression model, there are p independent variables:

$$y_i = \beta_1 x_{i1} + \beta_2 x_{i2} + \cdots + \beta_p x_{ip} + \varepsilon_i, \tag{4}$$

where x_{ij} is the ith observation on the jth independent variable, and where the first independent variable takes the value 1 for all i (so β_1 is the regression intercept).The least squares parameter estimates are obtained from p normal equations. The residual can be written as

$$\varepsilon_i = y_i - \hat{\beta}_1 x_{i1} - \cdots - \hat{\beta}_p x_{ip}. \tag{5}$$

The normal equations are

$$\sum_{i=1}^{n} \sum_{k=1}^{p} X_{ij} X_{ik} \hat{\beta}_k = \sum_{i=1}^{n} X_{ij} y_i,\ j = 1, \ldots, p. \tag{6}$$

In matrix notation, the normal equations are written as

$$(\mathbf{X}^\top \mathbf{X}) \hat{\beta} = \mathbf{X}^\top \mathbf{Y}, \tag{7}$$

where the $(ij)^{th}$element of X is x_{ij}, the ith element of the column vector Y is y_i, and the jth element of $\hat{\beta}$ is $\hat{\beta}_j$. Thus X is n×p, Y is n×1, and $\hat{\beta}$ is p×1. The solution is

$$\hat{\beta} = (X^\top X)^{-1} X^\top Y \tag{8}$$

By using the regression co-efficient as in equation(8) further predict the output using equation(3). Based on the residual error obtained using equation(5) we can understand the difference between the observed value and estimated function value. Small residual sum indicates tight fit of the model to the data.

3 A Regression Model for Finding Indicative Functional Dependencies

We seek for the indicative functional dependencies of the relational model. In order to find the prominent attributes, the linear regression model is more appealing since attributes in relational databases usually have linear relationships. After preprocessing we generate a regression model as described in the algorithm in the subsection that follows. Based on the error obtained, the prominent attributes will be the ones which have error more than threshold value. From the prominent attributes, the functional dependencies associated with them will be found. For the discovery of functional dependencies of the prominent attributes FD_Mine algorithm[2] is used.

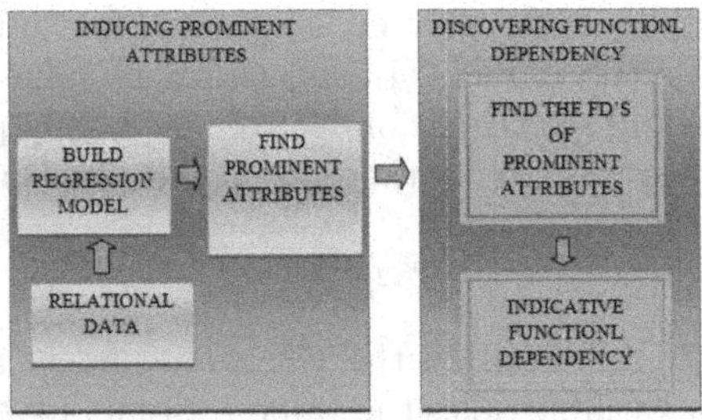

Fig. 1 Inducing Indicative Functional Dependencies

3.1 FD_ Mine for Discovering Approximate Functional Dependencies

The algorithm starts with finding the Nontrivial Closure of each candidate. Let F be a set of FDs over a dataset D and X be a candidate over D. The closure of candidate X with respect to F, denoted as Closure(X), is defined as { Y|X→Y can be deduced from F by Armstrong's axioms}. The nontrivial closure of candidate X with respect to F, denoted as Closure'(X), is defined as Closure'(X)=Closure(X)-X. After obtaining the NonTrivial Closure associated with the candidate, find the FDs and Keys. If there is a dependency like X_i →Closure[X_i], then add it to FD_ SET. If X_i∪Closure'[X_i]=R then X_i is a Key and add it to KEY_ SET.

By using the closure further find the equivalence among candidates. Let X and Y be two candidates of dataset D, Z = X∩Y, and let Closure(X) and Closure(Y) be the nontrivial closures of candidates X and Y respectively. If Closure(X)→Y-Z, and Closure(Y)→X-Z, then X→Y. Then we can say that X is equivalent to Y. It will take the candidates for the next level from the current set of candidates excluding the candidates which are elements in KEY_ SET and EQ_ SET. The pruning rules for deletion considered are detailed below. Former is, if X↔Y, then candidate Y can be deleted. It reduces the search space by eliminating redundant attributes and all their values from

database. In addition, if X→Y holds, then any superset YW of Y does not need to be examined. It also eliminates redundant candidates. For example, because XY↔WZ holds, if XYT→U holds, then WZT→U also holds. Latter, if X is a key, then any superset XY of X does not need to be examined.
Discovering the candidates for the next level is done mainly by two pruning rules. Former is, if Closure(X) and Closure(Y) are the nontrivial closures of attributes X and Y, respectively, then XY→Closure(X)∪Closure(Y) does not need to

Algorithm 1. Inducing Indicative Functional Dependencies(RM)

Input: RM : Relational Data with n columns with column type specifications and m number of instances
Output: Indicative Functional Dependencies

```
Preprocess(RM)
Set data RM as matrix $X^{mn}$ mn with m rows and n columns
for i:1 to n do
    Set the response matrix Y = $X^i$
    Set the Regression input as matrix excluding $i^{th}$ column
    Find the Regression coefficient using equation(8)
    Find the predicted output $Y^\star$ using equation(6)
    Find the residual_ error using equation(5)
    if residual_ error>Thrushold_ error then
        store $i^{th}$ column in prominent matrix P.
    end if
end for
FD_ Mine(P)
Display FD_ SET
```

Algorithm 2. Preprocess(RM)

Input: RM : Relational Datasource with T columns with column type specifications and I number of instances
Output: Preprocessed datasource

```
Let I be the number of instances in RM
for i:1 to T do
    if $i^{th}$ datatype is numerical then
        find the mean $\mu$ and stand deviation $\sigma$ of I numerical values
        for k:1 to I do
            find the normalized value using $\frac{x-\mu}{\sigma}$where x is the $k^{th}$ numeric value
        end for
    end if
    if $i^{th}$ datatype is boolean then
        assign 1,2 corresonding to boolean values.
    end if
    if $i^{th}$ datatype is ordinal then
        assign ASCII value corresponding to ordinal values.
    end if
end for
Return the preprocessed Data source RM
```

Algorithm 3. FD_ Mine(P)

Input: Database D and its attributes X1, X2, ... , Xm
Output: FD_SET

```
Initialization Step set R = {X_1, X_2, , , , X_m} set FD_ SET = φ
set EQ_ SET =φ, set KEY_ SET =φ set CANDIDATE_ SET = {X_1, X_2, , , , X_m} ∀X_1 ∈ CANDIDATE_ SET, set Closure[X_1] =φ
while CANDIDATE_ SET≠ φ do
   for i:1 to CANDIDATE_ SET do
      ComputeNonTrivialClosure(X_i)
      ObtaintFDandKey
   end for
   ObtainEQSet(CANDIDATE_ SET)
   PruneCandidates(CANDIDATE_ SET)
   GenerateNextLevelCandidates(CANDIDATE_ SET)
end while
Return(FD_ SET)
```

be inspected. This pruning rule is justified by one property, if Closure(XY) is the nontrivial closure of candidate XY, then Closure(X)∪Closure(Y)→Closure (XY) holds. It means that if X→Z and Y→W holds, then XY→WZ also holds. Latter is, let $X_1X_2..X_k \rightarrow X_{k+1}$ be a k-level FD. If any subsets $X_{i(1)}X_{i(2)}...X_{i(k-1)}$ of $X_1X_2...X_k$ satisfies $X_{i(1)}X_{i(2)}...X_{i(k-1)} \rightarrow X_{i(k)}$, then $X_1X_2..X_k \rightarrow X_{k+1}$ does not need to be inspected.

3.2 A Running Example

The following example will help to illustrate the working of our proposed method. Consider the data matrix corresponding to Table 1 after preprocessing.

$$\begin{pmatrix} St\_ID & SEM & LECTURE & Teaching\_Assistant & CLASSROOM \\ -0.904752163 & 1.1456439237 & 1.1326519639 & -1.4577379737 & 0.3330397768 \\ -0.5942742089 & 0.3818813079 & 0.3743591867 & -0.7717436331 & -1.4431723663 \\ 1.1909740269 & -1.1456439237 & -1.1498858907 & -0.0857492926 & -0.1110132589 \\ -0.904752163 & 1.1456439237 & 1.1632900559 & 1.2862393886 & 0.7770928126 \\ -0.5942742089 & 0.3818813079 & 0.3896782327 & 0.600245048 & -0.9991193305 \\ -0.5942742089 & 0.3818813079 & 0.3743591867 & -0.7717436331 & 0.3330397768 \\ 1.1715691548 & -1.1456439237 & -1.1498858907 & -0.0857492926 & -0.5550662947 \\ 1.2297837711 & -1.1456439237 & -1.1345668447 & 1.2862393886 & 1.6651988842 \end{pmatrix}$$

We can rename the attributes St_ ID, SEM, LECTURE, Teaching_ Assistant and CLASSROOM as A,B,C,D, and E respectively for simplicity. We model it as E=Aβ_1+Bβ_2+Cβ_3+Dβ_4. Each iteration of the linear programming have built a model and obtained the error corresponding to the attribute as shown below,

A= Bβ_1+Cβ_2+Dβ_3 ; error A 8.154760405719994
B= Aβ_1+Cβ_2+Dβ_3 ; error B 6.873863542250151
C= Aβ_1+Bβ_2+Dβ_3 ; error C 6.8686772519496415
D= Aβ_1+Bβ_2+Cβ_3 ; error D 6.345447645786461

From this, if a threshold value of 6.5 is taken, then A,B,C and E are the prominent attributes. Once the prominent attributes are obtained, FD_ Mine is applied. The Functional dependencies obtained after applying FD_ Mine on prominent attributes is shown here.

A→B, C→B, EA→B, EA→C, EB→A, EB→C, EC→B, EC→A

These Indicative Functinal Dependencies show every relationship corresponding to prominent attributes. On analyzing the data, it is clear that A→B and C→D are the minimal functional dependencies that must hold.

4 Experiments and Evaluation

We have done experiments on Bank Marketing dataset, Cancer dataset, Hepatitis dataset and German credit dataset. The Bank Marketing dataset is related with direct marketing campaigns of a Portuguese banking institution[17]. The dataset includes 45211 instances and it is described by 16 attributes(17th attribute is the target). Cancer dataset is one of three domains provided by the Oncology Institute that has repeatedly appeared in the machine learning literature[18]. This data set includes 201 instances of one class and 85 instances of another class. The instances are described by 9 attributes, some of which are linear and some are nominal. Hepatitis data contain 155 intances and 19 attributes[19] and German Credit dataset includes 1000 instances and 20 attributes[19].

Consider the error obtained after building the regression model for Hepatitis dataset in which the irrelevant attributes texted in bold is shown in table 2.

197 FDs obtained after doing FD_ Mine on prominent attributes.

Table 2 Error value obtained after building regression model for hepatitis dataset

ATTRIBUTES	ERROR
Y	14414.426748758575
A	279.5455395074454
B	6495.834383864317
C	172.1831644634682
D	235.56277521887853
E	289.3421243911014
F	209.12491356433384
G	247.10553932174213
H	277.1598256891142
I	275.0174665014448
J	236.1195779353524
K	273.77147334868874
L	250.71527562398734
M	284.23584048408014
N	285.83763885451185
O	237.94220598238593
P	17249.91434740195
Q	16250.255544219186
R	609.7656427628563
S	9183.104536055616

4.1 Comparison with FD_ Mine

Using the FD_ Mine algorithm there were 402 number of functional dependencies while using the proposed mathod, an enhanced result of 197 number of functional dependncies were obtained. Based on these 197 indicative functional dependencies it seems possible to have a better analysis of the data. Keeping all the 402 functional dependencies which contain both irrelevant and relevant functional dependencies, is not effective. If we have the functional dependencies of relevant attributes, we can use these functional dependencies as indicative functional dependencies and can effectively represent the important relationships existing in the data. The result obtained with other datasets are shown in figure.

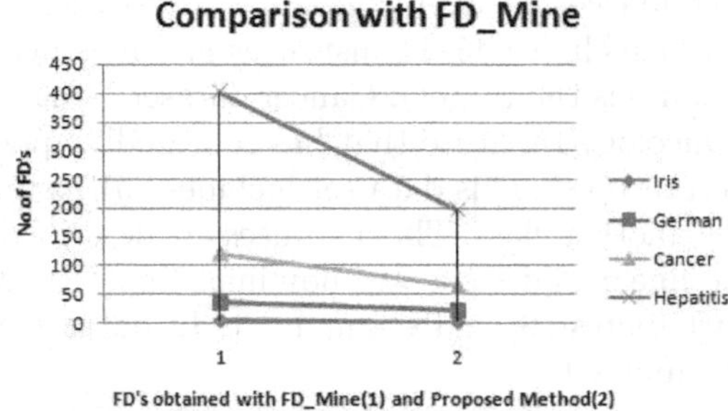

Fig. 2 Number of FDs for each dataset using FD_Mine(1) and the Proposed method(2)

4.2 Further Analysis

We have considered cancer dataset with 300 instances as training dataset. The error obtained after building regression model is shown below.

Table 3 Error value obtained after building regression model for training dataset

ATTRIBUTES	ERROR
A	58613.032162038995
B	3.839920591741155E8
C	1615.39366535798
D	1075.0571034117895
E	1127.1322769276765
F	1035.942936153549
G	1133.5622558866532
H	1467.9036038265647
I	1222.160756950302
J	1172.52687328578
K	659.0565812109907

In table 4 irrelevant attributes are texted as bold. We have obtained 64 functional dependencies as indicative functional dependencies after FD_ Mine. The residual sum obtained for cancer dataset with 300 instances as testing dataset is shown below.

Table 4 Error value obtained after building regression model for testing dataset

ATTRIBUTES	ERROR
A	148381.68681316302
B	3.292850305237574E8
C	1419.8966858442354
D	864.6887668065837
E	872.256176407442
F	860.0381271693109
G	895.476296981621
H	1064.4659326609094
I	963.1457589738627
J	879.2476545640903
K	451.34903553380616

The indicative functional dependencies obtained for test dataset is 63 based on the prominent attributes D,F,K and they are similar to the result obtained for training dataset. From this it is evident that the result obtained is accurate.

5 Conclusion

We presented a method to find the indicative functional dependencies for the relational databased with high number of attributes. Data from relational databases were processed to find the prominent attributes by building a regression model. Again we considered the dataset only with relevant attributes, further obtaining the indicative functional dependencies with the help of FD_Mine algorithm.

We have done experiments on different datasets and have seen that the result is accurate.

Acknowledgements. We thank Dr. M.R. Kaimal,Chairperson, Department of Computer Science and Engineering, Amrita School of Engineering for giving suitable suggestions for the successful completion of this work.

References

1. Synthesizing Third Normal Form Relations from Functional Dependencies philip a. bernstein, University of Toronto. ACM Transactions on Database Systems 1(4) (December 1976)

2. Yao, H., Hamilton, H.J., Butz, C.: FDMine Discovering Functional Dependencies in a Database Using Equivalences. In: International Conference on Data Mining 2002, pp. 729–732 (2002)
3. Mannila, H., Raiha, K.-J.: On the complexity of inferring functional dependencies. Discrete Applied Mathematics 40, 237–243 (1992)
4. Bra, P.D., Paredaens, J.: Horizontal decompositions for handling exceptions to functional dependencies. In: Gallaire, H., Minker, J., Nicolas, J.M. (eds.) Advances in Database Theory, vol. 2, pp. 123–141. Plenum Publishing Company, New York (1984)
5. Huhtala, Y., Krkkinen, J., Porkka, P., Toivonen, H.: TANE: An Efficient Algorithm for Discovering Functional and Approximate Dependencies. The Computing Journal 42(2), 100–111 (1999)
6. Flach, P.A., Savnik, I.: Database Dependency Discovery: A Machine Learning Approach. AI Communications 12(3), 139–160 (1999)
7. Wyss, C.M., Giannella, C.M., Robertson, E.L.: FastFDs: A Heuristic-Driven, Depth-First Algorithm for Mining Functional Dependencies from Relation Instances. In: Kambayashi, Y., Winiwarter, W., Arikawa, M. (eds.) DaWaK 2001. LNCS, vol. 2114, pp. 101–110. Springer, Heidelberg (2001)
8. He, Q., Link, T.W.: Extending Inferring Functional Dependencies in Schema Transformation. In: ACM 2004 (2004)
9. Montgomery, D.C.: Design and Analysis of Experiments, 5th edn. Wiley (2001)
10. Savnik, I., Flach, P.: Bottom-up induction of functional dependencies from relations. In: Piatetsky-Shapiro, G. (ed.) Knowledge Discovery in Databases, Papers from the 1993 AAAI Workshop (KDD 1993), Washington, DC, pp. 174–185. AAAI Press (1993)
11. Mannila, H., Raiha, K.-J.: Algorithms for inferring functional dependencies. Data & Knowledge Engineering 12, 83–99 (1994)
12. Bell, S., Brockhausen, P.: Discovery of Data Dependencies in Relational Databases. Technical Report LS-8 Report-14, University of Dortmund (1995)
13. Kivinen, J., Mannila, H.: Approximate dependency inference from relations. Theor. Comp. Sci. 149, 129–149 (1995)

Novel Research in the Field of Shot Boundary Detection – A Survey

Raahat Devender Singh and Naveen Aggarwal

Abstract. Segregating a video sequence into shots is the first step toward video-content analysis and content-based video browsing and retrieval. A shot may be defined as a sequence of consecutive frames taken by a single uninterrupted camera. Shots are the basic building blocks of videos and their detection provides the basis for higher level content analysis, indexing and categorization. The problem of detecting where one shot ends and the next begins is known as Shot Boundary Detection (SBD). Over the past two decades, numerous SBD techniques have been proposed in the literature. This paper presents a brief survey of all the major novel and latest contributions in this field of digital video processing.

Keywords: Shot boundary detection, scene change detection.

1 Introduction

The increased availability and usage of digital video obligates development of techniques that analyze video content automatically. Almost all content-analysis operations rely on automatic detecting of the boundaries between camera shots. Since SBD provides a foundation for almost all video abstraction and high-level segmentation tasks, it has a rich and vast history. The next section provides a brief and general introduction to the problem of detection of shot boundaries in a digital video and the various evaluative measures that are used to judge the performance of the various approaches that are suggested as a solution to this problem.

2 Introduction to Shot Boundary Detection

A shot is an unbroken sequence of frames taken from one uninterrupted camera. Shot Boundary Detection (SBD) is the detection and classification of transitions

Raahat Devender Singh · Naveen Aggarwal
UIET, Panjab University, Chandigarh, India
e-mail: {raahat3,navagg}@gmail.com

El-Sayed M. El-Alfy et al. (eds.), *Advances in Intelligent Informatics*,
Advances in Intelligent Systems and Computing 320, DOI: 10.1007/978-3-319-11218-3_41

between adjacent shots. There are a number of different types of transitions that can occur between shots which can be characterized as abrupt changes (hard cuts) or gradual transitions (dissolves, wipes, fades).

Hard cuts (or simply, cuts) represent the most common type of transition in videos where there is a clear distinction between the contents of two frames. Dissolves constitute the gradually disappearance of one frame and appearance of another. Disappearance of a frame into a dark frame (fade-out) or appearance of a frame out of a dark frame (fade-in) together are termed as FOI (Fade Out/In). In wipes appearing/disappearing shots coexist in different spatial regions of the transition frames [1].

2.1 Performance Evaluation

The comparison between the output of an algorithm and the ground truth is performed based on the number of missed and false detections. A missed detection is the one when a shot boundary actually existed but wasn't detected by the algorithm. On the other hand, a false detection occurs when the algorithm detects a transition where an actual boundary didn't exist. The basic evaluative measures in detection and retrieval problems are recall, precision and F-measure.

Recall estimates what proportion of the correct shot boundaries is detected, while precision appraises what proportion of the detected boundaries are correct. F-measure (or F1 score) reflects the temporal correctness of the detected results [2]. To evaluate the performance of their algorithms, researchers have been known to use either their own test sets or utilize standard test sets, like the ones provided by TRECVid [3], which makes it easier to compare their results with those of other techniques.

3 Survey of Novel Contributions to the Field of SBD

Despite the vastness of this field, most of the proposed approaches utilize a handful of techniques to tackle the problem of transition detection, which include pixel-based and histogram-based approaches, methods that focus on edge information, texture or color features of the video frames, methods that utilize motion or audio features, approaches based on statistical techniques or statistical classifiers etc. Consequently, various comprehensive surveys and reviews also focus on a limited and specific range of available techniques. Since this field of video processing has a long history, many past surveys do not include the latest additions to the pool of proposed methodologies.

This paper, however, presents all the major novel and latest contributions that utilize an amalgamation of concepts and techniques to achieve enhanced results for the daunting task of video shot boundary detection. For the sake of brevity, only a general insight into the working of these methods has been provided without going over any technical details of their functioning.

3.1 Techniques for Abrupt Transition Detection

This section summarizes the innovative techniques formulated to detect only the abrupt transitions (hard cuts) in video sequences.

3.1.1 Thresholding of Histogram Intersection

It's been observed that many SBD techniques use histogram to differentiate between transition and non-transition frames by applying thresholds to histogram intersections, which could be local, global or adaptive in nature. The following table summarizes the methods that fall under this category.

Table 1 Summarization of histogram intersection-based thresholding algorithms. R, P and F denote recall, precision and F-measure values, respectively.

Algorithm	Technique Used	Test Set	Results (%)	Salient Features
Audio-Assisted Video Segmentation Jiang et al. [4]	Integration of audio analysis and color tracking to cluster shots into semantically correlated groups	TV news broadcasts in MPEG-7 format	R=91.9 P=86.8	Includes an efficient audio segmentation and speaker change detection algorithm
Shot Connectivity Graph Javed et al. [5]	Exploitation of sematic structure of videos, segmentation of video into shots, construction of data-structure (graph) linking similar shots over time	4 episodes of a specific talk show	F=97.1	Robust due to lack of time constraint on segmentation, Good for videos with low spatial and temporal resolution
Colored Pattern Appearance Model Sze et al. [6]	Modelling of visual appearance of a small image block by the stimulus strength, the spatial and color pattern	MPEG-7 format home videos with strong noise, motion	R=95.2 P=35.4	Sufficiently invariant to strong noise and large object/camera motion, Adaptive threshold
Modified Phase Correlation Urhan et al. [7]	Use of Fourier transform to detect similarity between spatially sub-sampled frames, local and global thresholding of phase-correlation peaks, mean and variance tests for false detection removal	Archive B&W films MPEG-7 format hand-held camera videos	R=98.5 P=94.2	Reduced computational load due to spatial sub-sampling, Insensitive to global illumination changes, severe noise

All these methods use a limited and specific range of test videos, most of which contain a lot of noise and motion disturbances. Method suggested by Urhan et al. in [7] seems to provide the best results in this category.

Table 2 Summarization of hybrid approaches. R, P and F denote recall, precision and F-measure values, respectively. FAR stands for False Alarm Ratio.

Algorithm	Technique Used	Test Set	Results (%)	Salient Features
Constant False Alarm Ratio Approach Liu et al. [8]	Detecting cuts by calculating the current frame distance and counting the number of previous differences less than this one to guarantee a controllable false alarm ratio	MPEG-7 format movies, news videos	FAR=10	Controllable FAR, Computable misdetection ratio, High detection accuracy
Feature Tracking and Automatic Threshold Selection Whitehead et al. [9]	Measuring dissimilarity among frames using feature tracking metric by tracking fine details (corners, texture), determination of global threshold from PDF of difference values	Variety of videos with fast camera/ object motion, special effects	R=96.1 P=87.4 F=90.8	Reduced false alarm rate, Enhanced robustness against camera and object motion without additional computational cost
Best-Fitting Kernel Urhan et al. [10]	Identification of the boundary candidates based on the absolute frame difference between estimated and original image frames, classified using combination of local and global thresholds	Highly degraded B&W archive films	R=93.0 P=93.3 F=93.2	False detections because of intensive camera and object motion, Similar scene content causes missed cuts
Video Mosaicing and Montage Rules Chen et al. [11]	Use of video mosaicing for b/g reconstruction and film making aka montage rules for the extraction of spatio-temporal features from video sequences	MPEG-1 format home video, movies of various genres	R=77.3 P=77.5	Sufficiently invariant to strong noise, large object/camera motion
Modified Dugad Model Krulikovská and Polec [12]	Calculation of difference measure from successive frames within a sliding window Real-time video shot cut detection with the use of statistically adaptive threshold	Standard TRECVid test set	R=98.1 P=96.8 F=97.4	No computational delays since previously evaluated frames are used by the model to adapting itself to the sequence statistics

3.1.2 Hybrid Approaches

Instcad of simply applying threshold to histogram intersections, the following methods implement a hybrid technique to extract useful features from video frames and calculate frame dissimilarity measure (distance measure) and perform shot boundary detection.

It can be concluded from the table that the method proposed in [12] generates best results by operating on a benchmark dataset provided by TRECVid.

3.2 *Techniques for Gradual Transition Detection*

This section summarizes the innovative techniques formulated to detect only gradual transitions–GTs, like dissolves, fades and wipes in video sequences.

Table 3 Summarization of GT detection approaches. R, P and F denote recall, precision and F-measure values, respectively.

Algorithm	Technique Used	Test Set	Results (%)	Salient Features
Temporal Characteristics of Intensity Variation Truong et al. [13]	Application of two thresholds on the characteristic downwards curve formed by temporal intensity variation of frame pixels during a dissolve transition	Standard test set of variety of videos	Dissolves: R=82.2 P=75.1 Fades: R=92.8 P=89.6	Improver false alarm reduction via the use of two thresholds, Sufficiently invariant to noise and motion present in video
Morphological Filters and Radon Transform Nam and Tewfik [14]	Processing of temporal evaluation of intensity variation of pixels via morphological filters, detection of wipes using Radon transform of every frame	TV news, documentaries, music concerts and feature films	Dissolve: R=86.8 P=80.3 Wipes: R=77.7 P=76.3	Separate algorithms for different transitions, Effective distinction between fades and dissolves
Meta-Segmentation Approach Tsamoura et al. [15]	Color coherence, luminance center of gravity and intensity variation features extracted from frames and fed into a binary classifier: (Support Vector Machine-SVM)	TRECVid 2007 data set with intense motion and luminance variations	R=88 P=73	No need for threshold selection, Reduced sensitivity to local/global motion, illumination changes and noise in video
Audio-Visual Feature Extraction Sidiroupoulos et al. [16]	Incorporation of low and high level audio and visual cues (extracted from visual and auditory channels) into Generalized Scene Transition Graph	Set1-TRECVid dataset, Set2-movies, Set3-news clips	F(set1)= 88.9 F(set2)= 85.5 F(set3)= 78.7	Reduced computational cost, Diminished need of heuristic parameter setting, Generalizable to various genres of videos

Table 4 Summarization of statistical approaches to SBD. R, P, F denote recall, precision and F-measure, respectively.

Algorithm	Technique Used	Test Set	Results (%)	Salient Features
N-Distance Measure Model Huang and Liao [17]	Combination of intensity and motion features for scene change detection using a moving window and two thresholds	Action movies, news, sports music videos, commer-cials and sitcoms	Cuts: R= 97.7 P= 96.8 GTs: R= 93 P= 78	Relaxed threshold selec-tion, Efficient reduction in false alarm rate, Two thresholds help avoid false alarms due to camera zoom and pan operations
Motion and Illumination Estimation Li and Lai [18]	Exploitation of motion/ illumination estimation in a video via generalizable optical flow constraint, incorporation of motion vectors to encompass intensity changes	TRECVid data set videos	Cuts: R= 96.8 P= 92.9 GTs: R= 77.4 P= 81.4	Robust against translation, rotation, scaling, zooming and object motion, Simple data analysis, Easier global threshold selection
Change Detection Theory Lelescu and Schonfeld [19]	Modelling video sequenc-es as stochastic processes and scene changes as changes in the parameters of these processes, dimen-sionality reduction via PCA (performs additive (AMCD) and non-additive modeling (NAMCD))	MPEG-2 format news, sports and music videos, documen-taries	AMCD: R=92 P=72 NAMCD R=81 P=65	Single-pass, real time approach for compressed videos, No a priori assumptions: (limits scope of algo-rithm) No predefined windows, Invalidates alarms from camera zoom and pan
Multi-Stage Approach for Edit Detection Zheng et al. [20]	Cut detection via second-order difference method, FOI detection via mono-chrome frame identifica-tion, twin comparison and FSA for GT detection	Large data-base con-sisting of TRECVid 2003/ 2004 datasets	Cuts: R=92.1 P=91.2 F=92.6 GTs: R=91.7 P=71.4 F=80.3	Immune to abrupt illumi-nation change, large object/camera movement, Utilizes motion based self-adaptive threshold along with global and heuristically determined ones
Fuzzy Logic Based Cut Detection Fang et al. [21]	Combination of color, texture, edge variance, motion compensation features into a hybrid feature via their fuzzy membership, use of fuzzy rules and thresholds, analysis of change in pixel variance with Canny edge detector for GT detection	2 open test datasets	R=98.8 P=96.0 F=97.4	Easier assignment of importance to features, Easier selection of optimal thresholds, Fuzzification leads to a robust and reliable technique

Table 4 *(continued)*

Linear Transition Detector Grana and Cucchiara [22]	Iterative algorithm that exploits a linear transition model with additional capabilities to cope with non-linear factors such as camera/objects motions, color/luminance variations	Set 1: TRECVid data set Set 2: MPEG-7 format sports videos	Set1:Cuts: R= 95.7 P=82.3 GTs: R= 73.9 P= 65.4 Set2: Cuts: R=86, P=75 GTs: R=77.6 P=71.6	Requires only two parameters, thus makes the training process easier, Use of a precise model leads to more discriminative power than a general method, Assumes linearity of the transition effect
GED and Gaussian Functions Amiri and Fathy [23]	Definition of a distance metric from color histograms and GED to capture temporary color distribution of frame, metric demonstrated abrupt changes in hard cuts, semi-Gaussian behavior in GTs	Set 1: TRECVid 2006 data set Set 2: news, music, sports videos, cartoons, documentaries	Set1:Cuts: R=97, P=95 GTs: R=95, P=92 Set2: R= 96.8 P=94	Independent of video content, Cost effective, Less sensitive to object motion, color and illumination variations, camera zoom/pan, False alarms caused by flashlights and very high speed camera motion
Dual Detection Model Jiang et al. [24]	Incorporation of color histogram and pixel value differences (human visual features) in an adaptive binary search procedure, Feature matching using scale invariant feature transform (SIFT) algorithm for handling false alarms	TRECVid dataset in addition to videos from internet	Cuts: R=93.3 P=98.9 GTs: R=89.0 P=74.3	Simultaneous detection of cuts and GTs, Considerable reduction in computational complexity, Efficient removal of false alarms due to large camera/object motion

It's observed that the technique in [16] produces good results for TRECVid data set; however, fades are best detected by the method proposed in [13].

3.3 *Techniques for Abrupt and Gradual Transition Detection*

This section summarizes the novel techniques formulated to detect both hard cuts and gradual transitions in video sequences. All the techniques have been classified according to the dominant course of action followed by them. The categories are: statistical approaches, clustering and rule-based approaches, machine learning approaches and content based unprecedented approaches.

3.3.1 Statistical Approaches

The scope of this category ranges from approaches that utilize simple statistical techniques like sum of absolute difference of pixel intensities to those employing more elaborate statistical techniques such as Principal Component Analysis (PCA) and Finite State Automata (FSA).

On a standard test set (TRECVid in this case), the best results can be said to have been achieved by the method proposed in [23]. Fuzzy logic brings certain advantages to the field of transition detection [21] and is leads to generation of good results as well.

3.3.2 Clustering and Rule-Based Approaches

The following techniques either follow a specific modeling rule to detect boundary frames or try to distinguish between normal frames and transitions frames by judging similarities/dissimilarities among them and clustering them with or without the help of thresholds.

Table 5 Summarization clustering and rule-based approaches. R, P, F denote recall, precision and F-measure, respectively. C stands for cuts, FI means fade-in, FO means fade-out, D stands for dissolves.

Algorithm	Technique Used	Test Set	Results (%)	Salient Features
Video Edit Models Hampapur et al. [25]	A rule based model that studies video production processes to capture content and production style of every video individually	Cable TV data: news, sports and music videos, commercials and sitcoms	F= 91 (C) 75 (FI) 100 (FO) 73 (D)	Top-down approach: simpler computations, Robust against camera pan and zoom, Most errors occur due to object motion too close to camera
Histogram-Based Fuzzy C-Means Clustering Lo and Wang [26]	Amalgamation of threshold based and clustering based approaches to overcome the problem of correct threshold determination and accurate estimate of number of clusters (with the use of fuzzy logic)	Soap operas, commercials and movies of various genres	Hit Ratio: HR=91.2 False Ratio: FR=19.2	Both robust and gereralizable due to its fuzzy nature, Operates under several limiting assumptions which may be impractical to satisfy
Normalized Cut Criterion Damnjanovic et al. [27]	Partitioning of frames into clusters by focusing on their semantic structure with the help of adaptive thresholds	Videos showing different editing effects, news, music videos, documentaries, animations	R=91 P=89	Simultaneous detection of cuts and GTs, Not completely immune to illumination/color changes and extensive animation effects

On a wide range of test videos, the technique proposed by Damnjanovic et al. in [27] produces favorable results.

3.3.3 Machine Learning Approaches

In the methods that follow, features, after being extracted from the frames, are fed into supervised or supervised classifiers which then learn how to detect transitions from a training set and use this knowledge to detect transitions on actual test set, thereby avoiding the use of thresholds for decision making.

Table 6 Summarization of machine learning approaches to SBD. R, P, F denote recall, precision and F-measure, respectively.

Algorithm	Technique Used	Test Set	Results (%)	Salient Features
Foveation points Boccignone et al. [28]	Computation of consistency measure of the fixation sequences generated by an ideal observer looking at the video at each time instant, brightness, color, orientation information encapsulated into a Bayesian formulation	TRECVid 2001 dataset	Cuts: R=97 P=95 F=93 Dissolves: R=92 P=89 F=90	Single algorithm that detects both types of transitions (no needless computational overhead), Insensitive to small object motion, Slight computational bottleneck, requires a hardware solution in case of time-sensitive tasks
MV Filtration Scheme and RCH Han et al. [29]	Features in the form of dominant color mask projection and Motion Vectors (MVs) along with Region Color Histograms (RCH) integrated into SVM based classifiers	12 soccer videos with complex editing effects and sudden camera motion	Cuts: F=98 GTs: F=85	Enhanced capabilities to reduce false alarms and missed detections, Robust on a large data set, Immune to camera zoom/pan, Works for sports videos only
Supervised Learning Methodology Chasanis et al. [30]	Learning based technique that detects shot boundaries by capturing difference between the visual content of normal frames and transition frames, final classification performed via SVM	TV series, educational films, documentaries	Cuts: R=99.4 P=98.6 Dissolves: R=90.8 P=86.0	Threshold independent, Insensitive to content of video, No a priori knowledge required, Simultaneous detection of cuts & dissolves

Foveation point based approach in [28] uses a benchmark TRECVid dataset and is capable of generating very good results.

3.3.4 Content-Based and Unprecedented Approaches

This section summarizes methodologies that either depend highly on the underlying content of video sequences or utilize a key idea that had never been used for the task of transition detection. By focusing on the semantics of the video, these algorithms are able to diminish the effects of scaling, rotation and translation.

Table 7 Summarization of statistical approaches to SBD. R, P, F denote recall, precision and F-measure, respectively.

(*Note:-* Authors of [33] used a different set of evaluative measures: these were robustness (μ), missless error (Em), falseless recall (Rf) and gamma (γ). The best values obtained for these metrics were 0.11, 0.37, 0.51 and 0.80 (for cuts) and 0.11, 0.61, 0.43 and 0.69 (for flashes). Recall and error rates (α and β) were used to judge the fade detection results. For all fades recall and error were reported to be 0.95 and 0.30, however, for long fades, error of 0.05 was obtained.)

Algorithm	Technique Used	Test Set	Results (%)	Salient Features
Back-Ground Tracking Oh et al. [31]	Computation of background difference between frames, use of semantics of frame background to handle various kinds of camera motions	Cartoons, soap operas, sitcoms, music videos, science fiction	Cuts: R= 97.7 P= 96.8 GTs: R= 93 P= 78	Less sensitive to the predetermined threshold values, Good for large databases since it exhibits time/space efficiency, Intuitive and generic threshold makes the algorithm extendable
Spatial-Temporal Joint Probability Image Analysis Li et al. [32]	Combination of spatial and temporal characteristics of pixel intensities and use of joint probability for accurate characterization of the temporal evolution of the visual characteristics of videos	Sports and battlefield videos with fast camera and object motion and captions	Cuts: R=97, P=96 Cross dissolves: R=82, P=93 Dither dissolves: R=75, P=81	Efficient, fast, scalable and very suitable for real-time video segmentation, Poor performance in case of large object/camera motion
Virtual Rhythm Guimarães et al. [33]	Transformation of segmentation problem into pattern detection problem, use of morphological, topological, discrete geometry tools to segment videos without dissimilarity measures	Videos with zoom, tilt, pan effects, special effects, object motion	See 'Note' before table	Uses global information thus invariant to scaling /rotation/translation, Capable of accurately detecting flashes, Implementationally simple, Low processing cost, Reduced number of missed detections

Table 7 (*continued*)

Color Anglograms and Latent Semantic Indexing Zhao and Grosky [34]	Negotiation of gap between low-level features and high-level concepts by uncovering semantic correlation between frames via extracting underlying semantic structure of video frames	TV commercials, movie trailers and news videos	Cuts: R=91.4 P=82.9 GT: R=88.3 P=72.6	Invariant to scaling/ rotation/translation, Tolerant towards object/camera motion
Animate Vision Albanese et al. [35]	Computation of an arbitrary similarity measure between consecutive frames by utilizing mechanisms of visual attention (color, shape & texture) used by humans to recognize significant changes in the visual content	Movies and documentaries	Cuts; R= 100 P= 91.1 Dissolves R=98.0 P=87.7	Dynamic threshold for dissolve detection, Inefficient in presence of rapid camera/object movements
Visual Attention Model Mendi and Bayrak [36]	Utilization of color and luminance features of the frames to identify structural similarity between frames (motivated by the human visual system)	Two neurosurgical videos with cuts and dissolves	R=100 P=92	Immune to dissolve effect, Useful for key-frame extraction, Restricted data set used for evaluations

The animate vision approach of [35] produces best results in this category, but it's important to realize that the test videos used in the method (movies and documentaries) usually have specific production rules associated with them, which are somewhat helpful in detection of transitions. For a more general and elaborate test set, background tracking in [31] performs rather well.

3.4 *Conclusion*

This paper presented a brief survey of all the major novel and latest contributions in the field of shot boundary detection, along with the reported system performance statistics and special features of each methodology. This work can serve as a reference for working in SBD application.

References

1. Yuan, J., Wang, H., Xiao, L., et al.: A formal study of shot boundary detection. IEEE Trans. Circuits Syst. Video Technol. 17(2), 168–186 (2007)
2. Kasturi, R., Jain, R.: Dynamic vision. In: Kasturi, R., Jain, R. (eds.) Computer Vision: Principles. IEEE Computer Society Press (1991)

3. Smeaton, A.F., Over, P., Doherty, A.R.: Video shot boundary detection: Seven years of TRECVid activity. Comput. Vis. Image Underst. 114(4), 411–418 (2010)
4. Jiang, H., Lin, T., Zhang, H.: Video segmentation with the support of audio segmentation and classification. In: IEEE Int. Conf. Multimed. Expo., New York, vol. 3, pp. 1551–1554 (2000)
5. Javed, O., Khan, S., Rasheed, Z., et al.: A framework for segmentation of interview videos. In: Int. Conf. Internet Multimed. Syst. Appl., Las Vegas (2000)
6. Sze, K.W., Lam, K.M., Qiu, G.: Scene cut detection using the colored pattern appearance model. In: IEEE Int. Conf. Image Process., Spain, vol. 2, pp. 1017–1020 (2003)
7. Urhan, O., Güllü, M.K., Ertürk, S.: Modified phase-correlation based robust hardcut detection with application to archive film. IEEE Trans. Circuits Syst. Video Technol. 16(6), 753–770 (2006)
8. Liu, T.Y., Lo, K.T., Zhang, X.D., et al.: A new cut detection algorithm with constant false-alarm ratio for video segmentation. J. Vis. Commun. Image Represent. 15(2), 132–144 (2004)
9. Whitehead, A., Bose, P., Laganiere, R.: Feature based cut detection with automatic threshold selection. In: 3rd Int. Conf. Content Based Image Video Retr., Ireland (2004)
10. Urhan, O., Güllü, M.K., Ertürk, S.: Shot-cut detection for B&W archive films using best-fitting kernel. Int. J. Electron. Commun. 61(7), 463–468 (2007)
11. Chen, L.H., Lai, Y.C., Liao, H.M.: Movie scene segmentation using background information. Pattern Recognit. 41(3), 1056–1065 (2008)
12. Krulikovská, L., Polec, J.: Shot Detection using modified Dugad model. World Academy of Sci. Eng. Technol. 6 (2012)
13. Truong, B.T., Dorai, C., Venkatesh, S.: Improved fade and dissolve detection for reliable video segmentation. In: IEEE Int. Conf. Image Process., Vancouver, vol. 3, pp. 961–964 (2000)
14. Nam, J., Tewfik, A.: Detection of gradual transitions in video sequences using b-spline interpolation. IEEE Trans. Multimed. 7(4), 667–679 (2005)
15. Tsamoura, E., Mezaris, V., Kompatsiaris, I.: Gradual transition detection using color coherence and other criteria in a video shot meta-segmentation framework. In: IEEE Int. Conf. Image Process. Multimed. Image Retr., California, pp. 45–48 (2008)
16. Sidiropoulos, P., Mezaris, V., Kompatsiaris, I., et al.: Temporal video segmentation to scenes using high-level audio-visual features. IEEE Trans. Circuits Syst. Video Technol. 21(8), 1163–1177 (2011)
17. Huang, C.L., Liao, B.Y.: A robust scene-change detection method for video segmentation. IEEE Trans. Circuits Syst. Video Technol. 11(12), 1281–1288 (2001)
18. Li, W.K., Lai, S.H.: Integrated video shot segmentation algorithm. In: SPIE Conf. Storage Retr. Media Databases, California, pp. 264–271 (2003)
19. Lelescu, D., Schonfeld, D.: Statistical sequential analysis for real-time video scene change detection on compressed multimedia bitstream. IEEE Trans. Multimed. 5(1), 106–117 (2003)
20. Zheng, W., Yuan, J., Wang, H., et al.: A novel shot boundary detection framework. In: SPIE Vis. Commun. Image Process., vol. 5960, pp. 410–420 (2005)
21. Fang, H., Jiang, J., Feng, Y.: A fuzzy logic approach for detection of video shot boundaries. Pattern Recognit. 39, 2092–2100 (2006)
22. Grana, C., Cucchiara, R.: Linear transition detection as a unified shot detection approach. IEEE Trans. Circuits Syst. Video Technol. 17(4), 483–489 (2007)

23. Amiri, A., Fathy, M.: Video shot boundary detection using generelized Eigen value decomposition and Gaussian transition detection. Comput. Inform. 30(3), 595–619 (2011)
24. Jiang, X., Sun, T., Liu, J., et al.: An adaptive video shot segmentation scheme based on dual-detection model. Neurocomput. 116, 101–111 (2013)
25. Hampapur, A., Jain, R.C., Weymouth, T.: Production model based digital video segmentation. Int. J. Multimed. Tools Appl. 1(1), 9–46 (1995)
26. Lo, C., Wang, S.J.: Video segmentation using a histogram-based fuzzy C-means clustering algorithm. In: IEEE Int. Fuzzy Syst. Conf., vol. 3, pp. 920–923 (2001)
27. Damnjanovic, U., Izquierdo, E., Grzegorzek, M.: Shot boundary detection using spectral clustering. In: Eur. Signal Process. Conf., Poland, pp. 1779–1783 (2007)
28. Boccignone, G., Chianese, A., Moscato, V., et al.: Foveated shot detection for video segmentation. IEEE Trans. Circuits Syst. Video Technol. 15(3), 365–377 (2005)
29. Han, B., Hu, Y., Wang, G., et al.: Enhanced sports video shot boundary detection based on middle level features and a unified model. IEEE Trans. Consum. Electron. 53(3), 168–1176 (2007)
30. Chasanis, V., Likas, A., Galatsanos, N.: Simultaneous detection of abrupt cuts and dissolves in videos using support vector machines. Pattern Recognit. Lett. 30(1), 55–65 (2009)
31. Oh, J.H., Hua, K.A., Liang, N.: A content-based scene change detection and classification technique using background tracking. In: SPIE Conf. Multimed. Comput. Netw., California, vol. 3969, pp. 254–265 (2000)
32. Li, Z.N., Zhong, X., Drew, M.S.: Spatial temporal joint probability images for video segmentation. Pattern Recognit. 35(9), 1847–1867 (2002)
33. Guimarães, S.J.F., Couprie, M., de A Araújo, A., et al.: Video segmentation based on 2D image analysis. Pattern Recognit. Lett. 24(7), 947–957 (2003)
34. Zhao, R., Grosky, W.I.: Video shot detection using color anglogram and latent semantic indexing: From contents to semantics. In: Furht, B., Marques, O. (eds.) Handbook of Video Databases: Design and Applications, pp. 371–392. CRC Press (2003)
35. Albanese, M., Chianese, A., Moscato, V., et al.: A formal model for video shot segmentation and its application via animate vision. Multimed. Tools Appl. 24(3), 253–272 (2004)
36. Mendi, E., Bayrak, C.: Shot boundary detection and key frame extraction using salient region detection and structural similarity. In: 48th ACM Southeast Conf., Massachusetts (2010)

Fuzzy Based Approach to Develop Hybrid Ranking Function for Efficient Information Retrieval

Ashish Saini, Yogesh Gupta, and Ajay K. Saxena

Abstract. Ranking function is used to compute the relevance score of all the documents in document collection against the query in Information Retrieval system. A new fuzzy based approach is proposed and implemented to construct hybrid ranking functions called FHSM1 and FHSM2 in present paper. The performance of proposed approach is evaluated and compared with other widely used ranking functions such as Cosine, Jaccard and Okapi-BM25. The proposed approach performs better than above ranking functions in terms of precision, recall, average precision and average recall. All the experiments are performed on CACM and CISI benchmark data collections.

Keywords: Information retrieval, precision, recall, vector space model, ranking function.

1 Introduction

Information Retrieval (IR) means finding a set of documents that is relevant to the query. The ranking of a set of documents is usually performed according to their relevance scores to the query. The user with information need issues a query to the retrieval system through the query operational module.

Ashish Saini · Yogesh Gupta · Ajay K. Saxena
Electrical Engineering Department, Faculty of Engineering,
Dayalbagh Educational Institute, Uttar Pradesh, India
e-mail: {ashish7119,er.yogeshgupta,aksaxena61}@gmail.com

El-Sayed M. El-Alfy et al. (eds.), *Advances in Intelligent Informatics*,
Advances in Intelligent Systems and Computing 320, DOI: 10.1007/978-3-319-11218-3_42

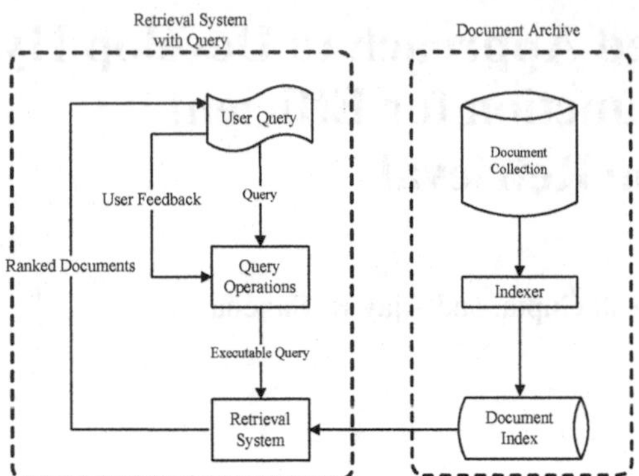

Fig. 1 A general IR system architecture

A general IR system architecture is shown in Fig. 1 [1]. In this figure, IR system is divided into two parts. First is the document archive, a set of textual units (often called document collection), and second is a retrieval system with query. A user of a retrieval system presents queries to describe which kinds of documents are desired. The retrieval system matches the queries against the documents in the textual archive. It then returns the user a list of sub-collection of the documents which are deemed as "best matches".

Traditionally, documents may be available in different forms e.g. full text, hypertext, administrative text, directory, numeric or bibliographic text [2]. It is very difficult to match these documents with queries to extract relevant information (documents). Therefore, IR models are used to represent these documents in an appropriate manner. The IR model gives the fundamental premises and forms the basis for ranking. Although different IR models are available in literature, but Vector Space Model (VSM) [3] is the best-known and most widely used IR model. This model creates a space in which both documents and queries are represented by vectors. A $/v/$- dimensional vector is generated for each document and query for a fixed collection of documents, where $/v|$ is the number of unique terms in the document collection. VSM has some advantages over other IR models such as it is simple model and can handle weighted terms. It also produces a ranked list as output that reduces administrative work.

Therefore, VSM is used as a base model in this paper. IR system computes the relevancy of the documents against the queries with the help of ranking function. Some efforts have made by different researchers to construct an effective ranking function for enhancing the performance of IR system in recent years. Fan et al. [4] proposed that ranking functions could be represented as trees. They also proposed a classical generational scheme for ranking function. Pathak et al. [5] proposed the linear combination of similarity measure and then used GA to optimize the weight of each similarity measure. This approach suffers two drawbacks. First, it is a time consuming process and second, it is not able to capture vagueness and uncertainty of query as well as of documents. Sahami [6] proposed a new approach to measure the similarity between short text snippets by leveraging web search results to provide greater context for the short texts. Faltings [7] also presented a novel approach that allowed similarities to be asymmetric while still using only information contained in the structure of the ontology. Torra et al. [8] used Fuzzy graphs to calculate similarity between words based on dictionaries. Chen [9] presented a new ranking function to handle the similarity problems of generalized fuzzy numbers based on the geometric mean averaging operator. Usharani et al. [10] used GA to calculate similarity of web document based on *cosine* similarity.

In the present paper, a new fuzzy hybrid ranking function called as *FHSM* is proposed and two ranking functions *FHSM1* and *FHSM2* are developed for vector space IR model by combining different ranking functions/similarity measures on the basis of optimal set of weights decided by fuzzy logic. The performance of proposed ranking functions are evaluated and compared with extensively used ranking functions *cosine*, *jaccard* and *okapi-BM25* on the basis of average precision and average recall. Precision-Query curves and Recall-Query curves are also used to compare the performance of these ranking functions.

The rest of present paper is organized as follows: section 2 describes proposed fuzzy based hybrid ranking function. Experiments and results are presented in section 3 and at last conclusion is drawn in section 4.

2 Proposed Fuzzy Based Hybrid Ranking Function

Fuzzy logic is used to determine appropriate weights for maximum retrieval efficiency in developing proposed hybrid ranking function. The *FHSM1* is the weighted sum of the scores returned by *cosine* and *jaccard* ranking functions and *FHSM2* is the weighted sum of the scores returned by *cosine* and *okapi-BM25* as given in (1) and (2) respectively.

$$\text{FHSM1 (D, Q)} = \text{w1* cosine + w2 * jaccard} \tag{1}$$

$$\text{FHSM2 (D, Q)} = \text{w1* cosine + w2 * okapi-BM25} \tag{2}$$

where D is the set of documents, Q is the set of queries and w_1 and w_2 are weights.

Mamdani [11] type FIS is used in proposed approach with the help of Matlab Fuzzy Logic Toolbox [12]. In this FIS, the values of two input variables such as the values of *cosine* and *jaccard/okapi-BM25* coefficients for each document are used to determine values of two output variables, weights w_1 and w_2 respectively. The range of input and output variables are represented by LOW, MEDIUM and HIGH. In this paper, triangular membership function is being used to map input space to a degree of membership of fuzzy set. The details of the membership functions for input and output variables of *FHSM1* and *FHSM2* are shown in Fig. 2.

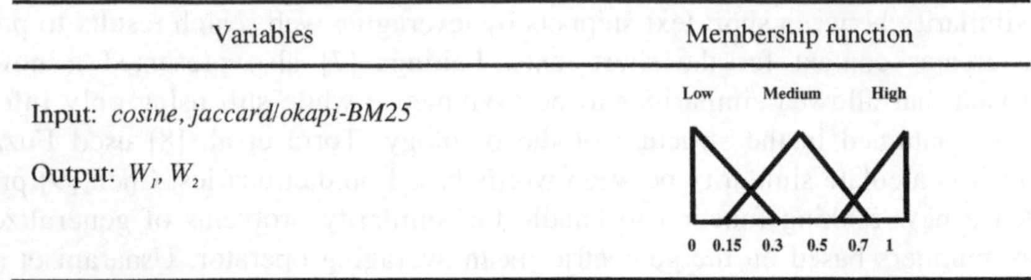

Variables	Membership function
Input: *cosine, jaccard/okapi-BM25* Output: W_1, W_2	Low Medium High 0 0.15 0.3 0.5 0.7 1

Fig. 2 Input and output membership functions

The following fuzzy rule base is constructed to calculate the value of w_1 and w_2.

1. If (*cosine* is Low) and (*jaccard/okapi-BM25* is Low) then (w_1 is Low) (w_2 is Low) (1).
2. If (*cosine* is Low) and (*jaccard/okapi-BM25* is Medium) then (w_1 is Low) (w_2 is Medium) (1).
3. If (*cosine* is Low) and (*jaccard/okapi-BM25* is High) then (w_1 is Low) (w_2 is High) (1).
4. If (*cosine* is Medium) and (*jaccard/okapi-BM25* is Low) then (w_1 is Medium) (w_2 is Low) (1).
5. If (*cosine* is Medium) and (*jaccard/okapi-BM25* is Medium) then (w_1 is Medium) (w_2 is Medium) (1).
6. If (*cosine* is Medium) and (*jaccard/okapi-BM25* is High) then (w_1 is Medium) (w_2 is High) (1).
7. If (*cosine* is High) and (*jaccard/okapi-BM25* is Low) then (w_1 is High) (w_2 is Low) (1).
8. If (*cosine* is High) and (*jaccard/okapi-BM25* is Medium) then (w_1 is High) (w_2 is Medium) (1).
9. If (*cosine* is High) and (*jaccard/okapi-BM25* is High) then (w_1 is High) (w_2 is High) (1).

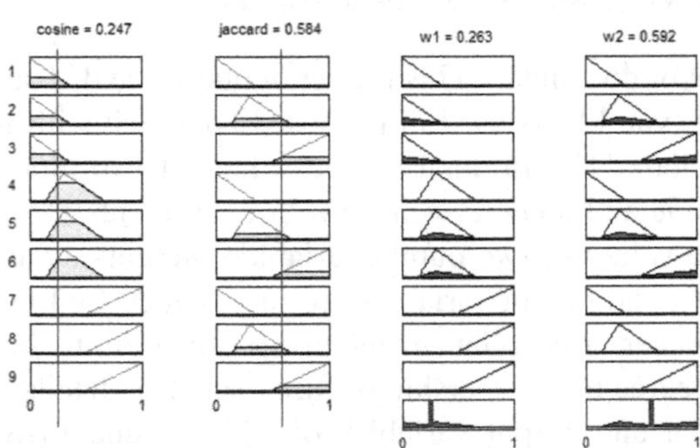

Fig. 3 Fuzzy inference engine for *FHSM1*

In order to interpret the entire fuzzy inference process at once, the fuzzy inference diagram [11] for *FHSM1* and *FHSM2* are presented in Fig. 3 and Fig. 4 respectively. In these figures, the four plots in each row represent the antecedent and consequent of the particular rule. Hence, each rule is a row of plots and each column is a variable. The first two column of plot show the membership functions referenced by the antecedent, or the if-part of each rule. The third and fourth columns of plots show the membership functions referenced by the consequent, or the then-part of each rule. The bottom most plots in third and fourth column represent the aggregate weighted decision values of w_1 and w_2. These figures may be used as a diagnostic for the performance of all nine fuzzy rules. For example, these show that which rules are active or how individual membership function shapes are influencing the results.

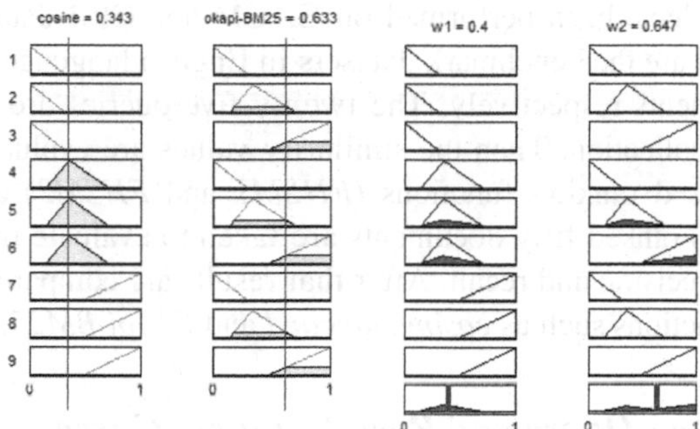

Fig. 4 Fuzzy inference engine for *FHSM2*

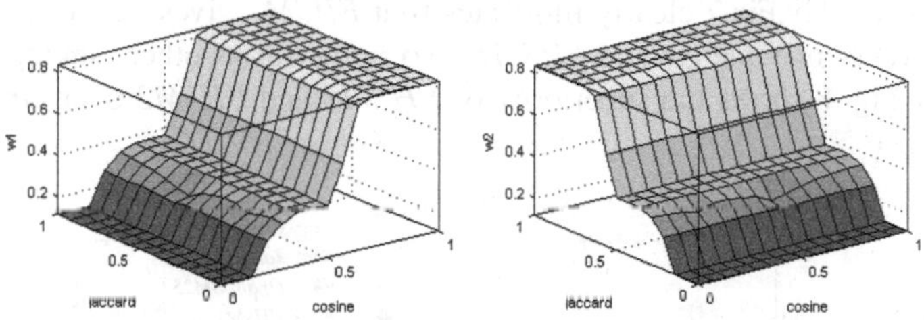

Fig. 5 Rule surface view of input and output variables of *FHSM1*

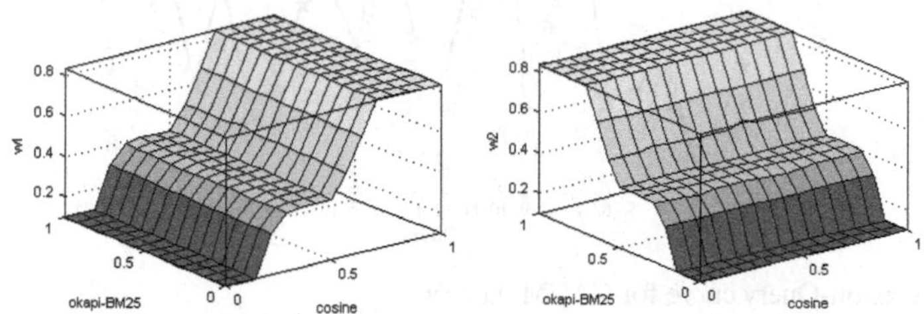

Fig. 6 Rule surface view of input and output variables of *FHSM2*

The fuzzy implication is modeled by Mamdani's minimum operator and the sentence connective is also interpreted as oring the propositions and defined by MAX operator. The AND operator is typically used to combine the membership values for each fired rule to generate the membership values for the fuzzy sets of output variables in the consequent part of the rule.

Further, the three dimensional rule surfaces [11] are plotted in Fig. 5 and Fig. 6 to display the dependencies of any one of two output variables (i.e. w_1 and w_2) on both the inputs variables (i.e. *cosine, jaccard* in case of *FHSM1* and *cosine, okapi-BM25* in case of *FHSM2*).

3 Experiments and Results

All experiments have been performed on CACM and CISI data collections [13]. CACM and CISI are the benchmark datasets in English language and contain 3204 and 1460 documents respectively. The twenty five queries are randomly chosen from each data collection. Then the similarity values are evaluated by using proposed Fuzzy based ranking functions (*FHSM1* and *FHSM2*) and ranking done accordingly. Top ranked fifty documents are taken to evaluate the performance of IR in terms of precision and recall. After that results are compared with the widely used ranking functions such as *cosine, jaccard* and *Okapi-BM25*.

3.1 Precision - Query and Recall - Query Curve

The precision-query curves for CACM and CISI datasets are shown in Fig. 7 and Fig.8 respectively.Fig.7 clearly illustrates that *FHSM2* gives better precision values for twenty one queries and *FHSM1* also outperforms other ranking functions. Similarly Fig. 8 shows the superiority of *FHSM1* and *FHSM2* over other ranking functions.

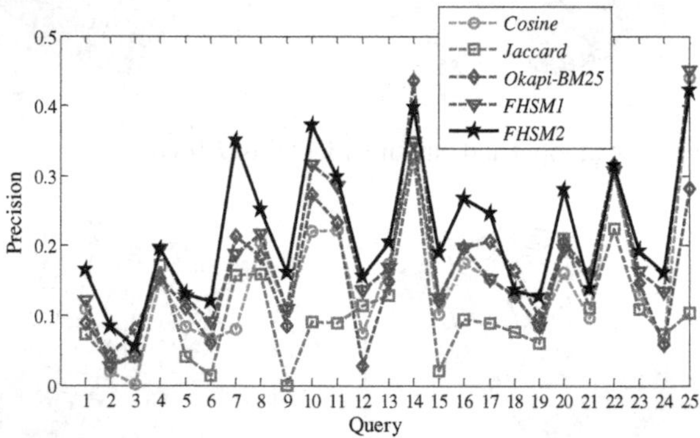

Fig. 7 Precision-Query curve for CACM dataset

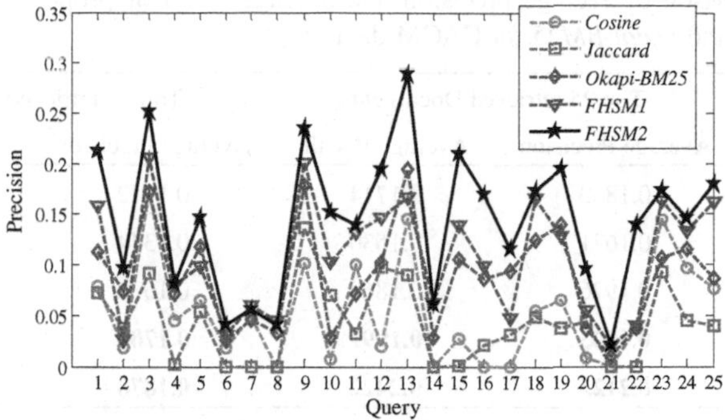

Fig. 8 Precision-Query curve for CISI dataset

The recall-query curves for CACM and CISI datasets are presented in Fig. 9 and Fig. 10 respectively. Fig.9 shows that *FHSM2* gives better recall values for twenty three queries out of twenty five and *FHSM1* also gets better recall values in comparison to other ranking functions. Fig.10 also concludes that *FHSM1* and *FHSM2* are superior to others.

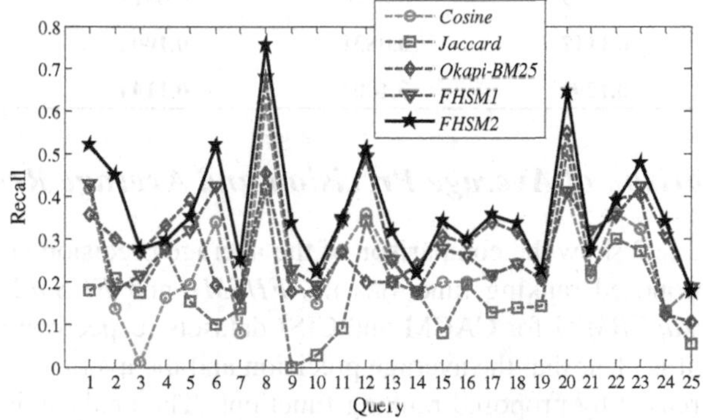

Fig. 9 Recall-Query curve for CACM dataset

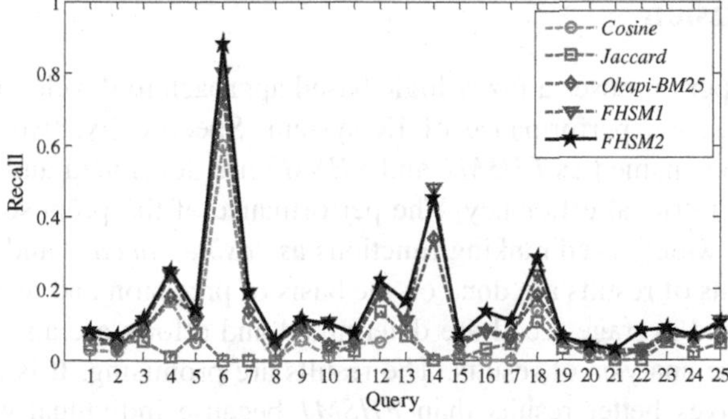

Fig. 10 Recall-Query curve for CISI dataset

Table 1 Comparison of average precision and average recall of proposed approach with *cosine, jaccard* and *okapi-BM25* for CACM dataset

Methods	Top 25 retrieved Documents		Top 50 retrieved Documents	
	Average Precision	Average Recall	Average Precision	Average Recall
Cosine	0.1880	0.1714	0.1472	0.2414
Jaccard	0.1671	0.1530	0.1344	0.2053
Okapi-BM25	0.1976	0.1868	0.1636	0.2818
FHSM1	**0.2023**	**0.1997**	**0.1701**	**0.2832**
FHSM2	**0.2125**	**0.2422**	**0.1878**	**0.3412**

Table 2 Comparison of average precision and average recall of proposed approach with *cosine, jaccard* and *okapi-BM25* for CISI dataset

Methods	Top 25 retrieved Documents		Top 50 retrieved Documents	
	Average Precision	Average Recall	Average Precision	Average Recall
Cosine	0.0546	0.0344	0.0531	0.0702
Jaccard	0.0383	0.0191	0.0416	0.0374
Okapi-BM25	0.1099	0.0718	0.0918	0.1178
FHSM1	**0.1112**	**0.0831**	**0.1006**	**0.1385**
FHSM2	**0.1292**	**0.1004**	**0.1144**	**0.1584**

3.2 Comparison of Average Precision and Average Recall

Table 1 and Table 2 show the comparison of the average precision and the average recall of the proposed ranking functions i.e. *FHSM1* and *FHSM2* with *Cosine, Jaccard* and *Okapi-BM25* for CACM and CISI datasets respectively. From Table 1 and Table 2, it is clear that the average precision and the average recall of the IR system are increased by proposed ranking functions. The analysis is done for top twenty five and top fifty retrieved documents.

4 Conclusion

The present paper proposes a fuzzy logic based approach to develop ranking functions to enhance the performance of IR system. Specifically, two fuzzy hybrid ranking functions named as *FHSM1* and *FHSM2* are developed and implemented to improve the retrieval efficiency. The performance of the proposed approach is compared with widely used ranking functions as *cosine, jaccard* and *okapi-BM25*. The comparisons of results are done on the basis of precision and recall. The average precision and average recall are determined and efforts are also made to perform query wise analysis of results. The results are promising. It is also observed that *FHSM2* gives better results than *FHSM1* because individually *okapi-BM25* (part of *FHSM2*) outperforms *jaccard* (part of *FHSM1*).

References

1. Liu, B.: Web Data Mining: Exploring Hyperlinks, Contents and Usage Data. Springer, Heidelberg (2007)
2. Gupta, Y., Saini, A., Saxena, A.K.: A Review on Important aspects of Information Retrieval. International Journal of Computer, Information Science and Engineering 7(12), 968–976 (2013)
3. Salton, G.: Automatic text processing: the transformation, analysis and retrieval of information by computer. Addison Wesley (1998)
4. Fan, W., Gordon, M., Pathak, P.: Automatic generation of a matching function by genetic programming for effective information retrieval. iN: America's Conference on Information System, Milwaukee, USA (1999)
5. Pathak, P., Gordon, M., Fan, W.: Effective information retrieval using genetic algorithms based matching functions adaption. In: Proceedings of 33rd Hawaii International Conference on Science (HICS), Hawaii, USA (2000)
6. Sahami, M., Heilman, T.: A Web-based Kernel Function for Measuring the Similarity of Short Text Snippets. In: 15th International Conference on World Wide Web, pp. 377–386 (2006)
7. Schickel, Z.V., Faltings, B.: OSS: A Semantic Similarity Function Based on Hierarchical Ontologies. In: : International Joint Conference on Artificial Intelligence, pp. 551–556 (2007)
8. Torra, V., Narukawa, Y.: Word Similarity from dictionaries: Inferring Fuzzy measures and Fuzzy graphs. International Journal of Computational Intelligence Systems 1(1), 19–23 (2008)
9. Chen, S.J.: Fuzzy information retrieval based on a new similarity measure of generalized fuzzy numbers. Intelligent Automation and Soft Computing 17(4), 465–476 (2011)
10. Usharani, J., Iyakutti, K.: A Genetic Algorithm based on Cosine Similarity for Relevant Document Retrieval. International Journal of Engineering Research & Technology 2(2) (2013)
11. Mamdani, E.H., Assilian, S.: An Experiment in Linguistic Synthesis with a Fuzzy logic controller. International Journal of Man–Machine Studies 7, 1–13 (1975)
12. Fuzzy Logic Toolbox User's Guide. The MathWorks Inc. (2004)
13. http://www.dataminingresearch.com/index.php/tag/dataset-2

Knowledge Transfer Model in Collective Intelligence Theory

Saraswathy Shamini Gunasekaran, Salama A. Mostafa, and Mohd. Sharifuddin Ahmad

Abstract. In a multi-agent environment, a series of interaction emerges that determines the flow of actions each agent should execute in order to accomplish its individual goal. Ultimately, each goal realigns to manifest the agents' common goal. In a collective environment, these agents retain only one common goal from the start, which is achieved through a series of communication processes that involve discussions, group reasoning, decision-making and performing actions. Both the reasoning and decision-making phases diffuse knowledge in the form of proven beliefs between these agents. In this paper, we describe the concept of discussions, group reasoning and decision making, followed by its corresponding attributes in proposing a preliminary Collective Intelligence theory.

Keywords: collective intelligence, decision-making, multi-agent systems, collaboration.

1 Introduction

Conceptually, an agent is sharply autonomous at understanding its environment in order to pursue its goal. These intelligent entities are robustly designed such that they are capable of learning and adapting themselves in group tasks as evident in many fields. Military and defense is one such field in which agents are deployed in unmanned aerial vehicles (UAVs) for the search and surveillance missions. As such, many of these multi-agent systems incorporate swarm-based algorithms that were introduced by Bonobeau and Meyer [1]. These algorithms somehow revolve around the idea of negotiation and game theories.

Agents in a multi-agent system negotiate with each other to achieve their individual goals [5][16]. The communication process that occurs between these

Saraswathy Shamini Gunasekaran · Salama A. Mostafa · Mohd. Sharifuddin Ahmad
College of Information Technology, Universiti Tenaga Nasional,
Putrajaya Campus 43000, Selangor Darul Ehsan, Malaysia
e-mail: {sshamini,sharif}@uniten.edu.my, semnah@yahoo.com

El-Sayed M. El-Alfy et al. (eds.), *Advances in Intelligent Informatics*,
Advances in Intelligent Systems and Computing 320, DOI: 10.1007/978-3-319-11218-3_43

agents is fundamental so as to coherently fulfill their group optimal goal. Sufficiency in terms of the communication process is reflected by the efficiency of a task being accomplished. Generally, the efficiency rate is subject to the problem domain.

In this paper, we propose our collective intelligence (CI) theory in which we mimic the actual communication process that occurs between two or more human entities and analyze the possible processes and their attributes in the pursuit of enhancing the efficiency of a collaborative goal attainment strategy. We discuss the processes that act as the driving force on the emergence of collective intelligence that is derived from an active collective discussion which occur amongst two or more human entities. Furthermore, we explain the attributes that are involved in the processes. The motivation of this work is inspired by the need to model such emergence which could be deployed in multi-agent systems for improved performance.

2 Research Objectives

Collective Intelligence (CI) systems have aroused the interest of many researchers due to its (i) adaptivity in uncertain environments, (ii) ability to organize themselves autonomously, and (iii) emergent behavior. Among others, multi-agent, adaptive, swarm intelligence, and self-organizing systems are considered to be CI systems. However, while the growth of research in CI systems continues, it has yet to lead to a systematic approach for model design of these kinds of systems [11]. Understanding the emergence of intelligent collective behaviors in social systems, such as norms and conventions, higher level organizations, collective wisdom and evolution of culture from simple and predictable local interactions has been an important research question since decades [12], [19]. Agent-based modeling of complex social behaviors by simulating social units as agents and modeling their interactions provides a new generative approach to understanding the dynamics of emergence of collective intelligent behaviors.

To achieve the aim of the research, the following objectives are proposed:

1. To investigate and identify the processes and attributes that instantiate the emergent intelligence of a collective interaction between human entities.
2. To organize these processes and attributes into adaptive sequential phases.

These objectives manifest a preliminary theory of the emergence of collective intelligence.

3 CI Mechanism

Embarking upon the theory of Collective Intelligence (CI) that emerges from the intellectual discussions amongst human entities reveals a higher rate of success in goal attainment[13][2][11][22][23]. CI, Swarm Intelligence, Collaborative Intelligence, Group Intelligence as the many terms deployed in the concept, the integrity between all of which often propose the idea of a group goal attainment. This is the basis that supports a team of actors to incorporate their individual diversity of knowledge and experience in the pursuit of a common goal [25][12], resulting in a dynamic process of intelligent group discussion, reasoning and decision-making.

Therefore to begin with, we observed and recorded nine separate meetings. Seven of these meetings were general business meetings and two were building design meetings. Interestingly, the observations lead us to three important discoveries.

1. Knowledge is the focal interaction attribute that is communicated between two or more entities.
2. Knowledge is transformed and cross-fertilized through a series of discussions, reasoning process and decision making.
3. Knowledge is communicated iteratively in order to manifest collective intelligence (e.g. goal attainment).

Figure 1 below describes the knowledge flow that is perceived from the nine meetings. The knowledge diffuses and transforms through four important phases which includes understanding problem, obtaining ideas, reasoning process and implementation and testing. The discoveries of the phases were carefully evaluated in [6].

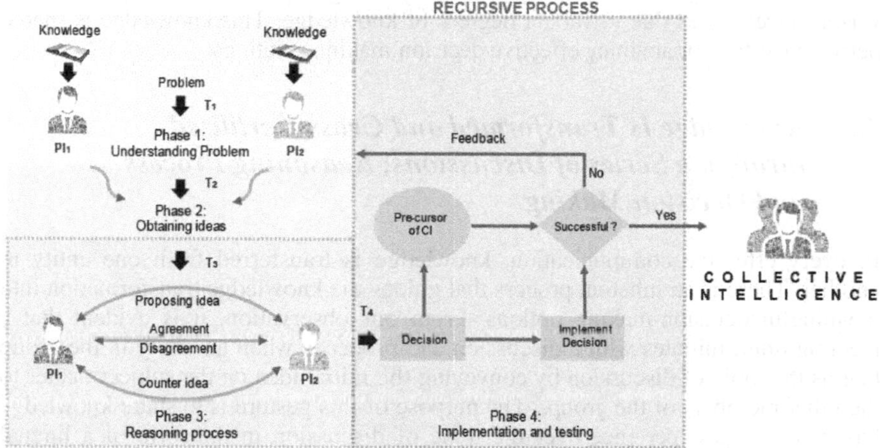

Fig. 1 Phases in the preliminary theory of Collective Intelligence

3.1 Knowledge as the Focal Interaction Attribute That Is Communicated between Two or More Entities

Communication is the key in any collective environment. It allows entities in a group to moderate information, ideas, and feelings as well as to create and share meaning. It is a two-way process of reaching mutual understanding [9][3]. As such, communication can be presented in a lingual and non-lingual form [14]. Lingual communication involves the characterization of words through verbal or written forms. Non-lingual communication involves the action reflected through a certain behavior. The effectiveness of a communication is often influenced by the ability of oneself to understand the motive or goal one has crafted and assimilating that within the environment it resides in [21]. Inevitably, different environments enforce different rules that govern the construction of a goal.

Seemingly, these rules are a reflection of information one has over the domain. This information is knowledge that one has gathered through education or experience in life [10]. For instance, a mother's knowledge on child upbringing may have differed from how it was a century ago or in terms of the various culture, ethnic, region or even countries we live in. A century ago, perhaps we would not have heard of vaccination for measles which at current time is a must in certain countries.

Therefore, knowledge is the root to the establishment of any form of communication [4]. Every entity in a group interaction is equipped with a certain degree of knowledge in order to communicate effective solutions in various fields [8] [17] [20] [24].

In our CI theory, we embark on the principle of representing each entity as having its own personal intelligence (PI). PI represents the mental consciousness one has over one's physical and neurological capability which is knowledge and enabling that to stimulate the social structure in order to achieve its goal. Success due to one's PI is often influenced by how one's knowledge is utilized in achieving a goal. Hence, as our CI preliminary theory dictates, each entity in a group interaction is governed by various levels of PI due to variant degrees of knowledge. This knowledge is shared between entities for attaining effective decision making solutions.

3.2 Knowledge Is Transformed and Cross-Fertilized through a Series of Discussions, Reasoning Process and Decision Making

In every effective communication, knowledge is transferred from one entity to another. There is an inherent process that guides the knowledge transformation into meaningful decision-making options. From our observation, it is evident that a meeting often initiates with a discussion and proceeds when an entity in the group begins the topic of discussion by conveying the initial idea on the subject matter to the other members of the group. The purpose of this gesture is to share knowledge based on one's experience over the topic of discussion in the form of a lingual proposal. However, during the course discussion, each member may argue the validity and effectiveness of the initial proposed idea and other corresponding ideas by elevating a legit reason.

Consequently, the interactions turn into argumentation when other members suggest counter proposals of fresh new ideas. Here, argumentation is the process of diminishing an idea with specific reasons that support its purpose. While some group reasoning may progress smoothly, most of these reasoning processes experience a string of arguments that are ultimately resolved through negotiation. Ultimately, this negotiation process ensures that an agreement is reached for decision-making purpose. This agreement progresses into action performance or re-discussed depending on its potential to solve the problem. A definition table ~~is~~ below summarizes the processes involved in the group interactions.

Table 1 Processes and its definition in knowledge transfer model in CI

Process	Definition
Discussion	The process of conveying ideas between members of a group. • The principal concept of a discussion is to be able to share knowledge based on the experience one has over the topic of discussion, which is in the form of a lingual proposals.
Argumentation	The process of diminishing an idea with specific reasons that support its purpose. • Argue the validity and effectiveness of the proposed idea, by elevating a legit reason.
Negotiation	The process of validating the proposed idea based on certain parameters. • Any argument has to reach a point of agreement or disagreement. In order to achieve either state, the proposed idea and its supporting reasons have to be validated.
Decision Making	Execution of an idea.

This observation indicates the existence of a knowledge transfer model in our CI theory that involves any group interaction. The knowledge transfer model describes the processes that represent knowledge transformation. It is important to note that during the reasoning process, the knowledge is transformed in an iterative manner between the argumentation process and negotiation process. This is because the nature of a group interaction often involves immense idea generation between various parties. These collections of ideas is refined for optimal solution requiring a continuous set of agreements, disagreements, proposals and counter proposals, Once an idea or a combination of ideas has been agreed upon, a decision is made to implement the idea. In this work, if the idea contributes to a successful action-performing outcome. We call the process of manifesting the outcome as collective intelligence (CI), otherwise the decisions of the outcome go through another cycle of discussion, agreement and negotiation.

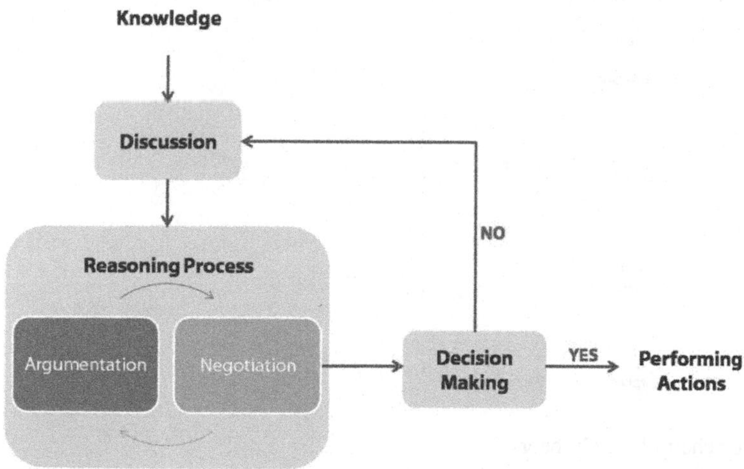

Fig. 2 Knowledge Transfer Model

The knowledge transfer model enables us to harvest a set of rules that yields the mechanism for our CI theory. This rule is vital in constructing the algorithm for the CI model in our research.

First, a problem is initiated for an appropriate construction of a common goal.
Then, each entity retrieves and realigns its existing knowledge concerning the problem.
One entity starts the discussion by sharing its knowledge on that problem.
The shared knowledge is either accepted or rejected by other members of the group.
When the knowledge is accepted, it diffuses into decision-making and actions-performing processes.
If the actions are successful in achieving the common goal, we have achieved CI else we start the discussion process again.
When the knowledge is rejected, it diffuses into a reasoning process.
In the reasoning process, that knowledge is argued until an agreement is reached during negotiation.
When the knowledge is accepted, it diffuses into decision-making and actions-performing processes.
If the actions are successful in achieving the common goal, we have achieved CI else we start the discussion process again.

Hence, from the rules, we construct the flowchart that represents the CI theory.

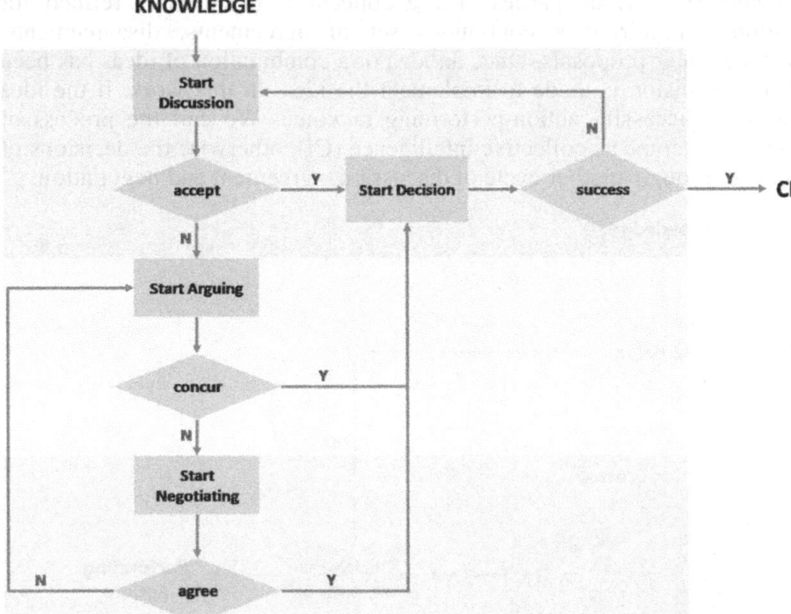

Fig. 3 Flowchart of the CI theory

Consequently, we convert this model into meta-rules that represents the CI preliminary theory.

```
Each entity retrieves its knowledge on the problem
Each entity aligns (assembles all known solutions) all knowledge it has about the problem
start Discussionprocess()
{An entity presents its knowledge
        If all entities share that knowledge it then becomes a common knowledge
        {start Decisionmakingprocesss()
        start Performingactionsprocess()}
        else{
        start Argumentationprocesss()
        if an entity conquers the meeting
        start Decisionmakingprocesss()
        start Performingactionsprocess()
        end()}
        else
        start Negotiationprocess()
        While ( there is no longer argumentation){
        start Decisionmakingprocesss()
        start Performingactionsprocess()}end()}
```

We further investigate the meta-rules as we identify that the processes in the knowledge transfer model only describe the knowledge flow that occurs between the collaborative entities, and there are attributes that represent the knowledge itself within those processes. These attributes are aligned according to the phases in the preliminary CI theory as well as the processes in the knowledge transfer model as shown in the table below.

Table 2 Phases, processes and attributes in CI

Phase	Process	Attributes
1	Discussion	1. Domain Identification 2. Domain Familiarity 3. Formation of a common Goal 4. Task Identification 5. Task Familiarity
2	Discussion	1. Idea Identification
3	Discussion	1. Propose Idea
	Argumentation	1. Disagreement 2. Counter Idea
	Negotiation	1. Idea Organization 2. Agreement
4	Decision Making	1. Idea execution 2. Idea storage

The first phase of problem understanding is crucial as it potentially establishes the inevitability of achieving a common goal. Here, domain identification is done to clarify the scope of the problem. Once this is achieved, domain familiarity is pursued to understand and identify the applicability of the domain with the existing knowledge content that each member of the group has. This continues with the formation of a common goal, which influences a collaborative mindset between the members of a group. Following this is the task identification, in which it is identifying tasks that would solve the problem Finally, in task familiarity, each member creates an association with its existing knowledge and experience. At this stage, each entity reflects solely on its personal intelligence (PI) to associate with the attributes mentioned above.

In the second phase, idea identification ensues. Here, solutions to the problem are postulated. As a norm, the entity that proposes the initial idea possesses higher authority, higher depth of knowledge and higher confidence level. The proposed idea goes through a series of agreement, disagreement and counter idea.

At this point, the third phase follows when reasoning is carried out on the ideas that are presented. Throughout this phase, idea organization is done to organize the task solutions based on the level of efficiency.

The last phase involves idea execution in which the accepted idea is implemented. Finally, all the ideas are stored in the memory of all the entities for future task execution. Another definition table is constructed to enable a clearer understanding on the attributes that represents the knowledge itself in the knowledge transfer model.

Table 3 Attributes definition

Attributes	Definition
Domain Identification	Determining and defining the scope of the problem.
Domain Familiarity	Understanding and identifying the applicability of the domain with the existing knowledge content.
Formation of a common Goal	Defining a universal aim to be achieved.
Task Identification	Defining the nature of the task by structuring the type of task that needs to be executed in order to achieve the common goal.
Task Familiarity	Identifying task similarity with existing knowledge content.
Idea Identification	Defining ideas that are associated with the execution of the task.
Propose Idea	Presentation of ideas.
Disagreement	Rejection of the idea that has been presented.
Counter Idea	Proposition of a new idea.
Idea Organization	Composition of all the ideas that are discussed and is organized based on the level of efficiency it has in accomplishing the task.
Agreement	Acceptance of the idea that has been presented.
Idea execution	Implementation of the accepted idea.
Idea storage	Saving the ideas that have been discussed and the idea that has been agreed upon for future references.

With these attributes, we could further enhance the meta-rules that represent the CI preliminary theory.

```
Each entity retrieves its knowledge on the problem
Each entity aligns( assembles all solution based on its success rate) all knowledge it has about the problem
start Discussionprocess(domainidentification, domainfamiliarity, goalformation, taskidentification, taskfamiliarity, ideaidentification, proposeidea ){
        An entity presents its knowledge
        If all entities share that knowledge it then becomes a common knowledge{
        start Decisionmakingprocesss(ideaexecution, ideastorage)
        start Performingactionsprocess()}else
        {start Argumentationprocesss(disagreement, counteridea)
        if an entity conquers the meeting
        start Decisionmakingprocesss(ideaexecution, ideastorage)
        start Performingactionsprocess()
        end()}
        else
        start Negotiationprocess(ideaorganization, agreement)
        While ( there is no longer argumentation)
        {start Decisionmakingprocesss(ideaexecution, ideastorage)
        start Performingactionsprocess()
        }end()
}
```

3.3 *Knowledge Is Communicated Iteratively to Achieve Collective Intelligence (Goal Attainment)*

In the preliminary theory of CI, the intention is to derive the intersection of knowledge of various entities in the group. As such, each entity is equipped with individual levels of PI. As the discussion, argumentation and negotiation process is an iterative process, knowledge is diffused from one entity to another transformed and cross-fertilized [7]. We term this process as knowledge intersection.

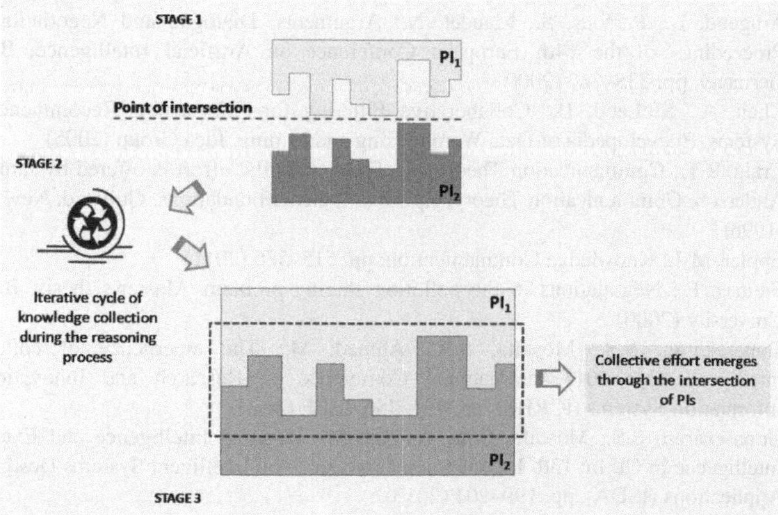

Fig. 4 Intersection of knowledge

The diffused knowledge is deliberated concisely and acts as an added value for current and future task execution. Hence, we propose that CI is the intersection of all the PIs, which can be represented as below:

$$CI = PI_1 \cap PI_2 \cap PI_3 \cap PI_i \cap \dots \dots \dots \cap PI_k$$

4 Conclusion and Further Work

Collective Intelligence in human entities is a valid and profound idea if the emergent intelligence resulting from discussions leads to a successful outcome. A successful outcome is demonstrated by the achievement of a common goal. The process of knowledge interactions and intersection represents the emergence process of collective intelligence in reaching a common goal.

A discussion involves a series of agreements and negotiation that is iterative in nature. This iterative process creates an intersection of knowledge that is diffused, transformed and cross-fertilized into intelligent outcomes. The four phases of understanding problem, obtaining ideas, reasoning and implementation and evaluation structures the flow of knowledge in the communication process.

In our future work we shall explore the argumentation and the negotiation parameters and the association of the BDI model with the CI theory.

Acknowledgments. This project is sponsored by the Malaysian Ministry of Higher Education (MoHE) under the Exploratory Research Grant Scheme (ERGS) No. ERGS/1/ 2012/STG07/UNITEN /02/5.

References

1. Amgoud, L., Parsons, S., Maudet, N.: Arguments, Dialogue and Negotiation. In: Proceedings of the 14th European Conference on Artificial Intelligence, Berlin, Germany, pp. 338–342 (2000)
2. Chen, A., McLeod, D.: Collaborative Filtering for Information Recommendation Systems. Encyclopedia of Data Warehousing and Mining. Idea Group (2005)
3. Craig, R.T.: Communication Theory as a Field. Asimilar effort is offered by James A. Anderson. Communication Theory: Epistemological Foundations. Guilford, New York (1996)
4. Eppler, M.J.: Knowledge Communication, pp. 515–526 (2011)
5. Freitsis, E.: Negotiations in the pollution sharing problem. Master's thesis, BarIlan University (2000)
6. Gunasekaran, S.S., Mostafa, S.A., Ahmad, M.: The emergence of collective intelligence. In: 2013 International Conference on Research and Innovation in Information Systems (ICRIIS), pp. 451–456. IEEE (2013)
7. Gunasekaran, S.S., Mostafa, S.A., Ahmad, M.: Personal Intelligence and Extended Intelligence in CI. In: 13th International Conference on Intelligent Systems Design and Applications (ISDA), pp. 199–204 (2013)

8. Gratton, L., Ghoshal, S.: Improving the quality of conversations. Organizational Dynamics 31, 209–223 (2002)
9. http://www.businessdictionary.com/definition/communication.html#ixzz30zXMJYkl
10. http://wiki.answers.com/Q/What_is_the_difference_between_knowledge_and_intelligence
11. Salminen, J.: Collective Intelligence. Collective Intelligence in Humans: A literature Review, 1Proceedings (2012)
12. Levy, P.: From Social Computing to Reflexive Collective Intelligence: The IEML Research Program. Information Sciences 180, 71–94 (2010)
13. Lévy, P.: Collective Intelligence: Mankind's Emerging World in Cyberspace. Plenum Press, New York (1997)
14. Lunenburg, F.C.: Communication: The process, barriers, and improving effectiveness, pp. 1–11 (2010)
15. Martijn, C.S.: On Model Design for Simulation of Collective Intelligence. Journal of Information Sciences 180, 132–155 (2010)
16. Jennings, N.R., Faratin, P., Lomuscio, A.R., Parsons, S., Sierra, C., Wooldridge, M.: Automated haggling: Building artificial negotiators. In: PacificRim International Conference on Artificial Intelligence, p. 1 (2000)
17. Rosenthal, U., 't Hart: Experts and Decision Makers in Crisis Situation. In: Knowledge: Creation, Diffusion, Utilization, vol. 12(4), pp. 350–372 (June 1991)
18. Singh, V.K., Gautam, D., Singh, R.R., Gupta, A.K.: Agent-Based Computational Modeling of Emergent Collective Intelligence. In: Nguyen, N.T., Kowalczyk, R., Chen, S.-M. (eds.) ICCCI 2009. LNCS (LNAI), vol. 5796, pp. 240 251. Springer, Heidelberg (2009)
19. Singh, V.K., Gupta, A.K.: Agent Based Models of Social Systems and Collective Intelligence. IAMA (2009)
20. Scarbrough, H.: Blackboxes, Hostages and Prisoners. Organization Studies 16(6), 991–1019 (1995)
21. Schank, R.C., Abelson, R.P.: Scripts, plans, goals, and understanding: An inquiry into human knowledge structures. Psychology Press (2013)
22. Malone, T.W.: Climate CoLab, Collective intelligence to address climate change - finalists chosen in world-wide contest. MIT Sloan Experts Blog (November 1, 2011)
23. Malone, T.: Can Collective Intelligence Save the planet. MIT Sloan Management Review (May 05, 2009)
24. Tsoukas, H.: The Firm as a Distributed Knowledge System: A Constructionist Approach. Strategic Management Journal 17, 11–25 (1996)
25. Wolley, et al.: Evidence for a collective intelligence factor in the performance of human groups (October 31, 2010), http://www.sciencmag.org

A Novel Way of Assigning Software Bug Priority Using Supervised Classification on Clustered Bugs Data

Neetu Goyal, Naveen Aggarwal, and Maitreyee Dutta

Abstract. Bug Triaging is an important part of testing process in software development organizations. But it takes up considerable amount of time of the Bug Triager, costing time and resources of the organization. Hence it is worth while to develop an automated system to address this issue. Researchers have addressed various aspects of this by using techniques of data mining, like classification etc. Also there is a study which claims that when classification is done on the data which is previously clustered; it significantly improves its performance. In this work, this approach has been used for the first time in the field of software testing for predicting the *priority* of the software bugs to find if classifier performance improves when it is preceded with clustering. Using this system, clustering was performed on *problem title* attribute of the bugs to group similar bugs together using clustering algorithms. Classification was then applied to the clusters obtained, to assign *priority* to the bugs based on their attributes *severity* or *component* using classification algorithms. It was then studied that which combination of clustering and classification algorithms used provided the best results.

1 Introduction

In large software development organizations, Bug Triaging is an important activity and an indispensable part of Software Testing. It helps in software bug management and taking critical decisions related to the fixing of software bugs. It involves things like finding out which bugs need immediate attention and which are not that important to be fixed, assigning appropriate developer to the bug who

Neetu Goyal · Maitreyee Dutta
National Institute of Technical Teachers Training & Research, Chandigarh, India

Naveen Aggarwal
Panjab University, Chandigarh, India

El-Sayed M. El-Alfy et al. (eds.), *Advances in Intelligent Informatics*,
Advances in Intelligent Systems and Computing 320, DOI: 10.1007/978-3-319-11218-3_44

could fix it etc. Although it being a significant part of software development, it is quite time consuming and tedious taking up considerable amount of time and resources. Hence there arises a need to develop an automated system for Bug Triaging so as to save developer's time and effort, ultimately leading to saving of the resources of the organization itself. This problem has been addressed by researchers [2, 14, 16, 20] by automating its various aspects like identification of duplicate bugs, assigning a bug to an appropriate developer who could fix it, classifying the *priority* and *severity* of the bugs, grouping similar bug reports etc. For achieving these tasks, they have used several methods, one of them being Data Mining techniques like classification. Although using these techniques have provided good results, but still there is a scope for improvement. This is because, in software development, one cannot afford to provide incorrect priority to the bugs. Assigning correct priority to the bugs plays a significant role in fixing the critical bugs on time. In order to achieve this, a new method has been proposed in this work. As found by few researchers [5, 26], when clustering is used as a pre-processing step in certain areas before other machine learning algorithms were applied on the data, such as classification, it can increase the performance of these machine learning algorithms. The same approach has been suggested in this work for automated classification of *priority* of the software bugs. It is being proposed to first cluster the software bugs based on their similarity instead of directly applying classification algorithms on them. Also it is proposed to classify the bugs based on *severity* or *component* attibute of the bugs to find which attribute out of these can provide better results. For this effect, various combinations of clustering and classification algorithms have been used in this work to study the effect of clustering on the performance of classifiers for bug priority classification.

2 Related Research Work

This section discusses that how various data mining techniques have been used by researchers in the area of Software Bug Triaging. For predicting suitable Developers, some researchers [1, 2, 3, 7, 11] have used classification techniques. While Čubranić and Murphy [7] used Bayesian learning approach, Anvik et al. [3] used SVM for classification. Ahsan et al. [1] used classifiers Decision Tree, Naïve Bayes, RBF Network, Random Forest, REPTree, Support Vector Machine and Tree-J48. In addition to this, Anvik et al. [2] developed a recommender for assigning appropriate component to a bug. For predicting the Severity of a bug, Lamkanfi et al. [15] have used Naïve Bayes classifier. Lamkanfi et al. [16] extended their work further by using Naïve Bayes Multinomial, K-Nearest Neighbour and Support Vector Machines classifiers as well. Chaturvedi and Singh [6] applyied several other machine learning algorithms in addition to these. To automate the bug priority classification, Yu at el. [28] used Artificial Neural Network (ANN). Kanwal and Maqbool [14] applied Naïve Bayes and SVM classifiers and found SVM to be better. Sharma et al. [25] worked across projects for bug priority prediction. For grouping related bug reports using classification technique, Nigam et al. [20] used Multi-Class Semi Supervised SVM with different Kernel functions. Hanchate et al. [12] applied Decision Tree for this whereas Neelofar et al. [19] had applied Multinomial Naïve Bayes algorithm. Whereas others [9, 17, 18, 22, 24] have resorted to clustering techniques for this purpose. For detecting duplicate bug reports, Runeson et al. [23] took the help of natural language information and Wang et al. [27] used natural language as well as execution information of a bug. In [8, 13], researchers had used clustring for detecting duplicate bug reports. Banerjee et al. [4] had used string matching to find longest common subsequence between bug reports for duplicate bug detection. Gegick et al. [10] had addressed an issue where Security Bug Reports (SBRs) are mislabelled as Non Security Bug Reports (NSBRs).

In [5, 26], researchers had explored that when clustering is used as a pre-processing step in certain areas before other machine learning algorithms are applied on the data, such as classification algorithms, it can increase the performance of these machine learning algorithms. This approach has been used in this work to improve the performance of classification algorithms.

3 Approach Used for Priority Classification

In this section, the approach used for automated priority classification of software bugs as part of software bug triaging has been presented. This system addresses the following two areas related to software bug triaging:

1. It clusters similar bugs based on their *problem title* using clustering technique.
2. In each cluster, *priority* is assigned to each bug using classification technique.

The data mining techniques used for automating the priority assignment of bugs are as follows:

1. **Information Retrieval** - In this work, the bug reports were clustered on the basis of similarity of their *problem title* fields. To compute this similarity using appropriate similarity measures, the bug reports need to be represented as feature vectors. Since all the words in the *problem title* text field are not significant enough to represent a bug report; therefore this text was pre-processed using text retrieval methods of Information Retrieval technique to determine the ones which were representative of this text. These retrieval techniques are as follows:
   a. Tokenization - In this step, those keywords were identified which could represent the *problem title* field. It involves converting the stream of characters of the given field into stream of words or tokens. In this process, capitals, punctuations, symbols such as brackets, hyphens and other non-alphabetic constructs etc are removed from the text.

b. Stop Word Removal - In this step, common words, like a, and, the, for, with etc that do not carry any specific information are removed.
c. Stemming - In this step, all the words left after performing step a and b are converted to their root forms. For example, words like connected, connecting, connection, and connections were converted to their root form connect.
d. Vector space model - In the end, Vector space model was used to represent the words obtained from the previous steps, as feature vector for each bug. Each word of this vector has a weight associated to it, which indicates the degree of importance of the particular attribute for the characterization of the Problem Title text for a particular bug.

2. **Clustering** - In this work, XMeans [21], Simple KMeans and Expectation Maximization clustering algorithms were applied on the data obtained from the previous steps. These algorithms were inputted with the pre-processed bug attributes from the previous step. Then using clustering process, similar bug reports were then grouped together in four different clusters.
3. **Classification** - Classification was applied in two ways. In one case, it was applied to each of the clusters obtained from the clustering algorithms mentioned in the previous section. In the second case, classification was directly applied to the data obtained from the Information Retrieval step mentioned in previous section. The classification was done to predict the priority of the bugs. It was done on the basis of the *severity* or *component* attributes of the bugs. The classification algorithms used in this work were Bayes Net, Random Forest and SMO [29].

4 Dataset and Resources Used

The dataset to be used for the proposed thesis work is the bug data of one of the software products of a software organization. This bug data is a collection of bug reports entered in the bug repository for the said product. These bug reports contain different fields or attributes providing the details of the bugs. Out of these fields, the ones which have been used in this work are: *defect id, problem title, severity, priority* and *component*. The dataset of software bugs for this work comprised of 36543 number of bug records. The clustering was applied on the whole dataset using the Classes to cluster mode of Weka. For classification, 66% of the dataset i.e. 24118 number of bug records were used for training the classifiers. Rest of the 34 % bug records i.e. 12425 number of bug records were used as test set for predicting the priority of the bugs. This dataset consisted of software bug reports for priority 0, 1, 2 and 3. Out of the total bug records, 1715 bug records were of priority 0, 5221 were of priority 1, 17648 were for priority 2 and 11959 records were of priority 3.

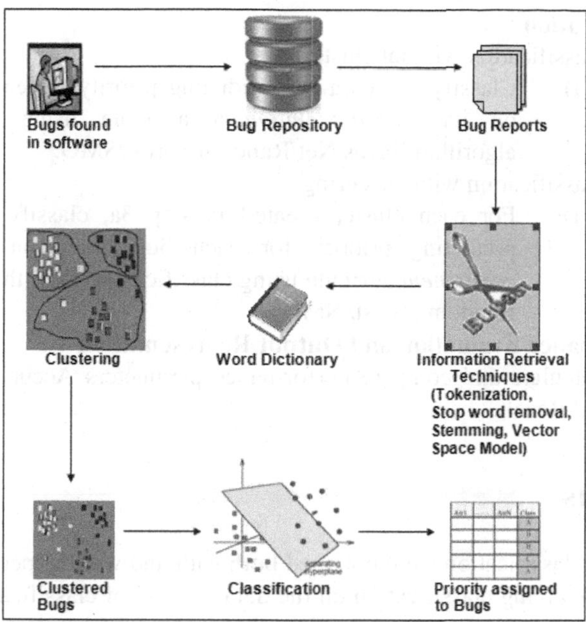

Fig. 1 Proposed system for priority classification of clustered software bugs

5 Algorithm

The automated system developed in this work has mainly 5 steps. An algorithm for this system has been presented as follows using those steps:

Algorithm for automated priority classification of clustered software bugs
Returns: Clustered and Priority classified instances of Bug Reports
Arguments: No. of clusters, Training data percentage, Test data percentage

1. Data Extraction

a. Extract data from the Bug database with only required bug attributes i.e, *defect id, problem title*, *category, component* and *priority*, ignoring other bug attributes in Bug Reports;

2. Feature Selection

a. Perform Tokenization on *problem title* attribute;
b. Remove Stop Words from terms obtained from previous step;
c. Apply Stemming on terms left from previous step;
d. Create a Feature Vector from the terms obtained from previous step using Vector Space Model.

3. Clustering

a. Group similar bugs together under 4 different clusters using clustering algorithm XMeans/Simple KMeans/EM.

4. **Classification**
   a. Classification without clustering
      i) Classify test data for predicting priority for each bug based on its *severity/component* attribute using classification algorithm Bayes Net/Random Forest/SMO.
   b. Classification with clustering
      i) For each cluster created in step 3a, classify test data by predicting priority for each bug based on its *severity/ component* attribute using classification algorithm Bayes Net/ Random Forest/ SMO.
5. **Performance Evaluation and Output Representation**
   a. Calculate and compare performance parameters Accuracy, Precision and Recall.

6 Results

The results of classification were obtained both with and without performing clustering before applying classification on the data. In case of classification of previously clustered data, the bug data was first clustered on the basis of *problem title*. Each cluster thus obtained was then classified separately based on the *severity* or *component* attribute of the bugs. The results comprised of details such as 'total number of instances' in each cluster, 'number of instances took as test data' out of the total number of instances, 'percentage of correctly classified instances' etc. Also the Precision and Recall of each cluster was computed. Finally the mean of all the clusters was calculated for the values of 'percentage of correctly classified instances', Accuracy, Precision and Recall.

The graphs for *severity* based classification with and without clustering have been presented in Figure 2 for all the combinations of clustering and classification algorithms used. Similarly graphs for *component* based classification for all such combinations have been presented in Figure 3. It can be noted from the graphs that in case of *severity* based classification, the results obtained for all the classifiers i.e. SMO, Bayes Net and Random Forest were the same with each other with or without clustering. Whereas in case of *component* based classification, these values slightly varied with each other.

From the graphs, it can be observed that, of all the clustering algorithms when applied before classification for both *severity* and *component* based classifications, the maximum performance improvement was obtained with XMeans. Simple KMeans proved to be the second best. While with Expectation Maximization, the performance improvement was the least. But it is to be noted that the overall increase in the performance was always there when classification was preceded by any of the above mentioned clustering algorithms.

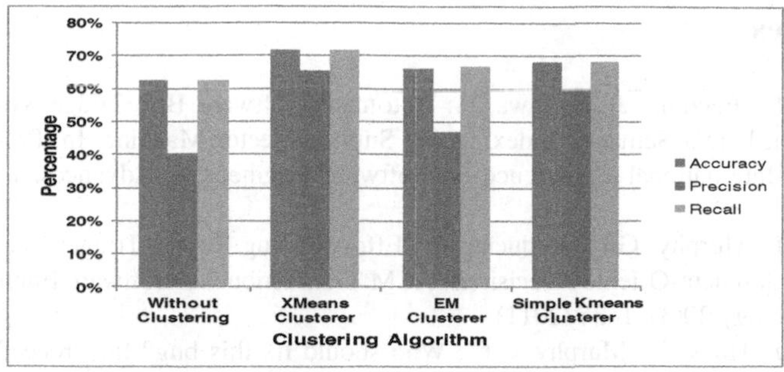

Fig. 2 Bar graph for Impact of Clustering on Accuracy, Precision and Recall of Bug Priority Classification based on Severity

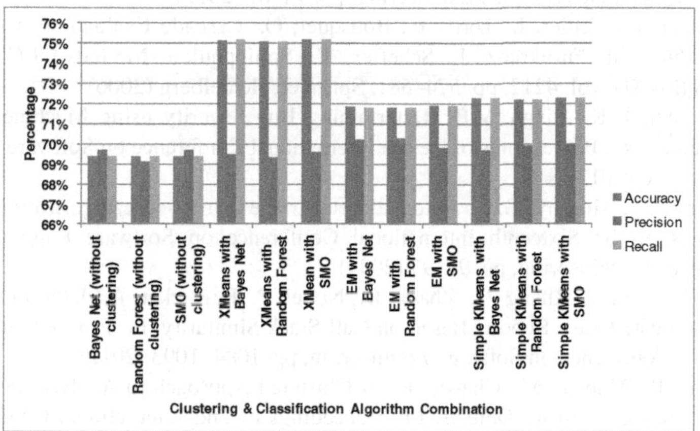

Fig. 3 Bar graph for Impact of Clustering on Accuracy, Precision and Recall of Bug Priority Classification based on Component

7 Conclusion

With XMeans, an overall performance improvement of 9.24% was achieved. It was followed by Simple KMeans with which an overall improvement of 5.67% was obtained. With Expectation Maximization, the improvement was 3.49%. Overall, the best results were obtained with XMeans clusterer, when used with the Bayes Net classifier. Hence it can be concluded that clustering improved the performance of the classifiers, when applied on the software bug data under consideration. The approach used in this work can be applied using other types of clustering and classification algorithms also in addition to the ones used in this work with other types of tools. Software bug data from other sources like Open data sources and data form other software organizations can be used to do the experiments.

References

1. Ahsan, S., Ferzund, J., Wotawa, F.: Automatic Software Bug Triage System (BTS) Based on Latent Semantic Indexing and Support Vector Machine. In: Proceedings of the 4th International Conference on Software Engineering Advances, pp. 216–221 (2009)
2. Anvik, J., Murphy, G.C.: Reducing the Effort of Bug Report Triage: Recommenders for Development-Oriented Decisions. ACM Transactions on Software Engineering and Methodology 20(3), 1–35 (2011)
3. Anvik, J., Hiew, L., Murphy, G.C.: Who should fix this bug? In: Proceedings of the 28th International Conference on Software Engineering, pp. 361–370 (2006)
4. Banerjee, S., Cukic, B., Adjeroh, D.: Automated Duplicate Bug Report Classification using Subsequence Matching. In: Proceedings of IEEE 14th International Symposium on High-Assurance Systems Engineering, pp. 74–81 (2012)
5. Candillier, L., Tellier, I., Torre, F., Bousquet, O.: Cascade Evaluation of Clustering Algorithms. In: Fürnkranz, J., Scheffer, T., Spiliopoulou, M. (eds.) ECML 2006. LNCS (LNAI), vol. 4212, pp. 574–581. Springer, Heidelberg (2006)
6. Chaturvedi, K.K., Singh, V.B.: Determining Bug Severity using Machine Learning Techniques. In: Proceedings of the 6th International Conference on Software Engineering, pp. 1–6 (2012)
7. Čubranić, D., Murphy, G.C.: Automatic bug triage using text categorization. In: Proceedings of the Sixteenth International Conference on Software Engineering and Knowledge Engineering, pp. 92–97 (2004)
8. Dang, Y., Wu, R., Zhang, H., Zhang, D., Nobel, P.: ReBucket: A Method for Clustering Duplicate Crash Reports Based on Call Stack Similarity. In: Proceedings of International Conference on Software Engineering, pp. 1084–1093 (2012)
9. Dhiman, P., Manish, M., Chawla, R.: A Clustered Approach to Analyze the Software Quality using Software Defects. In: Proceedings of 2nd International Conference on Advanced Computing & Communication Technologies, pp. 36–40 (2012)
10. Gegick, M., Rotella, P., Xie, T.: Identifying Security Bug Reports via Text Mining: An Industrial Case Study. In: Proceedings of 7th IEEE Working Conference on Mining Software Repositories, pp. 11–20 (2010)
11. Goyal, N., Aggarwal, N., Dutta, M.: A Novel Way of Assigning Software Bug Priority using Classification and Clustering. In: Proceedings of International conference on Computer Networks and Information Technology, pp. 535–547 (2014)
12. Hanchate, D.B., Sayyad, S., Shinde, S.A.: Defect classification as problem classification for Quality control in the software project management by DTL. In: Proceedings of the 2nd International Conference on Computer Engineering and Technology, pp. 623–627 (2010)
13. Jalbert, N., Weimer, W.: Automated Duplicate Detection for Bug Tracking Systems. In: International Conference on Dependable Systems & Networks, pp. 52–61 (2008)
14. Kanwal, J., Maqbool, O.: Bug Prioritization to Facilitate Bug Report Triage. Journal of Computer Science and Technology 27(2), 397–412 (2012)
15. Lamkanfi, A., Demeyer, S., Giger, E., Goethals, B.: Predicting the Severity of a Reported Bug. In: Proceedings of 7th IEEE Working Conference on Mining Software Repositories, pp. 1–10 (2010)

16. Lamkanfi, A., Demeyer, S., Soetens, Q.D., Verdonck, T.: Comparing Mining Algorithms for Predicting the Severity of a Reported Bug. In: Proceedings of the 15th European Conference on Software Maintenance and Reengineering, pp. 249–258 (2011)
17. Nagwani, N.K., Verma, S.: Software Bug Classification using Suffix Tree Clustering (STC) Algorithm. International Journal of Computer Science and Technology 2(1), 36–41 (2011)
18. Nagwani, N.K., Verma, S.: CLUBAS: An Algorithm and Java Based Tool for Software Bug Classification Using Bug Attributes Similarities. Journal of Software Engineering and Applications 5(6), 436–447 (2012)
19. Neelofar, M., Javed, Y., Mohsin, H.: An Automated Approach for Software Bug Classification. In: Proceedings of the 6th International Conference on Complex, Intelligent, and Software Intensive Systems, pp. 414–419 (2012)
20. Nigam, A., Nigam, B., Bhaisare, C., Arya, N.: Classifying the Bugs Using Multi-Class Semi Supervised Support Vector Machine. In: Proceedings of the International Conference on Pattern Recognition, Informatics and Medical Engineering, pp. 393–397 (2012)
21. Pelleg, D., Moore, A.W.: X-means: Extending K-means with Efficient Estimation of the Number of Clusters. In: Proceedings of the 17th International Conference on Machine Learning, pp. 727–734 (2000)
22. Podgurski, A., Leon, D., Francis, P., Sun, J., Wang, B., Masri, W., Minch, M.: Automated support for classifying software failure reports. In: Proceedings of the 25th International Conference on Software Engineering, pp. 465–476 (2003)
23. Runeson, P., Alexandersson, M., Nyholm, O.: Detection of Duplicate Defect Reports Using Natural Language Processing. In: Proceedings of 29th International Conference on Software Engineering, pp. 499–510 (2007)
24. Rus, V., Nan, X., Shiva, S., Chen, Y.: Clustering of Defect Reports Using Graph Partitioning Algorithms. In: Proceedings of the 21st International Conference on Software Engineering and Knowledge Engineering, pp. 442–445 (2009)
25. Sharma, M., Bedi, P., Chaturvedi, K.K., Singh, V.B.: Predicting the Priority of a Reported Bug using Machine Learning Techniques and Cross Project Validation. In: Proceedings of the 12th International Conference on Intelligent Systems Design and Application, pp. 539–546 (2012)
26. Vilalta, R., Achari, M.K., Eick, C.F.: Class Decomposition via Clustering: A New Framework for Low-Variance Classifiers. In: Proceedings of the 3rd International Conference on Data Mining, pp. 673–676 (2003)
27. Wang, X., Zhang, L., Xie, T., Anvik, J., Sun, J.: An approach to detecting duplicate bug reports using natural language and execution information. In: Proceedings of ACMIEEE 30th International Conference on Software Engineering, pp. 461–470 (2008)
28. Yu, L., Tsai, W.-T., Zhao, W., Wu, F.: Predicting Defect Priority Based on Neural Networks. In: Cao, L., Zhong, J., Feng, Y. (eds.) ADMA 2010, Part II. LNCS, vol. 6441, pp. 356–367. Springer, Heidelberg (2010)
29. Platt, J.C., Schlökopf, B., Burges, C., Smola, A.: Fast training of support vector machines using sequential minimal optimization. In: Advances in Kernel Methods—Support Vector Learning. MIT Press, Cambridge (1999)

Multi Objective Cuckoo Search Optimization for Fast Moving Inventory Items

Achin Srivastav and Sunil Agrawal

Abstract. The paper focuses on managing most important (class A) fast moving inventory items. A multi objective cuckoo search optimization is used to determine trade off solutions for continuous review order point, order quantity stochastic inventory system. A numerical problem is considered to illustrate the results. The results show a number of pareto optimal points are obtained using cuckoo search multi objective algorithm on a single run, which provides the practitioners flexibility to choose the optimal point.

1 Introduction

Inventory management is essential for survival and growth of any industry. Each distinct item of an inventory is known as stock keeping unit (SKU).Inventory items are classified in different classes as A (most important), B (moderate important) and C (least important). The class A constitutes 15 to 20 percent of stocks and account for 80 percent or more of total value. The B class constitutes 30 to 40 percent of stocks and account for 15 percent of total value. The C class constitutes 40 percent of stocks and account for 5 percent of total value. The above classification shows A class items are highly important, so we restrict our focus to them. The class A items can further be sub-classified on basis of lead time demand as fast moving (average demand during lead time 10 or more than 10), slow moving (average demand during lead time less than10) and very slow moving items (order quantity equal to one, case of expensive items). Here, we are focusing on A class fast moving inventory items. Silver et al. [8] has discussed fast moving A class inventory items and given an equation of expected total relevant cost for it. Silver et al. [8] has used simultaneous

Achin Srivastav · Sunil Agrawal
PDPM IIITDM Jabalpur, India

El-Sayed M. El-Alfy et al. (eds.), *Advances in Intelligent Informatics*,
Advances in Intelligent Systems and Computing 320, DOI: 10.1007/978-3-319-11218-3_45

approach to determine both lot size and safety stock factor to determine total relevant cost. The shortcoming with this approach is that there is single objective function is considered and only single optimal solution is obtained. In the inventory optimization, the cost and service level are conflicting objectives and optimizing a single objective can give unacceptable results with respect to other objectives. Therefore, it is essential to use multi objective optimization to obtain closer to realistic solutions of the inventory problems. There are number of evolutionary algorithm available for multi objective optimization . The most common are Genetic algorithm, Particle Swarm Optimization and Cuckoo search optimization. In Inventory optimization the researchers have used Ant colony optimization, GA and PSO. We have chosen Cuckoo search as it is more efficient and easy to implement than GA and PSO [11]. The use of Levi Flight provides more diversification than any other nature inspired algorithm. Secondly , number of parameter to be tuned are much less than any other nature inspired algorithm.

2 Literature Review

The literature on multi objective optimization of inventory system shows that work of Bookbinder and Chen [2], Agrell [1], Puerto and Ferandez [7], Mandal et al. [3] have used conventional preference based multi objective procedures [4]. Bookbinder and Chen [2] analyzed multi-echelon inventory systems using multi criterion decision making approach. Agrell [1] determined lot size and safety stock factor using interactive multi criteria framework. Puerto and Ferandez [7] obtained pareto optimal sets for stochastic inventory problem through advanced mathematical derivatives. Mandal et al. [3] developed optimal solution for fuzzy inventory model using geometric programming method.

Recent works on multi objective optimization on inventory problems are of Tsou [9], [10], Pasandideh et al. [6] and Park and Kyung [5]. Tsou [9] used multi objective particle swarm optimization (MOPSO) and technique for order preference in similarity to ideal solutions (TOPSIS) to determine optimal solutions for an inventory system. Tsou [10] used a improved particle swarm optimization algorithm to determine pareto front for continuous review inventory system. Pasandideh et al. [6] developed a bi-objective economic production quantity model for defective items using non dominated sorting genetic algorithm and multi objective particle swarm optimization. Park and Kyung [5] developed multi objective inventory model for obtaining optimal cost and fill rate using multi objective particle swarm optimization. The above literature shows that none of the paper has considered cuckoo search in solving multi objective inventory problem. Some of the prominent works on cuckoo search in other problems are [11], [12] and [13].

3 Notations

The notations are given below:

ETRC (k,Q) = expected total relevant cost
A = fixed set up cost per order
D = demand per year in units / year
σ_L= standard deviation of demand during lead time
k= safety factor
p(u≥ k) =stockout risk
s= order point
Q = order quantity
B_1= backorder cost in $/ stockout
r = inventory carrying charge, in $/$/year
v= unit variable cost in $/unit
C (k,Q) = expected cost in $
N(k,Q) = expected number of stockout

4 Multi Objective Cuckoo Search Approach for Stochastic (s, Q) Inventory System

Silver et al. [8] has given the following equation for Fast moving inventory items. It is combination of ordering cost, holding cost and stockout cost.

$$ETRC(k,Q) = \frac{AD}{Q} + \left(\frac{Q}{2} + k\sigma_L\right)vr + \frac{DB_1}{Q}\,p(u \geq k) \tag{1}$$

To obtain best solution both in terms of cost and service level, we formulate the two objective equations separately as below:

Minimize

$$C(k,Q) = \frac{AD}{Q} + \left(\frac{Q}{2} + k\sigma_L\right)vr \tag{2}$$

Minimize

$$N(k,Q) = \frac{D}{Q}\,p(u \geq k) \tag{3}$$

Subject to constraints:

$$0 \leq Q \leq D \tag{4}$$

$$0 \leq k \leq 4 \tag{5}$$

The equation (2) represents total cost and equation (3) calculates the frequency of stockout. Both are minimization functions. The equation (4) and (5) are constraint equation for order quantity and safety factor.

As discussed in introduction about benefits of cuckoo search of diversity and ease of implementation over PSO, GA and other mete heuristic evolutionary algorithms, we have used cuckoo search to solve the above inventory model.

Cuckoo search is population based stochastic search algorithm. It is developed by Yang and Deb [11]. It is based on life of cuckoo and exhibits levy flight distribution. It is widely used in non linear problems and quite useful for multi objective optimization problems. Cuckoos are known for their melodious voice and aggressive reproduction. In this work we have used weighted sum based cuckoo search optimization. Here we combine all objectives into a single objective by assigning weights to them where weights are generated randomly from a uniform distribution. We have varied the weights for improving diversity. We have used the following rules for cuckoo search [11]:

1. Each cuckoo lays one egg at a time and put it in randomly chosen nest.
2. The best nests with high quality egg is used to produce next generation.
3. The egg laid by a cuckoo is discovered by the host bird with a probability $p_a \in (0,1)$.In this case the host bird either throw the egg or simply abandon the nest so the fraction p_a of n host nests are replaced by new nests.

Based on the above mentioned assumption the basic steps of the Cuckoo search can be summarized in the flowchart mentioned in Fig. 1.In this process the global random walk is carried out by using levy flight distribution and new solution x^{t+1} for cuckoo *i* is generated as mentioned below:

$$x_i^{t+1} = x_i^t + \alpha L(s, \lambda) \tag{6}$$

The Levy flight essentially provides a random walk whose random step length (α) is drawn from a Levy distribution, which has an infinite variance with an infinite mean [11].

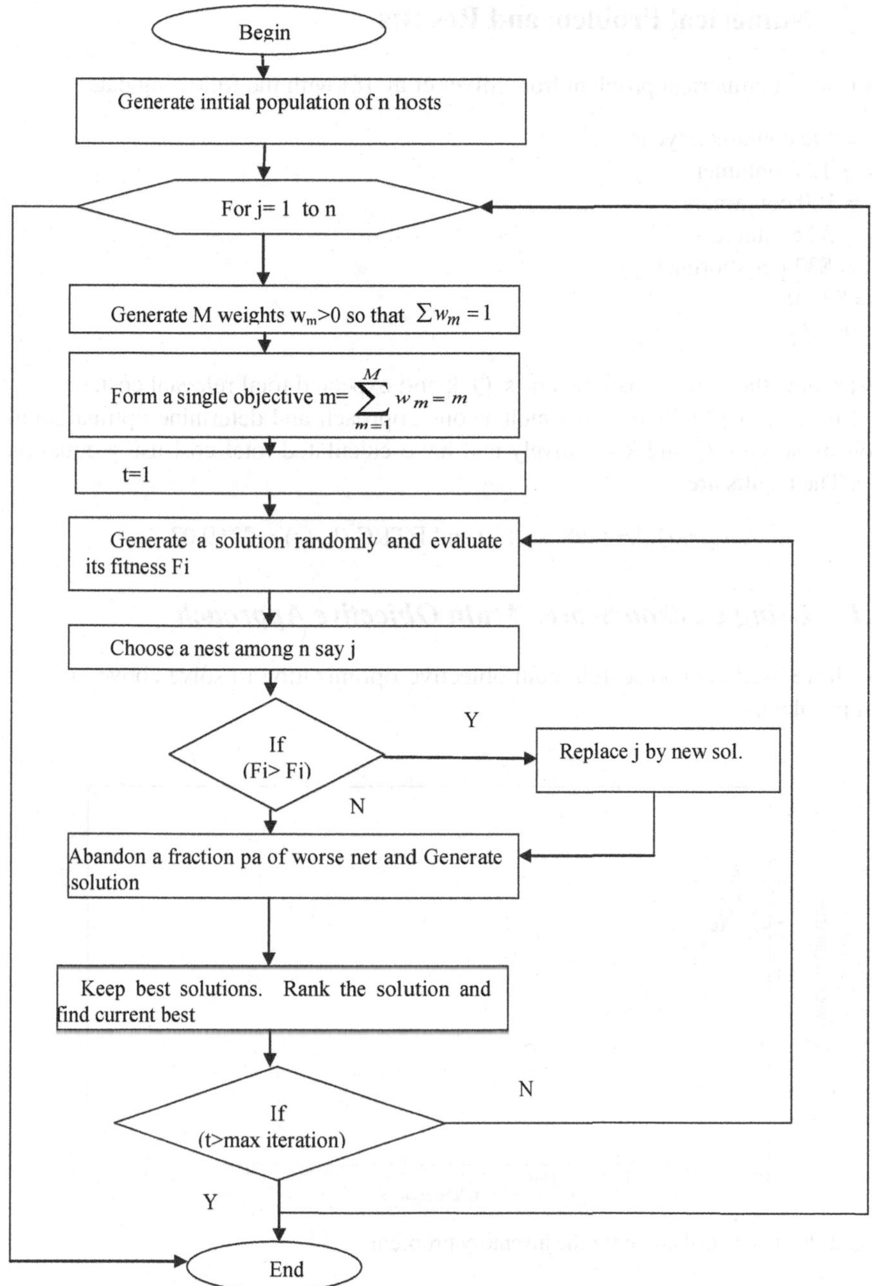

Fig. 1 Flowchart of Multi objective cuckoo search optimization

5 Numerical Problem and Results

Consider a numerical problem from Silver et al. [8] with the following data:

D = 700 containers/year
v= \$ 12 /container
x_L = 100 containers
σ_L = 30 containers
B_1 = \$32 per shortage
A= \$3.20
r= 0.24/ year

Determine the optimal parameters, s, Q, k and expected total relevant cost.

Silver et al. [8] have used simultaneous approach and determine optimal solution by solving Q and k iteratively and have calculated total cost using equation (1). The results are:

Q = 64, k= 0.98, s= 130 and ETRC (k, Q) = \$269.07.

5.1 *Using Cuckoo Search Multi Objective Approach*

We have used cuckoo search multi objective optimization to solve above numerical problem.

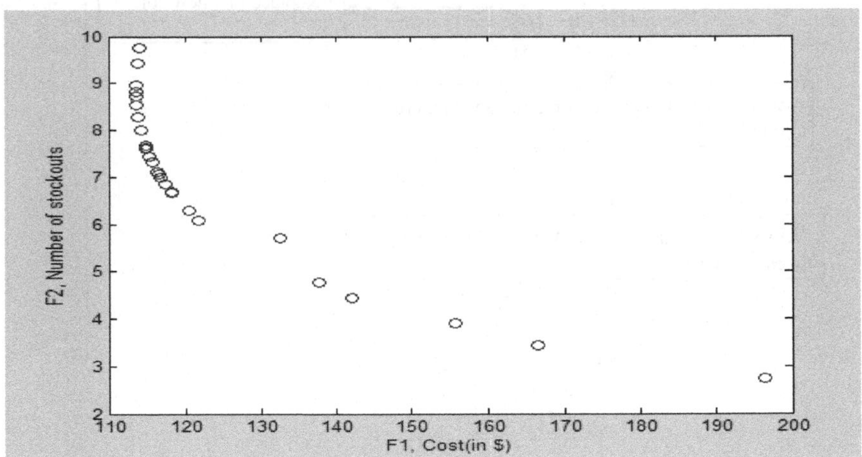

Fig. 2 Pareto optimal curve for the inventory problem

The Fig. 2 represents the pareto optimal curve which is obtained between f1 and f2.This shows with multi objective optimization we have obtained multiple optimal solution in a single run.

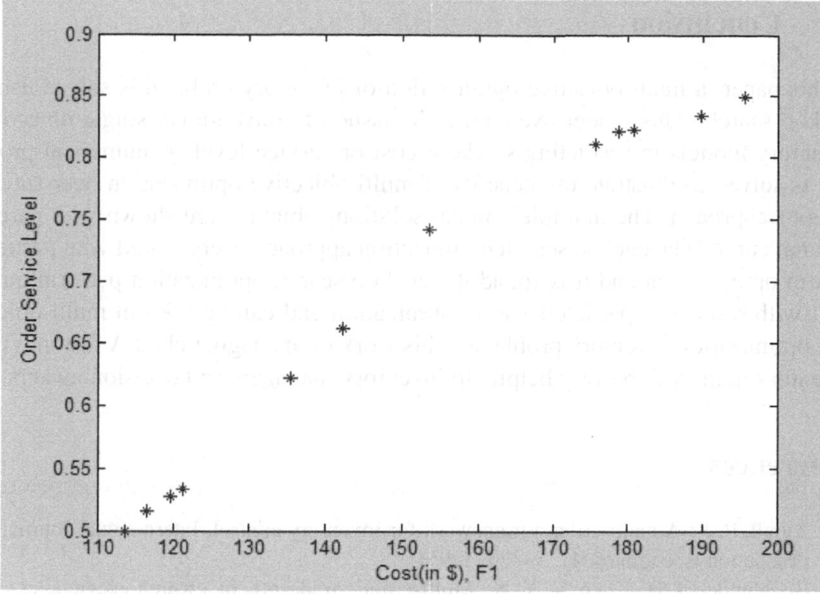

Fig. 3 Trade point curve between Cost and Order service level

The Fig. 3 represents the pareto curve which is obtained between cost and order service level. This curve will provide practitioners the flexibility to choose the desired service level and respective cost.

The assessment of performance of the multi objective cuckoo search algorithm is done on basis of value of metrics such as error ratio, spacing and spread. We have compared the results of multi objective cuckoo search algorithm with multi objective particle swarm algorithm on basis of the above three mentioned aspects. We have performed five runs of experiment to record the average value of the metrics.

Table 1 Comparison of performance of multi objective algorithms of cuckoo search and particle swarm optimization

Metrics	**Cuckoo Search**	**Particle Swarm Optimization**
Error ratio	0.1466	0.2920
Spacing	0.0233	0.0163
Max. Spread	0.4992	0.4013

The Table 1 shows that performance of multi objective cuckoo search algorithm is better than particle swarm optimization in error ratio and maximum spread. The small value of error ratio in cuckoo search algorithm suggests that number of non-dominated solutions which are not members of pareto front is quite low. The high value of spread shows that non dominated solutions are well diversified. The spacing value in cuckoo search algorithm is slightly more than particle swarm optimization. The good value of spacing in particle swarm optimization gets irrelevant as it has high error ratio.

6 Conclusion

In this paper, a multi objective optimization of inventory problem is solved using cuckoo search. This model overcomes the issue of conventional single objective inventory models in predicting stockout cost or service level. A numerical problem is solved to illustrate the benefits of multi objective optimization over simultaneous approach. The multiple optimal solutions obtained are shown with pareto optimal curve. The cuckoo search optimization approach is compared with particle swarm optimization and it is found the cuckoo search optimization perform quite well with respect to particle swarm optimization and can be used in multi objective optimization inventory problems. This work on managing class A fast moving inventory items will be very helpful to inventory managers and decision makers.

References

1. Agrell, P. J.: A multi-criteria framework for inventory control. International Journal of Production Economics, 41, 59–70 (1995)
2. Bookbinder, J. H., Chen, V. Y. X.: Multi-criteria trade-offs in a warehouse/retailer system. Journal of Operational Research Society, 43(7), 707–720 (1992)
3. Mandal, N. K., Roy, T. K., Maiti, M.: Multi-objective fuzzy inventory model with three constraints: A geometric programming approach. Fuzzy Sets and Systems, 150, 87–106 (2005)
4. Moslemi, H., Zandieh, A.: Comparisons of some improving strategies on MOPSO for multi objective (r,Q) inventory system 38(10), 12051–12057 (2011)
5. Park, K.J., Kyung, G.: Optimization of total inventory cost and order fill rate in a supply chain using PSO. International Journal of Advanced Manufacturing Technology, vol 70, 1533-1541 (2014)
6. Pasandideh, S.H.R., Niaki, S.T.A., Sharafzadeh, S.: Optimizing a bi-objective multi-product EPQ model with defective items, rework and limited orders: NSGA-II and MOPSO algorithms. Journal of Manufacturing Systems, 32, 764-770 (2013)
7. Puerto, J., Fernandez, F. R.: Pareto-optimality in classical inventory problems. Naval Research Logistics, 45, 83–98 (1998)
8. Silver, E. A., Pyke, D. F., Peterson, R.: Inventory Management and Production Planning and Scheduling. New York, John Wiley and Sons (1998)
9. Tsou, C. S.: Multi-objective inventory planning using MOPSO and TOPSIS. Expert Systems with Applications, 35, 136–142 (2008)
10. Tsou, C. S.: Evolutionary Pareto optimizers for continuous review stochastic Inventory systems. European Journal of Operational Research, 195, 364–371 (2009)
11. Yang, X. S., Deb, S.: Cuckoo search via L´evy flights Proceedings of World Congress on Nature and Biologically Inspired Computing (NaBIC 2009,India), IEEE Publications, USA, pp. 210-214 (2009)
12. Yang, X. S., Deb, S.: Engineering optimization by cuckoo search, International Journal of Mathematical Modelling and Numerical Optimization 1(4), pp. 330-343 (2010)
13. Yang, X. S., Deb, S.: Multi objective cuckoo search for design optimization, Computers and Operations Research 40 pp. 1616-1624 (2013)

An Empirical Study of Some Particle Swarm Optimizer Variants for Community Detection

Anupam Biswas, Pawan Gupta, Mradul Modi, and Bhaskar Biswas

Abstract. Swarm based intelligent algorithms are widely used in applications of almost all domains of science and engineering. Ease and flexibility of these algorithms to fit in any application has attracted even more domains in recent years. Social computing being one such domain tries to incorporate these approaches for community detection in particular. We have proposed a method to use Particle Swarm Optimization (PSO) techniques to detect communities in social network based on common interest of individual in the network. We have performed rigorous study of four PSO variants with our approach on real data sets. We found orthogonal learning approach results quality solutions but takes reasonable computation time on all the data sets for detecting communities. Cognitive avoidance approach shows average quality solutions but interestingly takes very less computation time in contrast to orthogonal learning approach. Linear time varying approach performs poorly on both cases, while linearly varying weight along with acceleration coefficients is competitive to cognitive avoidance approach.

Keywords: particle swarm optimization, community detection, social network analysis, k-means algorithm.

1 Introduction

Swarm intelligent (SI) techniques are stochastic search processes. Since last decade SI techniques have got widespread acceptance from the implementers of various fields. Extensively used SI techniques in various applications which include Particle Swarm Optimization (PSO) [2], Ant Colony Optimization

Anupam Biswas · Pawan Gupta · Mradul Modi · Bhaskar Biswas
Indian Institute of Technology (BHU), Varanasi, India

El-Sayed M. El-Alfy et al. (eds.), *Advances in Intelligent Informatics*,
Advances in Intelligent Systems and Computing 320, DOI: 10.1007/978-3-319-11218-3_46

(ACO) [1], Artificial Bee Colony (ABC) [7], Gravitational Search Algorithm (GSA) [8] and Intelligent Water Drop (IWD) [9]. All these approaches are initiated with a set of solutions called population, then in successive steps each candidate of the set learns collectively from other candidates and adapts itself in accordance to the solution space. Strategy incorporated and learning mechanism of these techniques mimic the natural facts and phenomenons. As far as PSO is concerned, it mimics the collective social behavior of birds flocking and fish schooling. Each candidate solution in the population is referred to as particle and population is referred to as swarm. Particles in the swarm share their knowledge with the neighbors. Comparatively better particle attracts other particles towards it. Besides, each particle's cognition about its own better state also has influence on its movement towards other better neighbors. Overall decision to move any particle is made through both particle's own experience as well as neighbor's experiences.

Let us consider that $x_i(t)$ denotes the position vector of particle i and $v_i(t)$ denotes the velocity vector of a particle in the search space at time step t, where t denotes discrete time steps. Let $x_{pb}(t)$ be the position vector of particle i's personal best position so far and $x_{gb}(t)$ be the position vector of global best particle so far. Then the velocity of particle i at time step $t+1$ is obtained as:

$$v_i(t+1) = v_i(t) + c_1 r_1 [x_{pb}(t) - x_i(t)] + c_2 r_2 [x_{gb}(t) - x_i(t)] \tag{1}$$

Here r_1 and r_2 are vectors of uniform random values in range $(0, 1)$. Parameters c_1 and c_2 are the cognitive acceleration coefficient and social acceleration coefficient respectively. Now, the position of particle i can be updated as given below:

$$x_i(t+1) = x_i(t) + v_i(t+1) \tag{2}$$

At each time step $t+1$ velocity and position of each particle i are updated from previous velocity and position at time step t using Equation 1 and Equation 2. This updating process continues until it reaches the maximum time step limit or acquires a solution with expected minimum error.

Numerous improvements to the basic PSO have been proposed in literature. Shi and Eberhart [3] introduced the concept of weighted velocity referred to as inertia weight (ω). With inertia weight, Equation 2 is modified as follows:

$$v_i(t+1) = \omega v_i(t) + c_1 r_1 [x_{pb}(t) - x_i(t)] + c_2 r_2 [x_{gb}(t) - x_i(t)] \tag{3}$$

Parameters of PSO algorithm such as inertia weight, cognitive acceleration coefficient (c_1) and social acceleration coefficient (c_2) have great impact on its performance. Proper and fine tuning of the parameters may result in faster convergence of the algorithm. In [4] time varying weight was introduced to balance exploration and exploitation of PSO. Ratnaweera et. al [5] proposed hierarchical self organization approach to PSO. Hybrid versions of PSO with other intelligent techniques are proposed in [6], [10].

In recent years Zhan et. al [11] proposed orthogonal learning method to PSO. Li et. al [12] proposed different learning strategy for each particle in the swarm. Pehlivanoglu [13] has proposed neural network based periodic mutation strategy into PSO. Biswas et. al [14] proposed cognitive avoidance approach to overcome misleading of particles.

In this paper we will study PSO as an application to Social Network Analysis (SNA). A social network is a network of people interacting with each other or related to each other via some inter-dependencies. Such social networks can be represented in the form of graphs and link between two nodes of the graph represents relation between them. An interesting problem in social networks is community detection, division of the network into groups which have strong intra-cluster dependencies and weak inter-cluster dependencies. Every aspect of SNA has link to identification of communities in the network. Hence detection of communities in the SNA has great importance.

Besides the graphical representation of a social network, it can also be viewed as a collection of feature vectors in which each feature is chosen specific to a target community. Such a representation of nodes in a social network provides the flexibility to use n-dimensional Euclidean geometry (where n is the cardinality of feature-set) for the similarity analysis of nodes. We have utilized PSO to exploit the vector representation of the network to accomplish the task of community detection. Our approach in detail can be found in section 3.

Rest of the paper is organized as follows: section 2 elaborates different variants that are considered for our experiment, section 3 describes our approach with social interpretation, section 4 consist of detailed analysis along with experimental setups and finally concluded in section 5.

2 Considered Variants

Over the decades several modifications had been proposed to the original PSO. We have selected these variants covering changes to PSO since its introduction to recent years on the basis of their popularity. We have considered four of them for this application centric study those are briefed below:

2.1 PSO-TVIW

Shi and Eberhart[4] introduced the concept of inertia weight to the original version of PSO, in order to balance the local and global search during the optimization process. Here, inertia weight is linearly varied weight over the generations.

$$\omega = \omega_{min} + (\omega_{max} - \omega_{min})\frac{(I_{max} - I_{current})}{I_{max}} \tag{4}$$

ω_{min} and ω_{max} are the minimum value and maximum value of weight respectively. I_{max} is the maximum number of iterations and $I_{current}$ is the present iteration number. Shi and Eberhart showed $\omega_{min} = 0.4$ and $\omega_{max} = 0.9$ gave better performance.

2.2 PSO-TVAC

Ratnaweera et.al [5] used very similar concepts as in PSO-TVIW. With extension to particle swarm optimization with time varying inertial weight they introduced linearly varying acceleration coefficients c_1 and c_2 along with ω. This additional linearity to PSO-TVIW becomes more effective to global search during beginning of the algorithm and to local search at the end.

$$c_1 = c_{1min} + (c_{1max} - c_{1min})(I_{max} - I_{current})/I_{max} \tag{5}$$

$$c_2 = c_{2max} + (c_{2min} - c_{2max})(I_{max} - I_{current})/I_{max} \tag{6}$$

$$\omega = \omega_{min} + (\omega_{max} - \omega_{min})(I_{max} - I_{current})/I_{max} \tag{7}$$

Here, c_{1min} is the minimum value of c_1, c_{1max} is the maximum value of c_1, c_{2min} is the minimum value of c_2 and c_{2max} is the maximum value of c_2. Ratanweera et.al [5] showed the algorithm performs better when c_1 varies from 2.5 to 0.5, c_2 from 0.5 to 2.5 and ω from 0.9 to 0.4.

2.3 PSOCA

The movement of the particle in the solution space is guided by the global and local best solution. During this course of movements certain unnecessary moves may also be encountered. In this version of PSO [14], a cognitive avoidance term has also been introduced to improve the efficiency of the algorithm, each particle maintains its worst value that it has attained so far along with the particle best and global best values to restraint particles move towards worst value.

2.4 OLPSO

One of the demerits of aforementioned variants of PSO is the failure to overcome the particle oscillations that often occur when it is trapped between the local and global best solutions. This reduces efficiency of the search due to delayed convergence. To remedy such issues, Zhan et. al [11] constructed a guidance vector P_o out of the global best and local best position vectors. It performs factor analysis for each particle to determine impact of each factor in the global best particle and local best position vectors. This has improved the speed of particle's movement towards the optimal solution.

3 Methodology

3.1 Social Interpretation

Our social life is full of choices. Everyone has different interests, likings and dis likings. For instance, a person may like Cricket and at the same he may dislike Hockey, or he may like to listen Hindi music over English and so on. Hence people can be attributed to various things based on their interests. For this particular example we have four attributes Cricket, Hockey, Hindi music and English music and we have two possibilities for each attribute, either it is liked or disliked by an individual. In real life situations there may be several instances of this kind. Suppose we represent liking or disliking with numerical values 1 and 0, the information is transformed into a boolean table where each row or instance represents a person's list of likings and dis likings and such instances can be represented with multi-labeled graph as shown in Fig. 1. By segregating the data into groups, we mean to identify network of people or communities having the most overlapping likes or dislikes. Such groups may consist mostly of people liking hindi music and cricket or a group which is comprised of hockey and cricket liking people and so on.

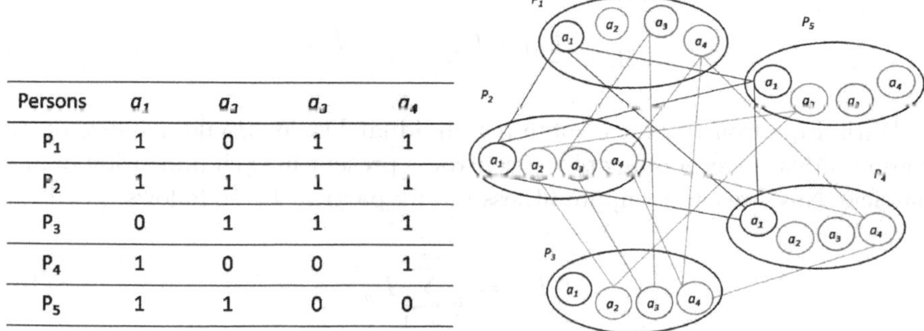

Persons	a_1	a_2	a_3	a_4
P_1	1	0	1	1
P_2	1	1	1	1
P_3	0	1	1	1
P_4	1	0	0	1
P_5	1	1	0	0

Fig. 1 Multi-label graph representation of instances with attributes. A person is connected with respect to an attribute to other if both have entries 1 in the table.

3.2 Utilization of PSO

To extrapolate PSO in our case, we have encoded particles in terms of instances. Suppose we have n instances containing m attributes and we want to group them into k communities. To do so we have used very similar approach to the k-means[15] algorithm in background of our approach. For measuring similarity between two instances considered Euclidean distances. Each centroid of k-means is a potential instance. We represent k dimensional particles with k centroids of k-means, but unlike k-means centroids are updated iteratively. Each particle in the population represents individual grouping of k communities. To determine fitness of a particle we have considered Davies-Bouldin index [16]. This index measures quality of a single community. As our particle is representative of k communities, to obtain fitness of a particle averaged Davies-Bouldin index of all communities.

Let $P_x = \{A_1, A_2, ..., A_k\}$ be a particle which represents a grouping of k communities, where $A_i = \{a_{i1}, a_{i2}, ..., a_{im}\}$ is any instance vector of all attribute values which has been chosen as the centroid for the ith community. Let $A_i = \{a_{i1}, a_{i2}, ..., a_{im}\}$ and $A_j = \{a_{j1}, a_{j2}, ..., a_{jm}\}$ be the ith and j^{th} instances respectively. Euclidean distance between these two instance can be computed as:

$$E_{ij} = \sqrt[2]{(a_{i1} - a_{j1})^2 + (a_{i2} - a_{j2})^2 + ..., (a_{im} - a_{jm})^2} \tag{8}$$

Let the community corresponding to the yth dimension of any particle P_x be $C_y = \{A'_1, A'_2, ..., A'_t\}$, the centroid of which is represented in the particle as A_y. This community is comprised of t instances. Davies-Bouldin index of this community can be obtained as follows:

$$B_y = \frac{E_{y1} + E_{y2} + ... + E_{yt}}{t} \tag{9}$$

With Equation 9 we can obtain individual Davies-Bouldin index of all communities corresponding to the centroids present in each dimension of any particle. Now we can compute fitness of any particle P_x as follows:

$$f(P_x) = \frac{1}{k} \sum_{y=1}^{k} B_y \tag{10}$$

Detail algorithm to implement our approach with PSO is explained in our previous work [17]. In our implementations of PSO variants, we specified an upper limit to the number of generations in order to terminate the execution and chosen the particle with best fitness from the final population for comparison.

4 Result Analysis

To perform correlative analysis of the four PSO variants with their implication to community detection, we carried out experiments on real world data sets. Comparative analysis of our approach has been done using variants of PSO to detect communities on three data sets. Two of them i.e. Blogger1 and Blogger2 are related to blogger networks and third one is standard Iris data. Though Iris data set is used mainly for feature selection, here we have considered it because it is related to feature and also very popular.

Blogger's data set 1 was collected from the UCI Machine Learning Database (http://archive.ics.uci.edu/ml/datasets/BLOGGER) [18]. The data set consists of a table of 100 bloggers with 6 attributes namely, Education, Political Caprice, Topics, Local media turnover and Local, political and social space. Iris data set [20], which contained 150 instances and 4 attributes, was also collected from the same source.

Blogger's data set 2 was collected from the Arizona State University [19]. The data file was processed and a csv file was made in which corresponding to an entry of a bloggers we had the tags associated with him. Some of these tags were "analysis / opinion", "apple", "hardware", "mods", "mac mini", "gaming", "software", "odds and ends" etc. This file was further processed to form a matrix in which the columns were the tags from the data set, and the rows represented the bloggers. Entries of the first 100 bloggers from this data set were used for further analysis in our experiments.

4.1 Experimental Setup

Each of the variants of PSO were executed with all three data sets. Parameter values that had been recommended by proposer of respective PSO variant is considered for our experiments, those values are also mentioned above.

4.2 Analysis

The results obtained with our approach from all four variants of PSO are presented in Fig. 2. In Blogger's data set 1 we found best results were generated for 2 communities. We found that the first community consisted predominantly entries of bloggers with a higher degree of education, few bloggers with medium level of education, and very few with low level of education. Most of the bloggers in the first cluster had entries for local media turnover (LMT) and local, political and social space (LPSS) as "YES". Bloggers in this cluster predominantly blogged about political topics. Second community predominantly had entries of blogger with medium level of education and few with low level of education and fewer with high level of education. Majority of bloggers in the second cluster blogged about news and there were fewer entries as compared to the first cluster where entries for LMT and LPSS was

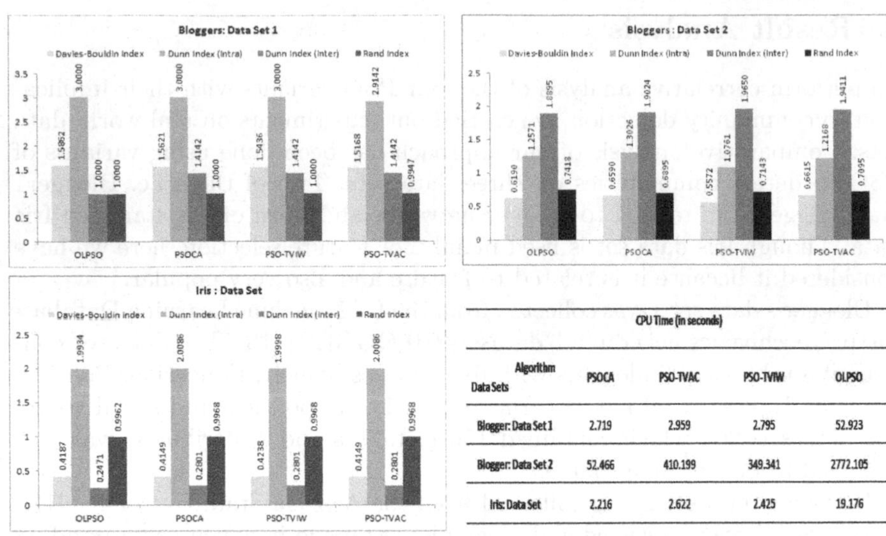

CPU Time (in seconds)

Algorithm / Data Sets	PSOCA	PSO-TVAC	PSO-TVIW	OLPSO
Blogger: Data Set 1	2.719	2.959	2.795	52.923
Blogger: Data Set 2	52.466	410.199	349.341	2772.105
Iris: Data Set	2.216	2.622	2.425	19.176

Fig. 2 Results of our approach to community detection on data sets

"YES". The data set was efficiently clustered into 2 sets viz. professional bloggers and seasonal (temporary) bloggers. With all variants of PSO we found Dunn index(intra) 3 and rand index 1. OLPSO has least Dunn index(inter) also obtained highest Davies-Bouldin index.

Blogger's data set 2 was grouped into 15 communities some of which were singleton sets, these singleton sets were justified because some bloggers in the data set blogged about topics different from all others. Characteristics of other clusters are as follows: The second cluster mostly consisted of bloggers tagged with "hardware" and "mods", the third cluster consisted of bloggers with common tags as "odds and ends", "internet tools", "productivity" and "software", fourth cluster had bloggers who had their blogs mostly tagged with "macworld" ,fifth cluster consisted of bloggers mostly tagged with "analysis opinion" and "apple" related words , sixth cluster had bloggers blogging about "operating systems" and "software" tags , ninth cluster had bloggers blogging about "ipod family", and "itunes", thirteenth cluster had bloggers tagged with "ipod family" and fourteenth cluster had bloggers tagged with "software". Again in this data set also OLPSO results in least Dunn index(inter) and highest rand index, but PSOCA results in highest Dunn index(intra). Besides detecting communities in a social network, we also tested our approach on the standard Iris data set. We found PSOCA and PSO-TVAC result in better rand index and Dunn index(intra), whereas OLPSO again results in least Dunn index(inter).

Overall PSO-TVIW outperforms all other variants in terms of isolation of communities formed. PSOCA and PSO-TVAC are competitive with respect to density within the communities with PSO-TVAC having a slightly better overall performance. OLPSO could not perform well on other quality measures but it yielded the results with highest rand index. If we look at the computation time of all variants, it seems OLPSO takes the highest time whereas, PSOCA takes the least time.

5 Conclusion

We have proposed an approach to detect communities in the social network using PSO and very similar approach to k-means algorithm in background. We have studied our approach on four variants of PSO. In this empirical study, we have evaluated these variants with three real world data sets. We have noticed PSO-TVAC and PSOCA shows almost similar results, but took more time. This time extra consumption is due to the slow convergence to optimal value. OLPSO outperforms in terms of distances between communities, but it takes higher computation time. Conclusion to our study is that if we are looking for more isolated communities OLPSO is better option, whereas if we are looking dense communities and also have constraints on computation time then PSOCA will be better option for detecting communities.

References

1. Dorigo, M.: Optimization, Learning and Natural Algorithms. PhD thesis, Politecnico di Milano, Italy (1992)
2. Kennedy, J., Eberhart, R.: Particle Swarm Optimization. In: Proceedings of IEEE Int. Conf. on Neural Networks IV, pp. 1942–1948 (1995)
3. Shi, Y., Eberhart, R.C.: A modified particle swarm optimizer. In: Proceedings of IEEE Int. Conf. on Evolutionary Computation, pp. 69–73 (1998)
4. Shi, Y., Eberhart, R.C.: Empirical Study of Particle Swarm Optimization. In: IEEE Int. Congr. Evolutionary Computation, pp. 101–106 (1999)
5. Ratnaweera, A., Halgamuge, S., Watson, H.C.: Self-organizing hierarchical particle swarm optimizer with time-varying acceleration coefficients. IEEE Trans. on Evolutionary Comp. 8(3), 240–255 (2004)
6. Poli, R., Chio, C.D., Langdon, W.B.: Exploring extended particle swarms: A genetic programming approach. In: Conference on Genetic Evolutionary Computation, pp. 33–57 (2005)
7. Karaboga, D.: An Idea Based On Honey Bee Swarm for Numerical Optimization. Technical Report-TR06, Erciyes University, Engineering Faculty, CED (2005)
8. Rashedi, E., Nezamabadi-pour, H., Saryazdi, S.: GSA: a gravitational search algorithm. Info. Science 179(13), 2232–2248 (2009)
9. Shah-Hosseini, H.: The intelligent water drops algorithm: a nature-inspired swarm-based optimization algorithm. International Journal of Bio-inspired Computation (IJBIC) 1(2), 71–79 (2009)

10. Epitropakis, M.G., Plagianakos, V.P., Vrahatis, M.N.: Evolving cognitive and social experience in Particle Swarm Optimization through Differential Evolution. In: IEEE Congr. on Evolutionary Comp (CEC), pp. 1–8 (2010)
11. Zhan, Z., Zhang, J., Li, Y., Shi, Y.: Orthogonal Learning Particle Swarm Optimization. IEEE Trans. on Evo. Comp. 15(6), 832–847 (2011)
12. Changhe, L., Yang, S., Nguyen, T.T.: A Self-Learning Particle Swarm Optimizer for Global Optimization Problems. IEEE Transactions on Systems, Man, and Cybernetics, Part B: Cybernetics 42(3), 627–646 (2012)
13. Pehlivanoglu, Y.V.: A New Particle Swarm Optimization Method Enhanced With a Periodic Mutation Strategy and Neural Networks. IEEE Trans. on Evolutionary Computation 17(3), 436–452 (2013)
14. Biswas, A., Kumar, A., Mishra, K.K.: Particle Swarm Optimization with Cognitive Avoidance Component. In: Inter. Conf. on Adv. in Computing, Communications and Informatics (ICACCI), pp. 149–154 (August 2013)
15. MacQueen, J.B.: Some Methods for classification and Analysis of Multivariate Observations. In: Proceedings of 5th Berkeley Symposium on Mathematical Statistics and Probability 1, pp. 281–297. University of California Press (1967)
16. Davies, D.L., Bouldin, D.W.: A Cluster Separation Measure. IEEE Trans. on Pattern Analysis and Machine Intelligence 1(2), 224–227 (1979)
17. Biswas, A., Gupta, P., Modi, M., Biswas, B.: Community Detection in Multiple Featured Social Network using Swarm Intelligence. In: International Conference on Communication and Computing (ICC 2014), Bangalore, June 12-14 (2014)
18. Ghaehchopogh, F.S., Khaze, S.R.: Data Mining Application for Cyber Space Tendency in Blog Writing: A Case Study. International Journal of Computer Applications (IJCA) 47(18), 40–46 (2012)
19. Agarwal, N., Liu, H., Tang, L., Philip, S.Y.: Identifying Influential Bloggers in a Community. In: 1st International Conference on Web Search and Data Mining (WSDM 2008), February 11-12, pp. 207–218 (2008)
20. Fisher, R.A.: The use of multiple measurements in taxonomic problems. Annual Eugenics 7(Pt. II), 179–188 (1936); also in "Contributions to Mathematical Statistics". John Wiley, NY (1950)

Evaluation of Data Warehouse Quality from Conceptual Model Perspective

Rakhee Sharma, Hunny Gaur, and Manoj Kumar

Abstract. Organizations are adopting Data Warehouse (DW) for making strategic decisions. DW consist of huge and complex set of data, thus its maintenance and quality are equally important. Using improper, misunderstood, disregarded data quality will highly impact the decision making process as well as its performance. The DW quality is depending on data model quality, DBMS quality and Data quality itself. In this paper, we have surveyed on two aspects of DW quality; one is how researchers have improved the quality of Data; and another is how data model quality is improved. The paper discusses that metrics are real quality indicators of DWs; they help the designers in obtaining good quality model that allows us to guarantee the quality of the DW. In this paper, our focus has been on surveying research papers with respect to quality of the multidimensional conceptual model of DW. Having surveyed various papers, we compared all the proposals concerning theoretical validation and empirical validation of conceptual model metrics for assessment of DW model quality.

Keywords: Data warehouse, multidimensional model, Object Oriented approach, metrics, Data quality, Data model quality, validation, UML, Conceptual model, Information quality, Framework, Quality indicator.

1 Introduction

Organizations have important and huge set of information. Thus to maintain it, efficiently they are adopting data warehouse (DW). Data Warehouse is a subject-oriented, integrated, time variant and non-volatile collection of data in support of management's decision making processes (Inmon, 1996). This huge set of data

---

Rakhee Sharma · Hunny Gaur · Manoj Kumar
Guru Gobind Singh Inderaprastha University, India

El-Sayed M. El-Alfy et al. (eds.), *Advances in Intelligent Informatics*,
Advances in Intelligent Systems and Computing 320, DOI: 10.1007/978-3-319-11218-3_47

has raised a new issue for the organizations: maintaining quality of data. If the data quality is not assured it leads to disastrous conditions, thus its quality is our prime aspect. Using improper, misunderstood, disregarded data quality will highly impact the decision making process as well as its performance. The research industry has taken many steps in improving the quality of DW so that the process of making decisions provides a new success mark to the organization. Several approaches have been proposed to improve the quality of data stored in DW, quality of the data model as well as the presentation of the data stored in DW. Sidi et al (Sidi et al., 2012) discussed about several data quality models and concluded about their suitability regarding DWs. The researchers also proposed a model, the combination of the data models surveyed in (Sidi et al., 2012). Chen et al in a paper (Chen et al., 2012) has included what data quality tools should be used in what problem.

Developing high quality DW is time consuming and expensive task, many projects, thus failed to deliver high data quality which leads to their failure. Hence, DW quality must be considered into whole DW development phase. Assuring quality at early stages of DW development can be achieved by improving the quality of data models used in designing. Data models are generally considered at three levels, conceptual, logical and physical. Serrano et al (Serrano et al., 2007) represented information quality as the quality of DW and presentation quality. Further, they have shown what actually must be considered to improve the quality of DW. It can be improved by improving database management system (DBMS) quality, data model quality or data quality. The research industries have mainly focused on improving data model quality. For improving quality DW modeling should be exposed at a higher level of design, increasing the understanding of users. This can be achieved by improving the quality of conceptual modeling phase of DW development. Nelson et al (Nelson et al., 2012) proposed a comprehensive conceptual modeling quality framework bringing together two well-known quality frameworks: the framework of Lindland, Sindre and Solvberg (LSS) and that of the Wand and Weber based on Bunge's ontology (BWW). They have defined several quality corner stones and that in turn defines quality dimension. Multidimensional (MD) model of DW is an early stage artifact and has been widely accepted as the foundation of DW, hence improving its quality will have a positive impact on DW quality. MD data models have been proposed at both logical and conceptual levels. At logical level star schema has been proposed by (Kimball and Ross, 2002). To develop the quality MD data model, some indicators are needed, namely metrics proposed in (Calero et al., 2001a). They defined valid metrics by following the metric proposal of (Calero et al., 2001b) as shown in fig1.Metrics are a way of measuring quality factor. These are validated through two methods proposed in (Calero et al., 2001a) Formal and empirical, although performed only formal validation using Zuse framework (Zuse, 1998). The empirical validation of the proposed metrics was done in (Serrano et al., 2002). This empirical validation have been further refined in (Serrano et al., 2003) and they have also proposed hypothesis a null and three alternative hypotheses. The validation has also been introduced at object oriented conceptual level modeling in (Serrano et al., 2004). They have proposed following metrics: Class scope, Star scope, Diagram scope. The validations of these metrics have been given by various researchers (Ali and Gosain, 2012; Gupta and Gosain, 2010; Gosain et al., 2010; 2011; 2013; Gosain and Mann, 2013; Kumar et al., 2013; Nagpal et al., 2012; 2013; Serrano et al., 2008) and result concludes that there exists some correlation between metrics and quality factors such as understandability, efficiency, completeness etc.

The remaining paper is structured as: section 2 presents impact of data quality on DW quality. Section 3 describes data model quality at conceptual level and also describes MD and object oriented MD approach. Section 4 discusses about metric definition and proposed metrics with theoretical and empirical validation in subsections 4.1 and 4.2, respectively, and the last section provides conclusion and future work.

Fig. 1 Steps for definition and validation of metrics (Calero et al., 2001b)

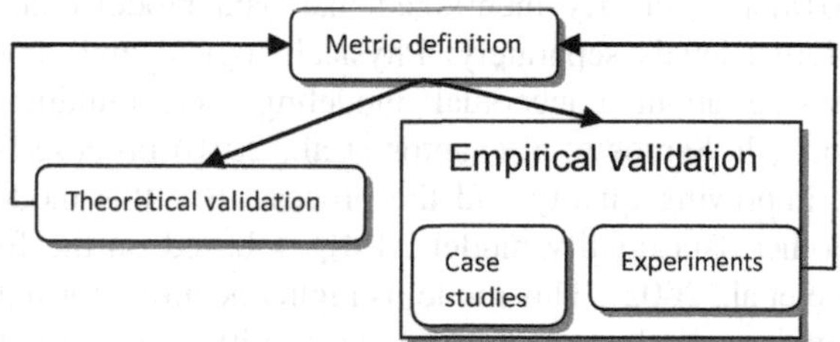

2 Data Quality

Data, mainly of DW plays important role in the decision making process. Its quality is major concerned area for researchers and organizations. The data quality degrades over time, if the processes and the information input qualities are not controlled, while data flows across information systems. In (Munawar et al., 2011) researchers have defined data quality of DW as business quality, information quality, technical quality. Fig 2 describes the data quality in detail. Business quality can be functional and non-functional; information quality can be quality of information content and quality of information access, technical quality deals with correctness, unambiguous, consistency and completeness. Developing high quality

DW is time consuming and expensive task, many projects, thus failed to deliver high data quality which leads to their failure. Data quality issues can be seen at any stage of development of DW. They must be tackled as soon as possible to improve the overall quality. Data quality can be measured using metrics and issues regarding data quality can be controlled at early stages.

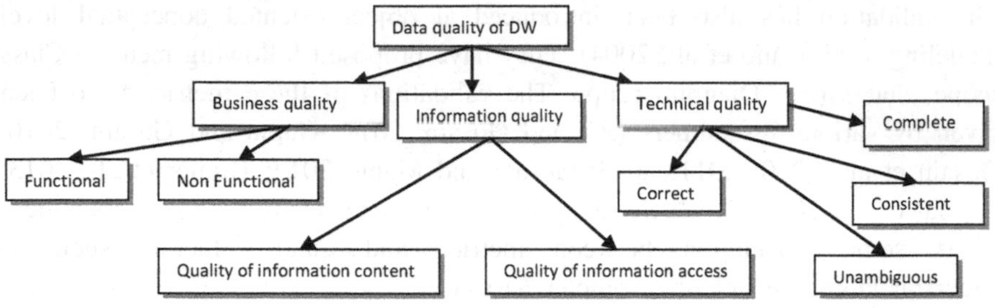

Fig. 2 Data Quality of DW (Munawar et al., 2011)

3 Data Model Quality

Data model quality is the clearest indicator of DW quality. In (Calero et al., 2001a) it is clearly mentioned that data model quality can be viewed at three different levels separately; Physical, logical and conceptual. In this section we describe about conceptual modeling, the multidimensional model and OO approach. Santos et al (Santos et al., 2010) proposed data quality model TDQM for improving quality and the process. But this model is not suitable for DWs. Another data quality model AIMQ is based on the foundation of model PSP/IQ (Lee et al., 2002). This model is highly accurate for making DW free of errors and provides completeness for product quality. Another data model quality framework is proposed by Moody et al (Moody and Shanks, 2003) and also introduces some quality factors such as simplicity, implementation, integration, understandability, flexibility and completeness which affect data model quality. For improving the quality DW modeling should be exposed at a higher level of design increasing the understanding of the user. This can be achieved through a conceptual modeling phase, which provides easy communication between the designer and the user, early detection of modeling errors and easily extendable schemas (Tryfona et al., 1999).

Many researchers (Gosain et al., 2011; Kumar et al., 2013; Nelson et al., 2012; Serrano et al., 2003; 2004; 2007; 2008) have worked on conceptual level data model quality. Conceptual modeling for DW can be tackled from MD modeling point of view. The structure of the multidimensional model depicts a star in the center and several outlying structures of data. MD models store information in facts and dimensions. A fact table is a structure that contains many occurrences of data and dimensions describe one important aspect of the fact table. The dimension table always defines the information that relates to the fact table. The fact table and dimension table are related with the existence of a common unit of data (Inmon, 1996). Dimensional data model is usually designed using the star schema modeling facility; which allows good response times and an easy understanding of data and metadata for both users and developers (Inmon, 1996). Serrano et al (Serrano et al., 2004) uses an extension of UML for the conceptual modeling of DW for defining metrics. The extension of UML uses the Object Constraint Language (OCL) for expressing the rules of new defined elements. The researchers adopted object oriented conceptual approach because it can represent the properties of DW easily at a conceptual level.

4 Metric Definition, Proposal and Validation

To guarantee the quality of multidimensional models for DWs, the subjective quality criteria or the design guidelines are not enough. Therefore, a set of formal and quantitative measures should be provided to reduce subjectivity and bias in evaluation and guide the designers in their work (Berenguer et al., 2005). These sets of measures that guide the designers in developing more precise DW are called metrics. Metrics should be defined in a procedural way and few steps must be followed to ensure their reliability. Steps to be followed in for defining the metrics are shown in figure 1 as proposed by (Calero et al., 2001a).Many researchers have been proposed the valid metrics as quality indicators for DW. These quality indicators help in evaluating the quality factors for DW such as understandability, efficiency, completeness etc. The first proposal was given by Calero et al (Calero et al., 2001a). They proposed a large set of metrics that can be applied at 3 levels, namely, table, star and schema. Table 1 represents the list of metrics used at all three levels.

Table 1 Proposed metrics for 3 levels (Calero et al., 2001a)

Table Metrics	Star Metrics	Schema Metrics
NA	NDT	NFT
NFK	NT	NDT
	NADT	NSDT
	NAFT	NT
	NA	NAFT
	NFK	NADT
	RSA	NASDT
	RFK	NA
		NFK
		RSDT
		RT
		RScA
		RFK
		RSDTA

Serrano et al (Serrano et al., 2007) proposed an initial set of metrics for the object oriented data warehouse conceptual model using UML. These metrics were also defined at three levels: class, star and diagram. The researchers focused mainly on star level metrics. Table 2 represents the proposed metrics at star level.

Table 2 Proposed metrics for Object Oriented DW conceptual model using UML (Serrano et al., 2007)

Metrics	Description
NDC	Number of dimension classes
NBC	Number of base classes
NC	Total number of classes
RBC	Ratio of base classes
NAFC	Number of FA attributes of fact class
NADC	Number of D and DA attributes of dimension classes
NABC	Number of D and DA attributes of base classes
NA	Total number of FA, D and DA attributes
NH	Number of hierarchy relationships
DHP	Maximum depth of hierarchy relationships
RSA	Ratio of attributes of star schema

Cherfi et al (Cherfi and Prat, 2003) proposed a metric approach based on UML and multidimensional schema quality evaluation framework (Si-Said et al., 2003) focusing on quality of multidimensional schemas considering specifically analyzability and simplicity criteria. The researchers focused mainly on specification view among the three views proposed in (Si-Said et al., 2003). The paper also addressed the problem of evaluation and maintaining balance between multidimensional schema's analyzability and simplicity and presented their associated metrics. Berenguer et al (Berenguer et al., 2005) proposed a framework for designing the metrics, being each metric a part of quality indicator. The researchers proposed metric design guidelines for UML packages in multidimensional modeling and metrics were proposed at three levels: level 1 defining one package for each star schema, level 2 defining on package for each dimension and fact and level 3 defining the whole content of both dimension and fact packages. Gosain et al (Gosain et al., 2011) proposed the metrics emphasizing on dimension hierarchies in multidimensional models. Table 3 represents the list of proposed metrics.

Table 3 Proposed metrics for dimension hierarchies (Gosain et al., 2011)

Metrics	Description
NMH	Number of multiple hierarchies
APMH	Average number of alternate paths in multiple hierarchies
ALMH	Average number of levels in multiple hierarchies
SMH	Number of shared multiple hierarchies
ADMH	Average number of dimensions participating in shared multiple hierarchies

4.1 Theoretical Validation

Defining metrics for DW is not sufficient to achieve a good quality multidimensional model for DW. They need to be validated to prove their correctness as quality indicators. Theoretical validation of metrics helps the designers to know that when to apply the metrics and in which manner. Calero et al (Calero et al., 2001a) performed the theoretical validation of some of proposed metrics chosen from all three levels using the Zuse framework (Zuse, 1998). This framework is a measurement theory based framework whose goal is to determine the scale to which a metric pertains (Calero et al., 2001a). The formal framework (Zuse, 1998) defines three mathematical structures, viz. the extensive structure, the independence conditions and the modified relation of belief. The result of the theoretical validation of metrics indicated that all the proposed metrics are either on the ordinal scale or on superior scale.

After this theoretical validation, the theoretical validation of proposed metrics (Serrano et al., 2007) was performed using Briand et al (Briand et al., 1996) framework in (Serrano, 2004) and using the DISTANCE framework (Poels and Dedene, 1996) in (Serrano et al., 2007). The validation result determined that all the metrics are characterized by the ratio scale and they are theoretically validated software metrics and are therefore rather serviceable. Gosain et al (Gosain et al., 2013) also performed the theoretical validation of metrics proposed in (Gosain et al., 2011) using Briand et al framework (Briand et al., 1996). The researchers observed that all the dimension hierarchies metrics considered in the paper satisfy all the properties that characterize the metric as either size measure or length measure. The result of the theoretical validation concluded that NLDH is a length measure metric while all other metrics are size measures which prove that the metrics are theoretically validated and contribute ominously towards structural complexity. Table 4 summarizes the theoretical validation work conducted by several researchers.

Table 4 Theoretical validation summarization

Paper Title	Framework Used	Metrics Used	Result
Towards Data Warehouse Quality Metrics (Calero et al., 2001a)	Zuse (Zuse, 1998)	Combination of star and schema level metrics (NA, NFK, NDT, NT, NADT, NAFT, NFT, NSDT, NASDT).	All the proposed metrics are either in the ordinal scale or in superior scale.
Metrics for Data Warehouse conceptual models understandability (Serrano et al., 2007)	Distance (Poels and Dedene, 1996)	Star Scope Metrics(NABC, RBC, NBC, NH, NDC, NC, NAFC, NA, RSA, DHP)	All the metrics are characterized by the ratio scale and they are theoretically validated software metrics and are therefore rather serviceable.
Validating dimension hierarchy metrics for the understandability of multidimensional model for data warehouse (Gosain et al., 2013)	Briand (Briand et al., 1996)	NMH(no. of multiple hierarchies) APMH (average no. of alternate paths in multiple hierarchies) ALMH (average no. of levels in multiple hierarchies) SMH (No. of shared multiple hierarchies) ADMH (average no. of dimensions participating in shared multiple hierarchies)	Metrics contribute significantly towards structural complexity of multidimensional models.

4.2 Empirical Validation

The objective of empirical validation is to prove the practical usefulness of the proposed metrics. Some ways of performing empirical validation are shown in figure 1. The empirical validation of metrics was first performed by Serrano et al (Serrano et al., 2002) by conducting a controlled experiment over proposed metrics using GQM (Goal Question Metric) approach (Basili and Weiss, 1984; Basili and Rombach, 1998). The researchers considered only schema level metrics for validation and the experiment was conducted by choosing 12 subjects (experts having experience in database design) to solve eleven DW schemas to measure the dependent variable understandability. They also formulate two hypotheses: one null and one alternate to find the correlation between the structural complexity measures and understandability of schemas. The result concluded that there seems to be a correlation between some metrics (NT, NFT, and NSDT) and complexity of the schema.The validation process was refined in (Serrano et al., 2003) by focusing only on two metrics NFT and NDT. The numbers of schemas were reduced to four and result indicated that NFT is a solid indicator of the complexity of the multidimensional models, but NDT is neither an indicator of complexity nor can it reduce the complexity.

Serrano et al (Serrano et al., 2004) empirically validates the proposed metrics for object oriented DW conceptual model using UML. The numbers of subjects taken were seventeen and the number of schemas was increased to ten for performing experiments. The outliers were also eliminated from collected data. The experiment verified two quality factors: understandability and modifiability and result concluded that there exist a high correlation between the metrics (NBC, NC, RBC, NABC, NA, NH and DHP) and understandability. And there is no correlation between metrics and modifiability. Serrano at al refined their experiment in (Serrano et al., 2007) with twenty five subjects and ten DW schemas. The quality factors considered in the experiment were understandability, efficiency and effectiveness. The researchers concluded that the metrics (NBC, NC, RBC, NABC, NA, NH and DHP) seems to be correlated with understandability of DW schema but with respect to efficiency and effectiveness, researchers found that a high inverse relationship exist between the metrics (NBC, NC, RBC and NH) and efficiency of subjects working with schemas. The empirical validation was replicated by Serrano et al in (Serrano et al., 2008) to confirm the results of previous experiments. The researchers conducted a four step analysis to check whether individually or collectively, the metrics are correlated with the understandability of schemas using statistical techniques. Only four metrics from schema level metrics were verified with thirteen schemas and eighty five subjects. The researchers found that all four metrics are good indicators of

DW quality. Gosain et al (Gosain et al., 2010) predicts the quality of the multidimensional model using machine learning technique: Neural Network Approach. The experiment was conducted on four metrics (NFT, NDT, NFK and NMFT) and ANN technique was applied to the data set provided in (Serrano et al., 2008). The results showed that technique proposed is able to predict the quality of the DW multidimensional model with acceptable accuracy.

The empirical validation procedure was also adapted by Gupta et al (Gupta and Gosain, 2010). The objective of the validation was to reduce the set of metrics to obtain the effective measures only and then performing empirical of these metrics to use them as quality indicators. The metrics proposed in (Calero et al., 2001a) were validated using Linear Regression technique and result predicted that NFT and NSDT are best quality indicators of DW schema. The Regression Analysis experiment was again conducted by Gupta et al (Gupta et al., 2011) using object oriented DW conceptual model using UML. The numbers of schemas taken were twenty solved by ten subjects. The metrics defined in (Serrano et al., 2007) were used for finding out the correlation between understandability and efficiency and complexity of the schema. Result of the experiment indicates that several metrics are correlated with the understandability of models as well as with efficiency of the subjects.

Gupta et al (Gupta and Gosain, 2012) extended their work of reducing the redundant metrics using the PCA (Principle Component Analysis) approach. The approach was applied on 41 schemas for the metrics proposed in (Calero et al., 2001a). Among seven metrics (NFT, NDT, NSDT, NAFT, NADT, NASDT and NFK), only three metrics (NFT, NFK and NAFT) are extracted as a principal component. The benefit of PCA approach is that it reduces the efforts required for further computation to get the precise quality indicators. Ali et al (Ali and Gosain, 2012) proposed a model based on fuzzy logic technique to find out the nonlinear relationship between the metrics and quality of object oriented multidimensional model, and performed empirical validation to assess the effectiveness of the proposed model. The results of fuzzy based models are compared with the controlled experiment conducted by researchers. The metrics used for empirical validation are NDC, NBC, NA and NC. The experiment was carried out by twenty subjects over eleven schemas. And result concluded that proposed model is a successful measure of the understandability of the multidimensional model for these four metrics.

Gosain et al (Gosain et al., 2013) extended their work (Gosain et al., 2011) by empirical validation the dimension hierarchy metrics for evaluating the correlation between quality factor understandability and complexity of the multidimensional schema. The validation was done with the help of controlled experiment with seventeen multidimensional schemas for thirty subjects (divided in two groups of fifteen subjects each). After experimenting, statistical techniques (Correlation and Linear Regression) were applied. The result indicates that the metrics are significant in predicting the quality of multidimensional models for DW.

Kumar et al (Kumar et al., 2013) performed the empirical validation of structural metrics proposed in (Serrano et al., 2004; 2008) by applying statistical methods (Logistic Regression) and machine learning techniques (Decision Trees, Naïve Bayesian Classifier) for predicting the effect of structural metrics on understandability. The researchers also represent the comparison of statistical techniques and machine learning techniques. The experiment was conducted for twenty schemas with eighteen subjects. The result of the experiment indicates that some metrics individually affects the understandability of the schema while some have combined effect on the understandability of multidimensional schema. The comparison result shows that Naïve Bayesian Classifier is a better method for prediction than Logistic Regression. Gosain et al (Gosain and Mann, 2013) used Linear Regression Analysis for predicting multidimensional quality. They have also evaluated understandability and efficiency through various combinations of metrics. The empirical validation was performed on metrics proposed in (Serrano et al., 2007). The experiment was performed using thirty three schemas by fifty subjects and the result indicates that all the metrics are solid indicators of predicting the quality for multidimensional schemas of DW. Table 5 summarizes the empirical validation work conducted by several researchers.

Table 5 Empirical validation summarization

Paper Title	Level Worked On	Metric Used	Hypotheses Formed	Technique Used	Quality Factors	Result
Validating metrics for Data Warehouse (Serrano et al., 2002)	Logical	Schema Level	One Null and One Alternate	Statistical	Understandability	There exist some correlation between some metrics and complexity of schema.
Experimental validation of multidimensional data model metrics (Serrano et al., 2003)	Logical	Schema Level	One Null and Three Alternate	Statistical	Understandability	NFT is a solid indicator of the complexity of the multidimensional models but NDT is neither an indicator of complexity nor can it reduce the complexity.
Empirical validation of metrics for conceptual models of Data Warehouse (Serrano et al., 2004)	Conceptual	Star Level	One Null and One Alternate	Statistical	Understandability and Modifiability	There exist high correlation between some metrics and Understandability. But no correlation between metrics and modifiability.
Metrics for Data Warehouse conceptual models understandability (Serrano et al., 2007)	Conceptual	Star Level	Three Null and Three Alternate	Statistical	Understandability, Efficiency and Effectiveness	Some metrics seem to be correlated with understandability and a high inverse relationship exists between metrics and efficiency of subjects.
Empirical studies to assess the understandability of Data Warehouse	Logical	Schema Level	Four Null	Statistical	Understandability	Four metrics of schema level are good indicators of Data Warehouse

Table 5 (*continued*)

Paper Title	Level Worked On	Metric Used	Hypotheses Formed	Technique Used	Quality Factors	Result
schema using structural metrics (Serrano et al., 2008)						quality.
Neural Network approach to predict quality of Data Warehouse multidimensional model (Gosain et al., 2010)	Logical	Schema Level	--	Machine Learning	Understandability	Technique proposed is able to predict the quality of Data Warehouse multidimensional model with acceptable accuracy.
Analysis of Data Warehouse quality metrics using LR (Gupta and Gosain, 2010)	Logical	Schema Level	--	Statistical	Understandability	Two metrics NFT and NSDT are best quality indicators of Data Warehouse schema.
Empirical validation of object oriented Data Warehouse design quality metrics (Gupta et al., 2011)	Conceptual	Star Level	One Null and One Valid	Statistical	Understandability and Efficiency	Several metrics are correlated with understandability of models as well as with efficiency of the subjects.
Validating Data Warehouse quality metrics using PCA (Gupta and Gosain, 2012)	Logical	Schema Level	--	Statistical	Understandability	Among seven metrics only three metrics are extracted as principal component.
Predicting the quality of object oriented multidimensional (OOMD) model of Data Warehouse using fuzzy logic technique (Ali and Gosain, 2012)	Conceptual	Star Level	--	Machine Learning	Understandability	The proposed model is a successful measure of the understandability of multidimensional model for metrics.
Validating dimension hierarchy metrics for the understandability of multidimensional models for Data Warehouse (Gosain et al., 2013)	Logical	Dimension Hierarchy	One Null and One Alternate	Statistical	Understandability	The metrics are significant in predicting the quality of Multidimensional models of Data Warehouse.
Empirical validation of structural metrics for predicting understandability of conceptual schemas for Data Warehouse (Kumar et al., 2013)	Conceptual	Star Level	One Null and Twelve Alternate	Statistical and Machine Learning both	Understandability	Some metrics individually affects the understandability of the schema while some have combined effect on the understandability of multidimensional schema.
Empirical Validation of metrics for object oriented multidimensional model for Data Warehouse (Gosain and Mann, 2013)	Conceptual	Star Level	Two Null and Two Alternate	Statistical	Understandability and Efficiency	All the metrics are solid indicators of predicting quality for multidimensional schemas of Data Warehouse.

5 Conclusion and Future Work

Data Warehouse stores large and complex set of information, so assuring its quality should be the prime concerned while designing DWs. The paper discusses about different aspects of data quality, data model quality and DW quality.

The paper focuses on surveying the research work on improving data warehouse quality. It discusses about the metrics proposed on logical and conceptual levels of DW. We have also highlighted the frameworks and techniques concerning theoretical and empirical validation of logical and conceptual model metrics for assessment of DW model quality. Researchers have performed empirical validations by considering different number of subjects and schemas; however, to achieve more precise results, empirical validation of metrics can be performed by considering more schemas and subjects.

References

Ali, K.B., Gosain, A.: Predicting the quality of object oriented multidimensional (OOMD) model of data warehouse using fuzzy logic technique. International Journal of Engineering Science & Advanced Technology 2(4), 1048–1054 (2012)

Basili, V.R., Weiss, D.: A methodology for collecting valid software engineering data. IEEE Transactionson Software Engineering 10(6), 728–738 (1984)

Basili, V., Rombach, H.: The TAME project: Towards improvement oriented software environments. IEEE Transactions on Software Engineering 14(6), 728–738 (1988)

Berenguer, G., Romero, R., Trujillo, J., Bilò, V., Piattini, M.: A set of quality indicators and their corresponding metrics for conceptual models of data warehouse. In: Tjoa, A.M., Trujillo, J. (eds.) DaWaK 2005. LNCS, vol. 3589, pp. 95–104. Springer, Heidelberg (2005)

Briand, L., Morasca, S., Basili, V.: Property based software engineering measurement. IEEE Transactions on Software Engineering 22(1), 68–86 (1996)

Calero, C., Piattini, M., Genero, M.: Metrics for controlling database complexity. In: Becker (ed.) Chapter III in Developing Quality Complex Database Systems: Practices, Techniques and Technologies. Idea Group Publishing (2001b)

Calero, C., Piattini, M., Pascual, C., Serrano, M.: Towards data warehouse quality metrics. In: Proceedings of the International Workshops on Design and Management of Data Warehouses, Interlaken, Switzerland (2001a)

Chen, M., Song, M., Han, J., Haihong, E.: Survey on data quality. IEEE (2012)

Si-said Cherfi, S., Prat, N.: Multidimensional schemas quality: Assessing and balancing analyzability and simplicity. In: Jeusfeld, M.A., Pastor, Ó. (eds.) ER Workshops 2003. LNCS, vol. 2814, pp. 140–151. Springer, Heidelberg (2003)

Gosain, A., Mann, S.: Empirical validation of metrics for object oriented multidimensional model for data warehouse. International Journal of System Assurance Engineering & Management (2013)

Gosain, A., Sabharwal, S., Nagpal, S.: Neural network approach to predict quality of data warehouse multidimensional model. In: Proceedings of International Conference on Advances in Computer Science. ACEEE (2010)

Gosain, A., Nagpal, S., Sabharwal, S.: Quality metric for conceptual models for data warehouse focusing on dimension hierarchies. ACM SIGSOFT Software Engineering Notes 36(4) (2011)

Gosain, A., Nagpal, S., Sabharwal, S.: Validating dimension hierarchy metrics for the understandability of multidimensional models for data warehouse. IET Software 7(2), 93–103 (2013)

Gupta, J., Gosain, A., Nagpal, S.: Empirical validation of object oriented data warehouse design quality metrics. In: Wyld, D.C., Wozniak, M., Chaki, N., Meghanathan, N., Nagamalai, D. (eds.) ACITY 2011. CCIS, vol. 198, pp. 320–329. Springer, Heidelberg (2011)

Gupta, R., Gosain, A.: Analysis of Data warehouse quality metrics using LR. In: Das, V.V., Vijaykumar, R. (eds.) ICT 2010. CCIS, vol. 101, pp. 384–388. Springer, Heidelberg (2010)

Gupta, R., Gosain, A.: Validating data warehouse quality metrics using PCA. In: Kannan, R., Andres, F. (eds.) ICDEM 2010. LNCS, vol. 6411, pp. 170–172. Springer, Heidelberg (2012)

Inmon, W.H.: Building the Data Warehouse. John Wiley & Sons (1996)

Kumar, M., Gosain, A., Singh, Y.: Empirical validation of structural metrics for predicting understandability of conceptual schemas for data warehouse. International Journal of System Assurance Engineering & Management (2013)

Kimball, R., Ross, M.: The data warehouse toolkit. Wiley, New York (2002)

Lee, Y.W., Strong, D.M., Kahn, B.K., Wang, R.Y.: AIMQ: A methodology for information quality assessment. Information and Management 40, 133–146 (2002)

Moody, D.L., Shanks, G.G.: Improving the quality of data models: Empirical validation of a quality management framework. Information System 28(6), 619–650 (2003)

Munawar, S.N., Ibrahim, R.: Towards data quality into the data warehouse development. In: IEEE Ninth International Conference on Dependable, Autonomic and Secure Computing (2011)

Nagpal, S., Gosain, A., Sabharwal, S.: Complexity metric for multidimensional models for data warehouse. In: CUBE, ACM (2012)

Nagpal, S., Gosain, A., Sabharwal, S.: Theoretical and empirical validation of comprehensive complexity metric for multidimensional models for data warehouse. International Journal of System Assurance Engineering & Management, 193–204 (2013)

Nelson, H.J., Poels, G., Genero, M., Piattini, M.: A Conceptual modeling quality framework. Software Quality Journal 20, 201–228 (2012)

Poels, G., Dedene, G.: DISTANCE: A framework for software measure construction. research report DTEW9937, Dept. Applied economics, Katholieke Universiteit Leuven, Belgium, p. 46 (1996)

Santos, G.D., Takaoka, H., de Souza, C.A.: An empirical investigation of the relationship between information quality and individual impact in organizations. In: AMCIS Proceedings (2010)

Serrano, M., Calero, C., Piattini, M.: Validating metrics for data warehouse. IEEE Proceedings 149(5) (2002)

Serrano, M., Calero, C., Piattini, M.: Experimental validation of multidimensional data model metrics. In: Proceedings of the 36th Hawaii International Conference on System Sciences (2003)

Bilò, V., Calero, C., Trujillo, J., Luján-Mora, S., Piattini, M.: Empirical validation of metrics for conceptual models of data warehouses. In: Persson, A., Stirna, J. (eds.) CAiSE 2004. LNCS, vol. 3084, pp. 506–520. Springer, Heidelberg (2004)

Serrano, M.: Definition of a set of metrics for assuring data warehouse quality. University of castilla, La Mancha (2004)

Serrano, M., Trujillo, J., Calero, C., Piattini, M.: Metrics for data warehouse conceptual models understandability. Information and Software Technology 49, 851–870 (2007)

Serrano, M., Calero, C., Sahraoui, H.A., Piattini, M.: Empirical studies to assess the understandability of data warehouse schema using structural metrics. Software Quality Journal 16, 79–106 (2008)

Sidi, F., Ramli, A., Jabar, M.A., Affendey, L.S., Mustapha, A., Ibrahim, H.: Data Quality Comparative Model for data warehouse. IEEE (2012)

Si-said Cherfi, S., Akoka, J., Comyn-Wattiau, I.: Conceptual modeling quality- from EER to UML schemas evaluation. In: Spaccapietra, S., March, S.T., Kambayashi, Y. (eds.) ER 2002. LNCS, vol. 2503, pp. 414–428. Springer, Heidelberg (2002)

Tryfona, N., Busborg, F., Christiansen, J.G.B.: starER: A conceptual model for data warehouse design. In: Proceedings of the Second ACM International Workshop on Data Warehousing and OLAP, pp. 3–8. ACM (1999)

Zuse, H.: A Framework of Software Measurement. Walter de Gruyter (1998)

Enhancing Frequency Based Change Proneness Prediction Method Using Artificial Bee Colony Algorithm

Deepa Godara and Rakesh Kumar Singh

Abstract. In the field of software engineering, during the development of Object Oriented (OO) software, the knowledge of the classes which are more prone to changes in software is an important problem that arises nowadays. In order to solve this problem, several methods were introduced by predicting the changes in the software earlier. But those methods are not facilitating very good prediction result. This research work proposes a novel approach for predicting changes in software. Our proposed probabilistic approach uses the behavioral dependency generated from UML diagrams, as well as other code metrics such as time and trace events generated from source code. These measures combined with frequency of method calls and popularity can be used in automated manner to predict a change prone class. Thus all these five features (time, trace events, behavioral dependency, frequency and popularity) are obtained from our proposed work. Then, these features are given as the input to the ID3 (Interactive Dichotomizer version 3) decision tree algorithm for effectively classifying the classes, whether it predicts the change proneness or not. If a class is classified into prediction of change prone class, then the value of change proneness is also obtained by our work.

1 Introduction

When adapting a system to new usage patterns or technologies, it is necessary to foresee what such adaptations of architectural design imply in terms of system quality [3]. Software systems are continuously subjected to changes. Changes are necessary to fix bugs, to enhance a product by adding new features, to adapt to a new environment and re-factor the source code [5]. Changes in software are made

Deepa Godara · Rakesh Kumar Singh
Uttarakhand Technical University, India

El-Sayed M. El-Alfy et al. (eds.), *Advances in Intelligent Informatics*,
Advances in Intelligent Systems and Computing 320, DOI: 10.1007/978-3-319-11218-3_48

due to a variety of reasons such as revision or adaptation, enhancements, fixing errors and perfective maintenance. Some parts of the object oriented software may be more prone to changes than others. Knowing which classes are change-prone can be very helpful; change-proneness may indicate specific underlying quality issues [8]. Handling change is one of those fundamental problems in software engineering. To deal with new technology and changing requirements evolutionary development has been proposed as the most efficient method [6]. Remedial actions can be taken if a maintenance process is able to identify which classes of the software are more change-prone. Evolution process of software can focus on such change prone classes [8].

Various software development projects fail in different ways but these all share common symptoms. Some of them are: changing requirements and specifications, unrealistic expectations, new technology, technology incompetence, unrealistic time frames, unclear objectives, devastating complexity [12]. A small change can unexpectedly disturb the system [13]. Change-prone classes in software require particular attention because they require effort and increase development and maintenance costs. Identifying and characterizing those classes can enable developers to focus preventive actions such as, peer-reviews, testing, inspections, and restructuring efforts on the classes with the similar characteristics in the future. As a result, developers can use their resources more efficiently and deliver higher quality products in a timely manner [2]. If faulty classes can be detected in advance, remedial measure such as focused inspection can be taken. To identify faulty classes prediction model using design metrics can be used [13].The accuracy of the predicted impact determines the accuracy of cost estimation and quality of project planning [11].

We believe that most of the researchers use static characteristics of the object oriented software to specify whether a design is good or not [14]. Behavioral Dependency Analysis (BDA) is a measure that determines the functional dependency between two entities. BDA can be divided into three groups based on the source of information: (1) execution trace based (2) model based and (3) code based. Sequence diagrams can be used to determine the behavioral dependency between two distributed objects [15].

UML is a general purpose modeling language that is widely accepted in software engineering community. It supports object-oriented designs which in turn encourage component reuse. It supports many diagrams to provide multiple views of the system under design [10]. The UML based design enabled us to apply formal verification and validation techniques [1]. The unified modeling language (UML) is a graphical language for visualizing, specifying, constructing, and documenting software-intensive systems. It helps to visualize design and communicate with others [12]. UML is an Object Management Group (OMG) standard. It is useful for modeling complex and large systems.OMG is proposing the UML specification for international standardization for information technology [9]. Due to vast acceptance of object-oriented design, we observe that polymorphism and inheritance are more frequently used to improve reusability [7].It offers many diagrams to model a software system such as sequence diagrams, class diagrams etc.

Section 2 enlightens the short notes for the proposed change proneness prediction methodology and the structure for the suggested methodology. At last, the conclusions are summed up in Section 3.

2 Proposed Methodology of Change Proneness Prediction

In our proposed work, the main intention is to develop an optimal frequency based change proneness prediction method using behavioral dependency and ABC algorithm. The input of our proposed work is an application. In our work, we have to predict the change proneness of any given application. In order to predict, at first, the features of every application need to be collected from the application. After the identification of these features, the change proneness in each class can be predicted by using ID3 Decision tree algorithm. If a class is classified as predicted class for change proneness, then the value of the change proneness prediction is also calculated.

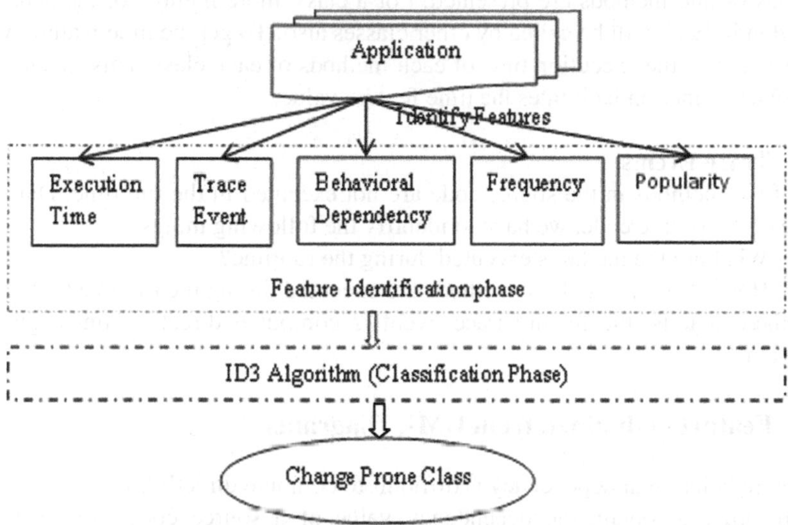

Fig. 1 Proposed change-proneness prediction method

2.1 Phase-I: Identification of Features

The features obtained from an application are time, trace events, behavioral dependency, frequency and popularity, which predict only the change proneness. These 5 features are derived from the application in the following three methods:-

(i) Features obtained directly from an application
(ii) Features obtained from UML Diagrams
(iii) Features obtained from optimal frequent item set mining and ABC

The features such as time and trace events of the methods are identified directly from the input easily. And the feature behavioral dependency is obtained from the UML diagrams. The dependency represents the behavior of each and every class in the software, which is calculated from the UML diagrams. And the frequency of interfaces and methods can be computed by using optimal frequent item set mining algorithm based on ABC. Using this frequency, we can find out the feature popularity.

2.1.1 Features Obtained Directly from an Application

In this way, only the two features time and trace events are obtained directly from the given input application. From the source code, we can directly find these features as given below.

2.1.1.1 Time

From the input source code, we can find this feature value, time. In source code, many classes and methods are presented. For a class, more number of methods is included in it. Each will be called by other classes also. To get the time feature, we have to calculate the execution time of each methods of each class. This execution time of every function facilitates the time feature value.

2.1.1.2 Trace Events

Some of the methods in the source code are not executed in the run time. During runtime of the source code, we have to identify the following things,

(i) What are the methods executed during the runtime?

(ii) How many times, these methods are executed during the runtime?

Using these details, the feature trace event is computed directly from a given application

2.1.2 Features Obtained from UML Diagrams

The feature behavioral dependency is difficult to compute directly from the source code. In order to obtain the dependency value of a source code, we need to generate UML diagrams for the source code. From these UML diagrams, we can find the behavioral dependency feature for the given application.

2.1.2.1 Generation of UML Diagrams

UML diagrams provide communication between the customer, system analysts and programmers, who write the source code. So, it acts as a mediator or information provider to find out the features. In general, 8 kinds of UML diagrams are generated. But for our purpose we need only 2 kinds of UML diagrams

UML Sequence Diagram: - It represents a sequence of interaction between the objects.

UML Class Diagram: - It represents the relationship between classes and interfaces.

2.1.2.2 Behavioral Dependency

Behavioral Dependency is one of the important features to predict the change proneness. In a source code, if one class gets changes, it also affects the other class. It can be found only through the dependencies between the classes or objects

The relationship between the sender object and receiver object is an example of the behavioral dependency, while sending a message between two objects. The changes in the class of the receiver object also affect the class of the sender object. The inheritance and polymorphism are also taken into account, during the measurement of behavioral dependency. The higher changes in behavioral dependency indicate the possibility of more changes to happen.

2.1.2.2.1 Measuring the Behavioral Dependency Using UML Models

To measure the behavioral dependency from the source code, we generate a UML Sequence and Class diagram for the source codes. From this, we then measure the behavioral dependency by using the following methods.

(i) Construct Object Dependency model
(ii) Construct System Object Dependency model
(iii) Form reachable path table
(iv) Calculate the weighted sum of reachable paths
(v) Calculate the Behavioral Dependency of a class

2.1.3 Features Obtained from Optimal Frequent Item Set Mining and ABC

The features frequency and popularity are obtained from the input by using optimal frequent item set and ABC algorithm.

Frequency

The frequency of the methods that are called by other methods is calculated using frequent item set mining method. Each function can be called by more number of functions. In this pattern, the rules for the method calls are generated. It is possible to happen that some of the methods cannot be called by other methods in the "changed" software. This will make changes in the frequency of the methods that are called by other methods. So, it is also one of the effective features to predict the change-proneness.

In our work, the method calls by different classes are considered as the set of instances. Each method call has a number of classes as the items.

The process of these frequent item set algorithm then identifies all the common sets of classes for the method calls, in which these set of classes have at least a minimum support. Minimum support specifies the minimum number of times the method calls exist in the common set of classes. Then the classes for the method calls are combined with parameters such as {1-length combination, 2-length combination, (n-1)-length combination} based on the number of classes for method calls. For example, if a set of classes is {a, b, c}, then 1-length combinations are {a}, {b}, {c} and the 2-length combinations are {a, b}, {b, c}, {a, c} taken from the common set. Likewise, according to the length of the set of classes, the combinations are taken by this frequent item set mining algorithm.

After the identification of a set of frequent item sets, association rules are produced. Every association rule has qualities such as support, confidence and lift.

After generating the rules, we have a number of rules to find the frequency of method calls for a particular class. To overcome this, we need to optimize these rules, for which Artificial Bee Colony (ABC) Algorithm is exploited in our work.

ABC Algorithm: - If we use only the frequent item set mining for getting the frequency feature values means, it is possible to produce more number of rules, in which many non-frequent item sets also presented. To avoid these non-frequent item sets, we use ABC algorithm within the frequent item set mining for optimizing the rules in this.

Then the frequency $(Freq_1)$ of the method calls is found using these optimized rules. This is the feature that obtained from frequent item set and ABC algorithms.

Popularity

This is the fifth feature that helps to predict the change proneness in the given input. By using the following equation and also the feature frequency, we can calculate this feature, popularity.

$$Popularity\ ,\quad PO = Freq_1 + Freq_2 \tag{1}$$

where, $Freq_2 = \dfrac{(MC_1).(MC_2)}{Total\ Number\ of\ methods\ used}$

MC_1 - Number of times a method called by other methods

MC_2 - Number of times a method calls other methods

Now all the features such as time, trace events, behavioral dependency, frequency and popularity are taken from the given application input.

2.2 *Phase-II: Classification of Classes for the Change-Proneness Prediction*

For classifying the classes which predict the change-proneness and do not predict the change-proneness, one of the classification algorithms named ID3 Algorithm is utilized. Based on the classification of classes, we can separate the classes which predict the change-proneness in classes from the classes which do not predict the change-proneness in the classes.

2.2.1 ID3 Algorithm

ID3 (Interactive Dichotomizer version 3) is a simple decision tree learning algorithm that is used for classification purpose. It builds decision trees based on greedy search in an up-down manner, which contains the set of nodes and edges that connect these nodes. According to our work, the result of classification is divided into two- (1) classes that predict the change-proneness (set of positive results - PR) and (2) classes that do not predict the change-proneness (set of negative results - NR). The total classes are represented as $T = PR \cup NR$. The number of sets of positive results (PR) is denoted as a and the number of sets of negative results (NR) is denoted as b. The probability of a class belongs to PR is $\frac{a}{a+b}$ and the probability of a class belongs to NR is $\frac{b}{a+b}$. The intermediate information required to make the decision tree as a source of the result PR or NR is given

$$\text{as, } I(a,b) = \begin{cases} -\frac{a}{a+b}\log_2\frac{a}{a+b} - \frac{b}{a+b}\log_2\frac{b}{a+b}, & \text{when } a,b \neq 0 \\ 0, & \text{otherwise} \end{cases} \quad (2)$$

If attribute (feature) X with values $\{v_1, v_2, ...v_N\}$ is used for the root of the tree, then it partitions T into $\{T_1, T_2, ...T_N\}$, in which T_i denotes the classes in T that have value v_i of X. Let T_i consists of a_i classes of PR and b_i classes of NR, then the intermediate information needed for the sub-tree for T_i is $I(a_i, b_i)$. The expected information EI needed for the decision tree with X as the root ($EI(X)$), is found as a weighted average as given below.

$$EI(X) = \sum_{i=1}^{N} \frac{a_i + b_i}{a+b} I(a_i, b_i) \quad (3)$$

From the above equation (10), the weight for the i th branches is the proportion of the classes T that belongs to T_i. Now, we can find the value of information gain IG by branching on X which is given below.

$$IG(X) = I(a,b) - EI(X) \quad (4)$$

All the features X are analyzed by ID3 algorithm and X, which gives maximum $IG(X)$ value is selected for constructing the decision tree. Then for the residual subsets $T_1, T_2, ...T_N$, to construct the tree, the same process recursively continues... For each $T_i, (i = 1,2,...N)$, if all the classes are positive, then it makes a decision with "YES" node and stops the process and if all the classes are negative, then it makes decision with "NO" node and stops the process; Otherwise ID3 chooses another attribute in the same way as given earlier.

Thus the classes which predict the change proneness and the classes which do not predict the change proneness are classified using ID3 algorithm.

2.2.2 Finding Change Proneness Prediction Value

Now from the classification results, the classes which predict change proneness are taken separately and the probability values for each class for the change proneness are found according to the following equation (12).

$$P(T_2) = P(T_2) + P(T_2 \mid T_1) * P(T_1) - P(T_2) * P(T_2 \mid T_1) * P(T_1) \qquad (5)$$

In the above equation (5), we have considered only two classes T_1 and T_2. Likewise, probability of all classes T_i is calculated as in equation (5) to find the change proneness value. Hence we can predict the change proneness of classes using the five features such as time, trace events, behavioral dependency, frequency and popularity.

3 Conclusion

In this paper, an effective change proneness prediction system was effectively constructed using frequent item set mining and ABC algorithm. At first, the features from this application were taken. The features taken were time, trace events, behavioral dependency, frequency and popularity. The features such as time, trace events were directly computed from the input application and the feature behavioral dependency was recognized from the UML diagrams of the input application. The feature frequency was estimated by means of frequent item set mining with ABC algorithm. Then the last feature popularity was assessed by treating the frequency as one of the inputs. Then the classes utilized in the application are categorized into two sets which specify whether the classes forecast the change-proneness or not by means of a decision tree algorithm, ID3. In future, the tests will be implemented on Java platform for authenticating any of the application regarding the fact whether it forecasts the change prone or otherwise, by means of the evaluation metrics such as precision, recall, FPR, FNR and accuracy. Further, we will be making comparisons of our work with other authors work to show that our approach gives more accurate results.

References

[1] Pataricza, A., Majzik, I., Huszerl, G., Várnai, G.: UML-based Design and Formal Analysis of a Safety-Critical Railway Control Software Module. In: Proceedings of the Conference on Formal Method for Railway Operations and Control Systems (2003)

[2] Güneş Koru, A., Liu, H.: Identifying and characterizing change-prone classes in two large-scale open-source products. Journal of Systems and Software 80(1), 63–73 (2007)

[3] Omerovic, A., Andresen, A., Grindheim, H., Myrseth, P., Refsdal, A., Stølen, K., Ølnes, J.: Idea: a feasibility study in model based prediction of impact of changes on system quality. In: Massacci, F., Wallach, D., Zannone, N. (eds.) ESSoS 2010. LNCS, vol. 5965, pp. 231–240. Springer, Heidelberg (2010)

[4] Glasberg, D., Emam, K.E., Melo, W., Madhavji, N.: Validating Object-Oriented Design Metrics on a Commercial Java Application. National Research Council (September 2000)

[5] Romano, D., Pinzger, M.: Using Source Code Metrics to Predict Change-Prone Java Interfaces. In: Proceedings of 27th IEEE International Conference on Software Maintenance, pp. 303–312 (2011)

[6] Arisholm, E., Sjøberg, D.I.K.: Towards a framework for empirical assessment of changeability decay. The Journal of Systems and Software 53(1), 3–14 (2000)

[7] Arisholm, E., Briand, L.C., Føyen, A.: Dynamic Coupling Measurement for Object-Oriented Software. IEEE Transactions on Software Engineering 30(8), 491–506 (2004)

[8] Bieman, J.M., Andrews, A.A., Yang, H.J.: Understanding Change-proneness in OO Software through Visualization. In: Proceedings of the International Workshop on Program Comprehension (2003)

[9] Thramboulidis, K.C.: Using UML for the Development of Distributed Industrial Process Measurement and Control Systems. In: Proceedings of IEEE Conference on Control Applications, pp. 1129–1134 (September 2001)

[10] Nguyen, K.D., Thiagarajan, P.S., Wong, W.-F.: A UML-Based Design Framework for Time-Triggered Applications. In: Proceedings of 28th IEEE International Symposium on Real-Time Systems, pp. 39–48 (2007)

[11] Lindvall, M.: Measurement of Change: Stable and Change-Prone Constructs in a Commercial C++ System. In: Proceedings of IEEE 6th International Software Metrics Symposium, pp. 40–49 (1999)

[12] Kušek, M., Desic, S., Gvozdanović, D.: UML Based Object-oriented Development: Experience with Inexperienced Developers. In: Proceedings of 6th International Conference on Telecommunications, pp. 55–60 (June 2001)

[13] Abdi, M.K., Lounis, H., Sahraoui, H.: A probabilistic Approach for Change Impact Prediction in Object-Oriented Systems. In: Proceedings of 2nd Artificial Intelligence Methods in Software Engineering Workshop (2009)

[14] Tsantalis, N., Chatzigeorgiou, A., Stephanides, G.: Predicting the Probability of Change in Object-Oriented Systems. IEEE Transactions on Software Engineering 31(7), 601–614 (2005)

[15] Garousi, V., Briand, L.C., Labiche, Y.: Analysis and visualization of behavioral dependencies among distributed objects based on UML models. In: Wang, J., Whittle, J., Harel, D., Reggio, G. (eds.) MoDELS 2006. LNCS, vol. 4199, pp. 365–379. Springer, Heidelberg (2006)

Formulating Dynamic Agents' Operational State via Situation Awareness Assessment

Salama A. Mostafa, Mohd. Sharifuddin Ahmad, Muthukkaruppan Annamalai, Azhana Ahmad, and Saraswathy Shamini Gunasekaran

Abstract. Managing autonomy in a dynamic interactive system that contains a mix of human and software agent intelligence is a challenging task. In such systems, giving an agent a complete control over its autonomy is a risky practice while manually setting the agent's autonomy level is an inefficient approach. This paper addresses this issue via formulating a Situation Awareness Assessment (SAA) technique to assist in determining an appropriate agents' operational state. We propose four operational states of agents' execution cycles; *proceed*, *halt*, *block* and *terminate*, each of which is determined based on the agents' performance. We apply the SAA technique in a proposed Layered Adjustable Autonomy (LAA) model. The LAA conceptualizes autonomy as a spectrum and is constructed in a layered structure. The SAA and the LAA notions are applicable to humans' and agents' collaborative environment. We provide an experimental scenario to test and validate the proposed notions in a real-time application.

Keywords: Autonomous agent, Multi-agent system (MAS), Layered adjustable autonomy (LAA), Dynamic environment, Decision-making.

Salama A. Mostafa · Mohd. Sharifuddin Ahmad · Azhana Ahmad ·
Saraswathy Shamini Gunasekaran
College of Information Technology, Universiti Tenaga Nasional, Putrajaya Campus
43000, Selangor Darul Ehsan, Malaysia
e-mail: semnah@yahoo.com, {sharif,azhana,sshamini}@uniten.edu.my

Muthukkaruppan Annamalai
Faculty of Computer and Mathematical Sciences, Universiti Teknologi MARA,
Shah Alam, Selangor Darul Ehsan, Malaysia
e-mail: mk@tmsk.uitm.edu.my

El-Sayed M. El-Alfy et al. (eds.), *Advances in Intelligent Informatics*,
Advances in Intelligent Systems and Computing 320, DOI: 10.1007/978-3-319-11218-3_49

1 Introduction

The main idea behind any intelligent autonomous systems is to minimize humans' intervention in the systems' operations [1]. In many domains, the applicability of such systems is highly constrained [2]. However, the multi-agent systems (MAS) paradigm reinforces the concept of autonomy to be much more significant [1] as autonomy is a core characteristic of software agents [3]. Nonetheless, the practical setting of the MAS requires that the behavior of the agents in the system subjects to each agent's local desire, as well as the global goal, which mandates dynamic distribution of the agents' autonomy [4]. This issue indicates the fact that agent autonomy is adjustable since interactions with others are required to achieve the goal [2].

The adjustable autonomy paradigm introduces many approaches that enable humans and agents to dynamically share control and initiative such as the mixed-initiative [5] and the sliding autonomy [4]. Some practical examples of agent-based systems that espouse these notions are: human-robots team [3, 6, 7], unmanned vehicles [8] and drones [9].

The Layered Adjustable Autonomy (LAA) model is an agent-based approach that offers solutions to some autonomy problems [10, 11]. The LAA encompasses modeling a spectrum of autonomy in a layered structure, where agents act at different layers of autonomy levels in order to fulfill the system's autonomy conditions. The LAA model aims to give the system explicit control over the agent's actions whenever needed by means of providing autonomy management model and measurement methods to ensure quality and robust decision-making [12].

In this paper, we propose a Situation Awareness Assessment (SAA) technique as a part of the LAA model. The SAA technique checks the agents' responses to the environmental changes and evaluates their performance. The LAA and the SAA feedbacks determine the agents' operational states; *proceed*, *halt*, *break* and *terminate*. The practical reason behind the proposed approach is to synchronize agents' actions by halting agents who are not qualified to operate and reduce system's disturbances by blocking and even terminating agents that operate erroneously.

2 Literature Review

Many researchers acknowledged the fact that autonomous agents might act undesirably due to different circumstances which might be beyond or under their control [2, 3, 4, 5, 7, 9, 14, 15]. Examples of possible situations are found in the cases of learning agents, coerced agents and manipulated agents [9, 13]. Different autonomy models are proposed in the literature including absolute autonomy [1], adjustable autonomy [3, 5, 9] and sliding autonomy [4, 6], each of which is built based on some arguments and offers solutions that resolve specific autonomy issues [7, 10]. Subsequently, situation awareness assessment privileges are frequently used in computer science, especially in computer automation aspects [16, 17, 18, 19, 20]. A significant outcome is the Situation Awareness Global Assessment Technique (SAGAT) proposed by Endsley [16] to measure the situation awareness level of a system. The aim of SAGAT is to assist a pilot of an air fighter to maintain high-level of situation awareness during a mission to enhance his/her decision-making quality and the cockpit's performance. Consequently, many different mechanisms are proposed to constrain agents' decisions via controlling their operational states, some of which are discussed in the following paragraphs.

Kapitan [13] comprehensively discussed agents' autonomy, freedom and responsibility of their actions. Kapitan considered an agent is retrospectively responsible of its action based on predefined conditions. For a given time, if an agent has the intention to do an action and performs the action based on self-obligation, then the agent is responsible for doing the action. He suggested including the motivation and omissions in responsibility determination. He determines the responsibility and the consequence of agent's choice for both doing and omitting an action.

Sierra and Schorlemmer [15] proposed a distributed mechanism to block or ostracize agents that violate certain norms. They experimented a cooperative norm scenario in which two options are possible; cooperate or defect. An agent's utility increases when it cooperates with cooperative agent and vice versa. The violator agents are categorized to unrestricted, semi-restricted and ostracized based on their violation records. The mechanism significantly reduces norm violations among the agents' community and increases their compliance to the cooperation norm.

3 The Layered Adjustable Autonomy

We propose the Layered Adjustable Autonomy (LAA) model to resolve some of the autonomy adjustment issues that concern the interactions of software and human agents to achieve common goals [10]. It enables global control of autonomous agents that guides or even withholds agents' actions whenever necessary. The LAA architecture separates the autonomy into a number of layers. The layers encapsulate an agent's decision process adjustment [12]. Each layer is attributed to deal with actions that correspond to the agent's autonomy level [11]. The LAA distributes the autonomy to a qualified agent that satisfies the knowledge (*know*) and/or authority (*can*) conditions to act on an event (i.e. agent: *know* (*event*) ∧ *can* (*event*) → *decide* (*event*)) [10]. This is achieved by formulating autonomy management structure, measurement criteria and intervention rules on the agents' decisions of actions [12].To give the agent the characteristics of intelligence and speed which are the core requirements for a dynamic interactive system, the agents in the LAA are designed to work through the Deliberation, Decision and Action modes. Consequently, the agents' interactions with its surrounding proceeds through five phases which are observe event, reason event, select action, obtain autonomy layer and process action as shown at the bottom of Figure 1 [21]. An Autonomy Analysis Module (AAM) and a Situation Awareness Module (SAM) are part of an agent's architecture. They cooperate to distribute the autonomy based on the LAA conditions. There are other uses of the AAM and SAM that are beyond the paper's scope[1]. The paper proposes a Situation Awareness Assessment (SAA) strategy to assess agents' performance and support the LAA in the autonomy distribution and adjustment. Figure 1 describes the proposed LAA model and its related parts including the SAA.

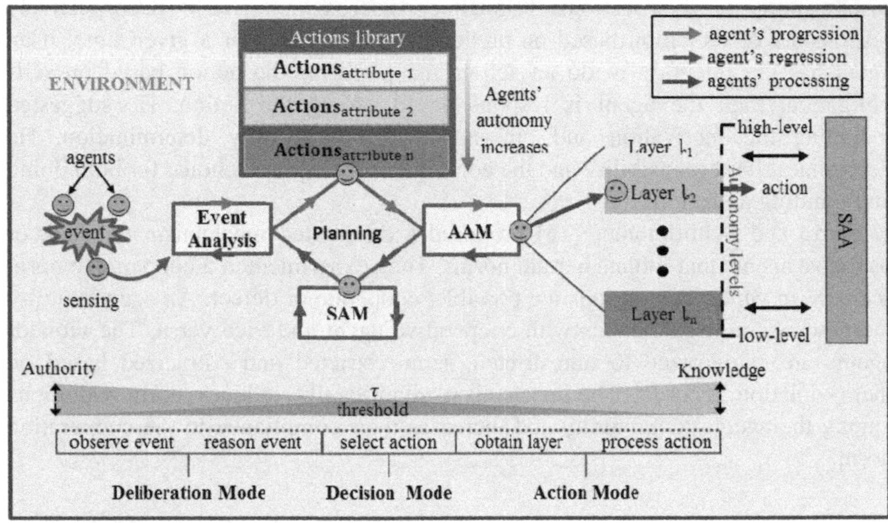

Fig. 1 The layered adjustable autonomy model

An autonomy layer can be viewed as a virtual entity that has specific autonomy properties and is attributed by specific sets of actions based on some conditions [11, 21]. We define an action, , as an organized sequence of activities that are represented via programmatic functions to accomplish discrete operations of the action. We regard the completion of a task, Ϯ, achieves the goal state. Task, Ϯ, comprises a set of discrete and/or repetitive actions, Ϯ = { , , ...}. The actions in the LAA are categorized based on some attributes. Each attributed actions category has a corresponding autonomy layer that has a set of autonomy properties [21]. When an agent selects an action, its autonomy is encapsulated based on the selected action's attribute and the autonomy layer's properties [10]. Figure 2 shows a generic model of the autonomy layers and the associations of the actions to their corresponding autonomy layers. The actions' categorization and the

[1] Only the basics of the AAM and SAM are described in this paper. The details can be found in [9], [10], [11] and [12].

autonomy layers' properties determination are domain specific issues. An example of possible application to the LAA model in a domain is detailed in Mostafa et al. [21]. Subsequently, the LAA model provides a mechanism that facilitates the agent to interact with a human as well as with other agents when it performs actions. The agent obtains a layer with a property that enables it to perform the interactions [21]. However, the interactions between the human and the agent are constrained via the actions' attributes [10]. Some actions are engineered to disable the agent from interacting with others.

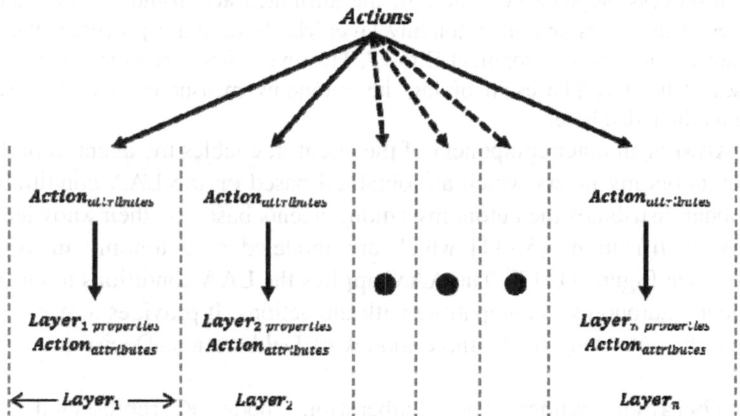

Fig. 2 A generic model of the autonomy layers

Definition 1. Let A be the set of all actions sorts, α be an action and $\alpha \in$ A; Ŧ be the possible tasks, Ŧ = {ŧ$_1$, ŧ$_2$, …} in which ŧ $\subset$ A;

Definition 2. Let ȹ be the *actions acquiring attribute function*, α̧ be the action after acquiring its attributes, ȹ: α $\rightarrow$ α̧ and Ą = {α̧$_1$, α̧$_2$, ...}; Ł be the possible autonomy layers and Ł={ƚ$_1$, ƚ$_2$, …}; ȸ be the *action and the layer association function*, α̧̇ be the action after inheriting its corresponding layer autonomy properties, ȸ: α̧ x ƚ $\rightarrow$ α̧̇ and Ą̇ = {α̧̇$_1$, α̧̇$_2$,...};

Definition 3. Let an agent ρ be defined as an entity which performs a set of actions A based on the current states of the environment, E, to produce a new set of environment E', ρ: α x E $\rightarrow$ E' and ρ functions based on the LAA model rules where ∀ α ∃ ƚ; ρ equipped with *observe*, *reason*, *select*, *obtain* and *process* functions to perform actions; assume that there exist an event *e* $\in$ E and ρ responds to *e* according to the LAA via the following phases, which are described by means of axioms;

Phase 1: $\exists\, \rho\, \exists\, E\, [(observe(\rho, E) \Rightarrow e]$

Phase 2: $\exists\, e\, [(reason(\rho, e) \Rightarrow ŧ]$

Phase 3: $\exists\, ŧ\, [((select(\rho, ŧ) \rightarrow \alpha) \rightarrow ꝑ(\alpha) \Rightarrow ą]$

Phase 4: $\exists\, Ł\, \exists\, ą\, [((obtain(\rho, Ł) \rightarrow ƚ) \rightarrow ȸ(ą, ƚ) \Rightarrow \dot{ą}]$

Phase 5: $\exists\, \dot{ą}\, [(process(\rho, \dot{ą}) \Rightarrow E'])))))$

The axioms describe the phases of an agent ρ exclusion cycle along with the LAA process. They outline the process of ρ performs action α ∈ ŧ in responses to event e. By reasoning over e, ρ selects an action α, ꝑ sets the action's attributes, ą; ρ obtains a layer ƚ to process the selected action, ą; the attributed action inherits the autonomy properties of its corresponding autonomy layer via ȸ, ą̇; and ρ processes the action; thus, transforming the environment E to E'. However, there are other intermediating processes of the five phases involving the autonomy measurement and distribution that are not included here.

The AAM is an inner component of the agent. It enables the agent to perform at different autonomy layers which are obtained based on the LAA conditions. The LAA model distributes the autonomy among agents based on their knowledge and authority qualifications [5, 11] which are modeled as autonomy measurement attributes (see Figure 1) [12]. The AAM applies the LAA conditions to ensure that each agent's autonomy is compatible with the actions. It provides a mechanism to distribute the autonomy to the three modes of Deliberation, Decision and Action as follows:

1. **Deliberation Mode:** The deliberation mode is represented by the observation and the reasoning phases. In this mode, the agent with the higher knowledge to deliberate on what it observes is more qualified than the agent with lesser knowledge. The outcome of this mode is a task that needs to be accomplished.
2. **Decision Mode:** In the action selection phase, both the knowledge and authority autonomy attributes are used to determine which agents to decide on actions as different agents might select different actions to perform the same task. The outcome of this mode is the selected action or actions to perform the task.
3. **Action Mode:** During the action execution phase, the agent's autonomy condition is built upon its authority to act. Hence, the agent that obtains an autonomy layer is able to access specific resources to perform the action. The outcome of this mode is the action execution.

The SAM is responsible for performing re-planning via checking the re-planning conditions of the agents after performing each action. The SAM uses a *plan validating function* (Ɣ) to investigate the validity of the plan which is a part of the agent deliberation mode. After each action processing, Ɣ checks the state of the beliefs that are responsible for motivating the processed desire or pre-desire (pre-D). If there is a match between the pre-beliefs (pre-B) and post-beliefs (post-B), in which both of them indicates states that match the pre-D, then the plan is valid. Otherwise, if there is a contradiction between the pre-B, the post-B and the pre-D, then the plan is no longer valid and plan reconsideration is required. The post-B might be combined with other transition states due to the environment's dynamism, hence, the pre-B and post-B are filtered via *filter function* before the comparison process. The re-planning might derives new desire (post-D) and causes the SAM to call the *options function* that considers four options; *operate*, continue with the plan; *repeat*, repeat the previous action; *insert*, insert new action; and *skip*, skip the current action.

4 The Situation Awareness Assessment Technique

The Situation Awareness Assessment (SAA) technique assesses agents' performance. The main objective of the SAA is to test the LAA model efficiency when agents are performing actions. The SAA identifies the successful and the unsuccessful agents' actions via comparing the actions' outcomes with predetermined and expected results. Consequently, the SAA measures the utility of the agents in order to adjust their operational states based on their performances. As mentioned earlier, each agent has its own AAM and SAM. The SAM provides a mechanism to interact with the SAA technique and process its feedback. The AAM applies the operational states' adjustment throughout the three modes of the agent's execution cycle. In addition, the agent has a log file, which contains records of the agent's activities including the operational states for every execution cycle. The following steps and Figure 3 illustrate the process of agents' engagements and data transmission with the LAA model and the SAA technique.

1. The LAA initial distributor retrieves the initial autonomy criteria.
2. The initial distributor sets the autonomy parameters of the layers and distributes the autonomy to the AAMs of agents.
3. The qualified agents that have *proceed* operational state perform actions.
4. The AAMs set the unqualified agents to the *halt* operational state.
5. The SAA checks the unsuccessful actions and notifies the SAMs.
6. The SAMs set the agents that cause unsuccessful action to the *block* state.
7. The agents update their respective log files.
8. The SAA retrieves all the log files' data of the agents.
9. The SAA measures the utility of the agents then decide on the *terminate* operational state of them.
10. The LAA dynamic distributor adjusts and distributes the autonomy to the AAMs of the agents based on the initial distributor and the execution cycle feedback.
11. The dynamic distributor updates the initial criteria file records for the next run.

After going through the 11 steps in the first round, in the proceeding rounds, the first and the second steps are not applicable.

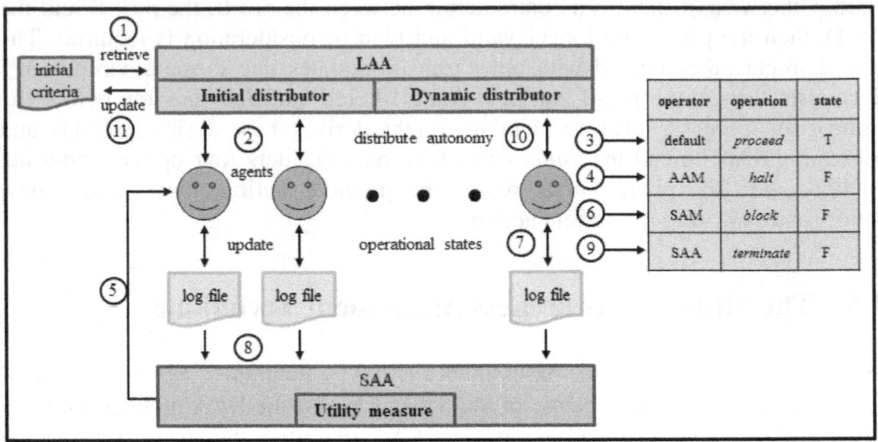

Fig. 3 Agents' engagement and data transmission

4.1 The Formulation of the Operational States

In this section, we detail the formulation of the proposed *proceed*, *halt*, *block* and *terminate* operational states and their corresponding procedures. We explain the proposed operational states configuration options. We also express the mechanism of agents' utility measurement and situation awareness assessment. The operational states are controlling conditions that determine the agents' execution cycle. The denotation of which are as follows:

1. *proceed*: The agent does a complete execution cycle if the operational states 2, 3 and 4 (given below) are invalid.
2. *halt*: The agent's execution cycle is interrupted due to some autonomy constraints. The AAM provides a mechanism that is responsible for the *halt* state activation condition.
3. *block*: The agent's execution cycle is interrupted due to some situation awareness constraints. The SAM provides a mechanism that is responsible for the *block* state activation condition.
4. *terminate*: The agent stops functioning as a result of occurrences of operational states 2 or 3. The SAA technique provides a mechanism that is responsible for the *terminate* state activation condition.

proceed operation: The agent that satisfies the LAA and the SAA conditions can carry out a complete execution cycle. Consequently, the agent that is under the *halt* or *block* state and does not terminate, its operational state reverts to the *proceed* state.

***halt operation*:** The AAM checks and distributes the autonomy three times in one execution cycle. The first autonomy checking and distribution is in the deliberation mode (phase 1 and 2), the second is in the decision mode (phase 3) and the third is in the action mode (phase 4 and 5). The *halt* operation might occur in any of the three modes. It occurs when the agent fails to fulfill the LAA conditions[2].

***block operation*:** The block operation is a consequence of an agent that performs unsuccessful action. After an agent performed an action, its SAM reasons the beliefs and the desires to investigate action successfulness (as explained in Section 3), then updates the corresponding agent's *block* operational state. Subsequently, the SAA assesses the agents' performances and notifies the SAMs to block the agents that performed unsuccessful actions. The SAA uses *performance update function* (ƥ), performance matrix (ℙ) and performance scheme matrix (Ṕ) to implement the assessment strategy and to keep track on all the agents. ℙ is updated via the ƥ based on Ṕ after each agent's run cycle as explained in Section 4.2. The *performance update function* ƥ process is defined as:

$$\text{ƥ}(\mathrm{P}_{i,j}) = \begin{cases} \mathrm{P}_{i,j} = \acute{\mathrm{P}}_{i,j}, & 1 \\ \mathrm{P}_{i,j} \approx \acute{\mathrm{P}}_{i,j} & 0 \\ \mathrm{P}_{i,j} \neq \acute{\mathrm{P}}_{i,j} & -1 \end{cases}$$

where $\mathrm{P}_{i,j}$ is the performance matrix, i is agent index, j is the action index $j = j + \sigma$ and σ is a plan shifting factor. The value of σ is changeable based on the planned actions checking points as it is difficult to check all the actions, especially in dynamic environment. 1 indicates successful actions, 0 indicates non-scheme action and $\sigma = 1$ and -1 indicates unsuccessful action. $\approx$ means checking the action is not required.

***terminate* operation:** It is valid when the performance of an agent has negative results. The *terminate* operation is measured by the SAA's *utility update function* (Џ). It is calculated after an agent encounters a *block* or *halt* operation. The *utility update function* uses three operational states which are *proceed* (S_p), *halt* (S_h) and *block* (S_b) as input and decides between the other two operational states which are S_p or *terminate* (S_t). The SAA uses a utility matrix (Ư) to implement the assessment strategy and to keep track on the agents' performance. The utility matrix is updated based on the three operational states *proceed* (S_p), *halt* (S_h) and *block* (S_b) occurrence after each agent's run cycle using the *utility update function* (Џ):

$$\text{Џ}(\text{Ư}_i) = \begin{cases} \text{Ư}_i = \text{Ư}_i + 1, & S_p \Rightarrow S_p \\ \text{Ư}_i = \text{Ư}_i - 1, & S_h or S_b \Rightarrow \begin{matrix} S_p, & (\neg\min(\text{Ư}) \wedge (\text{Ư}_i \geq 0)) \\ S_t, & (\min(\text{Ư}) \wedge (\text{Ư}_i < 0)) \end{matrix} \end{cases}$$

where $\min(\text{Ư})$ is a function that returns the minimum utility score of the agents, i is the agents index and $i = \{1, 2, \ldots\}$.

---

[2] The LAA model initial distributor of autonomy uses the data that is collected from the system's previous runs as initial criteria using probability formulae [12]. The data contains errors and failures history when agents implement actions and the impact of each. Some of the autonomy parameters are configured during testing based on the corresponding agent performance.

The utility measure procedure terminates the agents that perform unsuccessful actions. It ensures that there is at least one active agent in the system. Consequently, the SAA checks the efficiency of the LAA model as the LAA distributes the autonomy to the agents based on their performance.

4.2 Experimental Scenario

The above model is currently being applied to an online search application using a drone system. The drone performs surveillance mission and finds a number of objects within closed area. The agents are deployed to autonomously control the drone. The agents perform precompiled actions using an adjustable plan. The plan is a depth first search algorithm that represents a specific scheme and shared by the agents. The plan scheme is a grid that contains the required drone path and the objects' locations. The plan's scheme is used by the SAA technique to determine the agents' actions successes. The following diagram shows a visualize example of the plan scheme.

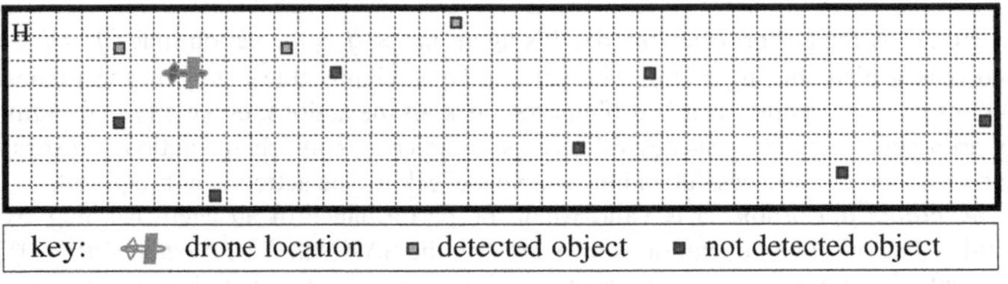

Fig. 4 The plan scheme

The SAA validates the agents' actions in the environment by assessing the drone performance based on the plan's scheme. The assessment includes drone path and location and the number of identified objects by the drone. In the experimental setting, we assume that the environment has fixed parameters each of which is observable and understandable by the agent. The experiment has two objectives; from the agents' utility measurement outcomes, we conclude the assessment of the overall performance, and ultimately, the accurate assessment leads us to reengineer the system in such a way that enhances its performance.

5 Conclusion and Future Work

In this paper, we present a generic representation to the Layered Adjustable Autonomy (LAA) in a multi-agent system. We propose a Situation Awareness Assessment (SAA) technique to test and evaluate agents' performance. The SAA outcomes are used to enhance agents' situation awareness. The motivation behind using *proceed*, *halt*, *block*, and *terminate* operational states is to synchronize agents' actions and reduce system disturbances. We provide an example scenario to prove the applicability of the proposed model.

The Layered Adjustable Autonomy (LAA) model and the SAA technique report the work-in-progress of our research and they are currently being applied to a drone system. In our future work, we shall report the results of the application. We argue that the proposed agents' operational states and the SAA technique can be further improved to manifest resilient measurement of autonomous systems.

Acknowledgments. This project is sponsored by the Malaysian Ministry of Higher Education (MoHE) under the Exploratory Research Grant Scheme (ERGS) No. ERGS/1/2012/STG07 /UNITEN/02/5.

References

1. Anderson, J., Evans, M.: Supporting Flexible Autonomy in a Simulation Environment for Intelligent Agent Designs. In: Proceedings of the Fourth Annual Conference on AI, Simulation, and Planning in High Autonomy Systems, pp. 60–66. IEEE Press, Tucson (1993)
2. Evans, M., Anderson, J., Crysdale, G.: Achieving Flexible Autonomy in Multi-Agent Systems using Constraints. Applied Artificial Intelligence: An International Journal 6, 103–126 (1992)
3. Johnson, M., Bradshaw, J.M., Feltovich, P.J., Jonker, C., van Riemsdijk, B., Sierhuis, M.: Autonomy and Interdependence in Human-agent-robot Teams. IEEE Intelligent Systems 27(2), 43–51 (2012)
4. Sellner, B., Heger, F., Hiatt, L., Simmons, R., Singh, S.: Coordinated Multi-agent Teams and Sliding Autonomy for Large-scale Assembly. Proceeding of the IEEE 94(7), 1425–1444 (2006)
5. Bradshaw, J.M., Feltovich, P.J., Jung, H., Kulkarni, S., Taysom, W., Uszok, A.: Dimensions of Adjustable Autonomy and Mixed-Initiative Interaction. In: Nickles, M., Rovatsos, M., Weiss, G. (eds.) AUTONOMY 2003. LNCS (LNAI), vol. 2969, pp. 17–39. Springer, Heidelberg (2004)
6. Brookshire, J., Singh, S., Simmons, R.: Preliminary Results in Sliding Autonomy for Coordinated Teams. In: Proceedings of the AAAI 2004 Spring Symposium, CA (2004)
7. Roehr, T.M., Shi, Y.: Using a Self-confidence Measure for a System-initiated Switch between Autonomy Modes. In: Proceedings of the 10th International Symposium on Artificial Intelligence, Robotics and Automation in Space, Sapporo, Japan, pp. 507–514 (2010)
8. Mercier, S., Tessier, C., Dehais, F.: Adaptive Autonomy for a Human-Robot Architecture. In: 3rd National Conference on Control Architectures of Robots, Bourges (2008)
9. Mostafa, S.A., Ahmad, M.S., Annamalai, M., Ahmad, A., Gunasekaran, S.S.: A Dynamically Adjustable Autonomic Agent Framework. In: Rocha, Á., Correia, A.M., Wilson, T., Stroetmann, K.A. (eds.) Advances in Information Systems and Technologies. AISC, vol. 206, pp. 631–642. Springer, Heidelberg (2013)

10. Mostafa, S.A., Ahmad, M.S., Annamalai, M., Ahmad, A., Gunasekaran, S.S.: A Conceptual Model of Layered Adjustable Autonomy. In: Rocha, Á., Correia, A.M., Wilson, T., Stroetmann, K.A. (eds.) Advances in Information Systems and Technologies. AISC, vol. 206, pp. 619–630. Springer, Heidelberg (2013)
11. Mostafa, S.A., Ahmad, M.S., Annamalai, M., Ahmad, A., Basheer, G.S.: A Layered Adjustable Autonomy Approach for Dynamic Autonomy Distribution. In: 7th International KES Conference on Agents and Multi-agent Systems Technologies and Applications (KES-AMSTA), vol. 252, pp. 335–345. IOS Press, Hue City (2013)
12. Mostafa, S.A., Ahmad, M.S., Ahmad, A., Annamalai, M., Mustapha, A.: A Dynamic Measurement of Agent Autonomy in the Layered Adjustable Autonomy Model. In: Badica, A., Trawinski, B., Nguyen, N.T. (eds.) Recent Developments in Computational Collective Intelligence. SCI, vol. 513, pp. 25–35. Springer, Heidelberg (2014)
13. Kapitan, T.: Autonomy and manipulated freedom. Noûs 34(s14), 81–103 (2000)
14. Bindiganavale, R., Schuler, W., Allbeck, J.M., Badler, N.I., Joshi, A.K., Palmer, M.: Dynamically altering agent behaviors using natural language instructions. In: Proceedings of the Fourth International Conference on Autonomous Agents, pp. 293–300. ACM (2000)
15. Sierra, A.P.D.P.C., Schorlemmer, M.: Friends no more: Norm enforcement in multi-agent systems (2007)
16. Endsley, M.R.: Situation Awareness Global Assessment Technique (SAGAT). In: Aerospace and Electronics Conference, NAECON 1988, Proceedings of the IEEE 1988 National, pp. 789–795. IEEE Press (1988)
17. Lili, Y., Rubo, Z., Hengwen, G.: Situation Reasoning for an Adjustable Autonomy System. International Journal of Intelligent Computing and Cybernetics 5(2), 226–238 (2012)
18. Endsley, M.R., Connors, E.S.: Situation Awareness: State of the Art. In: 2008 IEEE Power and Energy Society General Meeting-Conversion and Delivery of Electrical Energy in the 21st Century, pp. 1–4. IEEE Press, Pittsburgh (2008)
19. Mostafa, S.A., Ahmad, M.S., Tang, A.Y.C., Ahmad, A., Annamalai, M., Mustapha, A.: Agent's Autonomy Adjustment via Situation Awareness. In: Nguyen, N.T., Attachoo, B., Trawiński, B., Somboonviwat, K. (eds.) ACIIDS 2014, Part I. LNCS (LNAI), vol. 8397, pp. 443–453. Springer, Heidelberg (2014)
20. Mostafa, S.A., Ahmad, M.S., Ahmad, A., Annamalai, M.: Formulating Situation Awareness for Multi-agent Systems. In: Advanced Computer Science Applications and Technologies (ACSAT), pp. 48–53. IEEE Press, Kuching (2013)
21. Mostafa, S.A., Gunasekaran, S.S., Ahmad, M.S., Ahmad, A., Annamalai, M., Mustapha, A.: Defining Tasks and Actions Complexity-levels via Their Deliberation Intensity Measures in the Layered Adjustable Autonomy Model. In: The 10th International Conference on Intelligent Environments (IE 2014). IEEE Press, Shanghai (2014)

An Intelligent Modeling of Oil Consumption

Haruna Chiroma, Sameem Abdulkareem, Sanah Abdullahi Muaz,
Adamu I. Abubakar, Edi Sutoyo, Mungad Mungad, Younes Saadi,
Eka Novita Sari, and Tutut Herawan

Abstract. In this study, we select Middle East countries involving Jordan, Lebanon, Oman, and Saudi Arabia for modeling oil consumption based on computational intelligence methods. The limitations associated with Levenberg-Marquardt (LM) Neural Network (NN) motivated this research to optimize the parameters of NN through Artificial Bee Colony Algorithm (ABC-LM) to build a

Haruna Chiroma · Sameem Abdulkareem
Department of Artificial Intelligence, University of Malaya, Malaysia
e-mail: hchiroma@acm.org, sameem@um.edu.my

Haruna Chiroma
Federal College of Education (Technical), School of Science,
Department of Computer Science, Gombe, Nigeria
e-mail: hchiroma@acm.org

Sanah Abdullahi Muaz
Department of Software Engineering, University of Malaya, Malaysia
e-mail: samaaz.csc@buk.edu.ng

Adamu I. Abubakar
Department of Information System, International Islamic University Malaysia, Malaysia
e-mail: 100adamu@gmail.com

Edi Sutoyo · Mungad Mungad
Department of Information Systems, University of Malaya, Malaysia
e-mail: {edisutoyo,mungad}@um.edu.my

Younes Saadi
University Tun Hussein Onn Malaysia, Malaysia
e-mail: younessaadi@gmail.com

Eka Novita Sari
AMCS Research Center, Indonesia
e-mail: eka@amcs.co

Tutut Herawan
Universiti Malaysia Pahang, Malaysia
e-mail: tutut@um.edu.my

El-Sayed M. El-Alfy et al. (eds.), *Advances in Intelligent Informatics*,
Advances in Intelligent Systems and Computing 320, DOI: 10.1007/978-3-319-11218-3_50

model for the prediction of oil consumption. The proposed model was competent to predict oil consumption with improved accuracy and convergence speed. The ABC-LM performs better than the standard LMNN, Genetically optimized NN, and Back-propagation NN. The proposed model may guide policy makers in the formulation of domestic and international policies related to oil consumption and economic development. The approach presented in the study can easily be implemented into a software for use by the government of Jordan, Lebanon, Oman, and Saudi Arabia.

Keywords: Artificial Bee Colony, Neural Network, Levenberg-Marquardt, Oil Consumption, Prediction.

1 Introduction

A significant amount of energy supply to the world market comes from the Middle East. Yet, there are countries in the Middle East that significantly depend on oil importation for domestic consumption. Thus, we propose to model oil consumption in some selected Middle East countries. Jordan and Lebanon were selected due to their insufficient production of oil. Saudi Arabia was involved in the case study of its role in supplying significant quantity of oil to the global market and being a member of the Organization of the Petroleum Exporting Countries (OPEC). Oman attracted attention for involvement in the study as a result of its significant production of crude oil and the country is not a member of the OPEC [1].

There are studies on modeling of energy demand and consumption in the literature, for example, Assure *et al.* [2] applied Particle Swarm Optimization (PSO) and Genetic Algorithm (GA) to estimate oil demand in Iran based on socioeconomic indicators. The key indicators of energy demand in Turkey such as population, Gross Domestic Product (GDP), import and export were used for energy demand estimation in Turkey using Ant Colony Optimization (ACO) [3]. Chiroma *et al.* [4] applied Co-Active Neuro Fuzzy Inference System (CANFIS) because of its strengths over Fuzzy Neural Network (NN) and Adaptive Network-based Fuzzy Inference System (ANFIS). The CANFIS was modelled to predict crude oil price and comparative analysis shows that the CANFIS was found to perform better than the ANFIS. To improve the effectiveness of the CANFIS, Chiroma *et al.* [5] model the consumption of energy in Greece based on the hybridization of GA and CANFIS. It was found that the propose approach performs better than the comparison methods including CANFIS. Kaynar *et al.* [6] proposes NN and neuro–fuzzy system to predict natural gas consumption in Turkey. Results indicated that the prediction accuracy of the ANFIS was better than the Autoregressive Integrated Moving Average (ARIMA), multi-layer perceptron, and radial basis function networks. Malaysia oil production was estimated using ARIMA by [7]. However, the ARIMA typically assume normal distribution for input data which makes the ARIMA unsuitable for modeling energy because of the high level of energy uncertainty [8].

Applications of GA in training NN require lengthy local searches close to a local optimum. In GA, if population changes, the GA abolished the preceding knowledge of the problem [9]. The PSO has the possibility of being stuck in local optimal which can undermine its effectiveness [9]. Despite the effectiveness of NN in solving linear, nonlinear, and complex problems, the NN can possibly be trapped in local minima and over-fit the training data. There is no ideal framework for determining the optimal structure of the NN and the selection of the initial training parameters. Researchers typically employ the cumbersome trial-and-error technique to determine the optimal structure and selection of the NN parameters [10]. However, Karaboga and Basturk [11] reported that several comparative studies proved that the performance of the Artificial Bee Colony algorithm (ABC) outperforms GA, ACO, PSO, and Differential Evolution algorithms.

In this study, we propose to hybridize ABC and Levenberg-Marquardt NN (LM) due to its accuracy and convergence speed over other fast learning algorithms [12] and computational efficiency [13] in order to improve prediction accuracy and convergence speed in modeling oil consumption. The hybridization of ABC and LM (ABC-LM) is not a new concept, but applying the concept in the modeling of oil consumption in Jordan, Lebanon, Oman, and Saudi Arabia is the innovation of this study.

The rest of this paper is organized as follows. Section 2 describes the proposed method and data collection. Section 3 describes experimental results and discussion. Finally the conclusion and future works are presented in Section 4.

2 Proposed Method

2.1 Artificial Bee Colony

In the ABC algorithm, the colony of artificial bees (AB) involve three (3) major stages, including: employed bees, onlookers, and the scout. The first portion of the colony comprises of the employed AB, whereas the second portion contain onlookers. For each of the food source, only one employed bee exists. The employed bee of the discarded food source becomes the scout. The execution of ABC can be summarized in the following steps [11]:

Initialize population with random possible solutions

REPEAT

Step 1: Move the employed bees onto their food sources and determine their nectar amounts.

Step 2: Move the onlookers onto the food sources and determine their nectar amounts.

Step 3: Move the scouts for searching new food sources.

Step 4: Memorize the best food source found so far and continue

UNTIL (requirements are met and optimal solution found).

2.2 Data Collection

The ABC-LM is a data driven model, therefore, require data and careful design for successful modeling of the oil consumption. The oil referred to in this paper as reported by [1] comprised of liquefied petroleum gases, residual fuel, distillate oil, kerosene, jet fuel, motor gasoline, and other petroleum products. Our goal is to collect data for domestic oil consumption in Jordan, Lebanon, Oman, and Saudi Arabia. The oil consumption data were collected from the official website of the Energy Information Administration of the US department of Energy on a yearly basis for a period starting from 1981 to 2012. The data were found to be complete without missing points, outliers, and incomplete information. The oil consumption data are given in thousands of barrels. The independent variables responsible for domestic oil consumption are the installed power capacity and the residence yearly/electricity consumption [14]. The yearly ambient temperature, GDP, the amount of CO_2 pollution, number of air conditioners, electricity price, installation of renewable energy technologies and etc. [14]. According to Khazem [15], data availability is one of the criteria's for the selection of variables to be used in modeling. The data set for modeling in this study were not normalized to prevent the destruction of original patterns in the historical data [16]. The data were subjected to descriptive analysis in order to unveil their characteristics. The descriptive statistics of the data sets including maximum and minimum oil consumption are presented in Table 1. The standard deviation indicates that the dispersion of the oil consumption do not much deviated from each other except for Saudi Arabia.

Table 1 Descriptive statistics of the oil consumption

	Min.	Max.	Mean	Std. Dev.
Jordan	37.00	112.43	81.63	22.39
Lebanon	33.00	128.86	75.66	29.12
Oman	16.66	144.89	53.01	32.22
Saudi Arabia	610.0	2861.00	1418.79	547.30

2.3 Experiment Setup for Modeling ABC-LM for Oil Consumption

The complete modeling setup is represented in the flowchart shown in Fig. 1. The major steps involved in the modeling comprised of data collection and preprocessing, ABC and NN initialization, and evaluation of the proposed method.

The ABC-LM requires data for training and testing the efficiency of the model. Data partition in modeling has no specific universal acceptable partition ratio. Several data partition ratios were experimented and 80% for training and 20% for testing the effectiveness of the ABC-LM model was adopted. The objective of the

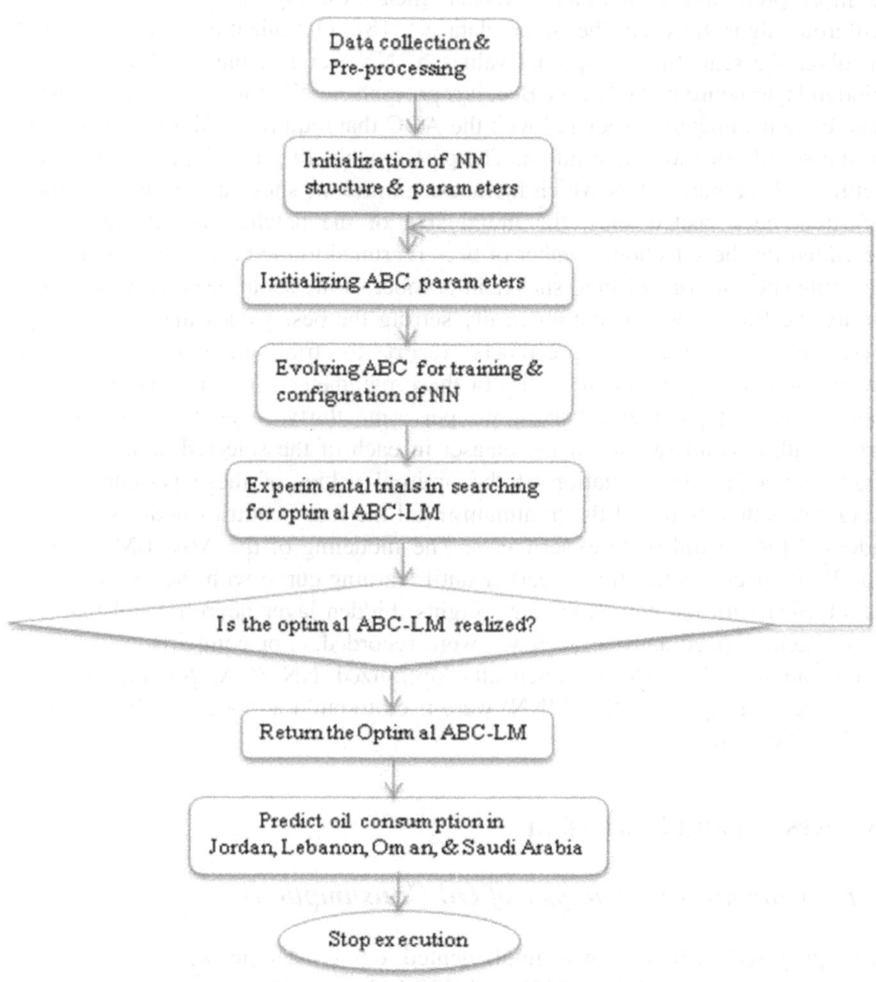

Fig. 1 The proposed method for intelligent modeling of the oil consumption

ABC is to optimize the NN weights, bias, and hidden neurons. The NN require initialization for the ABC optimization to start execution. The input neurons were fixed to five (5) since independent variables in the historical are five (5), and the dependent variable is the oil consumption because it depends on the independent variables. The hidden layer of the NN was set to one (1), though, the number of hidden layers can be more than one (1) but theoretical work of [17] argued that one hidden layer is sufficient for modeling any function. The activation functions of the hidden and output layer are sigmoid and linear, respectively, as recommended by [12]. The epoch for the training was set to 3000. The objective function used for the ABC in the experiment is Mean Square Error (MSE) because is more preferable than other statistical measures, especially when comparing different algorithms on the same dataset [18]. Optimization using the ABC involves the searching of optimal values of NN weights, bias, and a number of hidden layer neurons. For every Bee, it represents an NN for optimizing the MSE. There are parameters associated with the ABC that requires settings for efficiency in the search for the best solution. The performance of ABC depends on the best settings of the parameters which includes: number of sites chosen out of number of scout bees visited sites, the initial size of the patches, a number of bees recruited for the selection, number of bees recruited to the best sites, a number of the elite sites out of selected sites, and number of the scout bees. However, there is no idle framework for automatically setting the best values that can yield the best results. Typically, researchers resort to trial-and-error methods for determining the approximate values of these parameters. In our work, in order to obtain the best parameter values, we perform thirty seven (37) experimental trials with a small sample of the dataset in each of the selected countries under study for a fair determination of the optimal values of these parameters. The parameters that produced the minimum in all the four countries under study were adopted for the full scale experiments. The modeling of the ABC-LM continues for 3000 epochs as mentioned earlier until learning curve with the minimal MSE is returned with the corresponding weights, hidden layer neurons, and bias. The convergence speed and the accuracy were recorded. For comparison, the Back-propagation NN (BPNN), genetically optimized NN (GANN), and standard Levenberg-Marquardt NN (LMNN) were used to build a model for the prediction of the oil consumption.

3 Results and Discussion

3.1 Comparative Analysis of Oil Consumptions

The proposed ABC-LM was implemented on a machine with the following configurations: MATLAB 2012b and SPSS Version 20 on a personal computer (HP L1750, 4 GB RAM, 232.4 GB HDD, 32-bit OS, Intel Core 2 Duo CPU @ 3.00 GHz). The oil consumption for the countries selected for this study was analyzed using the statistical t-test for testing the significant difference between oil consumption of the comparison countries. The results of the analysis are presented in Tables 2-3. The oil consumption in Jordan was compared to that of Lebanon for fair comparison because both countries don't produce significant amount of oil and heavily depends on oil import for domestic use. The t-test analysis was performed under the assumption that the oil consumption in Jordan and Lebanon is equal. The t-test results for the two countries are shown in Table 2.

Table 2 T-test results *(p-value = 0.05, 95% interval)* for oil consumption in Jordan and Lebanon

	t	df	Sig.	Mean Difference
Jordan	20.628	31	0.000	81.63045
Lebanon	14.698	31	0.000	75.66600

The t-test results indicated in Table 2, shows that there is a significant difference between the oil consumption in Jordan and Lebanon. Therefore, we concludes that the oil consumption in Jordan is significantly higher than the oil consumption in Lebanon. Meaning that the GDP of Jordan is better than that of Lebanon since oil consumption is positively related to GDP [14]. Hence, the industrial and economic activities in Jordan could probably be higher compared to Lebanon as economic development is highly driven by oil consumption.

Table 3 T-test results *(p-value = 0.05, 95% interval)* for oil consumption in Oman and Saudi Arabia

	t	df	Sig.	Mean Difference
Oman	9.873	23	0.000	62.97208
Saudi Arabia	16.578	23	0.000	1619.72714

Revenue generation in both Oman and Saudi Arabia heavily depend on the sales of oil and they both produce significant amount of oil to the global oil market. The t-test results presented in Table 3 was performed under the assumption that the oil consumption in both Oman and Saudi Arabia is equal. Table 3 shows that there is a significant difference between oil consumption in Oman and Saudi Arabia. Therefore, the oil consumption in Saudi Arabia is significantly higher than that of Oman. The probable reason could be attributed to economic development in Saudi Arabia because oil consumption influences growth in GDP. The mean oil consumption for each of the countries as shown in Tables 2-3 indicated that Saudi Arabia has a significantly higher level of oil consumption than Jordan, Lebanon, Oman, and Saudi Arabia. Meaning that among the countries considered as our case study, the GDP of Saudi Arabia outperforms that of the other countries. This implies boom in economic development and industrial production. Thus, estimation of oil consumption become imperative for development planning in these countries.

3.2 ABC-LM for the Oil Consumption

The optimal parameters used for the modeling of oil consumption based on ABC-LM as described in section 2.4 are presented in Table 4. The parameters in Table 4 were realized after the best results were found at the end of the experimental trials in modeling the oil consumption.

Table 4 Parameters of the ABC-LM model

The optimal parameter	Setting
Number of sites chosen out of number of scout bees visited sites	32
Number of elite sites out of m selected sites	2
Number of scout bees	150
Number of bees recruited for the best e sites	50
Number of bees recruited for the selection	20
The initial size of the patches	0.5
Number of NN input neurons	5
Number of NN hidden neurons	7
NN hidden layer Activation function	Sigmoid
NN output layer activation function	Liner
NN Number of hidden layers	1
Number of output neurons	1
Epoch	3000

The best error convergence curve of the ABC-LM model is depicted in Fig. 2. The curve has no oscillation which signifies the smooth convergence to the best fitness function (MSE). The straight line at the end of the curve indicates convergence at a point where the optimum fitness could not improve. This phenomenon indicates successful convergence [19].

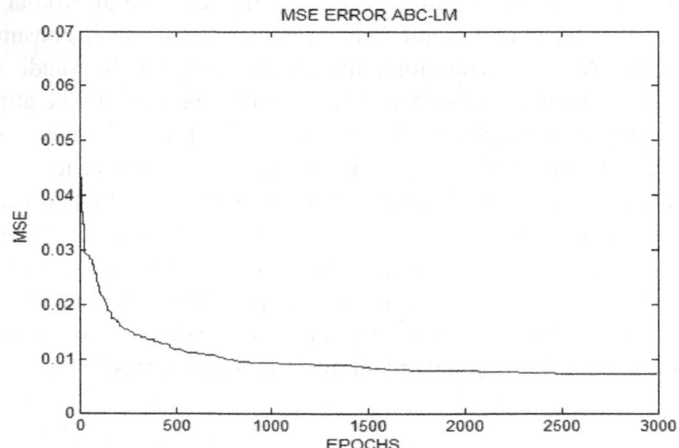

Fig. 2 Convergence of ABC-LM on the training dataset.

The ABC-LM model was used to predict oil consumption in the four countries under study. The oil consumption data from the 2007 to 2012 was reserved for testing the effectiveness of our proposed ABC-LM model. Figure 3 shows the prediction of oil consumption for each of the countries. The group bar charts indicate the actual and predicted oil consumption for each country. The actual oil consumption for Jordan is represented as Jordan, whereas the predicted oil consumption is represented as predict JOC. The actual oil consumption in Lebanon is represented as Lebanon, whereas the predicted oil consumption is represented as predict LOC. The actual oil consumption in Oman is represented as Oman whereas the predicted oil consumption is represented as predict OOC. The actual oil consumption in Saudi Arabia is represented as Saudi Arabia whereas the predicted oil consumption is represented as predict SOC.

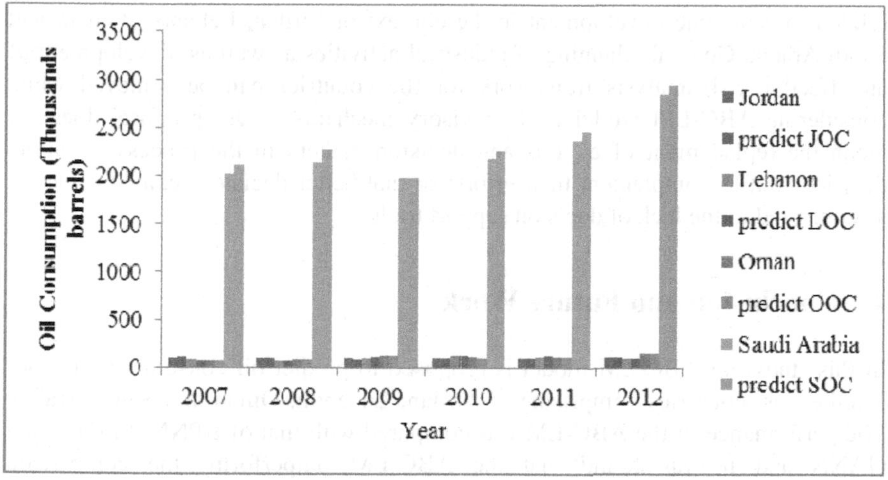

Fig. 3 ABC-LM model predicted oil consumption and actual oil consumption.

The performance of the ABC-LM model was compared with that of other soft computing techniques. The standard LMNN, GANN, and BPNN were used to predict oil consumption in all the countries under study. The best results were found in the Saudi Arabia oil market and reported in Table 5 for the purpose of comparing with our approach. The ABC-LM as shown in Table 5 is better than the comparison algorithms in terms of both accuracy and convergence speed which are the key indicators of measuring performance in soft computing, machine learning, and data mining. The performance of the ABC-LM could probably be attributed to the realistic search of the ABC to locate an exact solution and avoid the limitations of the LMNN.

Table 5 Comparing performance of ABC-LM, LMNN, GANN, and BPNN

Algorithm	CPU time	Accuracy (MSE)
LMNN	347	0.5185
GANN	238	0.1618
BPNN	593	0.7211
ABC-LM	192	0.0014

Accurate prediction of oil consumption can improve effective allocation of resources, distribution of oil in a country, tackling of an oil shortage. Also, it can guide in accurate formulation of international policy related to oil consumption. The ABC-LM model built for the prediction can be used by Jordan, Lebanon, Oman, and Saudi Arabia for the prediction of future oil consumption. The model proposes in this research can help in the formulation of oil consumption policy related to economic development in the context of Jordan, Lebanon, Oman, and Saudi Arabia. General, planning of industrial activities as well as development of an effective risk analysis framework for the countries can be achieved while considering ABC-LM model as an advisory mechanism. Our proposal does not mean the replacement of experts and decision makers in the process of taking decisions but to complement their efforts so that better decisions can be achieved as compared to the lack of decision support tools.

4 Conclusion and Future Work

In this study, an ABC-LM model is proposed to predict oil consumption in four Middle East countries comprising of Jordan, Lebanon, Oman and Saudi Arabia. The performance of the ABC-LM was compared with that of BPNN, LMNN, and GANN and it was found that the ABC-LM outperforms the comparison algorithms. The ABC-LM model developed in this research can constitute an alternative method for the prediction of oil consumption as well as a guide for the formulation of international and domestic policy related to oil consumption. Hence, sustainability can be enhanced for the development of economic growth, thereby, creates job opportunities for the unemployed population in the countries under study. The methodology documented in this study can easily be modified to predict oil consumption for other Middle East countries such as Iran, Iraq, Kuwait, Yemen, and etc. as well as modeling of oil consumption in OPEC countries.

In the future we will extend this research to include countries from North America and Europe. Our next direction is to find out whether oil consumption of OPEC countries is positively related? If so; can computationally intelligent algorithms be applied to predict oil consumption in an OPEC country based on oil consumption in other OPEC countries?.

Acknowledgment. This work is supported by University of Malaya High Impact Research Grant no vote UM.C/625/HIR/MOHE/SC/13/2 from Ministry of Higher Education Malaysia.

References

1. Energy Information Administration of the US Department of Energy (2013), http://www.eia.gov/
2. Assareh, E., Behrang, M.A., Assari, M.R., Ghanbarzadeh, A.: Application of PSO (particle swarm optimization) and GA (genetic algorithm) techniques on demand estimation of oil in Iran. Energy 35(12), 5223–5229 (2010)
3. Toksarı, M.D.: Ant colony optimization approach to estimate energy demand of Turkey. Energy Policy 35(8), 3984–3990 (2007)
4. Chiroma, H., Abdulkareem, S., Abubakar, A., Zeki, A., Gital, A.Y., Usman, M.J.: Co – Active Neuro-Fuzzy Inference Systems Model for Predicting Crude Oil Price based on OECD Inventories. In: 3rd International Conference on Research and Innovation in Information Systems (ICRIIS 2013), Malaysia, pp. 232–235. IEEE (2013)
5. Chiroma, H., Abdulkareem, S., Sari, E.N., Abdullah, Z., Muaz, S.A., Kaynar, O., Shah, H., Herawan, T.: Soft Computing Approach in Modeling Energy Consumption. In: Murgante, B., et al. (eds.) ICCSA 2014, Part VI. LNCS, vol. 8584, pp. 770–782. Springer, Heidelberg (2014)
6. Kaynar, O., Yilmaz, I., Demirkoparan, F.: Forecasting of natural gas consumption with neural network and neuro fuzzy system. In: EGU General Assembly Conference Abstracts, vol. 12, p. 7781 (2010)
7. Yusof, N.M., Rashid, R.S.A., Mohamed, Z.: Malaysia crude oil production estimation: An application of ARIMA model. In: 2010 International Conference on Science and Social Research (CSSR), pp. 1255–1259 (2010)
8. Su, F., Wu, W.: Design and testing of a genetic algorithm neural network in the assessment of gait patterns. Medical Engineering Physics 22, 67–74 (2000)
9. Dehuri, S., Cho, S.B.: A hybrid genetic based functional link artificial neural network with a statistical comparison of classifiers over multiple datasets. Neural Computing and Applications 19(2), 317–328 (2000)
10. Bishop, M.C.: Pattern recognition and machine learning. Springer, New York (2006)
11. Karaboga, D., Basturk, B.: On the performance of artificial bee colony (ABC) algorithm. Applied Soft Computing 5(1), 687–697 (2008)
12. Azar, A.T.: Fast neural network learning algorithms for medical applications. Neural Computing and Applications 23(3-4), 1019–1034 (2013)
13. He, K., Chi, X., Chen, S., Lai, K.K.: Estimating VaR in crude oil market: A novel multi–scale non-linear ensemble approach incorporating wavelet analysis and neural network. Neurocomputing 72, 3428–3438 (2009)
14. Ekonomou, L.: Greek long-term energy consumption prediction using artificial neural networks. Energy 35(2), 512–517 (2010)
15. Khazem, H.A.: Using artificial neural network to forecast the futures prices of crude oil. PhD dissertation, Nova south Eastern University, Florida (2007)

16. Jammazi, R., Aloui, C.: Crude oil forecasting: experimental evidence from wavelet decomposition and neural network modelling. Energy Economics 34, 828–841 (2012)
17. Pan, T.Y., Wang, R.Y.: State space neural networks for short term rainfall–runoff forecasting. Journal of Hydrology 297, 34–50 (2004)
18. Peter, G.Z., Patuwo, B.E., Hu, M.Y.: A simulation study of artificial neural networks for nonlinear time-series forecasting. Computers & Operation Research 28, 381–396 (2001)
19. Nawi, N.M., Rehman, M.Z., Khan, A.: Countering the problem of oscillation in Bat-BP gradient trajectory by using momentum. In: Herawan, T., Deris, M.M., Abawajy, J. (eds.) Proceedings of the First International Conference on Advanced Data and Information Engineering (DaEng 2013). LNEE, vol. 285, pp. 103–118. Springer, Heidelberg (2014)

Design and Implementation of a Novel Eye Gaze Recognition System Based on Scleral Area for MND Patients Using Video Processing

Sudhir Rao Rupanagudi, Varsha G. Bhat, R. Karthik, P. Roopa, M. Manjunath, Glenn Ebenezer, S. Shashank, Hrishikesh Pandith, R. Nitesh, Amrit Shandilya, and P. Ravithej

Abstract. In this modern era of science and technology, several innovations exist for the benefit of the differently-abled and the diseased. Research organizations worldwide, are striving hard in identifying novel methods to assist this group of the society to converse freely, move around and also enjoy those benefits which others do. In this paper, we concentrate on assisting people suffering from one such deadly disease – the Motor Neuron Disease (MND), wherein a patient loses control of his/her complete mobility and is capable of only oculographic movements. By utilizing these oculographic movements, commonly known as the eye-gaze of an individual, several day to day activities can be controlled just by the motion of the eyes. This paper discusses a novel and cost effective setup to capture the eye gaze of an individual. The paper also elaborates a new methodology to identify the eye gaze utilizing the scleral properties of the eye and is also immune to variations in background and head-tilt. All algorithms were designed on the MATLAB 2011b platform and an overall accuracy of 95% was achieved for trials conducted over a large test case set for various eye gazes in different directions. Also, a comparison with the popular Viola-Jones method shows that the algorithm presented in this paper is more than 3.8 times faster.

Keywords: Eye gaze Recognition, Video Processing, Video Oculography, Sclera based recognition, Otsu Algorithm, Motor Neuron Disease, Viola Jones.

Sudhir Rao Rupanagudi · Varsha G. Bhat · R. Karthik · P. Roopa · M. Manjunath · Glenn Ebenezer · S. Shashank · Hrishikesh Pandith · R. Nitesh · Amrit Shandilya · P. Ravithej
WorldServe Education, Bangalore

El-Sayed M. El-Alfy et al. (eds.), *Advances in Intelligent Informatics*,
Advances in Intelligent Systems and Computing 320, DOI: 10.1007/978-3-319-11218-3_51

1 Introduction

Over the past decade, several advancements in science and technology particularly related to fields such as artificial intelligence, gesture recognition and natural user interfaces have increased a hundred-fold. These developments prove to be very useful in easing the day to day life activities of mankind. For example, an automated vehicle capable of ferrying passengers without a driver [1], or a glass table being converted to a cost-effective touch screen [2] – these are just few of the several ways of how technology has changed the lives of people. It is also of great elation to note that this progress is also spreading towards the betterment of the differently-abled individuals as well. This includes meeting the two most important needs of such individuals – communication and locomotion. Scientists are finding various means to assist this section of the society to communicate freely. This is possible through various innovations such as Braille to speech convertors [3] or blink recognition systems [4], wherein blinks are converted to sentences.

With respect to locomotion, discoveries such as the "talking walking stick" for the blind [5] and the joystick controlled wheelchair [6] assist the differently abled in moving freely in public places and also in the safety of their homes with ease. Unfortunately, both these methodologies require the user to have their locomotory organs – hands or legs to be intact in order to use the same. In case of diseases such as the Motor Neuron Disease also known as the Lou Gherig's Disease [7], wherein the patient suffering from the same, loses his/her ability to move due to loss of nerve functionality, utilizing the aforementioned methods is clearly quite impossible.

According to statistics from the National Institute of Health (NIH), USA, approximately 25,000 individuals are currently suffering from the MND disease and an additional 4,500 patients get added to this toll every year [8]. Knowledge of the disease came into further prominence after the world's most eminent scientist Dr. Stephen Hawking contracted the disease in the late 1960's [9]. As mentioned previously, patients suffering from this disease lose all abilities to move and also in adverse cases lose their ability to speak as well. On the contrary, it is found that ocular movement – the ability to move their eyeballs, is still possible for these individuals [10]. Therefore, by tracking the gaze of these patients, certain activities including controlling the motion of a wheelchair remotely would certainly be possible.

Research on eye gaze recognition dates back to the year 1989, when Hutchinson et.al. [11] invented the eye gaze response interface computer aid - "Erica". The setup utilizes a camera and an infrared light aimed at the eyes of the user. Near infrared light is then reflected against the retina and the cornea of the eye, which is further captured by the camera. The position of the reflections obtained map to the direction of the eye gaze.

According to [12], eye gaze recognition techniques can be divided into two categories – model based and appearance based. A model based eye gaze recognition system utilizes features of the eyes such as contours or even reflections, similar to the method stated previously, in order to identify the eye region and the gaze. These methodologies require complex setups, high definition cameras and can be extremely cost ineffective in a common mans perspective to purchase [13]. Variations in lighting can also prove to be harmful in recognition, if considerable preprocessing is not performed. Appearance based systems on the other hand utilize simplified setups such as low resolution web cameras and use several comparison techniques of the captured eye image with a database of different eye gazes. Though cost effective, these methods require huge memories and a large training data set to identify the eye gaze [14]. Though immune to the lighting environment in which the setup is utilized, the system fails for unknown eye contours and also in case of individuals with squint. A major problem of concern for all the above methodologies is head tilt. Assumptions in all cases have been made that the head of the user is upright looking directly into the camera. In case of a slight variation in the angle of the head position, the above mentioned methods would fail and even if necessary corrections can be instantiated, the complexity of the algorithm would increase.

In this paper, we introduce a novel yet cost effective implementation for eye gaze recognition. The system utilizes a low cost computer camera fixed to a comfortable head mount in the form of a simple cap. The system is not affected by lighting or background variations and has been successfully tested for all types of individuals. The setup for the same has been elaborated in Sect. 2. The algorithm utilized for the gaze recognition has been explained in detail in Sect. 3.A display of the results obtained for varying individuals has been shown in Sect. 4. Sect. 5 deals with applications of our eye gaze recognition system, conclusion and also future scope of our research.

2 Setup

The setup used for our experiments is as shown in Fig. 1.

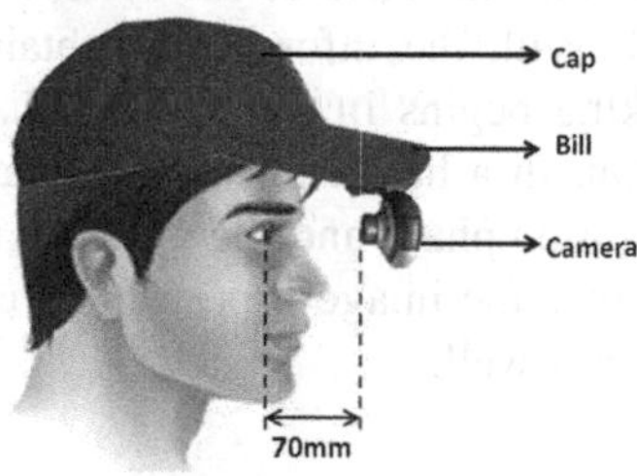

Fig. 1 An illustration of the setup used in our experiments

The camera utilized for our setup is an i-ball computer camera with a 2 Megapixel video Resolution. The video captures frames at a rate of 30 frames per second. This is in turn attached upright to the bill of a simple cap as shown in Fig. 1. Several setups which utilize electro-oculography [15] could be used, but ultimately a novel setup using a simple cap would be a better choice, due to its increased comfort and cost effectiveness for the user. A very important point to note is that since both eyes of a normal user point in the same direction while looking, the eye gaze of a single eyeball is therefore sufficient to track. In the case of a user suffering from squint, the camera can be focused on the normal eye. Therefore, it is due to this reason the camera is adjusted such that only one of the eyes is concentrated upon. Also, in order to ease detection of the eye, the camera is adjusted such that the left most corner of the eye, corresponds to the left most column of the image obtained. On an average, a 70 mm distance from the camera to the user's eye must be maintained. In the case of the "bill" not being long enough, it can be extended using cardboard or any other similar material.

The main advantage of the setup as explained above is that since the camera is placed close to the eye, the eye gaze can be tracked effortlessly. Also, since the background of the user is not taken into consideration, the setup provides a noise free environment. Since the camera is attached to a cap worn by the user, a movement in head position would not affect the eye gaze recognition since the camera would also move in the same direction as that of the user. In case of low illumination, soft LED lights which do not harm the eyesight of the user can be switched on for better results. Cameras with IR LED's could also be used.

3 Implementation

As explained in the above section, a live video feed which tracks the eye gaze of the user is obtained from the camera attached to the cap. This is further subjected to an eye gaze recognition algorithm, the resultant of which decides whether the user is looking straight, left, right or has closed his eyes. The flowchart of the algorithm utilized has been depicted in Fig. 2 and has been elaborated further below. The process of the eye gaze recognition can be divided into two phases – the first phase for calibration wherein the user provides initial training to the system as to how he/she looks straight. This ensures the system is robust and would work for any type of individual. With the information obtained from the calibration phase, the actual eye tracking begins in the second phase, after a certain delay delta required for calibration. In a hardware implementation perspective, it must be noted that both the calibration phase and the eye gaze recognition phase use the same methodologies and hence the image processing architectures can be reused, in turn proving area efficient as well.

3.1 Color Conversion

The first and foremost step in image processing is to convert the input color frames of the live feed into a color scheme suitable for segmentation. In this algorithm, conversion is performed into the YCbCr color space of which the intensity

(Y) component is utilized for further processing. Conversion from the RGB color scheme to the intensity component is represented by the equation (1) [16].

$$Y= + (0.299\times R) + (0.587\times G) + (0.114\times B) \quad (1)$$

where R, G and B are the Red, Green and Blue pixels respectively, of the color image. Another added advantage of color conversion is with respect to the memory required to store an image. In case of the RGB color space, 3 bytes of memory is required per pixel but in the case of the Y component, a minimal 1 byte is required, thereby reducing the memory requirement by 1/3rd. The Y component is then utilized for segmentation which is explained in detail next.

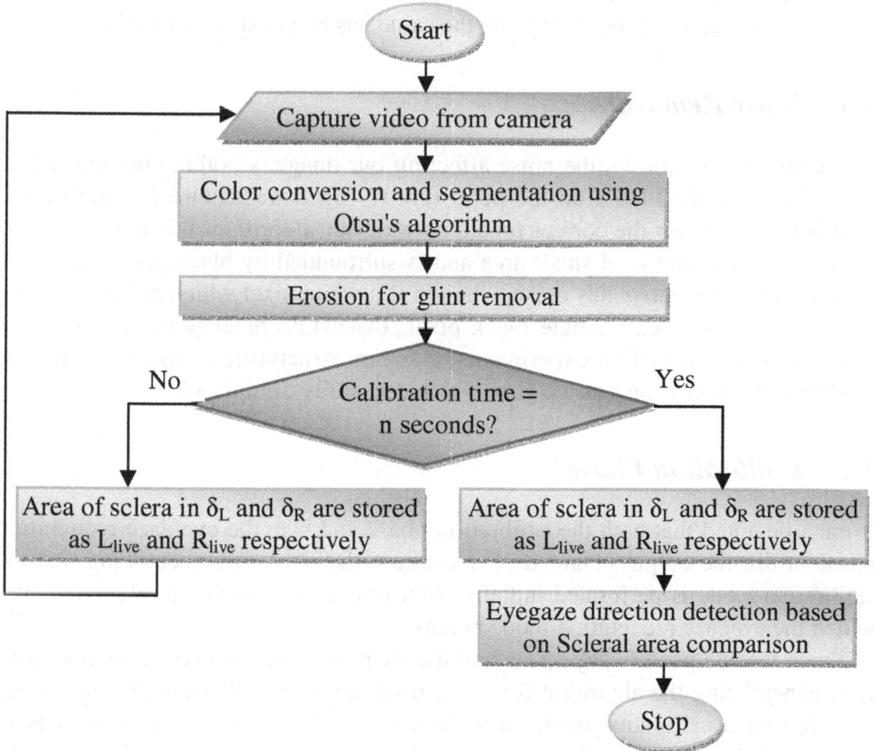

Fig. 2 Flowchart of the novel algorithm for eye gaze recognition using Scleral area of the eye

3.2 *Segmentation*

Once color conversion completes, the next major step is to extract and differentiate the various regions inside the eye i.e. the sclera and the pupil. This is an important step since the position of the pupil decides the direction of the gaze of the user. Amongst several thresholding algorithms available such as Watershed,

Histogram based segmentation, clustering e.t.c., the choicest algorithm to perform thresholding on a grayscale image was the Otsu's algorithm for segmentation. [17]. A very popular algorithm for automatic segmentation – the Otsu's algorithm was discovered by Otsu in 1991 and is utilized in a wide variety of applications such as text extraction, pest detection on leaves, face recognition and the like. The working of the same is elaborated in [18].

Upon application of this algorithm on the resultant acquired grayscale image, a binary image is obtained wherein the darker regions of the eye i.e. the pupil, eye-brows and eyelashes are converted to black (0) and the remaining regions such as the sclera, are converted to white(1). In special cases wherein the LED lights of the camera are switched on, a part of the pupil glows due to glint and this region of the pupil is also erroneously made white by the Otsu's algorithm. This noise is to be removed and its procedure to do the same has been explained further.

3.3 Noise Removal

As mentioned previously, the noise affecting our image is nothing but the reflection of light or the LED's on the pupil (glint). Since we require the pupil to be completely black for the correct functioning of our algorithm, this noise must be removed. Since glint is of small area and is surrounded by black pixels, a simple erosion [19] can remove this noise, wherein if a white pixel which belongs to glint is surrounded by even a single black pixel, that white pixel is made black in a resultant image. Based on experiments, a square structuring element of 5x5 was used to perform erosion and the glint was successfully diminished.

3.4 Calibration Phase

It must be noted that both the calibration phase and also the eye gaze recognition phase utilize the output obtained after erosion, the only difference being that the calibration mode is performed initially (first time use) to obtain a reference, after which the eyegaze recognition mode begins.

Since the eye position, eye gaze and the shape of eyes for every individual differs, generalizing the algorithm for every user becomes a difficult task and hence the need for a calibration mode. It is during this phase, the system is trained to understand the straight gaze of an individual. For this purpose the user is first made to look straight at the camera. A common method to identify a person's gaze is to store various images of the user looking straight, left or right and comparing the subsequent frames from the live camera feed, with the images stored. This method though accurate, leads to a high memory utilization and also slows down the algorithm considerably. Taking this disadvantage into account, a novel methodology to detect the eye gaze based on the area of the sclera region is utilized. First and foremost, through experimentation over a large image data set, the width (δ) of the human eye on an average, for a camera placed 70 mm away, was calculated. This is shown in Fig.3.

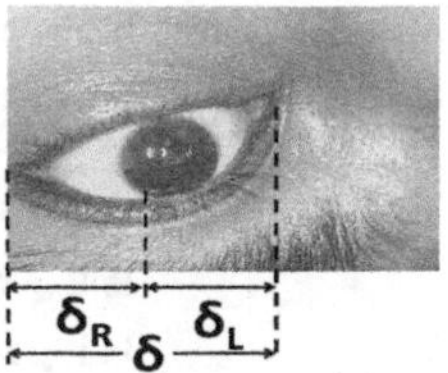

Fig. 3 Image depicting the width δ of the eye and also the width to the right and left of the pupil

The area of the sclera region to the left (δ_L) and the right (δ_R) of the pupil was then calculated. These in turn provide the reference areas – L_{ref} and R_{ref} of the sclera region to the left and right of the pupil, when a person looks straight. Utilizing these values as reference, a decision on which direction the user is looking can be taken later on in the eye gaze phase. In this way, instead of storing 3 reference images of 640x840, 2 bytes of reference data are sufficient to decide the eye gaze, in turn proving to be a very area efficient algorithm in a hardware perspective.

3.5 *Eye Gaze Recognition*

Once the reference area values for the sclera are obtained and calibration mode is complete, the left and right area values (L_{live} and R_{live}) of the sclera for every frame which subsequently follows is also calculated. It can be clearly seen from Fig. 4, that when the user looks right, the sclera area on the right side of the eye is more than the area present in the right half of the reference image. Similarly, it can be seen that the area of the sclera on the left side has reduced.

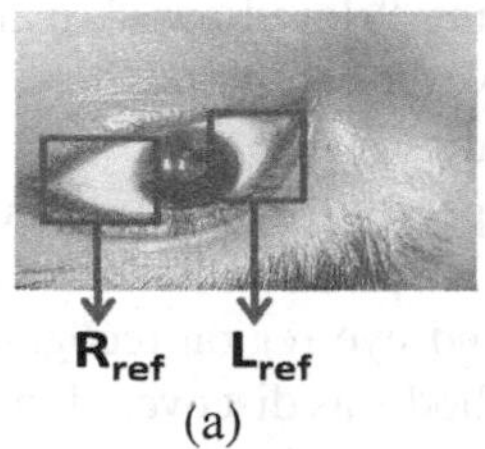

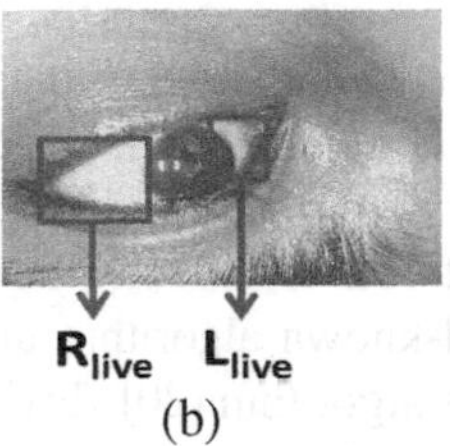

Fig. 4 Images showing the scleral areas enclosed in boxes in both the (a) Reference image as R_{ref} and L_{ref} and the (b) Live image as R_{live} and L_{live}

In this way, by performing a simple comparison of reference and live area values, the position of the eye can be deciphered. In the case of closed eyes, this can be easily detected since the reflection of the eyelid is detected white by the Otsu's algorithm and hence area will be much greater than the reference image. Table 1 shows the comparison chart between the areas of the reference and live images and its indicated meaning.

Table 1 Relation between direction of eye gaze and scleral area comparison

Sclera area comparison	Direction of eye gaze
Is $L_{ref} > L_{live}$ && $R_{ref} < R_{live}$	Looking right
Is $L_{ref} < L_{live}$ && $R_{ref} > R_{live}$	Looking left
Is $L_{ref} > L_{live}$ && $R_{ref} > R_{live}$	Eyes closed
Is $L_{ref} \approx L_{live}$ && $R_{ref} \approx R_{live}$	Looking straight

4 Results

The setup, as shown in Sect. 2, was used for our experiments and all algorithms were designed and developed on Simulink, a part of MATLAB – 2011b.

Fig. 5 shows the output obtained at every stage of the algorithm as elaborated in the previous section. Trials were conducted on over 50 individuals, of which three such individuals wearing the setup can be seen in Fig. 5a. The grayscale image of the eye region obtained and its corresponding segmentation into white and black using Otsu's thresholding can be seen in Fig. 5b and 5c respectively. Highlighting of the pupil area and removal of glint after performing the erosion operation can be seen in Fig. 5d, for all three individuals. Eroded images obtained for a user looking left, right and closed are also shown in Fig. 6.

As mentioned above, trials were conducted for over 50 individuals for various eye gazes. The percentage of accuracy of successfully identifying each eye gaze, as shown by the user, was tabulated and can be seen in Table 2. A slight reduction in accuracy can be seen in the case of recognizing "closed eyes", mainly due to non uniform illumination of the eyelid. This however can be rectified, by providing sufficient lighting on the face of the individual.

In order to further validate our results, the speed of our algorithm was compared in MATLAB 2011b (running on an Intel I3 processor of 2.5 GHz speed) with a well-known algorithm utilized for face and eye region recognition – The Viola-Jones algorithm [20]. The Viola-Jones method was discovered in 2001 by P. Viola and M. Jones and utilizes Haar-like features and ADABoost techniques [21]. It was found that our complete algorithm requires only 0.062 seconds to execute, whereas the Viola-Jones on the other hand requires 0.237 seconds to perform only the Haar transform step of the complete algorithm on a standard 640x480 image. This shows that the methodology shown in this paper is more than 3.8 times faster than the existing technique.

The only constraint of our algorithm is when the user wears spectacles and the experiments are conducted with the camera LED's switched on. Due to the reflection of the light on the spectacles worn, an accurate result is not obtained. This can be seen in Fig. 7. However future implementations can overcome this issue by utilizing a more intricate segmentation algorithm, though coming at a price of hardware area overhead.

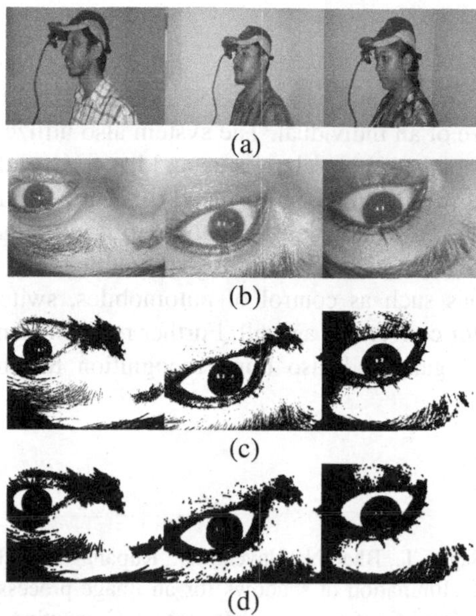

Fig. 5 (a) RGB images of three different users wearing the setup (b) The color converted images (c) The Otsu Thresholded images obtained and (d) Images after glint removal

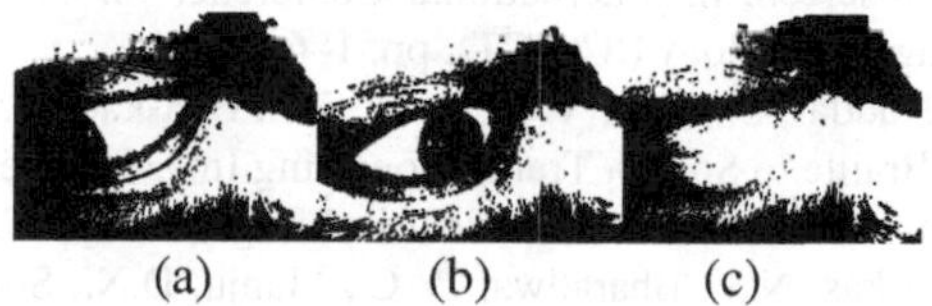

Fig. 6 Eroded images of a user (a) looking right (b) looking left and (c) closing his eyes

Table 2 Percentage of accuracy of successful identification of the eyegaze

Direction of Eyegaze	Accuracy of identification
Looking right	96%
Looking left	95%
Eyes closed	90%
Looking straight	98%
Total Percentage of Accuracy	95 %

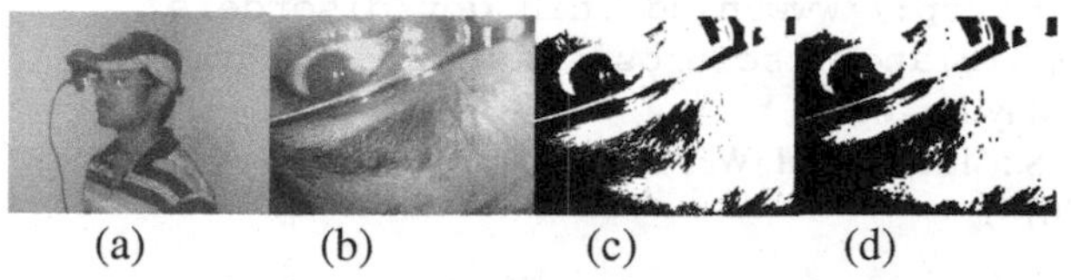

Fig. 7 Incorrect eye gaze recognition due to interference of light reflection from spectacles worn by the user

5 Conclusion

In this paper we have presented a novel and cost effective implementation for detecting the eye gaze of an individual. The system also utilizes a novel algorithm for eye gaze recognition based on scleral area and has an overall accuracy of 95%, without the interference of background or head-tilt. Further, the algorithm is also seen to be 3.8 times faster than the most popular Viola-Jones method of eye region recognition. The system, a boon for the differently-abled, can be used for a wide variety of applications such as control of automobiles, switching on electronic equipment and also for conversing as well. Further research can also be performed to integrate both eye gaze and also blink recognition for the benefit of these individuals.

References

1. Prashanth, C.R., Sagar, T., Bhat, N., Naveen, D., Rupanagudi, S.R., Kumar, R.A.: Obstacle detection & elimination of shadows for an image processing based automated vehicle. In: International Conference on Advances in Computing, Communications and Informatics (ICACCI), pp. 367–372 (2013)
2. Ravoor, P.C., Rupanagudi, S.R., Ranjani, B.S.: Detection of multiple points of contact on an imaging touch-screen. In: International Conference on Communication, Information & Computing Technology (ICCICT), pp. 1–6 (2012)
3. Rupanagudi, S.R., Huddar, S., Bhat, V.G., Patil, S.S., Bhaskar, M.K.: Novel Methodology for Kannada Braille to Speech Translation using Image Processing on FPGA. In: International Conference on Advances in Electrical Engineering, ICAEE 2014 (2014)
4. Rupanagudi, S.R., Vikas, N.S., Bharadwaj, V.C., Manju, D.N., Sowmya, K.S.: Novel Methodology for Blink Recognition using Video Oculography for Communicating. In: International Conference on Advances in Electrical Engineering, ICAEE 2014 (2014)
5. Kassim, A.M., Jaafar, H.I., Azam, M.A., Abas, N., Yasuno, T.: Design and development of navigation system by using RFID technology. In: IEEE 3rd International Conference on System Engineering and Technology (ICSET), pp. 258–262 (2013)
6. Tsai, M.-C., Hsueh, P.-W.: Synchronized motion control for 2D joystick-based electric wheelchair driven by two wheel motors. In: IEEE/ASME International Conference on Advanced Intelligent Mechatronics (AIM), pp. 702–707 (2012)
7. Corno, F., Farinetti, L., Signorile, I.: A cost-effective solution for eye-gaze assistive technology. In: Proceedings of IEEE International Conference on Multimedia and Expo (ICME 2002), vol. 2, pp. 433–436 (2002)
8. NINDS. Amyotrophic Lateral Sclerosis (ALS) Fact Sheet. NIH Publication No. 13-916 (June 2013), http://www.ninds.nih.gov/disorders/amyotrophiclateralsclerosis/detail_ALS.htm (accessed May 19, 2014)
9. Hawking, S.: The official Website, http://www.hawking.org.uk (accessed May 19, 2014)

10. Usakli, A.B., et al.: A hybrid platform based on EOG and EEG signals to restore communication for patients afflicted with progressive motor neuron diseases. In: Annual International Conference of the IEEE Engineering in Medicine and Biology Society, EMBC 2009, pp. 543–546 (2009)
11. Hutchinson, T.E., White Jr., K.P., Martin, W.N., Reichert, K.C., Frey, L.A.: Human-computer interaction using eye-gaze input. IEEE Transactions on Systems, Man and Cybernetics 19(6), 1527–1534 (1989)
12. Yuan, Z., Kebin, J.: A local and scale integrated feature descriptor in eye-gaze tracking. In: 4th International Congress on Image and Signal Processing (CISP), vol. 1, pp. 465–468 (2011)
13. Model, D., Eizenman, E.: An Automatic Personal Calibration Procedure for Advanced Gaze Estimation Systems. IEEE Transactions on Biomedical Engineering 57(5), 1031–1039 (2010)
14. Lu, H.-C., Wang, C., Chen, Y.-W.: Gaze Tracking by Binocular Vision Technology and PPBTF Features. In: International Conference on Intelligent Information Hiding and Multimedia Signal Processing (IIHMSP), pp. 705–708 (2008)
15. Al-Haddad, A., et al.: Wheelchair motion control guide using eye gaze and blinks based on bug 2 algorithm. In: 8th International Conference on Information Science and Digital Content Technology (ICIDT), vol. 2, pp. 438–443 (2012)
16. Dhruva, N., Rupanagudi, S.R., Sachin, S.K., Sthuthi, B., Pavithra, R., Raghavendra: Novel segmentation algorithm for hand gesture recognition. In: International Multi-Conference on Automation, Computing, Communication, Control and Compressed Sensing (iMac4s), pp. 383–388 (2013)
17. Otsu, N.: A Threshold Selection Method from Gray-Level Histograms. IEEE Transactions on Systems, Man and Cybernetics 9(1), 62–66 (1979)
18. Pandey, J.G., Karmakar, A., Shekhar, C., Gurunarayanan, S.: A Novel Architecture for FPGA Implementation of Otsu's Global Automatic Image Thresholding Algorithm. In: 27th International Conference on VLSI Design & 13th International Conference on Embedded Systems, pp. 300–305 (2014)
19. Gonzalez, R.C., Woods, R.E.: Digital Image Processing, 3rd edn. Prentice Hall, New Jersey (2009)
20. Phuong, H.M., et al.: Extraction of human facial features based on Haar feature with Adaboost and image recognition techniques. In: Fourth International Conference on Communications and Electronics (ICCE), pp. 302–305, 1-3 (2012)
21. Viola, P., Jones, M.: Rapid object detection using a boosted cascade of simple features. In: Proceedings of the 2001 IEEE Computer Society Conference on Computer Vision and Pattern Recognition (CPVR 2001), vol. 1, pp. I-511–I-518 (2001)

Correlation Based Anonymization Using Generalization and Suppression for Disclosure Problems

Amit Thakkar, Aashiyana Arifbhai Bhatti, and Jalpesh Vasa

Abstract. Huge volume of detailed personal data is regularly collected and sharing of these data is proved to be beneficial for data mining application. Data that include shopping habits, criminal records, credit records and medical history are very necessary for an organization to perform analysis and predict the trends and patterns, but it may prevent the data owners from sharing the data because of many privacy regulations. In order to share data while preserving privacy, data owner must come up with a solution which achieves the dual goal of privacy preservation as well as accurate data mining result. In this paper k-Anonymity based approach is used to provide privacy to individual data by masking the attribute values using generalization and suppression. Due to some drawbacks of the existing model, it needs to be modified to fulfill the goal. Proposed model tries to prevent data disclosure problem by using correlation coefficient which estimates amount of correlation between attributes and helps to automate the attribute selection process for generalization and suppression. The main aim of proposed model is to increase the Privacy Gain and to maintain the accuracy of the data after anonymization.

Keywords: Privacy Preserving Data Mining, Anonymization, Correlation, Data loss, Privacy Gain, Accuracy.

1 Introduction

Data mining is generally thought of as the process of extracting implicit, previously unknown and potentially useful information from databases. Utilizing this huge volume of data for decision making is very important for many organizations in today's many environments. P. Samarati [1] has mentioned that huge

Amit Thakkar · Aashiyana Arifbhai Bhatti · Jalpesh Vasa
Charotar University of Science and Technology, India

El-Sayed M. El-Alfy et al. (eds.), *Advances in Intelligent Informatics*,
Advances in Intelligent Systems and Computing 320, DOI: 10.1007/978-3-319-11218-3_52

amount of data and global networks are focusing on sharing of information which may be the most important feature for many organizations. They have given the solution to protect the data by anonymizing it and by removing the explicit identifiers such as name, phone number, etc. and converting the remaining data into anonymous data by using operations such as randomization, perturbation and cryptographic based techniques. De-identifying data by removing the explicit identifiers however, provides no guarantee of anonymity.

Released information may often lead to disclosure attack such linking attack which may be caused by linking the released data and publicly available data such as Voter Registration data. Such linking attacks are serious problem as they may reveal the identity of the person. Typical data published by some governmental agencies contained person identity like his/her name, address, city, date of birth contact number, and information about employment. These publicly distributed data can be used for linking identities with de-identified information, thus allowing re-identification of respondents. This situation has raised severe concern in the several areas where data are publicly available and have been subject to abuses, compromising the privacy of individuals.

In paper [12] author has shown an important method for privacy de-identification using the concept of k- anonymity. The idea in k-anonymity is to reduce the granularity of representation of the data in such a way that a given record cannot be distinguished from at least (k – 1) other records [12].

2 Background and Related Work

There are already some existing methods to find solutions for privacy problem. K-anonymity [4] can prevent the identity disclosure attack but not attribute disclosure attack. Another method, ℓ- diversity [2] method preserves the privacy against attribute disclosure attack. But, it is weaker in case of identity disclosure attack. T-closeness [7] method is good at attribute disclosure attack. It is computationally complex in achieving the privacy.

Sweeny [4] introduced k-anonymity as a property that each record is indistinguishable with at least k-1 records. This method has drawback that if there exist same value of sensitive attribute in equivalence class then privacy cannot be achieved as it can lead to homogeneity attack and background knowledge attack. In Enhanced P sensitive k-anonymity model [19], the modified micro data table T* satisfies (p+, α)-sensitive k- anonymity property if it satisfies k-anonymity, and each QI- group has at least p distinct categories of the sensitive attribute and its total weight is at least α. Compared to p sensitive k anonymity method, this method reduces the possibility of Similarity Attack and incurs less distortion ratio.

ℓ-diversity [2] method overcomes the drawbacks of k- anonymity. In this method, an equivalence class is said to have ℓ-diversity if there are at least ℓ "well-represented" values for the sensitive attribute. A table is said to have ℓ-diversity if every equivalence class of the table has ℓ- diversity. This method fails to preserve the privacy against skewness and similarity attacks.

In t-closeness method [7], an equivalence class is said to have t-closeness if the distance between the distribution of a sensitive attribute in this class and the distribution of the attribute in the whole table is no more than a threshold t. A table is said to have t-closeness if all equivalence classes have t-closeness. It preserves the privacy against homogeneity and background knowledge attacks.

All the above methods use generalization and/or suppression operation for masking the person specific data. Results shows that these operations causes considerable amount of information loss because higher the generalization hierarchy more information loss will be there. Also suppression causes elimination of entire row or attributes value. So we may loss important or relevant data that may be very useful to our data mining task. Proposed model eliminates these drawbacks by using correlation coefficient and two type of generalization hierarchy which reduces loss of important data and it also eliminates attacks such as background knowledge attack, homogeneity attack, similarity attack etc.

2.1 Problem Statement

There are various open issues are there in the current (Existing) System. Few of them are listed below. We are considering following drawbacks of the current system.

As we know that existing system uses generalization hierarchy for transforming the data into anonymized database, which causes considerable data loss. So Accuracy of data after Anonymization is decreased more. Also algorithm used for existing system such as k-anonymity and l-diversity have certain drawbacks such as k-anonymity can't handle homogeneity attack and background knowledge attack; and l-diversity is difficult to achieve. In existing system, user manually chooses quasi identifier attribute for generalization and suppression. So we have tried to overcome this problem in our proposed framework.

3 Proposed Framework

The proposed framework provides novel anonymization technique comprising correlation coefficient, generalization and suppression.

Identifier attributes: *Attributes such as Person ID and Name that can be used to identify a record are identifier attributes. Identifier attributes can lead to specific entity thus they present in initial data only.*

Key attributes or Quasi-identifier: *Attributes such as Zip Code and Age are key attributes or quasi identifier attributes which may be known by attackers. Key attributes are present in the anonymized data as well as in the initial data.*

Sensitive attributes: *Confidential attributes such as income medical information and Grade Pay that are assumed to be unknown to an intruder. Sensitive attributes are present in the anonymized data as well as in the initial data.*

To protect the data, the identifiers attributes are completely removed, and the key attributes are "perturbed", using disclosure control methods, in order to avoid the possibility of disclosure. We assume that the values for the sensitive attributes are not available from any external source. This assumption guarantees that an intruder cannot use the confidential attributes values to increase his/her chances of disclosure, and, therefore, their "masking" is unnecessary.

A. Flow Diagram for Proposed framework

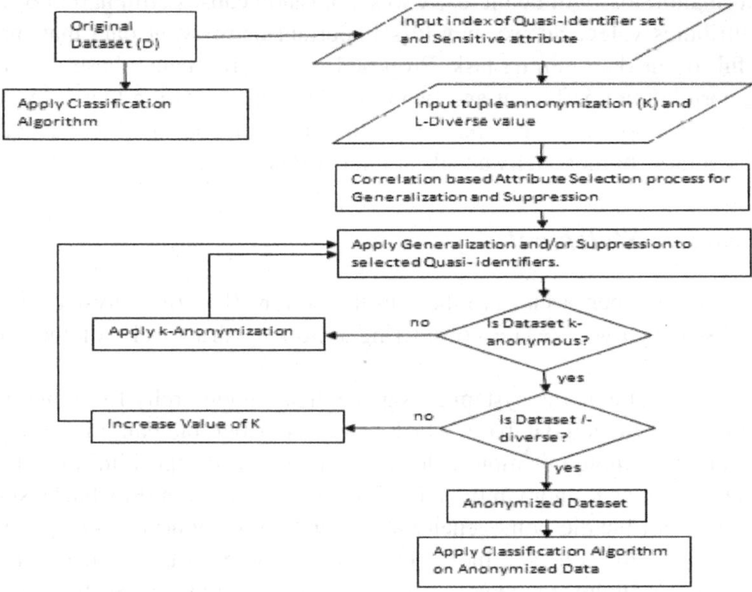

B. Simulation:

Step 1: Attribute Selection process for generalization and suppression:

In attribute selection process, we have used attribute correlation method which returns the correlation coefficient of the two attributes. Correlation coefficient is used to determine the relationship between two properties. For example, you can examine the relationship between an age and salary, between age and disease etc. If result of Correlation coefficient is greater than zero than two attributes are positively correlated means increase or decrease in one attribute value will correspondingly increase or decrease. In our proposed system, attribute having higher correlation will be subjected to generalization type 1 as it is positively correlated with other attributes and we want less information loss which results from generalization type 1, other attributes having lower correlation will be subjected to generalization type 2 or suppression . For example, we have age, zip-code and salary attributes. So, we have to calculate Correlation coefficient between them as below.

$$Correl(X,Y) = \frac{\sum(x - x')(y - y')}{\sqrt{\sum(x - x')^2 \sum(y - y')^2}}$$

Example:

We have used Numeric Dataset of Healthcare Medical System Database which contains information of all the patients, their disease, their personal data etc. Now, Suppose Medical Survey team wants' to make survey for disease outbreak and patterns. For that they can use the healthcare medical system but without revealing the identity of any patient.

Table 3.1 Original Patient Data

No.	Zip code	Age	Salary	Disease
3	29	47678	5000	Stomach cancer
4	36	47911	6000	Gastritis
5	52	47312	11000	Flu
6	36	47906	8000	Bronchitis
7	30	47677	7000	Bronchitis
8	36	47673	9000	Pneumonia
9	32	47678	22000	stomach cancer
10	25	47812	13000	Flu
11	36	47312	12000	Bronchitis
12	44	47614	35000	Stomach cancer
13	57	47333	50000	Pneumonia
14	23	47813	22000	Bronchitis
15	40	47678	25000	Gastric ulcer
16	59	47897	7000	Flu
17	39	47911	36000	Bronchitis
18	36	47312	22000	Gastritis
19	41	47865	18000	Pneumonia
20	19	47912	15000	Flu

So, Correl (zip-code, age) = -0.31519
Correl (zip-code, salary) = -0.24027
Correl (salary, age) = 0.371199

Table 3.2 Correlation coefficient for attributes

	Zip-code	Age	Salary	Sum
Zip-code	1	-0.31519	-0.24027	0.44454
Age	-0.31519	1	0.371199	1.056009
Salary	-0.24027	0.371199	1	1.130929

Table 3.3 Result after Generalization Type-1

No	Age	Zip code	Salary	Disease
5	52	47312	11000<=12000	Flu
11	36	47312	12000<=22000	Bronchitis
18	36	47312	12000<=22000	Gastritis

Table 3.4 Result after Generalization Type-2

No.	Age	Zip code	Salary	Disease
5	36-52	47312	11000<=12000	Flu
11	36-52	47312	12000<=22000	Bronchitis
18	36-52	47312	12000<=22000	Gastritis

Table 3.5 3-Anonymous Data

No	Age	Zip code	Salary	Disease
5	36-52	4****	11000<=12000	Flu
11	36-52	4****	12000<=22000	Bronchitis
18	36-52	4****	12000<=22000	Gastritis

Thus, we can apply generalization type 1 to attribute 'age' and generalization type 2 or suppression to other attributes.

Step 2: Generalization and Suppression on selected attributes.

Now, in the third step we apply generalization and/or suppression process to attributes, so attribute having highest correlation coefficient will be candidate for the generalization type 1. In generalization type 1, for each group find minimum and maximum value of the attribute and replace all the values of that attribute with [minimum-maximum] interval.

Now, we will apply generalization type 2 on attribute salary, so, find next nearest minimum and next largest value of each corresponding value of attribute. For example, salary=11,000, next nearest minimum value is itself as there as there is not any value which is less than 11,000 and next largest value is 12,000. So, we have interval 11000<=12000. So, we have above result in table 3.3.

For suppression, we first sort the dataset according to index of attribute having lowest correlation coefficient then we will match all the values of an attribute and replace unmatched digit with "*". For example, in above result, we can suppress last four digit with "*" i.e. 4****. We are not applying full domain suppression because it causes high data loss, so we keep first digit as it is and suppress the remaining digits.

Here, for above group in table 3.4, age attribute has three values 52, 36, 36. So, minimum value will be 36 and maximum value will be 52. Thus, we have [36-52] interval.

Step 3: Check for k-anonymity

Definition of k-anonymity [3] states that, a dataset complies with k-anonymity protection if each individual's record stored in the released dataset cannot be distinguished from at least k-1 individuals whose data also appears in the dataset.
For example,

Here, k=3 and above group is 3-Anonymous as for same zip-code and age value we have three different sensitive attribute value.

Step 4: Check for *l-diversity*

A data set satisfies *L-diversity* if it is divided into a partition and the sensitive attribute contains at least L different values in each group. *L-diversity* is needed in order to prevent published database from homogeneity attack and background knowledge attack.

Above example states that it is 3-diverse as there are three different values for same quasi-identifiers value.

Anonymized Data:

Table 3.6 3-Anonymized Data

No	Age	Zipcode	salary	Disease
1	[29<=30]	4767*	[3000-9000]	Gastric ulcer
2	[28<=44]	47***	[7000-50000]	Gastritis
3	[29<=32]	4****	[5000-25000]	Stomach cancer
4	[19<=39]	479**	[6000-36000]	Gastritis
5	[36<=52]	4****	[11000-22000]	Flu
6	[19<=39]	479**	[6000-36000]	Bronchitis
7	[29<=36]	4767*	[3000-9000]	Bronchitis
8	[30<=36]	4767*	[3000-9000]	Pneumonia
9	[29<=40]	4****	[5000-25000]	Stomach cancer
10	[23<=41]	478**	[13000-22000]	Flu
11	[36<=52]	4****	[11000-22000]	Bronchitis
12	[28<=57]	47***	[7000-50000]	Stomach cancer
13	[44<=57]	47***	[7000-50000]	Pneumonia
14	[23<=25]	478**	[13000-22000]	Bronchitis
15	[32<=40]	4****	[5000-25000]	Gastric ulcer
16	[39<=59]	479**	[6000-36000]	Flu
17	[36<=59]	479**	[6000-36000]	Bronchitis
18	[36<=52]	4****	[11000-22000]	Gastritis
19	[25<=41]	478**	[13000-22000]	Pneumonia
20	[19<=36]	479**	[6000-36000]	Flu

By Analyzing the Anonymized data, we conclude that there are fewer chances of Homogeneity attack and Background Knowledge attack which could be possible if intruder have some extra information. Here two type of generalization makes chances of attack to minimum.

4 Experimental Setup

We have used core-i3 processor with 4GB of RAM to evaluate our algorithm on different criterion. We also used weka tool to analyze our dataset on different classification algorithm. We have taken numerical dataset like Healthcare dataset, credit dataset to analyze proposed framework. Following parameters are measured.

Accuracy

Accuracy represents the percentage of correctly identified matches the engine detects from the possible matches. This measurement of accuracy is used on an overall basis for the entire dataset meaning correct matches include all classifier and null matches.

$$Accuracy = \frac{\#\,Correct\ matches}{Total\ \#\,possible}$$

Privacy Gain

Privacy gain can be defined as amount of Anonymization achieved after applying Anonymization process which results from generalization and suppression methods.

$$Privacy\ Gain\ \% = \left[1 - \left(\frac{Tn - Ta}{Ta}\right)\right] * 100$$

Where, Tn= Total no. of Instances

Ta= Total no. of instances anonymized

5 Results

Experiment-1: Compare performance of different database for different value of K and Q.

Table 5.1 Anonymization using Naïve Bayes Classifier(Q=3)

Data set	Instance Processed	Parameters	Original data sets	Anonymized data sets (Q=3)		
				K=4	K=6	K=10
Adult	33K	Correctly classified(%)	83.42	82.61	82.83	82.55
		Time taken to build model(sec)	0.09	0.08	0.07	0.06
		Privacy Gain(%)	-	90.26	90.91	-
Credit	10K	Correctly classified(%)	74.87	71.67	71.57	72.97
		Time taken to build model(sec)	0.01	0.05	0.02	0.01
		Privacy Gain(%)	-	37.2	48.4	65

Table 5.2 Anonymization using Naïve Bayes Classifier(Q=4)

Data set	Instance Processed	Parameters	Original data sets	Anonymized data sets (Q=4)		
				K=4	K=6	K=10
Adult	33K	Correctly classified(%)	83.42	82.4	82.5	82.29
		Time taken to build model(sec)	0.09	0.06	0.05	0.05
		Privacy Gain(%)	-	95.27	90.91	99.68
Credit	10K	Correctly classified(%)	74.87	71.77	71.57	73.17
		Time taken to build model(sec)	0.01	0.03	0.01	0.01
		Privacy Gain(%)	0	38.2	48.4	65

Experiment-2: Comparison of Result generated by Existing model, proposed model and ARX Tool after applying Homogeneity attack and Background knowledge attack.

Table 5.3 Original Data

No.	Zipcode	Age	Nationality	Condition
1	13053	28	Russian	Heart Disease
2	13068	29	American	Heart Disease
3	13068	21	Japanese	Viral Infection
4	13053	23	American	Viral Infection
5	14853	50	Indian	Viral Infection
6	14853	55	Russian	Heart Disease
7	14850	47	American	Viral Infection
8	14850	49	American	Viral Infection
9	13053	31	American	Cancer
10	13053	37	Indian	Cancer
11	13068	36	Japanese	Cancer
12	13068	35	American	Cancer

Table 5.4 Anonymization using existing l-diversity Model

No.	Zipcode	Age	Nationality	Condition
1	1305*	<=40	*	Heart Disease
2	1305*	<=40	*	Viral infection
3	1305*	<=40	*	Cancer
4	1305*	<=40	*	Cancer
5	1485*	>40	*	Cancer
6	1485*	>40	*	Heart Disease
7	1485*	>40	*	Viral infection
8	1485*	>40	*	Viral infection
9	1306*	<=40	*	Heart Disease
10	1306*	<=40	*	Viral infection
11	1306*	<=40	*	Cancer
12	1306*	<=40	*	Cancer

Table 5.5 Anonymization using proposed Model

No.	Zipcode	Age	Nationality	Condition
1	1****	[23-37]	Russian	Heart Disease
2	1****	[21-35]	American	Heart Disease
3	1****	[21-35]	Japanese	Viral Infection
4	1****	[23-37]	American	Viral Infection
5	1485*	[47-55]	Indian	Viral Infection
6	1485*	[47-55]	Russian	Heart Disease
7	1485*	[47-55]	American	Viral Infection
8	1485*	[47-55]	American	Viral Infection
9	1****	[23-37]	American	Cancer
10	1****	[23-37]	Indian	Cancer
11	1****	[21-35]	Japanese	Cancer
12	1****	[21-35]	American	Cancer

6 Conclusion and Future Extension

Privacy preserving data mining is a new era of data privacy in which private data has to be published and proposed system provides data security even after data has been published for any data mining application used by any organization. Proposed algorithm uses correlation based attribute selection method which automates the Anonymization process as user need to specify which attribute to select for different operation such as generalization and suppression. Also proposed algorithm uses two types of generalization to make database more secure as attacker is not able to apply homogeneity attack and background knowledge attack on published database. Proposed model increases considerable amount of Information gain after Anonymization which increases security of a data.

There are still many possible extensions in the future work. First, proposed approach uses only single sensitive attribute. So, in future work proposed approach for multiple sensitive attribute will be implemented. Second, proposed approach uses only numeric attributes as quasi identifiers. So, in future work one can adopt proposed approach for non-numeric or categorical quasi-identifier attributes. And if time permits this work can be extended to improve the performance measures such as running time, memory utilization, information loss etc.

References

[1] Ciriani, V., De Capitani di Vimercati, S., Foresti, S., Samarati, P.: K-Anonymous Data Mining: A Survey. In: Advances in Database Systems. Springer, US (2008)

[2] Machanavajjhala, A., Gehrke, J., Kifer, D., Venkitasubramaniam, M.: l-diversity: Privacy beyond k-anonymity. In: Proc. 22nd Intnl. Conf. Data Engg. (ICDE), p. 24 (2006)

[3] Samarati, P., Sweeney, L.: Protecting privacy when disclosing information: k-anonymity and its enforcement through generalization and suppression. Technical Report SRI-CSL-98-04, SRI Computer Science Laboratory (1998)

[4] Sweeney, L.: K-anonymity: A model for protecting privacy. Int. J. Uncertain. Fuzz. 10(5), 557–570 (2002)

[5] LeFevre, K., De Witt, D.J., Ramakrishnan, R.: Multidimensional K-Anonymity

[6] Samarati, P.: Protecting Respondents' Identities in Microdata Release. IEEE Transactions on Knowledge and Data Engineering 13(6) (2001)

[7] Li, N., Li, T., Venkatasubramanian, S.: t-Closeness: Privacy Beyond k-Anonymity and l-diversity. In: ICDE, pp. 106–115 (2007)

[8] Wu, Y., Ruan, X., Liao, S., Wang, X.: P-Cover K-anonymity model for Protecting Multiple Sensitive Attributes. IEEE (2010)

[9] Gionis, A., Tassa, T.: k-Anonymization with Minimal Loss of Information. IEEE Transactions on Knowledge and Data Engineering 21(2) (2009)

[10] Aggarwal, C.C., Yu, P.S.: Privacy-Preserving Data Mining: Models And Algorithms. Kluwer Academic Publishers

[11] Gionis, A., Tassa, T.: k-Anonymization with Minimal Loss of Information. IEEE Transactions on Knowledge and Data Engineering 21 (2009)

[12] Kedar, S., Dhawale, S., Vaibhav, W., Kadam, P., Wani, S., Ingale, P.: Privacy Preserving Data Mining. International Journal of Advanced Research in Computer and Communication Engineering 2(4) (2013)

[13] Malik, M.B., Ghazi, M.A., Ali, R.: Privacy Preserving Data Mining Techniques: Current Scenario and Future. In: 2012 IEEE Third International Conference on Computer and Communication Technology (2012)

[14] Mahesh, R., Meyyappan, T.: Anonymization Technique through Record Elimination to Preserve Privacy of Published Data. In: Proceedings of the 2013 International Conference on Pattern Recognition, Informatics and Mobile Engineering (2013)

[15] Maheshwarkar, N., Pathak, K., Chourey, V.: N-SA K-anonymity Model: A Model Exclusive of Tuple Suppression Technique. In: Third Global Congress on Intelligent Systems (2012)
[16] Liu, J., Luo, J., Huang, J.Z.: Rating: Privacy Preservation for Multiple Attributes with Different Sensitivity Requirements. In: 11th IEEE International Conference on Data Mining Workshops (2011)
[17] Wang, Q., Xu, Z., Qu, S.: An Enhanced K-Anonymity Model against Homogeneity Attack. Journal of Software 6(10) (2011)
[18] Mogre, N.V., Agarwal, G., Patil, P.: A Review on Data Anonymization Technique For Data Publishing. International Journal of Engineering Research & Technology (IJERT) 1(10) (2012) ISSN: 2278-0181
[19] Kisilevich, S., Rokach, L., Elovici, Y.: Member, IEEE, and Brach Shapira, Efficient Multidimensional Suppression for K-Anonymity. IEEE Transactions on Knowledge and Data Engineering 22(3) (2010)
[20] Shanthi, A.S., Karthikeyan, M.: A Review on Privacy Preserving Data Mining (2012)
[21] Ercan Nergiz, M., Clifton, C.: Thoughts on k-anonymization. Data & Knowledge Engineering 63, 622–645 (2007)
[22] Aggarwal, C.C.: On k-Anonymity and the Curse of Dimensionality. In: Proceedings of the 31st VLDB Conference, Trondheim, Norway (2005); International Journal of Information Security and Privacy 2(3), 28–44 (July-September 2008)
[23] Gal, T.S., Chen, Z., Gangopadhyay, A.: A Privacy Protection Model for Patient Data with Multiple Sensitive Attributes. International Journal of Information Security and Privacy 2(3), 28–44 (2008)
[24] Nargundi, S.M., Phalnikar, R.: k-Anonymization using Multidimensional Suppression for Data De-identificatio. International Journal of Computer Application 60(11), 975–8887 (2012)
[25] Wong, R.C.-W., Fu, A.W.-C., Wang, K., Pei, J.: Anonymization-Based Attacks in Privacy-Preserving Data Publishing. ACM Trans-actions on Database Systems 34(2) Article 8 (Publication date: June 2009)
[26] Woodward, B.: The computer-based patient record confidentiality. The New England Journal of Medicine 333(21), 1419–1422 (1995)

Combining Different Differential Evolution Variants in an Island Based Distributed Framework–An Investigation

Shanmuga Sundaram Thangavelu and C. Shunmuga Velayutham

Abstract. This paper proposes to combine three different Differential Evolution (*DE*) variants viz. *DE/rand/1/bin, DE/best/1/bin* and *DE/rand-to-best/1/bin* in an island based distributed Differential Evolution (*dDE*) framework. The resulting novel *dDE*s with different *DE* variants in each islands have been tested on 13 high-dimensional benchmark problems (of dimensions 500 and 1000) to observe their performance efficacy as well as to investigate the potential of combining such complementary collection of search strategies in a distributed framework. Simulation results show that *rand* and *rand-to-best* strategy combination variants display superior performance over *rand, best, rand-to-best* as well as *best, rand-to-best* combination variants. The *rand* and *best* strategy combinations displayed the poor performance. The simulation studies indicate a definite potential of combining complementary collection of search characteristics in an island based distributed framework to realize highly co-operative, efficient and robust distributed Differential Evolution variants capable of handling a wide variety of optimizations tasks.

Keywords: Differential Evolution, distributed Differential Evolution, Co-operative Evolution, Complementary Search Strategies.

1 Introduction

Since its inception in 1995, by R. Storn and K.V. Price, Differential Evolution (DE) has emerged as one of the powerful stochastic Evolutionary Algorithms (EAs) for continuous search domain [1, 2]. Like typical *EA*s, the differential evolution is a

Shanmuga Sundaram Thangavelu · C. Shunmuga Velayutham
Department of Computer Science and Engineering,
Amrita School of Engineering,
Amrita Vishwa Vidyapeetham, Ettimadai, Coimbatore, India
e-mail: {s_thangavel,cs_velayutham}@cb.amrita.edu

El-Sayed M. El-Alfy et al. (eds.), *Advances in Intelligent Informatics*,
Advances in Intelligent Systems and Computing 320, DOI: 10.1007/978-3-319-11218-3_53

population-based global optimizer that employs the iterative variation (mutation followed by crossover) and selection operations. However, the generation of offsprings from the scaled differences of random but distinct parent vectors in the current population (called the *differential mutation* operation) makes *DE* stand apart from other members of *EA* family. Depending on the way the offsprings are generated, through different parent vector perturbations, there exists multitude forms of differential mutation operations resulting in many *DE* variants.

Despite the existence of multitude of *DE* variants, interestingly no single variant of *DE* has displayed the best performance for a wide range of optimization problems as explained by no free lunch theorem [3]. Consequently, a trial-and-error selection of suitable *DE* variant becomes necessary for solving a given optimization problem. To alleviate this problem, recent research efforts employ ensemble learning with parameter and DE mutation strategy adaptation [4, 5]. These efforts, in effect, are an attempt to bring together a variety of search characteristics to effectively solve a diverse range of problems. This paper employs an alternative way to achieve a complementary collection of search characteristics by conceiving a distributed Differential Evolution algorithm which comprises different *DE* variants as subpopulations in different islands.

Towards this, the paper attempts a preliminary investigation to study about the potential interplay of different search characteristics in a distributed framework. Consequently, this paper employs three simple classical *DE* variants viz. *DE/rand/1/bin, DE/best/1/bin* and *DE/rand-to-best/1/bin* in an island based distributed *DE* framework to solve 13 high-dimensional (D = 500 and 1000) benchmark functions. Since high-dimensional optimization functions often pose severe challenges to the *EA*s in general, the collection of benchmark functions considered in this study would serve as a good test bed to analyze the potential, if there are any, of employing different *DE* variants in a distributed *DE* framework.

The rest of the paper is organized as follows. Section 2 describes the Differential Evolution and the proposed distributed Differential Evolution algorithms. Section 3 gives a brief overview of related research works. Sections 4 and 5 describe the design of experiments and the simulation results discussion respectively. Finally, Section 6 concludes the paper.

2 Distributed Differential Evolution Combining Multiple Search Strategies

Differential Evolution (DE) explores search space by starting with a uniformly randomized *D*-dimensional real-valued vectors $[x_{i,G}^1, x_{i,G}^2, x_{i,G}^3, \dots, x_{i,G}^{D-1} x_{i,G}^D]$. *NP* such vectors form the initial population (i.e. $G = 0$). At each generation, by employing the differential mutation, *DE* generates a mutant vector $V_{i,G}$ against each (target vector) $X_{i,G}$ in the current population using weighted differences of random distinct individuals of current generation. Being a preliminary investigation, this paper employs three simple mutation strategies viz.

1. *DE/rand/1*

$$V_{i,G} = X_{r_1^i,G} + F.(X_{r_2^i,G} - X_{r_3^i,G}) \quad (1)$$

2. *DE/best/1*

$$V_{i,G} = X_{best,G} + F.\left(X_{r_1^i,G} - X_{r_2^i,G}\right) \quad (2)$$

3. *DE/rand-to-best/1*

$$V_{i,G} = X_{r_3^i,G} + K.\left(X_{best,G} - X_{r_3^i,G}\right) + F.(X_{r_1^i,G} - X_{r_2^i,G}) \quad (3)$$

where K, F represent positive real scaling factors, $X_{best,G}$ denotes the best individual in the current population and r_1^i, r_2^i, r_3^i with $i \in 1, \ldots NP$ are mutually exclusive randomly generated indices anew for each mutant vector generation. The naming scheme employed (commonly in *DE* literature) is self explanatory: *rand, best, rand-to-best* represent the nature of individual being perturbed and *1* represents the number of parent pair(s) used in weighted difference calculation.

Following the differential mutation, *DE* typically employs either *binomial crossover* or *exponential crossover* that mixes a mutant vector $V_{i,G}$ with its corresponding target vector $X_{i,G}$ to generate a trial vector, $U_{i,G}$. Having observed the superiority of binomial crossover over the exponential crossover, in our earlier study [6], this paper employs the former. Finally, *DE* employs a one-to-one knockout competition between each target vector $X_{i,G}$ and its trial vector $U_{i,G}$ to decide the survivor for the next generation *G+1*.

The above mentioned mutation, crossover, selection cycle marks the end of one *DE* generation and this cycle is repeated generation after generation until a stopping criterion is satisfied.

The above considered *DE* variants viz. *DE/rand/1/bin, DE/best/1/bin* and *DE/rand-to-best/1/bin* (*/bin* represents the binomial crossover), by virtue of their differences in offspring generation, display different search characteristics. For example, while *DE/rand/1* is exploratory in nature, *DE/best/1* is greedy and tend to converge faster. This paper attempts to achieve a complementary collection of such search characteristics by employing the above discussed three *DE* variants as subpopulations in different islands thus resulting in a distributed Differential Evolution (*dDE*) framework. Further, this paper attempts a preliminary investigation to identify the potential, if there are any, of such complementary collection of search characteristics in a distributed framework to solve high-dimensional function optimization problems.

This investigation demands an extensive empirical analysis of the exhaustive combinations of the above identified three variants. Assuming an island size of 4, for the sake of easier analysis, 15 *dDE* variants arising out of combining the above said three variants in all possible combinations viz. *1:3, 2:2,* and *3:1* combinations of two of three variants; *2:1:1* combination of three variants and same variants in all islands (total 15 combinations) have all been implemented. Table 1 shows all such combinations of three variants in four islands, where *r_all, b_all* and *rb_all* respectively represents the case where the same variants *rand/1/bin, best/1/bin* and *rand-to-best/1/bin* occupy all four islands. The same notations with different proportions of combination represent the remaining 12 *dDE* variants. For the sake of brevity, this paper employs *r, b* and *rb* to represent respectively *DE/rand/1bin, DE/best/1/bin* and *DE/rand-to-best/1/bin* strategies. Figure 1 shows the algorithmic representation of the variant *brrb/1:1:2*, by way of an example and figure 2 shows the representation of few chosen *dDE* variants viz. *r_all, br/2:2, rrb/3:1* and *brrb1/1:1:2*.

Table 1 The 15 *dDE* variants arising out of combining *DE/rand/1/bin, DE/best/1/bin* and *DE/rand-to-best/1/bin* variants

S.No.	Variant Name	Description
1	*r_all*	*DE/rand/1/bin* in all four nodes (homogeneous mixing)
2	*br/1:3*	*DE/best/1/bin* in one node, *DE/rand/1/bin* in three nodes
3	*br/2:2*	*DE/best/1/bin* in two nodes, *DE/rand/1/bin* in two nodes
4	*br/3:1*	*DE/best/1/bin* in three nodes, *DE/rand/1/bin* in one node
5	*b_all*	*DE/best/1/bin* in all four nodes (homogeneous mixing)
6	*rrb/3:1*	*DE/rand/1/bin* in three nodes, *DE/rand-to-best/1/bin* in one node
7	*rrb/2:2*	*DE/rand/1/bin* in two nodes, *DE/rand-to-best/1/bin* in two nodes
8	*rrb/1:3*	*DE/rand/1/bin* in one node, *DE/rand-to-best/1/bin* in three nodes
9	*rb_all*	*DE/rand-to-best/1/bin* in all four nodes (homogeneous mixing)
10	*brb/3:1*	*DE/best/1/bin* in three nodes, *DE/rand-to-best/1/bin* in one node
11	*brb/2:2*	*DE/best/1/bin* in two nodes, *DE/rand-to-best/1/bin* in two nodes
12	*brb/1:3*	*DE/best/1/bin* in one node, *DE/rand-to-best/1/bin* in three nodes
13	*brrb/1:1:2*	*DE/best/1/bin in* one node, *DE/rand/1/bin* in one node, *DE/rand-to-best/1/bin* in two nodes
14	*brrb/2:1:1*	*DE/best/1/bin* in two nodes, *DE/rand/1/bin* in one node, *DE/rand-to-best/1/bin* in one node
15	*brrb/1:2:1*	*DE/best/1/bin* in one node, *DE/rand/1/bin* in two nodes, *DE/rand-to-best/1/bin* in one node

3 Related Works

Despite the fact that employing different *DE* variants as subpopulations in an island based *DE* is novel, this idea has been earlier attempted by researchers with respect to Genetic Algorithms (GAs). A gradual distributed real-coded genetic algorithms which employ different *GAs*, with different crossover operators, as subpopulations in an island based framework has been attempted by Herrera and Lozano in [7]. A two population island model combining binary coded GA and gray coded GA has been studied by Skolicki and De Jong in [8].

In *DE* literature, one of the efforts towards combining a variety of search strategy has resulted in self-adaptive (sequential) *DE* called *SaDE* [4] which employed *DE/rand/1/bin, DE/rand-to-best/2/bin, DE/rand/2/bin* and *DE/current-to-rand/1/bin* in a pool and let the successful strategy (based on the success probability in previous generations) to create the trial vector. This facilitated different variants to contribute in appropriate evolutionary phase during search displaying robust performance. It is worth noting that *SaDE* achieved a complementary collection of search characteristics in a Sequential *DE* framework. Mallipedi et. al.

extended this idea by including an ensemble of control parameters along with mutation strategies in the pool [5]. The above said extension has also been attempted by Wang et.al in their Composite DE (*CoDE*) [9]. Hybridization of *DE* with particle swarm optimization [10-13], ant colony optimization [14, 15], bacterial foraging optimization [16], artificial immune system [17] and simulated annealing [18, 19] have all been other interesting directions of research in *DE* literature to achieve a complementary collection of search strategy. However, unlike the above discussed works, this paper attempts to achieve a co-operative collection of search strategies in an island based distributed *DE* framework.

4 Design of Experiments

To investigate the potential of the above said framework, the 15 distributed DE variants (arising out of combining *rand/1/bin, best/1/bin* and *rand-to-best/1/bin* in various combinations in 4 islands) have been tested for their performance efficacy on a set of 13 benchmark problems of dimensionality 500 and 1000. Table 2 displays the 13 high-dimensional benchmark optimization problems. (adopted from [20]).

```
Initialize P_G = {X_{1,G}, ..., X_{NP,G}} with X_{i,G} = {X^1_{i,G}, ..., X^D_{i,G}}, i = 1, ..., NP; G = 0;
Compute {f(X_{i,G}), ..., f(X_{NP,G})}
Divide the population of size NP into 4 subpopulations of size NP/4
Scatter the sub populations to all nodes
Place DE/best/1/bin in 1 node, DE/rand/1/bin in 1 node
    and DE/rand-to-best/1/bin variant in 2 nodes,
WHILE stopping criterion is not satisfied DO
  FOR i = 1 to NP/4
  DO
    1. Mutation Step
    2. Crossover Step
    3. Selection Step
  END DO
  END FOR
  DO (For every migration frequency generation)
    1. Send nm candidates, selected by selection policy, to next node as per migration topology
    2. Receive the nm candidate from the previous node as per migration topology
    3. Replace the candidates selected by replacement policy by the received nm candidates
  END DO
  G = G + 1 ;
  Compute {f(X_{i,G}), ..., f(X_{NP,G})}
END WHILE
```

Fig. 1 The algorithmic representation of *brrb/1:1:2* variant

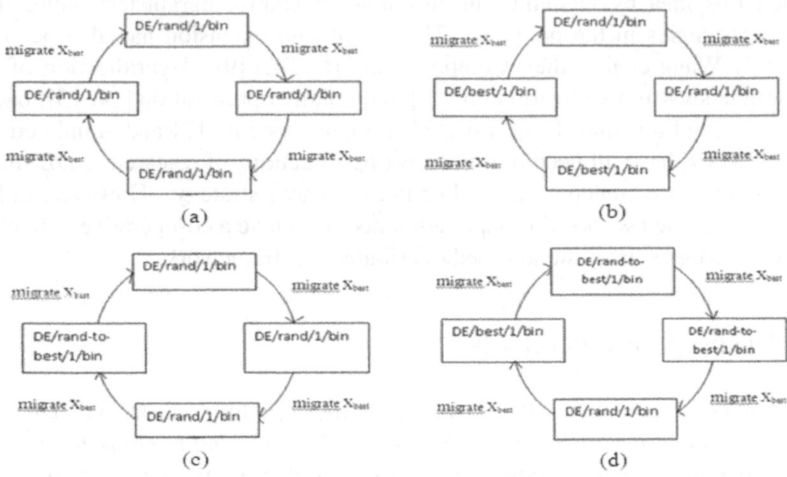

Fig. 2 Pictorial Representation of few *dDE mixing schemes* ((a) *r_all* (b) *br/2:2* (c) *rrb/3:1* (d) *brrb/1:1:2*) variants

Table 2 Details of the benchmark functions used in simulation experiments

Test Problem	Decision Space	Function Description
f_{01}- Ackley	$[-1,1]^n$	$-20+e+20exp\left(-0.2\sqrt{\frac{1}{n}\sum_{i=1}^{n}x_i^2}\right)-exp\left(\frac{1}{n}\sum_{i=1}^{n}cos(2\pi x_i)\right)$
f_{02}- Alpine	$[-10,10]^n$	$\sum_{i=1}^{n}\|x_i\, sin\, x_i + 0.1\, x_i$
f_{03}-Axis-parallel hyperellipsoid	$[-5.12,5.12]^n$	$\sum_{i=1}^{n} i\ x_i^2$
f_{04}- DeJong	$[-5.12,5.12]^n$	$\|\|x\|\|^2$
f_{05}- DropWave	$[-5.12,5.12]^n$	$-\frac{1+cos(12\sqrt{\|\|x\|\|^2})}{\frac{1}{2}\|\|x\|\|^2+2}$
f_{06}- Griewangk	$[-600,600]^n$	$\frac{1}{4000}\sum_{i=1}^{n}x_i^2-\prod_{i=0}^{n}cos\left(\frac{x_i}{\sqrt{i}}\right)+1$
f_{07}- Michalewicz	$[0,\pi]^n$	$-\sum_{i=1}^{n}sin\, x_i\left(sin\left(\frac{i x_i^2}{\pi}\right)\right)^{20}$
f_{08}- Pathological	$[-100,100]^n$	$\sum_{i=1}^{n-1}\left(0.5+\frac{sin^2\left(\sqrt{100x_i^2+x_{i+1}^2}-0.5\right)}{1+0.001*\left(x_i^2-2x_ix_{i+1}+x_{i+1}^2\right)^2}\right)$
f_{09}- Rastringin	$[-5.12,5.12]^n$	$\sum_{i=0}^{n}[x_i^2-10cos(2\pi x_i)+10n]$
f_{10}- Rosenbrock valley	$[-2.048,2.048]^n$	$\sum_{i=1}^{n-1}(100(x_{i+1}-x_i^2)^2+(1-x_i)^2)$
f_{11}- Schwefel	$[-500,500]^n$	$\sum_{i=1}^{n}\left(x_i sin(\sqrt{\|x_i\|})\right)$
f_{12}- Sum of powers	$[-1,1]^n$	$\sum_{i=1}^{n}\|x_i\|^{i+1}$
f_{13}- Tirronen	$[-10,5]^n$	$\sum_{i=1}^{n}\|x_i\|^{i+1}\ 3\,exp\left(-\frac{\|\|x\|\|^2}{10n}\right)-10\,exp(-8\|\|x\|\|^2)+\frac{2.5}{n}\sum_{i=1}^{n}cos\left(5(x_i+(1+i\, mod\, 2)cos(\|\|x\|\|^2))\right)$

The control parameters of the Differential Evolution algorithm viz. the population size (*NP*), Scale factors (*F* and *K*) and the crossover rate (C_r) have been set as follows. The population size has been fixed as 200 (with 50 as subpopulation size in each of 4 islands) and 400 (with 100 as subpopulation size in each of 4 islands) for the dimensions 500 and 1000 respectively. The scale factor (*F*) has been generated anew for each generation randomly in the range [0.3, 0.9]. In case of *DE/rand-to-best/1/bin*, the value of *K* has been set the same value as *F*. The crossover rate values $C_r \in [0,1]$ with step size 0.1 have been tested for all variants on each of the benchmark problems. Hundred independent runs have been performed for each variant–benchmark function–C_r value combination. A bootstrap test, to determine the confidence interval for the obtained mean objective function values, has been conducted. The C_r value, of each variant-test function combination, that corresponds to the best confidence interval (i.e. 95%) has been chosen to be used in our subsequent simulations. Table 3 shows the C_r values used by the three *DE* variants for each test function for both 500 and 1000 dimensions.

The control parameters of island based distributed framework have been set as follows. As has been said earlier, the simulation experiments assume an island size of 4 with 50 and 100 individuals in each island for 500 and 1000 dimensions cases respectively. Based on an earlier empirical analysis [21], the best solution in each island is allowed to migrate in a ring topology, once every 45 generations, replacing a random but best solution in the receiving island.

As *DE*s are stochastic in nature, each simulation experiment (involving each of 15 *dDE* on each test function) has been repeated 100 times (runs) with randomly initialized population for every run. Each simulation experiment will stop either if a maximum number of generations $G_{max} = 2500$ is reached or if a tolerance error value of 1×10^{-12} is reached.

Table 3 Crossover rate (C_r) values used for problems of Dimensions (D) 500 and 1000

D	Variant	f_{01}	f_{02}	f_{03}	f_{04}	f_{05}	f_{06}	f_{07}	f_{08}	f_{09}	f_{10}	f_{11}	f_{12}	f_{13}
500 /	*rand/1/bin*	0.9/0.9	0.8/0.9	0.7/0.9	0.7/0.8	1.0/1.0	0.7/0.8	0.0/0.9	0.0/0.0	0.9/0.9	0.7/0.0	0.9/1.0	0.0/0.0	0.1/0.1
1000	*best/1/bin*	0.3/0.3	0.2/0.3	0.1/0.1	0.1/0.1	0.2/1.0	0.1/0.1	0.2/0.2	0.0/0.0	0.2/0.3	0.1/0.2	0.8/0.6	0.2/0.3	0.1/0.1
	rand-to-best/1/bin	0.2/0.3	0.1/0.1	0.1/0.1	0.1/0.1	0.2/0.2	0.1/0.1	0.1/0.1	0.0/0.0	0.3/0.3	0.1/0.1	0.8/0.8	0.2/0.1	0.1/0.1

Considering the fact that the current work attempts to investigate the potential of combining different search strategies, the probability of convergence (P_c) and quality measure (Q_m) of different combinations have been reported instead of their mean performance. The probability of convergence [22] is the mean percent of number of successful runs (S_r) out of total number of runs (t_r) (t_r = 14 functions x 100 runs each = 1400) i.e. $P_c = s_r/t_r\,\%$. The quality measure [22] is the ratio of convergence rate of successful runs (C_s) and the probability of convergence (P_c) i.e. $Q_m = C_s/P_c$. The convergence rate of successful runs is the average number of function evaluations for successful runs i.e. $C_s = \sum_{j=1}^{S_r} FE_j/S_r$, where FE_j is the number of function evaluations in j^{th} successful run. A variant with high P_c value and lower number of function evaluations (i.e. low Q_m value) is considered to be a competitive variant.

Apart from the quality measure (Q_m), considering the fact that high-dimensional optimization functions pose severe challenges to the optimizer, a variant of Q_m calculation i.e. *Q-test* has also been reported as described in [20]. For each benchmark problem, the average final objective function value obtained by the best performing combination B has been considered along with the average fitness value at the beginning of the optimization b. The threshold value Thr, which represents 95% of the decay in fitness value of B, is calculated as $Thr = b - 0.95\ (B - b)$. Hence, a run is considered successful if a combination succeeds during a certain run to reach the Thr value. Using this, the Q_m has been recomputed to obtain Q-test values. The ∞ denotes the failure of a combination to reach the threshold (i.e. infinite time required to reach the threshold).

5 Simulation Results and Discussions

Tables 4 and 5 show the Q-test values obtained by the combinations of *DE* strategies for each function with 500 and 1000 dimensions respectively. As stated earlier, the entry ∞ indicates the failure of a particular combination to solve a function within G_{max} generations. For the sake of easier analysis, the 15 *dDE* variants are sorted in ascending order, based on their Q-test performance, in Tables 6 and 7 for 500 and 1000 dimensions respectively. In both Tables 6 and 7, *NA*s indicate the case where no other variants can be put as they have not been able to solve a particular problem.

In case of 500 dimensions, no variants could solve f_{11} and only 2 variants managed to solve f_5. Interestingly, in most cases viz. $f_1, f_2, f_3, f_4, f_6, f_7, f_{10}$ and f_{12}, the combinations of *rand/1/bin* and *best/1/bin* displayed a poor performance. This performance is contrary to the fact that *rand/1/bin* and *best/1/bin* are complementary strategies with the former exploratory in nature and the latter more greedy. However, in case of functions f_8 and f_9, *rand/1/bin* and *best/1/bin* combinations have displayed a slightly better performance. Interestingly *b-all, br/3:1, br/2:2* and *br/1:3* have displayed the top performance in case of f_{13}. The poor performance of *rand-best* combinations can be observed in 1000 dimensions too, in case of $f_1, f_2, f_3, f_4, f_6, f_8, f_9$ and f_{10}. While none of *rand-best* combinations could solve the function f_7, *rand-best* combinations *br/2:2, r-all* and *br/1:3* were the only variants (apart from *rrb/3:1*) to solve f_5. However, in case of functions f_{11}, f_{12} and f_{13}, *b-all* has displayed the best performance with *br/3:1, br/1:3* and *br/2:2* showing reasonably better performance.

On the contrary, the homogeneous *rb-all* has displayed the best performance in most of the cases ($f_2, f_3, f_4, f_5, f_6, f_7, f_{10}$ and f_{12} in the case of 500 dimensions; $f_1, f_2, f_3, f_4, f_6, f_7, f_8$ and f_{10} in case of 1000 dimensions). This behavior of *rand-to-best/1/bin* is quite understandable as the variant intrinsically combines both *rand* and *best* mutation strategies. Both in case of 500 and 1000 dimensions, between the best and poor performance, the intermediate performance have been displayed by the combinations of *rand* and *best* strategies with *rand-to-best* strategy.

Table 4 *Q*-Test values obtained (for each function and variant for 500 Dimensions)

Mixing	f_{01}	f_{02}	f_{03}	f_{04}	f_{05}	f_{06}	f_{07}	f_{08}	f_{09}	f_{10}	f_{11}	f_{12}	f_{13}
r_all	380559.2	332050	170000	183510	∞	184590	∞	488458.3	437836.4	160214	∞	142870	455710.4
b_all	318572.4	298092	151582	160908	∞	160964	492000	486950	∞	165434	∞	55176	436908.9
rb_all	310248.5	156976	60372	62934	437911	63276	293826	468222.7	394357.3	51410	∞	31260	466800
br/1:3	315123.7	265142	153296	164422	∞	163900	∞	486936.8	358600	164438	∞	72544	449139.7
br/2:2	334679.1	266706	150770	161218	∞	161340	487000	488294.7	369661.5	162276	∞	62886	443551.2
br/3:1	334683.3	276188	150198	160378	∞	160060	482400	488306.7	417331.3	161252	∞	57912	437330.1
rrb/3:1	230596	179038	67192	71014	∞	70670	331124	484170.2	296352	56934	∞	37224	466435.6
rrb/2:2	242580	168852	63476	66484	∞	66610	307568	478990.5	309688	54004	∞	34230	453291.9
rrb/1:3	272290	163826	61524	64534	∞	64018	297790	474079.5	325183.8	52306	∞	32592	467406.7
brb/3:1	309051.6	182798	67504	71006	∞	71480	379567.44	485587.5	392243	56320	∞	36114	445106.2
brb/2:2	300276.1	171612	63566	66538	∞	66058	365287.32	480829.1	385683	53728	∞	33742	455074.3
brb/1:3	296269.4	163844	61420	64458	472600	64448	338030.11	476260	384969.9	52148	∞	32386	460376
brrb/2:1:1	304239.1	183230	67384	71018	∞	70712	388581.82	487769.7	348662.5	56656	∞	35904	449810.8
brrb/1:2:1	277207.1	181020	67340	71190	∞	70960	357413.64	482170.7	324520	57010	∞	36498	449216.4
brrb/1:1:2	285765.2	170570	63388	66160	∞	66052	347000	482102.8	341577.1	53174	∞	33242	452660.4

Table 5 *Q*-Test values obtained (for each function and variant for 1000 Dimensions)

Mixing	f_{01}	f_{02}	f_{03}	f_{04}	f_{05}	f_{06}	f_{07}	f_{08}	f_{09}	f_{10}	f_{11}	f_{12}	f_{13}
r_all	∞	938325.4	297592	543076	200569.2	543616	∞	969573.3	∞	∞	∞	490788	882066.7
b_all	929346.7	720036	581188	613616	∞	615236	∞	982228.6	∞	327148	962776	123968	837565.2
rb_all	487208.1	396784	157708	164988	∞	165664	825828	934531.9	809272.3	125352	∞	170620	876600
br/1:3	981200	809012	292712	474652	205120	477648	∞	968577.8	∞	440136	986800	180220	866835.4
br/2:2	877444.4	673364	318568	489504	174000	489144	∞	973428.6	981600	359760	988974.4	147880	860163.9
br/3:1	919726.3	666157.6	365788	512252	∞	513936	∞	974782.6	∞	338776	979065.2	132440	859441.8
rrb/3:1	553052	443600	175960	189168	177800	189852	948800	966350	719051.5	139992	∞	201908	880016.2
rrb/2:2	513204	416848	165856	176796	∞	176228	893848.1	952464.5	774584	133472	∞	183676	884250.7
rrb/1:3	508056	406872	160172	171732	∞	171336	853127.3	951470	776084	130540	∞	180532	874813.6
brb/3:1	835756.3	482172	180256	190440	∞	189136	900990.5	962369.2	870711.1	139840	981774.5	129900	850289.7
brb/2:2	775329.3	436044	168392	176820	∞	175632	871708.1	952970.5	794717.5	132168	985085	130392	866240
brb/1:3	670484	411780	163624	170900	∞	171880	845994.7	956231.6	759153.2	130356	990302.7	145300	889154.3
brrb/2:1:1	844472.3	563684	179308	189648	∞	189828	908737.5	968547.8	894033.3	139588	989877.8	134460	848815.4
brrb/1:2:1	762262.9	526124	177132	189672	∞	189672	926411.4	967712.8	860148.1	139712	986400	146524	867604.9
brrb/1:1:2	721504.2	434204	166684	176684	∞	176336	882691.5	961253.3	807866.7	133076	983600	146504	877228.6

To get a still better picture of the performance efficacy of combining *rand* and *best* strategies with *rand-to-best* strategy, the Quality measure (Q_m) values have been calculated and are displayed in Tables 8 and 9 for 500 and 1000 dimensional cases respectively. Interesting patterns of performance efficacy can be observed in the case of 500 dimensions as shown in table 8.

The *rand* and *rand-to-best* strategy combinations have displayed the top performance in terms of Q_m values. This is closely followed by that of the *rand, best* and *rand-to-best* combination variants as well as the homogeneous *rb-all.* The *best* and *rand-to-best* combination variants have displayed a relatively moderate performance. As observed in case of Q-test, the *rand* and *best* combination variants have displayed poor performance in terms of Q_m. Interestingly, in case of 1000 dimensions, as can be seen in Table 9, the homogeneous *rb-all* has displayed the best performance. However, the other observations in 500 dimensions case largely hold good in 1000 dimensions too.

Table 6 *Q*-Test Results for 500 Dimensions (variants listed in sorted order based on the Q-Test values for each function)

f_{01}	f_{02}	f_{03}	f_{04}	f_{05}	f_{06}	f_{07}	f_{08}	f_{09}	f_{10}	f_{11}	f_{12}	f_{13}
rrb/3:1	rb_all	rb_all	rb_all	rb_all	rb_all	rb_all	rb_all	rrb/3:1	rb_all	NA	rb_all	b_all
rrb/2:2	rrb/1:3	brb/1:3	brb/1:3	brb/1:3	rrb/1:3	rrb/1:3	rrb/1:3	rrb/2:2	brb/1:3	NA	brb/1:3	br/3:1
rrb/1:3	brb/1:3	rrb/1:3	rrb/1:3	NA	brb/1:3	rrb/2:2	brb/1:3	brrb/1:2:1	rrb/1:3	NA	rrb/1:3	br/2:2
brrb/1:2:1	rrb/2:2	brrb/1:1:2	brrb/1:1:2	NA	brrb/1:1:2	rrb/3:1	rrb/2:2	rrb/1:3	brrb/1:1:2	NA	brrb/1:1:2	brb/3:1
brrb/1:1:2	brrb/1:1:2	rrb/2:2	rrb/2:2	NA	brb/2:2	brb/1:3	brb/2:2	brrb/1:1:2	brb/2:2	NA	brb/2:2	br/1:3
brb/1:3	brb/2:2	brb/2:2	brb/2:2	NA	rrb/2:2	brrb/1:1:2	brrb/1:1:2	brrb/2:1:1	rrb/2:2	NA	rrb/2:2	brrb/1:2:1
brb/2:2	rrb/3:1	rrb/3:1	brb/3:1	NA	rrb/3:1	brrb/1:2:1	brrb/1:2:1	br/1:3	brb/3:1	NA	brrb/2:1:1	brrb/2:1:1
brrb/2:1:1	brrb/1:2:1	brrb/1:2:1	rrb/3:1	NA	brrb/2:1:1	brb/2:2	rrb/3:1	br/2:2	brrb/2:1:1	NA	brb/3:1	brrb/1:1:2
brb/3:1	brb/3:1	brrb/2:1:1	brrb/2:1:1	NA	brrb/1:2:1	brb/3:1	brb/3:1	brb/1:3	rrb/3:1	NA	brrb/1:2:1	rrb/2:2
rb_all	brrb/2:1:1	brb/3:1	brrb/1:2:1	NA	brb/3:1	brrb/2:1:1	br/1:3	brb/2:2	brrb/1:2:1	NA	rrb/3:1	brb/2:2
br/1:3	br/1:3	br/3:1	br/3:1	NA	br/3:1	br/1:3	b_all	brb/3:1	r_all	NA	b_all	r_all
b_all	br/2:2	br/2:2	b_all	NA	b_all	br/2:2	brrb/2:1:1	rb_all	br/3:1	NA	br/3:1	brb/1:3
br/2:2	br/3:1	b_all	br/2:2	NA	br/2:2	b_all	br/2:2	br/3:1	br/2:2	NA	br/2:2	rrb/3:1
br/1:3	b_all	br/1:3	br/1:3	NA	br/1:3	NA	br/3:1	r_all	br/1:3	NA	br/1:3	rb_all
r_all	r_all	r_all	r_all	NA	r_all	NA	r_all	NA	b_all	NA	r_all	rrb/1:3

Table 7 *Q*-Test Results for 1000 Dimensions (variants listed in sorted order based on the Q-Test values for each function)

f_{01}	f_{02}	f_{03}	f_{04}	f_{05}	f_{06}	f_{07}	f_{08}	f_{09}	f_{10}	f_{11}	f_{12}	f_{13}
rb_all	rb_all	rb_all	rb_all	br/2:2	rb_all	rb_all	rb_all	rrb/3:1	rb_all	b_all	b_all	b_all
rrb/1:3	rrb/1:3	rrb/1:3	brb/1:3	rrb/3:1	rrb/1:3	brb/1:3	rrb/1:3	brb/1:3	brb/1:3	br/1:3	brb/3:1	brrb/2:1:1
rrb/2:2	brb/1:3	brb/1:3	rrb/1:3	r_all	brb/1:3	rrb/1:3	rrb/2:2	rrb/2:2	rrb/1:3	brb/3:1	brb/2:2	brb/3:1
rrb/3:1	rrb/2:2	rrb/2:2	brrb/1:1:2	br/1:3	brb/2:2	brb/2:2	brb/2:2	rrb/1:3	brb/2:2	brrb/1:1:2	br/3:1	br/3:1
brb/1:3	brrb/1:1:2	brrb/1:1:2	rrb/2:2	NA	rrb/2:2	brrb/1:1:2	brb/1:3	brb/2:2	brrb/1:1:2	brb/2:2	brrb/2:1:1	br/2:2
brrb/1:1:2	brb/2:2	brb/2:2	brb/2:2	NA	brrb/1:1:2	rrb/2:2	brrb/1:1:2	brrb/1:1:2	rrb/2:2	brrb/1:2:1	brb/1:3	brb/2:2
brrb/1:2:1	rrb/3:1	rrb/3:1	rrb/3:1	NA	brb/3:1	brb/3:1	brb/3:1	rb_all	brrb/2:1:1	br/1:3	brrb/1:1:2	br/1:3
brb/2:2	brb/3:1	brrb/1:2:1	brrb/2:1:1	NA	brrb/1:2:1	brrb/2:1:1	rrb/3:1	brrb/1:2:1	brrb/1:2:1	br/2:2	brrb/1:2:1	brrb/1:2:1
brb/3:1	brrb/1:2:1	brrb/2:1:1	brrb/1:2:1	NA	brrb/2:1:1	brrb/1:2:1	brrb/1:2:1	brb/3:1	brb/3:1	brrb/2:1:1	br/2:2	rrb/1:3
brrb/2:1:1	brrb/2:1:1	brb/3:1	brb/3:1	NA	rrb/3:1	rrb/3:1	brrb/2:1:1	brrb/2:1:1	rrb/3:1	brb/1:3	rb_all	rb_all
br/2:2	br/3:1	br/1:3	br/1:3	NA	br/1:3	NA	br/1:3	br/2:2	b_all	NA	br/1:3	brrb/1:1:2
br/3:1	br/2:2	r_all	br/2:2	NA	br/2:2	NA	r_all	NA	br/3:1	NA	rrb/1:3	rrb/3:1
b_all	b_all	br/2:2	br/3:1	NA	br/3:1	NA	br/2:2	NA	br/2:2	NA	rrb/2:2	r_all
br/1:3	br/1:3	br/3:1	r_all	NA	r_all	NA	br/3:1	NA	br/1:3	NA	rrb/3:1	rrb/2:2
NA	r_all	b_all	b_all	NA	b_all	NA	b_all	NA	NA	NA	r_all	brb/1:3

The relatively superior performance of *rand* and *rand-to-best* combination variants over the other combinations can be attributed to the fact that while *rand-to-best* strategy intrinsically combines *rand* and *best* strategies making it a very effective variant, the *rand* strategy facilitates in the search of former by maintaining the population variability unlike the greedy *best* strategy. In case of *rand, best* and *rand-to-best* combination variants, while *rand* strategy maintains population variability, the *best* strategy reduces it. Despite the fact that *rand* and *best* strategies are complementary in nature (both explorative and greedy), they lose to the intrinsic combinations in *rand-to-best/1/bin*.

Table 10 shows the execution time taken in a typical run (2000 generations) by *DE/rand/1/bin* on all benchmarks problems for both $D = 500$ and 1000 on Intel Core 2 Duo 1.86 GHz PCs with 2 GB DDR2 RAM in linux environment. It is worth noting that combining different *DE* variants, as proposed in this paper, has exactly the same operational scheme as that of a typical distributed *DE* (*dDE*). The only difference lies in the fact that the former employs different *DE* variants in each islands unlike the latter which employs same *DE* variant in all islands. Thus the proposed framework has similar processing time as its classical *dDE* counterpart.

Table 8 Q_m Results for 500 Dimensions (variants sorted on Q_m value)

Mixing	f_{01}	f_{02}	f_{03}	f_{04}	f_{05}	f_{06}	f_{07}	f_{08}	f_{09}
rrb/2:2	24258000	16885200	6347600	6648400	∞	6661000	30756800	30176400	30968800
rrb/3:1	23059600	17903800	6719200	7101400	∞	7067000	33112400	22756000	29635200
rrb/1:3	27229000	16382600	6152400	6453400	∞	6401800	29779000	41719000	32193200
brrb/1:2:1	23562600	18102000	6734000	7119000	∞	7096000	15726200	19769000	32452000
rb_all	30714600	15697600	6037200	6293400	35470800	6327600	29382600	45417600	35097800
brrb/1:1:2	26290400	17057000	6338800	6616000	∞	6605200	25678000	34711400	32791400
brb/1:3	29034400	16384400	6142000	6445800	3308200	6444800	31436800	38100800	35802200
brrb/2:1:1	27990000	18323000	6738400	7101800	∞	7071200	17097600	16096400	33471600
brb/2:2	27625400	17161200	6356600	6653800	∞	6605800	25935400	26445600	36254200
brb/3:1	28741800	18279800	6750400	7100600	∞	7148000	16321400	15538800	36478600
br/2:2	28782400	26670600	15077000	16121800	∞	16134000	1461000	9277600	33639200
br/1:3	23949400	26514200	15329600	16442200	∞	16390000	∞	9251800	35501400
br/3:1	28113400	27618800	15019800	16037800	∞	16006000	964800	7324600	27961200
b_all	27715800	29809200	15158200	16090800	∞	16096400	492000	3895600	∞
r_all	37294800	33205000	17000000	18351000	∞	18459000	∞	11723000	38529600

Mixing	f_{10}	f_{11}	f_{12}	f_{13}	SumFE	nc	P_c%	C_s	Q_m
rrb/2:2	5400400	∞	3423000	16771800	1.78E+08	1000	76.92308	178297.4	2317.866
rrb/3:1	5693400	∞	3722400	20989600	1.78E+08	992	76.30769	179193.5	2348.303
rrb/1:3	5230600	∞	3259200	14022200	1.89E+08	1017	78.23077	185666.1	2373.313
brrb/1:2:1	5701000	∞	3649800	27402200	1.67E+08	931	71.61538	179714.1	2509.434
rb_all	5141000	∞	3126000	7002000	2.26E+08	1081	83.15385	208795.7	2510.957
brrb/1:1:2	5317400	∞	3324200	23991000	1.89E+08	987	75.92308	191206.5	2518.424
brb/1:3	5214800	∞	3238600	23018800	2.05E+08	1021	78.53846	200364	2551.157
brrb/2:1:1	5665600	∞	3590400	33286000	1.76E+08	939	72.23077	187893.5	2601.295
brb/2:2	5372800	∞	3374200	31855200	1.94E+08	982	75.53846	197189.6	2610.453
brb/3:1	5632000	∞	3611400	36053600	1.82E+08	942	72.46154	192841.2	2661.29
br/2:2	16227600	∞	6288600	36371200	2.06E+08	881	67.76923	233883.1	3451.169
br/1:3	16443800	∞	7254400	32787200	2E+08	867	66.69231	230523.6	3456.525
br/3:1	16125200	∞	5791200	36298400	1.97E+08	851	65.46154	231799.3	3541
b_all	16543400	∞	5517600	39321800	1.71E+08	786	60.46154	217100.3	3590.717
r_all	16021400	∞	14287000	30532600	2.35E+08	877	67.46154	268418.9	3978.844

Table 9 Q_m Results for 1000 Dimensions (variants sorted on Q_m *value*)

Mixing	f_{01}	f_{02}	f_{03}	f_{04}	f_{05}	f_{06}	f_{07}	f_{08}	f_{09}
rb_all	48233600	39678400	15770800	16498800	∞	16566400	82582800	87846000	76071600
rrb/1:3	50805600	40687200	16017200	17173200	∞	17133600	84459600	76117600	77608400
rrb/2:2	51320400	41684800	16585600	17679600	∞	17622800	70614000	59052800	77458400
rrb/3:1	55305200	44360000	17596000	18916800	355600	18985200	19924800	46384800	69748000
brb/1:3	67048400	41178000	16362400	17090000	∞	17188000	64295600	72673600	71360400
brrb/1:1:2	69264400	43420400	16668400	17668400	∞	17633600	52078800	57675200	65437200
brb/2:2	76757600	43604400	16839200	17682000	∞	17563200	32253200	58131200	50067200
brrb/1:2:1	53358400	52612400	17713200	18967200	∞	18967200	32424400	37740800	23224000
brb/3:1	72710800	48217200	18025600	19044000	∞	18913600	18920800	37532400	7836400
brrb/2:1:1	70091200	56368400	17930800	18964800	∞	18982800	29079600	44553200	21456800
br/1:3	3924800	80901200	29271200	47465200	8204800	47764800	∞	34868800	∞
br/2:2	15794000	67336400	31856800	48950400	522000	48914400	∞	20442000	981600
br/3:1	17474800	65949600	36578800	51225200	∞	51393600	∞	22420000	∞
b_all	27880400	72003600	58118800	61361600	∞	61523600	∞	13751200	∞
r_all	∞	55361200	29759200	54307600	15644400	54361600	∞	29087200	∞
Mixing	f_{10}	f_{11}	f_{12}	f_{13}	***SumFE***	***nc***	P_c %	C_s	Q_m
rb_all	12535200	∞	17062000	43830000	4.57E+08	1037	79.76923	440381.5	5520.694
rrb/1:3	13054000	∞	18053200	51614000	4.63E+08	1038	79.84615	445783.8	5583.034
rrb/2:2	13347200	∞	18367600	59244800	4.43E+08	1008	77.53846	439462.3	5667.669
rrb/3:1	13999200	∞	20190800	65121200	3.91E+08	942	72.46154	414955	5726.555
brb/1:3	13035600	36641200	14530000	62240800	4.94E+08	1053	81	468797.7	5787.626
brrb/1:1:2	13307600	1967200	14650400	61406000	4.31E+08	968	74.46154	445431.4	5982.033
brb/2:2	13216800	78806800	13039200	69299200	4.87E+08	1020	78.46154	477705.9	6088.408
brrb/1:2:1	13971200	1972800	14652400	70276000	3.56E+08	854	65.69231	416721.3	6343.533
brb/3:1	13984000	92286800	12990000	73975200	4.34E+08	937	72.07692	463646.5	6432.663
brrb/2:1:1	13958800	35635600	13446000	66207600	4.07E+08	899	69.15385	452364.4	6541.421
br/1:3	44013600	986800	18022000	68480000	3.84E+08	760	58.46154	505135.8	8640.481
br/2:2	35976000	38570000	14788000	71393600	3.96E+08	765	58.84615	517026.4	8786.07
br/3:1	33877600	90074000	13244000	78209200	4.6E+08	824	63.38462	558794.7	8815.935
b_all	32714800	96277600	12396800	77056000	5.13E+08	836	64.30769	613737.3	9543.762
r_all	∞	∞	49078800	68801200	3.56E+08	645	49.61538	552560	11136.87

Table 10 Execution Time (in sec) of a variant *rand/1/bin* for a run (2000 generations) for 500 and 1000 dimensional problems

D	f_{01}	f_{02}	f_{03}	f_{04}	f_{05}	f_{06}	f_{07}	f_{08}	f_{09}	f_{10}	f_{11}	f_{12}	f_{13}
500	13.5	11.9	8.45	8.43	8.06	17.07	31.38	40.07	16.58	14.59	24.28	26.69	22.09
1000	42.8	39.54	26.73	26.92	26.33	57.01	110.29	136.54	56.23	46.6	85.42	84.91	73.76

The above observations regarding the combinations of *DE* variants provide insight about the definite potential of combining different strategies in an island based distributed framework. However, it is worth admitting that a thorough understanding of the performance efficacy of combining different *DE* variants in a distributed framework necessitates an extensive empirical performance analysis. A thorough theoretical and empirical understanding of the exploration-exploitation characteristics of *DE* variants would provide a much better insight in designing efficient and robust *dDE* variants suited for a wide variety of optimization tasks.

6 Conclusion

This paper proposed combining different Differential Evolution (DE) search strategies in an island based distributed framework to investigate the potential of such complementary collection of search strategies. Towards this, 15 distributed *DE* variants arising out of combining *DE/rand/1/bin, DE/best/1/bin* and *DE/rand-to-best/1/bin,* in an island size of 4, have all been implemented and tested on 13 high-dimensional (500 and 1000) optimization functions. The simulation results highlights the potential of *rand* and *rand-to-best* combination variants in terms of Quality measure (Q_m). The *best* and *rand-to-best* combination variants and *rand, best* and *rand-to-best* combination variants have displayed relatively moderate performance. This study hints both the potential of combining different *DE* variants in an island based distributed framework as well as the necessity of thorough empirical and theoretical analysis towards understanding the exploration-exploitation characteristics of different *DE* strategies. The insights from such analyses would provide us with definite guidelines in designing and building efficient and robust distributed *DE* variants (*dDE*) suited for solving a wide variety of optimization tasks. The implications of empirical parameter settings of both *DE* algorithm and *dDE* framework on the dynamics of search strategy combinations is also a potential direction for further research.

References

1. Storn, R., Price, K.: Differential Evolution – A Simple and Efficient Adaptive Scheme for Global Optimization Over Continuous Spaces. Technical Report TR-95-012, ICSI (1995)
2. Storn, R., Price, K.: Differential Evolution – A Simple and Efficient Heuristic Strategy for Global Optimization and Continuous Spaces. Journal of Global Optimization 11(4), 341–359 (1997)
3. Wolpert, D.H., Macreedy, W.G.: No Free Lunch Theorems for Optimization. IEEE Transactions on Evolutionary Computation 1(1), 67–82 (1997)
4. Qin, A.K., Huang, V.L., Suganthan, P.N.: Differential Evolution Algorithm with Strategy Adaptation for Global Numerical Optimization. IEEE Transactions on Evolutionary Computation 13(12), 397–417 (2009)
5. Mallipeddi, R., Suganthan, P.N., Pan, Q.K., Tasgetiren, M.F.: Differential Evolution Algorithm with Ensemble of Parameters and Mutation Strategies. Applied Soft Computing 11(2), 1679–1696 (2011)
6. Jeyakumar, G., Shunmuga Velayutham, C.: An Empirical Performance Analysis of Differential Evolution Variants on Unconstrained Global Optimization Problems. International Journal of Computer Information Systems and Industrial Management Applications (IJCISIM) 2, 77–86 (2010)
7. Herrera, F., Lozano, M.: Gradual Distributed Real-Coded Genetic Algorithms. IEEE Transactions on Evolutionary Computation 4(1), 43–63 (2000)

8. Skolicki, Z., De Jong, K.: Improving Evolutionary Algorithms with Multi-Representation Island Models. In: Yao, X., et al (eds.) PPSN VIII 2004. LNCS, vol. 3242, pp. 420–429. Springer, Heidelberg (2004)
9. Wang, Y., Cai, Z., Zhang, Q.: Differential Evolution with Composite Trial Vector Generation Strategies and Control Parameters. IEEE Transactions on Evolutionary Com-putation 15(1) (2011)
10. Hendtlass, T.: A Combined Swarm Differential Evolution Algorithm for Optimization Problems. In: Monostori, L., Váncza, J., Ali, M. (eds.) IEA/AIE 2001. LNCS (LNAI), vol. 2070, pp. 11–18. Springer, Heidelberg (2001)
11. Moore, P.W., Venayagamoorthy, G.K.: Evolving Digital Circuit using Hybrid Particle Swarm Optimization and Differential Evolution. International Journal of Neural Systems 16(3), 163–177 (2006)
12. Kannan, S., Slochanal, S.M.R., Subbaraj, P., Padhy, N.P.: Application of Particle Swarm Optimization Technique and its Variants to Generation Expansion Planning. Electric Power System Research 70(3), 203–210 (2004)
13. Omran, M.G.H., Engelbrecht, A.P., Salman, A.: Bare Bones Differential Evolution. European Journal of Operation Research 196(1), 128–139 (2009)
14. Chiou, J.P., Chang, C.F., Su, C.T.: Ant Direction Hybrid Differential Evolution for Solving Large Capacitor Placement Problems. IEEE Transactions on Power Systems 19, 1794–1800 (2004)
15. Zhang, X., Duan, H., Jin, J.: DEACO: Hybrid Ant Colony Optimization with Differential Evolution. In: Proceedings of the IEEE Congress on Evolutionary Computation, pp. 921–927 (2008)
16. Biswas, A., Dasgupta, S., Das, S., Abraham, A.: A Synergy of Differential Evolution and Bacterial Foraging Algorithm for Global Optimization. Neural Network World 17(6), 607–626 (2007)
17. He, H., Han, L.: A Novel Binary Differential Evolution Algorithm Based on Artificial Immune System. In: Proceedings of the IEEE Congress on Evolutionary Computation, pp. 2267–2272 (2007)
18. Das, S., Konar, A., Chakraborty, U.K.: Annealed Differential Evolution. In: Proceedings of the IEEE Congress on Evolutionary Computation, pp. 1926–1933 (2007)
19. Hu, Z.B., Su, Q.H., Xiong, S.W., Hu, F.G.: Self-Adaptive Hybrid Differential Evolution with Simulated Annealing Algorithm for Numerical Optimization. In: Proceedings of the IEEE Congress on Evolutionary Computation, pp. 1189–1194 (2008)
20. Weber, M., Neri, F., Tirronen, V.: Distributed Differential Evolution with Explorative-Exploitative Population Families. Genetic Programming and Evolvable Machines 10(4), 343–371 (2009)
21. Jeyakumar, G., Velayutham, C.S.: Empirical Study on Migration Topologies and Migration Policies for Island Based Distributed Differential Evolution Variants. In: Panigrahi, B.K., Das, S., Suganthan, P.N., Dash, S.S. (eds.) SEMCCO 2010. LNCS, vol. 6466, pp. 29–37. Springer, Heidelberg (2010)
22. Feoktistov, V.: Differential Evolution in Search of Solutions. Springer, USA (2006)

Word Sense Disambiguation for Punjabi Language Using Overlap Based Approach

Preeti Rana and Parteek Kumar

Abstract. Word Sense Disambiguation (WSD) is a concept for disambiguating the text so that computer would be able to interpret appropriate sense which is not difficult for a human to disambiguate. It is motivated by its use in many crucial applications such as information retrieval, information extraction, machine translation etc. WSD uses the punjabi WordNet which helps this approach to search the appropriate sense by providing information like synonyms, examples, concepts, semantic relation etc related to an ambiguous word. India is a multilingual country where people speak many different languages, which results in communication barrier. This acted as a motivation behind the building of IndoWordNet which has wordnets of major Indian languages. The expansion approach is used by many indian languages to develop their wordnets from the hindi wordnet. Millions of people know punjabi language in india but little computerized work has been done in the field for this language. It is therefore worthy to build up a punjabi lexical resource (WordNet) that can discover the richness of punjabi language. Word sense disambiguation uses lesk's algorithm in which context of ambiguous word is compared with the information concluded from WordNet and chooses the winner. The output will be the particular sense number designating the appropriate sense of ambiguous word. The evaluation has been done on the punjabi corpora and the results are encouraging.

1 Introduction

Word Sense Disambiguation (WSD) is well known and most interesting approach in a natural language processing which removes the ambiguity by identifying the most appropriate sense of a particular word according to its context.

Preeti Rana · Parteek Kumar
Thapar University, Patiala, India
e-mail: Preeti.11may@gmail.com, parteek.bhatia@thapar.edu

El-Sayed M. El-Alfy et al. (eds.), *Advances in Intelligent Informatics*,
Advances in Intelligent Systems and Computing 320, DOI: 10.1007/978-3-319-11218-3_54

Here, ambiguity means a word can have more than one meaning [2]. Human language is an ambiguous language so there is possibilities that many words can have multiple meanings these kinds of words are called polysemous word. For example, Punjabi sentences given in (1) and (2) illustrate the concept of ambiguity.Punjabi sentence:

ਉਸਨੇ ਮੇਰੇ ਤੇ **ਵਾਰ** ਕੀਤਾ। ...(1)

Transliterated sentence: *usnē mērē tē vār kītā|*
Equivalent English Sentence: He attacked at me.
Punjabi sentence:

ਮੈਂ ਉਸਨੂ ਕਈ ਵਾਰ ਫੋਨ ਕੀਤਾ। (2)

Transliterated sentence: maiṃ usnū kaī vār phōn kītā|
Equivalent English Sentence: I called her many times.

In example sentence 1 '**ਵਾਰ**' *vār* refers 'to attack' while in example sentence 2 '**ਵਾਰ**' *vār* refers to 'number of times'. From the above example, sentence 1 and 2 ambiguity is rarely a problem for human in day to day communication except in extreme cases but a computer has no basis of knowing which one is appropriate sense for a particular context. So, machine/computer needs software that can understand this natural language and identify the appropriate sense. Automatically identifying the correct sense of a word by a machine/computer is called Word Sense Disambiguation System. In this paper, overlap based approach of WSD has been presented. This approach uses Lesk's algorithm and punjabi WordNet as an important component of this implementation for the disambiguation.

This paper is divided into 7 sections. Section 2 describes applications of word sense disambiguation. Section 3 covers various learning methodologies used. Section 4 describes the implementation of Lesk's Algorithm. Section 5 Illustrate the proposed system with examples. Section 6 describes the evaluation process. Section 7 explains the future work and concludes the work presented in this paper.

2 Applications of Word Sense Disambiguation (WSD)

The applications where word sense disambiguation is used now a day are discussed below.

2.1 Machine Translation

Word sense disambiguation is very essential for the translation of words such as 'ਫਲ' *phal* depending on the context can be translated as 'fruit' or 'result'. Machine translation system must resolve two types of ambiguities. Firstly, ambiguity is in the source language when the meaning of a word is not apparent. Secondly, a word is not ambiguous in source language but after translation it has two different meanings. Hence, Word sense disambiguation system helps to improve the performance of machine translation system by resolving these sense ambiguities [6].

2.2 Text Processing

Word sense disambiguation is needed during text to speech translation. When the same word is pronounced in more than one way depending on its meaning. For example 'lead' can be a 'type of metal' or 'lead' can be 'in front off'.

2.3 Speech Processing

Word sense disambiguation is required when dealing with homophones words. Homophones words are those which are pronounced in the same manner but spelled differently. For example 'sealing' and 'ceiling'.

3 Learning Methodologies

Word sense disambiguation methods can be classified in two approaches i.e., Machine learning based approaches and Knowledge based approaches.

3.1 Machine Learning Based Approaches

These approaches rely on corpus. These approaches are supervised, unsupervised and hybrid [6].

3.1.1 Supervised Approach

Supervised approaches are very well known in every domain and also called inductive learningor classification in machine learning. This approach is based on the labeled training set automatically assigns the appropriate sense to an ambiguous word in the context. Supervised learning process: two steps

Learning (training): Learn a model using the training data.
Testing: Test the model using unseen test data to assess the model accuracy.

$$\text{Accuracy} = \frac{\text{Number of correct classifications}}{\text{Total number of test cases}}$$

This techniques uses machine learning approaches to assign the most appropriate sense to an ambiguous word suitable to its context. There is a classifier called word expert as shown in fig1 that assign suitable sense to its context. Training data set used in this approach consists of set of examples where the ambiguous word is manually tagged using machine readable dictionary. The supervised algorithm thus performed the ambiguous word sense disambiguation. For training each algorithm uses certain features associated with sense. This very fact forms the common thread of functionality of supervised algorithms. Different types of supervised approaches are decision tree, naive bayes, neural network, decision list etc.

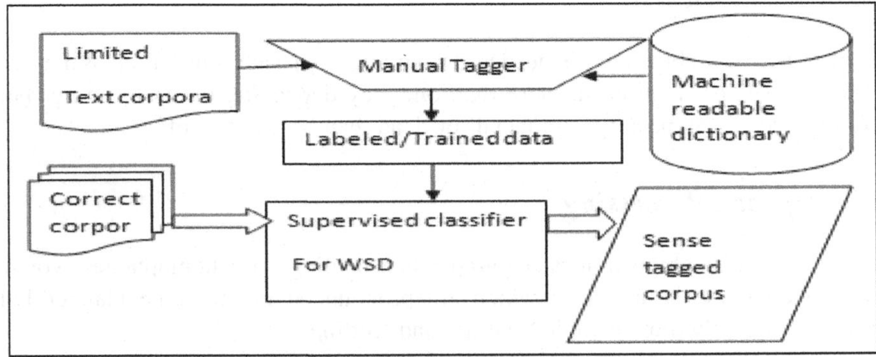

Fig. 1 Supervised learning approch of Word sense Disambiguation

3.1.2 Unsupervised Algorithms

This approach has no sense annoted data. So, there is a need to explore the data to find related information and for searching similar groups in data the technique named clustering is used which groups the data into clusters. Group of data that are similar to each other are placed in one cluster and unmatched group data are placed in other cluster. These clusters are then labeled by hand with known word senses.

3.1.3 Hybrid Approaches

This approach is combination of advantages of corpus based and knowledge based methods. The three algorithms come under hybrid approaches are sense learner, structural semantic interconnections (SSI) and iterative WSD.

3.2 Knowledge Based Approach

Knowledge based approaches works on overlap based method which uses machine readable dictionary WordNet for word sense disambiguation [5].

3.2.1 Introduction to WordNet

WordNet is a semantic lexicon (a reference book containing a list of words with them) for a language. It groups words into synonyms sets which are called synsets, provides general definitions (concept and example) and records semantic relations between these sets of synonyms. Each ambiguous word can have number of senses which have there *synset id's*. For each *synset id* we can find synonyms, concept and semantic relations. For an ambiguous word 'ਘਰ' *ghar*, *synset id's* are 1901 , 1902 , 1907 , 2761 , 4653 , 5036 , 28384 and for synset id 1901 synonyms are {ਮਕਾਨ *makān*, ਜਨਮਕੁੰਡਲੀ ਸਥਾਨ *janma kuṇḍlī sathān*, ਨਿਕੇਤਨ *nikētan*, ਸਵਦੇਸ਼ *savdēsh*, ਨਿਵਾਸ *nivās*, ਕੋਠਾ *kōṭhā*, ਹੋਮ *hōm*, ਅਸਥਾਨ *asathān*}.

Concept is ਮਨੁੱਖ ਦੁਆਰਾ ਛੱਤਿਆ ਹੋਇਆ ਉਹ ਸਥਾਨ ਜੋ ਕੰਧਾਂ ਨਾਲ ਘੇਰ ਕੇ ਰਹਿਣ ਦੇ ਲਈ ਬਣਾਇਆ ਜਾਂਦਾ ਹੈ *manukkh duārā chattiā hōiā uh sathān jō kandhāṃ nāl ghēr kē rahiṇ dē laī baṇāiā jāndā hai* and

Example is ਇਸ ਘਰ ਵਿਚ ਪੰਜ ਕਮਰੇ ਹਨ *is ghar vic pañj kamrē han.* Punjabi WordNet design is inspired by the famous Hindi WordNet which is again inspired by the famous English WordNet [1]. The basic semantic relations are given in table1.

Table 1 Illustrating the nature of relations in WordNet [1]

Relation	Meaning
Hypernymy/Hyponymy	Is-A(Kind-Of)
Entailment/Troponymy	Manner-of(for verbs)
Meronymy/Holonymy	Has-A(part whole)

We have the hypernymy relation (Is-A) links to {ਨਿਵਾਸਸਥਾਨ *nivās sathān*, ਰਿਹਾਇਸ਼ *rihāish*}. Its meronymy relation (Has-A) links to {ਸੌਣਵਾਲਾ ਕਮਰਾ *sauṇvāla kamrā*, ਬੈੱਡਰੂਮ *baiḍḍ rūm*} and holonymy relation (Is-A) links to{ਘਰ *ghar*, ਮਕਾਨ *makān*, ਨਿਕੇਤਨ *nikētan*,}as depicted in fig 2. Similarly we can have synonyms, concepts, examples and semantic relations for rest of all the *synset ids*.

3.2.2 Overlap Based Approach

The idea of overlap based approach is to count overlap between all the words in the context (context bag) and the feature of different senses of an ambiguous word (sense bag). Features are example, gloss, hypernymys, sense, semantic relations etc. Overlaps are counted for each sense and on the basis of overlaps sense with maximum overlaps is selected as appropriate sense to that context. Overlap based approaches uses Lesk's algorithm for the disambiguation.

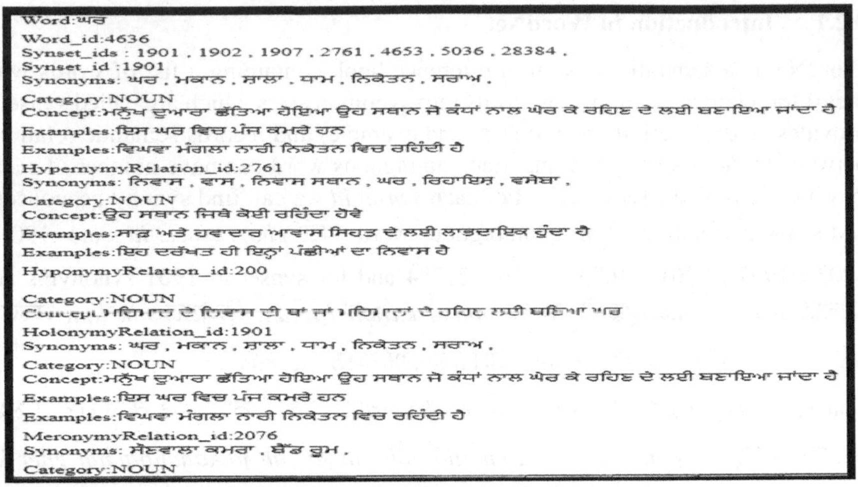

Word:ਘਰ
Word_id:4636
Synset_ids : 1901 , 1902 , 1907 , 2761 , 4653 , 5036 , 28384 ,
Synset_id :1901
Synonyms: ਘਰ , ਮਕਾਨ , ਸ਼ਾਲਾ , ਧਾਮ , ਨਿਕੇਤਨ , ਸਰਾਅ ,
Category:NOUN
Concept:ਮਨੁੱਖ ਦੁਆਰਾ ਛੱਤਿਆ ਹੋਇਆ ਉਹ ਸਥਾਨ ਜੋ ਕੰਧਾਂ ਨਾਲ ਘੇਰ ਕੇ ਰਹਿਣ ਦੇ ਲਈ ਬਣਾਇਆ ਜਾਂਦਾ ਹੈ
Examples:ਇਸ ਘਰ ਵਿਚ ਪੰਜ ਕਮਰੇ ਹਨ
Examples:ਵਿਧਵਾ ਮੰਗਲਾ ਨਾਰੀ ਨਿਕੇਤਨ ਵਿਚ ਰਹਿੰਦੀ ਹੈ
HypernymyRelation_id:2761
Synonyms: ਨਿਵਾਸ , ਵਾਸ , ਨਿਵਾਸ ਸਥਾਨ , ਘਰ , ਰਿਹਾਇਸ਼ , ਵਸੇਬਾ ,
Category:NOUN
Concept:ਉਹ ਸਥਾਨ ਜਿਥੇ ਕੋਈ ਰਹਿੰਦਾ ਹੋਵੇ
Examples:ਸਾਫ਼ ਅਤੇ ਹਵਾਦਾਰ ਆਵਾਸ ਸਿਹਤ ਦੇ ਲਈ ਲਾਭਦਾਇਕ ਹੁੰਦਾ ਹੈ
Examples:ਇਹ ਦਰੱਖਤ ਹੀ ਇਨ੍ਹਾਂ ਪੰਛੀਆਂ ਦਾ ਨਿਵਾਸ ਹੈ
HyponymyRelation_id:200
Category:NOUN
Concept:ਮਹਿਮਾਨ ਦੇ ਨਿਵਾਸ ਦੀ ਥਾਂ ਜਾਂ ਮਹਿਮਾਨਾਂ ਦੇ ਰਹਿਣ ਲਈ ਬਣਿਆ ਘਰ
HolonymyRelation_id:1901
Synonyms: ਘਰ , ਮਕਾਨ , ਸ਼ਾਲਾ , ਧਾਮ , ਨਿਕੇਤਨ , ਸਰਾਅ ,
Category:NOUN
Concept:ਮਨੁੱਖ ਦੁਆਰਾ ਛੱਤਿਆ ਹੋਇਆ ਉਹ ਸਥਾਨ ਜੋ ਕੰਧਾਂ ਨਾਲ ਘੇਰ ਕੇ ਰਹਿਣ ਦੇ ਲਈ ਬਣਾਇਆ ਜਾਂਦਾ ਹੈ
Examples:ਇਸ ਘਰ ਵਿਚ ਪੰਜ ਕਮਰੇ ਹਨ
Examples:ਵਿਧਵਾ ਮੰਗਲਾ ਨਾਰੀ ਨਿਕੇਤਨ ਵਿਚ ਰਹਿੰਦੀ ਹੈ
MeronymyRelation_id:2076
Synonyms: ਸੌਣਵਾਲਾ ਕਮਰਾ , ਬੈੱਡ ਰੂਮ ,
Category:NOUN

Fig. 2 Complete information of Punjabi ambiguous word ਘਰ using Punjabi WordNet

4 Implementation

Lesk's algorithm is a technique for assigning correct senses to words. An ambiguous word is assigned a correct sense by following steps.

1. The information from the context which contains the ambiguous word is collected in context bag.
2. The information in Punjabi WordNet like various senses, gloss, examples etc associated with a ambiguous word are collected in sense bag for the identification of correct sense.
3. The overlap between these two pieces of information take place and winner sense will be the sense with maximum overlaps.

4.1 Lesk's Algorithm: Finding the Word's Appropriate Sense

1. For the disambiguation of an ambiguous word ***w***, let the collection ***c*** be, the context bag which contains the sentence containing ambiguous word.
2. A word can have many possible senses. For each sense ***s*** of ambiguous word ***w***, do the following [3].

   – Let ***b*** be the sense bag of words obtained from the

Synonyms (similar senses), Examples (example Punjabi sentences), Glosses (concept), Hypernyms, Glosses of Hypernyms, Example of Hypernyms, Hyponyms, Glosses of Hypernyms, Example of Hypernyms, Meronyms, Glosses of Meronym, Example Sentences of Meronyms

– Measure the overlap between context bag ***c*** and sense bag ***b*** using the itersection similarity measure of both.

3. Output is the sense with maximum overlap is most appropriate sense suitable to its context. The idea behind this approach is there will be high overlap between the words present in the context and the all the related words found from the WordNet synsets, glosses, examples, relation like hyponymy, hypernymy, meronymy etc. using intersection similarity measure and the sense with maximum overlaps will be the winner sense [3].

5 Illustration of Proposed System with Example Sentences

1. Let us take two Punjabi sentence examples with same target word and explain the working of lesk's algorithm using these examples. The example Punjabi sentence (3) and example Punjabi sentence (4) illustrates concept of ambiguity having same ambiguous word 'ਘਰ' *ghar* but with different meanings.

   Punjabi sentences:

   ਉਸ ਦੇ ਛੇਵੇਂ ਘਰ ਵਿੱਚ ਸੂਰਜ ਦੀ ਦਸ਼ਾ ਹੈ| ... (3)

   Transliterated sentence: us dē chēvēṃ ghar vicc sūraj dī dashā hai.
   Equivalent English Sentence: Sun planetary period is in his sixth house.
   Punjabi sentences:

   ਉਸ ਨੂੰ ਮੇਰਾ ਘਰ ਭੁਲ ਗਿਆ | ... (4)

   Transliterated sentence: *us nūṃ mērā ghar bhul giā* ।
   Equivalent English Sentence: She forgot my home.

   For example sentence given in 3 and 4 an ambiguous word 'ਘਰ' disambiguation, the sentence containing the ambiguous word is taken as context bag ***c*** which contains all surrounding words of an ambiguous word .

   Context bag for example sentence 3: ਉਸ *us* ਦੇ *dē* ਛੇਵੇਂ *chēvēṃ* ਘਰ *ghar* ਵਿੱਚ *vicc* ਸੂਰਜ *sūraj* ਦੀ *dī* ਦਸ਼ਾ *dashā* ਹੈ *hai*

   Context bag for example sentence 4: ਉਸ *us* ਨੂੰ *nūṃ* ਮੇਰਾ *mērā* ਘਰ *ghar* ਭੁਲ *bhul* ਗਿਆ *giā* ।

2 For each sense ***s*** of ambiguous word ***w***, do the following.
   - Let Sense Bag ***b*** be the bag of words prepared from synonyms, glosses, example sentences, hypernyms, glosses of hypernyms, examples sentences of hypernyms, holonyms and glosses of holonyms etc. Sense bag is created for all the possible senses of an ambiguous word and each sense has a *synset ids*. Here, *synset id* means id of a sense. Each sense bag is represented by its *synset id*. For example , In Punjabi sentence 3 ambiguous word '**ਘਰ**' ghar, has 7 senses with *synset id's* a 1901, 1902, 1907, 2761, 4653, 5036, 28384 and first sense bag for punjabi sentence 3 is represented as SENSE1901 as shown in fig 3. Sense bags are created for all 7 senses. Sense bag SENSE1902 with maximum overlaps is discussed below.

Synonyms: ਜਨਮਕੁੰਡਲੀ ਸਥਾਨ *janmakuṇḍlī sathān*, ਕੁੰਡਲੀ ਸਥਾਨ *kuṇḍlī sathān*, ਘਰ *ghaī*

Concept: ਜਨਮਕੁੰਡਲੀ ਵਿਚ ਜਨਮਕਾਲ ਦੇ ਗ੍ਰਹਿਆਂ ਦੀ ਸਥਿਤੀ ਸੂਚਿਤ ਕਰਨ ਵਾਲੇ ਸਥਾਨਾਂ ਵਿਚੋਂ ਹਰੇਕ / *janmakuṇḍlī sathān janamkāl dē grahiāṃ dī sathitī sūcit karan vālē sathānāṃ vicōṃ harēk*

Example: ਜਨਮਕੁੰਡਲੀ ਸਥਾਨ ਤੋਂ ਗ੍ਰਹਿਆਂ ਦੀ ਦਸ਼ਾ ਦਾ ਪਤਾ ਲੱਗਦਾ ਹੈ। *janmakuṇḍlī sathān tōṃ grahiāṃ dī dashā dā patā laggdā hai*

ਤੁਹਾਡੀ ਜਨਮ ਪੱਤਰੀ ਵਿਚ ਸੂਰਜ ਨੌਵੇਂ ਘਰ ਵਿਚ ਹੈ। *tuhāḍī janam pattrī vic sūraj nauvēṃ ghar vic hai*

Hypernymy Relation:

Synonyms: ਖਾਨਾ *khānā*

Concept: ਰੇਖਾਵਾਂ ਆਦਿ ਨਾਲ ਘਿਰਿਆ ਹੋਇਆ ਸਥਾਨ। *rēkhāvāṃ ādi nāl ghiriā hōiā sathān*

Example: ਉਹ ਆਪਣੀ ਉੱਤਰ ਪੱਤਰਿਕਾ ਵਿਚ ਖਾਨੇ ਬਣਾ ਰਹੀ ਹੈ |*uh āpṇī uttar pattrikā vic khānē baṇā rahī hai*

Hyponymy Relation:

Concept: ਜਨਮਕੁੰਡਲੀ ਦੇ ਬਾਰਾਂ ਘਰਾਂ ਵਿਚੋਂ ਚੌਥਾ*janmakuṇḍlī dē bārāṃ gharlī gharāṃ vicōṃ cauthā*

Holonymy Relation:

Synonyms: ਜਨਮਕੁੰਡਲੀ ਸਥਾਨ *janmakuṇḍlī sathān*, ਕੁੰਡਲੀ ਸਥਾਨ *kuṇḍlī sathān* , ਘਰ *ghaī*

Concept: **ਜਨਮਕੁੰਡਲੀ ਵਿਚ ਜਨਮਕਾਲ ਦੇ ਗ੍ਰਹਿਆਂ ਦੀ ਸਥਿਤੀ ਸੂਚਿਤ ਕਰਨ ਵਾਲੇ ਸਥਾਨਾਂ ਵਿਚੋਂ ਹਰੇਕ** | janmakuṇḍlī sathān janamkāl dē grahiāṃ dī sathitī sūcit karan vālē sathānāṃ vicōṃ harēk

Example: **ਜਨਮਕੁੰਡਲੀ ਸਥਾਨ ਤੋਂ ਗ੍ਰਹਿਆਂ ਦੀ ਦਸ਼ਾ ਦਾ ਪਤਾ ਲੱਗਦਾ ਹੈ।** janmakuṇḍlī sathān tōṃ grahiāṃ dī dashā dā patā laggdā hai ,**ਤੁਹਾਡੀ ਜਨਮ ਪੱਤਰੀ ਵਿਚ ਸੂਰਜ ਨੌਵੇਂ ਘਰ ਵਿਚ ਹੈ** tuhāḍī janam pattrī vic sūraj nauvēṃ ghar vic hai

Meronymy Relation:

Synonyms: **ਗ੍ਰਹਿ** grahi, **ਖਗੋਲੀ** khagōlī, **ਅਕਾਸ਼ੀ ਪਿੰਡ** akāshī piṇḍ , **ਖਗੋਲੀ ਪਿੰਡ** khagōlī piṇḍ

Concept: **ਉਹ ਖਗੋਲੀ ਪਿੰਡ ਜਿਹੜਾ ਸੂਰਜ ਦੀ ਪ੍ਰਕਰਮਾ ਕਰਦਾ ਹੈ** |uh khagōlī piṇḍ jihṛā sūrajdīprakramākaradāhai.

Example: **ਪ੍ਰਿਥਵੀ ਇਕ ਗ੍ਰਹਿ ਹੈ** prithvī ik grahi hai.

Spelling of word **ਵਿਚ** is different in context bag and sense bag so not overlapped. Similarly all the sense bags are created for other 6 senses of an ambiguous word **w** as shown in fig 3. The same procedure is repeated for Punjabi sentence 4 as shown in fig 4.

- Compute the overlap between context bag and sense bags using the intersetion. Total counts tell the number of overlaps between sense bag and context bag for all 7 senses. All the underline words in the sense bag are the words which are overlapped with the context bag. For the sense bag SENSE1902 of ambiguous word present in example sentence 3 total counts are 23. Similarly sense bags are created for each sense and total counts of each sense is compared to find most probable sense suitable to its context as shown in fig 3. Fig 3 and fig 4 are showing each sense bag with its *id*, concept of the sense and total counts.

PUNJABI SENTENSE:ਉਸ ਦੇ ਛੇਵੇਂ ਘਰ ਵਿੱਚ ਸੁਰਜ ਦੀ ਦਸ਼ਾ ਹੈ
WORD TO BE DISAMBIGUATED :ਘਰ
CONTEXT_BAG: ਉਸ ਦੇ ਛੇਵੇਂ ਘਰ ਵਿੱਚ ਸੁਰਜ ਦੀ ਦਸ਼ਾ ਹੈ
TOTAL NUMBER OF SYNSET :7

SENSE 1901
CONCEPT: ਮਨੁੱਖ ਦੁਆਰਾ ਛੱਤਿਆ ਹੋਇਆ ਉਹ ਸਥਾਨ ਜੋ ਕੰਧਾਂ ਨਾਲ ਘੇਰ ਕੇ ਰਹਿਣ ਦੇ ਲਈ ਬਣਾਇਆ ਜਾਂਦਾ ਹੈ
TOTAL COUNT : 15
SENSE 1902
CONCEPT: ਜਨਮਕੁੰਡਲੀ ਵਿਚ ਜਨਮਕਾਲ ਦੇ ਗ੍ਰਹਿਆਂ ਦੀ ਸਥਿਤੀ ਸੂਚਿਤ ਕਰਨ ਵਾਲੇ ਸਥਾਨਾਂ ਵਿਚੋਂ ਹਰੇਕ
TOTAL COUNT : 23
SENSE 1907
CONCEPT: ਆਪਣਾ ਦੇਸ਼
TOTAL COUNT : 17
SENSE 2761
CONCEPT: ਉਹ ਸਥਾਨ ਜਿਥੇ ਕੋਈ ਰਹਿੰਦਾ ਹੋਵੇ
TOTAL COUNT : 18
SENSE 4653
CONCEPT: ਗੀਟੀ ਵਾਲੇ ਖੇਡ ਵਿਚ ਗੀਟੀ ਚਲਾਉਣ ਦੇ ਲਈ ਕਾਗਜ਼,ਲੱਕੜੀ ਆਦਿ ਦੇ ਉੱਪਰ ਬਣਿਆ ਹੋਇਆ ਵਿਭਾਗ
TOTAL COUNT : 18
SENSE 5036
CONCEPT: ਜੋ ਘਰ ਜਾਂ ਰਹਿਣ ਦੀ ਥਾਂ ਨਾਲ ਸੰਬੰਧਿਤ ਹੋਵੇ
TOTAL COUNT : 19
SENSE 28384
CONCEPT: ਉਹ ਸਥਾਨ ਜਿੱਥੇ ਤੁਸੀ ਰਹਿੰਦੇ ਜਾਂ ਟਿਕੇ ਹੋਵੇ ਅਤੇ ਜਿੱਥੋਂ ਉਦੇਸ਼ਾਂ ਦੀ ਸ਼ੁਰੂਆਤ ਅਤੇ ਸਮਾਪਤੀ ਹੁੰਦੇ ਹੋਵੇ
TOTAL COUNT : 16
WINNER WITH MAXIMUM MATCH:23

Fig. 3 Outputs Showing Correct Sense

PUNJABI SENTENSE: ਉਸ ਨੂੰ ਮੇਰਾ ਘਰ ਭੁਲ ਗਿਆ
WORD TO BE DISAMBIGUATED :ਘਰ
CONTEXT_BAG: ਉਸ ਨੂੰ ਮੇਰਾ ਘਰ ਭੁਲ ਗਿਆ
TOTAL NUMBER OF SYNSET :7

SENSE 1901
CONCEPT: ਮਨੁੱਖ ਦੁਆਰਾ ਛੱਤਿਆ ਹੋਇਆ ਉਹ ਸਥਾਨ ਜੋ ਕੰਧਾਂ ਨਾਲ ਘੇਰ ਕੇ ਰਹਿਣ ਦੇ ਲਈ ਬਣਾਇਆ ਜਾਂਦਾ ਹੈ
TOTAL COUNT : 6
SENSE 1902
CONCEPT: ਜਨਮਕੁੰਡਲੀ ਵਿਚ ਜਨਮਕਾਲ ਦੇ ਗ੍ਰਹਿਆਂ ਦੀ ਸਥਿਤੀ ਸੂਚਿਤ ਕਰਨ ਵਾਲੇ ਸਥਾਨਾਂ ਵਿਚੋਂ ਹਰੇਕ
TOTAL COUNT : 4
SENSE 1907
CONCEPT: ਆਪਣਾ ਦੇਸ਼
TOTAL COUNT : 5
SENSE 2761
CONCEPT: ਉਹ ਸਥਾਨ ਜਿਥੇ ਕੋਈ ਰਹਿੰਦਾ ਹੋਵੇ
TOTAL COUNT : 3
SENSE 4653
CONCEPT: ਗੀਟੀ ਵਾਲੇ ਖੇਡ ਵਿਚ ਗੀਟੀ ਚਲਾਉਣ ਦੇ ਲਈ ਕਾਗਜ਼,ਲੱਕੜੀ ਆਦਿ ਦੇ ਉੱਪਰ ਬਣਿਆ ਹੋਇਆ ਵਿਭਾਗ
TOTAL COUNT : 4
SENSE 5036
CONCEPT: ਜੋ ਘਰ ਜਾਂ ਰਹਿਣ ਦੀ ਥਾਂ ਨਾਲ ਸੰਬੰਧਿਤ ਹੋਵੇ
TOTAL COUNT : 6
SENSE 28384
CONCEPT: ਉਹ ਸਥਾਨ ਜਿੱਥੇ ਤੁਸੀ ਰਹਿੰਦੇ ਜਾਂ ਟਿਕੇ ਹੋਵੇ ਅਤੇ ਜਿੱਥੋਂ ਉਦੇਸ਼ਾਂ ਦੀ ਸ਼ੁਰੂਆਤ ਅਤੇ ਸਮਾਪਤੀ ਹੁੰਦੇ ਹੋਵੇ
TOTAL COUNT : 3
WINNER WITH MAXIMUM MATCH:6

Fig. 4 Outputs Showing Correct Sense

3 Output of this sense disambiguation is Punjabi sentence 3 has winner sense with maximum overlap 23 which shows **'ਘਰ'** *ghar* is related to astrology suitable to its context whereas Punjabi sentence 4 has winner sense with maximum overlaps are 6 which shows **'ਘਰ'** *ghar* is related to place to live. So, the same word is showing different results with respect to their context by giving appropriate sense of word regarding the context using lesk's algorithm for word sense disambiguation.

6 Evaluations

We use the Punjabi corpora (Punjabi text document) to calculate the accuracy. Various senses are calculated using Lesk's algorithm implemented in proposed system. Words which results appropriate senses are marked as correct and words which results wrong sense are marked as incorrect.

- Text Document

ਪਿੰਰਸ ਜਦੋਂ ਵੀ ਆਪਣੇ ਗੁਆਂਢੀ ਸੁਰਿੰਦਰ ਅੰਕਲ ਦੇ ਘਰ ਵੱਲ ਦੇਖਦਾ ਤਾਂ ਉਸ ਦਾ ਧਿਆਨ ਉਨ੍ਹਾਂ ਦੇ ਘਰ ਦਰੱਖਤ 'ਤੇ ਲੱਗੇ ਅਮਰੂਦਾਂ ਤੇ ਅਨਾਰਾਂ ਵੱਲ ਚਲਾ ਜਾਂਦਾ ।ਇਹ ਦੋਵੇਂ ਫਲ ਉਸ ਦੇ ਮਨਪਸੰਦ ਦੇ ਸਨ। ਉਸ ਦਾ ਦਿਲ ਉਨ੍ਹਾਂ ਨੂੰ ਤੋੜਨ ਨੂੰ ਕਰਦਾ ਪਰ ਉਹ ਬਹੁਤ ਉੱਚੇ ਲੱਗੇ ਹੋਏ ਸਨ ।ਉਹ ਸ਼ਰਮਾਉਂਦਾ ਹੋਇਆ ਉਨ੍ਹਾਂ ਤੋਂ ਮੰਗਦਾ ਵੀ ਨਹੀਂ ਸੀ ।ਇਕ ਦਿਨ ਉਸ ਨੇ ਇਕ ਵੱਡਾ ਸਾਰਾ ਅਨਾਰ ਦੇਖਿਆ ।ਉਸ ਦੇ ਮੰਹ ਵਿਚ ਪਾਣੀ ਆ ਗਿਆ। ਉਸ ਨੇ ਇਕ ਵੱਡਾ ਸਾਰਾ ਪੱਥਰ ਚੁੱਕਿਆ ਤੇ ਅਨਾਰ 'ਤੇ ਨਿਸ਼ਾਨਾ ਲਾਇਆ ਪਰ ਪੱਥਰ ਅਨਾਰ 'ਤੇ ਨਹੀਂ ਲੱਗਾ, ਸਗੋਂ ਘਰ ਦੀ ਖਿੜਕੀ ਦੇ ਸ਼ੀਸ਼ੇ 'ਤੇ ਵੱਜਾ। ਖਿੜਕੀ ਦਾ ਸ਼ੀਸ਼ਾ ਟੁੱਟ ਗਿਆ। ਸੁਰਿੰਦਰ ਅੰਕਲ ਵੀ ਸ਼ੀਸ਼ਾ ਟੁੱਟਣ ਦੀ ਆਵਾਜ਼ ਸੁਣ ਕੇ ਬਾਹਰ ਆ ਗਏ। ਉਨ੍ਹਾਂ ਨੇ ਪਿੰ੍ਰਸ ਨੂੰ ਦੌੜਦਿਆਂ ਦੇਖਿਆ ਤੇ ਪਿੰਰਸ ਨੂੰ ਆਵਾਜ਼ ਲਗਾਈ। ਪਿੰਰਸ ਡਰਦੇ ਮਾਰੇ ਦੌੜ ਗਿਆ। ਉਹ ਉਨ੍ਹਾਂ ਦੇ ਬੁਲਾਉਣ 'ਤੇ ਨਹੀਂ ਆਇਆ। ਉਹ ਸਮਝ ਗਏ ਕਿ ਇਹ ਕੰਮ ਪਿੰਰਸ ਦਾ ਹੀ ਹੈ। ਥੋੜ੍ਹੇ ਦਿਨਾਂ ਦੇ ਬਾਅਦ ਪਿੰਰਸ ਦਾ ਫੇਰ ਦਿਲ ਕੀਤਾ ਕਿ ਉਹ ਅਨਾਰ ਤੇ ਅਮਰੂਦ ਖਾਏ। ਉਹ ਅਨਾਰ ਦੇ ਦਰੱਖਤ ਕੋਲ ਖਲੋ ਗਿਆ ਪਰ ਦਰੱਖਤ ਉੱਤੇ ਅਨਾਰ ਉੱਚੇ ਲੱਗੇ ਸਨ ।ਅਮਰੂਦ ਵੀ ਨਾਲ ਦੇ ਦਰੱਖਤ 'ਤੇ ਉੱਚੇ ਲੱਗੇ ਹੋਏ ਸਨ। ਉਹ ਸੋਚ ਰਿਹਾ ਸੀ ਕਿ ਉਹ ਉਨ੍ਹਾਂ ਨੂੰ ਕਿਵੇਂ ਤੋੜੇ। ਉਹ ਉਨ੍ਹਾਂ 'ਤੇ ਪੱਥਰ ਨਹੀਂ ਮਾਰਨਾ ਚਾਹੁੰਦਾ ਸੀ, ਕਿਉਂਕਿ ਉਸ ਦਿਨ ਵੀ ਖਿੜਕੀ ਦਾ ਸ਼ੀਸ਼ਾ ਟੁੱਟ ਗਿਆ ਸੀ। ਖਿੜਕੀ ਦਾ ਸ਼ੀਸ਼ਾ ਟੁੱਟਣ ਦਾ ਸ਼ੱਕ ਸੁਰਿੰਦਰ ਅੰਕਲ ਨੂੰ ਉਸ ਉੱਪਰ ਹੀ ਸੀ। ਉਹ ਸੋਚ ਹੀ ਰਿਹਾ ਸੀ ਕਿ ਐਨੇ ਨੂੰ ਕਿਸੇ ਨੇ ਆਪਣੀ ਕਾਰ ਸੜਕ 'ਤੇ ਦਰੱਖਤ ਦੇ ਥੱਲੇ ਹੀ ਖੜ੍ਹੀ ਕਰ ਦਿੱਤੀ। ਪਿੰਰਸ ਨੇ ਜਦੋਂ ਕਾਰ ਦਰੱਖਤ ਦੇ ਥੱਲੇ ਦੇਖੀ ਤਾਂ ਉਹ ਖੁਸ਼ ਹੋ ਕੇ ਮਨ ਵਿਚ ਕਹਿਣ ਲੱਗਾ ਹੁਣ ਤਾਂ ਕੰਮ ਬਣ ਗਿਆ ਸਮਝੋ। ਪਿੰਰਸ ਨੇ ਹੌਸਲਾ ਕੀਤਾ ਤੇ ਕਾਰ ਉੱਪਰ ਚੜ੍ਹ ਗਿਆ। ਉਸ ਦਾ ਹੱਥ ਅਸਾਨੀ ਨਾਲ ਅਮਰੂਦਾਂ ਤੇ ਅਨਾਰਾਂ ਤੱਕ ਪਹੁੰਚ ਗਿਆ ।ਹੁਣ ਉਹ ਮਨ ਹੀ ਮਨ ਬਹੁਤ ਖੁਸ਼ ਸੀ। ਉਸ ਨੇ ਅਮਰੂਦ ਅਤੇ ਅਨਾਰ ਤੋੜੇ ਅਤੇ ਆਪਣੀ ਪੈਂਟ ਦੀ ਜੇਬ ਵਿਚ ਪਾ ਲਏ। ਉਸ ਨੇ ਚੰਗਾ ਮੌਕਾ ਦੇਖ ਕੇ ਹੋਰ ਵੀ ਅਨਾਰ ਤੋੜਨ ਦਾ ਮਨ ਬਣਾ ਲਿਆ। ਉਹ ਅਨਾਰ ਤੋੜ ਹੀ ਰਿਹਾ ਸੀ ਕਿ

ਅਚਾਨਕ ਉਸ ਦਾ ਹੱਥ ਭੂਡਾਂ ਦੀ ਖੱਖਰ 'ਤੇ ਜਾ ਕੇ ਲੱਗ ਗਿਆ, ਜੋ ਉਸ ਨੇ ਕਾਹਲ ਵਿਚ ਦੇਖਿਆ ਹੀ ਨਹੀਂ ਸੀ। ਭੂਡਾਂ ਦੀ ਖੱਖਰ 'ਤੇ ਜਿਉਂ ਹੀ ਉਸ ਦਾ ਹੱਥ ਪਿਆ, ਉਸ ਉੱਪਰ ਭੂਡਾਂ ਨੇ ਹਮਲਾ ਕਰ ਦਿੱਤਾ। ਕੋਈ ਭੂਡਾ ਉਸ ਦੇ ਮੂਹ 'ਤੇ ਲੜਿਆ, ਕੋਈ ਬਾਂਹ 'ਤੇ ਲੜਿਆ। ਉਹ ਉੱਚੀ-ਉੱਚੀ ਚੀਕਾਂ ਮਾਰਨ ਲੱਗਿਆ।ਉਸ ਦੀਆਂ ਚੀਕਾਂ ਦੀ ਆਵਾਜ਼ ਸੁਣ ਕੇ ਬਹੁਤ ਸਾਰੇ ਲੋਕ ਇਕੱਠੇ ਹੋ ਗਏ। ਸੁਰਿੰਦਰ ਅੰਕਲ ਵੀ ਬਾਹਰ ਨਿਕਲ ਆਏ ਤੇ ਉਨ੍ਹਾਂ ਨੇ ਵੀ ਪਿੰਰਸ ਦੀ ਹਾਲਤ ਦੇਖੀ। ਭੂਡਾਂ ਦੇ ਲੜਨ ਕਰਕੇ ਉਸ ਦਾ ਸਰੀਰ ਬੁਰੀ ਤਰ੍ਹਾਂ ਨਾਲ ਸੁੱਜ ਗਿਆ ਸੀ। ਉਹ ਉਸ ਨੂੰ ਫਟਾਫਟ ਡਾਕਟਰ ਕੋਲ ਲੈ ਗਏ। ਜਦੋਂ ਪਿੰਰਸ ਦੀ ਮੰਮੀ ਗੀਤਾ ਨੂੰ ਪਤਾ ਲੱਗਿਆ ਤਾਂ ਉਸ ਨੇ ਸੁਰਿੰਦਰ ਅੰਕਲ ਨੂੰ ਫੋਨ ਕਰਕੇ ਡਾਕਟਰ ਬਾਰੇ ਪੁੱਛਿਆ। ਪਿੰਰਸ ਦੀ ਮੰਮੀ ਵੀ ਡਾਕਟਰ ਕੋਲ ਪਹੁੰਚ ਗਈ। ਡਾਕਟਰ ਨੇ ਪਿੰਰਸ ਦਾ ਇਲਾਜ ਸ਼ੁਰੂ ਕਰ ਦਿੱਤਾ।ਕੁਝ ਦੇਰ ਬਾਅਦ ਪਿੰਰਸ ਕੁਝ ਠੀਕ ਹੋ ਗਿਆ ਤਾਂ ਉਸ ਨੇ ਆਪਣੀ ਗ਼ਲਤੀ ਮੰਨੀ ਤੇ ਉਸ ਦਿਨ ਸ਼ੀਸ਼ਾ ਤੋੜਨ ਵਾਲੀ ਗੱਲ ਵੀ ਮੰਨ ਗਿਆ ਸੁਰਿੰਦਰ ਅੰਕਲ ਨੇ ਹੱਸਦਿਆਂ ਕਿਹਾ, 'ਬੇਟੇ, ਉਸ ਦਿਨ ਤਾਂ ਭੂਡਾਂ ਖਿੜਕੀ ਦਾ ਸ਼ੀਸ਼ਾ ਤੋੜ ਕੇ ਭੱਜ ਗਿਆ ਸੀ ਪਰ ਅੱਜ ਭੰਡਾਂ ਨੇ ਅਸਲੀ ਅਮਰੂਦ ਤੇ ਅਨਾਰ ਚੋਰ ਫੜਾ ਹੀ ਦਿੱਤਾ।

Results obtained from this particular document are shown in table 2.

Table 2 Results obtained from the test document [3]

Sno	word	Synonyms	Comment
1	ਜਦੋਂ	ਜਿਸ_ਸਮੇਂ, ਜੇਕਰ	Correct
2	ਵੀ	ਨਾਲ ਜਾਂ ਸਿਵਾ,20	Correct
3	ਘਰ	ਮਕਾਨ , ਸ਼ਾਲਾ, ਜਨਮਕੁੰਡਲੀ ਸਥਾਨ	Correct
4	ਵੱਲ	ਸਮਰਥੱਕ, ਵੱਟ, ਸਲੀਕਾ, ਰੁਖ ਹੋਣਾ	Correct
5	ਧਿਆਨ	ਚੇਤਾ , ਖਿਆਲ, ਚਿੰਤਨ ਇਕਾਗਰਤਾ	Incorrect
6	ਅਨਾਰ	ਦਾੜੂ, ਪਟਾਕਾ	Incorrect
7	ਫਲ	ਫਰੂਟ, ਪਰਿਣਾਮ	Correct
8	ਪਰ	ਖੰਭ, ਪਰੰਤੂ	Incorrect
9	ਵੱਡਾ	ਪੂਰਵ ਜਨਮਿਆ, ਵਿਸਤ੍ਰਿਤ, ਲਾਰਜ, ਲੰਬਾ	Correct
10	ਬਾਹਰ	ਬਾਹਰਲੇ, ਅਧਿਕਾਰ ਤੋਂ ਪਰੇ	Correct
11	ਦੌੜ	ਰੇਸ, ਨਸਣਾ, ਰੰਨ	Correct
12	ਫੇਰ	ਦੁਬਾਰਾ, ਮਗਰੋਂ, ਚੱਕਰ, ਗੇੜਾ	Incorrect

Table 2 (*continued*)

13	ਕੋਲ	ਨਜ਼ਦੀਕ,ਅਧਿਕਾਰ ਵਿਚ	Correct
14	ਵੀ	20, ਨਾਲ ਜਾਂ ਸਿਵਾ	Correct
15	ਉੱਪਰ	ਉਤਾਂਹ , ਉਚਾਈ ਤੇ, ਉੱਤੇ, ਬਹੁਤਾ	Incorrect
16	ਕਾਰ	ਧੰਦਾ , ਗੱਡੀ	Incorrect
17	ਥੱਲੇ	ਅਧੀਨਤਾ ਵਿਚ,ਹੇਠਾਂ	Correct
18	ਕਰ	ਲਗਾਨ,ਸਿਕਰੀ,ਹੱਥ, ਕੰਮ ਕਰਨਾ	Incorrect
19	ਮਨ	ਚਿਤ,ਬੇਸਨ ਦੀ ਰੋਟੀ	Correct
20	ਬਣ	ਬਣਨਾ , ਜੰਗਲ	Incorrect
21	ਚੰਗਾ	ਸੱਭਿਅ,ਵਧੀਆ ਖਰਾ ਨੇਕ ਸ਼ੁੱਭ	Correct
22	ਤੋੜ	ਇਲਾਜ,ਉਪਾਅ,ਤੋੜਨਾ	Correct
23	ਲੋਕ	ਜਨਤਾ,ਤੱਲ	Correct
24	ਵਾਰ	ਹਮਲਾ,ਦਿਨ,ਦਫ਼ਾ	Incorrect

This way we tested the system on documents from various domains. The accuracy of the system has been calculatedby comparing the results of the proposed system with the manualy tagged corpus by the linguist. It has been observed that approximately 60% of the results of proposed system matched with senses marked in the manual tagged corpus. Table 2 summarizes the results.

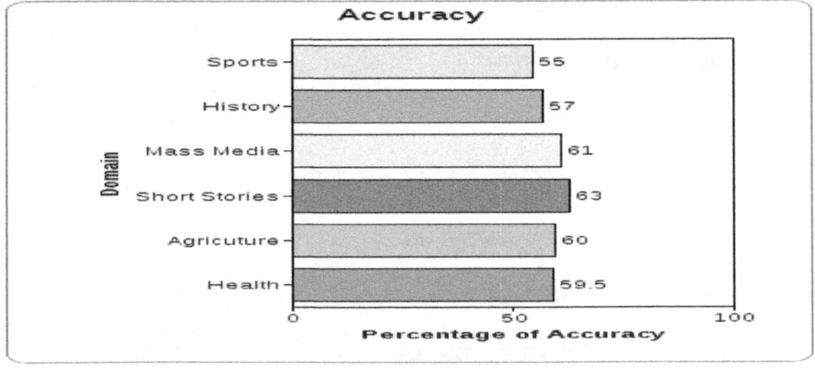

Fig. 5 Showing the WSD accuracy across domains for Punjabi Words [3]

Table 3 showing the percentage of accuracy achieved by proposed system in various domains.

Table 3 WSD accuracy across domains for Punjabi words [3]

Domain	Percentage
Health	59.5%
Agriculture	60%
Short Stories	63%
Mass Media	61%
Sports	55%
History	57%

7 Conclusion and Future Work

In this paper, Lesk's algorithm has been implemented for word sense disambiguation (WSD) of Punjabi text and the proposed system has been tested on approximately 500 Punjabi sentences and 100 Punjabi text corpora. To our knowledge, this is the first attempt at automatic WSD for a Punjabi language. However Rekha and Parteek had implemented Punjabi WordNet for creation of bilingual dictionaries. In this paper Punjabi WordNet has been implementad for WSD. From table 3 we depict that our accuracy ranges from about 55% to about 65%. In future this algorithm will be tested on more Punjabi text documents.

References

[1] Brent, M.R.: From grammar to lexicon: Unsupervised learning of lexical Syntax. J. Computational Linguistics 19(2), 1–20 (1993)

[2] Banerjee, S., Pedersen, T.: An Adapted Lesk Algorithm for Word Sense Disambiguation Using WordNet. In: Proceedings of Third International Conference on Intelligent Text Processing and Computational Linguistics, Mexico City, pp. 1–10 (2002)

[3] Sinha, M., Kumar, M., Pande, P., Kashyap, L., Bhattacharyya, P.: Hindi Word Sense Disambiguation, pp. 1–7. Department of Computer Science and Engineering Indian Institute of Technology, Bombay (2004)

[4] Sharma, R., Kumar, P., Sharma, R.K.: Word Sense Disambiguation (WSD). Master's thesis. Computer Science and Engineering Department, Thapar University, Patiala (2008)

[5] Narang, A., Kumar, P., Mohapatra, S.K.: Punjabi WordNet, Master's thesis. Computer Science and Engineering Department, Thapar University, Patiala (2012)

[6] Singh, V., Kumar, P.: Word Sense Disambiguation for Punjabi Language. PhD thesis. Computer Science and Engineering Department, Thapar University, Patiala (2013)

Flipped Labs as a Smart ICT Innovation: Modeling Its Diffusion among Interinfluencing Potential Adopters

Raghu Raman

Abstract. Smart ICT innovation like flipped classroom pedagogy is freeing up face-to-face in-class teaching system for additional problem based learning activities in the class. But the focus of flipped classrooms is more on the theory side with related lab work in science subjects further getting marginalized. In this paper we are proposing Flipped Labs - a method of pedagogy premeditated as a comprehensive online lab learning environment outside the class room by means of tutorials, theory, procedure, animations and videos. Flipped labs have the potential to transform the traditional methods of lab teaching by providing more lab time to students. An ICT educational innovation like flipped labs will not occur in isolation in an environment where two interrelated potential adopters namely teachers and students influence each other and both have to adopt for the innovation to be successful. In this paper we provide the theoretical framework for the diffusion and the adoption patterns for flipped labs using theory of perceived attributes and take into account the important intergroup influence between teachers and students. The results of this analysis indicated that Relative Advantage, Compatibility, Ease of Use, Teacher Influence and Student Influence were found to be positively related to acceptance of flipped labs.

Keywords: Flipped classroom, Online Labs, Simulations, chemistry, ICT, innovation.

1 Introduction

Smart ICT innovation like flipped classroom method of teaching is freeing up face-to-face class oriented teaching system for additional problem based learning

Raghu Raman
Amrita School of Business, Amrita Vishwa Vidyapeetham, Coimbatore, Tamilnadu, India
e-mail: raghu@amrita.edu

El-Sayed M. El-Alfy et al. (eds.), *Advances in Intelligent Informatics*,
Advances in Intelligent Systems and Computing 320, DOI: 10.1007/978-3-319-11218-3_55

activities. Also known as 'inverted classroom', or 'reverse instruction method', the flipped classroom method is one in which what is traditionally done in class is switched for what is traditionally done for homework [1]. This means, that rather than listening to lectures in school and doing their homework at home, students watch lecture videos at home and do their "homework" in the school, under the guidance of the teacher. therefore, rather than the teacher giving synchronous in-class group teaching, students are looked forward to make use of the video materials offered, by the side of additional resources to gain knowledge of concepts and finish assignments on their own at their own speed and at place fitting to the student [2]. In flipped classroom method of teaching, if any student is trailing, the teacher has additional time to endow with personalized facilitation when class time is on, even as other classmates are engaged in their problem based learning activity

The National Focus Group on "Teaching of Science", suggested prevention of marginalization of lab work in school science curriculum. Investment in this regard is required for improving school labs to promote experimental culture. But there seem to be two principal difficulties. Firstly, experiments require a certain minimum infrastructure – a lab with some basic equipment and consumables on a recurring basis. Learners get access to the physical lab for only a short period of time, which is often not sufficient to allow them to try different scenarios and hence limits the learning cycle. Secondly, assessment of practical skills in science in a sound and objective manner is by no means an easy task. The difficulty multiplies manifold if assessment is to be carried out for a class of 50 students which is very common in schools in India.

The traditional method of conducting classes using information from textbooks and lecture notes gives very little motivation or incentive to students to attend classes [3]. In fact, most of the times student only attend lectures to pass time and do not actually learn anything through them [4]. On the other hand flipped classrooms were found to promote higher levels of inventiveness and coordination among the students, thereby indicating the need to introduce flipped classroom practices for the future generations [5]. The need of the hour is to achieve maximum output from the students. For this, several studies have been conducted to determine how flipped classrooms are more advantageous than traditional classrooms. A study conducted by [6] shows that flipped classrooms helped students get higher grades and better achievements as compared to traditional classrooms. Another advantage with flipped classroom is the increase in the time available for problem solving and hands-on activities [7]. Flipped classrooms are noted for their use of modern technology as part of their teaching methods. Many studies have found that the use of technology in teaching "helps in reduction of attrition, increasing the outcomes and improving student satisfaction." [8]. This modern method of teaching has been found to be especially useful to students pursuing higher studies as the use of technology helps them to review their notes anytime they need to, even when they are not sitting in a classroom [9]. Perhaps the biggest advantage of a flipped classroom system is that it gives the teachers more time in getting the students to master a particular topic which is otherwise difficult to achieve in the limited time available during classroom teaching [10].

Educational innovations generally have two major groups of potential adopters – teachers and students (Fig. 1), whose ultimate benefit stems from interacting with each other. Since the two potential-adopter groups are interdependent, one group of potential-adopter decision to adopt the innovation could have a positive/negative effect on the other group. For the educational innovation to succeed both kinds of potential adopters must adopt it. Lack of feedback and interaction can lead to equilibrium where cost-savings are not realized, since no-one adopts to a critical mass point. In this paper, we consider instead interinfluence effects in the context of two groups of heterogeneous potential adopters where the users may realize gains by interacting with one another (Fig. 2)

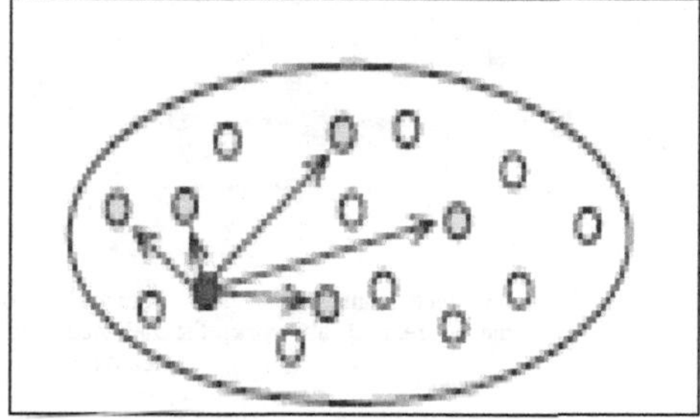

Fig. 1 Single group of potential-adopters – teachers or students

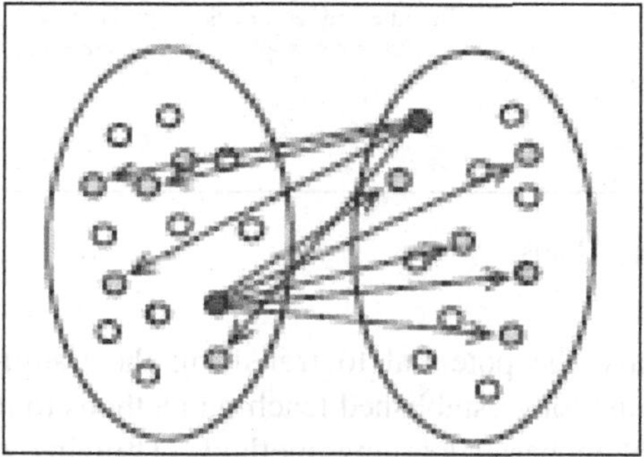

Fig. 2 Two groups of potential-adopters – teachers and students and their interinfluence

The adoption decision in an interrelated system of potential-adopters is a choice affected by three factors

1. diffusion rate of first group (teachers)
2. diffusion rate of second group (students)
3. interdependent influence diffusion rate (teachers-students-teachers)

2 Flipped Labs

Online Labs (OLabs) for science experiments is an ICT innovation based on the idea that lab experiments can be taught using the internet, more efficiently, less expensively, and offered to students who do not have access to physical labs [11, 12. 13]. It was developed to supplement the traditional physical labs. OLabs as flipped classroom method of pedagogy is premeditated as a comprehensive learning environment by means of tutorials, theory, procedure, animations and videos outside the classroom whereas the assessments takes the form of theoretical, experimental, practical and reporting skills. We call this approach of using OLabs in a flipped mode as flipped labs (Fig. 3).

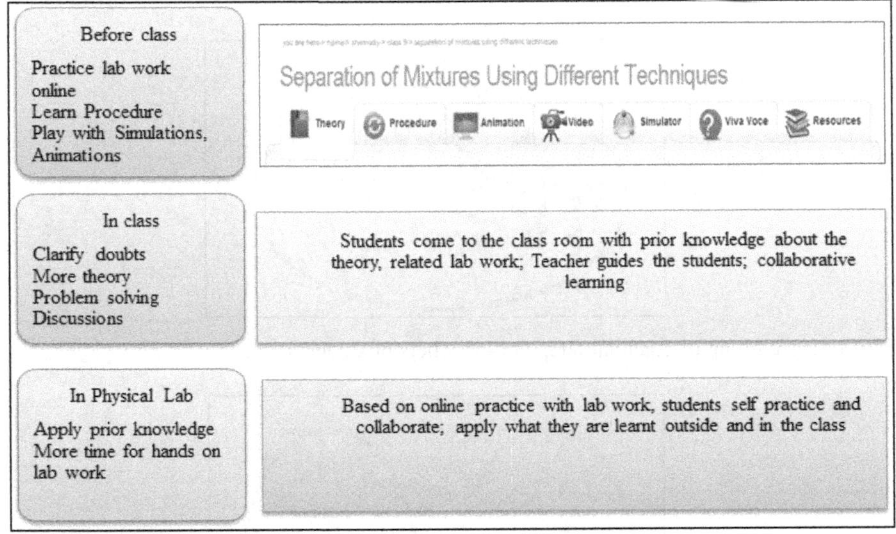

Fig. 3 Model of Flipped Labs

Flipped labs have the potential to transform the traditional methods of lab teaching by upturning long-established teaching methods to involve students in the practical and problem based learning methods. Utilizing flipped labs, teachers move out of the classroom and deliver online as a way to free up precious lab time for students to practice lab work outside the physical labs (Fig. 4). Flipped labs are not a substitute to traditional physical labs but it helps students familiarize with new concepts and acquire some pre knowledge before coming to the classroom. Today flipped classrooms are mostly focusing on the theory side and giving lecture notes to students via video, audio etc. However both theory and lab work are very important for conceptual understanding. But the real problem is that the lab time is limited. Sometimes the teacher will demonstrate the experiment or students will do the experiment in a group. Then there is this issue of lag between the time theory is taught in the classroom and the time student actually gets to perform the related lab work.

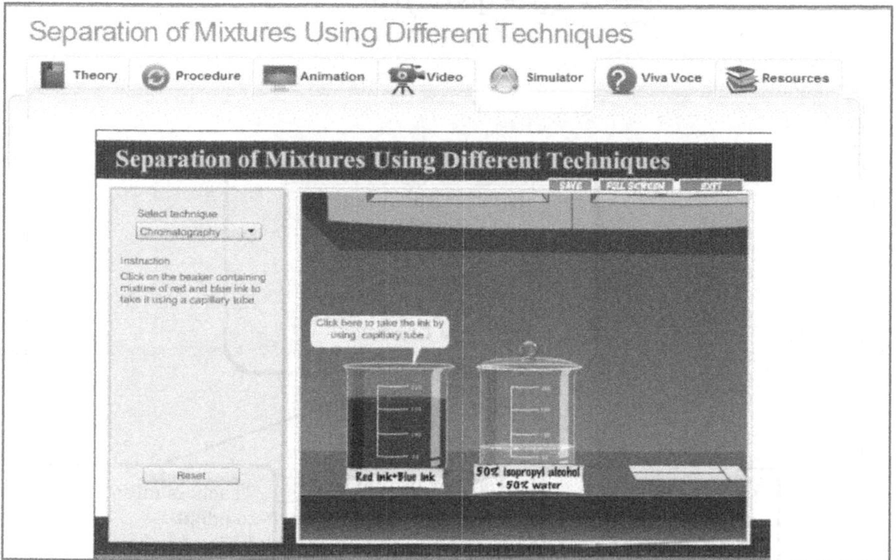

Fig. 4 Sample Flipped Labs showing simulation for students to practice outside class

3 Research Model

Flipped Labs is an innovation if it is 'perceived as new by an individual or other unit of adoption'. An ICT educational innovation like Flipped Labs will not occur in isolation in an environment where two interrelated potential adopters namely teachers and students influence each other and both have to adopt for the innovation to be successful. More importantly this paper provides the theoretical framework for the diffusion and the adoption patterns for Flipped labs using [14] theory of perceived attributes and takes into account the important intergroup influence between teachers and students. The Flipped Labs rate of adoption was investigated by assessing two groups of characteristics, which were the independent variables - innovation characteristics and environment characteristics (Fig. 5). We also considered the interinfluence between teachers and students.

Employing [14] framework, [15] proposed mathematical model as a nonlinear differential equation for diffusion of an innovation in a group of size M. In such a scenario [16] adoption of innovation is due to two influences viz. external influence (mass media) which is a linear mechanism and internal influence (word-of-mouth) which is a non-linear mechanism. The differential equation giving the diffusion is

$$\frac{dN(t)}{dt} = (p + qN(t))(M - N(t)) \tag{1}$$

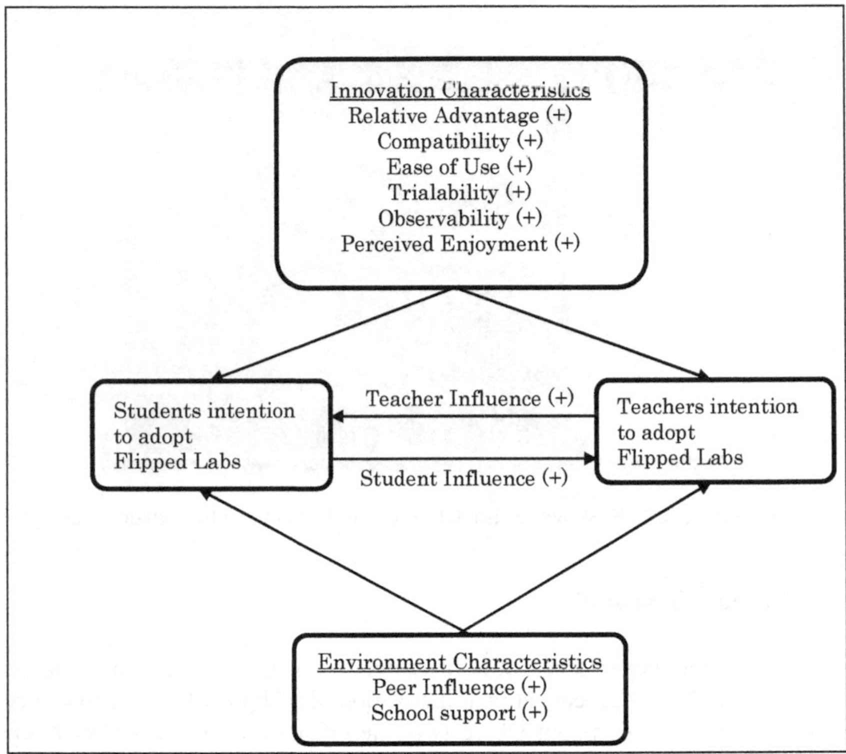

Fig. 5 Research model for diffusion of Flipped labs among students and teachers

Where N(t) is the cumulative number of adopter-students who have already adopted by time t, M is total number of adopter-students who will eventually use the innovation, p is the coefficient of external influence and q is the coefficient of internal influence

In terms of the fraction F(t) of potential adopter-students

$$F(t) = \frac{N(t)}{M} \tag{2}$$

the Bass model can be rewritten as

$$\frac{dN(t)}{dt} = (p + qMF(t))(1 - F(t)),\ F(t = 0) = F_0 \quad (3)$$

Equation (3) yields the S-shaped diffusion curve. It is assumed that the carrying capacity M of the adopter-students remains constant.

Now we extend the Bass model to account for the interinfluence between teachers and students. We define the following terms.

- p_1 external influence for the adopter-teachers
- q_1 internal influence for the adopter-teachers
- F total number of adopter-teachers who will eventually adopt the innovation
- f cumulative number of adopter-teachers who have already adopted by time t
- p_2 external influence for the adopter-students
- q_2 internal influence for the adopter-students
- S total number of adopter-students who will eventually adopt the innovation
- s cumulative number of adopter-students who have already adopted by time t
- m total number of adopters who will eventually adopt the innovation
- c_1 teacher influence on students ($c_1 > 0$, $c_1 < 0$, $c_1 = 0$)
- c_2 student influence on teachers ($c_2 > 0$, $c_2 < 0$, $c_2 = 0$)
- α proportion of teachers in the total population of potential adopters $(0 \leq \alpha \leq 1)$

The differential equation giving the diffusion for teachers (f) is

$$\frac{df}{dt} = f(t) = (p_1 + q_1 f + c_1 s)(F - f) \quad (4)$$

The differential equation giving the diffusion for students (s) is

$$\frac{ds}{dt} = s(t) = (p_2 + q_2 s + c_2 f)(S - s) \quad (5)$$

The equation (4), (5) takes into account the interinfluence of teachers and students

The differential equation giving the diffusion for the combined population of potential adopters (m) (students and teachers) is

$$\frac{dm}{dt} = m(t) = \alpha f(t) + (1 - \alpha)s(t) \quad (6)$$

4 Innovation Characteristics

Relative Advantage: [14] defines relative advantage as — 'the degree to which an innovation is perceived as being better than the idea that it supersedes'. We hypothesize that Relative advantage of Flipped Labs positively affects intention to adopt it.

Compatibility: [14] defines compatibility as — 'the degree to which an innovation is perceived as consistent with existing values, past experiences, and needs of potential adopters'. We hypothesize that Compatibility of Flipped Labs positively affects intention to adopt it.

Complexity/Ease of Use: Any innovation quickly gains a reputation as to its ease or difficulty of use [14]. We hypothesize that Complexity of Flipped Labs negatively affects intention to adopt it.

Trialability: Trialability is "the degree to which an innovation may be experimented with on a limited basis" [14]. Innovations that potential adopter can play with on a trial basis are more easily adopted as it presents less risk to the potential adopter. We hypothesize that Trialability of Flipped Labs positively affects intention to adopt it.

Observability: Another aspect of [14] is the degree to which the results of an innovation are visible to others. If potential adopters can see the benefits of an innovation, they will easily adopt it. We hypothesize that Observability of Flipped Labs positively affects intention to adopt it.

Perceived Enjoyment: According to [14] Perceived enjoyment is the 'degree to which using an innovation is perceived to be enjoyable in its own right and is considered to be an intrinsic source of motivation'. We hypothesize that Perceived Enjoyment of Flipped Labs positively affects intention to adopt it.

School Support: More often teachers and students are motivated to consider technology decisions that are sanctioned by the school management since those will have adequate support resources including the necessary IT infrastructure. We hypothesize that School support for Flipped Labs positively affects intention to adopt it.

Peer influence: Interpersonal influence appears to be extremely important in influencing potential adopters, as is demonstrated by the fact that the opinions of peers significantly affect the way in which an individual feels pressures associated with adoption of the innovation. We hypothesize that Peer influence for Flipped Labs positively affects intention to adopt it.

Teacher Influence: Since teachers play a pivotal role in implementing innovations, their perception of the innovation will strongly influence their students thinking. We hypothesize that Teacher support for Flipped Labs positively affects student's intention to adopt it.

Student Influence: Students perception of the innovation will strongly influence their teacher's thinking. We hypothesize that student support for Flipped Labs positively affects teacher's intention to adopt it.

5 Research Methodology

In our study, 81 students and 32 teachers participated. A five point Likert scale based questionnaire was administered to understand students' and teachers' perceptions on the factors that influence the adoption of flipped labs pedagogy. The survey consisted of nine independent research variables hypothesized to be a factor affecting the adoption. The independent research variables used in the study were Relative Advantage, Compatibility, Ease of Use, Observability, Trialability, Teacher Influence, Student Influence, Peer Influence and School Support.

In our study reliability of the attributes had values ranging from 0.72 to 0.86 for students and 0.69 to 0.89 for teachers which is in the acceptable range. Regression analysis was conducted for all nine adoption variables on the dependent variable and hypothesis results calculated (Table 1). There is strong support for attributes like Relative Advantage, Compatibility, Ease of Use and Teacher Influence. The regression equation for students was statistically significant ($p < .0001$) and explained approximately 78% of the variation ($R^2 = .78$).

Table 1 Summary of Hypothesis results (students)

Attributes	Mean	SD	t test	p-value	Result
Compatibility*	24.91	4.55	7.318	0.003451	Accepted
Ease Of Use*	10.03	1.94	-1.9245	0.00272	Accepted
Observability	6.63	1.79	-0.2441	0.4036	Rejected
Perceived Enjoyment	11.42	2.40	-1.0942	0.1372	Rejected
Peer Influence	9.26	2.65	-1.1231	0.1308	Rejected
Relative Advantage*	22.30	3.80	1.438	0.007534	Accepted
School Support	14.53	3.25	2.4526	0.2932	Rejected
Trialability	7.47	1.69	-0.9276	0.3769	Rejected
Teacher Influence*	14.36	3.06	3.6516	0.00287	Accepted

*p < 0.01

Interestingly for teachers also there is strong support for attributes like Relative Advantage, Compatibility, Ease of Use and student influence along with an additional attribute School support (Table 2). The regression equation for teachers was statistically significant ($p < .0001$) and explained approximately 76% of the variation ($R^2 = .76$).

Table 2 Summary of Hypothesis results (teachers)

Attributes	Mean	SD	t test	p value	Result
Compatibility*	32.72	4.76	-1.3302	0.009835	Accepted
Ease Of Use*	7.62	1.85	-2.1597	0.002084	Accepted
Observability	6.93	1.51	0.1642	0.4357	Rejected
Perceived Enjoyment	12.52	2.54	0.7958	0.2159	Rejected
Peer Influence	10.70	2.30	-1.5723	0.06491	Rejected
Relative Advantage*	20.48	3.69	-1.6358	0.00575	Accepted
Student Influence*	11.74	2.19	0.1567	0.00187	Accepted
School Support*	16.72	2.86	-1.6641	0.00158	Accepted
Trialability	7.50	1.83	-0.7514	0.2296	Rejected

*p < 0.01

6 Conclusions

This research has provided a deeper investigation into adoption of smart ICT innovation like flipped labs guided by the framework of Rogers' theory of perceived attributes. In this study we had proposed that certain characteristics of flipped labs as a smart ICT innovation could account for the degree of acceptance of innovation by teachers and students. The results of this analysis indicated that Relative Advantage, Compatibility, Ease of Use, Teacher Influence and Student Influence were found to be positively related to acceptance of innovation. For the first time we have empirical evidence to show that the adoption of an ICT innovation is directly impacted by the interinfluence between teachers and students. When interinfluence plays a key role in the diffusion of innovation, the adoption decisions of a teacher/student often goes beyond their own decisions. We posit that a teacher's adoption decision comes not only from her own experiences but also from student's influence. Likewise the students' adoption decisions are influenced by the teachers' perceptions. The results revealed that for innovation attributes like Relative Advantage, Ease of use, Compatibility and Interinfluence student and teacher perceptions were similar.

References

1. Lage, M.J., Platt, G.J., Treglia, M.: Inverting the classroom: A gateway to creating an inclusive learning environment. Journal of Economic Education 31(1), 30–43 (2000), doi:10.1080/00220480009596759
2. Tucker, B.: The Flipped Classroom. Education Next 12(1) (2012)

3. Mazur, E.: Peer Instruction (2012), `http://mazur.harvard.edu/research/detailspage.php?rowid=8` (retrieved)
4. Clark, D.: 10 Reasons to Dump Lectures (December 18, 2007), `http://donaldclarkplanb.blogspot.ca/2007/12/10reasons-to-dump-lectures.html` (retrieved)
5. Strayer, J.F.: The effects of the classroom flip on the learning environment: A comparison of learning activity in a traditional classroom and a flip classroom that used an intelligent tutoring system (2008)
6. Day, J.A., Foley, J.D.: Evaluating a Web Lecture Intervention in a Human– Computer Interaction Course. IEEE Transactions on Education 49(4), 420–431 (2006), doi:10.1109/TE.2006.879792
7. Toto, R., Nguyen, H.: Flipping the Work Design in an Industrial Engineering Course. In: Proceedings of the 39th ASEE/IEEE Frontiers in Education Conference, San Antonio, Texas, USA (October 2009)
8. Revere, L., Kovach, J.V.: Online Technologies for Engaging Learners: A Meaningful Synthesis for Educators. Quarterly Review of Distance Education 12(2), 113–124 (2011)
9. Copley, J.: Audio and video podcasts of lectures for campus-based students: production and evaluation of student use. Innovations in Education & Teaching International 44(4), 387–399 (2007), doi:10.1080/14703290701602805
10. Guskey, T.R.: Closing Achievement Gaps: Revisiting Benjamin S. Bloom's "Learning for Mastery". Journal of Advanced Academics 19(1), 8–31 (2007)
11. Nedungadi, P., Raman, R., McGregor, M.: Enhanced STEM learning with Online Labs: Empirical study comparing physical labs, tablets and desktops. In: 2013 IEEE Frontiers in Education Conference, pp. 1585–1590. IEEE (2013)
12. Raman, R., Nedungadi, P., Ramesh, M.: Modeling Diffusion of Tabletop for Collaborative Learning Using Interactive Science Lab Simulations. In: Natarajan, R. (ed.) ICDCIT 2014. LNCS, vol. 8337, pp. 333–340. Springer, Heidelberg (2014)
13. Karmeshu, R.R., Nedungadi, P.: Modeling diffusion of a personalized learning framework. Educational Technology Research and Development 60(4), 585–600 (2012)
14. Rogers, E.M.: Diffusion of innovations, 5th edn. Free Press, New York (2003)
15. Bass, F.M.: A new product growth for model consumer durables. Management Science 15, 215–227 (1969)
16. Karmeshu, Goswami, D.: Stochastic evolution of innovation diffusion in heterogeneous groups: Study of life cycle patterns. IMA Journal of Management Mathematics 12, 107–126 (2001)

Inquiry Based Learning Pedagogy for Chemistry Practical Experiments Using OLabs

Prema Nedungadi, Prabhakaran Malini, and Raghu Raman

Abstract. Our paper proposes a new pedagogical approach for learning chemistry practical experiments based on three modes of inquiry-based learning namely; structured, guided and open. Online Labs (OLabs) is a web-based learning environment for science practical experiments that include simulations, animations, tutorials and assessments. Inquiry-based learning is a pedagogy that supports student-centered learning and encourages them to think scientifically. It develops evidence based reasoning and creative problem solving skills that result in knowledge creation and higher recall. We discuss the methodology and tools that OLabs provides to enable educators to design three types of inquiry-based learning for Chemistry experiments. The integration of inquiry-based learning into OLabs is aligned with the Indian Central Board of Secondary Education (CBSE) goal of nurturing higher order inquiry skills for student centered and active learning. Inquiry-based OLabs pedagogy also empowers the teachers to provide differentiated instruction to the students while enhancing student interest and motivation.

Keywords: Inquiry-based learning, IBL, Structured inquiry, Guided inquiry, Open inquiry, Virtual Labs, Online Labs, olabs, Simulations, Chemistry, Chemical Sciences.

1 Introduction

Inquiry based learning is an approach to learning wherein the students' quest for knowledge, their thinking, opinions and observations form the core of the learning process. The underpinning postulation for inquiry-based learning is that both

Prema Nedungadi · Prabhakaran Malini · Raghu Raman
Amrita Center for Research in Advanced Technologies for Education, Amrita Vishwa Vidyapeetham, Amritapuri, Kerala, India
e-mail: {prema,malinip,raghu}@amrita.edu

El-Sayed M. El-Alfy et al. (eds.), *Advances in Intelligent Informatics*,
Advances in Intelligent Systems and Computing 320, DOI: 10.1007/978-3-319-11218-3_56

teachers and students are mutually accountable for learning. Faculties and teachers need to actively participate in the process and foster a culture wherein students are encouraged to express their ideas which are reverently challenged, verified, refined and improvised. This enables the children to deepen their understanding of the concept through interaction and questioning [1]. For students, this teaching methodology comprises open-ended explorations into a concept or a problem. In this process educators play the role of guides helping the students to explore their ideas and take their inquiry forward to make logical conclusions. Thus, it is a co-authored knowledge creation practice wherein educators and students work in tandem to foster enduring learning. This may also foster learning for all in the class and generate significant concepts and viewpoints [2].

Technology has transformed this learning methodology in manifold ways and more and more schools and colleges are resorting to technology centric inquiry-based learning. Educators believe that this learning methodology is highly effective in the fields of science, technology, engineering and math (STEM). These subjects are critical for students to excel in the 21st century [3], [4] and [5]. Research has proven that inquiry and inquiry-based teaching pedagogy is instrumental in developing student inclination towards difficult subjects such as science and maths and to nurture their curiosity and motivate them to learn right from the elementary school level. Comstock et al or Reference [6] analysed the web-based, knowledge-building tool, "Inquiry Page" in the National Science Foundation GK-12 Fellowship Program wherein scientists and educationists collaborate to integrate computer-based modelling, scientific visualization and informatics in the learning of science and mathematics. This allowed teachers and students to collaborate on the inquiry path and establish new knowledge and teaching resources together. A remarkable example of technology based learning for science practical skills is the Online Labs (OLabs). The OLabs Project was introduced to facilitate laboratory-learning experiences for school students throughout India who may not have access to adequate laboratory facilities or equipment [7]. OLabs encourages students to engage in scientific studies and get acquainted with scientific inquiry methodologies through simulations, animations, tutorials and assessment. Another commendable initiative is the Virtual Labs, a multi-institutional OER (Open Educational Resources) which is solely intended to provide lab experiments for undergraduate engineering students. This project developed an extensive OER repository comprising 1650 virtual experiments [8] and [9]. This paper compares three levels of inquiry learning; structured learning, guided learning and open learning and shows how educators may use the Chemistry OLabs to support these three modes of inquiry-based learning.

2 Literature Review

Spronken-Smith and Walker or Reference [10] have advocated inquiry-based learning as a student -centred approach which reinforces the nexus between teaching and research. Inquiry can be understood as the practice wherein students undertake investigative learning about science, conduct experiments, derive inferences from the evidence, evaluate their authenticity and then communicate their findings with due justification [11]. We consider three approaches towards inquiry-based instructions. Firstly, structured inquiry is the methodology in which students work upon an investigative process on a teacher-formulated question within the prescribed framework. The requisite material is also facilitated by the instructor. Teachers are aware of the outcomes but do not disclose them. Students have to discover for themselves through working out relationships between the variables. These are similar to cookbook activities but with fewer instructions and directions. This is the fundamental step suitable for elementary level students. Secondly, the guided inquiry, wherein teachers facilitate the material required for solving the problem at hand and students have to concoct their own procedure for problem resolution. This methodology is suited for secondary school students where framework still needs to be given but there is room for further experimentation. Thirdly, is the open inquiry, similar to guided inquiry but students get the room to formulate their own problem which they intend to investigate. This methodology is parallel to research work in science. Here, the instructor only provides the theoretical framework. Science fair activities are based on open inquiry. This pedagogy is best suited for senior school or college level students [12]. Open inquiry generates greater satisfaction amongst students as they get a sense of accomplishment while conducting the investigation, whereas guided inquiry helps students to produce better documentation for their findings within the provided framework. Thus, open inquiry is suitable for teaching pedagogy where higher autonomy, out-of the-box thinking and in-depth understanding of scientific concepts are warranted [12].

The acceptance of inquiry-based learning is rising and more and more educational institutes at both K-12 and higher education levels are desirous of adopting and integrating the pedagogy with the available technology. However, it is not easy and there are many challenges and barriers in implementing the techniques. The most prominently perceived barriers include the central or state controlled curricula, inadequate time for inquiry, student expectancies and competencies, infrastructure and technology accessibility, resistance from teachers and staff [13] and [14]. Duman and Guvan [15] highlighted the inability of the virtual learning systems to accurately assess the acumen levels of children and help them learn accordingly. This also includes children with special needs. Also, frequently, children are unable to generalize the learning of a session in spite of the session being interesting. However, the biggest challenge is aligning an inquiry-based approach with the mandated time-bound curriculum [16].

3 Inquiry Based OLabs Pedagogy

Inquiry-based strategies in science help students engage in active inquiry, problem solving and decision making and thus discover the significance of science to their lives. There are three different levels based on the manner in which inquiry is facilitated.

1) Structured inquiry: in which the question as well as the procedure for the inquiry are provided by the teacher and the results are to be discovered by the students.

2) Guided inquiry: in which the teacher provides students with only the research question and the materials. Students are encouraged to design an inquiry procedure and then generate the explanation.

3) Open inquiry: in which students generate questions on topics of interest, design investigations to answer the question, conduct the investigations, and communicate their results.

OLabs helps students propose their own research focus, supports in carrying out inquiry-based activities, produce their own data, and continue their inquiry as new questions arise. OLabs helps understanding by facilitating different methods to investigate the same problem.

3.1 Structured, Guided and Open Inquiry Based OLabs

OLabs can be designed to support three modes of inquiry-based learning. Observations and analysis are involved in each mode of inquiry. For each mode of inquiry, there is an instructional plan.

Teachers can design the experiment for structured, guided and open inquiry based on configuring the tabs in OLabs. (Fig.1a, Fig.1b, Fig.1c)

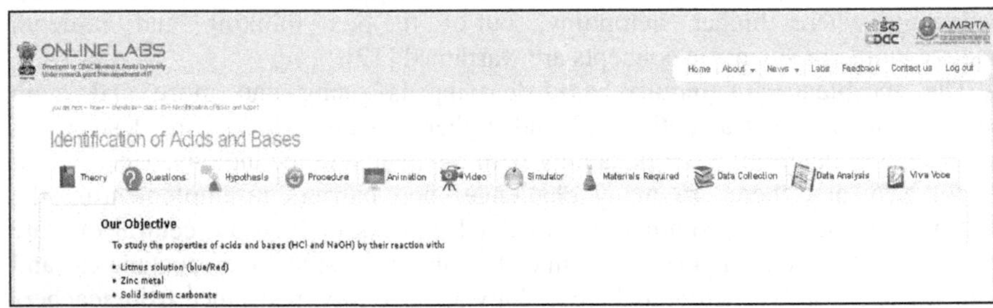

Fig. 1a. OLabs based on structured inquiry-based learning

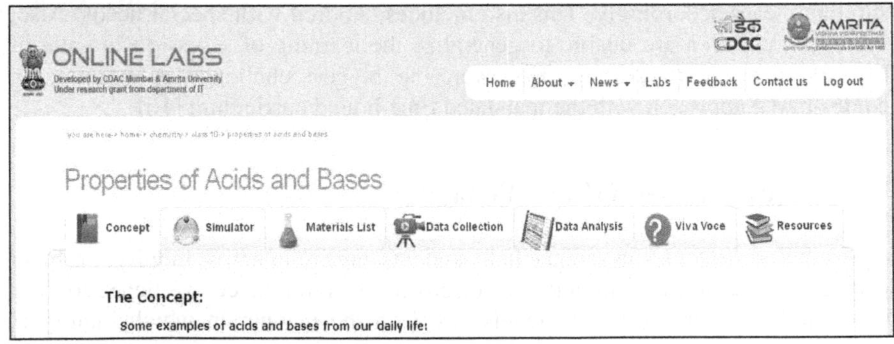

Fig. 1b. OLabs based on guided inquiry-based learning

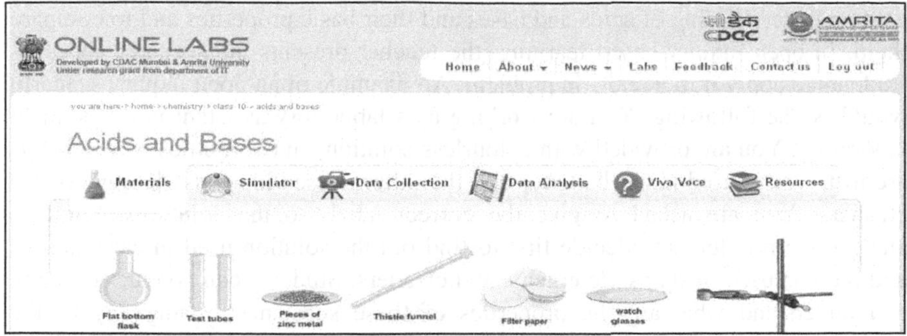

Fig. 1c. OLabs based on open inquiry-based learning

Generally speaking, OLabs may have one or more of the following tabs.

- **Theory** - Consists of the main objectives, concept and learning outcomes of the experiment.
- **Procedure** - Consists of the materials required for the experiment, real lab procedure, simulator procedure and the precautions.
- **Animation** - From the animation, students can clearly understand the methodology to do the experiment in the real chemical laboratory.
- **Simulation** – Using simulations, students can perform the experiments by changing variables and can observe the effect of the changes made. Students have the opportunity to do the experiment repeatedly using simulations until they clearly understand the concepts and the methodology.
- **Video** - Contains video of real laboratory experiments.
- **Viva Voce** - It consists of several questions of multiple choice categories. Students can verify the results of their experiment by answering these questions.
- **References** – Contains references, both book and websites links.

We describe how a Chemistry experiment, "Properties of Acids and Bases" can be configured under OLabs for Structured, Guided and Open Inquiry-based learning methods.

3.2 *Selection of Topic*

The first step of the OLabs inquiry process is selection of the topic and presentation of the scenario.

In structured inquiry-based learning, the teacher selects and presents the experiment, 'Identification of acids and Bases' from OLabs (amrita.olabs.co.in) along with the scenario based on the theory. The teacher can also use the lab videos to explain the methodology and the fundamental principles behind the concept. In the guided inquiry-based learning the teacher presents the scenario to investigate and understand the properties of acids and bases. After presenting this scenario, the students are then encouraged to read the Concept part of OLabs to gain an understanding of acids and bases and their basic properties and to compare them. In open inquiry-based learning, the teacher presents the scenario and asks students to apply it to a real-life problem. An example of an open inquiry scenario would be the following: You are working as a laboratory assistant in a Chemical laboratory. You are provided with colourless solutions in two bottles whose labels are lost. One is used in car batteries and the other is used in soaps, detergents and cleaners. You are asked to give the correct labels to the solutions. Students analyze the problem and decide first to find out the solution used in car batteries and the one used in soaps, detergents and cleaners. Students can conduct research and understand what are the properties of these substances, using any of the OLabs tabs or go to other websites and find out the basic principle of the properties of acids and bases.

3.3 Generating Questions

The next step in the inquiry process is generating the question to start the investigation. A question for our experiment might be, "How will you identify an acid or a base based on their properties?" The student needs to understand the concepts required for this main question.

In structured inquiry, the teacher states the questions for students, based on the objectives of the experiment's theory part. The teacher also explains the theoretical concepts. In guided inquiry, the teacher states the questions for students, such as, "Study What are the properties of acids and bases?". Furthermore, the teacher suggests the concepts that need to be learnt but does not explain the concepts or point to the answers. The students must search the Concept tab or references given in the Resource tab and understand the concepts for themselves. In open inquiry there is no involvement of teacher in the investigation and it is completely student-driven. The student formulates the questions and acquires basic knowledge about the topic. Students can either use the references given in Resource Tab or they can refer to other internet resources. Their findings may take them far beyond the scope of the original problem. The possible questions that students may consider are:

a) Which substance is used in car batteries?
b) Which substance is used in soaps, detergents and cleaners?
c) What are the properties of those substances?

3.4 Background Research

In structured inquiry, the teacher does the background work before designing the investigation. In the guided inquiry mode, students perform the background work before designing the investigation using references in the Resources tab for doing background work. In the open inquiry mode, students perform the background work using any internet resources for doing background work.

3.5 Constructing Hypothesis

In the structured inquiry mode, the teacher discusses the possible hypotheses and the method of investigation. The students write the possible outcomes of the hypothesis and then carry out the investigation as instructed by the teacher. In structured inquiry, the teachers talk about the possible hypothesis to students, based on the theory given in the Theory tab and students examine the hypothesis after carrying out the experimentation. In the guided inquiry, students write about the expected results based on the concept given in the Concept tab before conducting the investigation. Students can verify the hypothesis after conducting the experiment. In the open inquiry, the students write about the expected results before carrying out the experiment. In some cases, the expected results may be wrong. Students can examine the results after the experimentation.

3.6 Designing the Methodology

In structured inquiry mode, the real lab procedure and the procedure for carrying out the simulation are given in the Procedure tab. The teacher can design the investigation for the real laboratory experiment by viewing the animation given in the Animation tab. Students can also understand the real lab procedure from the Animation. Students can further their understanding of the real lab experiment by viewing the lab video given in the Video tab. In the guided learning mode, the teacher provides the list of materials and the students design the methodology for doing the experiment in a chemical laboratory based on the listed materials. The listed materials are given in the Materials list tab. Students also design the procedure for doing the simulation based on the listed materials in the side menu. If students face any difficulty during the design process, the teacher can assist them. In open inquiry, the design methodology is completely student driven. Students first choose the material required for the experiment from the number of materials given in the Materials tab. They have to write the methodology for the real laboratory experiment and the procedure for doing the simulation based on the required materials. In the simulator, a number of materials are given in the side menu.

3.7 Materials and Conducting the Investigation

In structured inquiry, the materials required for the experiment are given in the Materials required part of the Procedure tab. The teacher explains the materials for students based on this Materials required part. All the steps required for the simulation are given by clicking on Simulator Procedure in Procedure tab. Teacher explains the steps to students based on the Simulator Procedure. Students can carry out the experiment using the OLabs simulation. The teacher encourages use of the 'HELP' button to see the instructions. In the guided inquiry, the teacher provides the materials. The students write the entire step by step process for doing

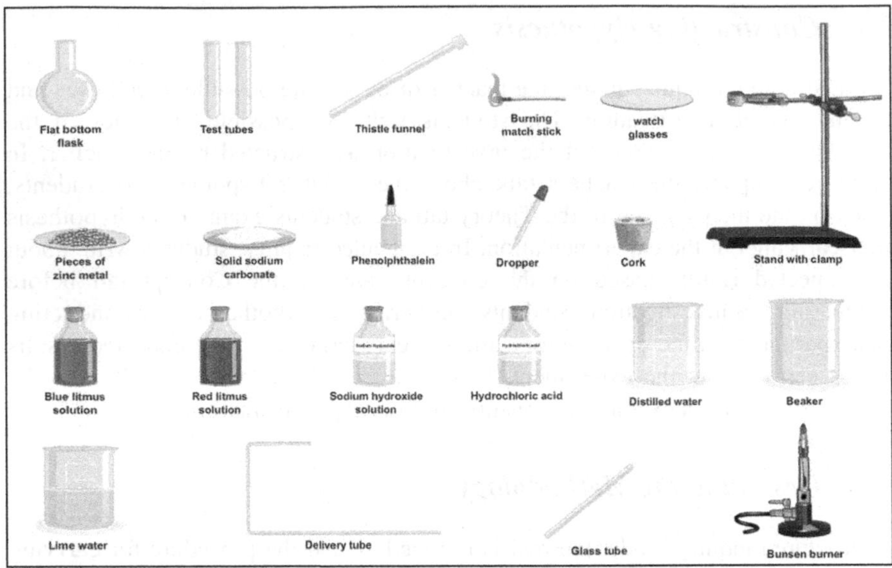

Fig. 2 Materials required for the experiment in structured inquiry

the simulation based on the materials provided and the background research. The listed material for doing real laboratory experiment is given in the Materials list tab. In the open inquiry, the basic methodology of the experiment has to be written by the students, based on the selected material of the experiment. Students can select the materials for doing the experiment from the Materials tab. The methodology should be designed based on the concept behind the experiment. Students can carry out the experiment based on the designed methodology.

3.8 Collecting and Analyzing Data

In structured inquiry, teachers guide students to collect and analyse data by giving specific instructions on the methods of tabulation and the type of graph. (Table 1)

Teacher tells the students to verify the results by answering questions in the Viva Voce tab. In guided inquiry, students formulate the observations, tabulation and graph of data. Teacher can guide them if there are any mistakes. In open inquiry, Students have to design the tabulation of the data observed.

Table 1 Tabulation of data in structured inquiry learning

Sample	Colour change on	
	Red litmus solution	Blue litmus solution
Sample 1		
Sample 2		

3.9 *Drawing Conclusions*

In structured inquiry, based on the analysis and data recorded students can identify which samples are acids and which are bases and then compare their conclusions with the observations and inferences given in the Procedure tab. In guided inquiry, based on the analysis and data recorded, students can identify which samples are acids and which are bases but make their own conclusions. In open inquiry, based on the analysis and observations students can identify which substance is used in car batteries which one is used in soaps, detergents and cleaners. They discuss in their conclusion whether their hypothesis was correct and record the summary of the answers for the questions raised.

4 Conclusions

We discuss the pedagogy of various types of inquiry-based learning such as structured, guided and open inquiry in the context of STEM skills. We present a detailed framework for using OLabs in each of the three inquiry models. We demonstrate the methodology and tools that OLabs provides to enable educators to create the three type of inquiry for science practical skills. The integration of inquiry-based learning methodology and OLabs is aligned to the Central Board of Secondary Education (CBSE) goal of nurturing higher order thinking skills for student-centered active learning. Inquiry-based OLabs empowers the teacher to provide differentiated learning to the students while enhancing student interest and motivation.

References

1. Quigley, C., Marshall, J., Deaton, C., Cook, M., Padilla, M.: Challenges to inquiry teaching and suggestions for how to meet them. Science Educator 20(1), 55–61 (2011)
2. Fielding, M.: Beyond student voice: Patterns of partnership and the demands of deep democracy. Revista de Educación 359, 45–65 (2012)
3. Nedungadi, P., Raman, R., McGregor, M.: Enhanced STEM Learning with Online Labs: Empirical study comparing physical labs, tablets and desktops. In: 43rd ASEE/IEEE Frontiers in Education Conference (FIE 2013), USA (2013)
4. Bybee, R.W.: Inquiry is essential. Science and Children 48(7), 8–9 (2011)
5. Lemlech, J.K.: Curriculum and Instructional Methods For the Elementary and Middle School, 7th edn. Pearson, New Jersey (2010)
6. Comstock, S., et al.: Fostering inquiry-based learning in technology-rich learning environments: The Inquiry Page in the GK-12 Fellows Program. In: World Conference on Educational Multimedia, Hypermedia and Telecommunications, pp. 340–341. Association for the Advancement of Computing in Education, Chesapeake (2002)
7. Nedungadi, P., Raman, R., McGregor, M.: Enhanced STEM Learning with Online Labs: empirical study comparing physical labs, tablets and desktops. In: 43rd ASEE/IEEE Frontiers in Education Conference (FIE 2013), USA (2013)

8. Raman, R., Achuthan, K., Nedungadi, P., Diwakar, S., Bose, R.: The VLAB OER experience: Modeling potential-adopter students' acceptance. IEEE Transactions on Education (in press, 2014)
9. Achuthan, K., et al.: The VALUE @ Amrita Virtual Labs Project. In: IEEE Global Humanitarian Technology Conference Proceedings, pp. 117–121 (2011)
10. Spronken-Smith, R., Walker, R.: Can inquiry-based learning strengthen the links between teaching and disciplinary research? Studies in Higher Education 35(6), 723–740 (2010)
11. Vreman-de Olde, G.C., de Jong, T., Gijlers, H.: Learning by Designing Instruction in the Context of Simulation-based Inquiry Learning. Educational Technology & Society 16(4), 47–58 (2013)
12. Sadeh, I., Zion, M.: Which Type of Inquiry Project Do High School Biology Students Prefer: Open or Guided? Research in Science Education 42, 831–848 (2012)
13. Colburn, A.: An Inquiry Primer. Science Scope 2, 42–44 (2000), `http://www.experientiallearning.ucdavis.edu/module2/el2-60-primer.pdf`
14. Trautmann, N., MaKinster, J., Avery, L.: What Makes Inquiry So Hard (And Why Is It Worth It?). In: National Association for Research in Science Teaching (NARST) 2004 Annual Meeting, Vancouver (2004)
15. Guven, Y., Duman, H.G.: Project based learning for children with mild mental disabilities. International Journal of Special Education 22(1), 77–82 (2007)
16. Parker, D.: Planning For Inquiry It's Not An Oxymoron. National Council of Teachers of English, Urbana (2007)

New Approach for Function Optimization: Amended Harmony Search

Chhavi Gupta and Sanjeev Jain

Abstract. Harmony search (HS) algorithm is an emerging population oriented stochastic metaheuristic algorithm, which is inspired by the music improvisation process. This paper introduces an Amended harmony search (AHS) algorithm for solving optimization problems. In AHS, an enhanced approach is employed for generating better solution that improves accuracy and convergence speed of harmony search (HS). The effects of various parameters on harmony search algorithm are analyzed in this paper. The proposed approach performs fine tuning of two parameters bandwidth and pitch adjustment rate. The proposed algorithm is demonstrated on complex benchmark functions and results are compared with two recent variants of HS optimization algorithms, improved harmony search (IHS) and highly reliable harmony search (HRHS). Results suggested that the AHS method has strong convergence and has better balance capacity of exploration and exploitation.

Keywords: Harmony search, Metaheuristic algorithm, Improved harmony search.

1 Introduction

Optimization is the process which is executed iteratively for finding the value of variables for which objective function or fitness function can be either minimize or maximize by satisfying some constraint . For a given problem domain, the main goal of optimization is to provide the mode of obtaining the best value of objective function. [1-2]. Problem domain can be optimized by either classical approaches or by advance optimization techniques. The classical approaches are based on pure mathematical methods like linear programming, graph theory etc. However for real time application, these methods cannot find global solution due to computationally expensive. These methods often trapped in local minima. Therefore, advance metaheuristic optimizations are designed in order to obtain the global optimum solution.

Chhavi Gupta · Sanjeev Jain
Madhav Institute of Technology & Science, Gwalior

El-Sayed M. El-Alfy et al. (eds.), *Advances in Intelligent Informatics*,
Advances in Intelligent Systems and Computing 320, DOI: 10.1007/978-3-319-11218-3_57

Meta means "high" and heuristic means to "guide" hence these methods provide the high level framework for any optimization problem. The purpose of metaheuristic optimization technique is to find near optimal solution by exploring the search space efficiently. Some of the popular methods such as Genetic Algorithm (GA), Simulated Annealing (SA), Ant Colony Optimization (ACO), Particle Swarm Optimization (PSO) etc. All these metaheuristic techniques mimic the natural phenomena, such as GA [3], simulate bio-inspired computation, SA [4] is motivated by an analogy to statistical mechanics of annealing in metallurgy, ACO [5] mimic social behavior of ant colony and PSO [6] is inspired by forging behavior of bird flock.

In the same way, Harmony search (HS) is a metaheuristic technique that is inspired by improvisation process of music players [7]. These techniques are widely used in numerous engineering problems such as resource allocation, scheduling, decision making. Various improvements have been proposed in literature to make HS more robust and convergent [8-15]. These improvements mostly cover two aspects [17]: (1) improvements in terms of hybridizing HS components with other metaheuristic algorithms and (2) improvements in terms of parameters setting. Taherinejad [8] modified HS by stimulating the SA's cooling behaviour of metallurgy. Omran & Mahdavi [9] designed an algorithm, which uses the global best particle concept of PSO for updating the bw parameter. Geem[10] presented the same idea of pbest of PSO in harmony memory consideration operator (HMCR) to get better the selection process. This new mechanism chooses the decision variable from the best harmony vector stored in harmony memory (HM) not randomly as in standard HS.

Coelho & Bernert [11] modified pitch adjustment rate (PAR) equation by introducing dynamic PAR that uses a new performance differences index (grade) in their approach. Geem [12] used fixed parameter values such as fixed number of iteration, PAR etc. while bandwidth (bw) was set to a range from 1 to 10% of the total value data range. Mahdavi et al. [13] gave the notion for determining the lower and upper bound of bandwidth automatically. They proposed the method that updates PAR dynamically at every iteration. Omran & Mahdavi [13] proposed Global-best harmony search (GHS) which utilized PSO concept of the global best particle that is the best particle in terms of fitness among all particles in the flock. Their method overcomes the drawback of IHS as they described. Wang and Huang [14] presented a new improvement to HS algorithm that modifies bw and PAR on the basis of maximum and minimum values in HM. Chakraborty et al. [15] proposed a new variant of HS by employing Differential Evolution (DE) mutation operator to HS. Yadav et al [17] introduced Intelligent Tuned Hamony Search which automatically estimates the value of PAR on the basis of its Harmony Memory.

In 1977, Wolpert & Macready introduced "No Free lunch theory" and concluded that every metaheuristic algorithm has different searching abilities and has its own advantage to deal with problem domain. So, no single algorithm is able to offer satisfactorily results for all problems. In other words, a specific algorithm may show very promising results on a set of problems, but may show poor performance on a different set of problems. Hence, there is always a possibility of development of new metaheuristic algorithm.

This chapter presents new population oriented stochastic algorithm called Amended harmony search (AHS) for updating PAR and bw dynamically and effectively. Results are evaluated on standard benchmark function and compared with IHS, HRHS and Section 2 gives the overview of basic HS, IHS and HRHS. Section 3 presents the proposed algorithm AHS. Experimental analysis on different test function is provided in Section 4. Finally conclusion is presented in Section 5.

2 Overview of Harmony Search

2.1 *Basic Harmony Search*

Harmony search (HS) [16] is recently developed random search algorithm which is motivated by musical improvisation process where musician improvise the pitches of their instruments to achieve better harmony. The working procedure of HS is narrated in the next five steps which are as follows [16]:

2.1.1 Initialize the Problem and Algorithm Parameters

In Step 1, the optimization problem is described as follows:

$$Minimize:\ f(x), \qquad Subjected\ to: x_i \in X, i = 1,2,.., d \tag{1}$$

where F(x) is an objective function that is to be minimized, x is the set of possible solution vector, d represent the dimension of solution vector. Initialize all the parameters harmony memory size (HMS) that represent the number of solution vectors in the harmony memory (HM), harmony memory considering rate (HMCR), bandwidth (bw), pitch adjusting rate (PAR) and the maximum number of iteration (T_{max}) or stopping criterion. The HM is a memory location where all the possible solution vectors are stored. [3].

2.1.2 Initialize the Harmony Memory

In Step 2, initialize the HM matrix with randomly generated solution vectors of size equal to HMS.

$$HM = \begin{bmatrix} x_1^1 & \cdots & x_d^1 \\ \vdots & \ddots & \vdots \\ x_1^{HMS} & \cdots & x_d^{HMS} \end{bmatrix} \tag{2}$$

2.1.3 Improvise a New Harmony

In Step 3, HMCR, PAR is applied to each variable of the new harmony/solution vector. In this step, a new solution vector $x_i' = (x_1', x_2',..., x_d')$ is created on the basis of three rules: (1) memory consideration, (2) pitch adjustment and (3) random selection and this generation of a new harmony is called 'improvisation' [16]. The HMCR is the rate of choosing one value from the stored values in the HM, while (1-HMCR) is the rate of randomly selecting one value from the possible range of values. Its value lies between 0 and 1.

$$x_i' = \begin{cases} x_i' \in \{x_i^1, x_i^2, ..., x_i^{HMS}\} \text{with probabilty } HMCR \\ x_i' \in \{X_i\} \quad \text{with probabilty } (1 - HMCR) \end{cases} \tag{3}$$

For every solution vector that comes from HM is required to find out whether it should be pitch-adjusted or not. To evaluate, newly created solution is pitch adjusted or not, following mechanism is used. This mechanism uses the rate of pitch adjustment i.e. PAR parameter which is given as:

$$\text{Pitch Adjustment Rate} = \begin{cases} YES \quad \text{with probabilty } PAR \\ NO \quad \text{with probabilty } (1 - PAR) \end{cases} \tag{4}$$

If the pitch adjustment decision for x_i' is No then do nothing otherwise the pitch adjustment decision is Yes then x_i' is replaced as follow:

$$x_i' = x_i' \pm r \times bw \tag{5}$$

where bw is an arbitrary distance bandwidth and r is a random number lies between 0 and 1.

2.1.4 Update Harmony Memory

If the fitness value corresponding to new solution vector is better than best fitness then the old solution vector is replaced with the new solution vector and included in the HM. The fitness corresponding to new solution vector should also be stored.

2.1.5 Check Stopping Criterion

If the stopping criterion is met, algorithm is terminated. Otherwise, Steps 3 and 4 are repeated.

2.2 *Improved Harmony Search (IHS)*

The basic HS algorithm was discussed in the last subsection and in this subsection variants of HS are presented. In basic HS, the value of parameters PAR and bw are initially adjusted i.e. fixed during the entire algorithm which results in slow convergence and also unable to offer the optimal solution to the problem. If low values of PAR, high values of bw are used then HS will converge slowly and require more iterations to give optimal solution [13]. If high PAR values, low bw values are adjusted then HS will converge fast but the solution may scatter around some potential optimal solution. Hence, such approach is required which adjust both the parameters dynamically in order to achieve near optimal solution. Mahadavi et al. [13] developed IHS which uses dynamic adjustment approach proposed for PAR and bw. They recommended that bw should decrease exponentially and PAR should increase linearly in every iteration so that the stage reached when both the parameters will create harmony i.e. the optimal solution with better convergence. So, PAR and bw change dynamically with iteration using equation (6) and (7) respectively:

$$PAR(t) = PAR_{min} + \frac{(PAR_{max} - PAR_{min})}{T_{max}} \times t \tag{6}$$

where PAR(t) is pitch adjusting rate for every iteration PAR_{min} is minimum pitch adjusting rate, PAR_{max} is maximum pitch adjusting rate, T_{max} is maximum iteration and t is current iteration.

$$bw(t) = bw_{max} \exp\left(\frac{Ln\left(\frac{bw_{min}}{bw_{max}}\right)}{T_{max}} \times t \right) \tag{7}$$

where bw(t) is bandwidth for each iteration , bw_{min} is minimum bandwidth and bw_{max} is maximum bandwidth.

In [8] Taherinejad proposed variant of HS i.e. highly reliable harmony search (HRHS) in which only PAR is updated with dynamic strategy. They adopted the concept from dispersed particle swarm optimization for dynamically updating the value of parameter PAR in each iteration. In this case, PAR is updated as follows:

$$PAR(t) = PAR_{max} - \frac{(PAR_{max} - PAR_{min})}{T_{max}} \times t \tag{8}$$

3 Proposed Amended Harmony Search (AHS)

In the traditional HS method, PAR and bw values are set during initialization of parameters therefore cannot be changed during iterations. PAR and bw affects the performance of HS algorithm. Mahdavi et al. [13] has proposed improvements in bw and PAR, which are adaptive to iteration. We are proposing two improvements in bw and PAR respectively. Firstly, bandwidth is made adaptive with respect to fitness value rather than iterations. Secondly, PAR is made adaptive to iteration but in different manner as given in eq. (8). We have used concept of SA which states that at initial time bad solutions may also be accepted. So rather than incrementing PAR value with respect to iteration, we have used a decrementing approach. As per iteration, best fitness of that iteration is used. Our proposed approach employs linear decrement to PAR. The PAR is given as follows:

$$PAR(t) = PAR_{max} - \frac{(PAR_{max} - PAR_{min})}{T_{max}} \times t \tag{9}$$

Our proposed approach employs exponential amendment to bw. The adaptive bw is given as follows:

$$c = \frac{best\,(objective\,) - mean\,(objective\,)}{best\,(objective\,) - worst\,(objective\,)} Ln(\frac{bw_{\min}}{bw_{\max}}) \tag{10}$$

$$bw = bw_{\min} (1 - e^{c}) \tag{11}$$

Although small bw values in final iterations increase the fine-tuning of solution vectors, but in early iterations bw must take a bigger value to enforce the algorithm to increase the diversity of solution vectors. Furthermore large PAR values with small bw values usually cause the improvement of best solutions in final iterations that converged to optimal solution vector [13].

4 Experiment

In our experimental study, 20 test functions have been demonstrated to analyze the performance of the proposed algorithm. Table1 represents the benchmark functions along with their range and dimension, d is the dimension of function and S represents the range of function respectively. F(x*) is the minimum value of the functions as given in Table1.

Table 1 Benchmark functions

S.No	Test Function Name	F(x*)	d	S
F1	Rosenbrock Function	0	2	$[-5,10]^d$
F2	Rastrigin Function	0	2	$[-5.12, 5.12]^d$
F3	Sum squares Function	0	4	$[-10,10]^d$
F4	Zakharov Function	0	2	$[-5,10]^d$
F5	Griewank Function	0	4	$[-600, 600]^d$
F6	Sphere Function	0	2	$[-5.12, 5.12]^d$
F7	Schwefel Function	0	4	$[-500,500]^d$
F8	Ackley Function	0	4	$[-32.768,32.768]^d$
F9	Hump Function	0	2	$[-5,5]^d$
F10	Trid Function	50	6	$[-36,36]^d$
F11	Cross in Tray Function	-2.06261	2	$[-10,10]^d$
F12	Schaffer N.2 Function	0	2	$[-100, 100]^d$
F13	Schaffer N.4 Function	.292579	2	$[-100,100]^d$
F14	Rotated Hyper-Ellipsoid Function	0	4	$[-65.536,65.536]^d$
F15	Booth Function	0	2	$[-10, 10]^d$
F16	Matyas Function	0	2	$[-10,10]^d$
F17	Levy N.2 Function	0	2	$[-10,10]^d$
F18	Colville Function	0	4	$[-10,10]^d$
F19	Michalewicz Function	-1.8013	2	$[0,\pi]^d$
F20	Sum of Different Powers Function	0	2	$[-1,1]^d$

AHS has applied to above mentioned minimization functions and results are compared with IHS as well as HRHS. Parameter settings for all the three algorithms are same. Harmony memory size is 150, number of iteration is 100, bandwith varies from 0.2 (bw_{min}) to 0.5 (bw_{max}) and pitch adjustment rate varies from 0.3(PAR_{min}) to 0.9 (PAR_{max}). All the algorithms use same parameter values for all the benchmark functions. As these are stochastic algorithm so each algorithm ran independently for 30 runs. The reported results are average and the best fitness value has been calculated from all 30 runs.

5 Results and Discussion

There are various performance measures of optimization algorithm such as fitness value obtained by it, stability obtained by it, best fitness value and convergence of algorithm. These parameters are explained in next three sub sections:

5.1 Mean Fitness Value

This section presents the experimental results tested to run for a fixed number of iterations, that is, 100. Table2 shows the mean fitness value examined by all the algorithms. AHS obtained optimum value of mean fitness value in most of the cases. All the algorithms are random in nature so to analyze the robustness and stability of each algorithm, standard deviation of fitness value for 30 runs is reported in Table2 along with mean fitness. The stability is measured in terms of standard deviation of algorithm. AHS obtained minimum value of standard deviation in most of the cases i.e. the more stable algorithm.

5.2 Best Fitness Value

Table3 provides the best fitness value obtained by each of the algorithm. Results demonstrate that the proposed AHS offer the best fitness value in most of the cases. From the results, it can be concluded that minimization function from F1 to F18 acquire the minimum best fitness when exploited using AHS. However in some cases, IHS performs well in comparison to HRHS.

5.3 Convergence

Fig. 1-4 shows the convergence graph of function F1, F2, F5 and F9 respectively. In the convergence graph, fitness value is plotted against number of iteration. The convergence graph of Rosenbrock test function between IHS, HRHS and AHS is shown in Fig.1. AHS has obtained fitness value f(x) = 0.04 at 33th iteration where as IHS has achieved this fitness value at 100th iteration. AHS has obtained fitness value f(x) = 0.00009 at 72th iteration where as HRHS has achieved this fitness value at 100th iteration.

The convergence graph of Rastrigin test function between IHS, HRHS and AHS is shown in Fig.2. AHS has obtained fitness value f(x) = 0.007 at 12th iteration where as IHS has achieved this fitness value at 100th iteration. AHS has obtained fitness value f(x) = 0.00025 at 67th iteration where as HRHS has achieved this fitness value at 100th iteration.

The convergence graph of Hump test function between IHS, HRHS and AHS is shown in Fig.4. AHS has obtained fitness value f(x) = 0.0001 at 12th iteration where as IHS has achieved this fitness value at 100th iteration. AHS has obtained fitness value f(x) = 3×10^{-7} at 37th iteration where as HRHS has achieved this fitness value at 100th iteration. It clearly indicates better convergence of proposed AHS.

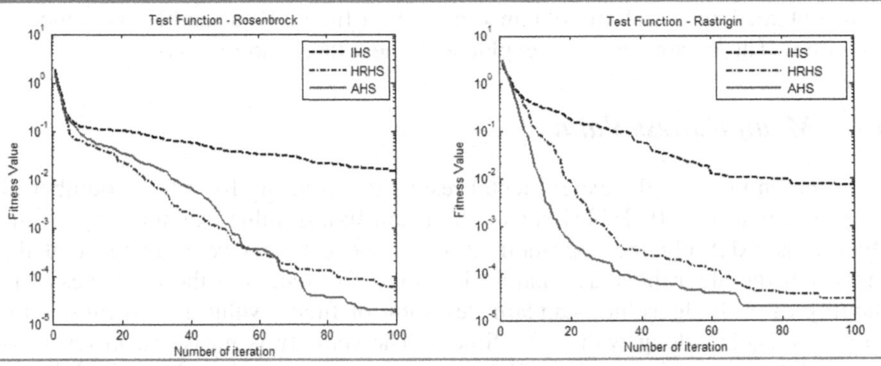

Fig. 1 Comparison of convergence of AHS, HRHS and IHS for function F_1-Rosenbrock

Fig. 2 Comparison of convergence of AHS, HRHS and IHS for function F_2-Rastrigin

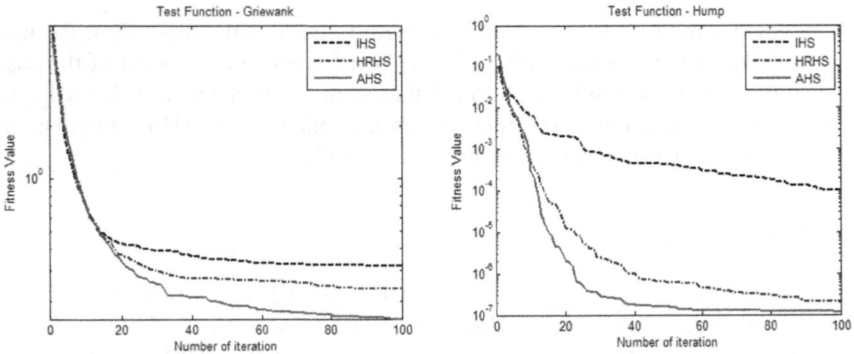

Fig. 3 Comparison of convergence of AHS, HRHS and IHS for function F_5- Griewank

Fig. 4 Comparison of convergence of AHS, HRHS and IHS for function F_9-Hump

Table 2 Mean fitness value among 30 runs

S. No	AHS	HRHS	IHS
F1	1.677803e-005±2.393e-005	6.008880e-005±8.415e-005	1.447735e-002±1.984e-002
F2	1.912802e-005±6.751e-005	2.866343e-005±4.171e-005	7.193735e-003±1.162e-002
F3	9.316768e-004±6.978e-004	1.479245e-003±7.826e-004	3.217780e-003±4.942e-003
F4	2.441712e-007±4.275e-007	1.482716e-006±3.811e-006	1.757589e-003±3.774e-003
F5	1.584128e-001±1.443e-001	2.347300e-001±1.773e-001	3.169654e-001±1.788e-001
F6	5.874454e-008± 1.437e-007	1.162043e-007 ± 1.759e-007	5.070051e-005± 1.097e-004
F7	8.749933e-005±5.653e-005	1.175002e-004±4.069e-005	1.295103e-001±3.248e-001
F8	1.644280e-002±0.021e-003	3.295896e-002±1.077e-002	1.469015e-001±1.605e-001
F9	1.149567e-007±1.743e-007	2.098885e-007±3.567e-007	8.075865e-005±1.459e-004
F10	-3.224073e+001±8.315	-2.931891e+001±8.250	-2.768043e+001±10.52
F11	-2.062612±2.739e-008	-2.062612±3.052e-008	-2.062607±6.816e-006
F12	2.980482e-004±6.520e-004	1.208354e-003±2.133e-003	2.686890e-003±3.955e-003
F13	5.000000e-001±0	5.000924e-001±2.922e-006	5.000970e-001±3.315e-006
F14	8.963265e-005±1.003e-004	3.497825e-004±2.602e-004	6.002708e-002±1.015e-001
F15	3.507000e-002±5.212e-002	3.314261e-002±3.329e-002	7.330346e-002±9.207e-002
F16	9.556852e-008±1.347e-007	2.729560e-007±3.264e-007	1.691262e-003±2.579e-003
F17	2.964354e-006±8.654e-006	2.204531e-006±3.139e-006	1.219548e-002±2.270e-002
F18	3.531193e-001±1.489	4.683879e 001±1.655	1.648905e+000±3.089
F19	-1.800385±9.412e-004	-1.799046±2.848e-003	-1.799513±2.225909e-003
F20	5.314191e-008±8.021e-008	2.812455e-008±6.277e-008	1.076526e-006±1.822e-006

Table 3 Best fitness value among 30 runs

S. No	AHS	HRHS	IHS
F1	2.855862e-009	1.343856e-008	2.053935e-006
F2	9.600143e-011	2.164715e-008	1.198169e-006
F3	5.815747e-005	1.764631e-004	3.118140e-005
F4	1.958559e-013	1.821665e-010	4.802143e-008
F5	3.663776e-007	1.093478e-005	3.607597e-003
F6	7.177261e-012	6.693637e-011	1.782266e-009
F7	5.185013e-005	5.779162e-005	5.700247e-005
F8	1.528496e-004	8.287648e-003	4.526073e-004
F9	7.067063e-015	1.839398e-009	1.655465e-007
F10	-4.668195e+001	-4.578741e+001	-4.441405e+001
F11	-2.062612	-2.062612	-2.062612
F12	1.634026e-012	1.547612e-011	7.103484e-012
F13	5.000000e-001	5.000874e-001	5.000909e-001
F14	5.090154e-006	1.838769e-005	3.497472e-005
F15	2.652623e-006	5.029626e-004	1.662573e-003
F16	5.522403e-011	1.398796e-009	1.470492e-007
F17	1.941169e-012	2.776362e-009	1.599821e-006
F18	4.729099e-005	2.915411e-003	3.596181e-004
F19	-1.801281	-1.801295	-1.801296
F20	1.370422e-012	1.212585e-012	8.114018e-010

6 Conclusion

In this paper, a novel Amended harmony search (AHS) algorithm is presented to improve the performance and efficiency of the harmony search algorithm. The AHS uses dynamic bandwidth (bw) and pitch adjustment rate (PAR) to maintain the proper balance between convergence and population diversity. In order to evaluate the performance of proposed AHS algorithm, we have applied it on a set of standard benchmark test functions. The results obtained by AHS shows that AHS has good searching ability and strong convergence. Results also suggested the AHS algorithm outperform the other two methods improved harmony search (IHS) and highly reliable harmony search (HRHS) in terms of fitness value. For the future scope, this method could be further applied to various engineering optimization problem.

References

1. Bandyopadhyay, S., Saha, S.: Unsupervised Classification. Springer, Heidelberg (2013), doi:10.1007/978-3-642-324512_2
2. Rao, R.V., Savsani, V.J.: Mechanical Design Optimization Using Advanced Optimization Techniques. Springer Series in Advanced Manufacturing. Springer, London (2012)
3. Kirkpatrick, S., Gelatt, C.D., Vecchi, M.P.: Optimization by Simulated Annealing. Science 220 (1993)
4. Man, K.F., et al.: Genetic Algorithms: Concepts and Applications. IEEE Transcations on Industrial Electronics 43(5) (1996)
5. Dorigo, M., et al.: The Ant System:Optimization by a colony of cooperating agents. IEEE Transactions on Systems, Man, and Cybernetics–Part B 26(1), 1–13 (1996)
6. Kennedy, J., Eberhart, R.C.: Particle Swarm Optimization. In: Proc. IEEE Int. Conf. on Neural Networks (1995)
7. Geem, Z.W., Kim, J.H., Loganathan, G.: A new heuristic optimization algorithm: Harmony search. Simulation 76(2), 60–68 (2001)
8. Taherinejad, N.: Highly reliable harmony search algorithm. In: Circuit Theory and Design, ECCTD, pp. 818–822 (2009)
9. Omran, M.G.H., Mahdavi, M.: Global-best harmony search. Appl. Math. Comput. 198(2), 643–656 (2008)
10. Ingram, G., Zhang, T.: Overview of applications and developments in the harmony search algorithm. In: Geem, Z.W. (ed.) Music-Inspired Harmony Search Algorithm. SCI, vol. 191, pp. 15–37. Springer, Heidelberg (2009)
11. Santos Coelho, L.D., de Andrade Bernert, D.L.: An improved harmony search algorithm for synchronization of discrete-time chaotic systems. Chaos Solitons Fractals 41(5), 2526–2532 (2009)
12. Geem, Z.W.: Improved harmony search from ensemble of music players. In: Gabrys, B., Howlett, R.J., Jain, L.C. (eds.) KES 2006. Part I. LNCS (LNAI), vol. 4251, pp. 86–93. Springer, Heidelberg (2006)
13. Mahdavi, M., Fesanghary, M., Damangir, E.: An improved harmony search algorithm for solving optimization problems. Appl. Math. Comput. 188(2), 1567–1579 (2007)
14. Wang, C.M., Huang, Y.F.: Selfadaptive harmony search algorithm for optimization. Expert Syst. Appl. 37(4), 2826–2837 (2010)
15. Chakraborty, P., Roy, G.G., Das, S., Jain, D., Abraham, A.: An improved harmony search algorithm with differential mutation operator. Fundam. Inform. 95, 1–26 (2009)
16. Lee, K.S., Geem, Z.W.: A new meta-heuristic algorithm for continues engineering optimization: Harmony search theory and practice. Comput. Meth. Appl. Mech. Eng. 194, 3902–3933 (2004)
17. Yadav, P., Kumar, R., Panda, S.K., Chang, C.S.: An Intelligent Tuned Harmony Search algorithm for optimization. Information Sciences 196, 47–72 (2012)

Multiobjective Mixed Model Assembly Line Balancing Problem

Sandeep Choudhary* and Sunil Agrawal

Abstract. The main objective of this paper is to improve the performance measures of a multi-objective mixed model assembly line. The motivation of our work is from the paper by Zhang and Gen [1] in which the assembly line problem is solved by using genetic algorithm. Mathematical solutions of their work show balance efficiency (E_b) of 86.06 percent, cycle time (C_T) of 54 minutes, work content (T_{wc}) is 185.8 minute, production rate (R_p) is 1.11E, where E is line efficiency. When the same mathematical model is reconstructed by changing decision variables (changing variables that hold the relationship between the task-station-models to task-worker-model) without changing the meaning of constraints and solved using the branch and bound (B&B) method using Lingo 10 software, there is a significant improvement in performance factors of assembly line. The mathematical solutions are obtained as follows. The balance efficiency (E_b) increases by 3.86 percent, cycle time (C_T) decreases by 25.55 percent, work content (T_{wc}) decrease by 22.17 percent and production rate (R_p) decease by 34.23 percent.

Keywords: Assembly line balancing, Branch & Bound, Genetic Algorithm, Multiobjective mixed model.

1 Introduction

In this very competitive world, no manufacturing organization wants to produce the products in large volume with low variety. Companies are now shifting towards the idea of producing more variety that attracts consumers. The products having soft variety can be produced on single assembly line known as mixed model assembly

Sandeep Choudhary · Sunil Agrawal
PDPM Indian Institute of Information Technology,
Design and Manufacturing Jabalpur, 482005, India
e-mail: {sandeep10165,sa}@iiitdmj.ac.in

* Corresponding author.

El-Sayed M. El-Alfy et al. (eds.), *Advances in Intelligent Informatics*,
Advances in Intelligent Systems and Computing 320, DOI: 10.1007/978-3-319-11218-3_58

line. In mixed model assembly line, there is a sequence of stations interconnected to each other through conveyor system. The speed of conveyor depends on the cycle time of bottleneck station. In order to design mixed model assembly line, there are number of criteria (objective functions) used by researchers.

These objectives are enlisted as: (1) to minimize the idle time at the workstations, (2) to reduce the delay and avoid waste of production, (3) to improve quality of assembled products, (4) maximizing utilization of manpower by reducing idle time, (5) minimizing the number of workstations, and (6) maintaining the smooth flow of production by eliminating the bottleneck station.

In literature attempts have been made by different authors to solve the problem of assembly line balancing. Chakravarty and Shtub [2] studied mixed-model line while considering work-in-process inventory with objectives of minimizing the labour, set-up cost and inventory. Rahimi Vahed *et al.* [3] introduced a new multiobjective scatter search for the mixed model assembly line (MMAL) which search for locally pareto-optimal frontier with the objectives to minimize: (1) total utility work, (2) total production variation, and (3) total setup cost. Whereas, Zhang and Gen [1] solved multiobjective mixed model assembly line balancing problem (MMALBP) by minimizing the following objective functions: cycle time minimization, workload variation minimization, and total cost minimization. The problem with three objectives was solved using genetic algorithm (GA). In the literature, the assembly line balancing problem has also been solved by methods like branch and bound (B&B), simulation annealing (SA), tabu search, ant colony optimization (ACO), and particle swarm optimization (PSO).

2 Mathematical Model

2.1 Previous Mathematical Model

This section discusses the model given by Zhang and Gen [1]. The improved version of the same is presented in the next subsection.

Indices

j,k : indices of task $(j,k = 1,2.....n)$
i : index of station $(i = 1,2.....m)$
w : index of worker $(w = 1,2.....m)$
l : index of model $(l = 1,2..... p)$

Decision variables

$$x_{jil} = \begin{cases} 1, \text{if task } j \text{ of model } l \text{ is assigned on station } i. \\ 0, \text{otherwise.} \end{cases}$$

$$y_{iw} = \begin{cases} 1, \text{if worker } w \text{ is working on station } i. \\ 0, \text{otherwise.} \end{cases}$$

Parameters

n : number of tasks.
m : number of stations, number of workers.
p : number of models.
r_l : expected demand ratio of model l during the planning period.
t_{jlw} : processing time of task j for model l processed by worker w.
d_{jw} : cost of worker w to process task j.
u_{il} : utilization of the station i for model l.
$u_{il} = t_l(S_i) / c_T$
$t_l(S_i)$: processing time at station i for model l.

$$t_l(S_i) = \sum_{j=1}^{n}\sum_{w=1}^{m} t_{jw} x_{jil} y_{iw}, \quad \forall i$$

c_T *: cycle time for combination of models.*

$$c_T = \sum_{l=1}^{p}\left\{ r_l \max_{1 \le i \le m}\left\{ \sum_{j=1}^{n}\sum_{w=1}^{m} t_{jlw} x_{jil} y_{iw} \right\}\right\}$$

Objective functions

$$Min \quad c_T = \sum_{l=1}^{p}\left\{ r_l \max_{1 \le i \le m}\left\{ \sum_{j=1}^{n}\sum_{w=1}^{m} t_{jlw} x_{jil} y_{iw} \right\}\right\} \tag{1}$$

$$Min \quad v = \sum_{l=1}^{p}\left\{ r_l \sqrt{\frac{1}{m}\sum_{i=1}^{m}(u_{il} - \overline{u_l})^2} \right\} \tag{2}$$

$$Min \quad d_T = \sum_{i=1}^{m} \sum_{j \in s_i} \sum_{w=1}^{m} d_{jw} y_{iw} \tag{3}$$

Constraints

$$s.t. \sum_{i=1}^{m} i\, x_{ijl} \ge \sum_{i=1}^{m} i\, x_{kil}, \quad \forall k \in \mathrm{Pre}(j), \forall j = 1...n, l = 1...p \tag{4}$$

$$\sum_{i=1}^{m} x_{jil} = 1, \quad \forall j = 1...n, l = 1..p \tag{5}$$

$$\sum_{w=1}^{m} y_{iw} = 1, \quad \forall i = 1...m \tag{6}$$

$$\sum_{i=1}^{m} y_{iw} = 1, \qquad \forall\, w = 1 \ldots m \tag{7}$$

$$x_{jil} \in \{0, 1\} \quad \forall\, j,i,l \tag{8}$$

$$y_{iw} \in \{0, 1\} \quad \forall\, i, w \tag{9}$$

The first expression in the objective function (1) of the model is to minimize the cycle time of the assembly line based on demand ratio of each model. The second expression in the objective function (2) is to minimize the variation of workload. The third expression in the objective function (3) is to minimize the total worker cost. Inequity (4) is the precedence relationships between tasks, that all predecessor of task j must be assigned to a station, which is in front of or the same as the station that task j is assigned in. Equation (5) ensures that task j must be assigned to only one station. Equation (6) ensures that only one worker can be allocated to station i. Equation (7) ensures that worker w can be allocated to only one station.

2.2 *Improved Mathematical Model*

The mathematical model presented in subsection 2.1 is modified as follows. In this improved mathematical model the intermediate relationship between task-station-model is changed to task-worker-model of the decision variable. With this small change, there is a significant improvement in the performance parameters of assembly line. Based on the numerical problem considered in this paper (section 3), the improvement in the performance measures like balance efficiency (E_b) improved by 3.86 percent and cycle time (C_T) improved by 25.55 percent can be reported here.

In the previous model given by Zhang and Gen [1], decision variable $x_{\ldots}$ is defined as follows.

$$x_{jil} = \begin{cases} 1, \text{ if task } j \text{ of model } l \text{ is assigned on station } i. \\ 0, \text{ otherwise.} \end{cases}$$

Here the indices are *jil* respectively represent task-station-model relationship. In this paper the modified index of decision variable $x_{\ldots}$ is defined as follows.

$$x_{jwl} = \begin{cases} 1, \text{ if task } j \text{ of model } l \text{ is assigned to worker } w. \\ 0, \text{ otherwise.} \end{cases}$$

The modified decision variable consisting of indices *jwl* respectively represents task-worker-model relationship. Rest of the mathematical model will remain same except replacing all x_{jil} with x_{jwl} in all the above equation.

3 Numerical Illustration

The modification proposed in the subsection 2.2 has been verified by considering a numerical problem taken from Zhang and Gen [1]. This problem considers two different models to be assembled on the single assembly line. There are four workstations on this assembly line with manning level one. The precedence diagram of two models is show in Fig. 1 and Fig. 2 respectively. Fig. 3 shows the combined precedence diagram of both the models. Further, task time taken by four workers to perform each task for model 1 and model 2 is given in table 1 and table 2 respectively. Task time of the combined model and the cost charged by each worker to perform a particular task is given in table 3. The demand ratio of model 1 is 0.60 and model 2 is 0.40. The problem is to distribute the tasks among four workers at four different stations such that the combined cycle time, variation of workload and cost charged by worker for both the models is minimized.

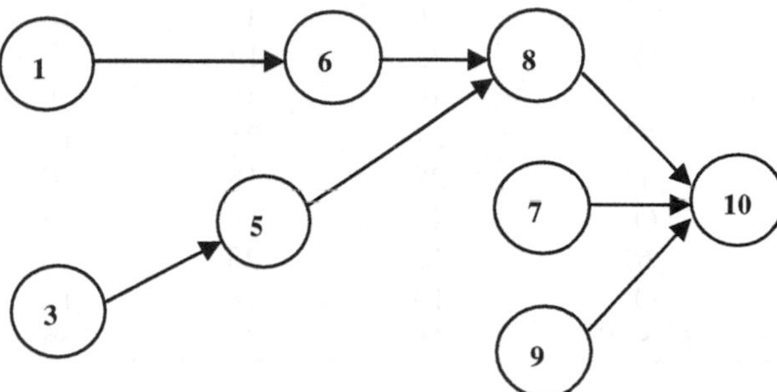

Fig. 1 A precedence diagram of model 1

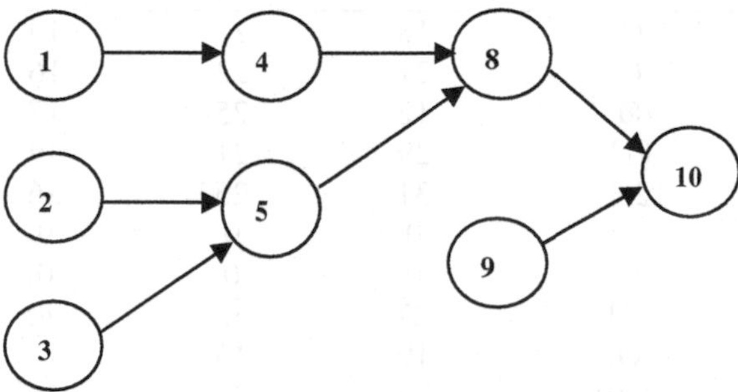

Fig. 2 A precedence diagram of model 2

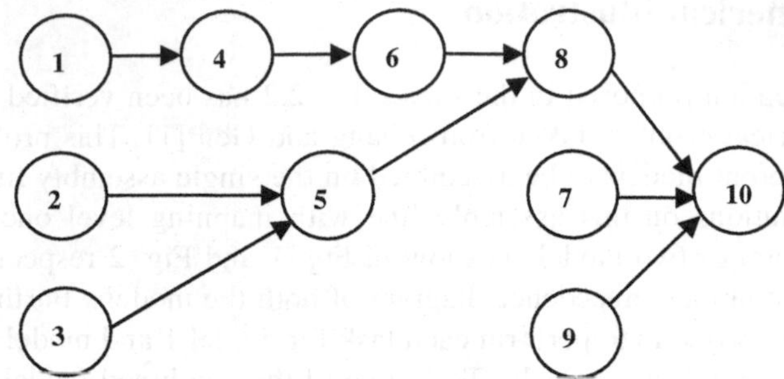

Fig. 3 Combined precedence diagram of both the models

Table 1 Data set of model 1

Task *j*	Precedence	*w1*	*w2*	*w3*	*w4*
1	Ø	17	23	17	13
2	-	0	0	0	0
3	Ø	22	15	27	25
4	-	0	0	0	0
5	{3}	21	25	16	32
6	{1}	28	18	20	21
7	Ø	42	28	23	34
8	{6}	17	23	40	25
9	Ø	19	18	17	34
10	{7,8,9}	16	27	35	26

Table 2 Data set of model 2

Task *j*	Precedence	*w1*	*w2*	*w3*	*w4*
1	Ø	18	22	19	13
2	Ø	21	22	16	20
3	Ø	12	25	17	15
4	{1}	29	21	19	16
5	{2,3}	31	25	26	22
6	-	0	0	0	0
7	-	0	0	0	0
8	{4}	27	33	40	25
9	Ø	19	13	17	34
10	{8,9}	26	27	35	16

Table 3 Data set of the combined model (combined time; worker cost)

Task j	Precedence	$w1$	$w2$	$w3$	$w4$
1	Ø	17.4; 83	22.6; 78	17.8; 81	13.0; 87
2	Ø	8.4; 79	8.8; 78	6.4; 84	8.0; 80
3	Ø	18.0; 88	19.0; 75	23.0; 73	21.0; 85
4	{1}	11.6; 71	8.4; 79	7.6; 81	6.4; 84
5	{2,3}	25.0; 69	25.0; 75	20.0; 74	28.0; 78
6	{4}	16.8; 72	10.8; 82	12.0; 80	12.6; 79
7	Ø	25.2; 58	16.8; 72	13.8; 77	20.4; 66
8	{6}	21.0; 73	27.0; 67	40.0; 60	25.0; 75
9	Ø	19.0; 81	16.0; 87	17.0; 83	34.0; 66
10	{7,8,9}	20.0; 74	27.0; 73	35.0; 65	22.0; 74

4 Results and Discussion

Comparison is done based on three solution approaches. First approach for task assignment is given by Zhang and Gen [1] after solving the mathematical model with G.A, second approach for task assignment is obtained when the mathematical model is solved by B&B method, and the third approach for task assignment is obtained when decision variable is changed and modified mathematical model (given in subsection 2.2) is solved by B&B method. Assembly line performance indices like balance efficiency, delay and production rate are used for comparison. Table 4 shows assignment of tasks with three different approaches and table 5 shows comparison of results with respect to various performance measures of assembly line.

Table 4 Task assignments with different approaches

	First approach				Second approach				Third approach			
Station	1	2	3	4	1	2	3	4	1	2	3	4
Model 1 Task	1,7	9,6	3,5	8,10	1,3,5	6,7,9	8,10	-	1,6	3,9	5,7	8,10
Model 2 Task	2,1	9,4	3,5	8,10	-	-	-	1,2,3,4,5,8,9,10	1,8	4,9	2,5	3,10
Worker no	4	1	3	2	2	1	4	3	4	2	3	1
Station time (min)	41.4	47.4	43	54	37.8	53.4	30.6	75.6	36	33.4	40.2	35

When the mathematical model with the decision variable x_{jil} is solved using G.A, mathematical result shows balance efficiency (E_b) of 86.06 percent, delay (d) of 0.139 and cycle time (C_T) of 54 minutes [1]. When same mathematical model given by Zhang and Gen [1] is solved by using B&B approach, balance efficiency (E_b) decrease to 65.27 percent, delay (d) increase to 0.347 and cycle time (C_T) increase to 75.6 minutes. Further, when the decision variable is changed to x_{jwl}

Table 5 Comparison of various results

Sr No	Description	Solved by GA when x_{jil} (Zhang and Gen, 2011)	Solved by B&B when x_{jil}	Solved by B&B when x_{jwl}
1	Work content time (T_{wc})	185.8 min	197.4 min	144.6 min
2	Cycle time (C_T)	54 min	75.6 min	40.2 min
3	Number of work-stations (m)	4	4	4
4	Number of workers (m)	4	4	4
5	Total cost	744	1346	952
6	Balance Efficiency (E_b)	86.06%	65.27%	89.92%
7	Delay (d)	0.139	0.347	0.100
8	Production rate (R_p)	1.11E	0.79E	1.49E

E: Line Efficiency

(as suggested in subsection 2.2 of the paper), there is significant improvement in the performance parameters of assembly line, i.e., balance efficiency (E_b) increases to 89.92 percent, delay (d) decreases to 0.100 and cycle time (C_T) decrease to 40.2 minutes.

Based on the results obtained, we can conclude that the first approach and third approach are comparable. In the third approach, the cycle time decreases by 13.8 min but cost get increased by 208 unit. Therefore, to reduce the cycle time of the assembly line by 1 minute, the cost of assembly line increases by 15 cost unit (called as crashing cost). The manufacturing firm has to decide which option is suited for them and accordingly they have to adopt any one of the option available to him. Few firms may go for adopting third approach if cost aspect is least weighted as compared to increase in production some may adopt first approach if they prefer to reduce cost and also maintaining production at average level.

5 Conclusions

There are significant changes in the performance parameters of assembly line when the same set of problem is solved with different approaches. It is not necessary that the performance parameters of assembly line will improve by adopting any other optimization technique to solve the assembly line problem. As in this paper, there is a worst result as shown by the performance parameters of assembly line when the same mathematical model given by author Zhang and Gen [1] is solved by B&B method. At same time when the decision variable is changed as proposed in the paper in subsection 2.2 (without changing the meaning of constraints) and solved by B&B method, there are significant improvements in the performance parameters of assembly line. For proper assignment of tasks to workers on stations which govern the performance parameters of assembly line, indices of the decision variables play important role in finding out the optimum solution.

References

1. Zhang, W., Gen, M.: An efficient multi-objective genetic algorithm for mixed- model assembly line balancing problem considering demand ratio based cycle time. J. Intell. Manuf. 22, 367–378 (2011)
2. Chakravarty, A.K., Shtub, A.: Balancing mixed modellines with in-process inventories. Management Science 31, 1161–1174 (1985)
3. Rahimi-Vahed, A.R., Rabbani, M., Tavakkoli-Moghaddam, R., Torabi, S.A., Jolai, F.: A multi-objective scatter search for a mixed model assembly line sequencing problem. Adv. Eng. Inform. 21, 85–99 (2007)
4. Beckerand, C., Scholl, A.: A survey on problems and methods in generalized assembly line balancing. European Journal of Operational Research 168, 694–715 (2006)
5. Bowman, E.H.: Assembly-line balancing by linear programming. Operations Research 8, 385–389 (1960)
6. Salveson, M.E.: The assembly line balancing problem. Journal of Industrial Engineering 6, 18–25 (1955)
7. Held, M., Karp, R.M., Shareshian, R.: Assembly line balancing-Dynamic programming with precedence constraints. Operations Research 11, 442–459 (1963)
8. Liu, S.B., Ng, K.M., Ong, H.L.: Branch-and-bound algorithms for simple assembly line balancing problem. Int. Journal Adv. Manufacturing Technology 36, 169–177 (2008)
9. Monden, Y.: Toyota production system. Institute of Industrial Engineers Press, Atlanta (1983)
10. Macaskill, J.L.C.: Production line balances for mixed-model lines. Management Science 19, 423–434 (1972)

Author Index